Handbook of Food Process Design

Handbook of Food Process Design

Volume II

Edited by

Jasim Ahmed, PhD
Kuwait Institute for Scientific Research
Safat, Kuwait

Mohammad Shafiur Rahman, PhD
Sultan Qaboos University
Muscat, Sultanate of Oman

WILEY-BLACKWELL
A John Wiley & Sons, Ltd., Publication

Blackwell Publishing was acquired by John Wiley & Sons in February 2007. Blackwell's publishing program has been merged with Wiley's global Scientific, Technical and Medical business to form Wiley-Blackwell.

Registered office: John Wiley & Sons, Ltd, The Atrium, Southern Gate, Chichester, West Sussex, PO19 8SQ, UK

Editorial offices: 9600 Garsington Road, Oxford, OX4 2DQ, UK
The Atrium, Southern Gate, Chichester, West Sussex, PO19 8SQ, UK
2121 State Avenue, Ames, Iowa 50014-8300, USA
111 River Street, Hoboken, NJ 07030-5774, USA

For details of our global editorial offices, for customer services and for information about how to apply for permission to reuse the copyright material in this book please see our website at www.wiley.com/wiley-blackwell.

Library of Congress Cataloging-in-Publication Data
Handbook of food process design / edited by Jasim Ahmed, M.Shafiur Rahman.
p. cm.
Includes bibliographical references and index.
ISBN-13: 978-1-4443-3011-3 (hardback)
ISBN-10: 1-4443-3011-X (hardback)
1. Food processing plants–Design and construction. 2. Food processing machinery.
3. Food industry and trade. I. Rahman, Shafiur.
TH4526.H36 2011
664'.02–dc23
2011022689

A catalogue record for this book is available from the British Library.

Wiley also publishes its books in a variety of electronic formats. Some content that appears in print may not be available in electronic books.

Set in 8/10 pt Trump Mediaeval by Toppan Best-set Premedia Limited
Printed and bound in Singapore by Markono Print Media Pte Ltd

1 2012

Contents

Preface

Generally, a process is defined as a sequence of events that transforms the biological materials of food products, via biochemical changes, into stable forms with added value. This can create new products or modify existing ones. *Process design* refers to the design of food processes and manufacturing methods, while *plant design* refers to the design of the whole processing plant. The processing of food is no longer as simple or straightforward as in the past. Food process design is an interdisciplinary science that is highly regarded by the food industry. The architecture of food process engineering is based on the solid foundations of chemical and mechanical engineering, together with the basics of microbiology, chemistry, nutrition, and economics. Other related disciplines, including instrumentation, computer science and mathematics, complete the discipline. Process design is the core of food engineering, and frequently begins with a concept and eventually ends in fabrication. Many types of documentation are involved in the process to test theories, display results, and organize data.

Today, the food industry is one of the largest manufacturing industries in the world and the significant contribution of food engineers to the industry is well recognized. A professional food engineer should be well versed in the basic principles, processes, flow diagrams, instrumentation and process control. The *Handbook of Food Process Design* has been developed primarily for fulfilling these expectations and is intended to be used by students in undergraduate and graduate courses in food process engineering/food technology/biochemical engineering, as well as by professionals working in the food industry. It could also be used by graduates in other disciplines, such as chemical and/or mechanical engineering.

The editors of this book have vast experience in teaching, research, and extension activities related to the food industry and have long realized the need for such a handbook on process equipment design to fill the current gap in the basic and applied fields of food engineering. They have endeavoured to gather eminent academics and professionals from across the globe and have succeeded in securing their participation in this book. All the contributors have diverse backgrounds, ranging from electronic engineering to food science.

The book contains 46 chapters in two volumes, with chapters grouped according to their similar subject matter. Chapters 1–12 are devoted to the basic principles, starting with units and dimensions, moving on to thermodynamics and reaction kinetics pertaining to foods, and followed by sensors and instrumentation involved in process automation. The handbook is well balanced by its coverage of unit operations involved in conventional and novel processing technologies to be used by the food industry.

Each chapter is intended to provide concise up-to-date descriptions of fundamentals, applications, solved problems, and methods of cost analysis. Chapters 13–18 cover heating and cooling systems used in food processing, including pasteurization, sterilization, refrigeration, and freezing. Drying is considered one of the most successful unit operations used in the food industry. Process design related to the drying of food materials is covered in Chapters 19–22.

Some important process designs, such as crystallization, extrusion, aseptic processing, baking, and frying, are well discussed in Chapters 23–28. Chapters 29–32 cover mechanical operations related to food process industries, including mixing/agitation, size reduction, and extraction and leaching processes. Chapters 33–40 focus on novel process designs, including pulsed light, ultrasound, ohmic heating, pulsed electric field, high pressure, and irradiation. Food packaging is discussed in Chapters 41–44, while quality systems and cost analysis are covered in Chapters 45 and 46.

The editors are confident that this handbook will prove to be interesting, informative, and enlightening to readers in the field. They would appreciate receiving new information and comments to assist in future development of the next edition.

Jasim Ahmed
Mohammad Shafiur Rahman

Acknowledgements

We would like to thank Almighty Allah for giving us life and the opportunity to gain knowledge to write this important book. We wish to express our sincere gratitude to the Sultan Qaboos University, Polymer Source Inc. and Kuwait Institute for Scientific Research for providing the opportunity and facilities to execute such an exciting project, and for supporting us in research and other intellectual activities around the globe.

We sincerely acknowledge the sacrifices made by our parents during our early education. Appreciation is due to all our teachers in the course of our careers. Special thanks to our colleagues and other research team members for their support and encouragement. We are grateful to our contributors for their wonderful cooperation and, finally, we are indebted to our families for their continued support and patience throughout the project.

About the Editors

Jasim Ahmed

Jasim Ahmed, Research Scientist, at Kuwait Institute for Scientific Research, Kuwait, is the author or co-author of over 150 technical articles including 95 refereed journal papers, 40 conference papers, 18 book chapters, 20 popular articles, and 4 books. He has edited several books including *Novel Food Processing: Effects on Rheological and Functional Properties* and *Starch-based Polymeric Materials and Nanocomposites: Starch Chemistry, Processing and Applications* published by CRC Press, Boca Raton, Florida, and *Handbook of Vegetables and Vegetables Processing* and *Handbook of Tropical and Subtropical Fruits Processing and Packaging* published by Wiley-Blackwell, NJ.

Dr Ahmed has served as an editor of the *International Journal of Food Properties* for more than 5 years. Furthermore, he has served as special editor for a number of other journals. He is also associated with the editorial boards of three international journals. In 2010, he was invited to serve as a sub-panel member for the Food Processing and Packaging Section of the Institute of Food Technology (IFT), Chicago, USA.

Dr Ahmed is a professional member of the Institute of Food Technology (IFT) and a life member of the Association of Food Scientists and Technologists (AFST), India. He has been involved in many professional activities, such as organizing international conferences, industrial training and workshops. He received the BTech (Food and Biochemical Engineering) in 1991 and MTech (Food and Biochemical Engineering) in 1993 from Jadavpur University, Kolkata, India, and PhD in Food Technology in 2000 from GND University, India. He worked as Visiting Professor and Research Director at McGill University, and Polymer Source Inc., Montreal, Canada, before moving to Kuwait.

Dr Ahmed was awarded a gold medal by Jadavpur University, India for securing the top position in the M.Tech degree. He has received several grants from various funding agencies to carry out his research during his academic career. He received a best reviewers' award by Elsevier in the area of food engineering in 2009.

Dr Ahmed has been involved in food processing teaching, research and industry over 18 years and has proved himself an active scientist in the area of food engineering. He has worked on food product development, food rheology and structure, novel food processing and the thermal behaviors of foods. His current research focus is on biopolymer and starch-based nanocomposites. Dr Ahmed's work has been well recognized globally: there are more than 1000 citations of his work and his h-index is 20.

Mohammad Shafiur Rahman

Mohammad Shafiur Rahman, Professor at the Sultan Qaboos University, Sultanate of Oman, is the author or co-author of over 250 technical articles including 90 refereed journal papers, 87 conference papers, 58 book chapters, 34 reports, 12 popular articles, and seven books. He is the author of the internationally acclaimed and award-wining *Food Properties Handbook*, published by CRC Press, Boca Raton, Florida, which was one of CRC's bestsellers in 2002. The second edition is now released under his editorship. He was also the editor of the popular book *Handbook of Food Preservation* published by CRC Press, Boca Raton, Florida. The first edition was one of CRC's bestsellers in 2003, and the second edition is now on the market. He was invited to serve as one of the associate editors for the *Handbook of Food Science, Engineering and Technology*, and as one of the editors for the *Handbook of Food and Bioprocess Modeling Techniques* published by CRC Press.

Professor Rahman initiated the *International Journal of Food Properties* (published by Marcel Dekker) and has served as the founding editor for more than 10 years. He also serves on the editorial boards of eight international journals. He is a member on the Food Engineering Series Editorial Board of Springer Science, New York, and serves as a section editor for the Sultan Qaboos University Journal of Agricultural Sciences. In 1998 he was invited to serve as a Food Science Adviser for the International Foundation for Science (IFS) in Sweden.

Professor Rahman is a professional member of the New Zealand Institute of Food Science and Technology and the Institute of Food Technologists, a member of the American Society of Agricultural Engineers and the American Institute of Chemical Engineers, and Member of the Executive Committee of the International Society of Food Engineering (ISFE). He has been involved in many professional activities, such as organizing international conferences, training workshops and other extension activities related to the food industry. He has been a keynote/plenary speaker at many international conferences. He received the BSc Eng. (Chemical) in 1983 and MSc Eng. (Chemical) in 1984 from Bangladesh University of Engineering and Technology, Dhaka, MSc in food engineering in 1985 from Leeds University, England, and PhD in food engineering in 1992 from the University of New South Wales, Sydney, Australia.

Professor Rahman has received numerous awards and fellowships in recognition of his research and teaching achievements, including the HortResearch Chairman's Award, the Bilateral Research Activities Program (BRAP) Award, CAMS Outstanding Researcher Award 2003, SQU Distinction in Research Award 2008, and the British Council Fellowship. In 2008 Professor Rahman ranked among the top five leading scientists and engineers of 57 OIC member states in the agroscience discipline.

Professor Rahman is an eminent scientist and academic in the area of food processing. He is recognized for his significant contributions to the basic and applied knowledge of food properties related to food structure, engineering properties and food stability. His total SCOPUS citation is more than 1200 and his h-index is 20, which indicates the high impact of his research in the international scientific community.

Contributors

Mohammad Shafiur Rahman, PhD
Professor
Department of Food Science and Nutrition, Sultan Qaboos University, Muscat, Oman
email: shafiur@squ.edu.om

Jasim Ahmed, PhD
Research Scientist
Food and Nutrition Program, Kuwait Institute for Scientific Research, Safat, Kuwait
email: jahmed2k@yahoo.com and jaahmed@kisr.edu.kw

Alik Abakarov, PhD
Visiting Professor
Technical University of Madrid, Higher Technical School of Agricultural Engineering, Madrid, Spain
email: alik.abakarov@upm.es

Tarif Ali Adib, PhD
Research Scientist
Water Quality Observation Laboratory, Lattakia, Syria
email: aliadib@hotmail.fr

Rajib Ul Alam Uzzal, PhD
Powertrain Control Engineer
Chrysler Technical Center
Chrysler Drive, Auburn Hills, Michigan, USA
email: rajiboic@gmail.com

Tanweer Alam, PhD
Associate Professor (Joint Director)
Indian Institute of Packaging, Mumbai, India
email: amtanweer@rediffmail.com

Eva Almenar, PhD
Assistant Professor
School of Packaging, Michigan State University, East Lansing, Michigan, USA
email: ealmenar@msu.edu

Sergio Almonacid, PhD
Associate Professor
Department of Chemical and Environmental Engineering, Universidad Técnica Federico Santa María, Valparaíso, Chile.
email: sergio.almonacid@usm.cl

Santanu Basu, PhD
Associate Professor
Department of Food Engineering
National Institute of Food Technology, Entrepreneurship and Management, Kundli, Haryana, India
email: santbasu@gmail.com

Pinaki Bhattacharya, PhD
Professor
Department of Chemical Engineering, Jadavpur University, Kolkata, India
email: pinaki_che@yahoo.com

Timothy J. Bowser, PhD, PE
Associate Professor
Department of Biosystems and Agricultural Engineering, and the Robert M.Kerr Food and Agricultural Products Center
Oklahoma State University, Stillwater, Oklahoma, USA
email: bowser@okstate.edu

Juan A. Chávez, PhD
Associate Professor
Sensor Systems Group, Department of Electronic Engineering, Universitat Politècnica de Catalunya, Barcelona, Spain
email: juan.antonio.chavez@upc.edu

Bijan Choudhury, PhD
Assistant Professor
Department of Biotechnology, Indian Institute of Technology, Roorkee, India
email: bijanfbs@iitr.ernet.in

M.A.A. Shoukat Choudhury, PhD
Assistant Professor
Department of Chemical Engineering, Bangladesh University of Engineering and Technology, Dhaka, Bangladesh
email: shoukat@che.buet.ac.bd

Rod Chu, BS
Retired Scientist
Ottawa, Ontario, Canada
email: rodchu2@yahoo.ca

Amalia Conte, PhD
Researcher
Department of Food Science, University of Foggia and Istituto per la Ricerca e le Applicazioni Biotecnologiche per la Sicurezza e la Valorizzazione dei Prodotti Tipici e di Qualità – BIOAGROMED, Foggia, Italy
email: a.conte@unifg.it

Maria Valeria De Bonis, PhD
Post-doctoral teaching assistant
CFDfood-DITEC, Università degli Studi della Basilicata, Potenza, Italy
email: mv.debonis@gmail.com

Matteo Alessandro Del Nobile, PhD
Associate Professor
Department of Food Science, University of Foggia and Istituto per la Ricerca e le Applicazioni Biotecnologiche per la Sicurezza e la Valorizzazione dei Prodotti Tipici e di Qualità – BIOAGROMED, Foggia, Italy
email: ma.delnobile@unifg.it

Ali Demirci, PhD
Professor
Department of Agricultural and Biological Engineering, Pennsylvania State University, University Park, Pennsylvania, USA
email: demirci@psu.edu

Kirk Dolan, PhD
Associate Professor
Department of Food Science and Human Nutrition, Department of Biosystems and Agricultural Engineering, Michigan State University, East Lansing, Michigan, USA
email: dolank@msu.edu

Ferruh Erdogdu, PhD
Professor of Food Process Engineering
Department of Food Engineering, University of Mersin, Mersin, Turkey
email: ferruherdogdu@yahoo.com

Ali Esehaghbeygi, PhD
Associate Professor
Department of Farm Machinery, College of Agriculture, Isfahan University of Technology, Isfahan, Iran
email: esehaghbeygi@cc.iut.ac.ir

E. Özgül Evranuz, PhD
Professor of Food Process Engineering
Food Engineering Department; Faculty of Chemical & Metallurgical Engineering; Istanbul Technical University, Maslak, İstanbul, Turkey
email: evranuz@itu.edu.tr

John J. Fitzpatrick, PhD
Assistant Professor
Department of Process and Chemical Engineering, University College, Cork, Ireland
email: j.fitzpatrick@ucc.ie

Federico Gómez Galindo, PhD
Research Scientist
Department of Food Technology, Engineering and Nutrition, Lund University, Sweden
email: Federico.Gomez@food.lth.se

Miguel J. García-Hernández, PhD
Professor
Sensor Systems Group, Department of Electronic Engineering, Universitat Politècnica de Catalunya, Barcelona, Spain
email: miguel.j.garcia@upc.edu

Gurbuz Gunes, PhD
Associate Professor
Food Engineering Department College of Chemical and Metallurgical Engineering, Istanbul Technical University, Istanbul, Turkey
email: gunesg@itu.edu.tr

M.N.A. Hawlader, PhD
Professor
Department of Mechanical Engineering, Faculty of Engineering
International Islamic University Malaysia, Jalan Gombak, Kuala Lumpur, Malaysia
email: mehawlader@iium.edi.my

Pär Henriksson
Managing Director
Arc Aroma Pure AB, Lund, Sweden
email: info@arcaromapure.se

K.A. Jahangeer, MEng, MIES
Professional Officer (Academic Research)
Department of Mechanical Engineering, National University of Singapore, Singapore
email: mpejkah@nus.edu.sg

Kazi Bayzid Kabir, PhD
Assistant Professor
Department of Chemical Engineering, Bangladesh University of Engineering and Technology, Dhaka, Bangladesh
email: kazibayzid@che.buet.ac.bd

Dattatreya M. Kadam, PhD
Senior Scientist
Central Institute of Post-Harvest Engineering and Technology (CIPHET), Ludhiana, Punjab, India
email: kadam1k@yahoo.com

Chinnadurai Karunanithy, PhD
Research Associate
Agricultural and Biosystems Engineering Department, South Dakota State University, Brookings, South Dakota, USA
email: Karunanithy.Chinnadurai@sdstate.edu.

Nene M. Keklik, PhD
Assistant Professor
Department of Food Engineering, Cumhuriyet University, Sivas, Turkey
email: meltemkeklik@gmail.com

Mohd. Kaleem Khan, PhD
Assistant Professor
Department of Mechanical Engineering, Indian Institute of Technology, Patna, India
email: mkkhan@iitp.ac.in

Meral Kılıç-Akyılmaz, PhD
Associate Professor
Food Engineering Department of Chemical and Metallurgical Engineering, Istanbul Technical University, Istanbul, Turkey
email: kilicmer@itu.edu.tr

Celale Kirkin, MS
Graduate student
Food Engineering Department College of Chemical and Metallurgical Engineering, Istanbul Technical University, Istanbul, Turkey
email: kirkin@itu.edu.tr

Tatiana Koutchma, PhD
Research Scientist
Agriculture and Agri-Food Canada, Guelph, Ontario, Canada
email: koutchmat@agr.gc.ca

Prabhat Kumar, PhD
R&D Scientist
Research and Development, Frito Lay North America, Plano, Texas, USA
email: prabhat116@gmail.com

Gerard La Rooy, Dip Mgt, MBA (Otago)
Round Earth Business Process Improvement, Havelock North, New Zealand
email: gerard.round.earth@xtra.co.nz

Frank P. Lucien, PhD
Associate Professor
School of Chemical Engineering, University of New South Wales, Sydney, Australia
email: f.lucien@unsw.edu.au

Francesco Marra, PhD
Professor
Department of Industrial Engineering, Section of Chemical and Food Engineering, University of Salerno, Fisciano, Italy
email: fmarra@unisa.it

Amin Allah Masoumi, PhD
Associate Professor
Department of Farm Machinery, College of Agriculture, Isfahan University of Technology, Isfahan, Iran
email: masoumi@cc.iut.ac.ir

Marcella Mastromatteo, PhD
Post-doc
Istituto per la Ricerca e le Applicazioni Biotecnologiche per la Sicurezza e la Valorizzazione dei Prodotti Tipici e di Qualità – BIOAGROMED, Foggia, Italy
email: ma.mastromatteo@unifg.it

Crispin Merkel, BS
Graduate student
School of Packaging, Michigan State University, East Lansing, Michigan, USA
email: merkelcr@gmail.com

Dharmendra K. Mishra, PhD
Thermal Process Expert
Nestle Nutrition, Fremont, Michigan, USA
email: dharmendra.mishra@rd.nestle.com

I.M. Mujtaba, PhD
Professor
School of Engineering, Design and Technology, University of Bradford, Bradford, UK
email: I.M.Mujtaba@brad.ac.uk

Kasiviswanathan Muthukumarappan, PhD
Professor
Agricultural and Biosystems Engineering Department, South Dakota State University, Brookings, South Dakota, USA
email: Kas.Muthukum@sdstate.edu

Tomás Norton, PhD
Senior Lecturer
Harper Adams University College, Newport, Shropshire, UK
email: tnorton@harper-adams.ac.uk

Helena Núñez
Research Associate
Department of Chemical and Environmental Engineering, Universidad Técnica Federico Santa María, Valparaíso, Chile

T. Koray Palazoglu, PhD
Associate Professor
Department of Food Engineering, University of Mersin, Mersin, Turkey
email: koray_palazoglu@yahoo.com

Tapobrata Panda, PhD
Professor
Biochemical Engineering Laboratory, Indian Institute of Technology Madras, Chennai, Tamilnadu, India
email: panda@iitm.ac.in

R.T. Patil, PhD
Director
Central Institute of Post-Harvest Engineering and Technology (CIPHET), Ludhiana, Punjab, India
email: ramabhau@yahoo.com

Q. Tuan Pham, PhD
Professor
School of Chemical Engineering, University of New South Wales, Sydney, Australia
email: tuan.pham@unsw.edu.au

Martín Picón-Núñez, PhD
Professor
Department of Chemical Engineering, University of Guanajuato, Guanajuato, Mexico
email: picon@quijote.ugto.mx

Emmanuel Purlis, PhD
Research Scientist
Centro de Investigación y Desarrollo en Criotecnología de Alimentos (CIDCA – CONICET La Plata), Facultad de Ciencias Exactas, UNLP, La Plata, Argentina
email: emmanuel@cidca.org.ar

Cristina Ratti, PhD
Professor
Food Engineering Program, Soil Science and Agri-Food Engineering Department, Institute for Nutraceuticals and Functional Foods, Université Laval, Québec, Canada
email: Cristina.Ratti@fsaa.ulaval.ca

Rohit Ruhal, MSc
Research Scholar
Department of Biotechnology, Indian Institute of Technology, Roorkee, India
email: rohitmicrobe@gmail.com

Gianpaolo Ruocco, PhD
Professor
CFDfood-DITEC, Università degli Studi della Basilicata, Potenza, Italy
email: gianpaolo.ruocco@unibas.it

Jordi Salazar, PhD
Associate Professor
Sensor Systems Group, Department of Electronic Engineering, Universitat Politècnica de Catalunya, Barcelona, Spain
email: jorge.salazar@upc.edu

K.P. Sandeep, PhD
Professor
Department of Food, Bioprocessing, and Nutrition Sciences, North Carolina State University, Raleigh, North Carolina, USA
email: kp_sandeep@ncsu.edu

Ilkay Sensoy, PhD
Assistant Professor
Department of Food Engineering, Middle East Technical University, Ankara, Turkey
email: isensoy@metu.edu.tr

U.S. Shivhare, PhD
Professor
University Institute of Chemical Engineering and Technology, Panjab University, Chandigarh, India
email: usshiv@yahoo.com

Muhammad Siddiq, PhD
Associate Professor
Food Science and Human Nutrition, Michigan State University, East Lansing, Michigan, USA
email: siddiq@anr.msu.edu

Ricardo Simpson, PhD
Professor
Department of Chemical and Environmental Engineering, Universidad Técnica Federico Santa María, and Centro Regional de Estudios en Alimentos Saludables, Valparaíso, Chile
email: ricardo.simpson@usm.cl

Josip Simunovic, PhD
Research Associate Professor
Department of Food, Bioprocessing, and Nutrition Sciences, North Carolina State University, Raleigh, North Carolina, USA
email: josip_simunovic@ncsu.edu

Siddhartha Singha, MTech
Research Scholar
Biochemical Engineering Laboratory, Indian Institute of Technology Madras, Chennai, Tamilnadu, India
email: sidsin2k4@gmail.com

Paul Takhistov, PhD
Associate Professor
Department of Food Science, Rutgers University, Newark, New Jersey, USA
email: takhistov@AESOP.Rutgers.edu

Nantawan Therdthai, PhD
Associate Professor
Department of Product Development, Faculty of Agro-Industry, Kasetsart University, Bangkok, Thailand
email: faginwt@ku.ac.th

Brijesh Tiwari, PhD
Lecturer
Manchester Food Research Centre, Manchester Metropolitan University, Manchester, UK
email: b.tiwari@mmu.ac.uk

Gary Tucker, PhD
Head of Baking and Cereals Processing
Campden BRI, Chipping Campden, UK
email: g.tucker@campden.co.uk

Antoni Turó, PhD
Associate Professor
Sensor Systems Group, Department of Electronic Engineering, Universitat Politècnica de Catalunya, Barcelona, Spain
email: antoni.turo@upc.edu

M. Reza Zareifard, PhD
Food Engineering Researcher
Agriculture and Agri-Food Canada, Food Research and Development Centre, Ste Hyacinthe, Quebec, Canada
email: MohammadReza.Zareifard@agr.gc.ca

Weibiao Zhou, PhD
Professor
Food Science and Technology Programme, National University of Singapore, Singapore
email: chmzwb@nus.edu.sg

26

Baking Process Design

Emmanuel Purlis

Introduction

Baking is the final and most important step in bakery production, and can be defined as the process which transforms dough, basically flour and water (plus other ingredients such as sugars, fat, egg, leavening agent, and other additives depending on the specific product), into a food with unique sensorial features. In particular, bread is a staple food and its production is of great importance from a commercial point of view, besides its cultural relevance.

White or French bread is the most popular type of bread, and is distinguished for having a crunchy and brown crust, a sponge and light crumb with soft texture and intermediate moisture, and a typical flavour. All these qualities are the result of a series of physical and chemical changes that occur during baking. Basically, dough is transformed into crumb due to starch gelatinisation and protein denaturation, and thermal expansion of carbon dioxide (produced by leavening agents) and water vapour; crust is subsequently formed as a result of water evaporation, cross-linking reactions and browning development, which is associated with formation of flavour and harmful compounds (Mondal and Datta, 2008).

Despite technological advances and process automation in the food industry, baking is a traditional food process that still largely depends on the experience of skilled technologists. In addition, since no microbiological risks are involved a priori

Handbook of Food Process Design, First Edition. Edited by Jasim Ahmed and Mohammad Shafiur Rahman.

(assuming that good manufacturing practices are applied), unlike other food processes such as pasteurisation or sterilisation, the baking process is assessed mainly by quality aspects (sensorial attributes) that are critical in the acceptance of the product consumers, i.e. surface colour together with texture and flavour. Nevertheless, knowledge of the relationship between process variables, such as process time and energy consumption, and material properties and operating conditions is of key interest to design engineers and equipment users.

With this in mind, the concept of process design appears to be a powerful tool. Formerly developed as a component of chemical engineering, process design is based on unit operations, transport phenomena, process control and process economics. Furthermore, the recent trend is to apply fundamental physical, chemical and engineering principles, use of computer modelling and process simulations. In this sense, systematic process design is being adopted instead of empirical approaches for the design and optimisation of food processes. As in chemical engineering, food process design seeks to reduce production costs. However, the overall quality of products, including sensorial, nutritional and safety aspects, is essential to any food process design (Maroulis and Saravacos, 2003).

Baking is a good example of the application of process design to food engineering. In recent years, efforts have been made to study the process from an integrated perspective, considering the transport phenomena and quality changes that occur during the process. Different approaches have been applied, mainly in the context of process optimisation. On the one hand, experimental studies allow empirical models (e.g. polynomial functions) to be developed by regression of experimental data that describe the variation of state variables (e.g. temperature, weight loss) or quality attributes (e.g. surface colour, texture parameters) as a function of operating conditions (e.g. oven temperature, baking time, heat transfer coefficient). Afterwards, optimisation can be performed using different methods, for example response surface methodology has been used to develop and improve new baking technologies for bread and cake (Demirekler *et al.*, 2004; Sevimli *et al.*, 2005), while (classical) nonlinear programming is another possibility (Therdthai *et al.*, 2002; Dingstad *et al.*, 2004).

On the other hand, a systematic approach can be used to design and optimise the baking process. This involves the use of transport models that describe transformations of the product (e.g. heat and mass transfer model coupled with quality kinetic models) as the starting point for process design. Using this concept, Hadiyanto *et al.* (2007, 2008a,b, 2009) developed a series of optimisation algorithms for quality-driven process design to improve bakery production.

In bread (and other products) baking, with regard to process design, process optimisation or direct technological application, it is clearly necessary to define certain parameters based on empirical information associated with sensorial attributes, i.e. subjective values. For instance, even for a multi-objective process optimisation based on sophisticated algorithms, it is necessary to use weight factors and target values (e.g. end point of the process) to establish the global objective function, based on previous experience and preference of consumers. In addition, because there is a variety of products or specifications related to different cultures and regulations, it is difficult to develop an objective and unique methodology that can design or optimise the baking process or determine a general heating strategy.

This chapter presents a review of the baking process from an engineering point of view, focusing on bread as product. The main objective is to deliver technological considerations based on a global perspective of the process, i.e. by simultaneously analysing transport phenomena and quality aspects occurring in the product during baking. In addition, baking equipment design is discussed in order to give a more extensive framework for baking process design. Process modelling and simulation appear to be very useful and powerful tools, following the principles of modern process design.

The Baking Process

Baking of bread (white or French type, made from wheat flour) is taken as an example for analysing the baking process. Throughout this section, experimental results will be used to understand the major mechanisms occurring in bread during baking, which is essential for process design. A conventional process using an electrical batch oven is studied, where natural or forced convection can be applied (see section 'Baking equipment' for description of baking oven). Although new technologies have been developed to improve traditional baking, e.g. microwave and infrared heating, mainly to reduce process time and energy consumption (Keskin *et al.*, 2004; Sumnu *et al.*, 2005), traditional baking represents a reference point.

Heat and Mass Transfer

From a transport phenomena point of view, bread baking is a simultaneous heat and mass transfer process occurring in a porous medium. In addition, a phase change (i.e. water evaporation) occurs in a moving front, which determines the three characteristic zones of the product:

1. **Crumb**: wet inner zone, where temperature does not exceed 100 °C and dehydration does not occur.
2. **Crust**: dry outer zone, where temperature increases above 100 °C and drying takes place.
3. **Evaporation front**: between the crumb and the crust, where temperature is about 100 °C and water evaporates.

At the beginning of the process, dough (with uniform temperature and water content) is placed in the oven at high temperature (>150 °C) and low relative humidity. Consequently, the food surface increases its temperature and drying starts because of the difference in water vapour pressure with respect to the surrounding air. In other

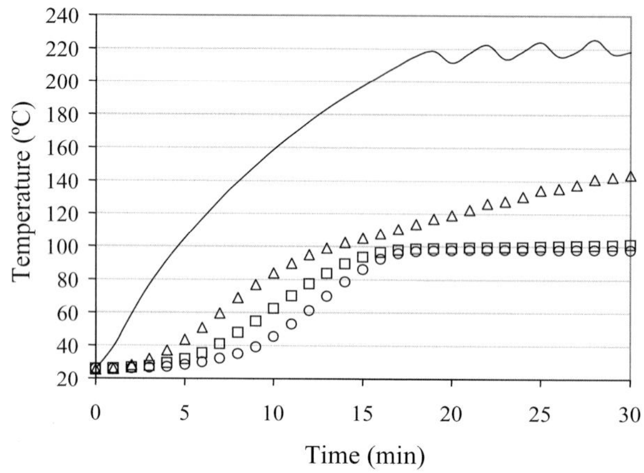

Figure 26.1 Typical temperature profiles during bread baking (220 °C under forced convection). Symbols indicate different zones of bread: surface (triangles), intermediate (squares), core (circles). Line corresponds to oven air. (From Purlis and Salvadori, 2010, courtesy of Elsevier.)

words, both temperature and water content gradients are established, and then simultaneous heat and mass transfer inside the bread begins, which can be characterised by temperature and water content profiles, and related variables (e.g. weight loss). The following representative experimental results are presented in order to understand the transport phenomena occurring during the process (details about materials and methods can be found in Purlis and Salvadori, 2009a).

Temperature variation in different regions of bread during baking is depicted in Figure 26.1. The temperature at the bread surface (crust zone) follows the behaviour of oven air temperature (set-up profile, non-steady regime), increasing continuously during the analysed period (30-min process). It is worth noting that in some situations a short plateau at 100 °C can be observed, which accounts for the rapid water evaporation at the surface. On the other hand, in the core (crumb zone), temperature starts rising after thermal gradient is established, reaching 100 °C asymptotically and showing a characteristic sigmoid trend. This rapid heating of the bread core has been explained by an evaporation–condensation mechanism (Sluimer and Krist-Spit, 1987; de Vries *et al.*, 1989). Finally, in the intermediate zone (i.e. underneath the surface but not at the core), the temperature increase shows hybrid behaviour: like the core, temperature does not exceed 100 °C but its variation before reaching the plateau is similar that of the surface region.

The typical variation in water content in bread during baking is presented in Figure 26.2, where two different processes can be clearly distinguished. The outer zones of the bread suffer dehydration throughout the process, leading to the formation of the crust. At the end of baking, values for water content of crust are generally in the range 5–10% (wet basis), depending on process conditions, i.e. high oven temperature and forced convection produce more dehydration. The crumb zone beneath the crust

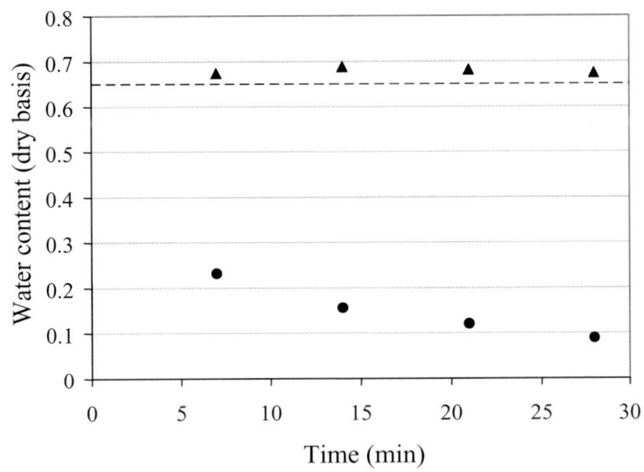

Figure 26.2 Typical water content (kg water/kg dry matter) profiles during bread baking (220 °C under forced convection). Symbols indicate different zones of bread: surface (triangles), core (circles). Dashed line corresponds to initial water content (raw dough).

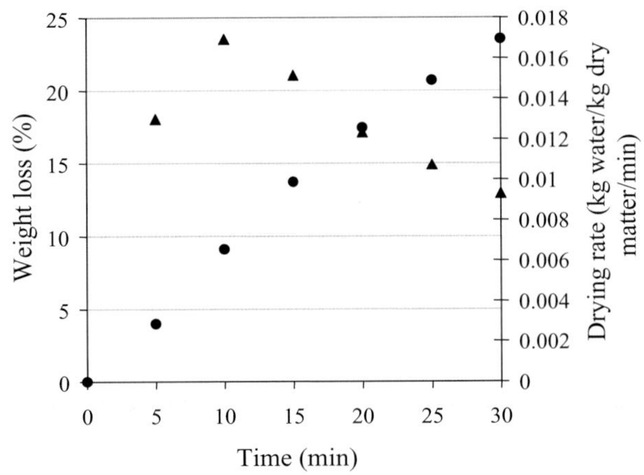

Figure 26.3 Typical variation of weight loss (circles) and drying rate (triangles) of bread during baking (220 °C under forced convection). (From Purlis and Salvadori, 2009a, courtesy of Elsevier.)

maintains the moisture content of the unbaked dough throughout the baking process. Furthermore, a slight increase can be detected in water content compared with initial conditions, also explained by the evaporation–condensation mechanism (Wagner *et al.*, 2007; Purlis and Salvadori, 2009a).

Weight loss continuously increases during baking, which is in agreement with the variation in water content at the crust zone, although the drying rate is not constant (Figure 26.3). According to Hasatani *et al.* (1991), the drying rate increases up to when the crust is formed and the evaporation front is established, and then begins to

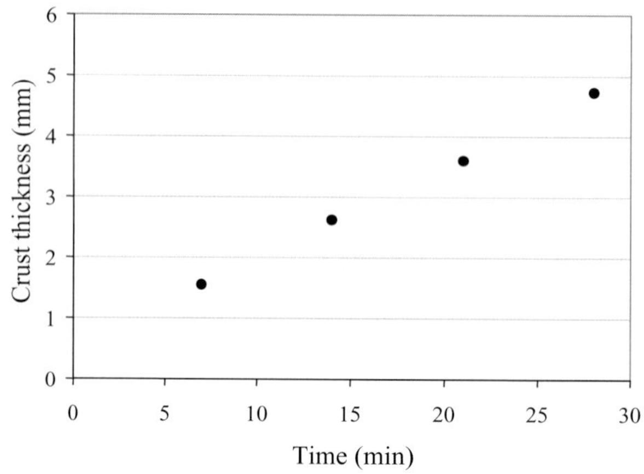

Figure 26.4 Enlargement of bread crust (i.e. outer dry zone) during baking (220 °C under forced convection). (From Purlis and Salvadori, 2009a, courtesy of Elsevier.)

decrease due to the enlarging crust thickness (Figure 26.4). In this way, formation of crust avoids further dehydration of the bread, the crust acting as a barrier to mass transfer to the oven ambient (Wählby and Skjöldebrand, 2002).

Finally, the volume change of leavened products (e.g. bread, cake) during baking is a distinguishing feature of this process. During baking, the dough firstly undergoes a volume increase because of thermal expansion of carbon dioxide (produced by leavening agents) and water vapour, until dough/crumb transition is reached. Secondly, slight shrinkage occurs due to final crust formation and setting (Sommier *et al.*, 2005). Subsequently, texture and other quality aspects are affected by these structural changes (Scanlon and Zghal, 2001).

Volume change depends on heat and mass transfer occurring during baking, and simultaneously change in volume may modify temperature and water content gradients due to changes in the characteristic dimensions of the system. Nevertheless, it is a complicated aspect of the process that depends on several variables, including product formulation, and it should be further studied. Purlis and Salvadori (2010) simulated the bread-baking process, including the volume change, in their mathematical formulation and found that it could be neglected from a technological point of view, after experimental validation using heat and mass transfer data.

Quality Aspects

For bread and other bakery products, although the various typical quality features are related to each product, surface colour and texture and flavour are the main consumer attributes because they are associated with level of satisfaction. However, it is first necessary to establish a minimum requirement for considering a product as baked. In bread baking, complete starch gelatinisation in dough should be considered as the first

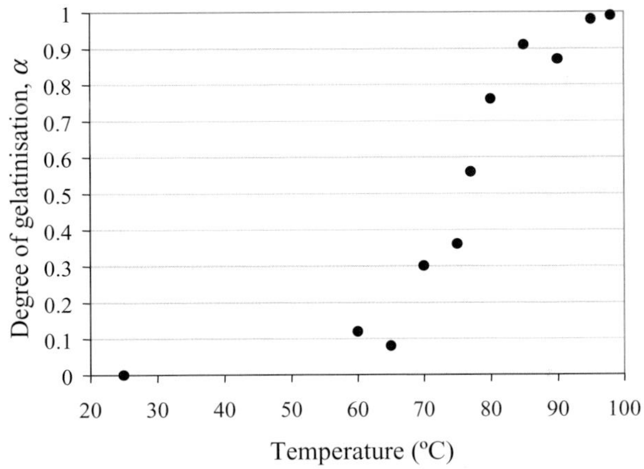

Figure 26.5 Extent of starch gelatinisation of bread crumb as a function of temperature during baking. (From Zanoni *et al.*, 1995, courtesy of Elsevier.)

quality index, i.e. the sensory acceptability of the product is not guaranteed if complete starch gelatinisation is not achieved (Zanoni *et al.*, 1995). Starch gelatinisation (together with protein denaturation) is responsible for the transition of dough (raw material) into bread (product) and starts at about 50 °C. The standard procedure for evaluating the degree of starch gelatinisation is by differential scanning calorimetry (DSC), which measures the temperature and enthalpy of this endothermic process (Fennema, 1996). Figure 26.5 shows experimental values for bread baking obtained by Zanoni *et al.* (1995), where the extent of starch gelatinisation (α) has been defined as $[1 - (Q/Q_{max})]$, where Q and Q_{max} are heat uptakes for partially baked and unbaked dough, respectively. Therefore, at initial condition, $\alpha = 0$, i.e. $Q = Q_{max}$ (unbaked dough).

Once the acceptability of the product is ensured, consumer preference should be the objective of baking process design (i.e. product quality). In this regard, two aspects are relevant: sensorial attributes and nutritional value of food. In bakery products, surface colour is an important attribute associated with aroma, taste, appearance, and with the overall quality of food, and has an important effect on consumer judgement: colour influences the anticipated oral and olfactory sensations because of the memory of previous eating experiences (Abdullah, 2008).

The formation of typical colour in bakery products during baking is widely known as *browning*. The development of browning is the result of the Maillard reaction and caramelisation of sugars. The ingredients of baked foods such as bread, cake, and biscuit (i.e. carbohydrates, proteins and water) are actually the reactants for these chemical reactions, which are catalysed by a low to medium level of moisture and high temperature obtained at the product surface during baking (Fennema, 1996). Browning is mainly influenced by temperature and water activity of the system, and results from the production and accumulation of coloured compounds during baking, principally hydroxymethylfurfural (HMF) and melanoidins. Browning can be followed

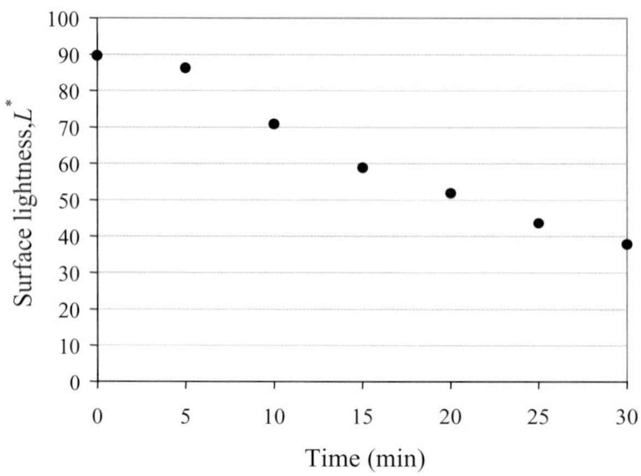

Figure 26.6 Typical variation of surface lightness (as representative variable of browning development) of bread during baking (220 °C under forced convection). (From Purlis and Salvadori, 2009b, courtesy of Elsevier.)

by measuring the concentration of reaction products or, alternatively, by reactant consumption. On the other hand, the concept of lightness is commonly used to describe the variation of colour during baking, since lightness is a parameter of the CIE $L^\star a^\star b^\star$ colour space ($L^\star$ ranges from 100 to 0, white to black), an international standard for colour measurement. The later approach is preferred in industrial applications, since the involved methods can be automated, rapid and non-destructive, e.g. computer vision (Abdullah, 2008).

There are certain minimum requirements for the initiation of colour formation during baking of bakery products. In general, browning is detected once the temperature exceeds 105–120 °C and water activity decreases to 0.4–0.7 (Purlis, 2010). Under such conditions, only the surface can show a significant change in colour during baking (see Figure 26.1 and previous discussion). Figure 26.6 depicts a representative case of browning development during bread baking, measured using a computer vision system (Purlis and Salvadori, 2009b). Standard or target values for lightness (or HMF concentration) are dissimilar because a great range of bakery products and operating conditions, besides consumer preference, is involved; some typical values of L^* for various products are shown in Table 26.1. Finally, when very high temperature and low water activity are achieved at the product surface, caramelisation takes place, producing more coloured compounds in addition to Maillard reaction products; this drastic condition is responsible for a burnt appearance characterised by low lightness of products, e.g. $L^\star < 60$ in bread (Purlis and Salvadori, 2009b).

The development of browning also produces important effects on the nutritional properties of bakery products (Purlis, 2010). A major concern is that the Maillard reaction is associated with the formation of acrylamide, which is probably a carcinogenic compound (Mottram *et al.*, 2002; Stadler *et al.*, 2002). Production of acrylamide is

Table 26.1 Some typical values of lightness (L^*) in bakery products for various baking conditions.

Product	L^*	Operating conditions
Biscuit	40–50	19 min, 200 °C
	55.7–14.4	6 min, 240–330 °C
	57.1	90 min, 180 °C
Fermented dough, ~10% sucrose	65.6	8–10 min, 220 °C
White bread	84.1, 77.2	50 min, 200 °C
	81.6	60 min, 200 °C
	81.9, 82.1	30 min, 210 °C
	83.0	16 min, 235 °C
Bread crisp	80.73	40 min, 140 °C
	72.40	34 min, 160 °C
	63.48	25 min, 180 °C
Bun	52.13	8 min, 225 °C
Muffin	83.9 ± 2.8	(Commercial, unknown)

Source: adapted from Purlis (2010).

strongly correlated with baking temperature and time and with content of asparagine and reducing sugars, and apparently starts at 120–130 °C so could be found only in the crust of bakery products. In addition, acrylamide formation is highly correlated with colour development (Ahrné *et al.*, 2007), and this represents a technological advantage for controlling its occurrence during the baking process.

Baking Design Based on Process Modelling and Simulation

Understanding the main mechanisms of the (bread) baking process, i.e. heat and mass transfer, and the quality changes occurring in the product are very useful for the development of accurate mathematical models that aim to simulate the process. Instead of performing experimental tests to analyse the process, numerical simulation allows working under standardised operating conditions, thus minimizing the uncertainties associated with such a complex process as (bread) baking. This section presents a summary of the results obtained from simulations of bread baking, with the objective of providing technological considerations about the process. Such conclusions are also of great importance for the design of baking equipment, major aspects of which are discussed in the next section.

Process Modelling and Simulation

The following discussion is based on the comprehensive study reported by Purlis (2011): process simulation was performed using a simultaneous heat and mass transfer model (extensively validated against experimental data) that considered bread baking as a moving boundary problem with phase change occurring in a porous medium. Major assumptions of the model are as follows:

- Bread is homogeneous and continuous; the porous medium concept is included through effective or apparent thermophysical properties.
- Heat arriving by convection and radiation to the bread surface is transported by conduction inside the product according to Fourier's law, but an effective thermal conductivity is used to incorporate the evaporation–condensation mechanism in heat transfer.
- Only liquid diffusion in the crumb and only vapour diffusion in the crust are assumed to occur, and are balanced by convective flux at the surface.
- Volume change is neglected.

The mathematical model consists of a system of partial differential equations involving heat and mass balance equations with the corresponding boundary and initial conditions (a worked example can be found in the Appendix). For a detailed description of the transport model, including thermophysical properties, the reader is referred to Purlis and Salvadori (2009a,c, 2010). In addition, kinetic models for describing product quality changes, i.e. starch gelatinisation (Zanoni *et al.*, 1995) and surface browning (Purlis and Salvadori, 2009b), during the process were coupled to the transport model. Input variables for process simulation were oven temperature (180–240 °C), heat transfer coefficient (5–25 $W \cdot m^{-2} \cdot K^{-1}$) and bread radius (0.025–0.035 m, characteristic length of an infinite cylinder); these ranges are in concordance with reported data for conventional baking ovens and common industrial practice (Purlis, 2011).

At this point, it is important to state that there are other mathematical models developed for baking, since the study and modelling of transport phenomena for this process still represent a challenge for food engineers. In this sense, it is not the intention of this chapter to discuss the modelling of baking, but to propose a systematic approach to designing the process. In this context, process modelling and simulation are useful tools for a global objective.

Technological Considerations of the Baking Process

For practical implications, it is useful to condense all the information obtained from process simulation into technological or industrial variables, i.e. variables that determine major changes in the product, and which can be effectively monitored during (or at some stage of) the process. Figure 26.7a shows the variation of core and surface temperatures and weight loss of product during (simulated) baking. It should be noted that similar profiles have been previously described from experimental data (Figures 26.1 and 26.3); quantitative differences are due to different operating conditions.

Regarding the quality aspects of the process, Figure 26.7b depicts the development of browning at the surface and the extent of starch gelatinisation at the core (coldest point). Kinetics models of these quality indices depend on temperature and water

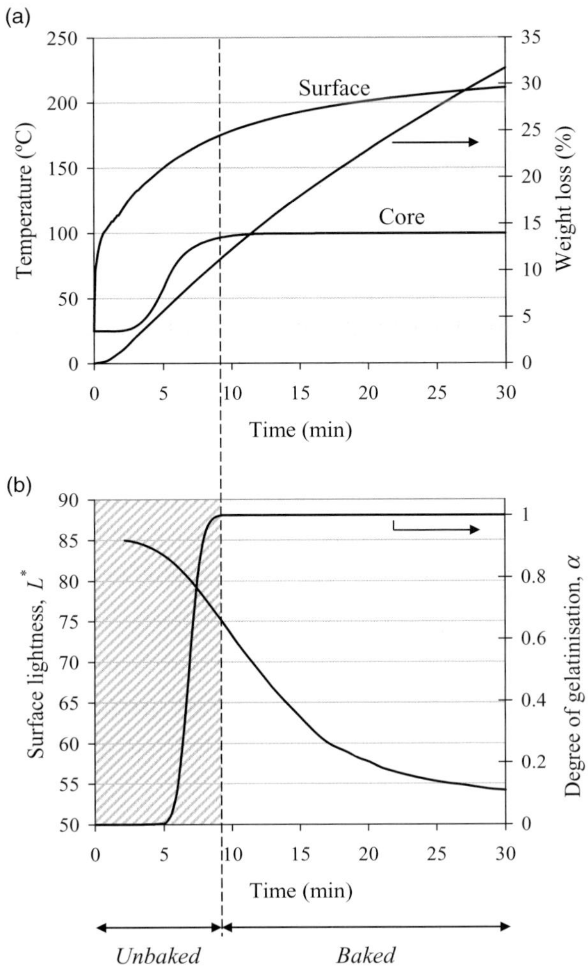

Figure 26.7 Variation of (a) core and surface temperature and weight loss, and (b) surface lightness and degree of gelatinisation at core of bread during (simulated) baking. Operating variables for simulation: oven temperature, 240 °C; heat transfer coefficient, 25 W·m^{-2}·K^{-1}; product radius, 0.03 m. Dashed line indicates $\alpha = 0.98$. (From Purlis, 2011, courtesy of Elsevier.)

activity (Purlis and Salvadori, 2009b) and temperature (Zanoni *et al.*, 1995), respectively. According to process design principles, it is important that quality changes can be described by transport phenomena, in order to control the process via the operating conditions.

It has already been established that complete starch gelatinisation in dough should be considered the first quality index, i.e. it is a minimum requirement to consider the product as baked; such a criterion is indicated in Figure 26.7. As can be seen, only acceptable products with L^* below 75 can be obtained under such operating conditions. Thus if less browned products are desired, the magnitude of heat and mass fluxes should be diminished in some way in order to allow the completion of baking before the target browning degree is accomplished. In addition, the increase in the characteristic length of product favours this situation, where certain values of browning cannot be obtained. This is because browning is mainly a superficial phenomenon (it only occurs when temperature is above 120°C) and transition of dough into crumb is assessed in the coldest point of the product. If the development of browning is accelerated, for example increasing heat transfer coefficient (h) and oven temperature, and the thermal gradient is diminished, for example increasing the characteristic length of product, the time required to achieve a low decrease in L^* is not enough to generate complete starch gelatinisation at the bread centre. Consequently, when slightly browned products are sought, it is recommended that there should not be a high driving force, for example $h > 15\,W{\cdot}m^{-2}{\cdot}K^{-1}$ and oven temperature > 220°C (Purlis, 2011).

In this context, where unacceptable products could be obtained, a control variable should be established to overcome this problem, i.e. achieving the target value of surface lightness without complete baking. This is because the degree of gelatinisation cannot be monitored directly during the process. One possible solution is to establish a minimum value of 95–96°C for the core temperature of bread at the end of the process (Purlis, 2011).

On the other hand, when minimum requirements are accomplished, it is useful to analyse the relationships between different process variables, e.g. baking time, and quality aspects. Figures 26.8, 26.9 and 26.10 show variation of baking time, surface

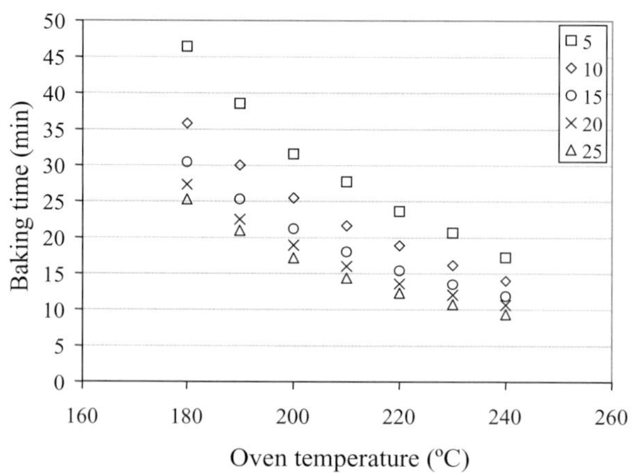

Figure 26.8 Variation of baking time as a function of oven temperature for different values of heat transfer coefficient (symbols, $W{\cdot}m^{-2}{\cdot}K^{-1}$) obtained by process simulation. The end point of the process corresponds to $L^* = 75$ for bread with radius 0.03 m. (From Purlis, 2011, courtesy of Elsevier.)

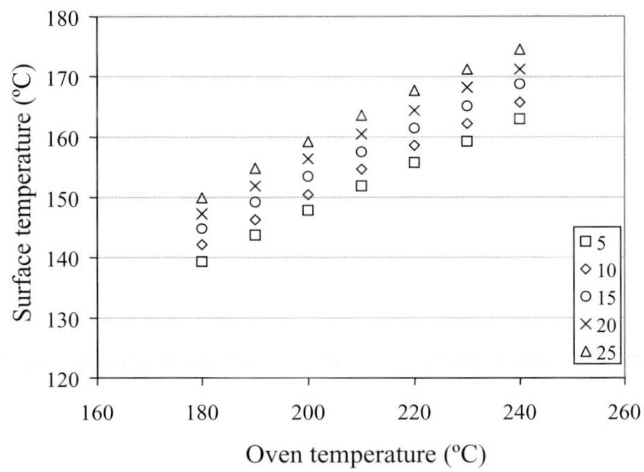

Figure 26.9 Variation of surface temperature of bread as a function of oven temperature for different values of heat transfer coefficient (symbols, $W{\cdot}m^{-2}{\cdot}K^{-1}$) obtained by process simulation. The end point of the process corresponds to $L^* = 75$ for bread with radius 0.03 m. (From Purlis, 2011, courtesy of Elsevier.)

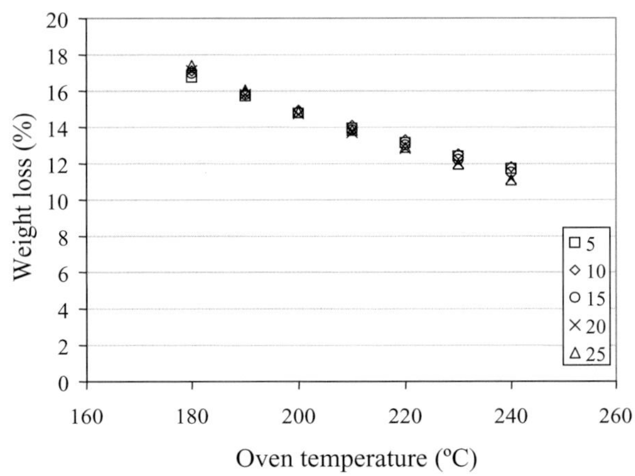

Figure 26.10 Variation of weight loss of bread as a function of oven temperature for different values of heat transfer coefficient (symbols, $W{\cdot}m^{-2}{\cdot}K^{-1}$) obtained by process simulation. The end point of the process corresponds to $L^* = 75$ for bread with radius 0.03 m. (From Purlis, 2011, courtesy of Elsevier.)

temperature and weight loss, respectively, with oven temperature for one representative condition (i.e. final $L^* = 75$) that was used to determine the end point of the process.

Firstly, baking time decreases when oven temperature and heat transfer coefficient are increased, showing an exponential trend (Figure 26.8); this is consistent with transport phenomena theory. On the other hand, for $h > 15\,\text{W·m}^{-2}\text{·K}^{-1}$, diminution of process time is produced in a slower manner. In this sense, when forced convection is applied, the cost of increasing the value of h (e.g. increasing oven fan velocity) would not be directly translated into reduction in baking time, i.e. the strategy of increasing h to diminish process time loses efficiency at high values of h. This can be explained by the relationship between internal and external resistance to heat transfer (i.e. Biot number): as heat transfer coefficient increases, the external resistance to heat transfer becomes negligible and all resistance is due to thermal conductivity of the product (low in foods).

The situation described above has a negative impact on the process, mainly from a nutritional point of view: high temperatures at the bread surface can be achieved when using high values of heat transfer coefficient and oven temperature, since surface temperature increases almost constantly with these two operating variables (Figure 26.9). Though browning and gelatinisation constraints are achieved in this case, the pathway for accomplishing the target L^* can produce a major detriment to bread quality due to the formation of harmful compounds such as acrylamide. In this way, it would be desirable to reduce the surface temperature of the product during baking as much as possible.

Secondly, the weight loss of bread decreases, approximately following linear behaviour, as oven temperature is augmented (Figure 26.10). This is because shorter times are required to achieve the final L^* value for increasing baking temperature, as heat flux is augmented. Nevertheless, it can be seen that weight loss is almost independent of heat transfer coefficient. This behaviour is probably due to the criterion used to establish the end point of baking, i.e. surface lightness; experimental data support this observation in the tested range of operating conditions (Purlis and Salvadori, 2007).

Baking Equipment

As baking is the most important step in bakery production, the oven is the major piece of equipment in this food process. In a baking oven, the generated heat is transferred to the food by radiation, convection and conduction, transforming the dough into product. As previously discussed, operating conditions determine food quality as well as process aspects. In particular, baking is an energy-intensive process due to the evaporation of water in the product (latent heat of vaporisation of water, $2.257\,\text{MJ·kg}^{-1}$ at $100\,°\text{C}$). The energy demand for a conventional baking process is around $3.7\,\text{MJ·kg}^{-1}$, although it can be higher (up to $7\,\text{MJ·kg}^{-1}$) depending on specific products and operating conditions. In this sense, baking is similar to drying, both demanding a high amount of energy in comparison with chilling, freezing and canning, which need less than $1\,\text{MJ·kg}^{-1}$ (Le Bail *et al.*, 2010).

Ovens are often operated in an empirical way, using trial and error, since information about manipulating oven settings for optimum production is still lacking and poorly understood (Broyart and Trystram, 2002). As a result, inconsistency in the quality of bakery products is common in most industrial-scale processes. Moreover, since quality is assessed at the end of baking, unacceptable products will have to be discarded because baking is a non-reversible process. Obviously, this is economically unfavourable (Wong *et al.*, 2007).

In summary, baking equipment design is of great importance for improving product quality and reducing energy consumption. Baking equipment (ovens) and related aspects are discussed in the following sections. It is worth noting that other devices are used for making bread and other products, before and after the baking process. Mixers, fermentation chambers, and divider, moulder and rounder devices are used to prepare dough from raw materials. After baking, mainly on industrial scale, chilling and packaging systems are required. In addition, freezing equipment is used for partially ('part') baked frozen products.

Baking Ovens

Baking ovens can be classified according to heat generation into direct and indirect heating ovens. In the former, heat is directly transferred by combustion gases from a clean gas fuel, such as natural gas or liquefied petroleum gas (LPG), or from a microwave power source. Indirect heating occurs when air in the baking chamber is heated through the oven walls, from steam tubes, or by electrical resistance. In general, direct heating ovens are more energy efficient and have shorter processing times (Maroulis and Saravacos, 2003).

In addition, baking ovens can be divided into batch and continuous ovens with respect to operation mode. In batch (or discontinuous) equipment, the product is placed directly onto decks (deck oven) or put in baking supports (trays, lids, pans, etc.) that are placed within a mobile rack or trolley (ventilated or rack oven) (Le Bail *et al.*, 2010). Forced convection is used to provide a more uniform airflow to the product and thus minimises variations in quality due to spatial distribution inside the oven (Anishaparvin *et al.*, 2010). Different configurations are designed with the aim of providing uniform baking to all product inside the oven, e.g. rotary and revolving ovens. Figure 26.11 shows some typical configurations of batch ovens, used mainly for small/medium-scale production.

In continuous equipment, baking is performed in tunnel ovens, where dough is transported on a conveyor band through the baking oven (Figure 26.12). Tunnel ovens of varying lengths and widths are common in most industrial-scale bakeries. In this type of oven, the baking time is defined by conveyor speed (i.e. residence time). According to specific products, the conveyor band can be made from steel, steel mesh, or 'stone' (granite, concrete, etc.). In general, the baking chamber is divided into several zones, where different baking conditions can be established in order to perform an efficient process (Figure 26.12a). As in batch equipment, tunnel ovens are equipped with fans at different locations to recirculate the hot air. A U-movement configuration

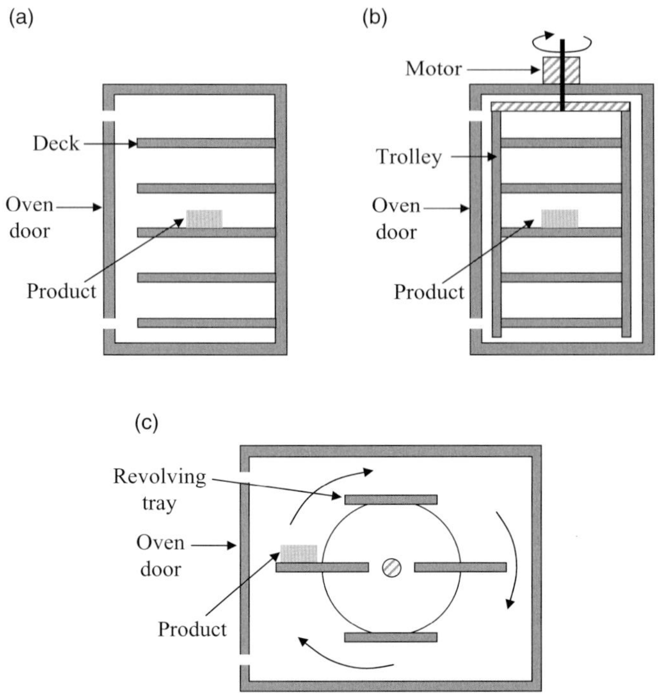

Figure 26.11 Schematic diagrams of batch ovens with different configurations: (a) deck oven; (b) rotary rack oven; (c) revolving oven.

is used to ensure uniform heating, increased energy efficiency and reduced length of tunnel (Figure 26.12b).

In a baking oven, the heat is transferred to the product by radiation from the oven heating surfaces, by convection from the hot air, and by conduction from surfaces in direct contact with the food (deck, tray, conveyor band). The proportion of each individual mode of heat transfer depends on oven design, configuration and operation, although heat supply by radiation is more difficult to control than that by convection (Zareifard *et al.*, 2009). Also, moisture control in the baking chamber is important for crust formation, whether required or not. For instance, in the case of crispy bread rolls (baguettes), steam is injected at the beginning of baking to plasticise the dough surface by steam condensation, which facilitates expansion of the bread (oven rise). Afterwards, the bread surface starts to dry and the typical crisp and browned crust is obtained (Le Bail *et al.*, 2010).

This short review of baking equipment shows that many oven configurations are available, depending on product specifications and production scale. Baking oven design (or selection) and operation should take into account the quality of bakery products and production costs (energy consumption, process time) in order to maximise the economic benefits of the process.

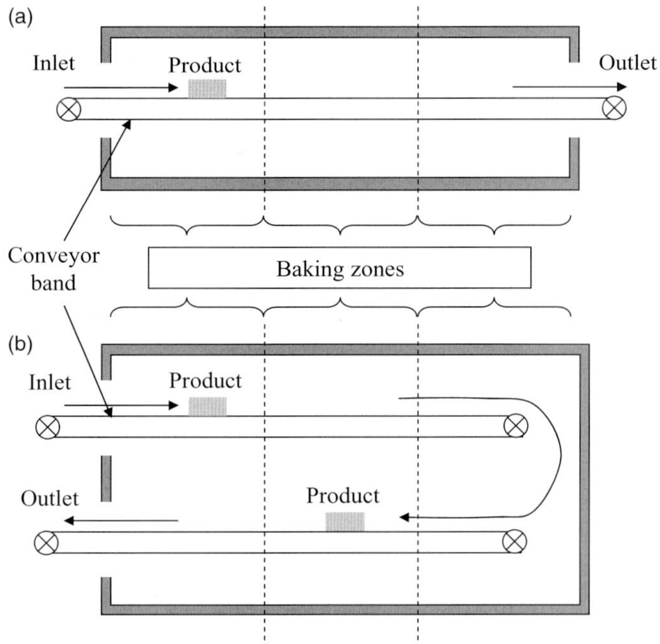

Figure 26.12 Schematic diagrams of tunnel ovens: (a) conventional configuration; (b) U-movement configuration.

Trends in Baking Technology

Food process engineers are continually developing new technologies and methodologies to improve the (conventional) baking process. With regard to baking technology, alternative or non-traditional heating sources have been investigated with the aim of reducing time and energy consumption of the process. Microwave heating has potential in this regard, since it involves internal heat generation and therefore the heating rate can be very high. Microwave energy is distributed throughout the product, and the oven is at ambient temperature. In this way, browning and crust formation are not promoted, and products with unacceptable texture, high moisture loss and rapid staling are obtained (Demirekler *et al.*, 2004). To solve these quality issues, different alternatives have been proposed in combination with microwave heating, leading to combination or hybrid ovens. Air impingement in combination with microwaves can produce acceptable browning, but products with low volume and firm texture are obtained (Li and Walker, 1996).

More recently, halogen lamp/microwave combination heating has been developed for baking oven design. Halogen lamp heating provides near-infrared radiation that is near the visible light region of the electromagnetic spectrum, with high frequency and low penetration depth. This radiation thus affects only the surface of foods, providing the required temperature values for browning development in bakery products (Keskin *et al.*, 2004). Very good results for bread and cake baking have been reported using a halogen lamp/microwave combination oven; products have comparable quality with conventionally baked items, while process time was reduced by 60–80% (Demirekler *et al.*, 2004; Sevimli *et al.*, 2005; Sumnu *et al.*, 2005). Further work will be necessary to implement this promising technology in practical or industrial situations.

As baking equipment design involves the study of fluid flow (momentum transfer) coupled with heat and mass transfer, computational fluid dynamics (CFD) is a powerful and versatile tool for process design. Basically, CFD is a simulation tool that applies numerical methods to model fluid flow situations for the prediction of (simultaneous) momentum, heat and mass transfer in a given process. This methodology reduces the amount of experimentation and empiricism associated with process design and optimisation. In addition, CFD allows handling a large amount of information, which is not possible with sensors or measurement devices during baking. For more information about CFD and its application in food processing, the reader is referred to Norton and Sun (2006) and Xia and Sun (2002). With regard to the baking process, CFD is rapidly becoming a common tool for food engineers to study and design the baking process and equipment under different operating conditions (Therdthai *et al.*, 2003, 2004; Mirade *et al.*, 2004; Wong *et al.*, 2007; Anishaparvin *et al.*, 2010; Boulet *et al.*, 2010).

Conclusions

Baking is a very complex process that involves many transport mechanisms and variables, involving both product quality and operating aspects. Knowledge of transport phenomena and chemical reactions occurring within the product as well as in the baking oven is essential to design, control and optimise the baking process. It is thus very useful to carry out simulations based on transport models coupled with (kinetic) models describing sensory and nutritional changes in the product, as a function of state variables and operating conditions. The results and discussion presented in this chapter lead to some technological considerations about bread baking that can be used as a starting point for process design when considering other products.

- Although the end point of baking may be determined by sensory attributes related to consumer preference (e.g. surface colour), a control variable associated with the complete process should be established in order to ensure the acceptability of the product. In bakery products, such a variable could be core temperature, with a lower limit value of 95–96 °C, which ensures complete starch gelatinisation.
- Intense heating should be avoided. High values of heat transfer coefficient and oven temperature may produce unbaked foods (although target parameters related to superficial phenomena are achieved). In addition, the baking time is not substantially decreased because of the low thermal conductivity of product (internal resistance to heat transfer).

- An advantageous strategy involves a low- to medium-intensity baking process. High-quality products are obtained since lower values of surface temperature are achieved, which avoids the generation of harmful compounds.
- It is important to promote the production and consumption of only slightly or minimally browned products, since the development of browning reactions is associated with the accumulation of toxic compounds. High-quality food will be obtained, and the avoidance of such reactions will also reduce the weight loss of products and energy consumption during baking, generating economic benefits.

There are still challenges to overcome: food engineers and equipment designers need to develop more efficient processes and new technologies for baking that help to reduce process time and energy consumption without degradation of product quality. Knowledge of transport phenomena occurring during the process and the use of simulation tools will certainly help to achieve this aim.

Appendix: Worked Examples

The following four examples illustrate the concepts discussed in this chapter. It is worth noting that the proposed solutions, mainly in equipment design, are rough estimations that should be considered only as starting points, bearing in mind the complexity of the studied process and specific construction issues.

Example A1: A Mathematical Model for Bread Baking

The major assumptions of a mathematical model for bread baking were presented earlier in this chapter. Here we develop the governing equations and the corresponding boundary and initial conditions for such a model, considering bread as an infinite cylinder of radius R. Then we assume that bread is baked as an individual piece on a perforated tray or steel mesh, without a tin or container. In this sense, a one-dimensional model can be formulated by assuming axial symmetry, i.e. only radial direction is considered (Figure 26.A1).

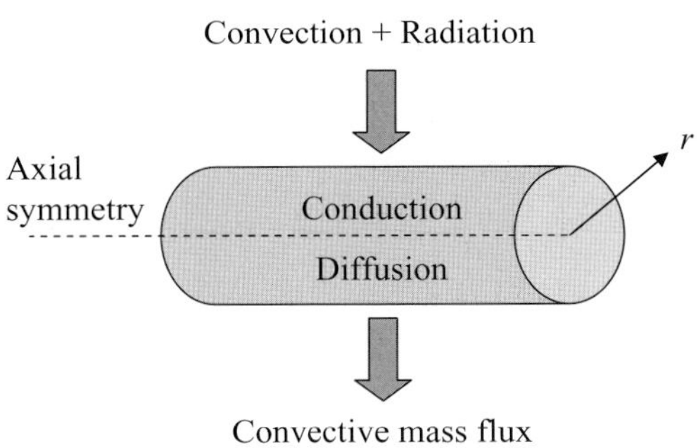

Figure 26.A1 Schematic description of the bread baking model.

Heat and mass balance equations are based on Fourier's law and Fick's diffusion law, respectively:

$$\rho C_p \frac{\partial T}{\partial t} = \frac{1}{r}\frac{\partial}{\partial r}\left(rk\frac{\partial T}{\partial r}\right) \tag{26.1}$$

$$\frac{\partial W}{\partial t} = \frac{1}{r}\frac{\partial}{\partial r}\left(rD\frac{\partial W}{\partial r}\right) \tag{26.2}$$

The corresponding boundary conditions at product surface ($r = R$) can be written respectively as:

$$-k\frac{\partial T}{\partial r} = h(T_s - T_\infty) + \varepsilon\sigma(T_s^4 - T_\infty^4) \tag{26.3}$$

$$-D\rho_s \frac{\partial W}{\partial r} = k_g\,[P_s(T_s) - P_\infty(T_\infty)] \tag{26.4}$$

where $P_s = a_w P_{sat}(T_s)$ and $P_\infty = (RH/100)P_{sat}(T_\infty)$.

At the centre of bread, i.e. $r = 0$:

$$\frac{\partial T}{\partial r} = 0 \tag{26.5}$$

$$\frac{\partial W}{\partial r} = 0 \tag{26.6}$$

For initial conditions, uniform temperature and water content are assumed. Note that thermophysical properties are temperature and/or water content dependent, so a coupled system of partial differential equations is obtained. This is called a simultaneous heat and mass transfer model.

Finally, the proposed model incorporates the enthalpy jump corresponding to phase change (evaporation front) by defining equivalent thermophysical properties. This is a physical approach used to formulate moving boundary problems such as bread baking (Purlis and Salvadori, 2009a,c, 2010).

Example A2: Power Estimation of a Batch Oven

In the design of baking equipment, it is essential to determine the power required for a given process. As an example, we estimate the power of a batch oven that can be initially loaded with 75 kg of bread dough (individual pieces with cylindrical shape, radius 3 cm). Let assume that the heat transfer coefficient of the oven is $15\,\mathrm{W{\cdot}m^{-2}{\cdot}K^{-1}}$, and that a target value of surface lightness ($L^* = 75$) has been established for bread.

According to the technological considerations in this chapter, a low to medium oven temperature will be used, e.g. 180 °C. From Figures 26.8 and 26.10, the baking time and weight loss for the analysed product can be estimated approximately as 30 min and 17%, respectively.

On the other hand, it has been reported that about 20% of the energy is used for water evaporation in the product during baking (Le Bail *et al.*, 2010). Therefore, a rough estimate of the oven power can be calculated as follows, taking one batch as basis:

$$\text{Mass of water evaporated} = \text{Initial mass of dough} \times \text{Weight loss} = 75 \times 0.17 = 12.75\ \text{kg}$$

$$\begin{aligned}&\text{Energy consumption due to water evaporation}\\&\quad = \text{Mass of water evaporated} \times \text{Latent heat of vaporisation of water (at 100 °C)}\\&\quad = 12.75 \times 2.257 \times 10^6\ \text{J}\cdot\text{kg}^{-1} = 28.777 \times 10^6\ \text{J}\end{aligned}$$

$$\begin{aligned}&\text{Power required for water evaporation}\\&\quad = (\text{Energy consumption due to water evaporation})/\\&\quad (\text{Time per batch or baking time})\\&\quad = (28.777 \times 10^6\ \text{J})/(1800\ \text{s}) \approx 16\ \text{kW}\end{aligned}$$

$$\begin{aligned}\text{Total power} &= (\text{Power required for water evaporation})/(\text{Proportion of energy}\\&\quad \text{used for water evaporation}) = (16\ \text{kW})/0.2 = 80\ \text{kW}\end{aligned}$$

So the oven will need approximately 80 kW of power for the analysed baking process. This basic procedure can also be applied to the situation frequently encountered by food engineers: estimating the production rate or energy distribution for different products given a particular oven with specified power.

Example A3: Operation of a Tunnel Oven

An important operating variable of a continuous tunnel oven is the speed of the conveyor band, which is determined by the baking time and the length of the baking zone. Suppose that bread loaves of cylindrical shape with radius 3 cm will be baked in a tunnel oven having a baking zone of 20 m length. If the target value for surface lightness is selected as $L^* = 75$, Figure 26.8 together with the length of the baking zone determine the speed of the conveyor band. An operating map for the specified product and oven is shown in Figure 26.A2 for a wide range of baking conditions.

Although this is a simple example, it is important to note how the operating variables should be changed according to product specifications. In this case, the more intensive the baking process (high values of oven temperature and/or heat transfer coefficient), the higher the speed of the conveyor band, and the lower the residence

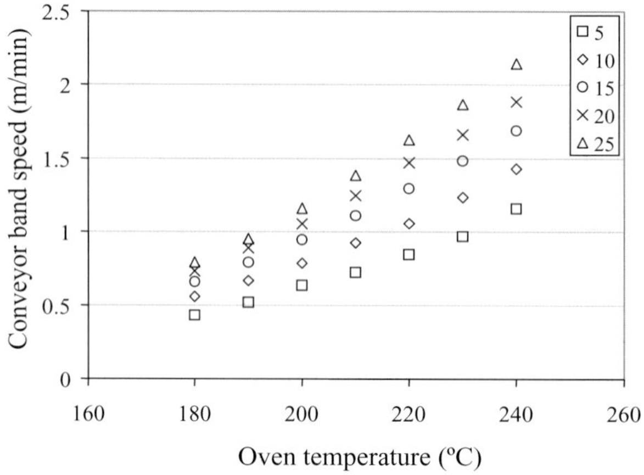

Figure 26.A2 Variation of speed of the conveyor band in a tunnel oven (baking zone length, 20 m), as a function of oven temperature for different values of heat transfer coefficient (symbols, $W{\cdot}m^{-2}{\cdot}K^{-1}$) obtained by process simulation (see Figure 26.8). The end point of baking corresponds to $L^* = 75$ for bread with radius 0.03 m.

time of product. Finally, if the speed of the conveyor band is not controlled accurately, unacceptable products will be obtained.

Example A4: Thermal Insulation of an Oven

Insulation of the oven is an important aspect of baking equipment design, since a well-insulated oven has minimum energy losses. Since the surrounding ambient of an oven is at room temperature, a heat flux is established from the baking chamber (or combustion chamber) through the oven walls towards the exterior of the equipment, producing energy losses. Suppose a batch oven with a total area of $20\,m^2$ (sum of all walls) and 70 kW power can operate at a maximum temperature of 350 °C with a heat transfer coefficient of $25\,W{\cdot}m^{-2}{\cdot}K^{-1}$ in the baking chamber.

Rockwool ($k_{ins} \approx 0.06\,W{\cdot}m^{-1}{\cdot}K^{-1}$) is commonly used as insulating material in oven construction; a thick layer of this material is placed between the internal and external walls of the ovens. Consider that a maximum energy loss of 5% has been established for the described equipment, i.e. 3.5 kW or $175\,W{\cdot}m^{-2}$. We need to determine the thickness of the insulating layer required for this design situation (Figure 26.A3).

Firstly, we consider that heat flux occurs by steady-state conduction through plane walls in series. Then we assume negligible thermal resistance of internal and external walls of the oven, due to the high conductivity of steel or similar construction materials. Therefore, the heat flux per unit area can be expressed as (Bird *et al.*, 2006):

$$q = \frac{T_{\text{oven}} - T_{\text{room}}}{1/h_{\text{oven}} + \delta/k_{ins} + 1/h_{\text{room}}} \tag{26.7}$$

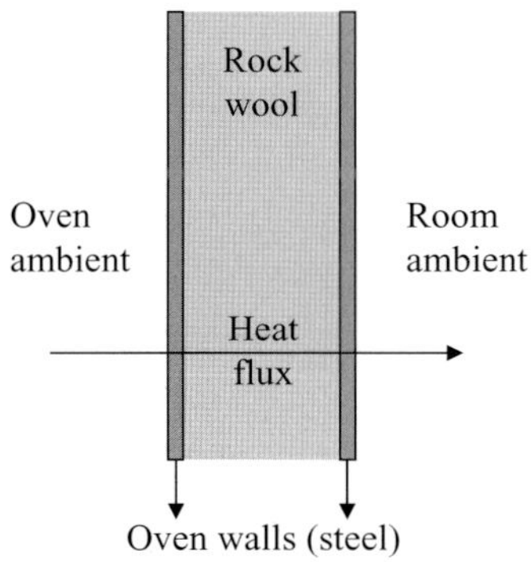

Figure 26.A3 Schematic description of heat flux through oven walls in a baking oven.

If the room temperature is 30 °C and the corresponding heat transfer coefficient is $10\,W{\cdot}m^{-2}{\cdot}K^{-1}$, the thickness of the insulating layer δ will be approximately 100 mm. This procedure can also be applied to the selection of an insulating material given a fixed thickness, e.g. for insulation replacement due to a change in power supply.

Nomenclature

a_w Water activity
C_p Specific heat ($J{\cdot}kg^{-1}{\cdot}K^{-1}$)
D Water (liquid or vapour) diffusion coefficient of product ($m^2{\cdot}s^{-1}$)
h Heat transfer coefficient ($W{\cdot}m^{-2}{\cdot}K^{-1}$)
k Thermal conductivity ($W{\cdot}m^{-1}{\cdot}K^{-1}$)
k_g Mass transfer coefficient ($kg{\cdot}Pa^{-1}{\cdot}m^{-2}{\cdot}s^{-1}$)
L^* Lightness
P Water vapour pressure (Pa)
q Heat flux per unit area ($W{\cdot}m^{-2}$)
R, r Radius (m)
RH Relative humidity (%)
T Temperature (K)
t Time (s)
W Water (liquid or vapour) content ($kg{\cdot}kg^{-1}$)

Greek Symbols

α Degree of starch gelatinisation
ε Emissivity
ρ Density ($kg{\cdot}m^{-3}$)
σ Stefan–Boltzmann constant ($5.67 \times 10^{-8}\,W{\cdot}m^{-2}{\cdot}K^{-4}$)

Subscripts

∞ ambient
s solid or surface
sat saturated

References

Abdullah, M.Z. (2008) Quality evaluation of bakery products. In: *Computer Vision Technology for Food Quality Evaluation* (ed. D.-W. Sun). Academic Press, Burlington, VT, pp. 481–522.

Ahrné, L., Andersson, C.-G., Floberg, F., Rosén, J. and Lingnert, H. (2007) Effect of crust temperature and water content on acrylamide formation during baking of white bread: steam and falling temperature baking. *LWT Food Science and Technology* 40: 1708–1715.

Anishaparvin, A., Chhanwal, N., Indrani, D., Raghavarao, K.S.M.S. and Anandharamakrishnan, C. (2010) An investigation of bread-baking process in a pilot-scale electrical heating oven using computational fluid dynamics. *Journal of Food Science* 75: E605–E611.

Bird, R.B., Stewart, W.E. and Lightfoot, E.N. (2006) *Transport Phenomena*, 2nd edn. John Wiley & Sons, New York.

Boulet, M., Marcos, B., Dostie, M. and Moresoli, C. (2010) CFD modeling of heat transfer and flow field in a bakery pilot oven. *Journal of Food Engineering* 97: 393–402.

Broyart, B. and Trystram, G. (2002) Modelling heat and mass transfer during the continuous baking of biscuits. *Journal of Food Engineering* 51: 47–57.

Demirekler, P., Sumnu, G. and Sahin, S. (2004) Optimization of bread baking in a halogen lamp–microwave combination oven by response surface methodology. *European Food Research and Technology* 219: 341–347.

de Vries, U., Sluimer, P. and Bloksma, A.H. (1989) A quantitative model for heat transport in dough and crumb during baking. In: *Cereal Science and Technology in Sweden* (ed. N.G. Asp). Lund University Chemical Centre, Sweden, pp. 174–188.

Dingstad, G.I., Egelandsdal, B., Mevik, B.-H. and Færgestad, E.M. (2004) Modelling and optimization of quality and costs on empirical data of hearth bread. *Lebensmittel-Wissenschaft und-Technologie* 37: 527–538.

Fennema, O.R. (1996) *Food Chemistry*, 3rd edn. Marcel Dekker, New York.

Hadiyanto, H., Asselman, A., van Straten, G., Boom, R.M., Esveld, D.C. and van Boxtel, A.J.B. (2007) Quality prediction of bakery products in the initial phase of process design. *Innovative Food Science and Emerging Technologies* 8: 285–298.

Hadiyanto, H., Esveld, D.C., Boom, R.M., van Straten, G. and van Boxtel, A.J.B. (2008a) Control vector parameterization with sensitivity based refinement applied to baking optimization. *Food and Bioproducts Processing* 86: 130–141.

Hadiyanto, H., Esveld, D.C., Boom, R.M., van Straten, G. and van Boxtel, A.J.B. (2008b) Product quality driven design of bakery operations using dynamic optimization. *Journal of Food Engineering* 86: 399–413.

Hadiyanto, H., Boom, R.M., van Straten, G., van Boxtel, A.J.B. and Esveld, D.C. (2009) Multi-objective optimization to improve the product range of baking systems. *Journal of Food Process Engineering* 32: 709–729.

Hasatani, M., Arai, N., Katsuyama, H. *et al.* (1991) Heat and mass transfer in bread during baking in an electric oven. In: *Drying '91* (eds A.S. Mujumdar and I. Filkova). Elsevier Science Publishers, Amsterdam, pp. 385–393.

Keskin, S.O., Sumnu, G. and Sahin, S. (2004) Bread baking in halogen lamp–microwave combination oven. *Food Research International* 37: 489–495.

Le Bail, A., Dessev, T., Jury, V., Zuniga, R., Park, T. and Pitroff, M. (2010) Energy demand for selected bread making process: conventional versus part baked frozen technologies. *Journal of Food Engineering* 96: 510–519.

Li, A. and Walker, C.E. (1996) Cake baking in conventional, impingement and hybrid ovens. *Journal of Food Science* 61: 188–191, 197.

Maroulis, Z.B. and Saravacos, G.D. (2003) *Food Process Design*. Marcel Dekker, New York.

Mirade, P.S., Daudin, J.D., Ducept, F., Trystram, G. and Clément, J. (2004) Characterization and CFD modelling of air temperature and velocity profiles in an industrial biscuit baking tunnel oven. *Food Research International* 37: 1031–1039.

Mondal, A. and Datta, A.K. (2008) Bread baking: a review. *Journal of Food Engineering* 86: 465–474.

Mottram, D.S., Wedzicha, B.L. and Dodson, A.T. (2002) Acrylamide is formed in the Maillard reaction. *Nature* 419: 448–449.

Norton, T. and Sun, D.-W. (2006) Computational fluid dynamics (CFD): an effective and efficient design and analysis tool for the food industry. A review. *Trends in Food Science and Technology* 17: 600–620.

Purlis, E. (2010) Browning development in bakery products: a review. *Journal of Food Engineering* 99: 239–249.

Purlis, E. (2011) Bread baking: technological considerations based on process modelling and simulation. *Journal of Food Engineering* 103: 92–102.

Purlis, E. and Salvadori, V.O. (2007) Bread browning kinetics during baking. *Journal of Food Engineering* 80: 1107–1115.

Purlis, E. and Salvadori, V.O. (2009a) Bread baking as a moving boundary problem. Part 1: Mathematical modelling. *Journal of Food Engineering* 91: 428–433.

Purlis, E. and Salvadori, V.O. (2009b) Modelling the browning of bread during baking. *Food Research International* 42: 865–870.

Purlis, E. and Salvadori, V.O. (2009c) Bread baking as a moving boundary problem. Part 2: Model validation and numerical simulation. *Journal of Food Engineering* 91: 434–442.

Purlis, E. and Salvadori, V.O. (2010) A moving boundary problem in a food material undergoing volume change: simulation of bread baking. *Food Research International* 43: 949–958.

Scanlon, M.G. and Zghal, M.C. (2001) Bread properties and crumb structure. *Food Research International* 34: 841–864.

Sevimli, K.M., Sumnu, G. and Sahin, S. (2005) Optimization of halogen lamp–microwave combination baking of cakes: a response surface methodology study. *European Food Research and Technology* 221: 61–68.

Sluimer, P. and Krist-Spit, C.E. (1987) Heat transport in dough during the baking of bread. In: *Cereals in a European Context* (ed. I.D. Morton). Ellis Horwood, Chichester, UK, pp. 355–363.

Sommier, A., Chiron, H., Colonna, P., Della Valle, G. and Rouillé, J. (2005) An instrumented pilot scale oven for the study of French bread baking. *Journal of Food Engineering* 69: 97–106.

Stadler, R.H., Blank, I., Varga, N. *et al.* (2002) Acrylamide from Maillard reaction products. *Nature* 419: 449–450.

Sumnu, G., Sahin, S. and Sevimli, M. (2005) Microwave, infrared and infrared-microwave combination baking of cakes. *Journal of Food Engineering* 71: 150–155.

Therdthai, N., Zhou, W. and Adamczak, T. (2002) Optimisation of the temperature profile in bread baking. *Journal of Food Engineering* 55: 41–48.

Therdthai, N., Zhou, W. and Adamczak, T. (2003) Two-dimensional CFD modelling and simulation of an industrial continuous bread baking oven. *Journal of Food Engineering* 60: 211–217.

Therdthai, N., Zhou, W. and Adamczak, T. (2004) Three-dimensional CFD modelling and simulation of the temperature profiles and airflow patterns during a continuous industrial baking process. *Journal of Food Engineering* 65: 599–608.

Wagner, M.J., Lucas, T., Le Ray, D. and Trystram, G. (2007) Water transport in bread during baking. *Journal of Food Engineering* 78: 1167–1173.

Wählby, U. and Skjöldebrand, C. (2002) Reheating characteristics of crust formed on buns, and crust formation. *Journal of Food Engineering* 53: 177–184.

Wong, S.-Y., Zhou, W. and Hua, J. (2007) CFD modeling of an industrial continuous bread-baking process involving U-movement. *Journal of Food Engineering* 78: 888–896.

Xia, B. and Sun, D.-W. (2002) Applications of computational fluid dynamics (CFD) in the food industry: a review. *Computers and Electronics in Agriculture* 34: 5–24.

Zanoni, B., Peri, C. and Bruno, D. (1995) Modelling of starch gelatinization kinetics of bread crumb during baking. *Lebensmittel-Wissenschaft und-Technologie* 28: 314–318.

Zareifard, M.R., Boissonneault, V. and Marcotte, M. (2009) Bakery product characteristics as influenced by convection heat flux. *Food Research International* 42: 856–864.

27

Membrane Separation and Design

Rohit Ruhal and Bijan Choudhury

Introduction

In recent years, the use of membranes in the food processing industry has become widespread, with major applications reported in the dairy and beverage industries. In the 1960s, the dairy industry was the first to implement membrane technology and the development of newer membrane materials and design has resulted in wider utilization of membranes in this sector. Concentration and fractionation are two major areas of interest to the membrane specialist in the dairy industry (Daufin *et al.*, 1998). The use of membranes in the dairy industry has resulted in better-quality products with minimum waste generation. In the case of concentration, membrane separation processes replaced thermal processes, resulting in energy efficiency. Similarly, fractionation with membranes resulted in minimum waste generation and recycling of byproduct. A major motivation for the acceptance of membrane processes in the food processing industry is their energy efficiency, retention of food quality, and minimum waste generation (Bruschke, 1995). In recent years it has been reported that membranes can also be used for cold sterilization of liquid food materials, with improved shelf-life and retention of flavor and nutrients. Membrane technology has also been more openly accepted in the beverage industry and has improved the quality of product along with better economics (Memtec Ltd, 1998). Some of the current examples

Handbook of Food Process Design, First Edition. Edited by Jasim Ahmed and Mohammad Shafiur Rahman.

include water desalination and purification and cold sterilization of beverages; recovery and fractionation of proteins from whey; clarification of fruit juice, wines and beer; removal of bacteria from water; effluent treatment for removal of heavy metals and organic materials; separation of oil and water emulsions; and removal of volatile organic compounds from air.

Membrane technology can be subdivided into four different processes depending on the size of membrane pore: microfiltration, ultrafiltration, reverse osmosis, and nanofiltration. Microfiltration is generally used for separation of macromolecules and particles that range in size from 1 to 10 μm. Ultrafiltration separates macromolecules by molecular mass, generally from 5 to 500 kDa. A membrane with a cut-off of 5 kDa can retain 90% of macromolecules with molecular mass of 5 kDa. Ultrafiltration finds wider application in whey protein concentration and fractionation, while reverse osmosis finds application in the desalination of water. Nanofiltration is a recent addition to membrane technology and its area of application lies between that of ultrafiltration and reverse osmosis.

The material comprising the membrane plays an important role and today commercial membranes are mostly made of polymers or inorganic material. Although inorganic membranes have the advantage of strength and steam sterilizability, their cost plays an important role in their suitability in commercial processes. A number of membrane modules are used in the commercial food processing industry: hollow-fiber module, spiral wound module, plate and frame module, and tubular module. Among these, the spiral wound module has the highest specific area per unit volume but has limited application with a solution containing suspended particles. In the process flow-sheet, any membrane module can be adapted without much difficulty and the requirement for less sophisticated control also favors easy replacement of traditional separation with membrane technology. However, the phenomenon of concentration polarization (see section Microfiltration) results in low acceptability of membrane technology in continuous processes. Thus design parameters have a huge influence on the successful implementation of membrane processes in the commercial food processing industry. This chapter discusses the issues involved in the design and implementation of membrane technology in food processing operations.

Process Flow-sheet for Membrane Operation

A membrane process can be represented by the flow-sheet shown in Figure 27.1. In commercial membrane processes, permeate or retentate can be a product and, depending on the requirement, the operation flow-sheet needs to be modified. In most commercial processes, tangential filtration is carried out, where flow of liquid is parallel to the membrane surface. To improve the filtration rate, high cross-flow velocity is maintained and this results in reduced thickness of the concentration polarization layer, which limits flux through the membrane.

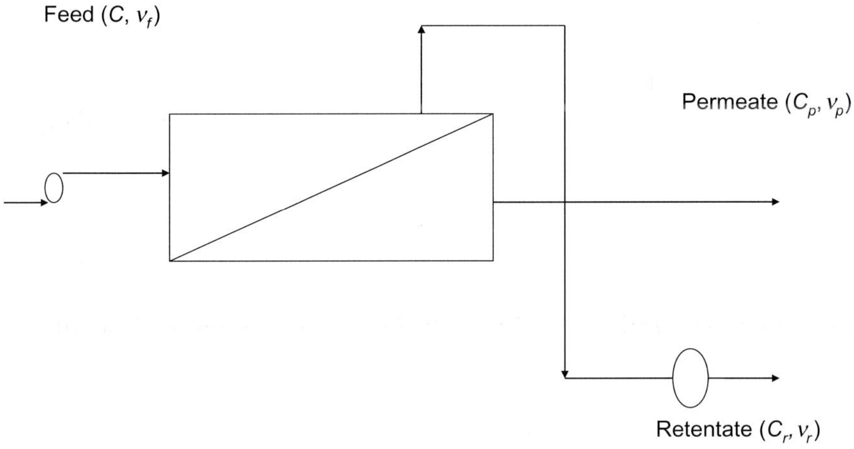

Figure 27.1 Process flow-sheet for membrane separation process.

Material Balance

If a protein is concentrated in a membrane having feed concentration C in a batch operation, then we can write:

$$v_r + v_p = v_f \tag{27.1}$$

$$C_r v_r + C_p v_p = C v_f \tag{27.2}$$

where v_r is volumetric flow rate of retentate ($m^3 \cdot s^{-1}$), v_p volumetric flow rate of permeate ($m^3 \cdot s^{-1}$), v volumetric flow rate of feed ($m^3 \cdot s^{-1}$), C_r concentration of protein in retentate ($g \cdot L^{-1}$), C_p concentration in permeate ($g \cdot L^{-1}$), and C concentration of protein in solution ($g \cdot L^{-1}$). Now if we define concentration factor (F_c) and recovery factor (Q) as follows:

$$F_c = C_r / C \tag{27.3}$$

$$Q = vP / vF \tag{27.4}$$

then we can write

$$F_c = \frac{1}{1-Q}\left[1 - \frac{C_p Q}{C}\right]$$

which correlates concentration factor and recovery factor. Depending on the desired concentration, the process can be shut down at any point. However, due to the formation of the concentration polarization layer, there is a limit on the achievable final concentration of protein. In the case of a semi-continuous process, part of the retentate can be recycled to maintain a pseudo-steady flux across the membrane.

Basic Theoretical Principle, Membrane Operation Mode and Membrane Materials

Transmembrane pressure is the major driving force for the movement of solution across a membrane. Because of its larger size, the solute is retained by the membrane; as the process continues, the concentration of solute increases and this mass of solute is deposited on the membrane. Therefore there is a concentration gradient on the feed side of the membrane, with maximum concentration at the membrane surface. This build-up of solute cake (in membrane filtration) or solute layer (in ultrafiltration) results in higher membrane resistance and thus decline in permeate flux across the membrane. The flow of liquid is tangential to the membrane surface, and this flow attempts to reduce the thickness of the cake. However, the increase in solute concentration may also increase viscosity, which may affect transmembrane flux. Even if a pseudo-steady state is achieved, operation may not continue for long as there is a possibility of irreversible fouling of the membrane and eventual shutdown of the operation. Further cleaning and back-flushing of the membrane may result in restoration of the original flux. For solutes that are soluble, the concentration adjacent to the surface may reach the level of insolubility and precipitation of solute may result in higher membrane resistance and decline in permeate flux. In recent years, efforts have been made to reduce fouling by optimizing process parameters, design parameters, and membrane materials.

Diffusion-based membrane processes include diafiltration and pervaporation, where concentration difference across the membrane is the driving force. Membrane separation is generally operated in batch or continuous modes. In batch mode, the volume of solution decreases and concentration of solute increases with time. However, there is an upper limit on the maximum concentration of solute achievable with membrane concentration processes. In batch mode, steady state is never achieved and this mode of operation is suitable for smaller-scale filtration. In continuous mode, the concentration of solute in retentate remains constant. This mode of operation is suitable for large-scale membrane separation processes. In other modes of operation, bleeding of retentate is carried out to maintain a constant level of solute concentration in the feed. By maintaining a constant concentration, steady-state flux can be achieved at a constant transmembrane pressure. Commercially, batch mode is more adaptable on the process flow sheet.

Membrane materials include natural polymers like cellulose acetate, synthetic polymer membranes, and composite or ceramic membranes. The first commercial membranes used in reverse osmosis and ultrafiltration were composed of cellulose acetate, which provides high permeate flux and is easy to manufacture but with the disadvantages that it breaks down at high temperature and is sensitive to chemicals. Recently, polymer membranes made from polyamide, polyacrylonitrile, nylon, and polysulfones have been used, especially in ultrafiltration, and these have better chemical resistance. Ceramic or composite membranes made from carbon, zirconium oxide, or alumina have the advantage of cleaning and sanitation due to their inert nature.

Membrane Modules

Four membrane modules are generally used in industrial applications: tubular module, hollow fiber module, spiral wound module, and plate and frame module.

Tubular membranes are located inside a tube (Figure 27.2). Tubular membranes have a diameter of about 5–15 mm. Because of the size of the membrane surface, plugging of tubular membranes is not likely. A drawback of tubular membranes is that the packing density is low, which results in high capital cost per module. Ceramic membranes are generally assembled in tubular modules. Some of the advantages of tubular module are easy replacement of membrane and cleaning of membrane surface and no need for any costly pretreatment of feed. Disadvantages of this module are high energy cost per unit of liquid treated, low surface area per unit volume, and necessity of high liquid flow rate to maintain turbulence in the module.

A hollow fiber module (Figure 27.3) consists of a large number of small-diameter fibers sealed at the ends with epoxy resin, polyurethanes and silicone rubber. Membranes are asymmetric in nature, the feed side having lower pore size, and are self-supporting. They have very high specific area per unit volume compared with tubular membranes, up to 30 000 $m^2 \cdot m^{-3}$. It is preferable to use hollow fiber modules with clean solution, and thus are most used in gas separation or pervaporation. Hollow fiber membranes have low diameter so the chances of plugging are very high.

Plate and frame modules (Figure 27.4) were initially used for commercial purposes but their low specific area is unattractive. Plate and frame modules have been adapted into a conventional filter press. In this module two membranes are sandwiched with

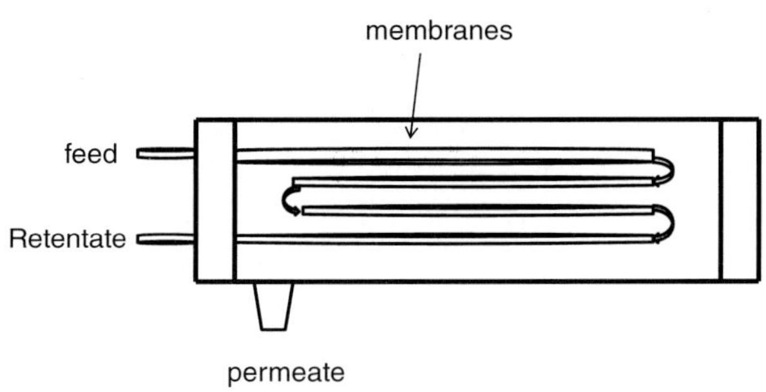

Figure 27.2 Tubular module. (From EPA Research & Development.)

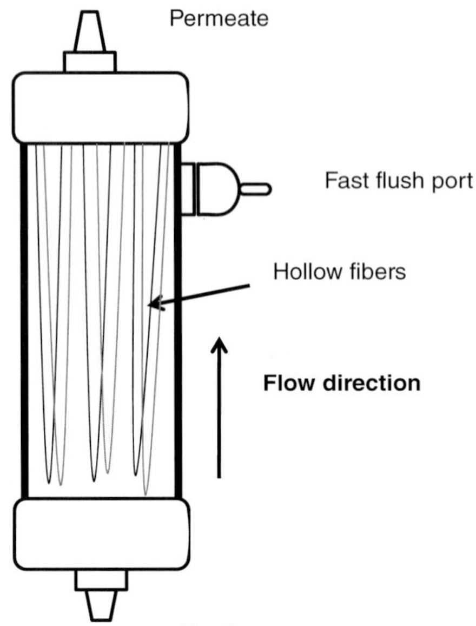

Figure 27.3 Hollow fiber module. (From EPA Research & Development.)

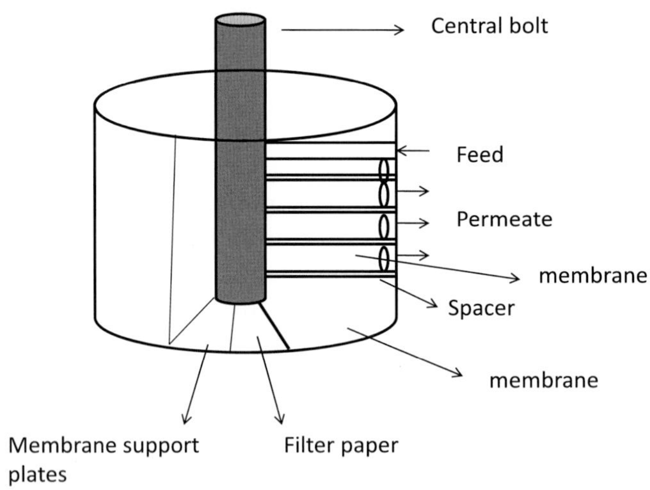

Figure 27.4 Plate and frame module. (From USEPA, 2005. Membrane Filtration Guidance Manual (EPA 815-R-06-009). Office of Water.)

the feed sides facing each other. A spacer is placed between the feed and permeate sides. To achieve the desired membrane area, a number of such sets are placed in parallel. Membrane replacement and cleaning is easier and there is no need for costly pretreatment of feed.

Spiral wound module (Figure 27.5) is an adaptation of plate and frame module, where membrane sets are wrapped around a central line for collecting permeate. It has the

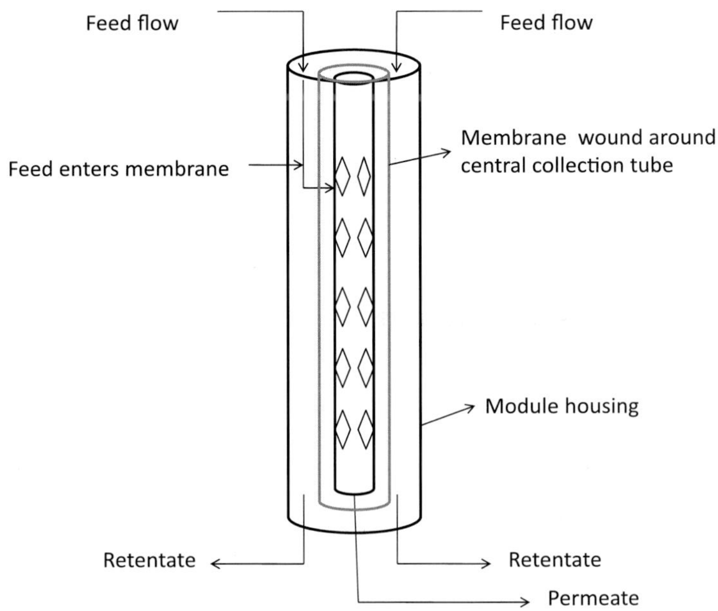

Figure 27.5 Spiral wound module. (From EPA Research & Development.)

advantages of compact structure with high pressure durability. It has the highest specific surface area among all the modules but feed needs to be free of particles.

Types of Membrane Process

A membrane is a selective barrier and this selectivity is important for separation (Reis and Zydney, 2007). Various selectivity parameters such as diffusivity, shape, size, electrostatic charge, physicochemical interactions, volatility, and solubility can be used to influence product concentration and removal of solute from solvent (e.g desalting, fractionation, gas separation, pervaporation). Membrane materials include organic polymers like polysulfone, polyether sulfone, cellulose acetate, polyamides, and polyacrylonitrile and inorganic material such as borosilicate glass, pyrolyzed carbon, and zirconia carbon. The driving forces of membrane processes include concentration gradient, transmembrane pressure, chemical potential, osmotic pressure, electric field, magnetic field, and partial pressure. For ultrafiltration, microfiltration, and reverse osmosis, membranes are classified on the basis of molecular size of pores and transmembrane pressure, whereas pervaporation is based on separation driven by partial pressure and selectivity that depends on solubility and diffusivity of species in the membrane.

Reverse Osmosis

Reverse osmosis is essentially a pressure-driven membrane diffusion process for separating dissolved solvents. During the process, feed solution at high pressure is passed over the feed side of the membrane. The operating pressure is kept higher than the osmotic pressure of the feed solution so that water will flow from the more concentrated solution to the more dilute solution through the membrane. Reverse osmosis is used for desalination of seawater. The attractive features of the process are that there is no no requirement of phase change and it is a relatively low-energy requiring process. Normally, solvent (e.g. water) flows across a semipermeable membrane from the dilute solution to the more concentrated solution until equilibrium is reached. Thus applying high pressure at the high-concentration side will cause this process to reverse. This application of high pressure at the feed side results in solvent migrations from the concentrated solution, leaving a higher concentration of solute. In process application, the waste stream containing salts flows past the membrane, resulting in solvent permeation (e.g. water) through the membrane and the remaining solutes (e.g. organic or inorganic components) get concentrated on the feed side of the membrane.

Most reverse osmosis membranes are made of polymers like cellulose acetate, and aromatic polyamides, and are rated at 96–99% NaCl rejection. These membranes are generally of two types: asymmetric or skinned membranes and thin-film composite membranes. The support material is commonly polysulfones, while the thin film is made from various polyamines and polyurea. Reverse osmosis membranes have the smallest pore structure, with pore diameter ranging from about 5 to 15 Å (0.5–1.5 nm). The extremely small size of these pores allows only the smallest organic molecules and uncharged solutes to pass through the semipermeable membrane along with the water. Greater than 95–99% of inorganic salts and charged organics get rejected due to the repulsion of similar charge present at the membrane surface at process conditions.

The major advantage of reverse osmosis is requirement of low power for recovery of salts and chemicals from process effluents. No latent heat of vaporization is involved for affecting separations; the main energy requirement is a high-pressure pump. It also requires low floor space (compact high-capacity units) as compared to traditional unit operation. The major problem of reverse osmosis is susceptibility to fouling. This is caused due to particulate and colloidal matter that is present at high concentration at the feed side of the membrane surface. Fouling of membranes by soluble components or colloids present in solution causes failures. Thus pretreatment is used to remove particulate matter from the feed water. A system operating at a high permeate flux is more likely to experience high fouling rates and thus require intermittent chemical cleaning. Low operational temperature range is another limitation of reverse membrane process. For cellulose acetate membranes, the temperature operation range is 18–30 °C; while lower temperatures result in decreased fluxes and higher temperatures increase the rate of membrane hydrolysis. These membranes are not chemically inert.

Strongly acidic or basic solutions, strong oxidizing agents, solvents, and other organic compounds should be avoided as these agents cause dissolution of the membrane. Some concentrated solutions may have high initial osmotic pressures and these solutions are economically unattractive for reverse osmosis process.

Ultrafiltration

Ultrafiltration is a membrane separation process that operates under a pressure gradient and it can fractionate components of a liquid with respect to their solvated size, molecular weight, and structure. Even electrostatic charges play a role in fractionation of solutes. Ultrafiltration membranes have a pore diameter of 0.01–0.1 mm and typically remove high-molecular-weight substances, colloidal materials, and organic and inorganic polymeric molecules, whereas microfiltration membranes remove all bacteria and some viral contamination, even though viruses are smaller than the pores of a microfiltration membrane (Bailey *et al.*, 2000). This is because viruses can attach themselves to bacterial bio film. Ultrafiltration membranes can also filter viruses. Permeability generally depends on pore size but other factors like electrostatic charge on membrane surface and solute also play a role. Ultrafiltration also depends on applied pressure, nature of feed material, temperature of operation, feed concentration, and flow rate of feed stream across the membrane.

Ultrafiltration membranes are produced from variety of synthetic polymers and have high thermal stability and chemical resistivity but are unable to tolerate harsh cleaning chemicals. Polyethersulfone is most preferably used as an ultrafiltration membrane material as it has high rigidity, creep resistance, and good thermal and dimensional stabilities. The choice of membrane is guided by molecular weight cutoff, chemical and thermal resistance. However in recent times membrane surface properties like hydrophilicity have also become important criteria for selection of ultrafiltration membrane. Hollow fiber, flat sheet cassettes, spiral wound cartridges, and tubular modules are generally used as the ultrafiltration module in the commercial process. Some of the advanced techniques, especially for protein, use charged ultrafiltration membranes, electro-ultra filtration, and ultrafiltration in the presence of an ultrasonic field. Ultrafiltration is most suitable for concentrating protein in whey, cheese-making process, reduction of salt concentration in process stream, and fractionation of whey protein. Ultrafiltration membrane finds widespread application in the dairy industry.

Microfiltration

Microfiltration is a technique that allows concentration of the liquid containing insoluble particle, and the liquid passing through the filter is known as microfiltrate. Pore diameter of microfiltration membrane ranges from 0.1 to 10 mm. Cross-flow microfiltration is used for applications like removal of bacteria and micelle casein. The material properties of the membrane determine its efficiency. Previously polysulfone and polycarbonate were used but their performances were not satisfactory. Polyethersulfone is now the standard material of microfiltration membranes used in biological applications. The pore size of commercially available microfiltration ceramic membranes ranges from 0.1 to 20 mm. Ceramic membranes are most advantageous in the dairy and food industry because of their strong mechanical resistance and wide pH tolerance. The operation of microfiltration leads to a higher solute concentration closer to the membrane surface than in the bulk feed stream; this phenomenon is called concentration polarization and is caused by diffusive back flow of solute to the bulk feed.

Microfiltration needs to be started with care as it can avoid rapid fouling. In the dairy industry, microfiltration membranes are first filled with warm water and then milk (Saboya and Maubois, 2000). Other dairy applications of microfiltration include removal of microorganisms from milk to produce the sterilized raw material used to prepare fluid milks, cheeses or long- storage dairy products. The reason for this cold sterilization process is that milk collected from farms is contaminated with various pathogenic bacteria.

Microfiltration offers an economically interesting alternative to heat treatment. In microfiltration, skim milk heated to 50 °C is circulated at a velocity of $7.2\,m \cdot s^{-1}$ along the membrane with uniform transmembrane pressure around 0.5 bar (0.05 MPa). Another application is the selective separation of micellar casein and selective fractionation of globular milk fat. Microfiltration is also used in microbial fermentation (upstream and downstream processing of biomass) and concentration of other metabolites in batch and continuous process operations.

Microfiltration is commonly used to recover macromolecules and retain suspended colloidal particles and is being integrated into both upstream and downstream processes. Large numbers of MF applications are reported, including removal of small molecules from bigger protein molecules, clarification of suspensions for cell harvesting, and sterilization of heat sensitive liquids to remove viruses and bacteria (Saxena *et al.*, 2009). Another important role is protein separation and purification, for example recovery of extracellular proteins from biomass. Other applications in the food industry include processed meat, nutritional beverages, infant formulas and dairy products.

Pervaporation

Liquid mixtures can be separated on nonporous polymer membranes by partial evaporation. The procedure is called pervaporation because the substance crossing the membrane changes phase state (Marriot *et al.*, 2001). The substance encounters the membrane as a liquid and leaves it as a vapor, vacuum pressure on the permeate side causing the vapor to desorb. The membrane used in pervaporation can be hydrophilic or organophilic depending on its composition and is generally nonporous. Pervaporation is used in the wine industry because it can produce a more concentrated alcohol permeate and thus the wine becomes less dense compared to membrane filtration process where only the consistency of the wine product changes slightly. In pervaporation, temperature is a very important parameter and choice of optimum temperature is the main goal for optimum process design. In comparison with other membranes, pervaporation has higher temperatures leading to higher flux and less membrane surface demand, which makes it economical because it increases efficiency and production.

Flux Equations

Membrane filtration can be divided into two zones, a pressure dependent zone and a pressure-independent zone (Ghosh, 2006). In the pressure-dependent zone transmembrane pressure is the driving force, whereas in the pressure-independent zone mass transfer is the only limiting mechanism. In the case of a clean membrane, with pure solvent, permeate flux (J_v) is only dependent on transmembrane pressure and this can be represented as:

$$J_v = \frac{\Delta p}{R_m} \tag{27.5}$$

where Δp is transmembrane pressure (Pa) and R_m membrane resistance (Pa·s·m^{-1}). In the case of ultrafiltration of solute, the additional resistance of the concentration polarization layer (R_c, Pa·s·m^{-1}) reduces permeate flux compared with pure solvent flux at the same hydrodynamic conditions and transmembrane pressure:

$$J_v = \frac{\Delta p}{(R_m + R_c)} \tag{27.6}$$

In the case of the concentration polarization layer, driving force is also reduced due to osmotic pressure (Π) of concentrated solute around the membrane, so permeate flux can be represented as:

$$J_v = \frac{(\Delta p - \Delta \Pi)}{(R_m + R_c)} \tag{27.7}$$

The van't Hoff equation gives the osmotic pressure of such a concentration polarization layer as:

$$\Pi = MRT \tag{27.8}$$

where Π is osmotic pressure (kPa), M molar concentration (mol·m^{-3}), T absolute temperature (K), and R universal gas constant ($8.314\,\text{J·mol}^{-1}\text{·K}^{-1}$). In the case of ultrafiltration, the concentration polarization layer is a problem that leads to increased resistance and lower permeate flux, so the above equation is limited to a restricted range of pressures in the pressure-dependent zone. Hence in the pressure-independent zone with concentration polarization layer resistance, mass transfer in ultrafiltration can be described by a diffusion equation (assuming pseudo-steady state) (Earle and Earle, 2004):

$$J_v = k' \ln(C_w / C) \quad (27.9)$$

where C_w and C are the solute concentrations at the interface and in the bulk solution respectively. The effect of the physical properties of the material can be predicted from known relationships for the mass transfer coefficient k' (m·s^{-1}). This equation has been found to predict, with reasonable accuracy, the effect on the mass transfer of changes in the physicochemical properties of the solution. This is done through well-established relationships between diffusivity D, mass transfer coefficient k', and other properties such as density (ρ), viscosity (μ), and temperature (T) giving:

$$(k'd/D) = a[(dv\rho/\mu)m] \times (\mu/\rho D)n \quad (27.10)$$

$$\text{or } (\text{Sh}) = (k'd/D) = a(\text{Re})m(\text{Sc})n \quad (27.11)$$

where d is the hydraulic diameter, (Sh) the Sherwood number ($k'd/D$), (Sc) the Schmidt number ($\mu/\rho D$), and a, m, and n are constants. In the case of microfiltration, permeate flux is represented by the following equation:

$$J_v = \frac{\Delta P}{\mu(R_m + R_C)} \quad (27.12)$$

where μ is permeate viscosity (Pa·s), R_m membrane resistance (m^{-1}), and R_C cake resistance (m^{-1}) = rV_c/A_m where r is specific cake resistance (m^{-2}):

$$r = [180(1-\varepsilon)/\varepsilon^3](1/d_s^2) \quad (27.13)$$

where ε is cake porosity, d_s particle diameter, and A_m membrane area (m^2). In the case of dialysis, the driving force is the concentration difference across the membrane and this can be represented as:

$$J_v = \Delta C/R_m \quad (27.14)$$

In the case of dialysis, two solvents pass through the feed side and permeate side (dialyzate side) and thus overall flux of solute depends on membrane resistance (R_m), feed side mass transfer resistance (R_f) and dialyzate side mass transfer resistance (R_d):

$$J_v = \frac{\Delta C}{R_m + R_f + R_d} \tag{27.15}$$

Sieving Coefficient in Ultrafiltration

It has long been predicted that the ratio of the diameter of the macromolecule to the diameter of the membrane pore is the only criterion for selecting an ultrafiltration membrane with appropriate molecular weight cut-off. However, it was experimentally observed that sieving of macromolecules through an ultrafiltration pore is also dependent on environmental parameters (pH, ionic strength, etc.), hydrodynamic parameters, and membrane properties. Thus to select an appropriate membrane, a new parameter, intrinsic sieving coefficient, was introduced and is defined as (Ghosh, 2006):

$$\psi = C_p/C_w \tag{27.16}$$

The difficulty in determining C_w has resulted in the introduction of a new parameter known as apparent sieving coefficient, ψ_a, defined as:

$$\psi_a = C_p/C \tag{27.17}$$

It was observed that sieving coefficient and apparent sieving coefficient are related by the following mathematical expression:

$$\ln[\psi_a/(1-\psi_a)] = J_v/k + \ln[\psi/(1-\psi)] \tag{27.18}$$

Thus by determining the apparent sieving coefficient at various values of permeate flux (J_v), intrinsic sieving coefficient (ψ) and mass transfer coefficient can be calculated for a particular solute. However, it is worthwhile noting that ψ needs to be constant in the permeate flux range studied. It has been reported that ψ also depends on J_V, but at low permeate flux and high permeate flux is independent of J_v. Apparent sieving coefficient can be used for determining the selectivity of a membrane for fractionation of two macromolecules:

$$\alpha = \psi_{a1}/\psi_{a2} \tag{27.19}$$

where ψ_{a1} is apparent sieving coefficient of more transmitted solute and ψ_{a2} is apparent sieving coefficient of less transmitted solute.

Mode of Operation

Membrane operation is generally carried out in two modes: batch operation (Figure 27.6) and continuous mode (Figure 27.7). In large-scale operations, continuous mode

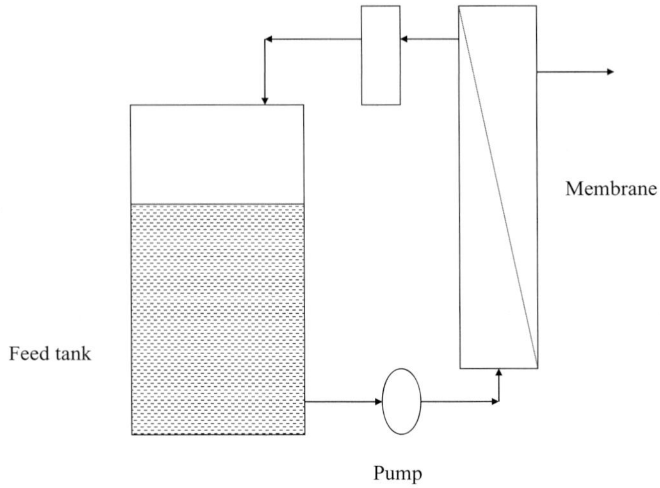

Figure 27.6 Batch operation.

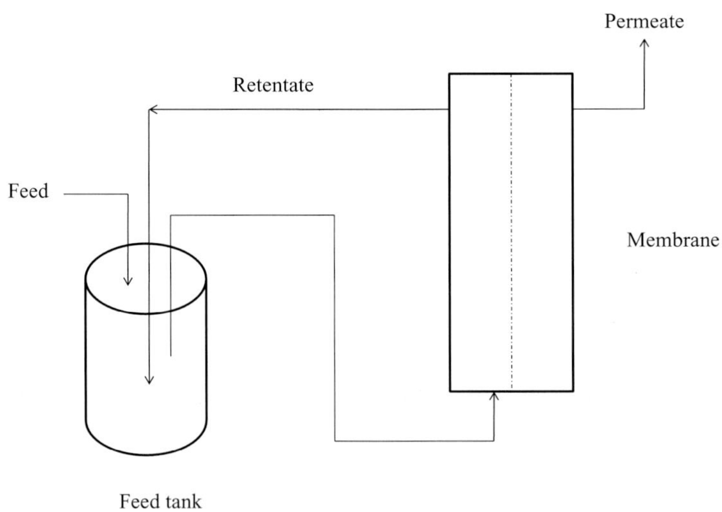

Figure 27.7 Continuous mode.

of operation is important as low membrane area in batch operation cannot provide desired permeate flux.

Design of Membrane

The calculation of membrane area is the desired objective using known variables like recovery factor, feed flow and composition, temperature, transmembrane pressure, permeate flow, permeability for solvent flow, and rejection factor. The concentration polarization layer effect is not considered in this method. The steps involved in determining the area of membrane are as follows.

1. Calculation of the concentration of the permeate stream:

$$C_p = 2C(1-R)/(2-Q) \tag{27.20}$$

assuming nearly complete rejection of solute, where R is rejection of solute.

2. Calculation of the flow of retentate (v_r) from the global balance Equation 27.1 and Q value:

$$v_r = v_f(1-Q)$$

3. Calculation of the concentration of the retained stream (C_r) from Equation 27.2 using previously determined C_p.
4. Calculation of the mean concentration of the concentrated solution:

$$C_m = (v_r C_r + v_f C)/(v_r + v_f) \tag{27.21}$$

5. Determination of the concentration of the permeate stream:

$$C_p = C_m(1-R) \tag{27.22}$$

6. Steps 3, 4, and 5 are repeated using this value of the concentration of the permeate stream to obtain a new value of such concentration. If this final value coincides with the C_p value calculated in Equation 27.19, then the process is finished; otherwise steps 3, 4, and 5 are repeated in an iterative way until the values coincide.
7. Calculation of the osmotic pressure of the concentrated solution using as concentration value that of the weighed mean and Equation 27.8.
8. Determination of the solvent flux from the following equation:

$$J_v = \varepsilon(\Delta P - \Delta\Pi)$$

where ε is permeability of membrane for solvent.

9. Calculation of the membrane area needed to carry out the separation is given by $A = v_p/J_v$.

Another approach for designing membranes has been proposed by Lombardi and Moresi where the concentration polarization layer is included (Ibraz and Barbosa-Cánovas, 2002).

Fouling of Membrane in Ultrafiltration and Microfiltration

Fouling of the ultrafiltration/microfiltration membrane occurs due to of adsorption of the membrane surface and leads to increased resistance to flow, which can reduce filtration flux rate and process economy, efficiency and recovery. Membrane fouling refers to the irreversible alteration in membrane properties caused by feed stream components. Fouling may be due to formation of a gel layer due to concentration polarization, adsorption or deposition, and pore blocking. Generally, fouling is characterized by measuring clean water flux decline at constant pressure.

Fouling in microfiltration membranes, particularly those used for purification of proteins, has been overcome with the introduction of microsieves, which have well-structured morphology and controlled porosity that improve separation behavior and which show high feed flow rates. These lead to improved permeate flux, low operating pressures, and improved process economics. Other anti-fouling methods include surface modification by chemical methods to, increase surface hydrophilicity, coating, and graft polymerization over membrane surface. The membrane material can also influence fouling: in ultrafiltration, surface-modified polyvinylidene fluoride (PVDF) membranes have been found to have low fouling properties.

Cleaning and Sanitation

Cleaning of membrane equipment depends on the nature of the membrane material and availability of high quality water. Water should be free of suspended colloidal particles or microorganisms that are likely to cause irreversible blockage. A commonly used membrane cleaning cycle includes the following steps: (i) rinse with heating at 50 °C; (ii) alkaline cleaning with ternary detergents at 65–70 °C for 20 min; acid cleaning with HNO_3 (0.5%) for 20 min at 50 °C followed by rinse with water at 50 °C.

Cost

Membrane separation is a highly cost-effective unit operation as as observed with the production of apple juice, in which the gross profit and return on investment were higher using a membrane process with or without pervaporation or reverse osmosis in comparison with conventional operation. The annual manufacturing cost was also low in the membrane process while payback time was also low compared with conventional operation (Cuperus, 1998).

Applications

Membranes in their various configurations, including spiral, hollow fiber, and plate and frame, are used throughout the food industry. The two main benefits are improved production process and the recovery of valuable products (Slater, 1989). Improved production processes include consistent high-quality permeate and retentate, reduced operating cost, low maintenance, low pressure drop, chemical resistance, and longevity of membrane (Strathmann, 1981). Microfiltration is used for clarification and sterilization, both of which are the best options compared to centrifugation and heat sterilization. Thus microfiltration can be used for the efficient removal of suspended particles, fat globules, and high-molecular-weight proteins. Ultrafiltration is used for fractionation, concentration, and purification, for example concentrated skimmed milk has more calcium and protein content compared to conventionally treated milk, while in fruit juices ultrafiltration removes polysaccharides, microbes, viruses, and yeasts that make the final product unstable. Reverse osmosis has the advantages of high-quality product separation without heat damage, reduced waste treatment volume, and hence lower capital investment. Unique applications of membranes in the food industry include concentration of oil emulsions, recovery of proteins, clarification of wines, concentration of fruit juices, and treatment of food processing waste products. Potential applications of various membrane systems in food processes are wide, for example in sugar mills ultrafiltration has helped in the production of colorless sugar and improved the sugar extraction process.

Membrane-based techniques have a fundamental role in the separation and purification of proteins. Ultrafiltration, microfiltration, and nanofiltration have greatest application in protein purification. Microfiltration is used mainly for the recovery of extracellular protein produced by the fermentation of bacteria but also for removal of microbes and viruses in preparation of therapeutic proteins. Much effort has been made to improve mass transfer in ultrafiltration and microfiltration membranes. Some examples include rotating disk filters, cylindrical taylor vortex devices, and conically shaped rotors (Saxena *et al.*, 2009). A material like polyethersulfone has been designed for selective permeation and protein recognition/purification and is used in the purification of soya protein. Electrically enhanced membrane filtration (EMF) is an advanced technique that utilizes an electrical field for filtration. An electric field has been found to have good impact on protein transmission and selectivity (Nagarale *et al.*, 2006). In one approach ultrafiltration in the presence of ultrasonic waves is used to separate proteins, which can solve the problem of polarization and fouling (Saxena *et al.*, 2009).

Glucose syrup can be used as a food ingredient for biscuits, soft drinks, jams, and food preservatives. Ultrafiltration is used to eliminate suspended solids and proteins from wheat hydrolysate, and does not even affect dextrose equivalent in filtered sugars (Pinelo *et al.*, 2009). In the case of vegetable proteins and their various derivatives used for a variety of beverages, processing with membranes has the advantages of reduction of transport cost and better manufacture of intermediate products. Membranes are also used for separation and harvesting of enzymes and microorganisms. In fermentation, cells are harvested by centrifugation but for economic reasons the harvesting must be continuous, which is implemented using continuous cross-flow membrane filtration.

In the beer industry, membranes are used to increase brightness (Le, 1987) and clarity, while filtration removes solids and particles from malt, hops and yeast. Rough beer is filtered in order to eliminate yeast and colloidal particles responsible for haze (Takács *et al.*, 2007). Microfiltration of beer can ensure microbiological stability in a single operation. In the milk and dairy industry, membranes can be used to control microbial growth instead of heat treatment. Membrane filtration is used for globular milk fat fractionation (Daufin *et al.*, 2001). Microfiltration is used in an integrated protein extraction process for the manufacture of micelle casein products and whey protein isolates. Membranes are also used in the treatment of dairy effluents and waste stream. Membrane processes should contribute to process simplification, improve competitiveness, and be environmentally beneficial.

In juice production, use of membranes leads to better process economics and product quality. Enzymatic membrane reactor is used for the clarification and maceration of juice, reverse osmosis for concentration, and pervaporation for recovery of aromas (Girard and Fukumoto, 2000). By choosing a suitable membrane configuration, it is also possible to achieve direct pulping and clarification in one step. Membranes are also used for agrowaste treatment. Membranes are used in processes especially developed for the treatment of persistent compounds like chlorobenzene, tetrachloroethylene, and trichloroethylene. In water these components are mostly present at a very low concentrations, which makes conventional processing inefficient.

Membrane systems are extensively used throughout the various biological and chemical industries to control variety of products (Pal *et al.*, 2009). These membrane processes are successful because they are effective and economically implemented at the very large scale required for industrial application. A combination of different membrane processes gives interesting benefits not achieved by single operation. On an industrial scale, the integration of various membrane operations can be beneficial because it leads to low environmental impacts, low energy consumption, and high-quality products. Dairy proteins, food additives, nutraceuticals and therapeutics processing by membrane integration can be advantegous. An example is shown in Figure 27.8.

Conclusions

Most of the membrane technology used in the food industry originated in the dairy industry. Ultrafiltration is the most widely used process in the dairy industry. Microfiltration is used for removing microorganisms and casein micelles. Membranes play an important role in the fractionation of milk. Current membrane processes are limited by their low capacity due to fouling. Microfiltration reduces the amount of bacteria and spores without affecting the taste and flavors of milk. It can be used for the pretreatment of skim milk for production of raw milk cheeses. Ultrafiltration has found major application in the production of cheese. Whey contains dissolved salts,

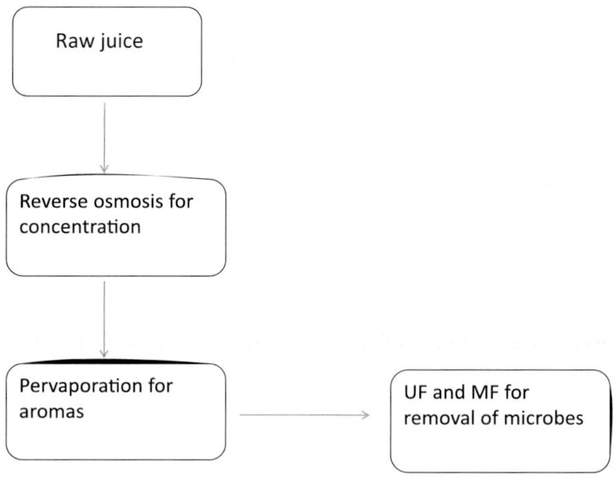

Figure 27.8 Integration of various membrane systems for fruit juice processing.

sugars and 25% of the original protein, and thus ultrafiltration of whey allows these additional foods to be recovered. Another role of membrane separation is in fractionation of buttermilk, an application where both ultrafiltration and microfiltration can be used. In the food and beverage industry, safety, microbiological quality, functionality, texture, and flavors can be improved by membrane processes. Clarification of fruit juices can be done by microfiltration or ultrafiltration and deacidification by electrodialysis. Significant developments have been made in membrane technology in the last decade but the potential is limitless and there is a great need for research in all these aspects. Lower cost, lower fouling rates, and lower energy consumption will improve commercialization.

References

Bailey, A.F.G., Barbe, A.M., Hogan, P.A., Johnson, R.A. and Sheng, J. (2000) The effect of ultrafiltration on the subsequent concentration of grape juice by osmotic distillation. *Journal of Membrane Science* 164: 195–204.

Bruschke, H. (1995) Industrial application of membrane separation processes. *Pure and Applied Chemistry* 67: 993–1002.

Cuperus, F.P. (1998) Membrane process in agro-food: state of the art and new opportunities. *Separation and Purification Technology* 14: 233–239.

Daufin, G., Escudier, J.P., Carrère, H., Bérot, S., Fillaudeau, L. and Decloux, M. (2001) Recent and emerging application of membrane processes in food and dairy industry. *Transactions IChemE* 79A: 1–14.

Earle, R.L. and Earle, M.D. (2004) Unit Operations in Food Processing, 4th edn. New Zealand Institute of Food Science and Technology. Available at www.nzifst.org.nz/unitoperations

Ghosh, R. (2006) *Principles of bioseparations engineering*. World Scientific, Singapore: pp 199–242.

Girard, B. and Fukumoto, R.L. (2000) Membrane processing of fruit juices and beverages: a review. *Critical Reviews in Biotechnology* 20: 109–175.

Ibraz A., and Barbosa-Cánovas, G.V. (2002) *Unit operations in food engineering*. CRC Press, Boca Raton, FL.

Le, M.S. (1987) Recovery of beer from tank bottoms with membranes. *Journal of Chemical Technology and Biotechnology* 37: 59–66.

Marriott, J.I., Sørensen, E. and Bogle, I.D.L. (2001) Detailed mathematical modelling of membrane modules. *Computers and Chemical Engineering* 25: 693–700.

Memtec Ltd (1998) The use of cross flow membranes to clarify and stabilize beer. *Filtration and Separation* 35: 860–861.

Nagarale, R.K., Gohil, G.S. and Shahi, V.K. (2006) Recent developments on ion-exchange membranes and electro-membrane processes. *Advances in Colloid and Interface Science* 119: 97–130.

Pal, P., Sikder, J., Roy, S. and Giorno, L. (2009) Process intensification in lactic acid production: a review of membrane based processes. *Chemical Engineering and Processing* 48: 1549–1559.

Pinelo, M., Jonsson, G., Anne, S. and Meyer, A.S. (2009) Membrane technology for purification of enzymatically produced oligosaccharides: molecular and operational features affecting performance. *Separation and Purification Technology* 70: 1–11.

Reis, R.V. and Zydney, A. (2007) Bioprocess membrane technology. *Journal of Membrane Science* 297: 16–50.

Saboya, V.L. and Maubois, J.-L. (2000) Current developments of microfiltration technology in the dairy industry. *Lait* 80: 541–553.

Saxena, A., Tripathi, B.P., Kumar, M. and Shahi, V.K. (2009) Membrane-based techniques for the separation and purification of proteins: an overview. *Advances in Colloid and Interface Science* 145: 1–22.

Slater, C.S. (1989) Instruction in membrane separation processes. *Journal of Membrane Science* 44: 265–272.

Strathmann, H. (1981) Membrane separation processes. *Journal of Membrane Science* 9: 121–189.

Takács, L., Vatai, G. and Korány, K. (2007) Production of alcohol free wine by pervaporation. *Journal of Food Engineering* 78: 118–125.

28

Food Frying Process Design

Ferruh Erdogdu and T. Koray Palazoglu

Introduction

Frying is a significant and highly versatile process in the food industry due to significant sales and availability of a wide range of fried products (Saguy and Dana, 2003). There are different shapes and forms of fried products available in the market, including chips, crisps, doughnuts, and battered and breaded foods (Bouchon, 2006). Significant increase in market volume of fried products is the result of customer preference for the taste, texture and appearance of fried food products (Rimac-Brncic *et al.*, 2004).

Frying is an old and popular process that originated and developed around the Mediterranean area (Bouchon, 2006). It is defined as a process of immersing a food product in edible oil or fat at a high temperature, typically 150–200 °C (Farkas *et al.*, 1996). The most common temperature range for the process is between 170 and 190 °C (Singh, 1995), where a higher rate of heat is transferred to the product and water is evaporated through the surface (Hubbard and Farkas, 1999). Shallow frying and deep frying are ways of producing the fried product under atmospheric conditions, and deep fat frying is mostly preferred in bulk production of fried food products. In addition to traditional atmospheric frying, where shallow and deep frying options are used, vacuum frying is another type of applied frying process (Garayo and Moreira, 2002). Vacuum frying, which has less adverse effects on oil quality (Shyu *et al.*, 1998), is

Handbook of Food Process Design, First Edition. Edited by Jasim Ahmed and Mohammad Shafiur Rahman.

considered to be an option for lower oil content fried products, with desirable texture and flavor characteristics (Garayo and Moreira, 2002).

Besides traditional atmospheric and vacuum frying processes, high-pressure frying, where the boiling point of the fried product is determined by the pressure generated by vapor released from the product, is another approach for frying food (Rao and Delaney, 1995). Pressure frying, leading to higher heat transfer rates (Erdogdu and Dejmek, 2010) is also known to yield juicier and tender products (Ballard and Mallikarjunan, 2006; Innawong *et al.*, 2006). This may be explained by the faster crust formation during the earlier stages of a frying process due to higher heat transfer rates occurring via the effect of higher heat transfer coefficients (Erdogdu and Dejmek, 2010). In pressure frying processes, either nitrogen gas or the vapor released from fried food to the medium is used to build up the desired pressure level. Increase in pressure level is also noted to lead to more moisture retention in the fried product (Ballard and Mallikarjunan, 2006; Sahin and Sumnu, 2009). Microwave frying has also been applied recently to reduce the frying time via additional heat generation in the fried product and to decrease oil absorption by the product during frying, with its disadvantages of nonuniform heating of the product and deterioration of oil at high temperatures (Sahin and Sumnu, 2009).

Frying comprises simultaneous heat transfer (convection between oil and surface of the food product and conduction within the food product) and mass transfer (removal of water vapor, formed via the effect of high temperature, as a result of concentration and pressure gradients). It is also considered to be a moving boundary problem, where moisture in the form of vapor is removed at the evaporation front, and this front moves towards the center as the frying process continues (Costa *et al.*, 1999; Singh, 2000). In this intensive heat transfer process, significant evaporation and spatially varied pressure generation are produced as a function of porous structure of the fried product (Ni and Datta, 1999). The core region of the product stays at the boiling temperature of water for a longer period without increasing beyond this, while crust region temperature rises as the moisture content decreases due to rapid evaporation at the surface (Figure 28.1). As also explained by Singh (1995), core region temperature is unaffected by oil temperature because of the presence of the boundary at the evaporating front of moisture. On the other hand, the influence of oil temperature on the rise in temperature of the crust section is clearly seen even in the early stages of a frying process (Figure 28.1).

Frying is a process of cooking and drying through contact with hot oil, and it is intended to make the product more palatable (Oreopoulou *et al.*, 2006). Frying creates unique flavors and textures in the fried product (Farinu and Baik, 2005), and typical crust and core regions are distinctive of fried products. The crust region, nonexistent prior to frying, develops on the porous surface and becomes thicker as the core region shrinks with frying time (Farinu and Baik, 2005). The crust of the fried product provides mostly texture, color, and flavor while the core region provides the bulk of the moisture and nutrients, contributing to the texture (Farkas *et al.*, 1996; Farinu and Baik, 2005). A soft and moist interior along with a crispy crust provides increased

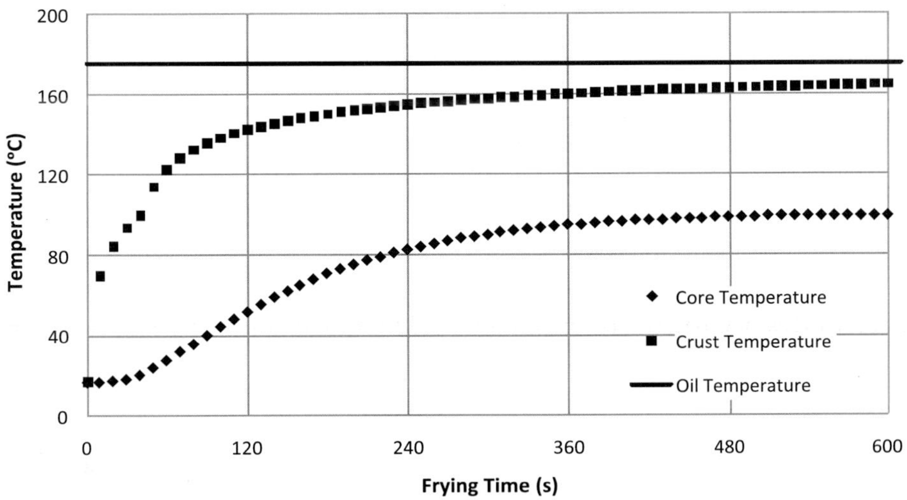

Figure 28.1 Center and surface temperature changes of a fried potato.

palatability to the fried product (Innawong *et al.*, 2006). Texture formation is noted to be the result of modification of viscoelastic properties and mechanical strains (Parkash and Gertz, 2004) while desirable surface color and flavor are due to the activation of nonenzymatic browning, Maillard reactions (Achir *et al.*, 2008; Alvis *et al.*, 2009), and toasting and caramelization (Alvis *et al.*, 2009) triggered by the simultaneous heat and mass transfer process. In addition to the formation of surface color and flavor, these reactions also contribute to the formation of potentially toxic and mutagenic products, such as acrylamide, furfurals, and polycyclic aromatic hydrocarbons (Achir *et al.*, 2008). Surface color and flavor formation might also become a significant process design and optimization criterion with the formation kinetics of these products of health concern (Erdogdu and Sahmurat, 2007). As stated by Achir *et al.* (2008), monitoring these reactions that occur above the boiling point of water (as in a frying process) at industrial scale can be used to control the formation of the undesirable substances, as applied in similar circumstances to coffee and peanut roasting (Davidson *et al.*, 1999; Hernandez *et al.*, 2008).

A frying process consists of four stages (Farkas *et al.*, 1996; Farinu and Baik, 2005; Oreopoulou *et al.*, 2006):

1. **Initial heating** (nonboiling phase of the frying process): the product surface is heated from its initial temperature to the boiling temperature of water at the given condition. Natural convection dominates heat transfer from the frying medium in this stage; this usually takes a short time and a negligible amount of water is lost.
2. **Surface boiling**: there is rapid loss of moisture from the surface with apparent bubble formation with nucleate boiling.

3. **Falling rate**: the longest stage, with temperature of the core region increasing to the boiling point and crust formation with its increasing thickness.
4. **Bubble end point**: the final stage of frying, describing the apparent end of moisture loss from the product.

Following initial heating and the start of surface boiling, bubbling increases to its maximum (Costa *et al.*, 1999). The following stage is characterized by a significant increase in heat transfer coefficient, where the surface conditions change from free convection to boiling with the start of crust formation. During this stage, random fluctuations in the temperature of the oil are associated with turbulent motion of the fluid, resulting in an increase in the convective heat transfer (Farinu and Baik, 2007). In the subsequent stages, heat transfer rate is decreased because the increase in the crust section (with its lower thermal conductivity, $0.1\,W{\cdot}m^{-1}{\cdot}K^{-1}$) serves as an insulating layer (Farkas and Hubbard, 2000) and mass transfer is reduced via the reduction in moisture removal (Farkas *et al.*, 1996).

For better process control and design and optimization purposes, it is important to determine the thermal parameters (thermal conductivity and specific heat of fried product) and physical parameters (shape, size, surface area, volume and density of food product) that dominate the frying process (an extensive review of thermal and physical properties of fried products is given by Mittal, 2009), because development of a mathematical model is required in order to determine first the temperature changes inside the product and then the formation of the health-concerning chemicals. Thus reducing the formation of these chemicals is the objective function while the required degree of frying for texture and flavor formation is the implicit constraint for a frying process within a certain time, with oil temperature as the explicit constraint. In addition, physical and chemical changes occurring in the fat or oil can also be used as optimization parameters since an economic and superior-quality fried product is produced under optimum frying conditions (Blumenthal and Stier, 1991).

According to Erdogdu and Dejmek (2010), coupled heat and mass transfer with bubbling make frying a difficult process to model and further optimize. Changes occurring in the oil create extra challenges when trying to optimize a given frying process for different objectives such as fried product and frying medium quality. In addition, the complex flow field around the fried product requires that the boundary conditions for the thermal transport equations utilize the convective heat transfer coefficient to characterize the interface between fluid (oil) and solid (fried product) (Hubbard and Farkas, 1999). The vigorous movement of bubbles across this solid–fluid interface causes considerable mixing in the oil (Oreopoulou *et al.* 2006) that affects the heat transfer coefficient by increasing turbulence across the interface (Costa *et al.*, 1999; Vitrac *et al.*, 2000). The heat transfer coefficient was reported to be the most significant parameter, besides the thermal and physical properties of the product itself, affecting surface temperature change and this needs to be controlled for when simulating reactions resulting in the formation of flavor and health-concerning chemicals during the frying process (Achir *et al.*, 2008). Hence, when modeling the simultaneous heat and mass transfer process during frying, the first step is to determine the characteristics of heat transfer and fluid flow across the interface (Hubbard and Farkas, 1999, 2000).

Based on the description of frying of food products, this chapter illustrates developments in the area of food frying from an engineering and food chemistry point of view and discusses possible ways to improve process design based on simultaneous heat and mass transfer phenomena.

Fried Products

The potato is probably the most popular food for frying (Rossell, 2001). Potatoes are used to make French fries and potato chips, the most popular fried food products in the world followed by fried chicken and fried fish. As indicated, poultry and seafood are also fried but commonly after being coated with a batter or breading. Once fried, this coating provides a crispy texture, a juicy interior, and aids in browning (Moreira *et al.*, 1999). Frying has also been used as an important processing step in the production of frozen coated (breaded or battered) products, which are usually par-fried prior to freezing and indeed require another frying step during final preparation.

Other widely consumed fried products include tortilla chips, which have become increasingly popular over the last two decades. Although frying is not the only cooking step in tortilla chip processing (it follows a baking step), the unique characteristics of tortilla chips only develop after the frying step. Many different dough products are also prepared by frying, of which doughnuts, produced on a large scale, are probably the most popular.

Quality Attributes of Fried Products

Frying is a process that imparts unique characteristics to foods, for example the desirable taste, crispy texture, and golden color of fried potato products are developed during the frying process. Maillard reaction between reducing sugars and amino acids is responsible for desired color and flavor formation. The quality of fried food products depends on the quality of the frying oil and type of fried product. In general, moisture and oil content, color and flavor, texture, nutritional properties and shelf-life stability of the food product are the characteristics that determine overall quality (Moreira *et al.*, 1999). Among these, moisture content is a significant factor for quality maintenance and shelf-life stability. Especially in industrial-scale fried potato chip production, on-line sensors are preferred to maintain quality (Brescia and Moreira, 1997). These sensors use the near-infrared reflectance (NIR) technique, and surface moisture is easily measured with careful calibration to determine the efficiency of the production line.

Color is the first sensory attribute perceived by consumers, and it is used to judge the acceptability of a food product. It is closely associated with food quality. If the

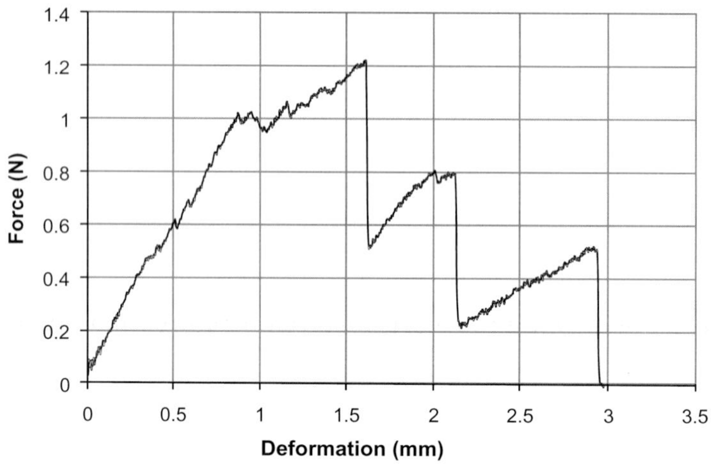

Figure 28.2 A typical force–deformation curve for potato chip.

color is unacceptable, the two other major quality attributes, flavor and texture, are unlikely to be judged at all (Claybon and Barringer, 2002). Color is not just a physical characteristic of the fried product but rather represents a range of input signals perceived by the consumer (Chen *et al.*, 2001). For example, in French fries, while a dark color might suggest poor quality, a golden yellow color indicates the use of proper methodology in the frying process. Although, on-line colorimeters are easily used for color measurement and control of the frying processes, appropriate sensory evaluation techniques should also be implemented to validate the instrumental results.

Texture is a multi-parameter characteristic of foods (Bourne, 1978) and is defined as the properties arising from structural elements and physiological senses (Szczesniak, 1963). Crispness, denoting freshness and high quality (Moreira *et al.*, 1999), describes the textural characteristic of fried food products. Crispness is the initial slope of the force–deformation curve obtained instrumentally. Figure 28.2 shows a typical force–deformation curve obtained during texture measurement of a potato chip sample. A highly jagged plot is desirable for a potato chip, as jaggedness (force fluctuations) is associated with crispy texture. Puncture tests are also reported to provide discrimination between the mechanical properties of the crust and can be used for products of different sizes and shapes since they are less sensitive to the cross-sectional geometry of the product (Moreira *et al.*, 1999).

The oil content of fried products is another parameter of concern, especially with the widespread anxieties about the Western diet. Frying oil becomes a major component of the final product, ranging from 35 to 45% by weight in potato chips. Although oil is known to reinforce the fried taste and pleasant feeling in the mouth (Keijbets, 2001), there has been a great deal of interest in reducing the fat content of fried products since a high-fat diet is known to increase the risk of obesity and cardiovascular disease. The need for alternative processes to produce low-fat fried products is increasing as health-conscious consumers become more concerned about the fat and calorie content of food. The American Heart Association and other health organizations have been encouraging consumers to limit their fat intake to less than 30% of total calories. However, producing low-fat fried products that also taste good poses a unique challenge. With regard to the oil absorption of fried products, there is a general understanding that oil uptake mostly occurs during the cooling stage of the process.

Fried products have public health significance not only for their high fat content but also because of their high level of acrylamide, a potential carcinogen that forms during the processing of certain foods at high temperatures and low moisture, conditions associated with frying, roasting and baking. Fried potato products such as potato chips and French fries are among the foods with the highest amounts of acrylamide, probably due to the relatively high levels of suspected acrylamide precursors (asparagine and reducing sugars) present in potato. Composition of the food product and frying conditions have been shown to influence the quality attributes as well as the formation of acrylamide. The main pathway of acrylamide formation in foods is linked to the Maillard reaction, a reaction that is also responsible for the development of the desirable characteristic color, texture, flavor, and aroma of fried products. Hence it is extremely important to determine the effect of factors related to composition and processing conditions on acrylamide formation during frying. In addition, monitoring the evolution of the Maillard reaction during a frying process might be helpful in controlling and optimizing the frying process.

Frying Oils

The oil is the most critical but also the most variable component of a frying system (Blumenthal and Stier, 1991) since each oil affects the fried food in a particular manner, and all are affected by heat, light, oxygen, hydrolysis, and components of the fried food (Blumenthal and Stier, 1991). A wide range of oils and fats, including vegetable oils (sunflower oil, soybean oil, cottonseed oil, canola oil, palm oil, almond and peanut oils), animal fats (animal tallow and lard) or a mixture of these, can be used in a frying process (Blumenthal and Stier, 1991; Bouchon, 2006). These range from nonhydrogenated refined fats and oils to hydrogenated products (Oreopoulou *et al.*, 2006). Among these, oils are mostly preferred because of their convenience in handling. As explained in the previous sections, oil uptake by fried products is a significant issue, especially with regard to health concerns. Most frying oil confined to the surface of the fried product is absorbed during the cooling stage. While Gamble and Rice (1987) suggest that vacuum formation due to steam condensation results in the absorption of oil in the cooling stage, Moreira and Barrufet (1998) explain this mechanism in terms of capillary forces. In addition to these theories, Ufheil and Escher (1996) proposed a hypothesis involving surface phenomena, where equilibrium between adhesion and drainage of oil is involved on removing the fried product from the oil. Type of oil (due to the effect of reduced interfacial tension via the formation of surfactants; Blumenthal and Stier, 1991) and quality of oil are expected to significantly influence oil absorption (Bouchon, 2006).

Selection criteria for frying oils are influenced by stability to oxidation (long frying stability), cost, flavor stability of fried product, long frying fluidity, smoke formation, low tendency to foam, low tendency to polymerize, and melting point in the case of fats (frequently the higher the melting point, the lower the tendency to oxidation) (Bouchon, 2006; Oreopoulou *et al.*, 2006). Among these, the most important factors, excluding cost, are inherent stability to oxidation, because of its direct effect on quality of the fried product, and smoke point, because of its effect on long-term use at higher frying temperatures. Sunflower oil is the most commonly used oil in frying due to its high smoke point and high content of linoleic acid, an essential multi-unsaturated fatty acid (Oztop *et al.*, 2007). In contrast, olive oil has oxidative stability, with its high content of oleic acid, but has a smoke problem because of its nontriglyceride constituents (Oreopoulou *et al.*, 2006).

Physical changes that occur in oils during the frying process include increased viscosity, darkening in color, and increased foaming with decrease in smoke point (Warner, 2004). In addition, because of the high temperatures used during the frying process, because components of the fried products leach into the oil, and because of thermal and hydrolytic breakdown of the oil on interaction with oxygen, the oil transforms into a mixture of different compounds. Eventually, this results in variation in the heat transfer characteristics of the oil due to changes in heat transfer coefficient. This variation is a function of the reduction in surface tension with formation of surfactants due to the increased contact between oil and fried product (Blumenthal and Stier, 1991). The increase in concentration of surfactants (soaps, phospholipids, monoglycerides, diglycerides, slightly polar thermal polymers, highly polar thermal polymers, etc.) results in increased degree of foaming and bubble size, leading to the accelerated oxidative changes (Blumenthal and Stier, 1991). As also stated by Blumenthal and Stier (1991), formation of polar materials also results in changes in the physical properties of the oil, for example increase in viscosity, thermal conductivity and specific heat and decrease in surface tension. All these changes affect the heat transfer coefficient, leading to changes in temperature distribution within the fried product with the formation of flavor and health-concerning components. Based on the formation of polymers and surfactants, an optimal quality fried product can be produced if the rate of polymer formation is limited and surfactant concentration is maintained in the range 35–75 ppm. This can be accomplished using active filters to remove surfactants and by establishing a uniform temperature gradient inside the fryer to control polymer formation (Blumenthal and Stier, 1991).

Based on these changes in the frying oils, it is extremely important to determine the variations in quality parameters of oils for better design and optimization of frying processes since proper monitoring would result in fried product of good quality (Lalas, 2009). Determination of the peroxide value and iodine value, changes in the fatty acid profile and smoke point, formation of conjugated dienes, polar compounds and polymers, changes in color and sensory evaluation are possible ways to follow change in quality of frying oils (Lalas, 2009). Among these, peroxide value can be conveniently used to monitor changes in frying oils in order to indicate the state of oil deterioration (Oreopoulou *et al.*, 2006).

Frying Equipment

A fryer is a pot or kettle with a means of heating the oil, but industrial-scale fryers might be somewhat larger sizes of batch (Figure 28.3) or continuous (Figure 28.4) construction. In batch fryers, used for small-scale production, oil can be heated directly by electrical resistance or by gas underneath the fryer itself. These fryers are capable of frying several hundred kilograms of product per batch (Gupta, 2009). In some installations, oil might also be heated in a shell and tube heat exchanger, with steam (180–200 psi) passing through the jacketed side of the heat exchanger (Gupta, 2009).

In large-scale processing conditions, continuous fryers are preferred: a frying vessel maintains the oil at the desired temperature, a conveyor feeds the product into the fryer and removes it at the opposite end, and an extraction system eliminates the fumes (Bouchon, 2006). In these systems, it is also possible to have multiple heating zones (multi-zone fryers) with their own temperature control units, enabling the use of optimal variable temperature control to improve product quality by applying

Figure 28.3 Batch fryer. (Courtesy of Heat and Control Company.)

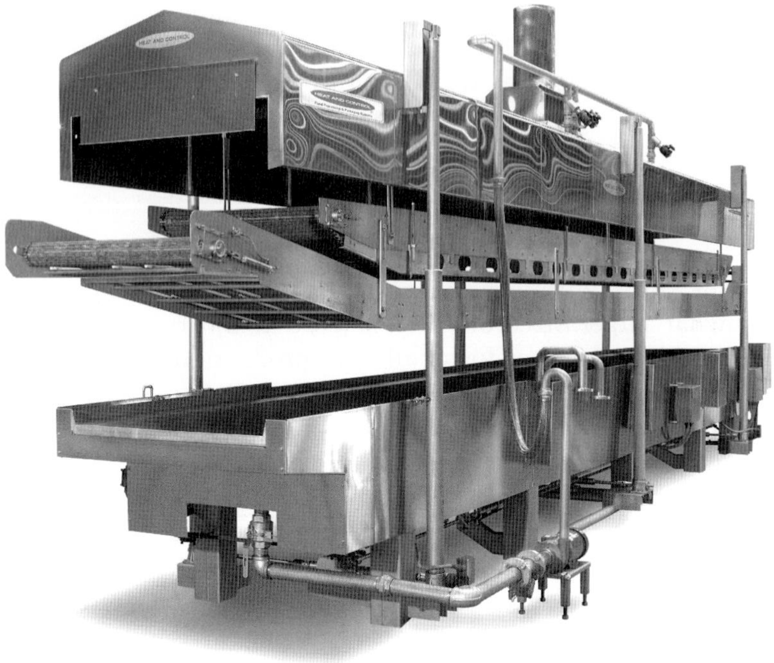

Figure 28.4 Continuous fryer. (Courtesy of Heat and Control Company.)

required optimization models. The heated oil might also enter these fryers at different entry points (Gupta *et al.*, 2004). In addition to multi-zone fryers, straight-through and horseshoe fryers are the other types of continuous fryers (Gupta, 2009).

In continuous fryers, oil absorbed by the fried product should be continuously replenished with fresh oil (oil turnover defined as the ratio of weight of oil in fryer to the weight of oil added per hour). Oil turnover represents the time required to replace the oil in the fryer, and a fast turnover is desired in order to maintain the quality of oil for longer processing conditions in continuous frying operations (Bouchon, 2006; Oreopoulou *et al.*, 2006). Gupta (2009) lists the oil turnover times for various fryers and fried products, for example 4–5 hours for extruded corn chips, 9.5–11 hours for potato chips. Based on fryer design, a turnover time of 5–12 hours is assumed to be normal in continuous production, and frying oil quality cannot be maintained with turnover times of 20 hours or more (Rossell, 2001).

In the calculation of turnover time, fryer capacity and oil content of the fried product are important parameters.

Example 1

In a fryer containing 2500 kg of oil for frying 1000 kg product per hour having 35% oil content, oil absorbed by the product and oil turnover time can be calculated as (Gupta, 2009):

$$\text{Oil absorbed by the product} = 0.35 \times 1000\ \text{kg/h} = 350\ \text{kg/h}$$

$$\text{Oil turnover time} = \frac{2500\ \text{kg}}{350\ \text{kg/h}} = 7.14\ \text{hours}$$

In addition to batch and continuous fryers, vacuum fryers (also counted as batch fryers) are used to produce chips of high quality with low levels of absorbed oil, since lower temperatures can be applied during the frying process. Vacuum fryers are preferred when frying fruits and vegetables. They are operated at pressures of 100 mmHg or less, where frying can be conducted at 121.1 °C (Gupta, 2009).

The important criteria when selecting or designing a suitable fryer are listed below (Gupta *et al.*, 2004; Gupta, 2009):

- desired product characteristics (texture, color and appearance);
- product buoyancy (directly affects conveyor design since it determines whether product sinks or floats in the oil);
- required production capacity;
- required thermal load, depending on moisture content of the feed and the finished product for the given production capacity;
- desired oil turnover time to minimize oil degradation during frying;
- oil filtration system and ease of fryer sanitation with maintenance of the system;
- direct or indirect mode of oil heating;
- technical support from the manufacturer.

Among these criteria, product buoyancy is a significant factor in selecting or designing a fryer. Different conveyor designs are available to handle the buoyancy of the food material. For example, for products that become buoyant very early in the frying process (e.g. potato chips), the frying system must include both main and submerger conveyors (Figure 28.5). On the other hand, a design similar to that shown Figure 28.6 should be used for products that undergo a buoyancy change during frying, such as

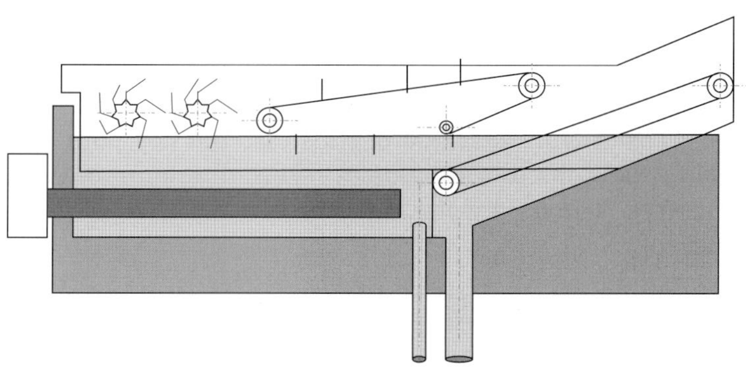

Figure 28.5 Fryer conveyor design for buoyant products. (Courtesy of Coat and Fry Technologies.)

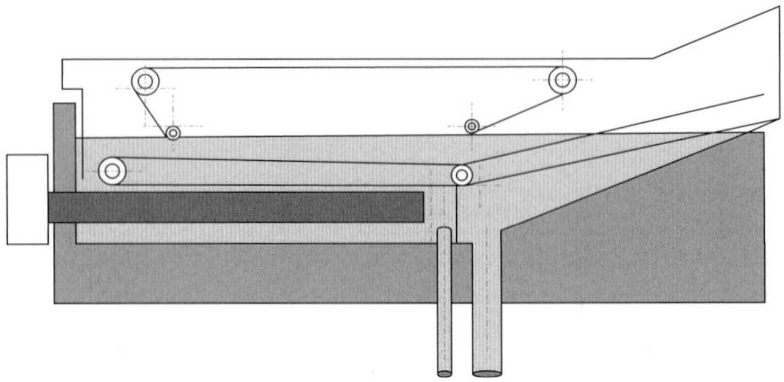

Figure 28.6 Fryer conveyor design for products that undergo a buoyancy change during frying. (Courtesy of Coat and Fry Technologies.)

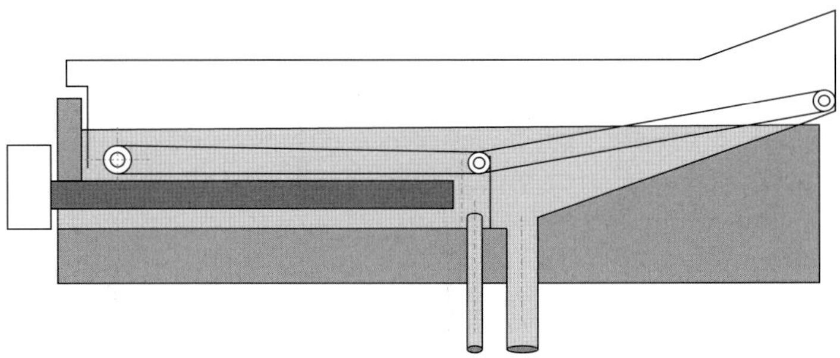

Figure 28.7 Fryer conveyor design for nonbuoyant products. (Courtesy of Coat and Fry Technologies.)

French fries. There are also products which naturally remain submerged throughout the frying process (e.g. breaded chicken); such products require only a main conveyor (Figure 28.7).

Fryer capacity is another important parameter and selection of a particular capacity depends on production volume required and various other factors including number of operating hours, required frying time, and frequency of fryer sanitation. Gupta *et al.* (2004) gives the following example to illustrate the key points (desired production rate, physical dimensions of the product, frying time required, and belt loading) in determining the fryer size for 1 hour of frying.

Example 2

Desired production rate, 1500 $kg{\cdot}h^{-1}$
Recommended product loading, 7.25 $kg{\cdot}m^{-2}$

$$\text{Cook area required} = \frac{1500\ \text{kg/h}}{7.25\ \text{kg/m}^2} \times \frac{1\ \text{h}}{60\ \text{min}} = 3.45\ \text{m}^2$$

Therefore fryer length (assuming 1 m of fryer width) = $3.45\,\text{m}^2/1\,\text{m} = 3.45\,\text{m}$.

Fryers also include exhaust systems to collect and dispose of the vapor generated during frying and hot gases in a safe and sanitary manner. Exhaust systems are significant components of a fryer. Because a sudden drop in oil temperature is observed after immersion of the food, temperature recovery is another important parameter when choosing a fryer for optimum process conditions. Also, the frying process deposits carbon on the fryer's heating elements (Blumenthal and Stier, 1991), resulting in longer recovery times and leading to increases in the immersion time with possible loss of quality. A uniform temperature distribution within the fryer is also important for ensuring uniform temperature distribution inside the fried product, since temperature distribution directly affects the convective heat transfer coefficient.

Energy consumption during frying is another important problem from an engineering and economic point of view (Rywotycki, 2002, 2003). Research on this problem is limited, and a model to design and describe a food drying system to lower thermal energy consumption is required for process optimization purposes. With this objective in mind, Rywotycki (2002) lists the required tasks as the design and construction of a fryer insulated with low thermal conductivity material, formulation of a mathematical model for calculating the thermal power consumption as a function of frying temperature, and validation of the model with actual situations. The required mathematical model should include (Rywotycki, 2003):

- heat requirement for heating and evaporating water from raw product;
- heat requirement for raw product heating and heating supplementary fat/oil;
- heat transmitted to the environment through fryer casing and fryer ventilation system;
- heat radiated into the environment.

Heat and Mass Transfer during Frying

Heat Transfer and Heat Transfer Coefficient

Frying is a very efficient heat transfer process not only because of high heat transfer coefficient values associated with the hot frying oil, but also because of the hydrodynamic conditions set up around the food material. During frying, heat and mass transfer occur simultaneously. As heat is transferred from hot oil to the surface of the food material, moisture is transferred from the interior to the surface. Heat transfer to the food material results in an increase in its internal energy, which in turn leads to the vaporization of water (boiling phase). Rapid escape of water as vapor bubbles creates turbulence around the food material, further enhancing the rate of heat transfer.

The heat transfer coefficient is a measure of how fast heat transfer occurs between a surface and a fluid. As in every convective heat transfer process, the heat transfer coefficient is a critical processing parameter that plays a vital role in frying. Although dimensionless convective heat transfer correlations exist for determining heat transfer coefficients, the boiling phase of frying requires the use of experimental methods. A common practice for determining the heat transfer coefficient during thermal processing of foods is to use the lumped capacitance method, where highly conductive metals are used. However, for thermal processes such as frying during which simultaneous heat and mass transfer take place, the lumped capacitance method may fail to yield an accurate heat transfer coefficient.

Singh (1995) and Sahin *et al.* (1999) have pointed out that use of a metal transducer to determine heat transfer coefficient during frying will yield a value quite different from what would be obtained by using the actual food product. When a cold lumped system is immersed in a hot medium, the object absorbs heat from its surroundings very rapidly and as a result the oil around the object cools substantially, resulting in an increase in its density. This creates a downward flow and hence convection currents around the object at a strength that would never exist for a food particle during frying. This may lead to over-calculation of the heat transfer coefficient, especially at the beginning of the frying process. As opposed to the cool-down effect of the lumped system, bubbles are present around the food being fried. As moisture within the food comes in contact with the surrounding hot oil, it suddenly boils and forms bubbles. This bubble formation may in fact become very violent and alter the conditions around the food. The presence of bubbles was reported to enhance the heat transfer process by creating turbulence (Costa *et al.*, 1999; Vitrac *et al.*, 2000; Budzaki and Seruga, 2005). This effect of bubbles on transport properties definitely needs to be taken into account when determining heat transfer coefficient.

Several researchers (Costa *et al.*, 1999; Ngadi and Ikediala, 2005) have determined heat transfer coefficients during deep-fat frying of foods using the lumped capacitance approach. Ngadi and Ikediala (2005) used drum-shaped models made of aluminum to estimate the heat transfer coefficient during deep-fat frying of chicken drums. They obtained heat transfer coefficient values ranging from 67 to 163 $W{\cdot}m^{-2}{\cdot}K^{-1}$ for different combinations of oil temperature, chicken drum size and oil viscosity. Costa *et al.* (1999) determined heat transfer coefficients during frying of French fries at 140 and 180 °C using a steel piece and employing the lumped system approach. They placed potato strips in the fryer before immersing the steel piece to mimic the hydrodynamic conditions of the actual frying process. They found that the heat transfer coefficient reached a maximum value of 443 $W{\cdot}m^{-2}{\cdot}K^{-1}$ at 140 °C and 650 $W{\cdot}m^{-2}{\cdot}K^{-1}$ at 180 °C. At the end of frying, as bubbling ceased, lower heat transfer coefficient values were reported (353 $W{\cdot}m^{-2}{\cdot}K^{-1}$ at 140 °C and 389 $W{\cdot}m^{-2}{\cdot}K^{-1}$ at 180 °C). The heat transfer coefficient was also determined from the surface temperature data obtained during frying of a potato strip at 140 °C. The maximum heat transfer coefficient value was found to be around 600 $W{\cdot}m^{-2}{\cdot}K^{-1}$, greater than that obtained with the steel piece at the same oil temperature (443 $W{\cdot}m^{-2}{\cdot}K^{-1}$).

Table 28.1 Heat transfer coefficient values (W·m^{-2}·K^{-1}) determined with different methods at different frying oil temperatures.

Oil temperature (°C)	Lumped system (without bubble source)	Lumped system (with bubble source)	Actual food material (proposed approach)
150	206.3 ± 3.8	587.8 ± 86.0	286.7 ± 15.4
170	223.9 ± 6.0	684.5 ± 62.8	227.3 ± 8.0
190	241.2 ± 8.9	727.2 ± 14.9	181.3 ± 6.5

In another approach proposed by Yıldız *et al.* (2007), the actual food material was used to determine heat transfer coefficient, while knowledge of thermocouple location within the potato strip was not required, eliminating the errors as a result of using unreliable surface temperature data. Heat transfer coefficient values were found to be 287, 227, and 181 W·m^{-2}·K^{-1} for oil temperatures of 150, 170, and 190 °C, respectively, indicating a decreasing trend with increasing oil temperature. Table 28.1 shows the heat transfer coefficient values determined by all three methods (lumped system, lumped system with bubbles, and actual food material) at different frying temperatures. According to this table, when the lumped system is used without a bubble source in the frying medium, heat transfer coefficient increases slightly with increasing oil temperature, probably due to the effect of reduced viscosity of the frying oil at higher temperatures. However, heat transfer is greatly enhanced when bubbles are present. The effect of bubbles appears to become more pronounced as the frying temperature increases, indicating that greater rate of water loss from the bubble source at higher temperatures favors the heat transfer process.

When the actual food material is used, water vaporization plays a primary role in determining internal temperature increase, and hence heat transfer coefficient determined from the time–temperature data. As water vaporizes from the food material, it extracts a large amount of the incoming energy, reducing the energy available for temperature increase. Therefore, heat transfer coefficient determined using this method incorporates the effect of the energy utilized for vaporizing the water in the food being fried, making the term "apparent heat transfer coefficient" more appropriate. The difference between heat transfer coefficient values determined with the lumped system in the absence of bubbles and with the actual food material is not large. However, increasing the temperature of the frying oil results in opposite effects in these two methods: heat transfer coefficient decreases with increasing frying temperature when the actual food material is used. It is thought that the increasing rate of water vaporization at higher oil temperatures absorbs increasingly more of the incoming energy, resulting in a reduction in the apparent heat transfer coefficient.

Another experimental method of determining heat transfer coefficient is based on obtaining the moisture loss rate during frying (Hubbard and Farkas, 1999). In this method, an energy balance is set up between the food and the frying medium (Equation 28.1). Erdogdu and Dejmek (2010) provide a detailed discussion on the energy balance methodology.

$$h \cdot A(T_{\text{oil}} - T_{x=L}) = -\frac{dm}{dt} \Delta H_{\text{vaporization}}$$
$$h = -\frac{dm}{dt} \frac{\Delta H_{\text{vaporization}}}{A(T_{\text{oil}} - T_{x=L})} \tag{28.1}$$

where h is the heat transfer coefficient ($W \cdot m^{-2} \cdot K^{-1}$), A the surface area of the product (m^2), $\Delta H_{\text{vaporization}}$ the heat of vaporization of water at the prevailing conditions (e.g. $2257 \times 10^3 J \cdot kg^{-1}$ at 100 °C), dm/dt the rate of moisture loss from the food ($kg \cdot s^{-1}$), and $T_{x=L}$ and T_{oil} are surface temperature of the food and oil temperature (°C), respectively. The major assumption when writing Equation 28.1 is that there is negligible temperature gradient dT/dt in the core region of the food being fried. This requires heating the food to the boiling temperature of water prior to immersion in the frying oil (Hubbard and Farkas, 2000). When this is not done, Equation 28.1 should include the internal energy increase term to account for the temperature change within the core region (Farinu and Baik, 2007):

$$h \cdot A(T_{\text{oil}} - T_{x=L}) = -\frac{dm}{dt} \Delta H_{\text{vaporization}} + m \cdot c_p \frac{dT}{dt} \tag{28.2}$$

where m is mass (kg) and c_p is the specific heat capacity ($J \cdot kg^{-1} \cdot K^{-1}$).

Blending different oils to provide lower levels of viscosity, and hence higher heat transfer coefficient values at frying temperatures, has recently been suggested for improving the quality of fried products and reducing costs associated with frying (Debnath *et al.*, 2010). However, one should remember that the bottleneck of heat transfer during frying is the conduction of heat within the food material. Therefore, increasing heat transfer coefficient or oil temperature may easily result in near-surface regions reaching higher temperatures during frying, which will have a major effect on increased overall nutrient degradation and excessive browning, not to mention the increased levels of acrylamide in some products.

Mass Transfer

During frying, heat transferred from hot oil initially causes physical and chemical changes in the product, such as starch gelatinization, protein denaturation, crust formation, and vaporization of water from the surface. Mass transfer during frying is characterized by the movement of water in vapor form from the fried product into the oil. During the production of potato crisps, for example, moisture content was reported to reduce from 80% to below 2% in the final product (Gamble *et al.*, 1987). In the early stages of the frying process surface water is lost by evaporation, resulting in intense bubbling over the surface. In the latter stages, via the effect of the evolving moisture concentration gradient, the moisture may be assumed to diffuse from the inside to the evaporation interface from where it is removed by the final evaporation.

The rate of mass transfer is defined as the ratio of driving force over resistance (Rice and Gamble, 1989): the driving force is provided by the conversion of moisture into vapor, and the resistance to mass transfer is characterized by mass diffusion and surface resistance of the product (Farinu and Baik, 2005).

In the literature, as also reported by Singh (1995), there have been various empirical relationships describing moisture loss during frying. Kozempel *et al.* (1991) applied Fick's second law and solution to describe the moisture loss and oil gain in French fries. Yildiz *et al.* (2007) also used this approach to determine both diffusion coefficient for moisture and mass transfer coefficient using the time–moisture data of the fries samples obtained at different time intervals. They found that mass transfer coefficient increased from 1.12 to $2.07 \times 10^{-5}\,m^2{\cdot}s^{-1}$ while moisture diffusion coefficient increased from 9.2 to $18.2 \times 10^{-9}\,m^2{\cdot}s^{-1}$ with an increase in frying temperature from 150 to 190 °C. This method enables determination of both mass transfer coefficient and diffusion coefficient using a single set of data. A detailed discussion of this method is also given by Erdogdu (2005).

There is strong evidence that most of the oil is confined to the surface region of the fried product (Bouchon, 2006) and is mostly absorbed after the frying process is completed and during the cooling period due to condensation of steam creating a vacuum effect (Gamble and Rice, 1987). Ufheil and Escher (1996) suggested that oil uptake is a surface phenomenon, while Moreira and Barrufet (1998) explain this mechanism in terms of capillary forces. Farinu and Baik (2005) summarized the factors affecting oil uptake of fried products as the replacement of moisture where oil is entering the voids left by moisture removal, crust formation resulting in adsorption of oil, interfacial tension between the product and frying oil, capillary rise and porosity formation (Pinthus and Saguy, 1994).

Process Control

In the case of fried potato products, since the raw material is a natural product, production must be carefully planned (Nikolaou, 2006). Potato tubers contain 18–28% dry matter, of which 60–80% is starch (Varns and Sowokinos, 1974). Since starch is directly related to final product texture and fat content, potato tubers destined for frying must have relatively high starch level. In contrast to starch, the levels of reducing sugars in potatoes must be low.

Reducing sugars play a critical role in development of color and formation of acrylamide in potato chips and French fries. A high level of reducing sugars causes products to turn brown early in the frying process. Selection of suitable varieties is therefore an essential step for successful production. Although an upper limit has not been specified for potato chip production, varieties with reducing sugars content of less than $3\,g{\cdot}kg^{-1}$ fresh weight are recommended by CIAA (2009). In Switzerland, potatoes with a reducing sugars concentration as low as $0.5\,g{\cdot}kg^{-1}$ are desired by potato chip manufacturers (Foot *et al.*, 2007). Storage conditions play an important role in controlling reducing sugars in potato tubers, since temperatures below 8–10°C are known to promote liberation of reducing sugars from starch (Grob *et al.*, 2003).

Frying conditions also have a profound effect on product quality attributes. A high frying temperature may cause browning of the surface before the interior is cooked, while a low frying temperature results in more oil absorption and a soggy product. Acrylamide formation is also strongly dependent on processing conditions. Therefore, from a process control point of view, it is important to understand the effect of compositional and processing variables on quality attributes and acrylamide formation during frying.

Continuous monitoring of product moisture content is an effective way of controlling the frying process. For example, the moisture content of potato chips must be reduced to 2% by weight or less (Baumann and Escher, 1995) in order to obtain a crispy texture and an extended shelf-life. Therefore, the moisture content of potato chips can be a good indicator of process performance, and can easily be measured in real time using on-line moisture sensors positioned at the fryer exit.

The quality of oil is of paramount importance with regard to fried food quality. The fact that frying oil becomes a major component of fried food (15–20% in doughnuts and breaded or battered fish and chicken, 10% in French fries, and 35–45% in potato chips) further emphasizes this importance (Rossell, 2001). It should be noted that as soon as the frying oil comes into contact with water, it begins to degrade irreversibly. Its free fatty acid concentration increases, leading to off-flavors in the final product. The free fatty acid content of frying oil can be controlled by continuously adding fresh oil to the fryer. Oil turnover in a fryer must be 8 hours or less in order to avoid free fatty acid build-up (Bennett, 2001). Using the fryer at its designed capacity and removing crumbs from the fryer are also important with regard to maintaining oil quality.

Conclusions and Future Needs

Frying is a simultaneous heat and mass transfer process, and therefore design and optimization of a frying process should consider quantification of the changes in both the fried product and the frying medium (due to the irreversible degradation process with the start of frying) via the known thermal, physical, and kinetic-related parameters. For example, in addition to knowledge of heat and mass transfer parameters, the formation kinetics of the substances involved with the frying system should also be known in order to improve design of a frying process. Inclusion of the changes occurring in the frying medium brings additional positive benefits when designing a better frying process.

In addition to a design and optimization study, a future need lies in reducing the oil absorption of fried products. Producing low-fat fried products with the required quality attributes poses a unique challenge, and a comprehensive study is expected to provide additional information about the mass transfer of oil uptake. Fontes *et al.*

(2011) describe the optimization of the deep-fat frying process of sweet potato chips in palm olein and stearin where the objective was to minimize the oil content incorporated in the food product by determining the optimum frying temperature.

References

Achir, N., Vitrac, O. and Trystram, G. (2008) Simulation and ability to control the surface temperature history and reactions during deep fat frying. *Chemical Engineering and Processing* 47: 1953–1967.

Alvis, A., Velez, C., Rada-Mendoza, M., Villamiel, M. and Villada, H.S. (2009) Heat transfer coefficient during deep-fat frying. *Food Control* 20: 321–325.

Ballard, T.S. and Mallikarjunan, P. (2006) The effect of edible coatings and pressure frying using nitrogen gas on the quality of breaded fried chicken nuggets. *Journal of Food Science* 71: S259–S264.

Baumann, B. and Escher, F. (1995) Mass and heat transfer during deep-fat frying of potato slices I. Rate of drying and oil uptake. *LWT Food Science and Technology* 28: 395–403.

Bennett, R.M. (2001) Managing potato crisp processing. In: *Frying: Improving Quality* (ed. J.B. Rossell). CRC Press, Boca Raton, FL, chapter 10.

Blumenthal, M.M. and Stier, R.F. (1991) Optimization of deep-fat frying operations. *Trends in Food Science and Technology* 2: 144–148.

Bouchon, P. (2006) Frying. In: *Food Procesing Handbook* (ed. J.G. Brennan). Wiley-VCH Verlag GmbH and Co. KGaA, Weinheim, Germany, chapter 8.3.

Bourne, M. (1978) Texture profile analysis. *Food Technology* 32: 62–72.

Brescia, L. and Moreira, R.G. (1997) Modeling and control of a continuous frying process: a simulation study. Part I. Dynamic analysis and system identification. *Food and Bioproducts Processing* 75: 3–11.

Budzaki, S. and Seruga, B. (2005) Determination of convective heat transfer coefficient during frying of potato dough. *Journal of Food Engineering* 66: 307–314.

Chen, C-S., Chang, C-Y. and Hsieh, C-J. (2001) Improving the texture and color of fried product. In: *Frying: Improving Quality* (ed. J.B. Rossel). CRC Press, Boca Raton, FL, chapter 13.

CIAA (2009) A toolbox for the reduction of acrylamide in fried potato products/potato crisps. Confederation of the Food and Drink Industries of the EU. Available at www.ciaa.eu/documents/others/crisps-EN-final.pdf (accessed September 15, 2010).

Claybon, K.T. and Barringer, S.A. (2002) Consumer acceptability of color in processed tomato products by African-American, Latino and prototypical consumers. *Journal of Food Quality* 25: 487–498.

Costa, R.M., Oliveira, F.A.R., Delaney, O. and Gekas, V. (1999) Analysis of the heat transfer coefficient during potato frying. *Journal of Food Engineering* 39: 293–299.

Davidson, V.J., Brown, R.B. and Landman, J.J. (1999) Fuzzy control system for peanut roasting. *Journal of Food Engineering* 41: 141–146.

Debnath, S., Vidyarthi, S.R. and Singh, R.P. (2010) Impact of blending of frying oils on viscosity and heat transfer coefficient at elevated temperatures. *Journal of Food Process Engineering* 33: 144–161.

Erdogdu, F. (2005) Mathematical approaches for use of analytical solutions in experimental determination of heat and mass transfer parameters. *Journal of Food Engineering* 68: 233–238.

Erdogdu, F. and Dejmek, P. (2010) Determination of heat transfer coefficient during high pressure frying of potatoes. *Journal of Food Engineering* 96: 528–532.

Erdogdu, F. and Sahmurat, F. (2007) Mathematical fundamentals to determine the kinetic constants of first order consecutive reactions. *Journal of Food Process Engineering* 30: 407–420.

Farinu, A. and Baik, O.-D. (2005) Deep fat frying of foods: transport phenomena. *Food Reviews International* 21: 389–410.

Farinu, A. and Baik, O.-D. (2007) Heat transfer coefficients during deep fat frying of sweet potato: effects of product size and oil temperature. *Food Reviews International* 40: 989–994.

Farkas, B.E. and Hubbard, L.J. (2000) Analysis of convective heat transfer during immersion frying. *Drying Technology* 18: 1269–1285.

Farkas, B.E., Singh, R.P. and Rumsey, T.R. (1996) Modeling in heat and mass transfer in immersion frying. I, model development. *Journal of Food Engineering* 29: 211–226.

Fontes, L.C.B., Oliveira, F.G. and Collares-Queiroz, F.P. (2011) Optimization of the deep-fat frying process of sweet potato chips in palm olein or stearin. *American Journal of Food Technology* 6: 348–361.

Foot, R.J., Haase, N.U., Grob, K. and Gondé, P. (2007) Acrylamide in fried and roasted potato products: a review on progress in mitigation. *Food Additives and Contaminants* 24: 37–46.

Gamble, M.H. and Rice, P. (1987) Effect of pre-fry drying on oil uptake and distribution on potato crisp manufacture. *Journal of Food Science and Technology* 22: 535–548.

Gamble, M.H., Rice, P. and Selman, J.D. (1987) Relationship between oil uptake and moisture loss during frying of potato slices from c.v. Record U.K. tubers. *International Journal of Food Science and Technology* 22: 233–241.

Garayo, J. and Moreira, R. (2002) Vacuum frying of potato chips. *Journal of Food Engineering* 55: 181–191.

Grob, K., Biedermann, M., Biedermann-Brem, S. *et al.* (2003) French fries with less than 100 μg/kg acrylamide. A collaboration between cooks and analysts. *European Food Research and Technology* 217: 185–194.

Gupta, M.K. (2009) Industrial frying. In: *Advances in Deep-Fat Frying of Foods* (eds S. Sahin and S.G. Sumnu). CRC Press, Boca Raton, FL, chapter 12.

Gupta, M.K., Grant, R. and Stier, R.F. (2004) Critical factors in the selection of an industrial fryer. In: *Frying Technology and Practices* (eds M.K. Gupta, K. Warner and P.J. White). AOCS Press, Champaign, IL, chapter 7.

Hernandez, J.A., Heyd, B. and Trystram, G. (2008) On-line assessment of brightness and surface kinetics during coffee roasting. *Journal of Food Engineering* 87: 14–322.

Hubbard, L.J. and Farkas, B.E. (1999) A method for determining the convective heat transfer coefficient during immersion frying. *Journal of Food Process Engineering* 22: 201–214.

Hubbard, L.J. and Farkas, B.E. (2000) Influence of oil temperature on convective heat transfer during immersion frying. *Journal of Food Processing and Preservation* 24: 143–162.

Innawong, B., Mallikarjunan, P., Marcy, J. and Cundiff, J. (2006) Pressure conditions and quality of chicken nuggets fried under gaseous nitrogen atmosphere. *Journal of Food Processing and Preservation* 30: 231–245.

Keijbets, M.J.H. (2001) The manufacture of pre-fried potato products. In: *Frying: Improving Quality* (ed. J.B. Rossell). CRC Press, Boca Raton, FL, chapter 9.

Kozempel, M.F., Tomasula, P.M. and Craig, J.C. Jr (1991) Correlation of moisture and oil concentration in French fries. *LWT Food Science and Technology* 24: 445–448.

Lalas, S. (2009) Quality of frying oil. In: *Advances in Deep-Fat Frying of Foods* (eds S. Sahin and S.G. Sumnu). CRC Press, Boca Raton, FL, chapter 4.

Mittal, G.S. (2009) Physical properties of fried products. In: *Advances in Deep-Fat Frying of Foods* (eds S. Sahin and S.G. Sumnu). CRC Press, Boca Raton, FL, chapter 16.

Moreira, R.G. and Barrufet, M.A. (1998) A new approach to describe oil absorption in fried foods: a simulation study. *Journal of Food Engineering* 35: 1–22.

Moreira, R.G., Castell-Perez, M.E. and Barrufet, M.A. (1999) *Deep-Fat Frying: Fundamentals and Applications.* Aspen Publishers, Gaithersburg, MD, chapter 4.

Ngadi, M. and Ikediala, J.N. (2005) Natural heat transfer coefficients of chicken drum shaped bodies. *International Journal of Food Engineering* 1(3), Article 4.

Ni, H. and Datta, A.K. (1999) Moisture, oil and energy transport during deep-fat frying of food materials. *Transactions in Chemical Engineering* 77(C): 194–204.

Nikolaou, M. (2006) Control of snack food manufacturing systems. *IEEE Control Systems Magazine* 26: 40–53.

Oreopoulou, V., Krokida, M. and Marinos-Kouris, D. (2006) Frying of foods. In: *Handbook of Industrial Drying*, 3rd edn (ed. A.S. Mujumdar). CRC Press, Boca Raton, FL, chapter 52.

Oztop, M.H., Sahin, S. and Sumnu, G. (2007) Optimization of microwave frying of potato slices by using Taguchi technique. *Journal of Food Engineering* 79: 83–91.

Parkash, S. and Gertz, C. (2004) New theoretical practical aspects of the frying process. *European Journal of Lipid Science and Technology* 106: 722–727.

Pedreschi, F. and Zuniga, R.N. (2009) Kinetics of quality changes during frying. In: *Advances in Deep-Fat Frying of Foods* (eds S. Sahin and S.G. Sumnu). CRC Press, Boca Raton, FL, chapter 5.

Pinthus, E.J. and Saguy, I.S. (1994) Initial interfacial tension and oil uptake by deep-fat fried foods. *Journal of Food Science* 59: 804–807, 823.

Rao, V.N.M. and Delaney, A.M. (1995) An engineering perspective on deep-fat frying of breaded chicken pieces. *Food Technology* 49: 138–141.

Rice, P. and Gamble, H. (1989) Modeling moisture loss during potato slice frying. *International Journal of Food Science and Technology* 24: 183–187.

Rimac-Brncic, S., Lelas, V., Rade, D. and Simundic, B. (2004) Decreasing of oil absorption in potato strips during deep-fat frying. *Journal of Food Engineering* 64: 237–241.

Rossell, J.B. (2001) Factors affecting the quality of frying oils and fats. In: *Frying: Improving Quality* (ed. J.B. Rossell). CRC Press, Boca Raton, FL, chapter 7.

Rywotycki, R. (2002) The effect of fat temperature on heat energy consumption during frying of food. *Journal of Food Engineering* 54: 257–261.

Rywotycki, R. (2003) Food frying process control system. *Journal of Food Engineering* 59: 339–342.

Saguy, I.S. and Dana, D. (2003) Integrated approach to deep fat frying: engineering, nutrition, health and consumer aspects. *Journal of Food Engineering* 56: 143–152.

Sahin, S. and Sumnu, G.S. (2009) Alternative frying technologies. In: *Advances in Deep-Fat Frying of Foods* (eds S. Sahin and S.G. Sumnu). CRC Press, Boca Raton, FL, chapter 13.

Sahin, S., Sastry, S.K. and Bayindirli, L. (1999) The determination of convective heat transfer coefficient during frying. *Journal of Food Engineering* 39: 307–311.

Shyu, S., Hau, L. and Hwang, L.S. (1998) Effect of vacuum frying on the oxidative stability of oils. *Journal of the American Oil Chemists' Society* 75: 1393–1398.

Singh, R.P. (1995) Heat and mass transfer in foods during dep-fat frying. *Food Technology* 49: 134–137.

Singh, R.P. (2000) Moving boundaries in food engineering. *Food Technology* 54: 44–48, 53.

Szczesniak, A.S. (1963) Classification of textural characteristics. *Journal of Food Science* 28: 385–389.

Ufheil, G. and Escher, F. (1996) Dynamics of oil uptake during deep-fat frying of potato slices. *LWT Food Science and Technology* 29: 640–644.

Varns, J.L. and Sowokinos, J.R. (1974) A rapid micro-starch quantitation method for potato callus and its application with potato tubers. *American Journal of Potato Research* 51: 383–392.

Vitrac, O., Trystram, G. and Raoult-Wack, A.-L. (2000) Deep-fat frying of food: heat and mass transfer, transformations and reactions inside the frying material. *European Journal of Lipid Science and Technology* 102: 529–538.

Warner, K. (2004) Chemical and physical reactions in oil during frying. In: *Frying Technology and Practices* (eds M.K. Gupta, K. Warner and P.J. White). AOCS Press, Champaign, IL, chapter 7.

Yildiz, A., Palazoglu, T.K. and Erdogdu, F. (2007) Determination of heat and mass transfer parameters during frying of potato slices. *Journal of Food Engineering* 79: 11–17.

29

Mechanical Separation Design

Timothy J. Bowser

Definition and Purpose

Mechanical separation of raw agricultural materials into foods has been practiced by humans for centuries. Winnowing is an ancient method of mechanical separation that uses air currents to remove chaff from grain. The Bible frequently mentions a "winepress," an invention used to separate juice from grapes. Egyptian tomb inscriptions dating to 1450 BC show a sedimentation device employed to remove solids from drinking water (Baker and Taras, 1981), a primary ingredient in processed foods. In the fifth century BC, Hippocrates invented the "Hippocrates Sleeve," consisting of a cloth bag for filtering rainwater (Columbia Electronic Encyclopedia, 2007). The Industrial Revolution brought about large-scale mechanization of separation processes in the food industry, resulting in profound changes. Today's modern food plant contains a significant amount of specialized equipment dedicated to separation processes.

Mechanical separation is any operation that divides a mixture of ingredients into two or more fractions by the use or manipulation of physical force. Ingredients may be of plant or animal origin, from naturally occurring materials, fine chemicals, or waste products occurring in any of three phases (solid, liquid and gas) or combinations of these. The goal of mechanical separation of foods is to economically produce useful products or ingredients. Common examples of mechanical separation in food processing include sifting, filtering, clarifying, rinsing, winnowing and dewatering. The

Handbook of Food Process Design, First Edition. Edited by Jasim Ahmed and Mohammad Shafiur Rahman.

Table 29.1 A sampling of mechanical separation techniques and food products they are used on (listed alphabetically by mechanical separation technique).

Mechanical separation technique	Food products undergoing mechanical separation	Description of mechanical separation technique	Element(s) separated from food product
Centrifuge	Oils	Spin at high speeds	Unwanted phase(s)
Clarify or settle	Sugar syrup, oils, fats, juices, liquid foods	Settle	Solids, dense phase, light phase
Cyclone	Spray dried or powdered foods like milk, juices, coffee, butter	Air flotation and gravity	Air
Dewater	Fruits, vegetables, meats, seeds, nuts and more	Shake, centrifuge, air blow, absorb	Free water
Filter	Juices, oils, pumpable liquids	Porous media barrier used to remove particles	Particles
Peel	Fruits, vegetables	Powered knife or steam	Surface skin of fruit or vegetable
Pit	Fruit	Pitting machine	Seed of fruit
Press	Oilseeds, fruits, vegetables	Squeeze	Liquid from solid
Rinse	Fruits, vegetables, meats, seeds, nuts, and more	Rinse with water	Debris and foreign materials
Sift	Whole or ground solids like seeds, flour	Porous media and mechanical agitation	Debris, foreign material, size classification of base material
Sort	Fruits, vegetables, grain, nuts, eggs	Screens, controlled gaps or spaces	Different sizes or shapes of same material
Trim	Meat	Powered knife	Unwanted materials like hair, tissue, fat, debris
Winnow or scalp	Grains, nuts, seeds, fruits, vegetables	Air flotation and gravity	Debris and foreign materials

purpose of this chapter is to introduce the unit operation of mechanical separation and its applications in the food industry.

Food Products Processed by Mechanical Separation

Mechanical separation is a unit operation that is nearly ubiquitous in food processing. Table 29.1 lists examples of mechanical separation techniques and the food products they are used on.

Theoretical Principles of Mechanical Separation

Differences in the physical properties of materials, such as particle size, surface area, texture, density, color, melting point, phase, shape, viscosity, vapor pressure, polar charge, magnetism (Seader and Henley, 1998), and wettability (Answers.com, 2010), can be exploited to achieve mechanical separation of mixtures. There are five general categories of mechanical separation (Seader and Henley, 1998): (i) phase creation, (ii) phase addition, (iii) barrier, (iv) solid agent, and (v) force field. In the remainder of this section, the five categories are explained and examples of each given.

Phase Creation

A second phase that can be separated from the feed phase is created by a sudden change in pressure, or the addition of energy through heat or mixing. Flash vaporization is a common mechanical separation process used to remove water from food products. Milk pasteurized by direct steam injection adds water (condensate) in the process. Added water is removed by flash vaporization when the line pressure of the hot liquid milk is suddenly reduced and water instantly evaporates (Bowser, 2003). Spray drying of liquid foods like broths, milk, and juices relies on the addition of heat to create a second phase (gas) that is gently separated from the liquid and solid phases of the product.

The mass and energy balance technique described by Crapiste and Rotstein (1997) is practical for the design of phase creation processes. Materials composition is defined on a dry basis where water, W, removed from the solid is given by:

$$W = S(x_i - x_f) \tag{29.1}$$

where S is the flow rate of the bone dry solid product and x the moisture content of the product. Subscripts i and f indicate initial and final conditions. When water is transferred to the air (spray drying), W is given by:

$$W = A(y_f - y_i) \tag{29.2}$$

where A is the flow rate of dry air and y the moisture content of the air (kg vapor/kg dry air). For the case of a solid product or air, the energy balance can be represented by the following equations:

$$Q_S = SC_{pf}(T_{pf} - T_{pi}) + W[\Delta H - C_w(T_{pi} - T_r)] \tag{29.3}$$

$$Q_A = A(C_{gi} + C_{vi}y)(T_{ai} - T_{af}) + WC_v(T_{ai} - T_r) \tag{29.4}$$

where Q_S and Q_A are net energy transferred from the solid product and to the air, respectively, and $Q_S = Q_A$; C is specific heat, T is temperature and ΔH is the latent heat of vaporization of water at a reference temperature. Subscripts a, g, p, r, and v denote air, dry air, solid product, reference temperature, and water vapor, respectively.

Phase Addition

A second phase that can be separated from the feed phase is added to dissolve or remove or physically change a portion of the feed material. Water washing or rinsing is a case of separation by phase addition. Water may carry away or dissolve a residue on the surface of the feed or fractions of the feed mixture. Dissolved air flotation (DAF) uses air bubbles to float particles to the surface of a liquid mixture (Britz *et al.*, 2006) for subsequent removal via another mechanical separation method such as skimming. Ferreira-Leitao *et al.* (2010) tested the use of CO_2 and steam in a sugar cane bagasse pretreatment process to increase sugar yields for ethanol production. Both CO_2 and steam were added phases that served to enhance the separation process. Barrington *et al.* (2002) described the use of limestone dust to precipitate phosphorus from dairy manure. A 3.0% dosage of limestone dust was incorporated into a manure slurry, resulting in recovery of 90% of phosphorus.

Barrier

A physical barrier such as a screen, filter, or membrane is used to control or enhance the movement of particles or species with respect to the feed material. Reverse osmosis is commonly used to purify water used in food processing (American Water Works Association, 2007). Membrane systems have been used extensively to process fruit juices, sugar solutions, and dairy products (Saravacos and Kostaropoulos, 2002). Screens may be employed to clean, grade, sort, and separate fruits, vegetables, nuts, seeds and other food materials (Henderson *et al.*, 1997). Moving screens have an optimum frequency, amplitude, and angle for each application. Residence time of product on the screens and sieve construction must also be considered. Some standards for product-specific sieve analysis are available (e.g. American National Standards Institute, ANSI, and American Society of Agricultural and Biological Engineers, ASABE). In the absence of a standard, sieving parameters must be determined experimentally (Retsch, 2010).

In filtration systems, the rate of fluid filtration is determined by examining flow as described by Earle (1983). Filtration rate is equivalent to the driving force divided by the resistance to flow:

$$\frac{dV}{dt} = (A\Delta P)/R \tag{29.5}$$

where A is the filter area, ΔP the driving force or pressure difference across the filter, and R the resistance to flow that develops from the filter material and the cake of solids that accumulate on it. Total resistance is proportional to fluid viscosity:

$$R = \mu r(L_c + L) \tag{29.6}$$

where μ is the viscosity of the fluid, r the specific resistance of the filter cake, and L_c and L the thicknesses of the filter cake and filter material, respectively. For constant-pressure filtration systems, ΔP is constant and the thickness of the filter cake is equal to the volume of solids in the product to be retained by the filter divided by the surface area of the filter:

$$L_c = wV/A \tag{29.7}$$

where w is solids retained per unit volume of liquid, V the volume of liquid to be filtered, and A the surface area of the filter. Integrating Equation 29.6 from $V = 0$ to $V = V$ from time $t = 0$ to $t = t$ and substituting into Equation 29.7, we obtain:

$$\frac{tA}{V} = \left(\frac{\mu rw}{2\Delta P}\right)(V/A) + \frac{\mu rL}{\Delta P}\mu r \tag{29.8}$$

Equation 29.8 can be used to predict filter performance by experimentally determining the slope ($\mu rw/2\Delta P$) of the line.

Solid Agent

A solid agent (normally loose or packed granular substance) has surface-active sites that separate by adsorption or chemical reaction with components of the feed mixture. Water is softened by ion exchange as the calcium and magnesium ions in the water are replaced by sodium ions (Nunn, 1997) as shown by the following reaction for calcium removal:

$$Ca^{2+}_{(aq)} + 2NaR_s \leftrightarrow CaR_{2(s)} + 2Na^{+}_{(aq)} \tag{29.9}$$

where R is the residual substance on the ion exchanger. The process is reversible and the solid agent can be reused until excessively fouled. Water softening is often preceded and followed by other mechanical separation techniques like filtering and reverse osmosis (barrier category) and sedimentation (force field). Chromatography is another example of separation by solid agent.

Force Field

Separation by force field involves the use of an imposed field, like gravity, centrifugal, magnetic or electrical, over the feed mixture. For example, gravity separation is a critical processing step in the manufacture of specialized cheeses (Ma and Barbano, 2000). Often the force field is established with a gradient; in the case of a spinning disk, the centrifugal effects are higher at the circumference, causing lighter feed components to migrate to the axis of the disk and heavier fractions to move toward the perimeter. Thermal gradients have been used to improve the purity of rice bran oil fractions (Dunford and King, 2001) and glyceride mixtures (King *et al.*, 1997). A magnetic field can be used to remove ferromagnetic particles from a mixture. A centrifuge develops a pressure gradient that separates fluid mixtures by molecular weight (Seader and Henley, 1998) and particles from solution based on their shape, density, and the viscosity of the fluid medium (IQS, 2010). The measurement of pressure applied to a sample in a centrifuge is termed "relative centrifugal force" (RCF) and is measured in units of gravity as (Geankoplis, 1978):

$$\mathrm{RCF} = \frac{r(2\pi N)^2}{g} \tag{29.10}$$

where g is the gravitational constant, r the radius of rotation, and N the speed of rotation.

Equipment Used for Mechanical Separation

Much of the equipment available for mechanical separation has been developed over the years based on experience (Saravacos and Kostaropoulos, 2002) and need. Some of the most successful pieces of mechanical separation equipment exploit a number of the unique physical properties of the fractions to be divided. For example, a centrifuge can make use of phase, density, and viscosity and a simple water rinse can take advantage of solubility, relative density, fluid transport, porosity, wettability, and texture. Table 29.2 provides examples of equipment used in the five categories of mechanical separation and lists some advantages and disadvantages of each.

This remainder of this section lists, in alphabetical order, some of the important equipment used in mechanical separation of foods, providing a brief description and some of the physical parameters and empirical design information used to specify and install the equipment. Equipment manufacturers' websites are provided for the reader to access images and additional information.

Centrifuge

The centrifuge is used to separate liquids and particles suspended in a carrier fluid by centrifugal force (force field). Important process variables include percent suspended solids, volumetric slurry throughput, solids throughput, and product consistency at point of discharge. Incompressible materials of 45 μm or larger are normally suited to filtration-type centrifuges and finer materials to sedimentation-type centrifuges (Broadbent, 2001). For coarse solids, a low residual liquid content is achievable. Substantial volumetric throughputs can be achieved relative to the size of a centrifuge, which may help offset the high capital and operating expense (Purchas, 1981).

Throughput of a sedimenting centrifuge is proportional to the square of the smallest particle size to be separated; therefore halving the throughput will permit separation of a particle size that is decreased by a factor of $1/\sqrt{2}$. Temperature is an important process variable since it affects density and viscosity of the liquid. Increased tempera-

Table 29.2 Examples of equipment used in mechanical separation and advantages and disadvantages of each.

Category	Equipment	Advantage	Disadvantage
Phase creation	Flash vaporizer (e.g. milk cooling)	Compact rapid process Precise control possible	Rough treatment of product (high shear) Requires pressurization of product stream
	Gun puffing (e.g. cereal puffing)	Compact rapid process Precise control possible	Pressure vessel required Requires intensive heat transfer
	Steam peeling (e.g. potato peeling)	Rapid process Low waste in comparison with other peeling processes No chemicals	Pressure vessel and steam source required
Phase addition	Classifier, wet and dry (e.g. winnowing of chaff [skin] from grain, water flotation of trash from produce)	Inexpensive Quick and efficient Continuous Few moving parts	Will not remove particles of similar size and density Uniform physical properties of infeed material required
	Dissolved air flotation (e.g. removal of fats from waste water stream)	Relatively inexpensive Effective	Flocculation aids may be required and may contaminate byproducts Requires large tanks Low efficiency Additional separation equipment needed (e.g. skimming)
	Leaching (e.g. vanilla flavoring from beans)	Inexpensive Effective Few or no moving parts	Inefficient Slow Difficult to control
	Rinsing/washing	Inexpensive Quick Efficient	Water residue may be undesirable May remove desirable components Water treatment and disposal costs
Barrier	Filter (e.g. liquid foods such as juices, fats and solutions)	Inexpensive Effective Rapid and compact Well understood	Clogs over time Difficult to clean Eventually filter material must be replaced Pressure differentials
	Membrane (e.g. liquid foods such as juices, fats and solutions)	Rapid and compact	Clogs over time Difficult to clean Membrane eventually requires replacement Pressure differentials
	Press (e.g. expel liquid from fruit)	Inexpensive Rapid Continuous or batch	Inefficient Not preferential Pressing aid sometimes required
	Screens (e.g. sift flour, dewater produce)	Inexpensive Effective and rapid Well understood	Difficult to clean Multiple passes may be required Pore clogging Vibration may be needed to increase effectiveness

(*Continued*)

Table 29.2 (*Continued*)

Category	Equipment	Advantage	Disadvantage
Solid agent	Chromatography (e.g. protein isolation) (Clark, 2009)	Superior product High efficiency Compact	New technology Intellectual properties agreements
	Water softener	Effective and rapid Relatively inexpensive compared with other softening methods	Requires regeneration Produces regeneration waste Frequent maintenance required Cleanability issues
Force field or gradient	Centrifuge (filtration and sedimentation types)	Compact Effective and rapid Continuous High capacity	Numerous moving parts with high maintenance Expensive compared to settling Set-up and controls may be difficult
	Clarifier	Inexpensive No moving parts Simple controls	Large space and time requirements
	Electrical (e.g. electrodialysis; used to demineralize dairy whey)	Compact, effective, rapid	Used in tandem with membrane (barrier) separation Fouling Cost
	Magnet (removal of ferrous metal from grain)	Compact Effective and simple Rapid	Removal of ferrous material and cleaning of magnet surface

ture tends to decrease liquid density and viscosity, which increases throughput and reduces the size of the smallest particle that can be separated (Purchas, 1981). Tarleton and Wakeman (2007) list nine basic types of commercial centrifuge that are available with many variations and options that result in an extensive array of choices. Consulting with equipment manufacturers or an experienced engineer is often required to determine the best centrifuge for a particular job. Images and videos of a variety of centrifugal separators may be viewed at www.wsus.com.

Chromatograph

Chromatography is a mechanical separation technique that has recently extended from analytical instrumentation to production. It is used to separate molecules based on their differences in structure and/or composition. A mobile phase is forced through a column containing stationary solids or immobilized liquids (solid agent) that adsorb substances from the mobile materials and/or cause differences in the rate of migration of the mobile materials through the column. Flow rate, pressure, and mobile phase composition are important factors in system design (Hadden *et al.*, 1971). Etzel (2004) reports the use of chromatography in an inexpensive commercial separation process for dairy proteins. Images of commercial equipment and diagrams of principles of operation can be viewed at www.upfront-dk.com.

Clarifier

A clarifier employs gravity to settle solids out of liquids (force field). Clarification should be used where possible (especially as an upstream process prior to other separation steps such as filtration) because of low cost and high efficiency. Clarification can remove particles down to 0.1 μm, while filtration is normally capable of removing particles of 0.45 μm (Riffer, 2000) with solids loadings from 4 to 40% w/w (Tarleton and Wakeman, 2007). Clarifiers have no moving parts and are simple to control and operate. Liquid flow rate and solids load are important design criteria. Some clarifiers have a back-washing facility to enable surfaces where solids accumulate to be cleaned. A typical solids removal rate is 80–90% (Salah, 2010). Visit www.hydroflotech.com to view inclined plate clarifiers and application information.

Classifier, Air

Streams of particles are divided according to their terminal velocity in air (Senden and Tels, 1978), which depends on the physical characteristics of the product such as projected area, density, mass, volume, and drag coefficient (Henderson *et al.*, 1997). Particles in the range 2–40 μm are commonly separated using air (Saravacos and Kostaropoulos, 2002). Ferrari *et al.* (2009) optimized air classification techniques in a process to separate β-glucan-enriched barley flours. Wilhelmi *et al.* (2009) studied air classification for separation of field peas used in ethanol production. Photos and animated figures of air classification equipment can be viewed at www.sweco.com.

Classifier, Water

Hydrocyclones are used to separate solid particles from a fluid (water). They operate under a wide range of flow rates and pressures. The fluid flows in a spiral path and particles are separated by centrifugal force and the difference in specific gravity between the solid and liquid (force field). The square root of the diameter of the hydrocyclone is proportional to the minimum-sized particle that can be separated; thus small-diameter hydrocyclones are required to separate fine particles. Small diameters result in reduced throughput, which is incompatible with most process needs. This issue is generally resolved by staging cyclones in parallel (Purchas, 1981). Minimum effective particle size is about 10 μm (Seo and Lee, 2009). Sand and silt removal from wash-water is a common application. Fluid classifiers generally contain no moving parts and are simple to operate and maintain. Photos of equipment and animated figures of water classifiers are available at www.lakos.com.

Cyclone

In a cyclone separator, centrifugal and gravitational forces act on particles to separate them from air (force field). Air and entrained solids tangentially enter a conical tank and descend in a spiral motion that can be visualized by imagining a vortex. Uses of cyclone separators include removing dust from air and air from product in conveying and spray drying systems. Particles under 5 μm are not effectively separated in cyclones. Although the theory of cyclone separation is well known, most designs are based on trial and error. Akulich and Lustenkov (2007) optimized the operating and design parameters of a combined dust separator that incorporates a cyclone separator and bag filter. Zenz (2001) provides design tips for cyclone separators. See www.sweco.com for equipment photos and animations.

Deaerator

A deaerator uses vacuum or steam to remove air or other dissolved or trapped gases from a solid or liquid ingredient (phase creation and/or addition). Deaeration is often used in boiler feedwater treatment to remove oxygen and carbon dioxide, which cause corrosion of heat exchange surfaces. Oxygen may also be removed from food products to prevent oxidation and to establish uniform density and texture or viscosity. In a pressurized deaerator, the liquid is maintained at its saturation temperature, reducing the solubility of gas in the liquid to zero (Spirax Sarco, 2010). Deaeration has been shown to improve flavor, increase shelf-life of fruit juices (Marshall, 1951), and reduce thermal resistance of microorganisms in foods undergoing pasteurization processes (HRS Spiratube, 2010). Important design variables are pressure (differential), temperature, flow, and product level. Ferro *et al.* (2002) successfully developed a deaeration model for a seawater desalination process that helped to reduce the cost of the deaerator equipment and operation. Images and a drawing showing product flow in a deaerator are given at www.cornellmachine.com.

Dissolved Air Flotation Unit

DAF systems use chemical agents to flocculate (or aggregate) particles to be removed from water. Fine air bubbles are introduced to the water and attach to the flocculated particles, lifting them to the water surface (phase addition). A skimmer removes the floating particles. DAF systems are relatively compact and operate under loading rates as high as $7\,\mathrm{L{\cdot}min^{-1}{\cdot}m^{-2}}$ (Leopold, 2010). Hanafy and Nabih (2007) reported the most appropriate design conditions for processing oil and water emulsions. Chemical flocculants may make recovery of waste streams, such as oil, difficult for agricultural or food use since many are considered adulterants. Design parameters include recycle pressure, saturation efficiency, recycle rate, tank detention, hydraulic loading rate, solids loading rate, and air/solids ratio (Ross *et al.* 2000). Images and an animated DAF system can be viewed at www.fbleopold.com.

Electrostatic Precipitator

Electrostatic precipitation is often used to remove airborne contaminants in process exhaust streams. Airborne particles are electronically charged and then captured by attraction to oppositely charged collection surfaces (force field). Design information includes airflow, contaminant loading, contaminant electrical charge properties, viscosity, and density. Electrostatic precipitators have been studied as a means to achieve even and rapid application of smoke to food products such as meats (Baron *et al.*, 2008). Particles may be collected for recycling or reuse. See www.uasinc.com for equipment images and specific details.

Filter

Filters remove particles from a fluid stream using porous materials such as wire mesh, solids beds, cloth, and sintered metal (barrier). The porous medium catches the particles and allows the fluid to pass through. A pressure difference is always created across the porous barrier. The effluent from the barrier is known as the filtrate. The two classes of filters are (i) cake and (ii) deep bed. Cake filters rely on a porous barrier with openings smaller than the particles to be filtered. Particles that accumulate on the barrier form a cake, which eventually may become the primary filter and flow resistance. Deep bed filtration uses fibers, sand, or diatomaceous earth as the filtering medium. They are useful for fluids that carry less than about 1% of suspended solids. Important design information includes fluid viscosity, pressure gradient, particle characteristics, and media fouling (Chou, 2000a). Filtration equipment can be viewed at www.komline.com.

Grader

Grading is the separation of mixtures of ingredients into two or more streams that may be defined based on their value and intended use (Henderson *et al.*, 1997). Grading is normally preceded by other separation steps like color sorting, classifying, screening, and washing. Egg grading is a typical example. Eggs are graded according to size, but if eggs are not cleaned and defects removed prior to sizing, the separated eggs may not make grade specification.

Magnet

Magnets are utilized to remove ferrous fractions from nonmagnetic materials (force field). Materials to be separated may be solids, liquids, or a combination. Important design variables include temperature, flow rate, flow characteristics, materials to be processed, magnet capacity, and cleaning operations. Efficiency of a magnetic separation process can be improved by maximizing the gradient and intensity of the applied field (Tarleton and Wakeman, 2007). Luk and Liu (2005) reported use of a magnetic coagulator for treatment of wastewater sludge. A good overview of magnet selection is given by Eriez Manufacturing Co. (2007). Photos of magnets can be viewed at www.eriez.com.

Membrane

Membrane filtration uses semipermeable barriers to remove particles from fluid streams similar to filters, but targets much smaller particles (sized from bacteria down to salts). Membrane filtration is divided into four categories, depending on the size of the particles to be removed (listed in order of decreasing size of particle removed): microfiltration (0.1–20 μm), ultrafiltration (0.001–0.1 μm), nanofiltration (molecular weight 200–1000), and reverse osmosis (molecular weight < 200). System operating pressure increases with decreasing size of particle to be removed. Important variables for system design include fluid viscosity, membrane thickness, fouling, and pressure difference across the membrane. Plugging and filter cake build-up severely reduce system effectiveness and must be alleviated by back-washing, prefiltration, and membrane surface cleaning (Saska, 2000; Tarleton and Wakeman, 2007). Information and images of membrane systems can be viewed at www.cuno.com and geafiltration.com.

Press

Pressing, or mechanical expression, is a means of separating juices and oils from agricultural products. Pressure applied to the product breaks open cell walls and tissue to release liquid contents. Expression depends on the following factors (Brennan *et al.*, 1990): mechanical pressure, yield stress of feed material, cake porosity, and viscosity of expressed liquid. Batch presses are configured as hydraulic or screw rams that compress the feedstock. Feed material may be contained in a cloth sack or by perforated materials (barrier). Continuous presses are preferred to batch presses because of less labor, higher throughput, and greater efficiencies. The four major types are screw, roller, belt, and reamer (Saravacos and Kostaropoulos, 2002). Images of batch presses are available at www.goodnature.com while continuous systems for food and sludge can be found at www.vincentcorp.com and www.komline.com respectively.

Puffing Gun

A puffing gun is a batch processing device that relies on a sudden change in pressure to vaporize moisture (phase creation) in a product (typically a wholegrain like rice or wheat), rapidly expanding it. The same effect can also be achieved on a continuous basis for dough-based products using an extruder. Important design considerations for gun and extrusion puffing include product throughput, pressure differential, moisture content, specific heat, and gelatinization temperature. The Puritan Puffing Machine can be viewed at www.purmfg.com. Wenger Manufacturing (www.wenger.com) has developed an extrusion process that utilizes carbon dioxide as the expansion agent as an alternative to water.

Rinser

Rinsing is a simple method used to remove residue and soils by dipping products in water or by pouring or spraying water on the product (phase addition). Separation of materials occurs by dissolving, floating, sinking, transporting, or dislodging materials from the surface of the primary feed fraction. A multitude of spray nozzles are available for this purpose (www.spray.com) and a wide variety of equipment has been designed to rinse products (www.key.net and heinzen.com). Chemicals may be added to the rinse water to enhance the separation process (www.ecolab.com). Rinse water may be cleaned using other separation process, such as filtration, and recycled to reduce waste. Important design considerations include temperature, impingement force, and physical properties of residues (e.g. solubility, density, stickiness, and particle size).

Screen

Screens or sieves are probably the most widely used mechanical separation device (barrier). Screening processes may be wet or dry, depending on process requirements and capabilities. Important design considerations for the product include particle size (range), density, shape, moisture content, throughput, flowability, and corrosiveness. The Association of Equipment Manufacturers (www.aem.org) offers a rich group of references for vibrating screen use, including an applications handbook. Screens can be mechanically agitated and/or inclined to improve throughput efficiencies. Many variants of screen materials, geometry, and weaves are available to optimize the separation process (Tarleton and Wakeman, 2007). Pressurized air or gas is often used as a combined separation technique and to dislodge particles captured in screens. Wedged particles may be removed by the addition of a screen self-cleaning device (Sweco, 2002), which incorporates elements that slide over or bump the upper or lower surfaces of the screen element. A wide range of screening equipment can be viewed at www.sweco.com and www.kason.com.

Sorter

Sorting is the separation of mixtures of ingredients into two or more streams that may be defined based on their size, shape, density, texture, and color (Henderson *et al.*, 1997). Mechanical sorting equipment includes a wide variety of screens, rollers, belts, weighing stations, and flotation and other devices. Sorting overlaps with many of the other separation processes. One very important sorting process is optical sorting, which makes use of sensors to determine shape, color, and moisture of particles. Food objects such as peas are dropped or projected through a system that rapidly takes a digital image of each particle and analyzes it. The object is diverted to the appropriate fraction by air jet, conveyor, gravity, or other means. Vision sorting technology has been successfully used to mechanically separate items that have similar density and color such as green glass particles and peas. Advanced vision technology used for sorting and grading can be viewed at www.key.net and www.bestsorting.com.

Steam Peeler

Steam peeling is a rapid process used to remove the outer layer, or skin, from fruits and vegetables via steam (phase creation). Heat is rapidly transferred by steam into the surface layer of a product that is held under pressure. The pressure is suddenly released, causing the heated moisture in and on the surface layer to instantly evaporate and expand, shredding the outer surface of the product. The application of steam peeling can be very effective, with yields as high as 93% using steam pressure of 20 bar (2 MPa) and steam time of up to 8 s (Odenberg Engineering, 2008). Efficiency of operation depends on many factors such as the cultivar, condition of the produce, intended use (final product), throughput, cost and availability of steam and waste handling capabilities of the plant. Garrote *et al.* (2000) reported the effects of heating on steam-peeled potatoes. Images and information about steam peelers is available at www.odenberg.com.

Water Softener

Water softeners are used to remove "hardness" from water caused by an excess of calcium and magnesium using an ion-exchange resin bed (solid agent). Sodium ions on the bed are traded for the calcium or magnesium ions in the water. Hardness can result in inefficient use of soap, soap scum residues, chalky deposits on fixtures and in pipes, and off-flavors in foods (Crittenden *et al.*, 2005). Water usage rates, peak demand, hardness characteristics of available water, water system pressure, and knowledge of the intended use of softened water are all needed to design softening systems. More information and images of industrial water softening systems can be viewed at www.reskem.com.

Design of Mechanical Separation Processes

Identification and definition of the mixture and fractions to be removed must occur early in the design process. The design engineer will reckon the physical force(s) to be manipulated and the materials properties to be exploited in the mechanical separation process. An understanding of the theoretical principles of the separation process will assist in selection of possible equipment choices and alternatives. A host of additional information must be collected and evaluated, including budget, site utilities, space available, ambient conditions, future processing plans, materials and waste

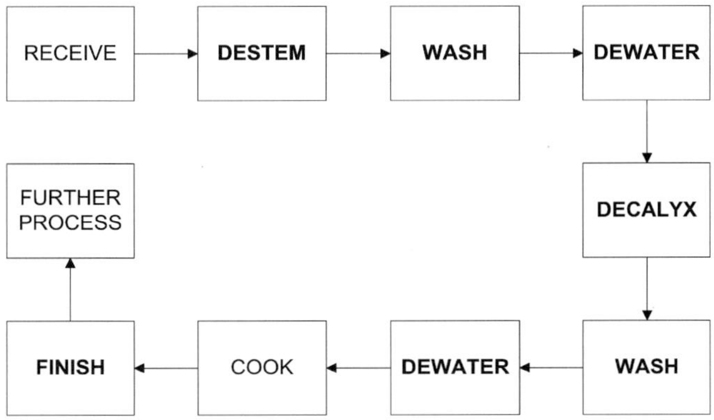

Figure 29.1 Process flow diagram of a portion of an apple sauce manufacturing process.

handling, maintenance and operations, quality standards, and safety. Safety of the mechanical separation process must be assessed early and continuously throughout the project. A brief section devoted to process hazard and safety analysis appears below.

Mechanical separation of food products requires unique considerations compared with separation of nonfoods. Separation processes must be cleanable, sanitary, and open to inspection or validation when applied to food products (Clark, 2003). "Cleanable" means that soils or process residues may be completely removed from the system during routine cleaning operations. "Sanitary" indicates the absence of filth, pathogens, and foreign materials. Inspection is necessary to prove that clean and sanitary conditions exist and confirmed evidence must be available to substantiate the efficacy of routine cleaning operations (Seiberling, 1999).

A process flow diagram (PFD) is often employed as an early step in designing a process (Bowser, 2007). Figure 29.1 shows a PFD of a portion of an apple sauce manufacturing process that utilizes five unique mechanical separation steps (shown in bold). The mechanical separation steps of Figure 29.1 are described in Table 29.3 and are employed to clean and prepare apples for processing and to separate cooked pulp (raw apple sauce) from stems, core, skin, and seeds. The use of the mechanical separation process "finish" (see Table 29.3 for description) facilitates a low-labor, high-throughput production scheme for a superior quality commercial product.

Material and energy flows in mechanical separation processes can be identified using a PFD. Figure 29.2 shows a simplified materials flow based on the apple sauce PFD in Figure 29.1. Table 29.4 uses information from Figure 29.2 to calculate the material flow rate of product entering and leaving each step of the given portion of the apple sauce manufacturing process.

In the *Handbook of Sugar Refining*, Chou (2000b) lists the basic principles of sugar cane refining, which are adapted here to form the 10 basic principles of mechanical separation of foods.

Table 29.3 Purpose and description of mechanical separation steps shown in Figure 29.1.

Mechanical separation step	Purpose	Description
Destem	Separate attached stems, leaves and twigs	Apples are transported over a bed of rollers that rotate the apples and in the process pinch and remove any attached twigs, stems or leaves
Wash	Separate soil and residues	The fruit is deluged in water which dislodges or dissolves soil and residues
Dewater	Separate excess water from fruit	Shaking sieve conveyor and forced air are used to remove surface moisture
Decalyx	Separate calyx from fruit	Calyx (dried sepals of the flower) must be separated at this point with an automated trim knife or they will be very difficult to remove later
Finish	Separate seeds, skin, stem and core from fruit pulp	Soft fruit pulp is fluidized under the pressure of rotating paddles which force it through a screen, leaving the tougher, denser parts of the fruit behind

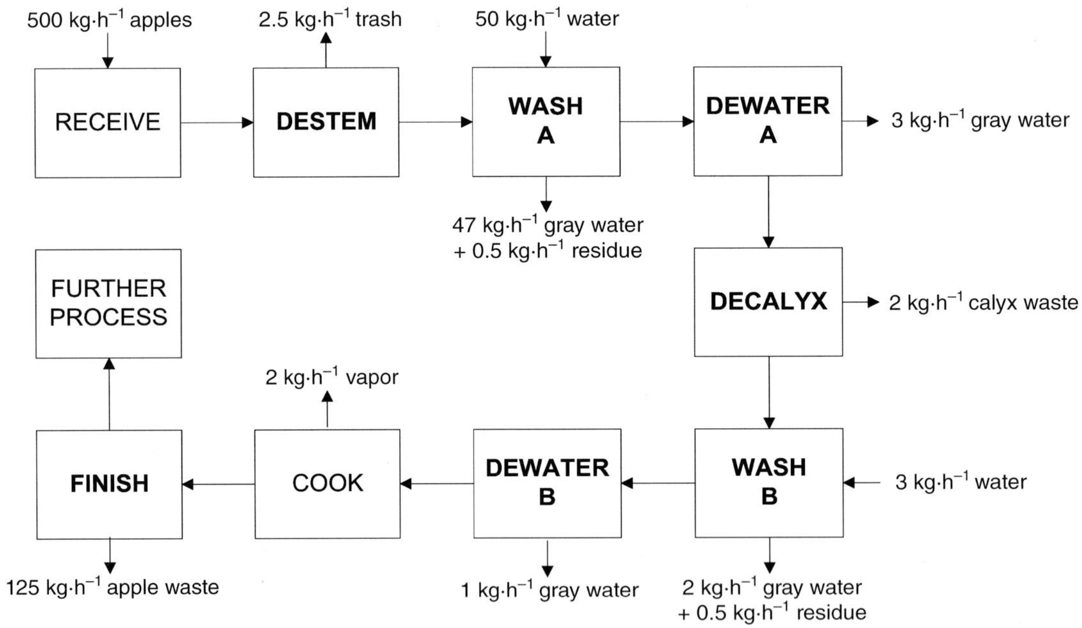

Figure 29.2 Materials flow for a portion of an apple sauce manufacturing process.

1. Once a separation has been made, the divided portions should not be recombined.
2. Each separation step should be completed to the highest feasible level.
3. Feedstock in process should be limited to the smallest quantity possible.
4. Separations should be carried out as quickly as possible.

Table 29.4 Materials flow rate between each process step of the PFD shown in Figure 29.2 (represented by connecting arrows between process steps).

Process step (see Figure 29.2)	Material input	Material output
Receive		500.0 $kg \cdot h^{-1}$ apples
Destem	500.0 $kg \cdot h^{-1}$ apples	497.5 $kg \cdot h^{-1}$ destemmed apples
Wash A	497.5 $kg \cdot h^{-1}$ destemmed apples	497.0 $kg \cdot h^{-1}$ washed apples
Dewater A	497.0 $kg \cdot h^{-1}$ washed apples	497.0 $kg \cdot h^{-1}$ washed dry apples
Decalyx	497.0 $kg \cdot h^{-1}$ washed dry apples	495.0 $kg \cdot h^{-1}$ decalyxed apples
Wash B	495.0 $kg \cdot h^{-1}$ decalyxed apples	494.5 $kg \cdot h^{-1}$ clean apples
Dewater B	494.5 $kg \cdot h^{-1}$ clean apples	494.5 $kg \cdot h^{-1}$ prepared apples
Cook	494.5 $kg \cdot h^{-1}$ prepared apples	492.5 $kg \cdot h^{-1}$ cooked apples
Finish	492.5 $kg \cdot h^{-1}$ cooked apples	367.5 $kg \cdot h^{-1}$ apple pulp
Further process	367.5 $kg \cdot h^{-1}$ apple pulp	

5. The number of separation steps should be minimized.
6. Separation processes should be simplified to the maximum feasible level.
7. Imposed quality standards should be surpassed, but not considerably exceeded.
8. Reprocessing of separated components should be reduced or eliminated.
9. Waste byproducts should be minimized or eliminated.
10. Separation equipment should be self-cleaning or easy to clean.

Process Control

Process control of mechanical separation systems is generally supplied by the equipment manufacturer. Additional controls may be needed to successfully install mechanical separation equipment in processes, especially for feed and discharge streams. Gravimetric and volumetric techniques are among the most simple and relatively accurate means of process control of bulk materials encountered in physical systems. Gravimetric control makes use of load cells or mechanical scales to measure ingredients or products based on weight. Weight measurement can be batch or continuous.

Batch weight may be obtained using load cells to weigh an entire vessel, or structure, including its contents. Net weight is obtained by subtracting the weight of the vessel and any attached hardware. An example of a continuous weight measurement is a force plate or weigh belt that measures loading and integrates it over time to compute total weight (Omega Engineering, 1998). Loss-in-weight systems incorporate load cells that support a hopper with an integral feeder. The feeder continuously empties the hopper and a control system modulates the rate of emptying based on a set point and feedback from the load cells (Siev *et al.*, 1993).

Volumetric control utilizes a known volume of space that is filled with the material being metered. The space is emptied and refilled periodically to establish a given volume displacement or a rate of successive displacements. A piston pump, rotary vane, and screw feeders are examples of continuous volumetric feed devices. Mass measurement and control systems, such as a coriolis meter, are extremely accurate since they measure mass directly. Changes in material density, viscosity, pressure, and temperature do not affect the measurement. Current coriolis meter designs include straight tubes with no flow obstructions (Micromotion, 2010), which are reliable and simple to clean.

Hazard and Safety Issues

Mechanical separation systems are prone to failure and human errors in application, operation, and maintenance are likely. Hazards and safety risks must be reduced to the lowest extent possible. Kletz (1998) offers excellent ideas for reducing process hazards, including an atlas of safety thinking, which lists five simple concepts to be applied: intensify, substitute, attenuate, simplify and change.

1. **Intensify** refers to the practice of reducing volumes of hazardous materials to less objectionable levels. Enclosure and shrinkage of separation processes that generate hazardous dust is an example of intensification.
2. **Substitute** means to use a less hazardous material or processing step. For example, supercritical CO_2 can be used as a solvent instead of hexane, and a wet separation can be used instead of dry to control dust.
3. **Attenuate** pertains to reducing the conditions that might cause hazards. For example, a separation process might be carried out at a lower ambient temperature to reduce the growth rate of unwanted microorganisms.
4. **Simplify** means to reduce process complexity and opportunities for error. Often processes are unnecessarily complicated to provide system flexibility or to accommodate existing facilities, equipment, and outdated specifications.
5. **Change** is needed as soon as hazards are identified. Early identification of hazards (during the design and review phases of a project) is critical.

Cleaning and Sanitation Methods

Features that enable the cleaning and sanitation of mechanical separation equipment make them suitable for processing foods. Two basic cleaning regimens are employed to clean mechanical separation equipment: dry and wet. Dry cleaning systems are suitable for low-moisture stable products like flours, grains, nuts, seeds, minerals, and powders. Brushes, air jets and vacuum systems may be used to substantially remove product residue from equipment. Periods between cleaning cycles may be lengthy for dry systems when product safety is not an issue.

Wet cleaning systems incorporate water-washing techniques, which might include rinsing, impingement, scrubbing, chemical additives, heating, steaming, and drying steps. Periods between water washing cycles are normally short and occur at least daily. Water washing removes food products and ingredients before they become hazardous due to microbial growth or other spoilage effects.

Corrosion resistance is a key factor for cleaning and sanitation, since many processed materials and cleaning agents cause corrosion of product contact surfaces. Some byproducts of corrosion are considered adulterants that will render the product unsaleable. Smooth, polished, impervious product contact surfaces without cracks or pockets make cleaning more effective. Seams and hollow sections in food contact areas are not permitted. Equipment design should allow for complete access and disassembly for access when necessary. Self-draining surfaces prevent puddling of materials. A concise summary of equipment design and construction considerations is given by Schmidt and Erickson (2005).

Capital and Operating Costs

Capital and operating costs of mechanical separation processes and associated equipment tend to be less expensive than other separation processes because they utilize existing forces of nature such as gravity and magnetic fields, and the innate physical properties of materials. A settling process to remove oil from water is less energy intensive and requires simpler equipment and controls than an evaporation or distillation process. Many pieces of mechanical separation equipment have few or no moving parts, further reducing costs. The requirements for cleanability and frequent need for corrosion-resistant materials increase capital costs compared with nonfood applications.

Most mechanical separation equipment runs automatically without the need for continuous, dedicated operator supervision. Significant operating costs include normal maintenance activities such as replacement of bearings, wear surfaces, belts, and seals; utilities like electricity, water, steam, and natural gas; cleaning and sanitation; and waste products disposal and handling.

Future Needs

Mechanical separation is an important operation in food processing that has ancient origins. Physical forces such as gravity and magnetism are used to separate mixtures based on differences in their physical properties. Mechanical separation processes are generally low cost and simple to implement. Future needs of mechanical separation processes in the food industry include improved cleanability of equipment, reduced waste, value-added processing, and improved control and separation methods. Cleanability is the chief requirement for all food processing equipment and rapid and effective cleaning methods will reduce costs and improve food safety. Mechanical separation processes should be designed to eliminate or reclaim waste streams. Valuable components in food and agricultural products must be identified and economic methods developed to separate them. Better measurement and control tools are needed to improve process efficiency and reliability. Finally, new separation methods should be developed to reduce or combine process steps.

References

Akulich, A.V. and Lustenkov, V.M. (2007) Optimization of the operating-design parameters of a combined dust separator. *Journal of Engineering Physics and Thermodynamics* 80: 358–363.

American Water Works Association (2007) *M46 Reverse Osmosis and Nanofiltration*, 2nd edn. AWAA, Denver, CO.

Answers.com (2010) Mechanical separation techniques. Available at www.answers.com/topic/mechanical-separation-techniques (accessed January 27, 2010).

Baker, M.N. and Taras, M. (1981) *The Quest for Pure Water. The History of the 20th Century*. American Water Works Association, Denver, CO.

Baron, R., Havet, M., Solliec, C., Pierrat, D. and Touchard, G. (2008) Numerical and experimental study of a continuous electrostatic smoking process. *IEEE Transactions on Industry Applications* 44: 1052–1058.

Barrington, S., Lesauteur, P., Shin, M. and Gelinas, J.B. (2002) Precipitation of swine and cattle manure phosphorus using limestone dust. *Journal of Environmental Science and Health B* 37: 613–623.

Bowser, T.J. (2003) Steam injection heating. In: *Encyclopedia of Agricultural and Food Engineering* (ed. D.R. Heldman). Marcel Dekker, New York, pp. 944–947.

Bowser, T.J. (2007) Food processing facility design. In: *Handbook of Farm, Dairy, and Food Machinery* (ed. M. Kutz). William Andrew Publishing, Norwich, NY.

Brennan, J.G., Butters, J.R., Cowell, N.D. and Lilly, A.E. (1990) *Food Engineering Operations*, 3rd edn. Elsevier Applied Science, London.

Britz, T.J., Schalkwyk, C.V. and Hung, Y.T. (2006) Treatment of dairy processing wastewaters. In: *Waste Treatment in the Food Processing Industry* (eds L.K. Wang, Y.T. Hung, H.H. Low and C. Yapijakis). CRC Press, Boca Raton, FL, pp. 1–28.

Broadbent, T. (2001) Centrifuges: the choice. *Filtration and Separation* 38: 30–33.

Chou, C.C. (2000a) Filtration processes. In: *Handbook of Sugar Refining* (ed. C.C. Chou). John Wiley & Sons, New York, pp. 155–168.

Chou, C.C. (2000b) Refinery design criteria. In: *Handbook of Sugar Refining* (ed. C.C. Chou). John Wiley & Sons, New York, p. 361.

Clark, N. (2009) King protein rules. Available at www.upfront-dk.com (accessed March 11, 2010).

Clark, P.J. III (2003) Mechanical separation systems design. In: *Encyclopedia of Agricultural and Food Engineering* (ed. D.R. Heldman). Marcel Dekker, New York, pp. 595–597.

Columbia Electronic Encyclopedia, 6th edn. 2007. Columbia University Press, New York.

Crapiste, G.H. and Rotstein, E. (1997) Design and performance evaluation of dryers. In: *Handbook of Food Engineering Practice* (eds K.J. Valentas, E. Rotstein and R.P. Singh). CRC Press, Boca Raton, FL, pp. 125–166.

Crittenden, J.C., Trussell, R.R., Hand, D.W., Howe, K.J. and Tchobanoglous, G. (2005) *Water Treatment: Principles and Design*, 2nd edn. John Wiley & Sons, Hoboken, NJ.

Dunford, N.T. and King, J.W. (2001) Thermal gradient deacidification of crude rice bran oil utilizing supercritical carbon dioxide. *Journal of the American Oil Chemists' Society* 78: 121–125.

Earle, R.L. (1983) *Unit Operations in Food Processing*. New Zealand Institute of Food Science and Technology. Available at www.nzifst.org.nz (accessed April 21, 2010).

Eriez Manufacturing Co. (2007) *How to choose and use magnetic separators*. Bulletin SB-95A. Erie, PA.

Etzel, M. (2004) Manufacture and use of dairy protein fractions. *Journal of Nutrition* 134: 996S-1002S.

Ferrari, B., Finocchiaro, F., Stanca, A.M. and Gianinetti, A. (2009) Optimization of air classification for the production of β-glucan-enriched barley flours. *Journal of Cereal Science* 50: 152–158.

Ferreira-Leitao, V., Perrone, C.C., Rodrigues, J., Franke, A.P.M., Macrelli, S. and Zacchi, G. (2010) An approach to the utilization of CO_2 as impregnating agent in steam pretreatment of sugar can bagasse and leaves for ethanol production. *Biotechnology for Biofuels* 3: 7.

Ferro, E., Ghiazza, E., Bosio, B. and Costa, P. (2002) Modelling of flash and stripping phenomena in deaerators for seawater desalination. *Desalination* 142: 171–180.

Garrote, R.L., Silva, E.R. and Bertone, R.A. (2000) Effect of thermal treatment on steam peeled potatoes. *Journal of Food Engineering* 45: 67–76.

Geankoplis, C.J. (1978) *Transport Process and Unit Operations*, 3rd edn. Prentice Hall, Englewood Cliffs, NJ.

Hadden, N., Baumann, F., MacDonald, F. *et al.* (1971) *Basic Liquid Chromatography*. Varian Aerograph, Palo Alto, CA.

Hanafy, M. and Nabih, H.I. (2007) Treatment of oily wastewater using dissolved air flotation technique. *Energy Sources, Part A: Recovery, Utilization, and Environmental Effects* 29: 143–159.

Henderson, S.M., Perry, R.L. and Young, J.H. (1997) *Principles of Process Engineering*, 4th edn. American Society of Agricultural and Biological Engineers, St Joseph, MI.

HRS Spiratube (2010) Deaeration system. Available at www.hrs-spiratube.com (accessed March 12, 2010).

IQS (2010) Industrial Quick Search Manufacturer Directory. Available at www.iqsdirectory.com/centrifuges (accessed March 10, 2010).

King, J.W., Sahle-Demessie, E., Temelli, F. and Teel, J.A. (1997) Thermal gradient fractionation of glyceride mixtures under supercritical fluid conditions. *Journal of Supercritical Fluids* 10: 127–137.

Kletz, T. (1998) *Process Plants: A Handbook for Inherently Safer Design*. Taylor and Francis, Philadelphia, pp. 199–202.

Leopold (2010) Operating principle of DAF. Available at www.fbleopold.com/ (accessed March 24, 2010).

Luk, G.K. and Liu, Z. (2005) Treatment of oily food industrial sludge with a magnetic coagulator. In: *Proceedings of the Annual Conference of Canadian Society of Civil Engineers, June 2005, Toronto, Ontario, Canada*.

Ma, Y. and Barbano, D.M. (2000) Gravity separation of raw bovine milk: fat globule size distribution and fat content of milk fractions. *Journal of Dairy Science* 83: 1719–1727.

Marshall, C.R. (1951) Oxidation in apple juice. II. Some observations on deaeration. *Journal of the Science of Food and Agriculture* 2: 321–327.

Micromotion (2010) Food and beverage solutions. Available at www.emersonprocess.com (accessed March 24, 2010).

Nunn, R. (1997) *Water Treatment Essentials for Boiler Plant Operation*. McGraw-Hill, New York.

Odenberg Engineering (2008) Odenberg peeling process technology delivers improved yield peeling line technology. Available at www.odenberg.com (accessed March 24, 2010).

Omega Engineering, Inc. (1998) *Transactions in Measurement and Control, Vol. 3, Force-related Measurements*. Pitman Publishing and Omega Press, New York, pp. 62–69.

Purchas, D.B. (1981) *Solid/Liquid Separation Technology*. Uplands Press, Croydon, England, pp. 610–619.

Retsch (2010) The basic principles of sieve analysis. Available at www.retsch.com (accessed March 24, 2010).

Riffer, R. (2000) Process selection. In: *Handbook of Sugar Refining* (ed. C.C. Chou). John Wiley & Sons, New York, pp. 363–378.

Ross, C.C., Smith, B.M. and Valentine, G.E. (2000) Rethinking dissolved air flotation (DAF) design for industrial pretreatment. In: *Water Environment Federation and Purdue University Industrial Wastes Technical Conference, St Louis, Missouri* (14): 43–156.

Salah, M.A. (2010) Cooling tower blowdown treatment using a Lamella® gravity settler. Available at www.iqsdirectory.com/centrifuges (accessed March 12, 2010).

Saravacos, G.D. and Kostaropoulos, A.E. (2002) *Handbook of Food Processing Equipment*. Plenum Publishers, New York.

Saska, M. (2000) Application of membrane technology in sugar manufacturing. In: *Handbook of Sugar Refining* (ed. C.C. Chou). John Wiley & Sons, New York, pp. 335–352.

Schmidt, R.H. and Erickson, D.J. (2005) *Sanitary design and construction of food equipment*. Document FSHN0409, Florida Cooperative Extension Service, Institute of Food and Agricultural Sciences, University of Florida, Gainesville.

Seader, J.D. and Henley, E.J. (1998) *Separation Process Principles*. John Wiley & Sons, New York.

Seiberling, D.A. (1999) CIP sanitary process design. In: *Handbook of Engineering Practice* (eds K.J. Valentas, E. Rotstein and R.P. Singh). CRC Press, Boca Raton, FL, pp. 582–583.

Senden, M.M.G. and Tels, M. (1978) Mathematical model of vertical air classifiers. *Resource, Recovery and Conservation* 3: 129–150.

Seo, Y. and Lee, S. (2009) *Effect of a hydrocyclone and a centrifugal separator in purifying rolling lubricant*. European Aerosol Conference, Karlsruhe, Abstract T077A01.

Siev, R., Mair, D.C. and Liptak, B.G. (1993) Solids flow meters and feeders. In: *Flow Measurement* (ed. B.G. Liptak). Chilton Book Company, Radnor, PA, pp. 127–141.

Spirax Sarco (2010) Pressurized deaerators. Available at www.spiraxsarco.com (accessed March 12, 2010).

Sweco (2002) *Superiority of Sweco Cloth*. Bulletin 457. Sweco, Division of M-I, LLC, Florence, KY.

Tarleton, E.S. and Wakeman, R.J. (2007) *Solid/Liquid Separation: Equipment Selection and Process Design*. Elsevier, Oxford, pp. 11–25.

Wilhelmi, A.J., Wiesenborn, D.P., Gustafson, C.R. and Pryor, S.W. (2009) Models for fractionation of field peas to supplement corn ethanol. *Applied Engineering in Agriculture* 25: 709–717.

Zenz, F.A. (2001) Cyclone design tips. *Chemical Engineering* 108: 60–73.

30

Mixing and Agitation Design

Siddhartha Singha and Tapobrata Panda

Introduction

Mixing and agitation are important unit operations in the food industry. Although it rarely serves as a stand-alone step in a processing operation, it is unquestionably one of the most important steps that determine the sensory, organoleptic, and nutritional quality of the end product. The food industry constantly tries to reduce production costs due to the strong competition of the global market and also has a strong drive to diversify product range. To meet the requirement of these two objectjves, mixing is considered an important unit operation. Much of the industry is trying to upgrade conventional mixing processes or to opt for advanced mixing solutions. Prior to the in-depth discussion in this chapter, it is necessary to distinguish the two apparently close terms "mixing" and "agitation." Agitation is technically well defined and controls pattern of flow in a fluid. Power consumption is the single most important criterion for characterizing agitation processes. On the other hand, mixing is defined by attainment of homogeneity in the end product and is not always well defined (Griskey, 2006). In this chapter, the term "mixing and agitation" defines an operation where at least uniform distributions of two or more components occur by promotion of flow. In the food process industry, the purpose of mixing and agitation is not only to increase homogeneity of the bulk material but also to achieve many process objectives. At times the objective is so complex that the equipment involved is not recognized as a

Handbook of Food Process Design, First Edition. Edited by Jasim Ahmed and Mohammad Shafiur Rahman.

simple mixer, but is named after the process. Equipment such as emulsifiers are basically mixers, but they cause uniform distribution of two or more components, i.e. stable dispersion of oil/aqueous droplets in a continuous aqueous/oil phase. Therefore, mixers often serve one or more purposes in many food processes. Operations may be classified as liquid or solid mixing and a few examples are listed below.

- **Liquid mixing**: blending (liquid whole and skim milk in tank), enhanced heat and mass transfer effect (chocolate mix heating), accelerating reaction (enzymatic hydrolysis of corn starch for HFCS production), change of texture (conching), size reduction (emulsification of fluids), dissolution of solids in liquid/blending of miscible liquids (flavor-syrup mixing), dispersion of gas/solids/liquids in liquid (carbonation/fruit pieces in yoghurt/emulsification), drying (concentrated milk), kneading (dough preparation).
- **Solid mixing**: blending (different types of cocoa bean mixing), mixing of active ingredients into bulk powder (vitamin fortification in flour), mixing of multi-component mixers as formulation (cereal mix), admixer of liquid(s) in solid material (oil addition in instant soup mix), particle coating (chocolate coating), powder coating on particle (spice on snacks), cohesive powder/paste mixing with particles (mixing of salad constituents).

This unit operation can appear in various stages of a food process and with different degrees of impact on final product quality. For example, in a typical dairy, mixing may be required in the milk receiving tank (mixer with approximately 4000 liters capacity capacity and mixing time about 10 min), milk standardization process (high-shear batch mixers, rotor–stator type, 4–15 kW power), solid ingredients premixing (ribbon mixers to mix powders like sugar, cocoa powder and carrageenan in 5–10 min), in-line solid–liquid blending (rotor–stator mixer with 2000 $L{\cdot}h^{-1}$ output), fermentation tank (agitated tank with low shear impeller), and yoghurt mixing (static mixer with output up to 10 000 $L{\cdot}h^{-1}$)

In subsequent sections theoretical discussion will follow description of different mixer types and their design principles, operational features and cost analysis of mixing processes.

Mixing and Agitation: Theoretical Principles

Food processes deal with raw materials with different physicochemical properties. For example, standardization of milk is a common unit operation in any dairy to attain proper process specification, and typically needs mixing of solid nonfat (cohesive solid powder), cream (highly viscous non-Newtonian liquid), and milk (liquid). The situation becomes more complex when the mixing operation accompanies physicochemical reactions and phase change. In complex processes, such as bakery dough preparation, flour, water, yeast, salt and other ingredients are mixed and worked to obtain a plastic mass. Therefore the dough mixer should not only uniformly distribute the multiple solid and liquid raw materials but also facilitate gluten development that eventually gives rise to a viscoelastic dough. Such complexity demands expansion of mixing theories developed traditionally for the chemical industries.

Assessment of Mixing Quality

Mixing in food processes can be classified, in relation to the continuous phase, as liquid mixing (follows laws of fluid mechanics) and solid mixing (follows laws of solid physics and statistics). The theory of fluid mixing is well developed due to its inherent simplicity and wide use across the process industries, whereas solid mixing is generally empirical. They also differ in assessment. When one or more component mixes, the element size of the component(s) (particulate solid or liquid) becomes smaller and smaller in order to attain a desired size. These arrange themselves throughout the available volume of the mixer or continuous phase during the mixing operation. At some point in such mixing, the uniformity can be quantified by two parameters, the *scale of segregation* and the *intensity of segregation*. The scale of segregation is the average dimension of separated regions, whereas the intensity of segregation is the different concentration of the minor components throughout the mixture. As the mixture becomes more homogeneous, both parameters should decrease. In practice, after some time, the scale of segregation becomes less than a minimum detectable level and sample variance may be preferred. In an ideal mixer, complete mixing can be confirmed when every sample analyzed at regular intervals during mixing as well as from different locations within the mixer will have the same composition of individual components as the whole mixture. For completely mixed samples, the variance (S^2) of concentration of minor component(s) should be zero. However, mixing components are too different in their properties and never attain such conditions in practice. Especially for solid mixing, where no intrinsic motion like molecular diffusion is present and mixing simply randomizes the particles, uniformity is measured statistically (Lacey, 1943). Thus to assess mixing of a particulate mass, *scale of scrutiny* becomes essential as it gives an idea about the best sample size. Then an acceptable value of variance (S_∞^2) needs to be fixed to determine the end point of mixing. In the case of miscible liquid–liquid mixing or solid dissolution, the mixture eventually becomes single phase, resulting in S_∞^2 being zero.

In a batch mixer, mixing time is determined by the mixing index (M), a dimensionless parameter. A few expressions for M are as follows:

$$S_\infty (S_0 - S) \Big/ (S_0 - S_\infty)_S \quad \text{(Beaudry, 1948)}$$

$$(S_0^2 - S^2) \Big/ (S_0^2 - S_\infty^2) \quad \text{(Lacey, 1954)}$$

$$\left[(\ln S_0^2 - \ln S^2) \Big/ (\ln S_0^2 - \ln S_\infty^2) \right]^{0.5} \quad \text{(Ashton and Valentin, 1966)}$$

for $0 < M < 1$. M at 0 and 1 indicates completely unmixed and fully mixed solids, respectively. S_∞^2, S_0^2 and S^2 are the variances at an accepted mixing level, at a completely segregated condition, and at any point during mixing, respectively. The choice of mixing index depends on a particular mixing operation and should correlate linearly with time. Using one of the above mixing indices, rate of change of variance (dS^2/dt) can be expressed in terms of the driving force, i.e. $(S^2 - S_\infty^2)$ by Equation 30.1, which defines mixing time for a particular level of mixing (Equation 30.2):

$$\frac{dS^2}{dt} = -k\left(S^2 - S_\infty^2\right) \tag{30.1}$$

$$\text{or } \left[\left(S^2 - S_\infty^2\right)/\left(S_0^2 - S_\infty^2\right)\right] = M = e^{-kt} \tag{30.2}$$

In the food industry, mixing is often undertaken with more than two components, as well as with multiple phases, and thus defining a "suitable" mixing index is a difficult task. It is also problematic to choose a local variable (e.g. concentration of a minor component) and then monitor it accurately in order to obtain a quantitative expression of the mixing index. However, the assessment of a food mixing operation is performed not only by measuring "degree of homogeneity" of the mixture in terms of one of the minor components but also by textural and/or organoleptic properties.

Mechanisms of Mixing

The selection of an appropriate mixer or optimization of a mixing process requires an understanding of the mechanism of mixing. Usually, in any liquid undergoing mixing, two regions are visible: a convective flow region that distributes mass throughout the space, and a high-shear region that increases contact area or reduces the degree of segregation. In liquid phase mixing, the mechanism depends on the flow pattern, i.e. laminar flow, turbulent flow, and transition flow, which is characterized by Reynolds number (Re = $dv\rho/\mu$) where ρ is density and μ viscosity of the liquid. On the other hand, characteristic length (d) and velocity (v) vary for different mixers. The ranges for Re for different set-ups are as follows: stirred tank, 10–10000; static mixer (SMX Sulzer Static Mixer), 300–1500/3000 (Hirech *et al.*, 2003); jet mixer, 100–2000 (Revill, 1992); and simple pipe, 2100–4000. Practically, when the viscosity of either of the components and the mixture is typically less than 100 mPa·s (www.komax.com/det-reynolds.html), flow in a mixer is expected to be turbulent (Hirech *et al.*, 2003). Liquid in such a flow contains fluid elements in swirling motion or eddies that exchange position in space. Larger eddies decay into smaller ones to produce a fast mixing effect (convective mixing) and eventually eddies decay to the molecular level (diffusive mixing). In a mixer, both mechanisms prevail simultaneously. In an agitated tank, diffusive mixing occurs mostly near the agitator whereas convective mixing prevails in the rest of the space. The dissolution of sugar in water and the blending of milk and water in a propeller-agitated tank are unique examples of turbulent mixing. When mixing highly viscous liquid(s), flow promotion and mixing becomes difficult. In laminar flow, the liquid phase moves in layers, each layer moving with different velocity. The interfacial shear acting between these adjacent layers is called laminar shear. For such viscous liquids, if velocity increases in the direction of motion, an idealized elongation flow takes place that deforms the fluid elements. Apart from these two mechanisms, the mixer element can generate a "slicing and replacing" effect that redistributes the components throughout the mixing space (distributive mixing). The three mechanisms discussed so far reduce the degree of segregation or increase the contact area between different components but the segregated elements finally decay by molecular diffusion. For a specific mixer the contribution of each mechanism may vary depending on a particular process and material (Lindley, 1991).

No molecular diffusion exists in powder mixing. The mixing is always assessed with respect to a predetermined scale of scrutiny at a coarser scale (Lindley, 1991). The spatial orientation of particles in a mixer is the result of two opposite phenomena, randomization and segregation. Solid particles not only need to be mixed carefully but the mixture requires careful handling, otherwise well-mixed particles can completely segregate during downstream processes (Ottino and Khakhar, 2000).

Food products encompass a variety of solids, for example free-flowing solids such as salt or cohesive solids such as milk powder. Differences in the properties of the particles to be mixed, i.e. size, shape, density, and resilience, causes them to segregate. Particle size determines cohesiveness or flowability of the material (Uhl and Gray, 1986). Smaller particles tend to stick together because of inter-particle forces such as van der Waals' force, electrostatic forces, and moisture bonding forces. On the other hand, bigger particles tend to separate due to gravity. As a rule of thumb, any particle larger than 75 μm will completely segregate. Particles of less than 10 μm in size show no tendency to segregate (Nienow *et al.*, 1992). Complex food particles quite often deviate from this generalization. In any case, increased cohesiveness restricts particle flow as well as segregation. A good mixing process needs to optimize between ease of flow and segregation. When two particulate materials are mixed, a group of particles can displace relative to each other and create homogeneity at a coarser scale for convection mixing. For diffusion mixing, individual particles move relative to each other and create mixing at finer scale. This is different from molecular diffusion in liquid phase caused by the Brownian motion of molecules. In another mechanism, shear mixing, solid particles slip along failure planes in the material. The contribution of these mechanisms to the mixing of solids depends on the process as well as the mixer design employed (Bayram and Göğüş, 2008). If very fine powder is mixed with coarse particles, the fine particles can stick to the larger particles because of surface properties and in this way segregation is avoided. In a liquid admixture process, such as mixing fat in dry soup mixes, micro-level coating of particles occurs. In agglomeration, liquid bridges (mostly moisture) hold particles together to produce coarser particles (Niranjan *et al.*, 1994).

Mixing Equipment: Mode of Operation and Comparative Analysis

Various types of mixers and agitators are used to perform complex food mixing. In industry, agitators (impellers, pumps, etc.) are often regarded as simple "processing aids" because they are an auxiliary part of many unit operations (heat transfer, crystallization, etc.). For simplicity, this chapter defines the terms "mixers" and "agitators" synonymously. In the food industry, mixing can be achieved (at least partially) with simple equipment (hoppers, dosing pumps, and open pipelines) or with specially designed equipment (static mixers or rotor–stator mixers). These can be classified by mode of operation (batch/continuous, in-line/in-tank), scale, application, or mixing intensity (high flow/low shear, low flow/high shear). There are two categories: liquid mixers deal with pure liquids, solutions or suspensions where the major phase is liquid of various consistencies (low viscosity, slurry, paste or plastic mass); and solid mixers handle powders with or without moisture, particulates with or without moisture and with various flow properties (free flowing to cohesive).

Mixers for Liquid Mixing

In the food industry, many liquid mixers are used for various operations. Discussion of all of them is outside the scope of this chapter, but it would be logical to investigate some of the most popular liquid mixers in order to examine their methods of operation, scope, and limitations. Liquid mixing operations can be in-line or in-tank. As the name suggests, in-line mixing does not require a dedicated vessel, unlike in-tank mixing. Each has its advantages and disadvantages. Often, choice between the two is not obvious. For some food mixing processes, a hybrid system utilizing both types may be advantageous.

In-line Mixing Equipment

There are many types of equipment which can be used to enhance mixing within pipeline, for example elbows, nozzles, orifice plates, Tee mixers, jet mixers, valves, mixing pumps, in-line static mixers, and in-line dynamic mixers. Apart from in-line static and dynamic mixers, the other devices have a very narrow range of application and are not discussed further. In a static mixer, a helical (or some other shape) mixing element is placed in a pipeline to restrict the flow and impart a mixing effect on the components. The pressure drop across the element is converted into energy for mixing. This is a good alternative to agitated vessel mixers because of low equipment cost, minimal space requirement, maintenance, power requirement, and rapid mixing time. Mixing can occur in both laminar and turbulent regimes depending on the process fluid. In laminar flow, mixing occurs because of flow division and redistribution. In turbulent flow, the mixing element magnifies the degree of turbulence and radial mixing compared with the same phenomenon in an empty pipeline. For a laminar-type static mixer, the greater the number of stationary elements, the greater the uniformity of the material, but the increased pressure drop leads to the requirement for a pump with higher capacity. Such mixers are useful for mixing soluble solids or miscible liquids in highly viscous liquids, or dispersion of gas or liquids in highly viscous liquids (e.g. sauce, margarine, jams, yoghurt). On the other hand, when blending soluble and less viscous liquids or dispersion of gas/liquids in low viscosity liquids, the stationary element promotes eddy formation. Smaller pipe length and fewer number of mixing elements is adequate to create homogeneity in a turbulent flow regime than in a laminar flow regime. Another class of in-line mixer employs a moving part to promote flow and generate shear. There are basically two types, rotor–stator type and extruder type. Both generate locally high shear forces that cause mixing (Etchells and Meyer, 2004).

In general, an in-line mixer is a good choice if a mixing process requires the following:

1. continuous operation;
2. handling a fluid with process viscosity <50 mPa·s;
3. solid in liquid mixing with uniform solid particle size and consistent concentration;
4. low residence time (process time being critical for loss of quality);
5. process pressure is too high or seal problem;
6. limited mixing duties like blending, dissolution, dispersion, heat transfer and reaction in plug flow mode (e.g. enzymatic reaction with starch solution);
7. limited space and access;
8. limited sanitation and maintenance;
9. precise mixed product quality;
10. low capital cost.

It is not feasible to compare the efficiency of mixing using this equipment for a given process. Each is case specific and beyond the scope of this chapter. Generally, static mixers and rotor–stator in-line mixers are quite useful in several mixing operations across the food industry, especially in laminar flow regime. Static mixers are attractive for food processors because of proven process capacities in bulk mixing as well as in speciality mixing.

In-tank Mixing Equipment

In-tank mixers comprise two major components: a vessel and a flow promotion system. The flow promoter may be a mechanical agitator or a forced flow of gas or liquid. Design of both the components influences the performance of the mixer. Theoretically, there are many tank geometries (cylindrical body with spherical/ASME dish/conical bottom, cuboid, etc.) and types (with or without baffle) that can be used for mixing. In the food industry, the choice of tank for tank mixers is limited by other operational constraints like ease of cleaning and sanitation. Dished or shallow bottom and unbaffled tanks are preferred for food mixing. Complementary to tank design, there is a range of impeller designs available to impart different mixing duties. In jet mixers, instead of an impeller a liquid jet generates the motion required for mixing. The various designs of in-tank mixer can be classified as general purpose or special purpose. The first category is simpler in design and operation than the second category, which is complex in design and operation and specific for a process or even a product

General-purpose Mixers

These types of mixers are generally used for simple mixing duties, for example blending of low-viscosity liquids, solid–liquid or liquid–liquid dispersions. Design and operation are simple as well as flexible and they may or may not be dedicated for a process.

Agitated Tanks

This is the most common type of mixer in the food industry, where an impeller mounted inside a cylindrical vessel moves the liquid along with other ingredients. Movement of the impeller creates a three-dimensional velocity field: a radial component (V_r) moves the liquid element towards the wall, an axial component (V_a) moves the liquid element along the axis of the impeller up or down, and a tangential component (V_t) rotates the liquid element in the horizontal plane. Axial and radial velocity components contribute more to mixing than the tangential velocity component. Three basic types of impeller (propeller, turbine, and paddle) predominantly generate axial flow, radial flow and tangential flow, respectively (Figure 30.1a–c).

Paddle impellers are large flat blade(s) fixed on a shaft and cover 50–70% of tank diameter. They produce more radial flow and very little axial flow at moderate operational speed (20–150 rpm). Such impellers improve heat transfer and handle crystallization, liquid blending, and dissolution of solids. However, they are expensive and often need baffles or other accessories for better mixing. When there is a need for a variety of mixing processes (blending, air dispersion and emulsion premixing with moderately viscous liquid), a turbine impeller is a better choice. This has at least four blades fixed on the same boss that cover 30–50% of the tank diameter. Turbine impellers, despite being expensive, can generate strong axial and radial components at 30–500 rpm. Propeller mixers are useful for low-viscosity high-volume mixing, solid dispersion and simple emulsification duties. They operate at very high speed (500 to a few thousand rpm) and promote rotational and axial flow. In contrast to the other two types, they have shorter blades that cover about 25% of tank diameter.

Hydrodynamically, in agitated vessels turbulent mixing is expected, where eddies can increase mixing intensity. In fact, a mixer contains both laminar and turbulent regions. There is more turbulence close to the impeller and baffles. The state of turbulence strongly depends on liquid rheology as well as impeller type. Impeller power number provides an indication of the expected turbulence. Depending on what kind

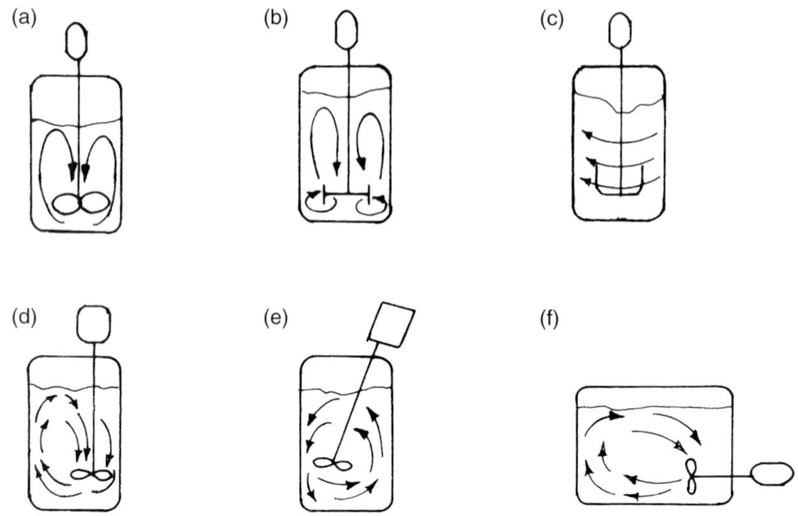

Figure 30.1 Schematic of agitated tank showing (a) axial flow impeller, (b) radial flow impeller, (c) tangential flow impeller, (d) off-centered axial flow impeller, (e) inclined axial flow impeller and (f) horizontally mounted axial flow impeller. Arrows indicate flow patterns generated by different types of impeller and different position of impeller.

of mixing is desired, various types of impeller are employed in the food industry. However, impeller manufacturing companies are constantly upgrading impeller designs for versatile application and improved efficiency. In addition to this, one can frequently encounter problems with such rotating impellers, for example vortex formation, centrifugal segregation, and dead space. To avoid this, baffles, alternative blade geometries (instead of flat blades), multiple blades, and different orientation of impellers and multiple impellers can be employed. Different orientation of impellers, for example off-centered, inclined and horizontal mounting (see Figure 30.1d–f), exhibit distinct flow patterns compared with a centrally located and vertically mounted impeller.

Portable mixers are a popular variant of the agitated vessel mixer for low-volume (<1000 L) and flexible mixing duties in the food industry. They come in various sizes and shapes, with impellers usually clamp-mounted or cup-mounted near the side of an open tank, usually without baffles. The impeller makes a narrow angle with the vessel axis for better volumetric flow and mixing (see Figure 30.1e). The type of impeller, shaft length, and angle of offset from the axis govern the relative magnitude of the axial, radial, and tangential velocity components at a given rotation. The relative magnitude of the velocity components is important for optimizing a portable mixer for a given type of mixing duty. Standard makes of such mixers have a propeller impeller with motor power between 0.25 and 3 hp, with or without a gear drive for multiple or single speed output.

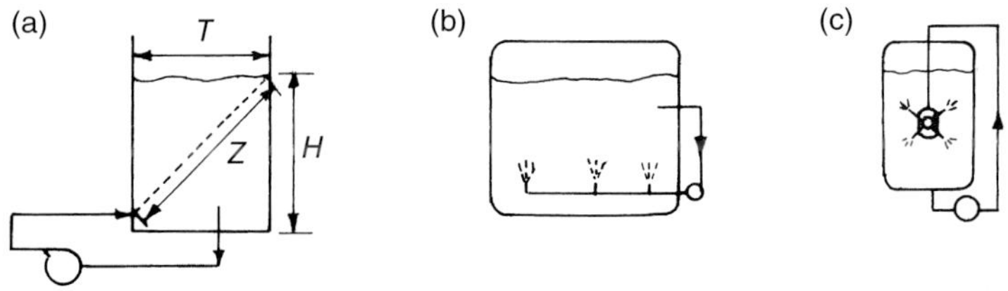

Figure 30.2 Different jet mixer configurations showing (a) single static nozzle, (b) multiple static nozzle and (c) four rotary head.

Jet Mixers

For low-viscosity processes, a submerged nozzle can be used to introduce a high-velocity jet of fluid into another slow-moving/static fluid to create mixing. Such mixers can be used in combination with agitated tanks or as a stand-alone mixing element in-tank/in-line (Revill, 1992). Available designs include static heads and rotary heads. In static-head designs, one or more nozzles is fixed and each nozzle creates a jet in one direction all the time (Figure 30.2a,b). In the rotary-head design (see Figure 30.2c), one or more rotating nozzles in a plane simultaneously generate a high-velocity jet for more efficient mixing than fixed-head jets (Nordkvist *et al.*, 2008). Although fixed head jet mixers are suitable for general purpose mixing in large tanks, the rotary head type jet mixers possess extended applications. This type of mixer is ideal where stringent sanitation is required. Rotary jet mixers can be used for mixing sugar syrups, liquid ingredients and aroma with water, mixing of "smoothies," deoxygenation and carbonization of water, beer, and soft drinks, and even for dissolution of solid material like food polymers. Advanced devices can be used for multiple roles such as mixing, deoxygenation, carbonization, and CIP.

Special-purpose Mixers

This class of mixers handles difficult powder dissolution, solid or liquid dispersion in liquid, emulsification with wide-viscosity liquids, and special food processes involving slurrys, pastes or viscoelastic dough. There are two types: high-shear agitated vessels and tanks with rotor–stator assembly.

High-shear Agitated Vessels

These are specially designed vessels for mixing highly viscous paste, slurry or dough. For highly viscous substances (>5000 mPa·s) in agitated tanks, liquid motion decreases gradually with distance from the impeller and turbulent mixing is not possible. Since for non-Newtonian liquids (viscosity depends on shearing time or shear rate) shear rate decreases from impeller to tank wall, different apparent viscosities exist inside the mixer. The situation can be even more complex where processing alters the rheology of the components, such as increase in viscosity, during mixing. This causes nonuniform mixing. The problem can be avoided by using larger impellers or multiple impellers to induce intimate contact between the mixing element and the process liquid. However, this is at the expense of high power. Often, viscoelastic food materials (e.g. dough) require specific mechanical work to be peformed to achieve a particular quality. Many subgroups of mixer belong to this category and mostly are custom made for a given food process. Some important and versatile designs are worth mentioning and five basic types are described in Table 30.1 and Figure 30.3. Because they execute

Table 30.1 Comparative analysis of various high-shear agitated vessels.

Type of mixer	Description	Limitation(s)	Purpose
Pan mixers or vertical mixers (Type I)	Stationary vessel with single or double impellers of 95% of tank diameter (anchor–gate type, multiple paddle type, helical, etc.) rotating at fixed place	High power requirement with increase in impeller size	Useful for batter mixing, low shear dispersions
Pan mixers or vertical mixers (Type II)	Stationary vessel with single or double impellers rotating in a planetary path (rotation speed 40–370 rpm). Different blades are used for different objectives: hooks for dough mixing, whisks for batter preparation	Low volume and batch operation	Useful for paste mixing, ingredients blending, dough mixing, wheeping and whisking at low power requirement
Horizontal trough mixers	Horizontal trough with one or two heavy blades rotating parallel to the vessel axis (at 35 and 70 rpm). Mixing capacity 45–1360 kg of dough. Two blades can rotate independently or in a coordinated manner. Blades (sigma or Z-type) move tangentially or intermeshing. Power input varies between 0.065 and 0.8 $kW{\cdot}kg^{-1}$. "Short dough" needs much less power	Precise control required for mixing of viscous material due to generated heat and local reaction	Can perform wide variety of mixing like kneading, dispersing masticating
High-speed mixers	Construction similar to vertical or horizontal type operating at higher agitator speed (300–1200 rpm) for higher output rate (3000–5400 $kg{\cdot}h^{-1}$). Better for small batches (220–385 kg) with high-moisture dough. Sometimes partial vacuum may assist mixing	Precise control system required	Best for faster dough process especially in continuous mode (mixing time ~3 min)
Screw conveyors or extruder	A single or double Archimedean screw moves materials to knead and mix through obstruction in a cylindrical barrel. The vessel wall can have tooth intermeshing with the screws to impart additional shear. Space between the screws as well as space between the screws and the wall is high-shear zone	Precise control required for mixing viscous material due to generated heat and local reaction situation	Kneading and mixing viscous pastes like margarine or butter

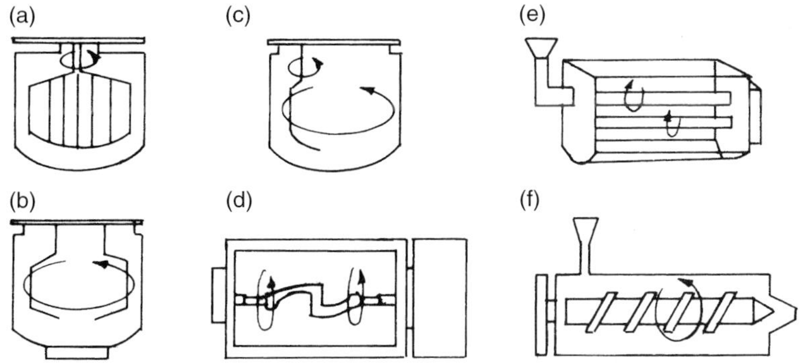

Figure 30.3 Simplified design of (a) a single impeller pan mixer (type I), (b) a double impeller pan mixer (type I) with bottom clearance, (c) a single impeller planetary mixer, (d) a horizontal trough mixer with Z-type blade, (e) a double roller high speed mixer and (f) an extruder.

specialized mixing duties, their design frequently includes automatic control systems, vacuum environment, and cooling arrangements.

Rotor-stator Mixers

The basic design of such a mixer comprises a high-speed rotor surrounded by a stationary perforated casing. The assembly draws liquid in and expels it at high velocity to create a circulatory motion as well as high shear. The liquid flow pattern for a typical rotor–stator head is shown in Figure 30.4(d). This is ideal for high-shear dispersions like gum dispersion in water or other viscous food mixture. There are various designs of rotor and stator assemblies which at different rotor speed can achieve a wide range of mixing duties (see Figure 30.4a–c).

Mixers for Solid Mixing

Compared with liquid mixers, solid mixers are relatively simple devices. Because solid mixing science is yet to be fully explored, the selection and operation of mixers are based on experience rather than theory. Six basic types of mixer are employed in the food industry: tumbler mixers, agitated mixers/convective mixers, high-shear mixers, pneumatic mixers, and gravity silo mixers. Often, more than one mixer can be used for the same application.

Tumbler Mixers

This is the simplest solid mixer, consisting of a covered vessel rotating about an axis that rolls over the particles inside the vessel. Common vessel geometries are double cone (Figure 30.5a) and V or Y shaped (Figure 30.5b). Typical speed of rotation for such

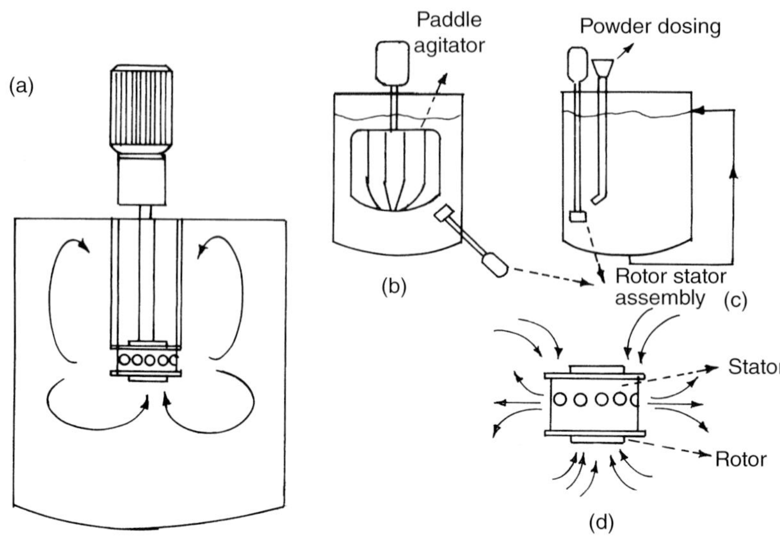

Figure 30.4 Schematic diagram of (a) a typical rotor–stator assembly in tank, (b) rotor–stator mixer assisted with a paddle agitator for highly viscous mixing, (c) rotor–stator mixer with recirculation system for difficult powder dissolution, and (d) rotor–stator head with circular aperture.

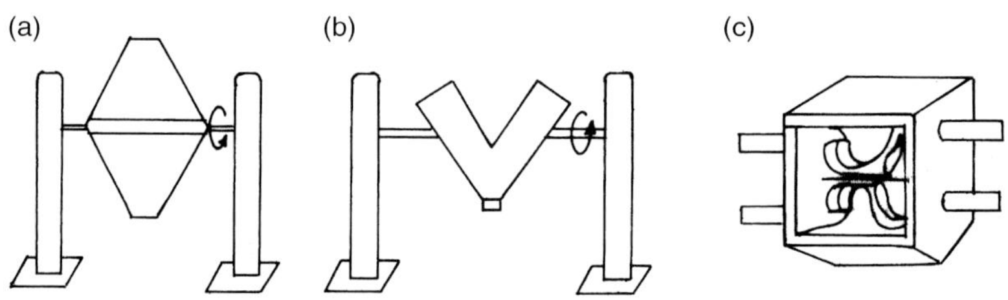

Figure 30.5 (a) Double cone tumbler mixer; (b) V-type tumbler mixer; (c) double arm high shear mixer.

a mixer is 15–60 rpm. They can be as big as 50 m^3 and are normally operated at under half capacity. Addition of internal impellers, application of vacuum and/or baffles can improve efficiency of mixing. Usual mixing time is about 15 min. These mixers are useful for coarse mixing operations, such as tea blending, sandwich fillings, meat, poultry and vegetable products.

Agitated Mixers/Convective Mixers

These mixers have a stationary vessel with an impeller. The impeller moves the mass of particles from one point to another and creates mixing. There are many impeller

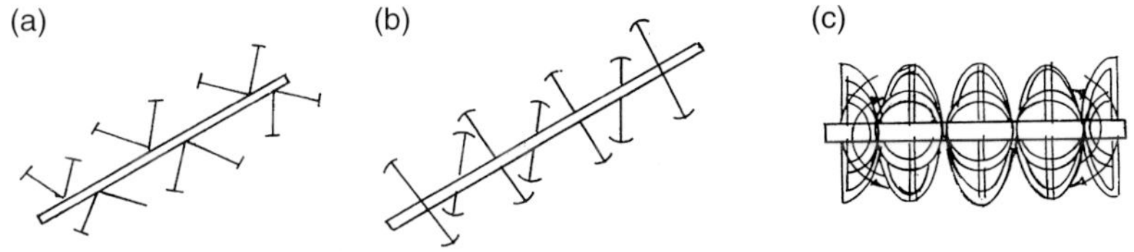

Figure 30.6 Simplified design of agitators: (a) plow type, (b) paddle type and (c) ribbon type.

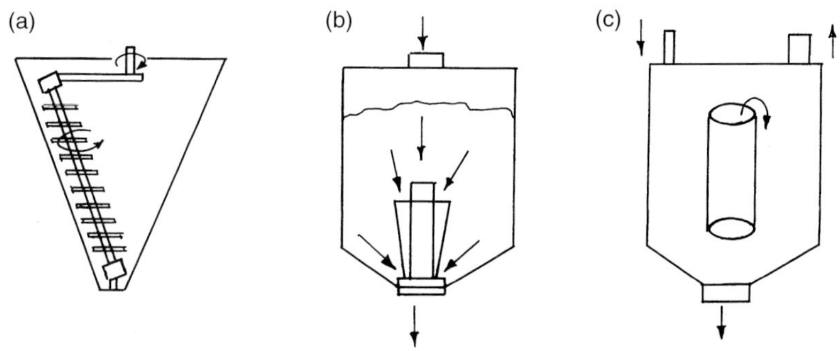

Figure 30.7 Schematic diagram of (a) orbital screw (Nauta) mixer, (b) gravity silo mixer and (c) modified air merge mixer.

designs, for example ribbon type, Z type, Archimedean screw, or paddle type (Figure 30.6). A twin impeller is used for heavy mixing duties. All designs should ensure mixing by sweeping the entire mixture volume or by maintaining natural circulation to force the entire mass through the impeller region. These mixers are particularly useful for powders with agglomerating tendencies (burger mixes, spices and dry ingredients). There are five common types of convective mixers: paddle mixer, plow mixer, ribbon mixer, orbital screw (Figure 30.7a), and sigma mixer (see Table 30.2 for comparative analysis).

High-shear Mixers

These are similar to convective mixers but can generate higher shear, resulting in size reduction and simultaneous mixing (see Figure 30.5c). They are ideal devices for converting improperly mixed cohesive powders by rupturing clumps. Unfortunately, because of low heat dissipation by solids, the heat generated by the high-shear conditions may sometimes cause heat accumulation. In such cases, a control mechanism is required to retain the organoleptic quality of the food products.

Table 30.2 Different types of convective solid mixers with their comparative features.

Parameters	Plow mixer	Paddle mixer	Ribbon mixer	Orbital screw (Nauta mixer)	Sigma mixer
Typical ingredients handled	Particle size up to 5 mm. Free-flowing to cohesive powder	Cohesive powder	Particle size up to 5 mm. Free-flowing to cohesive powder	Particle size up to 500 μm, from free-flowing to cohesive powder	Narrow application in dry mixing but suitable for thick viscous pastes or dough
Mode of operation	Batch/continuous	Batch/continuous	Batch/continuous	Batch	Batch
Structural features	Single or double shafts with plows at regular intervals Spray nozzles for agglomeration High-speed choppers for lump breaking	Single or double shafts with paddles Spray nozzles for agglomeration Heating and cooling arrangements	Single or double shafts with ribbons Choppers for lump breaking Heating and cooling arrangements	Direction of screws can be from above and below Satellite screw can be placed to increase efficiency Heating and cooling arrangements	Different shaped blades (sigma, spiral, masticator, etc.) are available for specific applications Spray nozzles Heating and cooling arrangements Grooved core for better shear
Mixing time (min)	<5 with random mixing	<6 with random mixing	30–120 (for fibrous materials)	<10	10–30
Limitations	For CIP/SIP proper seal needed Complete emptying is difficult	Properly sized bearing is required	Ribbon must withstand load Drive needs to be designed properly Clean up difficult	Seals are expensive Clean up not easy for sticky substances Friable products may degrade	Heat removal difficult Cleaning is difficult
Energy consumption ($kW \cdot m^{-3}$)	<150	<150	<12	<80	<80
Allowable fill of vessel size	30–70%	40–85%	40–85%	10–75%	40–65%
Vessel size (m^3)	Up to 40	Up to 40	Up to 50	Up to 50	Up to 60
Typical applications	Used for agglomeration, food flavoring blending, cake mix	Coffee mix, cocoa/sugar mix	Bakery premixes, instant coffee mix, spice blend, pet food, etc.	Bakery premixes, chocolate drink mixes	For very intense mixing, particularly pasty material

Source: adapted from Muzzio *et al.* (2004).

Gravity Silo Mixers and Pneumatic Mixers

Properly designed hoppers can mix solids by gravity when feeding ingredients into a unit operation (see Figure 30.7b). This is an auxiliary device for simple mixing duties. In pneumatic mixers for solids, gas or air is forced through the bed of particles to produce turbulence and mixing. They are applicable for mixing, coating, agglomeration, and drying for example. Apart from such versatility, mixing time is also very much less (2–3 min) compared with tumbler or convective mixers. However, their design and operation is complicated. Some variants of pneumatic mixer include air merge type (see Figure 30.7c), vertical screw, conical screw, Zeplin, Binsert and fluid bed.

Design Principles of Mixers in the Food Industry

When designing a new mixer, the process requirement decides the type of mixing to be used (agitated vessel, static mixer, rotor–stator, etc.) and the size of the mixer. Empirical design equations are used to select and size mixers for a given process. The same equations can be used to rate or optimize an existing mixer for a new process. The final conceptual design of the mixer is guided by the secondary requirements of the process, i.e. special operational requirements (variable speed for process versatility, operation in fill-up or draw-off condition, vacuum operation, etc.), capital cost, operational cost, management constraints (particular service life, low maintenance, lower manual labor, etc.), and regulatory constraints. The conceptual design should match engineering design standards and manufacturing tolerances. For challenging mixing processes (e.g. dough mixing, particulate batter mixing, hygroscopic powder mixing), where underlying scientific understanding is limited, a pilot plant study may be needed to scale up. Some selected mixer designs are discussed from the vast pool of mixers used in the food industry.

Design and Scale-up Principle: Liquid Mixing System

In terms of technology, there are five distinct mixing systems used in the food industry: agitated vessels, heavy-duty agitated vessels, jet mixers, static mixers, and rotor–stator mixers. These technologies are applied either in-tank or in-line.

In-tank Mixing Systems

Design of Agitated Tanks for Simple Mixing

Operationally such equipment can be of two types: off-line mixing tanks (e.g. milk storage tanks in dairies or oil storage tanks in vegetable oil processing plants) and on-line process vessels. For off-line mixing tanks, mixing time is greater than in on-line process vessels due to large volume as well as stagnancy and this has significant effects on design (Oldshue, 1983).

On-line mixing vessels can perform a variety of operations, for example solid dissolution/dispersion in liquid or mixing of miscible/immiscible liquids. The basic equipment comprises four major parts: tank, impeller, shaft bearing, and a motor that drives the shaft (with or without gears). Agitated vessels differ in impeller type, location of the impeller, and number of impellers. The impellers used in such mixers can be any of three basic types (turbine, propeller or paddle) or their variants. Tanks where liquid height is greater than tank diameter require multiple impellers; where tank diameter is greater than 10 m, side-entry impellers are preferred.

Mechanistically, impellers are like pumps that generate flow ($Q \propto ND^3$, circulating capacity) and shear ($\kappa \propto N^2D^2$, shear or velocity head) by utilizing the power supplied by the motor (Ludwig, 2002; Hemrajani and Tatterson, 2004). Power input (P) for an agitator can be calculated by Equation 30.3:

$$P = Q\kappa\rho = P_o\rho N^3 D^5 \tag{30.3}$$

where N is impeller speed, D impeller diameter, ρ liquid density, and P_o power number of the impeller, which is a function of impeller geometry and impeller Reynolds number [$\mathrm{Re}_{\mathrm{imp}} = (D^2N\rho)/\mu$]. There are many studies in the literature of several impeller–tank combinations and process conditions (blending, dispersion, etc.) but no universally accepted correlation exists between P_o and Re. The next section deals with process-specific design of agitated tank mixers using some of the most accepted correlations.

Irrespective of the process, the design of a new agitated vessel starts with the tank and then moves on to the impeller, motor drive, shaft and seal. The type of process and volume of charge decides tank geometry and size. Once the tank design is known, process requirements determine the type of impeller (shape and dimension) and its location (mounting). Impellers can be selected using experimental data from laboratory trials or by applying rules of thumb followed by the industry. For blending, depending on process volume, one of the three basic types of impeller can be used (turbine, up to 1850 m^3; propeller, up to 3700 m^3; paddle, up to 5550 m^3) with ratio of tank diameter to impeller diameter (T:D) varying from 3 : 1 to 6 : 1. Similarly for solid dispersion, the applicability of the three basic types of impeller depends on solids percentage (turbine, up to 80%; propeller, up to 50%; paddle, 65–92%) where T:D and H_{tank}:T vary from 2.0 : 1 to 3.5 : 1 and from 1 : 1 to 2 : 1, respectively. Ludwig (2002) has presented many such rules of thumb for impeller selection.

The motor drive is selected based on power requirement. Power consumption depends on the operational speed of the agitator. Calculation of the minimum operational speed can be done using empirical equations that correlate "mixing time" and "mixing quality." Mixing time is either a management decision or dictated by the specific food process (e.g. minimum time of spoilage or flavor loss). Mixing quality for a given mixing process can be expressed as percentage homogeneity (for blending), drop size (for liquid dispersion), or simply mixing index. There are many correlations available in the literature for calculating minimum speed of the impeller and scientific intuition is necessary to choose the most suitable one. After the calculation of impeller speed, motor power (Equation 30.3) can be obtained and a motor drive selected. In practice, impeller power (P) should be 80% of the rated power of a standard motor for safer operation. Along with the design of the motor, a gear or belt drive is selected at this stage. For variable speed operation, variable speed AC or DC drives are selected. Since impeller speed and power rating of the motor are known, torque on the shaft (λ) can be calculated using the relation $P = 2\pi N\lambda$. The torque determines the shaft length and type suitable for the agitator. At this step a proper seal needs to be designed, as closed tank mixing must be supported by a durable seal for containment. Eventually, cost estimation and further process optimization help finalize the actual design.

Successful design requires a process-specific design approach that depends on the mixing process (blending, solid dissolution or suspension in liquid, or improvement of reaction rate or heat transfer). The next three sections discuss the principles of design of an agitated tank mixing both miscible and immiscible liquids as well as solid dispersion in liquid. To define each process, the use of chemical engineering specifications is simpler than representing mixing quality in terms of organoleptic properties. For example, in the case of emulsification, a particular droplet size in the dispersed phase is a common mixing objective in the chemical industry. However, in the case of a food emulsion, droplet size is not the only aim: achieving a particular flavor and consistency is also important. Nevertheless, to make the design process easier, a basic chemical design scheme is followed with proper modification to suit food systems.

Liquid–Liquid Blending in Turbulent Flow or Transition Regime

When designing a new agitator to blend miscible liquids, two factors are predetermined: desired percentage homogeneity, i.e. mixing quality (z), and mixing time (θ). Then, depending on fluid properties (ρ and μ), the flow regime can be calculated and eventually minimum mixing speed (N). For Newtonian liquid blending using propeller or turbine impellers, if $\mathrm{Re}_{\mathrm{imp}}$ is above 200 the flow regime is definitely turbulent. However, near the turbulent transition boundary, it is necessary to determine the flow regime precisely using either of two parameters: Fourier number ($\mathrm{Fo} = \mu\theta_{95\%}/\rho T^2$) or $\left(P_o^{1/3}\,\mathrm{Re}\right)$. At the turbulent transition boundary, $1/\mathrm{Fo} = 1225$ and $P_o^{1/3}\,\mathrm{Re} = 6370$ (Grenville and Nienow, 2004). P_o is fairly constant in the turbulent transition regime for a given type of impeller and it is usually supplied by the manufacturer. Once the type of impeller is selected, the required operational speed (N_{cal}) of the impeller can be calculated using Equations 30.4 or 30.5 (Grenville, 1992).

$$\text{In turbulent regime, } P_o^{1/3}\,\mathrm{Re} = 5.2/\mathrm{Fo} \tag{30.4}$$

$$\text{In transition regime, } P_o^{1/3}\,\mathrm{Re} = 183/\sqrt{\mathrm{Fo}} \tag{30.5}$$

This speed is compared with output speed of standard gearboxes available in the market and the value closest to standard output speed is selected (N) (Hemrajani and Tatterson, 2004). Subsequently, power required (P_{cal}) by the agitator can be calculated by substituting N in Equation 30.3. Finally, a standard motor (rated power P) can be selected to provide the required power using an appropriate safety factor.

In Equations 30.4 and 30.5, Fo assumes mixing time for 95% homogeneity ($\theta_{95\%}$). However, any percentage homogeneity (z) can be investigated by replacing $\theta_{95\%}$ with mixing time for z% homogeneity ($\theta_{z\%}$), which can be calculated in terms of $\theta_{95\%}$ using Equation 30.6. (Grenville and Nienow, 2004),

$$\theta_z = \theta_{95} \ln[(100 - z)/100]/\ln 0.05 \tag{30.6}$$

When designing an agitator and motor for shear-thinning liquids (ketchup, sauces, etc.), the method described earlier changes because for such liquids viscosity decreases with shear. Since mixing becomes poorer as apparent viscosity increases, mixing will be good near the impeller and poor near the wall. Equations 30.4 and 30.5 need to be modified to ensure mixing even in the most difficult region, i.e. at the wall of the mixer where apparent viscosity is highest. Although apparent viscosity near the wall is a complex function of liquid properties, impeller speed, and impeller and tank geometry, to simplify the design relevant wall properties (suffix w for wall properties) can be calculated using Equations 30.7, 30.8, 30.9, and 30.10 (Grenville and Nienow, 2004). For all standard motor speeds (N), corresponding values of μ_w, Fo_w and Re_w can be calculated and used for determining flow regions (at turbulent transition boundary $1/\mathrm{Fo}_w = 1225$ and $P_o^{1/3}\mathrm{Re}_w = 6370$). Substituting the wall property values in Equation 30.4 or 30.5 and finally in Equation 30.11, calculated speed (N_{cal}) can be found and tabulated with N. When N and N_{cal} are the same, a particular N can be used for further design as used for blending of Newtonian liquids.

$$\text{Torque } \lambda = P_o \rho N^2 D^5 / 2\pi \tag{30.7}$$

$$\text{Wall shear stress } \tau_w = \lambda / 1.622 T^3 \tag{30.8}$$

$$\text{Wall shear rate } \gamma_w = (\tau_w / K)^{1/n} \tag{30.9}$$

$$\text{Wall apparent viscosity } \mu_w = K \gamma_w^{n-1} \tag{30.10}$$

$$N_{cal} = \left\{ \left(P_o^{1/3} \mathrm{Re}_w \right) \mu_w / P_o^{1/3} \rho D^2 \right\} \quad \text{(Grenville and Nienow, 2004)} \tag{30.11}$$

The design procedure changes significantly for extremely shear-thinning liquids like yoghurt, fermentation broth, and high solid suspensions (flow behavior index, $n \leq 0.3$). In these cases, mixing occurs locally near the impeller and a distinct zone of mixing (termed a "cavern") forms near the impeller (Figure 30.8a). Beyond this active region

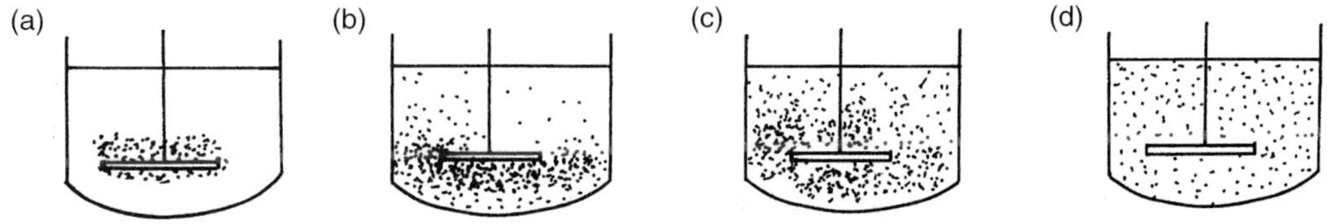

Figure 30.8 (a) Schematic diagram of a cavern-forming liquid. (b–d) Schematic diagrams of solid suspensions under different conditions: (b) on-bottom suspensions; (c) off-bottom suspensions; (d) uniform suspensions.

in the tank almost no mixing can be noticed. At the boundary of the active region shear stress equalizes with the yield stress of the liquid and when the torque applied by the impeller is large enough to expand the cavern to the wall, it starts rising along the wall boundary and ensures efficient mixing. Thus the shape of the cavern provides clues for the designer. The geometry of the cavern depends primarily on impeller type. For a given cavern dimension, minimum speed (N_{cal}) required for the mixing can be determined by Equation 30.12, involving P_o, fluid yield stress (τ_c),and tank dimension. For effective mixing, $D_c = T$ and H_c/D_c is generally supplied by the manufacturer of a particular impeller. Subsequently, standard gearbox speed (N) and motor rating can be selected as described elsewhere.

$$(D_c/D)^3 = P_o\rho N^2 D^2/\tau_c(H_c/D_c + 1/3)\pi^2 \quad \text{(Elson } et\ al.\text{, 1986)} \tag{30.12}$$

for $0.25 \leq D/T \leq 0.6$, where H_c and D_c are height and diameter of the cavern, respectively. One drawback of this correlation is accurate determination of yield stress makes difficult to use it.

Dispersive Mixing in Turbulent Regime

Food dispersions contain at least one component as dispersed phase (droplets for immiscible liquid–liquid mixing; solid particles for solid suspension in liquid) suspended in a continuous phase. During mixing the dispersed phase liquid(s) element (droplets or solid particles) breaks down to smaller size and spreads uniformly throughout the continuous liquid phase to achieve a particular quality (consistency and flavor). The resultant dispersion can be characterized by the following parameters.

Element size of the dispersed phase: Mean droplet sizes can be represented in many ways. One such representation is Sauter mean diameter (Sauter, 1926, 1928) for liquid dispersions:

$$d_{32} = \sum_i^m n_i d_i^3 / \sum_i^m n_i d_i^2,\ d_{32} = 6\phi/a_v \tag{30.13}$$

where n_i is the number of drops of size d_i.

where ϕ and a_v are volume and area of the droplet, respectively. Dispersion with droplet size less than 0.5 μm requires high-shear mixers (i.e. emulsifiers) and surface-active agents. With droplet size above 3 μm, dispersion becomes highly unstable and the material needs constant stirring or flow.

For solid dispersions, one way the mean particle diameter (d_p) can be represented is by mass-mean diameter:

$$(d_p)_{43} = \sum_i^m n_i d_i^4 / \sum_i^m n_i d_i^3 \quad \text{(Baldi } et\ al.\text{, 1978)} \tag{30.14}$$

where n_i is the number of particles of size d_i.

Ratio of dispersed phase and continuous phase: For liquid dispersions, it is indicated as the ratio of dispersed and continuous phase volume (φ). If φ is less than 0.01, the system is dilute, above which coalescing becomes important (e.g. milk). Common food dispersions are highly concentrated systems ($\varphi > 0.20$). For solid dispersions, the ratio is in terms of weight fraction of solid (X, kg solid/kg liquid).

Flow regime in the mixer: For liquid dispersions, Re_{imp} using average properties dictates the nature of the flow: for laminar regions, $Re_{imp} < 10$; for turbulent regions, $Re_{imp} > 10000$; for transition regions, $10 < Re_{imp} < 10000$ (Oldshue, 1983). For solid dispersions, particle Reynolds number ($Re_p = d_p v_l \rho_l / \mu$) indicates the flow pattern in a mixer with solid dispersed in liquid.

Weber number and dispersion stability: For liquid dispersions, two important parameters are Weber number (We = disruptive inertial force/cohesive surface force = $\rho_c N^2 D^3 / \sigma$, where ρ_c is the density of continuous phase and σ is the interfacial tension) and dispersion stability, indicative of time of phase separation (t_s). For $t_s < 1$ min, the system is fast coalescing; for $t_s > 2$ min, the dispersion is moderately stable with slow coalescence.

Solid suspensions are of three types (see Figure 30.8b–d). The rheology of suspension depends on the nature (especially shape) of the solid particle. Analogous to We, in solid suspensions the Froude number (Fr) (inertial force/gravity force) is a characteristic parameter:

$$\text{For immediate suspended solids, } Fr = \frac{N_{js}^2 D}{g_c}\left(\frac{\rho_l}{\rho_s - \rho_l}\right)$$

$$\text{For floating solids, } Fr = ND^2 / g_c$$

where N_{is} is the impeller speed in immediately suspended solids.

For the design of mixers using liquid–liquid or solid–liquid dispersions, the following correlations can be used to calculate required N. The design steps are similar to liquid–liquid blending in turbulent regime.

For dilute liquid–liquid dispersions ($\varphi < 0.01$):

For dispersed phase with lower viscosity (Chen and Middleman, 1967):

$$d_{32}/D = C_1 We^{-3/5} \tag{30.15}$$

For dispersed phase with higher viscosity:

$$d_{32}/D = 0.053\mathrm{We}^{-3/5}\left(1+0.92V_i^{0.84}\right)^{3/5} \tag{30.16}$$

$$\text{where } V_i = (\rho_c/\rho_d)^{1/2}\,\mu_d\,ND/\sigma \quad \text{(Calabrese } et\ al.\text{, 1986)} \tag{30.17}$$

Usually d_{32} is a predetermined quality parameter.

For dense dispersions ($\varphi > 0.2$), Skelland and Seksaria (1978) described Equation 30.18 with limited set of system values C_2 and a:

$$N_{\min}D^{0.5}/g^{0.5} = C_2\,(T/D)^a\,(\mu_c/\mu_d)^{1/9}\,(\Delta\rho/\rho_c)^{0.25}\,(\sigma/D^2\rho_c g)^{0.3} \tag{30.18}$$

where $\Delta\rho = |\rho_d - \rho_c|$

Suffix c denotes continuous phase in Equations 30.17 and 30.18.

Solid dispersions with immediate suspended condition is mathematically expressed by Zwietering (1958) as:

$$N_{js} = \psi v^{0.1}\left[\frac{g_c(\rho_s - \rho_l)}{\rho_l}\right]^{0.45} X^{0.13} d_p^{0.2} D^{-0.85} \tag{30.19}$$

for $0.02 < X < 0.15$, where ψ is a dimensionless number that is a function of impeller type and tank geometry, N_{js} impeller speed for just suspended condition (rps), g_c gravitational acceleration constant ($9.81\,\mathrm{m{\cdot}s^{-2}}$), and ρ density ($\mathrm{kg{\cdot}m^{-3}}$). Suffixes s and l indicate properties for solid particle and liquid, respectively. Values of ψ for various impellers and different configurations are available in the literature (Nienow, 1976; Armenante and Nagamine, 1998) or can be obtained experimentally for a defined process.

For dispersion with floating solids (Joosten *et al.*, 1977):

$$\mathrm{Fr} = 3.6\times10^{-2}\,(D/T)^{-3.65}\,[(\rho_c - \rho_s)/\rho_l]^{0.42} \tag{30.20}$$

Liquid Blending in Laminar Region

In laminar flow region (process viscosity >5000 mPa·s), design of an agitated vessel is complex and often process specific. It starts with the critical step of impeller selection. When Re is less than 50, a special type of impeller (helical ribbons or the like) is undoubtedly required. For intermediate-range Re, both propeller and special types of impeller may be applicable. Mixing in the laminar flow region comprises two distinct phases, the stagnation zone and its disappearance. $\mathrm{Re}_{\mathrm{Trans}}$ $[(1.8T/aD)^2]$ denotes the disappearance of the stagnant zone (Wichterle and Wein, 1981), where $a = 0.375P_o^{1/3}$. If $\mathrm{Re}_{\mathrm{Trans}}$ is between 100 and 200, a specially designed impeller is essential for mixing. Once the impeller is chosen, Equation 30.21 can be used to calculate the desired speed of rotation (N_{cal}) for a given time of mixing (θ):

Dimensionless blending time (Grenville and Nienow, 2004)

$$N\theta = 896\times10^3\,A^{-1.69} \tag{30.21}$$

where A is a function of impeller geometry (Rieger *et al.*, 1988).

For helical impellers,

$$A = 82.8(h/D)(c/D)^{-0.38}(p/D)^{-0.35}(w/D)^{0.20}n_b^{0.78} \tag{30.22}$$

where c = wall clearance, h = impeller height, p = pitch, w = blade width, and n_b = number of blades.

However, for mixing in laminar range, P_o (=A/Re) is not constant and is a function of Re. Therefore, power input cannot be calculated as usual. Subsequently using Equation 30.3, power requirement can be evaluated to select a standard motor.

A special case in this category is dough mixer design. During mixing of the ingredients (flour, sugar, salt, yeast, water and/or milk), the dough attains maximum viscosity, but after a certain period of time is converted into a sticky mass with low viscosity. Therefore an optimal shear force for a given time is essential for good-quality dough. Various designs are available to meet this objective, i.e. vertical change-can type, horizontal batch or continuous type, and with single or multiple agitators (hook, paddle, spiral, etc.). In most cases, selection of a mixer for high-viscosity materials is based on experience. The design is tested at pilot plant level for scale-up. Power ($P \propto N^{(1+n)}$) and torque ($\lambda \propto N^n$) where $0.25 < n < 0.5$ (Valentas *et al.*, 1997) are common parameters for comparing different dough mixers.

Design Calculation

Problem 1

A vegetable oil manufacturing company wants to produce fortified oil with ricinol in liquid form. The mixing time specified is 1 min. The process requires 99.9% homogeneity. Design a suitable impeller and mixer drive and calculate the shaft torque. Relevant information: tank diameter (T), 1.75 m; liquid height (H), 1.75 m; process viscosity,m 35 mPa·s; specific gravity, 0.94.

Solution

Vegetable oil behaves as a Newtonian fluid. General form of Fo and $P_o^{1/3}$ Re considers mixing time for 95% homogeneity whereas the present problem requires 99.9% homogeneity. Using Equation 30.6, mixing time can be calculated for 95% homogeneity:

$$\begin{aligned} \theta_{99.9\%} &= \theta_{95\%}[\ln(100-99.9)/100]/\ln 0.05 \\ \theta_{99.9\%} &= 2.306\theta_{95\%} \end{aligned} \tag{30.23}$$

Substituting $\theta_{99.9\%}$ in Fo expression, the flow regime condition is modified for turbulent transition boundary (1/Fo = 1225):

$$\begin{aligned} 1/\text{Fo} &= \rho T^2/\mu\theta_{95\%} = 1225 \\ 1/\text{Fo} &= \rho T^2/\mu\theta_{99.9\%} = 1225/2.306 = 531.2 \end{aligned} \tag{30.24}$$

In this case

$$1/\text{Fo} = (940\ \text{kg}\cdot\text{m}^{-3})\times(1.75\ \text{m})^2/(0.035\ \text{Pa}\cdot\text{s})\times(60\ \text{s}) = 1370.8$$

Since 1/Fo > 531.2, the flow condition is turbulent. A pitched-blade turbine will be suitable for such volumetric application, with P_o = 1.80 and D/T = 0.5. Using the data, minimal operational speed can be calculated using Equation 30.25 with modification for turbulent flow regime:

$$P_o^{1/3} N\theta_{99.9\%} D^2 / T^{1.5} H^{0.5} = 5.2\times 2.306 = 11.99 \quad (30.25)$$

or $N = (11.99/(0.5)^2) \times 60 \times (1.8)^{1/3} = 0.66$ rps = 39.6 rpm (since T = H = 1.75 m). This value is not a standard speed and the closest standard speed of 37 rpm (Hemrajani and Tatterson, 2004) can be selected. A higher speed will reduce mixing time but at the expense of power cost. So a lower speed drive is preferred for the process. For the gearbox speed output, power requirement can be calculated from Equation 30.3:

$$P = P_o \rho N^3 D^5 = 1.8\times 940\times\left(\frac{37}{60}\right)^3\times(1.75)^5 = 6512.41\ \text{W}$$
$$= 8.73\ \text{hp}.$$

Choosing the nearest standard motor drive of 10 hp is not a safe option (standard motor power rating can be found in Hemrajani and Tatterson, 2004) because the mixer will consume 87% of the rated power. Hence, the next highest standard motor of 15 hp is a better option. The torque (λ) on the shaft can be calculated using the equation (30.7). So the value of

$$\lambda = 1.8\times 940\times\left(\frac{37}{60}\right)^2\times\frac{(1.75)^2}{2}\times 3.14 = 1681.63\ \text{N.m.}$$

Scale-up Principle

In the design of large-scale mixing tanks for blending or dispersions, there are many scale-up techniques that consider geometry and kinematic and dynamic similarities. Suitable scale-up depends on the specific process. Two commonly used criteria for mixing in a turbulent transition flow regime are constant power per unit volume (P/V) and torque per unit volume (T_Q/V). Often T_Q/V is preferred because it has a direct impact on the overall size and cost of the mixer, including the gearbox (Hemrajani and Tatterson, 2004). For mixing in laminar flow, constant power input per kilogram is a useful scale-up criterion (Rudolph *et al.*, 2009).

Design of Jet Mixers

For liquids with low viscosity (<1000 mPa·s) and larger tank volumes (>1000 US gallons), jet mixing is an effective technique for in-tank mixing but has longer mixing time (typically hours). Many configurations of the nozzle are available for jet mixing in a tank. One simple design is bottom drawing and side-bottom entering jet. Tank and nozzle dimensions and nozzle angle determine the free path and mixing efficiency of a jet. Jet free path (Z) is the longest distance that a liquid jet travels before reaching the farthest point of the liquid boundary.

Designing the output of a jet mixer must consider nozzle diameter, volumetric flow rate ($Q = 0.25\pi U D_N^2$, where U is jet velocity), and pressure drop ($h_1 = 2.5U^2/2g$) for pump specification. Normally, the flow regime in a jet mixer is turbulent. In such condition, for a given blending time, Fo number and Equation 30.26 provides $\mathrm{Re}_{\mathrm{nozzle}}$. From $\mathrm{Re}_{\mathrm{nozzle}}$ [$= (D_N v\rho)/\mu$], nozzle diameter (D_N) can be calculated. The diameter should be the nearest standard pipe diameter (from Perry and Green, 1997):

$$\mathrm{Re}_{\mathrm{nozzle}} = B/\mathrm{Fo} \quad \text{(Greenville and Nienow, 2004)} \qquad (30.26)$$

for $B = 3.0$, when $\mathrm{Re} > 10000$ (turbulent flow), $0.2 < H/T < 2.0$, $0.178 < V < 1200\,\mathrm{m}^3$, $1.32 \times 10^{-2} < (UD_N/Z) < 13.7 \times 10^{-2}\,\mathrm{m{\cdot}s^{-1}}$, $86 < Z/D < 753$. The design calculation below gives some more details of jet mixers.

Problem 2

A food industry uses a storage tank for a dilute molasses solution. Design a jet mixing system for mixing old and new batches in 20 min. Relevant information: tank diameter (T), 8 m; liquid height (H), 15 m; process fluid viscosity, 1.5 mPa·s; specific gravity, 0.90.

Solution

For the system shown in Figure 30.2(a), jet free path length:

$$Z = \left[(H)^2 + (T)^2\right]^{1/2} = \left[(15)^2 + (8)^2\right]^{1/2} = (289)^{1/2} = 17\ \mathrm{m}$$

To determine the flow regime, calculation of $\mathrm{Re}_{\mathrm{nozzle}}$ is necessary and that can be found using Equation 30.26 and the expression for Fo:

$$\mathrm{Fo} = \mu\theta/\rho(Z)^2 = (0.0015\ \mathrm{Pa\cdot s})\times(1200\ \mathrm{s})/\left[(900\ \mathrm{kg\cdot m^{-3}})\times(17\ \mathrm{m})^2\right] = 6.92\times10^{-6}$$

$$\mathrm{Re} = 3/\mathrm{Fo} = 3/6.92\times10^{-6} = 4.33\times10^{5}\ (\mathrm{Re} > 104\ \text{turbulent regime})$$

From the definition of Re, nozzle diameter (D_N) can be calculated:

$$D_N = \mathrm{Re}\,\mu/\rho U = (4.33\times10^{5})(0.0015\ \mathrm{Pa\cdot s})/(900\ \mathrm{kg\cdot m^{-3}})(10\ \mathrm{m\cdot s^{-1}}) = 0.072\ \mathrm{m}$$

wher jet velocity (U) is assumed to be $10\,\mathrm{m{\cdot}s^{-1}}$. Nearest standard pipe diameter of 0.0725 m (3 in 80 schedule) can be used as nozzle diameter (refer to Perry and Green, 1997). Therefore, for the designed nozzle, volumetric flow rate and pressure head loss are:

$$Q = 0.25\pi UD^2 = 0.25(3.14)(10)(0.0725)^2 = 0.041\ \mathrm{m^3 \cdot s^{-1}} = 2.46\ \mathrm{m^3 \cdot min^{-1}}$$

$$h_1 = 2.5(U^2/2g) = 2.5(10)^2/(2)(9.81) = 12.74\ \mathrm{m}$$

Design of Rotor–Stator Mixers

Rotor–stator mixers for dispersive processes are quite useful in the food industry. Unfortunately, understanding of the science that underlies such mixing is lacking. Design is mostly dependent on the manufacturer's in-house technology and almost no correlation data are available in the literature. The same is true for in-line mixers involving rotor–stators. There are three basic types of rotor–stator mixer: (i) colloid mills and toothed devices, (ii) radial discharge impeller (Ross or Silverson type), and (iii) axial discharge impeller (Chemineer type). Standard mechanical design of rotor–stator mixers includes proper selection of the motor, design of mechanical or electrical drives for variable speed options, design of gearbox, seal and bearing, pressure tolerance of the unit, and selection of appropriate construction materials. Often, process-specific designs are supplied by the manufacturer. Customized design of rotor–stator mixers will employ pilot plant study and many manufacturers offer a rental service for such study.

Common scale-up criteria include maintaining a constant tip speed ($V_{tip} = \pi ND$) for pilot and industrial scale. Shear rate (Y) generated by the rotor–stator is the ratio of V_{tip} to the gap between rotor and stator (δ). It is common practice not to change δ during scale-up.

In-line Mixing: Static Mixer Design and Scale-up Principle

The design of static mixers involves two steps: selection of the mixer module and the sizing. Normally, selection of a mixer module is based on data from the manufacturer or from a similar process. An important selection criterion is the ratio of the pressure drop across the static mixer module versus that across empty pipe of the same length (K_T or K_L, $\Delta P_{staticmix} : \Delta P_{pipe}$, where K_T and K_L are the pressure ratios for a particular static mixer module and process in turbulent and laminar conditions, respectively). Examples of K_T and K_L values, respectively, are as follows: Kenics (KMS), 6.9 and 1.5; Koch-Glitsch (SMX), 37.5 and 500; Koch-Glitsch (SMF), 5.6 and 130. Generally, the best static mixer should impart the lowest pressure drop for the same mixing duty. Various methods are adopted by individual manufacturers for sizing and comparing different mixing elements.

Graphical Method of Sizing

Manufacturers can provide empirical plots I and II (Figure 30.9) for a given element. In plot I, the number of mixing elements is plotted against velocity of fluid. Each line

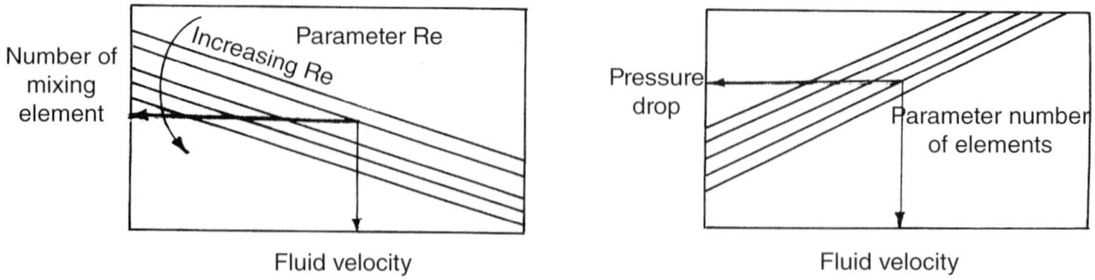

Figure 30.9 Empirical plots for designing static mixers (not drawn to scale). Concept based on the plots from the technical report of Komax Inc., USA. (With kind permission of Komax Systems Inc., Huntington Beach, CA, USA.)

in the family of straight lines represents a particular Re. The first step is to calculate velocity of the fluid and Re. Velocity can be calculated from the desired flow rate of fluid (Q) and cross-sectional area of pipe to be used. For each calculated value of Re, a line drawn from a particular velocity value on the x-axis will intersect the Re line and this then indicates the number of elements required on the y-axis. It is better to take the next higher value.

In plot II, pressure drop is plotted against fluid velocity. Each line in the family of straight lines represents a particular number of elements. From plot I, the number of elements is selected as the appropriate parameter in plot II. A line drawn from a known velocity value on the x-axis will intersect the selected number of element line and this then indicates the pressure drop on the y-axis. Since these empirical plots are based on a standard fluid, calculated pressure drop needs to be corrected for density and viscosity difference of the fluid under consideration. The design output comprises the number of elements of the mixer and pump head required. Another method of comparing static mixers for blending of liquids is by coefficient of pressure drop and by correlation of coefficient of variation (CoV). In a static mixer the mixing quality can be measured by measuring concentration of one of the smaller (or essential) components in samples drawn from different positions in the pipeline and calculating the variance (CoV). CoV correlates with pipe geometry and the type of static mixer element. This is used as the design equation (Etchells and Meyer, 2004).

The common scale-up strategies for static mixers involve identical static mixers connected in parallel or series and increased tube diameter by keeping geometric similarity or pressure constant. Scale-up factor (S' = throughput for full scale/throughput for pilot plant) in parallel connection can be calculated as the ratio of number of tubes in full scale and pilot scale, whereas for series connection S' can be calculated as the ratio of Reynolds numbers. In the case of geometric similarity or constant pressure-based scale-up in turbulent region both L/V and V^3/D is constant, but in the laminar region only L/V is constant (Thakur *et al.*, 2003).

Design and Scale-up Principle: Solid Mixing System

In terms of flowability, food solids can range from extremely cohesive to free-flowing. Neither extreme is good for mixing. A free-flowing solid will have a tendency to completely segregate irrespective of time of mixing, whereas a highly cohesive material will be difficult to agitate but does not pose segregation problems once mixed well. A quantitative measure of flowability is the flowability index (= σ_1/σ_c, where σ_1 is consolidation stress and σ_c compaction stress). According to flowability index, solids can be classed as free-flowing (>10), easy flowing (4–10), cohesive (2–4), very cohesive/not flowing (1–2), and hardened (<1) (Jenike, 1964; Tomas and Schubert, 1979). Many components can affect flowability of granular solids or solid powders: particle size and its distribution, shape, moisture, composition (e.g. presence of fat or flow conditioner), hygroscopic nature, and bulk density (Ganesan *et al.*, 2008). If the flow properties of the solids to be mixed are known and the change in flow properties during process requirement is also known, one can select a generic design of mixer. Depending on other constraints, more detailed design can be made. Pilot plant study followed by scale-up is often preferred for better mixing process.

Rules of thumb for the mixing of solids include the following:

1. Use of ingredients of similar size, shape and other properties improves mixing efficiency. This option is rare in the food industry.
2. Highly segregating solids (particle size >5 mm) can be packed in proportion instead of mixing.
3. If the weight percentage of any component to be mixed is less than 0.5%, premixing of the material is better. This is quite common in food mixing.
4. If the particles to be mixed have a wide range of sizes and size reduction is allowed, use of hammer or ball mills is a better choice. If addition of moisture is allowed, wet milling may be tried.
5. If the moisture content is more than 2% and the mixture behaves like paste, close clearance mixers like sigma blade or extruders are required.
6. For a free flowing solid with segregating tendency: shear-insensitive materials can undergo wet milling; shear-sensitive materials can undergo ribbon or plow mixing.
7. For a free-flowing solid without segregating tendency, many mixers are suitable (e.g. silo mixer, tumbler mixer, ribbon mixer, plowshare), but for cohesive solids (lump forming) special designs (e.g. high-intensity pan mixer, orbiting screw with lump breaker, plowshare with chopper) are required.
8. Fluidizable solids pose dusting problem, so pan mixer, orbiting tapered screw and plowshare are suitable varieties.
9. Continuous mixing operation is preferred for large process volumes (>5 m^3) or where a low number of components (<3) are to be mixed without much flexibility of operation. If the resultant mixture has no segregation problems, most of the solid food mixing processes are batch mode (Manjunath *et al.*, 2004).

Scale-up of tumbler mixer strongly depends on the properties of the materials to be mixed and does not vary with mixer variants (double cone, V-type, etc.). There are many scale-up criteria for agitated or convective type solid mixers. Parameters that are usually considered constant in both pilot scale and large scale include Froude number ($Fr = R\omega/g$, where ω is speed of rotation), peripheral velocity ($V = \pi ND$), mixing time [$t = (L/T)^2\ (T/v$ where T is the mixer tank diameter having mixer length of L) for $Fr < 3$ and $t \approx (L/T)^2(T/v)^3$ for $Fr > 3$], and power consumption. Constant power consumption is the most versatile scale-up criterion in solid mixing.

Advanced Process Design of Mixers

In the processed food sector, a wide variety of novel food products and ready-to-eat foods are introduced into the market periodically. The food industry is adopting faster, cleaner, and more economic processes to stay ahead in competition. Cutting-edge technology demands mixing devices for complex rheological systems (high solids content, 65–75%, chocolate mix), mixing with additional processing duties (incorporation of air in viscous liquid), cleaner mixer designs (especially for solid powder mixing or powder–liquid mixing), power economy (for power consuming dough or viscous liquid mixing), and flexible operation (flexibility of volume as well as recipe). Naturally, the design of a mixer (e.g. rotor–stator type or static mixer) and the mixing process (e.g. one-pot mixing instead of premixing step or a combination of mixers) look to exploit advanced technology. This is visible in three aspects of mixing, namely evaluation of mixing quality, modeling of mixing process (i.e. suitable composition of complex and often process-specific design equations), and design of mixing equipment.

Because the food industry commonly attempts to achieve a particular consistency or texture and flavor and the presence of multiphase systems, the evaluation of mixing quality is difficult and often subjective. To evaluate mixing efficiency in a mixer, the velocity field can be measured using particle image velocimetry (PIV) and laser doppler anemometry (LDA). To visualize blending, a nonintrusive technique such as laser-induced fluorescence (LIF) can be used. For faster and economic evaluation of hydrodynamics (shear, velocity profile) in a mixer, computational fluid dynamics (commercial CFD codes: Fluent, CFX Phoenix, Pam) (Xia and Sun, 2002) is routinely used. For solid mixers, near infrared (NIR) spectroscopy can be used for quality determination.

Traditional mathematical models of mixing processes are mostly derived for ideal systems with low volume of dispersed phase, binary mixtures, low viscosity levels, simple rheology, and nonhygroscopic solids with limited particle–particle interactions. Some examples of advanced models are discussed below.

A model of local shear distribution for dough mixing in a slow mixer with asymmetrical sigma blades using LDA technique is given by (Prakash and Kokini 2000):

$$f(\eta) = \frac{(b+1)\,r_1 r_2}{\{k\tan(\omega t+\phi)-\tan(1.5\omega t+\phi)\}(r_2-r_1)}\eta^{b} \qquad (30.27)$$

where η is average shear rate, b and k (properties of the fluid) are constants, and r_1 and r_2 are the distances of the nearest point with maximum shear rate and the nearest point with minimum shear rate, respectively from the longitudinal axis of the reference (slower) blade at the selected encoder position. These distances depend on fluid rheology. This design technique gives quantitative values of shear rate at various locations, which provide a picture of a physical phenomenon like gluten development at those locations. This information can be used to compare two mixers as well as optimize operation of a mixer.

Better power prediction for thixotropic fluids and Bingham-type liquids can be performed using Equation 30.28 (Maingonnat *et al.*, 2005):

$$A' = 78.7 + 88.3\exp(-0.029\mathrm{Bi})$$
$$\text{where Biot number, } \mathrm{Bi} = \frac{\tau_0(t)}{N\mu_{e0}} \tag{30.28}$$

where $\tau_0(t)$ and μ_{e0} are the intial values of time-dependent shear and viscosity.

Portillo *et al.* (2007, 2008) have investigated mixing in a continuous drum mixer using a compartment model to optimize parameters, i.e. feed rate, active concentration of products, and blender processing angle. Chaudhuri *et al.* (2006) have used the discrete element method (DEM) to simulate mixing in a blender.

For the mixing equipment manufacturer, computer-aided design (CAD) is currently a standard tool for conceptual design. CAD converts a process specification into detailed drawing of the mixing equipment, including for example a customized agitator (most probably with multiple impellers or multiple shafts for an agitated tank), an element for static mixer or a rotor–stator head. This improves not only the accuracy of design but also alows coordination between the food processor and the mixer manufacturer.

Operational Issues of Mixing Equipment

Process Control

In the food industry, conventional manual process control systems are being replaced by automated control systems due to increased scale and complexity of food processing. Consistent mixing ensures better utilization of resources, regulatory conformation, and process integration. Nevertheless, food mixing processes can be manual, semi-automatic or automatic depending on the tolerance level (determined by scale and complexity) of the mixing process. For a storage tank where mixing has a 5% tolerance level, a manual control system is adequate; however, for a premixing unit with allowable tolerance of 2%, a semi-automatic or automatic control system is best. In a mixing-intensive process such as mayonnaise production (emulsification) or a high-volume ready-to-cook powder blending unit, the quality of the mixture must not deviate by more than 1% from the desired specification. In such cases automatic systems with PID loop control systems and faster and accurate valves need to be employed despite higher cost (Lanham, 2003). For batch mixers, a dedicated control system comes with a semi-automated or automated (PLC) controller with appropriate sensors, valves, motors, and user-friendly human–machine interface. Typically, the control system can store a large number of recipes, has alarms to indicate process failures, and can record process history. They are quite flexible and cost-effective at small scale (typical productivity, 1200 $kg \cdot h^{-1}$). Currently, mixer manufacturers design control systems for mixers as stand-alone units or as attachable units for semi- continuous or continuous processes. However, continuous plants require stringent process control or mixing that is synchronized with other unit operations. Therefore, such plants employ distributed control systems or integrated control systems for the mixer, involving system and application software, computer and processing hardware, automation networks, controllers (multiple PLCs, e.g. Siemens TI545 controllers), process instruments and sensors and processing systems, and integration components.

Figure 30.10(a) is a schematic diagram of a simple control system for mixing in an agitated tank. Citric acid is being mixed with water to achieve a certain pH, indicated by a pH electrode. When it attains the set value provided by the operator, the controller will shut off the dosing valve for citric acid. The outlet has an additional pH electrode to record the pH history of the product. Figure 30.10(b) is an example of an in-line mixing control system, where two or more fluids (in the example confectionary batter and steam) pass through a mixer to attain a process objective (i.e. cooking). The sensor–controller senses and determines if the mixing objective is reached by measuring the temperature of the exit fluid. The mixing objective is correlated with the temperature of the product because the actual mixing objective, i.e. cooking, is difficult to quantify or measure directly. Hence, the target temperature or end point of mixing will be achieved by regulating the steam valve. However, such a simple feedback control system may not always be useful. Currently, the food industry is adopting practices of process integration, large-scale continuous production, predictive maintenance, and enterprise resource planning (ERP).

Maintenance

For dynamic equipment like mixers, maintenance occurs before installation (design) and during installation (Oldshue, 1983) not just after installation. Different mixing equipment needs different frequency and extent of maintenance. To minimize maintenance costs, a simple design is preferred and of course the design should consider operational fluctuations such as shock loading and overfill conditions. Corrosion protection should take account of both storage and operational conditions. Proper coating (e.g. vapor phase inhibitor, anti-rust compound) of internal parts is essential. During installation, inspection of the equipment for any damage and proper mounting and alignment reduces the chance of frequent failures of the motor due to vibration and other operational damage. It is better to use a checklist (e.g. electrical wiring) for installation of every mixer.

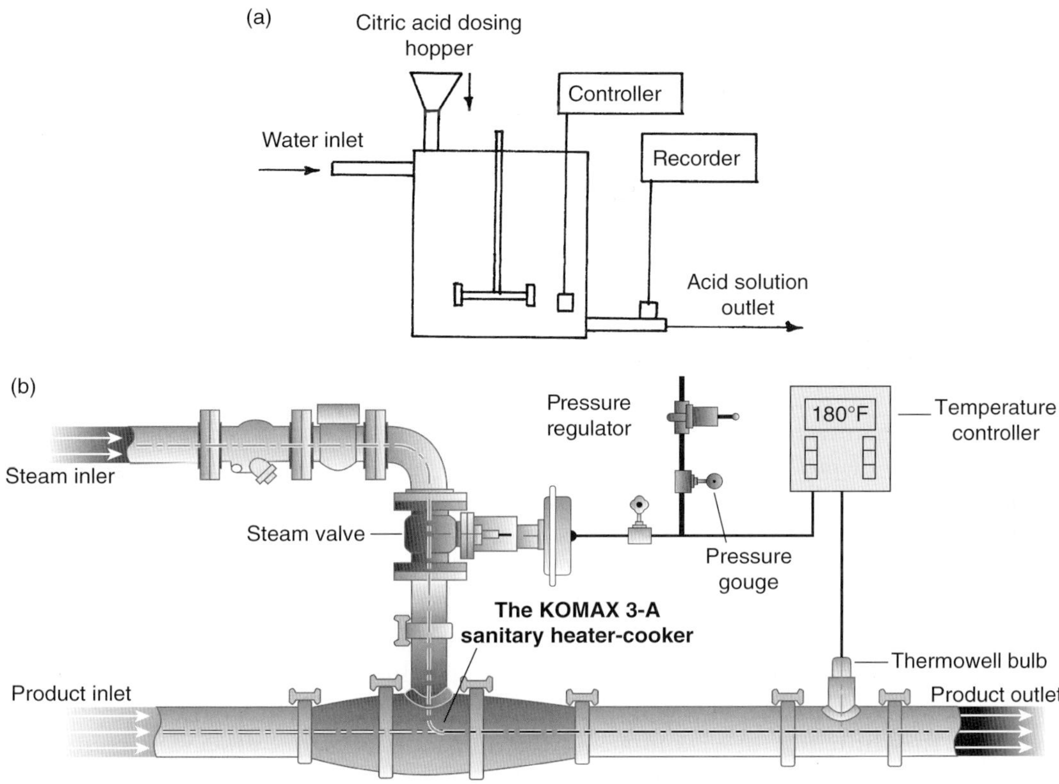

Figure 30.10 (a) An automatic control system for acidification of fruit juice. (b) In-line static mixer for continuous steam cooking. (With kind permission of Komax Systems Inc., Huntington Beach, CA, USA.)

The following mixers are listed in decreasing requirement for maintenance: liquid mixers, high-shear mixers > impeller agitated tanks > rotor–stator mixer > jet mixer and solid mixers, high-shear mixers > convective solid mixers with fluidization > convective solid mixers > tumbler mixer > gravity silo mixer. The food industry typically follows a standard operating procedure (developed in combination with the manual supplied by the manufacturer) when devising a maintenance program for the mixer element, tank, electrical equipment, and instrumentation. For mechanically agitated vessels, with a short-term (say weekly) lubrication system, agitator drives, tank accessories excluding impeller (e.g. seals, dosing systems, scrapers, pneumatic systems), drive motor, pumps and control accessories need inspection for physical damage, lubrication and calibration (wherever required), whereas agitators, tanks, motor coupling, seals, gear drives, and temperature control systems require inspection every month for physical damage, alignment, and efficiency. This is more elaborate than short-term inspection and may need corrective actions like replacement of seals, rubber parts and valves. Once or twice a year, components like the drive motor, lift or tilt system and gear drives of an agitated vessel may need overhauling.

For in-line static mixers, the mixing element requires minimum maintenance (one of the incentives of such mixers) while maintenance for the pump and the instrumentation is the same as for other in-tank mixers. For jet mixing, nozzles need additional maintenance because the nozzle can undergo erosion (operational wear and tear), accidental damage, and corrosion as well as clogging or deposition of solid layers inside the nozzle, which can reduce the efficiency of the nozzle. A preventive program should monitor pump flow rate, upstream pressure drop, and jet shape to check for the problems. Generally, high-shear mixers and solid mixers require more frequent maintenance because of the process load.

Cleaning and Sanitation Methods

Cleaning and sanitation procedures can be manual or automated depending on the scale and type of process. The parts of the mixer that are in contact with the food materials (mixing element, pipeline, tank, dosing systems like hoppers, motors, and valves) need to be cleaned and sanitized properly. Mixers can be cleaned and sanitized in place or by dismantling the unit. All in-line mixers have a cleaning-in-place (CIP) option. Except for high-viscosity mixing processes, CIP and sanitation can be used for tank mixers. Food processors follow a standard operating procedure for cleaning and sanitation that takes account of the processing schedule, i.e. before start-up, between batches, after last batch of the day. The complexity of mixing and mixer design increases the difficulty of cleaning and sanitation. Since this process accounts for much process time, man hours, power, and chemicals, mixer manufacturers always upgrade mixer designs for easier cleaning and sanitation. The philosophy behind such clean design is simple: there are fewer internal parts (i.e. less wet surface area), it cuts down multiple steps, there is faster intake, faster and complete discharge of the mix, no dead space, and reduced leakage.

High-viscosity mixers are generally considered the most difficult to clean and sanitize. Detachable agitators or change-can mixer design can improve the situation. Dry powder mixers can also create cleaning difficulties by dusting, with sticky powders especially difficult to clean. Dustproof seals (covered with neoprene/silicon gasket) and automatic dosing systems with pneumatically or hydraulically operated lids can be a good option to stop dusting. Use of scrapers and wide-cornered vessels can eliminate dead zones in paste and sticky powder mixers. Static mixers and spray nozzle mixer in-tank have inherent advantages for better CIP operations.

Capital and Operating Costs for Different-sized Equipment

In the food industry, the cost analysis of mixers includes (i) capital cost (inclusive of all dedicated auxiliaries), (ii) operational cost (cost of power mainly for the motor and other instrumentation and labor costs), (iii) maintenance and sanitation cost (materials as well as labor costs), and (iv) depreciation cost. It is subjective and difficult to perform a comparative cost analysis of all the different types of mixer suitable for a given mixing process but mixer selection is guided by both fixed and recurring costs. However, for different designs of a particular class of mixer, cost comparison is possible. For a given volume of agitated vessel with side-entering impeller (25 hp with 0.71 m diameter) and top-entering impeller (10 hp with 1.63 m diameter) the capital cost ratio would be about 2:5, whereas the operating cost ratio would be 5:2. Frequently, industry uses standard correlations for calculation of the capital cost of mixers. For a large-scale agitated vessel, the capital cost can be modeled in terms of torque and impeller diameter (Equation 30.29) (Oldshue, 1983):

$$\text{Capital cost} = aT^{0.8}D^{\frac{(1.6x+4)}{2.4}} \tag{30.29}$$

The value of x varies with the process ($0 > x > -3$). Operating cost equals motor power rating in kilowatts per unit electrical power cost.

Summary and Future Needs

From an industrial perspective, there are ample opportunities in the field of mixing in the food processing sector. In order to manage existing food industries (troubleshooting mixing problems) effectively and to commercially realize future mixing-intensive food processes (new food emulsions, preparatory food mixes, etc.), understanding of mixing science is essential. This requires the close cooperation of food processors, mixing equipment manufacturers and of course academia, and will not only enrich mixing science but also allow better mixing solutions.

From a food processors perspective, the challenges in mixing can be summarized as follows:

- Development of standards for comparing different mixer types.
- Better monitoring and control of mixing processes, particularly for food systems with memory (where physicochemical reactions initiated by mixing continue long after the cessation of mixer operation).
- Improvement in mixing for rheologically difficult fluids like shear thinning and viscoelastic types as well as liquids with foaming tendency.
- Conversion of batch processes to continuous mode.
- Scale-up of mixing of highly viscous liquids and solids.
- Reducing number of mixing steps to one-pot mixing.
- Mixing in vessels with unusual geometries (i.e. H/T away from 1).

To overcome these challenges, basic research in mixing processes is required as well as in various interdisciplinary areas such as particle properties, multiphase interaction (solid–liquid, liquid–liquid, solid–liquid–air, liquid–liquid–air), modeling and simulation techniques and control strategies.

References

Armenante, P.M. and Nagamine, E.U. (1998) Effect of low off-bottom impeller clearance on the minimum agitation speed for complete suspension of solids in stirred tanks. *Chemical Engineering Science* 53: 1757–1775.

Ashton, M.D. and Valentin, F.H.H. (1966) The mixing of powders and particles in industrial mixers. *Transactions of the Institution of Chemical Engineers* 44: 166–186.

Baldi, G. and Conti, R. and Alaria, E. (1978) Complete suspension of particles in mechanically agitated vessels. *Chemical Engineering Science* 33: 21–25.

Bayram, M. and Göğüş, F. (2008) Mixing: determining mixing parameters. In: *Experiments in Unit Operations and Processing of Foods* (eds M. Vieira and P. Ho). Springer, New York.

Beaudry, J.P. (1948) Blender efficiency. *Chemical Engineering* 55: 112–114.

Calabrese, R.V. Wang, C.Y. and Bryner, N.P. (1986) Drop break-up in turbulent stirred-tank contactors: II Correlations for mean size and drop size distribution. *AIChE Journal* 32: 677–681.

Chaudhuri, B., Mehrotra, A., Muzzio, F.J. and Tomassone, S. (2006) Cohesive effects in powder mixing in a tumbling blender. *Powder Technology* 165: 105–114.

Chen, H.T. and Middleman, S. (1967) Drop size distribution in agitated liquid–liquid systems. *AIChE Journal* 13: 989–995.

Elson, T.P., Cheesman, D.J. and Nienow, A.W. (1986) X-ray studies of cavern sizes and mixing performance with fluids possessing a yield stress. *Chemical Engineering Science* 41: 2555–2562.

Etchells, A.W. and Meyer, C.F. (2004) Mixing in pipelines. In: *Handbook of Industrial Mixing: Science and Practice* (eds E.L. Paul, V.A. Atiemo-Obeng and S.M. Kresta). John Wiley & Sons, Hoboken, NJ, pp. 391–477.

Ganesan, V., Muthukumarappan, K. and Rosentrater, K.A. (2008) Flow properties of DDGS with varying soluble and moisture contents using jenike shear testing. *Powder Technology* 187: 130–137.

Grenville, R.K. (1992) *Blending of viscous Newtonian and pseudo-plastic fluids*. PhD thesis, Cranfield Institute of Technology, England.

Grenville, R.K. and Nienow, A.W. (2004) Blending of miscible liquids. In: *Handbook of Industrial Mixing: Science and Practice* (eds E.L. Paul, V.A. Atiemo-Obeng and S.M. Kresta). John Wiley & Sons, Hoboken, NJ, pp. 507–542.

Griskey, R.G. (2006) *Transport Phenomena and Unit Operations: A Combined Approach*. John Wiley & Sons, New York.

Hemrajani, R.R. and Tatterson, G.B. (2004) Mechanically stirred vessels. In: *Handbook of Industrial Mixing: Science and Practice* (eds E.L. Paul, V.A. Atiemo-Obeng and S.M. Krestanew). John Wiley & Sons, Hoboken, NJ, pp. 345–390.

Hirech, K., Arhaliass, A. and Legrand, J. (2003) Experimental investigation of flow regimes in an SMX Sulzer static mixer. *Industrial and Engineering Chemistry Research* 42: 1478–1484.

Jenike, A.W. (1964) *Storage and flow of solids*. Bulletin 123. Engineering Experiment Station, University of Utah.

Joosten, G.E.H., Schilder, J.G.M. and Broere, A.M. (1977) The suspension of floating solids in stirred vessels. *Transactions of the Institution of Chemical Engineers* 55: 220–222.

Lacey, P.M.C. (1943) Chemical engineering research and design. *Transactions of the Institution of Chemical Engineers* 21: 53–59.

Lacey, P.M.C. (1954) Developments in the theory of particle mixing. *Journal of Applied Chemistry* 4: 257–268.

Lanham, R. (2003) Processing Magazine' Flow Handbook. Available at www.mixers.com/specifications/Flow%20Controls%20Story.pdf.

Lindley, J.A. (1991) Mixing processes for agricultural and food materials: 1. Fundamentals of mixing. *Journal of Agricultural Engineering Research* 48: 153s–170s.

Ludwig, E.E. (2002) *Applied Process Design for Chemical and Petrochemical Plants*. Gulf Professional Publishing, Houston, TX.

Maingonnat, J.F., Muller, L. and Leuliet, J.C. (2005) Modelling the build-up of a thixotropic fluid under viscosimetric and mixing conditions. *Journal of Food Engineering* 71: 265–272.

Manjunath, K., Dhodapkar, S., and Jacob, K. (2004) Mixing of particulate solids in the process industries. In: *Handbook of Industrial Mixing: Science and Practice* (eds E.L. Paul, V.A. Atiemo-Obeng and S.M. Krestanew). John Wiley & Sons, Hoboken, NJ, pp. 924–985.

Nienow, A.W. (1976) The effect of agitation and scale-up on crystal growth rates and on secondary nucleation. *Chemical Engineering Research and Design* 54: 205–207.

Nienow, A.W., Harnby, N. and Edwards, M.F. (1992) Introduction to mixing problems. In: *Mixing in Process Industries*, 2nd edn (eds A.W. Nienow, N. Harnby and M.F. Edwards). Butterworth-Heinemann, London, pp. 1–16.

Niranjan, K., Smith, D.L.O., Reilly, C.D., Lindley, J.A. and Phillips, V.R. (1994) Mixing processes for agricultural and food materials: 1. Review of mixer types. *Journal of Agricultural Engineering Research* 59: 145–161.

Nordkvist, M., Vognsen, M., Nienow, A.W., Villadsen, J. and Gernaey, K.V. (2008) Mixing by rotary jet heads: indications of the benefits of head rotation under turbulent and transitional flow conditions. *Chemical Engineering Research and Design* 86: 1454–1461.

Oldshue, J.Y. (1983) *Fluid Mixing Technology*. McGraw-Hill, NewYork.

Ottino, J.M. and Khakhar, D.V. (2000) Mixing and segregation of granular mixing. *Annual Review of Fluid Mechanics* 32: 55–91.

Perry, R.H. and Green, D.H. (1997) *Perry's Chemical Engineers' Handbook*. McGraw-Hill, NewYork, section 6.

Portillo, P.M., Muzzio, F.J. and Ierapetritou, M.G. (2007) Hybrid DEM-compartment modeling approach for granular mixing. *AIChE Journal* 53: 119–128.

Portillo, P.M., Muzzio, F.J. and Ierapetritou, M.G. (2008) Characterization of continuous convective powder mixing processes. *Powder Technology* 182: 368–378.

Prakash, S. and Kokini, J.L. (2000) Estimation and prediction of shear rate distribution as a model mixer. *Journal of Food Engineering* 44: 135–148.

Revill, B.K. (1992) Jet mixing. In: *Mixing in Process Industries*, 2nd edn (eds A.W. Nienow, N. Harnby and M.F. Edwards). Butterworth-Heinemann, London, pp. 159–180.

Rieger, F., Novak, V. and Havelková, D. (1986) Homogenization efficiency of helical ribbon agitators. *Chemical Engineering Journal* 33: 143–150.

Rudolph, L., Atiemo-Obeng, V., Schaefer, M. and Kraume, M. (2009) Power consumption and blend time of co-axial tank mixing systems in non-Newtonian fluids. Presented at 13th European Conference on Mixing, Kings College, London, 14–17 April 2009. Available at proceedings.europeanmixing13.org/data/Rudolph_et_al.pdf (accessed October 7, 2010).

Sauter, J. Die Grössenbestimmung der in Gemischnebeln von Verbren nungskraftmaschinen vorhandenen Brennstoffteilchen. *VDI-Forschungsheft* No. 279 (1926) und No. 312 (1928).

Skelland, A.H.P. and Seksaria, R. (1978) Minimum impeller speeds for liquid–liquid dispersion in baffled vessels. *Industrial and Engineering Chemistry Process Design and Development* 17: 56–61.

Thakur, R.K., Vial, C., Nigam, K.D.P., Nauman, E.B. and Djelveh, G. (2003) Static mixers in the process industries: a review. *Chemical Engineering Research and Design* 81: 787–826.

Tomas, J. and Schubert, H. (1979) Particle characterisation. In: *Partec 79*, pp. 301–319. Nurnberg, Germany.

Uhl, V.W. and Gray, J.B. (1986) *Mixing: Theory and Practice*. Academic Press, New York.

Valentas, K.J., Rotstein, E. and Singh R.P. (1997) *Handbook of Food Engineering Practice*. CRC Press, Boca Raton, FL.

Wang, C.Y. and Calabrese, R.V. (1986) Drop breakup in turbulent stirred-tank contactors: II. Relative influence of viscosity and interfacial tension. *AIChE J* 32: 667–676.

Wichterle, K. and Wein, O. (1981) Threshold of mixing non-Newtonian fluids. *International Chemical Engineering* 21: 116–120.

Xia, B. and Sun, D.W. (2002) Applications of computational fluid dynamics CFD in the food industry: a review. *Computers and Electronics in Agriculture* 34: 5–24.

Zhao, W.Q., Pu, B.Y. and Hartland, S. (1993) Measurement of drop size distribution in liquid/liquid dispersion by encapsulation. *Chemical Engineering Science* 48: 219–227.

Zwietering, T.N. (1958) Suspending of solid particles in liquid by agitators, *Chemical Engineering Science* 8: 244–253.

31

Extraction Process Design

Q. Tuan Pham and Frank P. Lucien

Introduction

In the food and chemical industries, extraction or solvent extraction is the process of separating a component substance (the solute) from a solid or liquid mixture by dissolving it in a liquid solvent. Depending on the phase of the mixture and the extraction agent, extraction can be divided into the following types:

- liquid–liquid extraction, where a solvent extracts a solute from a liquid phase;
- solid–liquid extraction, or leaching, where a solvent extracts a solute from a solid phase;
- supercritical extraction, where a fluid under supercritical conditions is used as the solvent.

Laboratory or commercial-scale applications in the food industry include the extraction or recovery of colours, flavours, proteins, amino acids, enzymes, vitamins, oils, sugars, starch, and other carbohydrates. Solvents commonly used in the food industries include water, alcohol, and other organic solvents, or supercritical CO_2.

Extraction is usually preceded by, or carried out simultaneously with, other operations to reduce the size of the solid (size reduction), to intimately mix the liquid or solid raw material with the solvent (mixing and agitation), and to separate the two

Handbook of Food Process Design, First Edition. Edited by Jasim Ahmed and Mohammad Shafiur Rahman.

phases (mechanical separation). Extraction may also take place along with digestion or solubilisation of insoluble components to produce solutes that can be extracted, such as the hydrolysis of proteins with the help of enzymes, acids, bases or salts.

Extraction results in a solution consisting mainly of solvent and solute, but one that may often also contain other impurities. The extracted component may then have to be separated from the solvent by another separation process such as distillation or crystallisation. For supercritical extraction this separation is easily achieved by reducing the pressure and turning the solvent into a gas. In this chapter we are concerned only with the extraction operation itself, and the reader is referred to other chapters for the accompanying processes. An extraction plant can be carried out by assembling the components necessary for each stage of the process, but there is also specialised proprietary extraction equipment designed to combine several of the necessary steps together into one unit.

Liquid–Liquid Extraction

What is Liquid–Liquid Extraction?

Liquid–liquid extraction (LLE) is the extraction of a solute A from a solution of A in a liquid carrier C, using a liquid solvent B. The solvent is chosen to be highly miscible with A but immiscible with C. Some A will move from the feed solution to the solvent until equilibrium between the two phases is achieved. The product may be the extract, the raffinate, or both.

LLE is a more complicated separation process than distillation due to the introduction of a solvent, which usually needs subsequently to be separated from the product as well as from the carrier streams, and recirculated to avoid wastage, using additional equipment such as distillation columns. LLE is more attractive than distillation when the components to be separated have close volatility; when the solute has low volatility (e.g. salts); when the product is heat sensitive (e.g. vitamins, antibiotics, food flavours); or when the desired product is the solvent–solute solution rather than pure solute. If the process is carried out in a single stage, the extraction efficiency is limited by the carrying capacity of the solvent, which depends on the equilibrium distribution ratio and the mass ratio of solvent to carrier. Therefore, like most other mass transfer operations, LLE is usually carried out in countercurrent cascades (e.g. spray columns, tray columns or packed beds). A recent review of applications of LLE is given by Batista *et al.* (2009).

Equipment

LLE equipment must be designed to carry out two basic operations, which have opposite objectives and requirements: mixing and separation. Good mixing is necessary to increase the mass transfer coefficient, increase contact area between the two phases and minimise the diffusion path of the solute molecules, by breaking up one or both liquid phases into small droplets suspended in the other phase. Separation means re-agglomerating these droplets into a continuous phase or at least into large enough drops, so that they can flow in the opposite direction to the other phase under gravity or centrifugal force and be separated from it. Since liquids that mix readily will be difficult to separate and *vice versa*, and the density difference between liquids is usually small (much less than between vapour and liquids, as in a distillation column), the retention time in a LLE unit is often significant and great care has to be paid to equipment design. Often, mixing and separation have to be mechanically assisted.

Mixer-Settlers

In the most basic set-up, a LLE stage will comprise a mixer followed by a separator. Any suitable type of mixer can be used, such as a pump, a mixing tank with agitator, a pressure nozzle or a static inline mixer. The reader is referred to the Chapter 30 for further information. The resulting mixture or emulsion is then separated by agglomeration and settling or centrifugation. Mixing and separation can be combined in a single vessel, termed a mixer-settler, possibly partitioned into a mixing compartment and a separation compartment (Figure 31.1).

Non-agitated Columns

In practice single-stage extraction is rarely used as the amount of extraction is usually limited. For increased extraction efficiency, several mixer-settlers may be connected to form a countercurrent cascade, with the light and heavy phases moving in opposite directions. Countercurrent operation can also be carried out in standard types of countercurrent contacting column, such as spray columns, packed columns or sieve plate columns. In spray columns (Figure 31.2) one phase is broken up into droplets

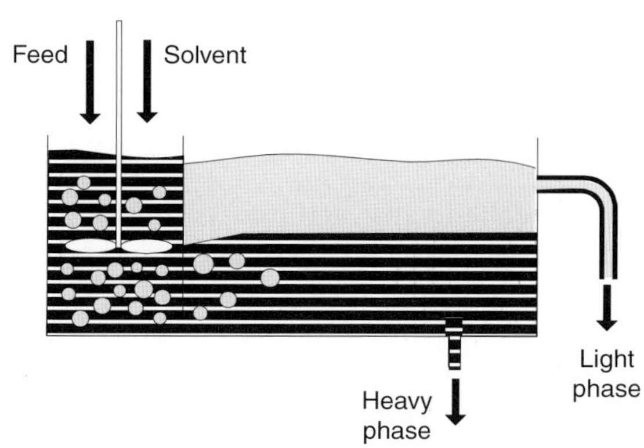

Figure 31.1 Mixer-settler.

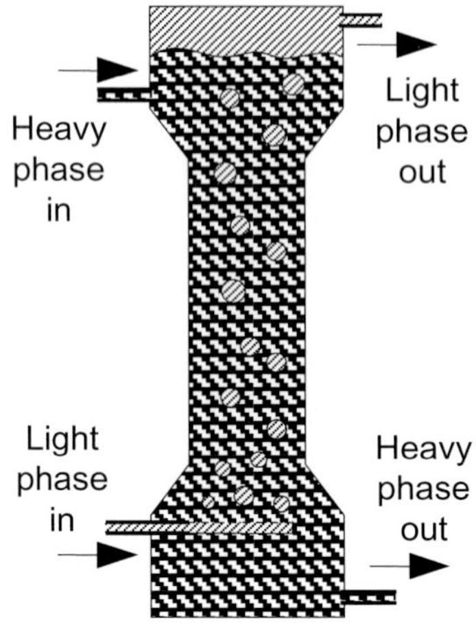

Figure 31.2 Spray column for liquid–liquid extraction (LLE).

which rise or fall countercurrently to the other continuous phase under gravity. In packed columns, the middle section is filled with packing and both phases may be broken up to some extent as each moves from one particle of packing to another. Many different types of packing have been designed to enhance mixing. In sieve plate columns, one phase rises through perforations in a plate where they are broken up into droplets, which coalesce and then are broken up again by the next plate. The second phase flows across each plate, then through a slot down to the next plate. For all column types, a settling section must be provided at each end to allow thorough separation.

The above types of columns are driven purely by gravity. They will only work well if the densities of the two phases are significantly different. If that is not the case, the droplets have to be reasonably large or the flow velocities slow enough to ensure that the droplets can rise or fall in the opposite direction to the other phase, both of which requirements lead to longer residence times and larger equipment sizes.

Mechanically Agitated Equipment

A number of proprietary mechanically agitated columns, an example of which is shown in Figure 31.3a, have been designed to enhance mixing (Reissinger and Schroter, 1978; Perry and Green, 1997).

- In *pulsed columns* the flows may be pulsed at 1.5–4 Hz to enhance mass transfer.
- *Rotary agitated columns* have a central shaft running the length of the column fitted with impellers. Most columns are divided into compartments, usually by annular plates, each compartment containing one impeller. The impellers help

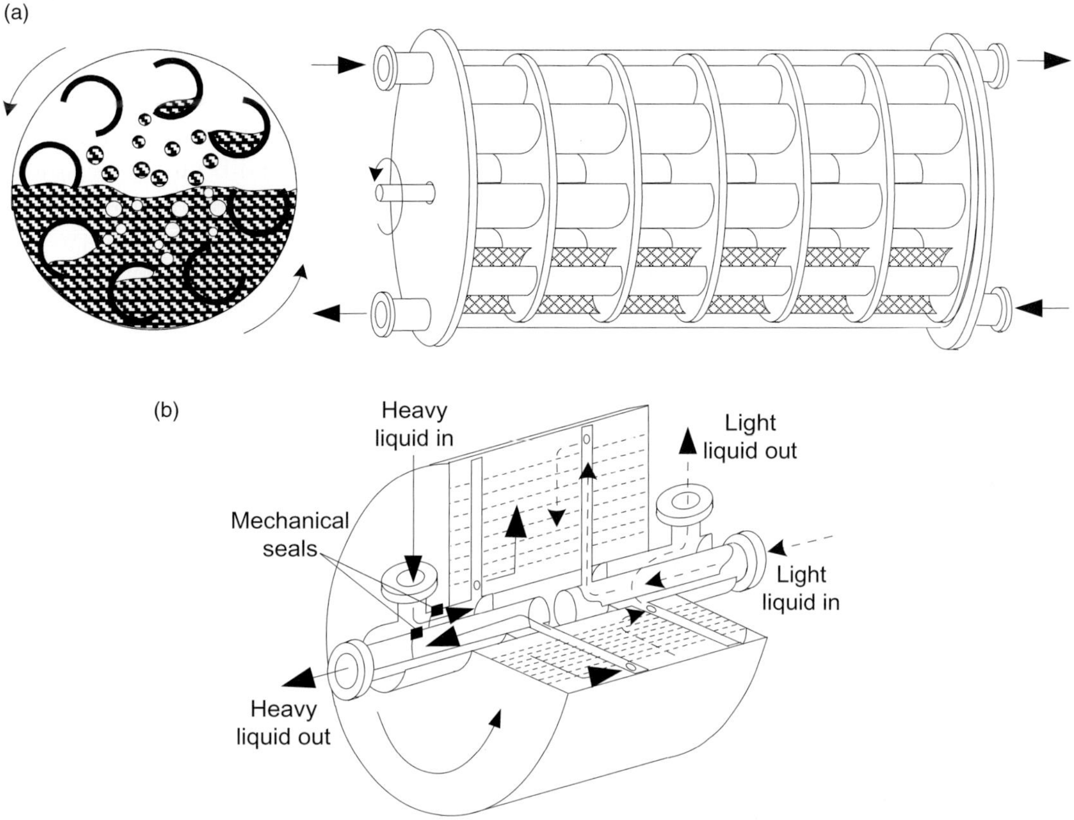

Figure 31.3 Mechanically agitated and centrifugal extractors: (a) Graesser raining bucket contactor. (b) Podbielnak centrifugal extractor (rotor only, casing not shown).

break up the liquids and enhance mass transfer. The best known type is the rotary disc contactor (Lo *et al.*, 1983). Others include the Scheibel column, the Oldshue–Rushton extractor, and the Kuhni column.

- *Reciprocating agitated columns* such as the Karr column use longitudinally reciprocating perforated plates.
- The *Graesser raining bucket contactor* (Figure 31.3a) has a disc and bucket assembly which rotates inside a horizontal shell. The light phase occupies the top half of the shell while the heavy phase occupies the lower half. The C-shaped buckets pick up the heavy liquid and release it into the light liquid, and *vice versa*. The discs divide the shell into compartments or stages, while an annular gap between the discs and the shell enables the liquids to flow countercurrently. Agitation is gentle; thus this equipment is suitable for liquids that tend to form emulsions.

Centrifugal Equipment

The major advantage of centrifugal equipment, an example of which is shown in Figure 31.3b, is efficient phase separation. Since phase separation occupies most of the equipment volume and takes up most of the total residence time, centrifugal equipment is very compact and can handle emulsions, which are difficult to separate. Furthermore, the rapid rotation allows the inclusion of efficient mixers, resulting in rapid mass transfer and the ability to handle liquids with high surface tension (which are difficult to break up). One possible drawback is the high pressure experienced by the fluids, up to 100 bars, which could adversely affect phase equilibrium by making the components more miscible.

- *Centrifugal extractors* are basically centrifugal separators similar to those used for separating a liquid, with the addition of a mixing impeller or mixing plate to intimately mix the feed and solvent.
- The *multistage centrifugal extractor* (Rousselet Robatel) is a multistage mixer-centrifuge. Each stage consists of a mixing chamber near the axis, where the two liquids are mixed into a fine dispersion by a stationary disc mounted on the central drum, which also performs a pumping action to move the fluids. The mixture then moves to a settling chamber where the two fluids are separated by centrifugal force and routed in opposite directions. The design ensures both efficient mass transfer and rapid separation.
- The *Podbielnek centrifugal extractor* (Figure 31.3b) is essentially a very fat rectangular sieve plate column that is rolled into a cylindrical shape and rotated rapidly. The heavy liquid is released near the axis and the light liquid near the rim of a centrifuge. Concentric perforated cylindrical plates break up the streams, in a manner similar to a conventional sieve plate column. The centrifugal force is much higher than gravity, leading to higher velocities, better separation, and smaller residence times. Emulsions can easily be separated into separate liquid streams.

Table 31.1, which summarises equipment characteristics, can be used for a preliminary screening. Reissinger and Schroter (1978) presented a simplified flowchart for selecting LLE equipment.

Process Design

Single-Stage Extraction

Let a feed stream (F) containing the solute to be extracted be intimately mixed with an immiscible solvent stream (S) in a mixer until the two phases are in equilibrium. The two phases are then allowed to separate and exit the settler as two single-phase streams, called the extract (E) and raffinate (R) streams. The former consists mainly of solvent and extracted solute, while the latter will be mainly the carrier fluid with some residual solute. However, both streams will in general contain all three components.

Table 31.1 Equipment characteristics summary. Note: any type of equipment can be connected into a multistage cascade.

Equipment type	Residence time	Height	Floor area	Handles emulsions	Pressure effect	Mixing	Separation	Multistage
Mixer-settler	Large	Small	Large	No	No	Poor	Poor	No
Non-assisted columns	Large	Large	Small	No	No	Poor	Poor	Yes
Pulsed columns	Large	Large	Small	No	No	Medium	Poor	Yes
Rotating disc columns	Medium	Large	Small	No	No	Medium	Poor	Yes
Reciprocating plate column	Medium	Large	Small	Yes	No	Medium	Poor	Yes
Graesser raining bucket	Medium	Small	Large	Yes	No	Medium	Poor	Yes
Centrifugal extractors	Small	Small	Small	Yes	Yes	Good	Good	No
Multistage centrifugal extractors	Small	Small	Small	Yes	Yes	Good	Good	Yes
Podbielnak centrifugal extractor	Small	Small	Small	No	Yes	Medium	Good	Yes

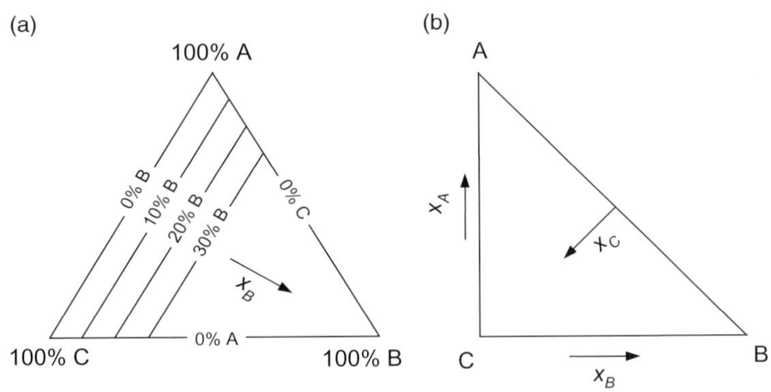

Figure 31.4 How compositions are represented on a ternary diagram. (a) Equilateral triangle. (b) Right triangle.

If the flows and compositions of F and S are known, we can determine those of the products E and R by using a ternary diagram (Figure 31.4), in which the composition of a mixture is represented by a point in a triangle. The vertices of the triangle represent pure components (A, B or C). The concentration of any component *i* is directly proportional to the distance from the point representing the mixture to the side opposite vertex *i*. For example, points on the side opposite vertex B represent binary

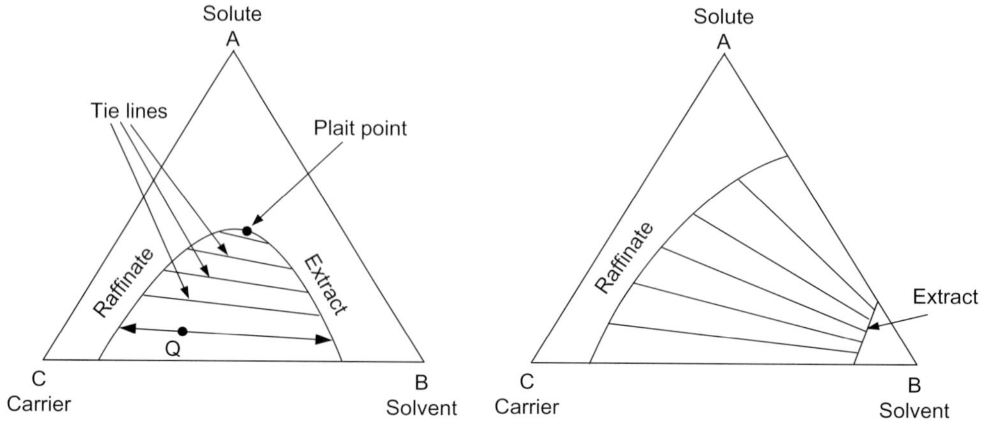

Figure 31.5 Ternary diagram with two liquid phases. On the left B and C form immiscible phases. On the right both B–C and A–B form immiscible phases.

mixtures containing no B. Lines parallel to this side represent mixtures of varying concentration of B – the closer a line is to vertex B, the higher the concentration of B.

Ternary diagrams do not have to be equilateral triangles. Triangles of any shape may be used, as long as the scale remains linear, i.e. the mass fraction of each component is proportional to the distance from one side of the triangle. Right-angled triangles (Figure 31.4b) are popular as they can be plotted using normal Cartesian graph paper or plotting software (the mass fractions of two components are plotted on the x and y axes respectively, that of the third being worked out from difference, although it could also be read on a diagonal scale).

If there are two immiscible phases, a mixture Q will spontaneously separate into two phases, raffinate (carrier-rich) and extract (solvent-rich), as shown in Figure 31.5. The line joining the two equilibrium concentrations is called a tie line. Different starting mixture compositions result in different tie lines and product compositions. The loci of the equilibrium concentrations (the tips of the tie lines) form the equilibrium curve, one half of which represents the extract and the other the raffinate. For some systems and at given temperature ranges, at a certain point the compositions of the extract and raffinate come together: this is the plait point. Beyond that point, the components are completely miscible and there is only one phase. Most systems become more miscible as temperature increases; thus the two-phase region will tend to shrink as temperature rises and may disappear altogether. For efficient extraction, the two-phase region must be as large as possible, so extraction is easier at lower temperatures.

Lever Rule

If two streams P and Q with flow rates p and q respectively are mixed together, the composition $\mathbf{x} = (x_A, x_B)$ of the mixture will be:

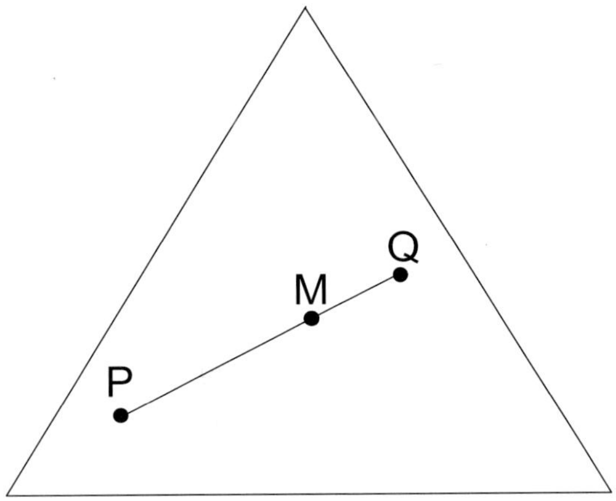

Figure 31.6 Lever rule. A mixture of one part P and two parts Q is represented by point M on PQ, which is twice as far from P as from Q.

$$\mathbf{x}_M = \frac{p\mathbf{x}_P + q\mathbf{x}_Q}{p+q} \tag{31.1}$$

which can be rearranged into:

$$\begin{gathered} p\mathbf{x}_M + q\mathbf{x}_M = p\mathbf{x}_P + q\mathbf{x}_Q \\ \frac{\mathbf{x}_M - \mathbf{x}_P}{q} = \frac{\mathbf{x}_Q - \mathbf{x}_M}{p} \end{gathered} \tag{31.2}$$

which shows that, on any graph with linear composition scales such as the ternary diagram, the mixture composition is represented by a point M (termed the mixing point) between P and Q on the line PQ, such that the distances PM and MQ are in inverse proportion to the flow rates (Figure 31.6):

$$\frac{\overline{PM}}{\overline{MQ}} = \frac{q}{p} \tag{31.3}$$

This graphical relationship, known as the lever rule, is readily used to determine graphically the separation performance of a single-stage extraction process (Figure 31.7):

1. Plot the entering feed and solvent concentrations on the ternary diagram.
2. Determine the mixture M of feed and solvent from the lever rule: draw a line joining the entering compositions and calculate the position of M on the line from Equation 31.3.
3. Draw a tie line (by interpolation) passing through M and determine its intersections with the equilibrium curves.

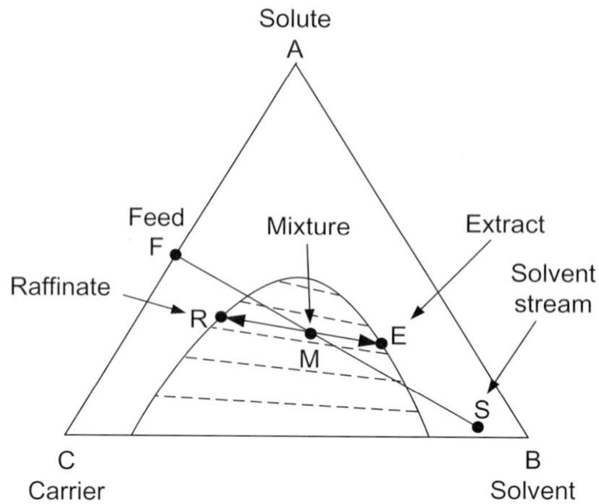

Figure 31.7 Single equilibrium stage extraction process. S refers to the entering solvent stream, which may be contaminated, while B refers to the (pure) solvent substance.

4. Read off the extract and raffinate compositions from these intersections.
5. Apply the lever rule once more to determine the product flow rates from the ratio RM/ME, given that their sum is equal to the sum of entering feed and solvent flow rates:

$$E = (F + S)\frac{\overline{RM}}{\overline{RE}}, \ R = (F + S)\frac{\overline{ME}}{\overline{RE}} \tag{31.4}$$

Note that although the ternary diagram and graphical method have been presented in terms of mass flow rates and mass fractions, the same methods can be used if flow rates and concentrations are measured in molar terms. The ternary diagram and the distribution ratios K_i will of course have to be expressed in terms of mole fraction. The same applies to other methods presented throughout this chapter.

Graphical Methods for Countercurrent Cascade

Extraction effectiveness can be increased by having the two phases flowing in opposite directions. There is no point having more than one equilibrium stage if the liquids flow in the same direction (co-current cascade), since the streams leaving the first stage are already in equilibrium and no further extraction will happen in the subsequent stages. Cross-current cascades, where the feed flows through a series of stages each with a supply of fresh solvent, are better than a co-current cascade, but less efficient than a countercurrent cascade.

In designing a countercurrent cascade (Figure 31.8) we usually know the compositions and flow rates of the two entering streams (feed F and solvent S), and also the required composition or at least the solute concentration of one of the product streams,

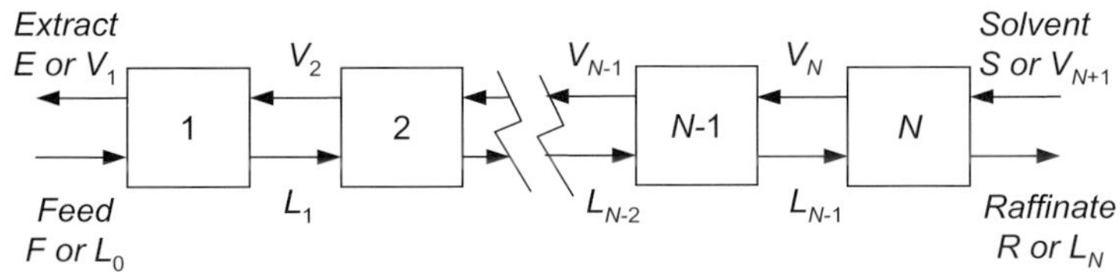

Figure 31.8 Countercurrent cascade of equilibrium extraction stages. V denotes the solvent-rich (extract) stream and L the carrier rich (raffinate) stream. Subscripts refer to the stage from which a stream originates.

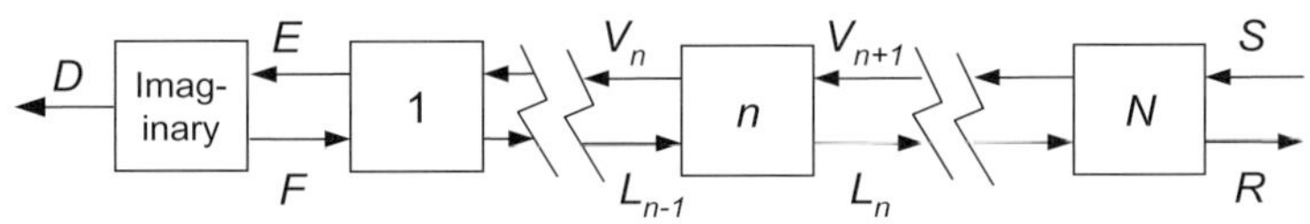

Figure 31.9 The difference stream D is an imaginary stream which, if mixed with an L stream anywhere along the cascade, will produce the V stream passing it at the same location.

either extract or raffinate. Our job is then to calculate the number of equilibrium stages, N.

Before we can describe the design method, we need to introduce a new concept: the *difference point*. Let us put an imaginary stage at the end of the cascade so as to obtain a single stream D coming out at that end (Figure 31.9). From the material balance over the imaginary stage, mixing D with the stream F will give E. Therefore, D can be termed the *difference stream*: it is the difference between F and E. On a ternary diagram, the lever rule dictates that D, F, and E lie on the same straight line. Similarly, by considering the material balance over the whole cascade, including the imaginary stage, we can see that D is also the difference of streams S and R, and therefore D, S, and R must be collinear. Thus, on the ternary diagram, D is simply the intersection of the straight lines FE and SR (extended as necessary) (Figure 31.10a). Do not be surprised that D lies outside the ternary diagram: it is only an imaginary stream and can have physically unrealistic compositions, such as mole or mass fractions outside the range 0–1.

What makes the difference point useful is that, by considering the mass balance between any interstage point along the cascade and the stream D (Figure 31.9), it can be seen that any pair of interstage streams (streams passing each other between two stages), such as (V_{n+1}, L_n) or (V_n, L_{n-1}), will also have the same difference stream D. Hence, on a ternary diagram, all lines connecting pairs of interstage compositions must pass through D. Together with the requirement that any stream leaving a stage must lie on the equilibrium line, this gives us the means of graphically stepping from one stage to the next on the ternary diagram. We are now ready to describe the Hunter–Nash graphical design method for an LLE cascade (Treybal, 1963).

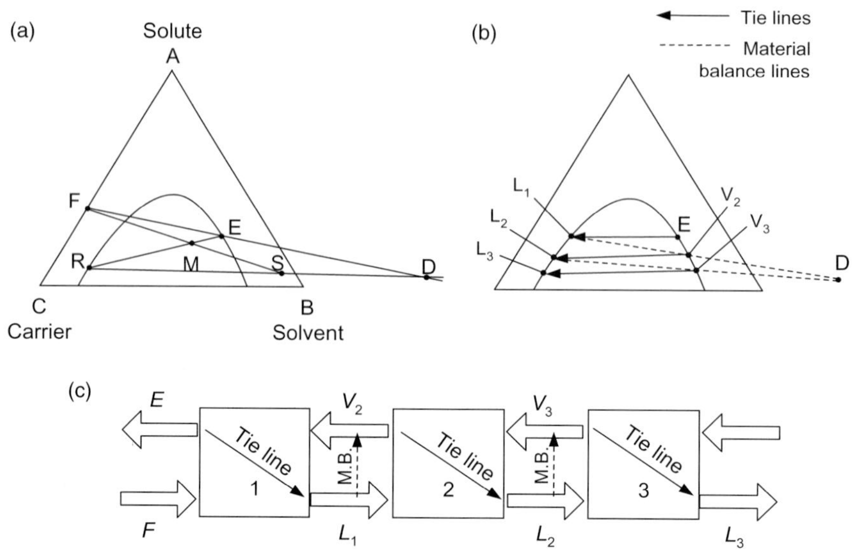

Figure 31.10 (a) Identifying inlet/outlet compositions, mixing point M and difference point P on a ternary diagram. (b) Stepping between tie lines and material balance lines to count stages. (c) Physical interpretation of stepping procedure. M.B., material balance line.

Step 1: Identify All Inlet and Outlet Compositions on the Diagram

For illustration's sake, let us assume that the composition of the extract E is known and that of the raffinate R remains to be determined. Since mixing these two exit streams would give the same mixture as mixing the entering streams F and S, these two pairs must have the same mixing point on the ternary diagram. Furthermore, we know that the compositions of both the leaving streams E and R must lie on the equilibrium curve, since each comes from an equilibrium stage. Thus, the unknown composition (in this case that of the extract) is simply obtained by locating the mixing point M of feed F and solvent S, then extending a straight line from the (known) extract R through M until it re-intersects the equilibrium curve at the raffinate R (Figure 31.10a).

Step 2: Identify the Difference Point (Operating Point)

Join F to E and R to S and extend these lines until they intersect (Figure 31.10a). The intersection is the difference point D, as shown earlier.

Step 3: Use the Equilibrium Relationship

We may start from either end of the cascade, but for illustrative purpose let us assume that we start with stage 1, where the extract E leaves. The stream L_1 leaving stage 1 is in equilibrium with E, and therefore E and L_1 are at the ends of a tie line (Figure 31.10b). This allows us to determine the composition of L_1 on the ternary diagram by drawing a tie line from E.

Step 4: Use the Material Balance

Since the stream V_2 passes the stream L_1 forming an inter-stage pair, the points representing their compositions on the ternary stage must be collinear with the difference point D, as already shown. Therefore, by joining point D to L_1, we obtain the composition of V_2 where the material balance line DL_1 intersects the equilibrium curve (Figure 31.10b).

By repeating steps 3 and 4 over and over until we reach or pass the composition of the desired raffinate at the other end, we can determine the necessary number of equilibrium stages. Each stage is represented by a tie line that we have drawn in Step 3 (Figure 31.10b,c).

From the graphical method just described, a necessary condition for any separation to be possible is that the mixing point M must be inside the two-phase region, since the extract and raffinate compositions must be on the equilibrium curve and on different sides of M. Thus, the line joining feed and entering solvent must pass through the two-phase region. Since the two-phase region tends to shrink with rising temperature due to increasing miscibility, separation by LLE is easier at lower temperature and product tends to be purer. Another condition for feasibility is that the difference point D must not be aligned with any tie line in the operating region. If this happens, we will go back and forth between the same L and V compositions in steps 3 and 4, i.e. an infinite number of stages will result. These inoperability conditions will occur if the solvent-to-feed ratio is smaller than necessary or if the solvent already contains more than a certain amount of solute.

Example 1

1 $kg \cdot s^{-1}$ of vegetable oil containing 8% fatty acid by mass is de-acidified by extracting with pure ethanol at 1.5·$kg \cdot s^{-1}$ in a countercurrent cascade. Equilibrium data are given in the ternary diagram in Figure 31.11, where A is oleic acid (solute), C is vegetable oil (carrier) and B is ethanol (solvent). Determine the number of equilibrium stages to bring the acid mass fraction to 1% or below.

Solution

1. Identify the points F (feed composition), S (fresh solvent) and R (desired raffinate). R must lie on the raffinate (feed) side of the equilibrium curve and have $x_A = 0.01$.
2. Use the lever rule to identify mixing point M: MF/SM = 1.5 or MF/SF = 1.5/(1 + 1.5) = 0.6.
3. Extend RM to re-intersect the equilibrium curve at E, the extract composition.

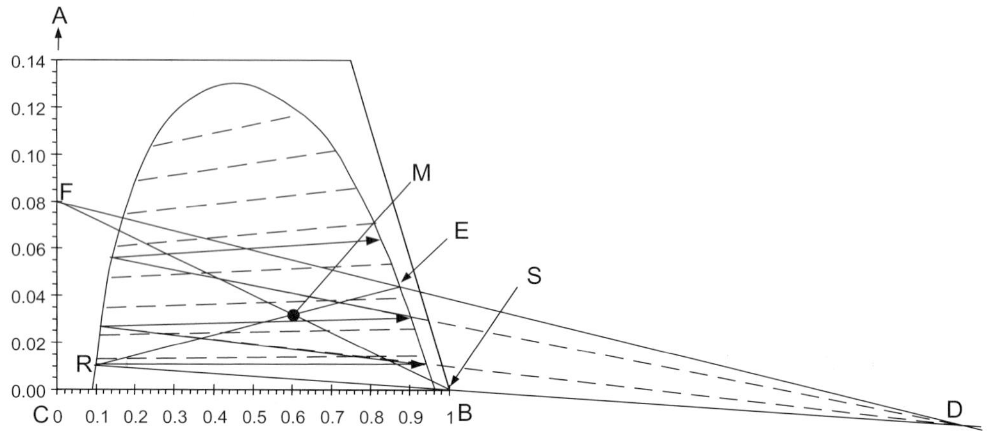

Figure 31.11 Graphical method for liquid–liquid extraction (LLE) example.

4. Using a ruler, RM/RE = 0.65; therefore 65% of the total mass goes to extract and 35% goes to raffinate. Thus, the raffinate flow is 0.35 × (1.0 + 1.5) = 0.875 kg·h^{-1}, the extract flow is 0.65 × (1.0 + 1.5) = 1.625 kg·h^{-1}.
5. Extend FE and RS to get difference point D.
6. Draw a tie line from R by interpolation, re-intersecting the equilibrium curve at the composition of the extract stream leaving the last stage, V_N.
7. Join D to this extract composition and extend to re-intersect the equilibrium curve. This is the composition of the next raffinate stream, L_{N-1}.
8. Repeat the previous two steps until the composition of the final extract is reached or passed. This should happen at between two and three equilibrium stages, say 2.5 stages. (Each stage is represented by a tie line.)

Approximate Methods for Countercurrent Cascade

An alternative approach that does not require graphical construction and is therefore more suitable for calculators and spreadsheets, is to use an approximate method such as Kremser's method (Kremser ,1930; Edmister, 1957). A proof of the method can be found in Seader and Henley (2006). Kremser's method is a rating method, i.e. it calculates the performance (product concentration) from a given cascade with known feed and solvent flows and concentrations. However, it can be used for design by iterating or using an equation solver tool in a spreadsheet. It is applicable when the flow rates V and L do not vary greatly along the cascade.

First we define the *extraction factor* E_i and its inverse, the *absorption factor* A_i:

$$E_i \equiv \frac{K_i V}{L},\ A_i \equiv \frac{L}{K_i V} \tag{31.5}$$

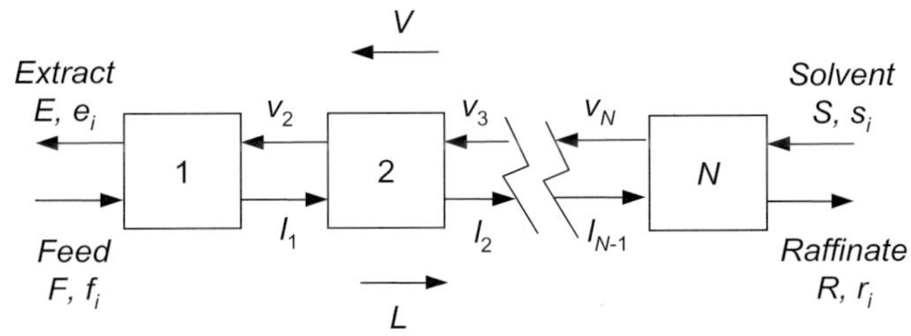

Figure 31.12 Notation for Kremser's method. Lower case symbols represent component mass flows.

Since $y_{i,n} = v_{i,n}/V$ and $x_{i,n} = l_{i,n}/L$, where $v_{i,n}$ and $l_{i,n}$ are the component flows of component i from stage n, the equilibrium relationship between streams leaving stage n can be written as:

$$v_{i,n} = E_i l_{i,n}, \; l_{i,n} = A_i v_{i,n} \tag{31.6}$$

We shall now derive Kremser's equations for a given component i. In the following derivation we will drop the component subscript i but retain the subscript for stage number when necessary. Assuming for the moment that the solvent is free of solute, i.e. $s = 0$, material balance over the whole column (Figure 31.12) gives:

$$f = e + r \tag{31.7}$$

Equilibrium for stage 1 (Equation 31.6) gives:

$$e = El_1$$

Component material balance for the section comprising stages 2 to N gives:

$$l_1 = v_2 + r$$

Continuing to alternate between material balance and equilibrium:

$$v_2 = El_2$$

$$l_2 = v_3 + r$$

$$\cdots$$

$$v_{N-1} = El_{N-1}$$

$$l_{N-1} = v_N + r$$

$$v_N = Er \tag{31.8}$$

By performing backwards substitutions from Equation 31.8 back to Equation 31.7 (substituting for v_N then l_{N-1} then v_{N-1} etc.) we eliminate all the intermediate variables l_n, v_n and obtain a relationship between f and r:

$$f = (E^N + E^{N-1} + \ldots + E + 1)r = \frac{E^{N+1} - 1}{E - 1} r \tag{31.9}$$

Hence:

$$r = \frac{E - 1}{E^{N+1} - 1} f = \Phi^E f \tag{31.10}$$

$$e = f - r = (1 - \Phi^E) f \tag{31.11}$$

where Φ^E, defined by Equation 31.10, is the unextracted fraction of component i entering with the feed.

Similarly, by considering a cascade where component i enters with the solvent but not with the feed, we obtain:

$$e = \frac{A - 1}{A^{N+1} - 1} s = \Phi^A s \tag{31.12}$$

where Φ^A is the unabsorbed fraction of component i entering with the solvent.

To get the equation for the extract flow for the general case (non-zero f and s) we combine the extracted fraction from Equation 31.11 and unabsorbed fraction from Equation 31.12:

$$e = \Phi^A s + (1 - \Phi^E) f \tag{31.13}$$

Summary of Kremser's Method

1. Determine the mean distribution ratios $K_i \equiv y_i/x_i$ for all components i, averaged over the cascade, where x_i is the concentration of i in the L phase (feed-raffinate) and y_i that in the V phase (solvent-extract).
2. Estimate mean values of total flows L and V over the cascade.
3. Calculate the extraction and absorption factors for component i from Equation 31.5.
4. Calculate the unabsorbed and unextracted fractions:

$$\Phi_i^E \equiv \frac{E_i - 1}{E_i^{N+1} - 1}, \quad \Phi_i^A \equiv \frac{A_i - 1}{A_i^{N+1} - 1} \tag{31.14}$$

5. Calculate the mass flow e_i of component i in the final extract from:

$$e_i = \Phi_i^A s_i + (1 - \Phi_i^E) f_i \tag{31.13}$$

6. Calculate the mass flow r_i of component i in the final raffinate from material balance over the cascade:

$$r_i = s_i + f_i - e_i \tag{31.15}$$

7. Repeat steps 3–6 for the other components.
8. Once the flows e and r of all components are determined, add them up to recalculate E and R.
9. Go back to step 2 if necessary (i.e. if L and V have changed significantly).

If L and V vary greatly, Kremser's method is not applicable and a graphical or rigorous method should be used.

Example 2

Use Kremser's method to verify the results obtained graphically in Example 1, i.e. verify that about 2.5 ideal stages will produce a raffinate containing 1% solute.

Solution

Read off the solute mass fractions (x_A, y_A) at the ends of a tie line and calculate $K_A = y_A/x_A$. For example, the bottom tie line gives $x_A = 0.011$, $y_A = 0.012$, hence $K_A = 1.09$. Do this for a few tie lines in the operating range and calculate the geometric mean to get $K_A = 1.114$.

The raffinate and extract curves are quite close to the sides of the ternary diagram, showing that the solvent and carrier are not very soluble in each other (this can be verified by calculating their K-values). Therefore, let us assume that L and V are constant at $L = 1.0\,\text{kg·s}^{-1}$, $V = 1.5\,\text{kg·s}^{-1}$. Now calculate the extraction factor:

$$E_A = \frac{K_A V}{L} = 1.115 \times 1.5/1.0 = 1.672$$

$$\Phi_A^E = \frac{E_A - 1}{E_A^{N+1} - 1} = \frac{1.672 - 1}{1.672^{2.5+1} - 1} = 0.133$$

The parameters A_A and Φ_A^A are not required since there is no solute in the entering solvent.

$$f_A = Fx_{A,feed} = 1.0 \times 0.08 = 0.08\ \text{kg} \cdot \text{s}^{-1}$$

From Equation 31.13:

$$e_A = (1 - \Phi_A^E) f_A = (1 - 0.133) \times 0.08 = 0.069\ \text{kg} \cdot \text{s}^{-1}$$

$$r_A = s_A + f_A - e_A = 0 + 0.08 - 0.069 = 0.011\ \text{kg} \cdot \text{s}^{-1}$$

The solute mass fraction in the raffinate is:

$$x_{A,raffinate} = r_A/L = 0.011$$

This is very close to the specification of 1% solute in the raffinate. In fact, setting $N = 2.6$ will give exactly 1% solute in the raffinate. We will leave the reader to carry out the iteration (steps 7–9).

Rigorous Methods for Countercurrent Cascade

Rigorous methods are methods that solve rigorously the material and energy balances and equilibrium equations over each stage in a cascade, using a purpose written computer program. Most rigorous methods involve the simultaneous solutions of the MESH equations (Material balances, Equilibria, Summation, enthalpy or H-balance), which apply to each stage. For each stage n, these equations consist of:

- one material balance for each component i:

$$(Lx_i + Vy_i)_{in} - (Lx_i + Vy_i)_{out} = 0 \tag{31.16}$$

- one equilibrium relationship for each component i relating the streams leaving the stage:

$$y_i = K_i x_i \tag{31.17}$$

- two summation equations:

$$\sum_i x_i = 1, \sum_i y_i = 1 \tag{31.18}$$

- enthalpy or H-balance:

$$(LH_L + VH_V)_{in} - (LH_L + VH_V)_{out} + \text{Heat input} = 0 \tag{31.19}$$

These equations are solved together with thermodynamic relationships for calculating K-values and specific enthalpies H_L and H_V, all of which vary with composition, temperature and pressure. LLE usually takes place under isothermal conditions so the enthalpy balance equation may be dropped. The equations are non-linear, with strong interactions between the parameters (such as the dependence of K-values on composition and *vice versa*); therefore an iterative solution method must be used. There are several different approaches to solving the MESH equations, such as the Sum Rate Method (Sujata, 1961; Friday and Smith, 1964), the Simultaneous Correction Method (Naphtali and Sandholm, 1971), the Inside Out Method (Boston and Sullivan, 1974), etc., differing in their approach to minimising the number of calculations and accelerating convergence. Some well-known commercial process simulation packages that implement one or more rigorous methods are Aspen, Hysys, Prosim, Chemcad, and ChemSep. Since readers are unlikely to write a program themselves and will make use of one of these packages, we will not go into the details of the computerised solution methods.

Activity Coefficient Models

In using rigorous methods, the most important choice for the user is that of a thermodynamic model, which must be used to calculate K-values from composition and temperature. Many such models have been proposed but only a few can accurately handle two-phase liquid systems such as those encountered in LLE. The condition for equilibrium between two liquid phases L and V is that the activity of component *i* is the same in each phase, or:

$$\gamma_i^L \xi_i^L = \gamma_i^V \xi_i^V \tag{31.20}$$

where γ_i are the activity coefficients and ξ_i the mole fraction of *i* in the L and V phases respectively. The molar K-values (mole fraction in V/mole fraction in L at equilibrium) are therefore given by:

$$K_i^{molar} \equiv \frac{\xi_i^V}{\xi_i^L} = \frac{\gamma_i^L}{\gamma_i^V} \tag{31.21}$$

while the mass fraction K-values (mass fraction in V/mass fraction in L at equilibrium) are given by:

$$K_i = \frac{y_i}{x_i} = \frac{M_i \xi_i^V / M_V}{M_i \xi_i^L / M_L} = \frac{M_L}{M_V} \frac{\gamma_i^L}{\gamma_i^V} \tag{31.22}$$

where M_L and M_V are the average molecular weights of each phase. Thus, the calculation of K-values is reduced to the problem of calculating activity coefficients.

The best known activity coefficient models for LLE calculations are the Non-Random Two-Liquid (NRTL) model (Renon and Prausnitz, 1968) and the UNIQUAC model (Abrams and Prausnitz, 1975). These models calculate activity coefficients in a liquid mixture from a set of empirical binary interaction parameters that are found by curve-fitting *x–y* equilibrium data. NRTL and UNIQUAC parameters for many systems have been published in the DECHEMA Chemical Data Series publications (Arlt *et al.*, 1979–1987). These data are available online in the DETHERM database (DECHEMA, 2009) and incorporated in most major commercial process simulation software packages.

When experimental data are unavailable, a Group Contribution Model can be used to predict binary interaction parameters from the functional groups on each molecule. Well-known group contribution models include the UNIFAC Model (Fredenslund *et al.*, 1975; Hansen *et al.*, 1991), the Modified UNIFAC (Dortmund) model (Weidlich and Gmehling, 1987; Gmehling *et al.*, 1998) and the ASOG model (Derr and Deal, 1969; Kojima and Tochigi, 1979). Functional group UNIFAC parameters for LLE were published by Magnussen *et al.* (1981) and ASOG parameters by Tochigi *et al.* (1990). Batista *et al.* (1999) presented improved values for both UNIFAC and ASOG parameters.

Commercial simulation software usually contains databases for interaction parameters that enable many systems to be modelled. Food engineers, however, often do not have precise knowledge of the molecular structure of components they are dealing with, hence equilibrium data often have to be obtained experimentally and curve-fitted to determine the binary parameters. Some process simulation software has facilities for calculating binary parameters from x–y data by regression.

Solid–Liquid Extraction (Leaching)

What is Solid–Liquid Extraction?

Solid–liquid extraction (SLE) is the removal of a soluble component A from a solid C by contact with a liquid solvent B. It is also called leaching, although this term is sometimes reserved for situations when the dissolution of A is caused or accompanied by a chemical reaction. In this chapter we will use the terms solid–liquid extraction and leaching interchangeably. An everyday example is the leaching of coffee from ground coffee beans with hot water. The desired product of leaching may be the solute (which will have to be separated from the solvent in the extract liquid by other means), the liquid extract (i.e. solute–solvent solution) or the depleted solid. Osmodehydration is the extraction of water using a low water activity solution (such as a concentrated sugar solution), accompanied by diffusion of other solutes into the solid. SLE is a very widely used process in the food industry and the number of applications is still growing. Table 31.2 lists some typical applications (Schwartzberg, 1980).

To accelerate the diffusion of solutes out of the solid, leaching is often preceded by some form of size reduction, such as grinding, breaking, cutting or flaking. It will be seen later that the required extraction time is proportional to the square of the particle size. Furthermore, grinding helps in breaking down the cell wall structure of many foods, which facilitates the diffusion process.

Equipment

Due to the difficulty of circulating solids, leaching is often carried out in batch fashion. Therefore, leaching equipment can be classified into batch extractors and continuous extractors.

Table 31.2 Examples of solid–liquid extraction in the food industry (Schwartzberg, 1980). If the solid product is listed, it is the desired product; if not, the solute is the product.

Solid product	Solute	Raw material	Solvent
	Sugar	Sugar beet, sugar cane	Water
	Vegetable oil	Oil seeds	Hexane
Softened corn kernels	Corn steep solids	Dry corn kernel	Water
	Instant coffee	Roasted ground coffee	Water
Collagen	Keratin, globulins, mucosopolysaccharides, albumins, elastins	Hide, ossein	Calcium hydroxide solution, dilute acid
	Gelatine	Collagen	Water or dilute acid
	Pectin	Apple pomace	Dilute acid
Tapioca	Cyanogenic glucosides	Manioc	Water
	Vanilla	Vanilla beans	35% ethanol, 65% water
	Carageenan	Kelp	Water
	Zein	Corn	90% ethanol, 10% water
	Iodine	Seaweed	H_2SO_4
Fish protein concentrate	Fish oil	Trash fish	Ethylene dichloride, ethanol, butanol, hexane
	Prune juice	Dried prunes	Water
	Fish oil	Fish	Ethylene dichloride, butanol, hexane
Alfalfa protein concentrate	Chlorophyll pigments, chlorogenic acid	Coagulated alfalfa proteins	Acetone, ethanol, butanol, isopropanol
	Flavours	Flowers, herbs	Hexane, benzene, fat, alcohol, petroleum ether
	Abthrocyanines, betanines	Chokeberries, grape skins, beets	Alcohol, water
	Grape juice solubles	Grape pomace	Water
	Citrus juice solids	Peel, press, pulp and rag	Water
Apple juice		Apple chunks	Water
Butter	Butyric acid, other low molecular weight organic acids	Rancid butter	Water
Apple pomace for pectin production	Sugar	Apple pomace	Water
Citrus peel for pectin production	Sugars, flavanoids, hesperidin	Citrus peel	Water
	Brewing worts	Malted barley, grain adjuncts	Water
	Hop-flavoured brewing wort	Hops	Brewing wort
Single-cell protein	Substrate residue	Single-cell protein	Water
Soy protein concentrate	Sugars, non-protein solids	Defatted soy flour	Ethanol–water at isoelectric point
Soy protein isolate	Non-protein solids	Defatted soy flour	pH 9 caustic solution
Kelp for algin production	Salt	Giant kelp	Dilute HCl
Partially dehydrated and salt-free pectin	Water, Al salts	Alcohol precipitated pectin	Isopropanol
Partially dehydrated fruits	Water, Al salts	Fruit	Sugar solution

Batch Extractors

Agitated vessels are often used for batch leaching of small particles that can be easily suspended in the liquid. Various types of impellers, propellers or paddles may be used. The reader is referred to Chapter 30 for further information. The leaching time depends on the size of the particles, the diffusivity of the solute in the solid matrix, and the mass transfer coefficient. The latter depends on the flow pattern and mechanical energy input to the mixer. When the desired residence time has been reached, agitation is stopped and the solid allowed to settle out of the liquid. The liquid is then decanted or filtered.

Percolators are another possibility, especially when the particle sizes are large or dense and difficult to keep in a suspended state. The solid is held in a vessel while the solvent is fed at the top and percolates down the bed, possibly under pressure to increase the flow rate. A well-known example is the espresso coffee machine.

Countercurrent Extractors

Batch extraction is not very efficient as the most that can be achieved in a batch unit is one equilibrium stage. As with LLE, higher extraction efficiencies require a countercurrent cascade with solid and solvent flowing in opposite directions. Batch percolating extractors can be operated as a countercurrent cascade in a semi-continuous manner. Several vessels are connected in a series and the solvent flows through the vessels sequentially, say from left to right. When the solid in the first (leftmost) vessel becomes depleted, it is emptied, filled with fresh solid, and shifted to the end of the cascade, while the second tank receives the fresh solvent. In practice, this rearrangement can be achieved simply by re-routing the fluid flow with a system of valves (Figure 31.13).

A countercurrent cascade can also be assembled from continuous mixers and separators, similar to the mixer-settlers used for LLE. Separators may include gravity settlers (clarifiers and thickeners), filters, hydrocyclones or centrifugal separators. Dedicated countercurrent leaching units contain several countercurrent stages within the same vessel. They differ mainly in the arrangement used to convey the solid from one stage to another. Schwartzberg (1980) gives a comprehensive review of equipment used in the food industry, although many types have become obsolete. Two examples are shown in Figure 31.14.

Belt and screw conveyors are readily converted to leaching equipment simply by adding a liquid circulation system (pump or gravity). In the perforated-belt extractor, a horizontal perforated belt conveys the solid from left to right. Solvent is introduced as a spray at the right end, collected under the belt, pumped to the next spray nozzle to the left, and so on, creating countercurrent contact. In screw extractors, the screw conveys the solid up the slope while the solvent percolates down the slope (Figure 31.14).

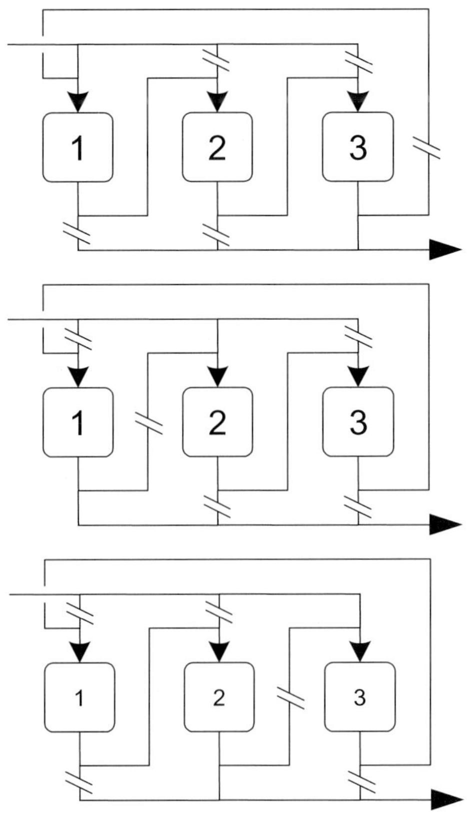

Figure 31.13 Using batch percolators in a countercurrent cascade.

Process Design

Single-Stage Batch Extraction

In a single-stage batch extractor, solid particles are mixed and suspended in the liquid for a given time, and then the two phases are separated. The main concerns are (i) to allow the solute sufficient time to diffuse out of the solid, (ii) to provide adequate agitation to suspend the solid particles and disperse the solute, and (iii) to ensure a sufficiently large liquid-to-solid ratio so that enough solute may be extracted, given the limits imposed by equilibrium. In the following treatment we shall assume that the solute is evenly dispersed in the liquid phase, i.e. there is no concentration gradient in the liquid near the solid's surface.

The solute's diffusion in a solid particle is usually described by Fick's second law of diffusion:

$$\frac{\partial c}{\partial t} = D_{AB} \nabla^2 c \qquad (31.23)$$

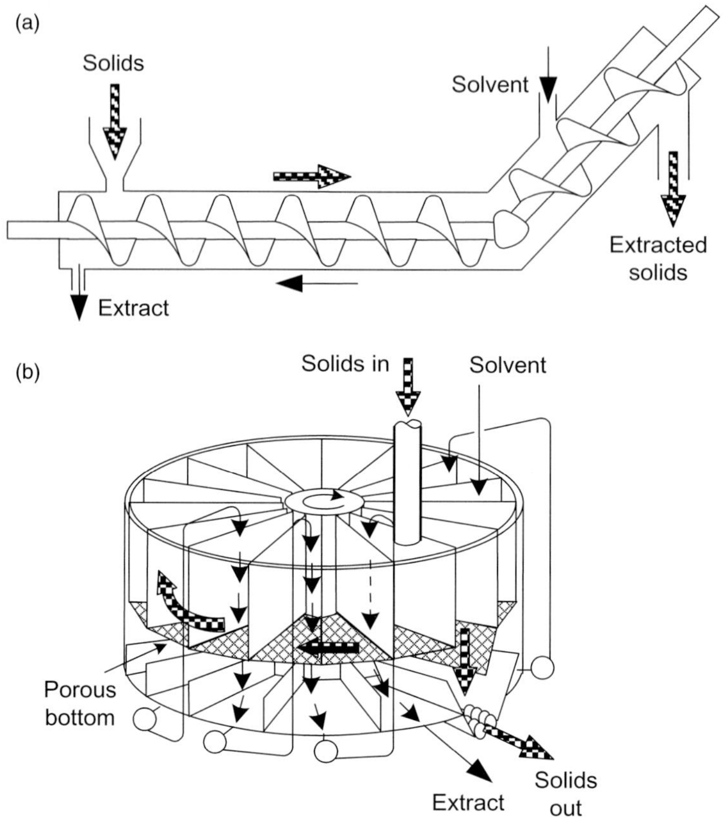

Figure 31.14 Some commercial solid liquid extractors. (a) Hildebrandt screw extractor. (b) Rotocel extractor.

where D_{AB} is the diffusion coefficient, or diffusivity. Values of diffusivity for many food systems are listed by Schwartzberg (1982). The above equation can be solved using the initial condition:

$$c(t=0, r \leq a) = c_0 \tag{31.24}$$

$$c^L(t=0) = c_0^L \tag{31.25}$$

where a is the half thickness or radius of the particle, c_0 the initial solid-phase solute concentration, c^L the liquid-phase solute concentration and c_0^L the initial liquid-phase solute concentration. The boundary condition is:

$$c(r=a, t>0) = c^L/K_c \tag{31.26}$$

$$c^L = R(c_0 - c) + c_0^L \tag{31.27}$$

where R is the solid-to-liquid volume ratio and K_c is the concentration distribution ratio of the solute, $(c^L/c)_{equilibrium}$. Due to diffusion, the outer part of the particle becomes depleted of solute faster, while the solute from the inner part diffuses out to replace it. The leaching rate gradually decreases as the solid becomes depleted and c^L gradually rises from the dissolved solute. Eventually, when the liquid and solid come to equilibrium, we obtain by material balance:

$$Rc_0 + c_0^L = Rc_{eq} + K_c c_{eq} \tag{31.28}$$

where the subscript eq indicates equilibrium values, hence:

$$c_{eq} = \frac{Rc_0 + c_0^L}{R + K_c} \tag{31.29}$$

Solutions for simple particle shapes (infinite slabs or flakes, infinite cylinders or spheres) have the form of infinite series of exponential terms. They are given in Crank (1975) and summarised by Schwartzberg (1982). Here the solution is expressed in terms of fractional distance from equilibrium:

$$\phi \equiv \frac{c_m - c_{eq}}{c_0 - c_{eq}} \tag{31.30}$$

where c_m is the mean solid-phase concentration at time t. ϕ is plotted against the dimensionless time or Fourier number Fo, defined by:

$$Fo \equiv \frac{D_{AB}t}{a^2} \tag{31.31}$$

As R/K_c increases (less liquid for given mass of solids), the approach to equilibrium is faster, but the equilibrium concentration also becomes higher according to Equation 31.29. For the case $R/K_c = 0$ (infinite liquid capacity), the liquid concentration remains constant as extraction proceeds. For this case, when enough time is allowed to bring the unleached fraction to less than about 0.7, the contributions of all exponential terms in the infinite series solution except the first become negligible, and ϕ decreases exponentially with time:

$$\phi = je^{-fFo} \tag{31.32}$$

The parameters j and f depend on the shape of the particles:

- *slabs (flakes)*: $j = 8/\pi^2 = 0.811$, $f = \pi^2/4 = 2.467$;
- *infinite cylinders*: $j = 0.692$, $f = 5.783$;
- *spheres*: $j = 6/\pi^2 = 0.608$, $f = \pi^2 = 9.870$.

It can be seen from Equation 31.32 that the required extraction time is approximately proportional to the square of the particle size; hence the importance of size reduction. The above solution depends on the following assumptions:

1. The diffusion coefficient D_{AB} is constant. Because most foods are inhomogeneous, this is often not true. However, if the variation is not great, an average value can be used.
2. The particles have uniform initial solute concentration.
3. The fluid is perfectly well agitated, so that the solute concentration in the fluid is uniform right up to the solid–liquid interface. In other words, the only resistance to mass transfer is in the solid.

Example 3

Solvent extraction is used to extract flavours from seeds in a batch stirred vessel. Solute diffusivity in the solid is $1.0 \times 10^{-10}\,\mathrm{m^2 \cdot s^{-1}}$ and $K_c = 2.0$. Calculate the time for 85% of the solute to be removed, assuming that the seeds are spherical, with average particle diameter of 3.0 mm, and the solid-to-liquid volume ratio is 0.25.

Solution

The equilibrium concentration is:

$$c_{eq} = \frac{Rc_0 + c_0^L}{R + K_c} = \frac{0.25 + 0}{0.25 + 2.0} c_0 = 0.111\, c_0$$

The final normalised mean concentration is:

$$\phi = \frac{c_m - c_{eq}}{c_0 - c_{eq}} = \frac{(1 - 0.85)c_0 - 0.111\, c_0}{c_0 - 0.111\, c_0} = 0.044$$

$$R/K_c = 0.25/2.0 = 0.125$$

Interpolating from Figure 31.15 we obtain $Fo = 0.25$. Hence:

$$t = Fo.a^2/D_{AB} = 0.25 \times (0.0015)^2/(1.0 \times 10^{-10}) = 5625\ \mathrm{s} = 94\ \mathrm{min}$$

Continuous Countercurrent Extraction

In continuous countercurrent equipment such as screw or belt conveyor extractors, the solid and liquid phases flow continuously past each other without reaching equilibrium at any point. Nevertheless, countercurrent flow will allow very high extraction efficiencies; in fact, if the entering solvent is free of solute and the solute-carrying

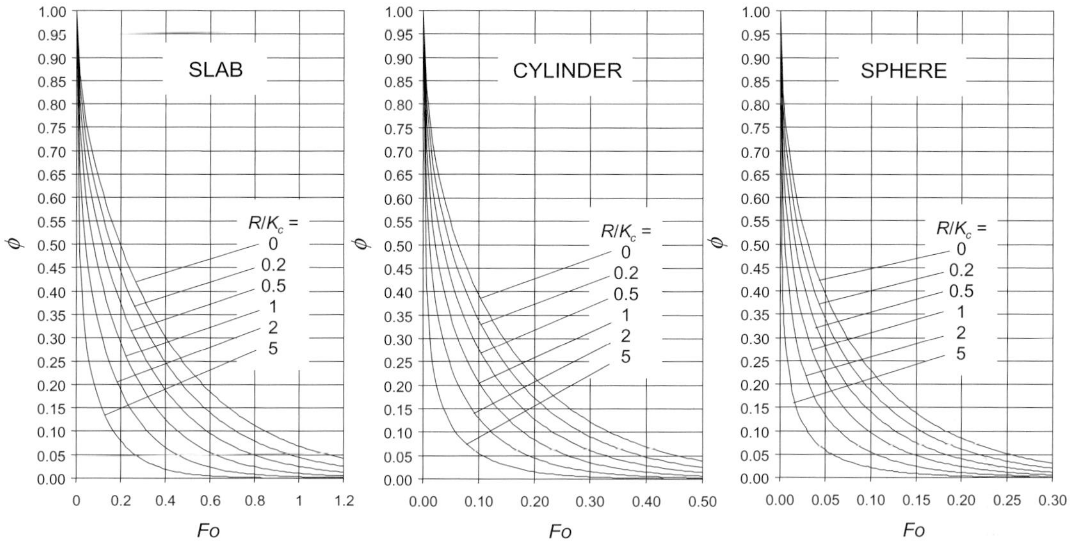

Figure 31.15 Fractional distance from equilibrium for batch leaching of simple shaped particles.

capacity of the liquid is at least equal to that of the solid, the unleached fraction can approach zero as closely as one wishes given enough contact time. Continuous countercurrent extraction is described by the same equations as single-stage batch extraction, as long as the factor R/K_c is given a negative sign (since the solid stream encounters solvent with decreasing solute concentration, as more and more solute is leached out of the solid). The solution for $R/K_c \geq -1$ is plotted in Figure 31.16 in terms of ln φ, where the fractional distance from equilibrium φ is defined as:

$$\varphi \equiv \frac{c_m - c_0^L/K_c}{c_0 - c_0^L/K_c} \tag{31.33}$$

For $R/K_c = 0$ (infinite liquid capacity) $\varphi = \phi$ and both graphs can be used.

Example 4

Using the data in Example 3, calculate the leached fraction when continuous countercurrent leaching is used, other conditions being the same.

Solution

From Figure 31.16, for abs$(R/K_c) = 0.125$ and $Fo = 0.25$, ln $\varphi = -3.0$, hence $\varphi = 0.05$, $1 - \varphi = 0.95$. Thus 95% of solute is removed by countercurrent leaching, compared to 85% in the batch leaching for the same contact time.

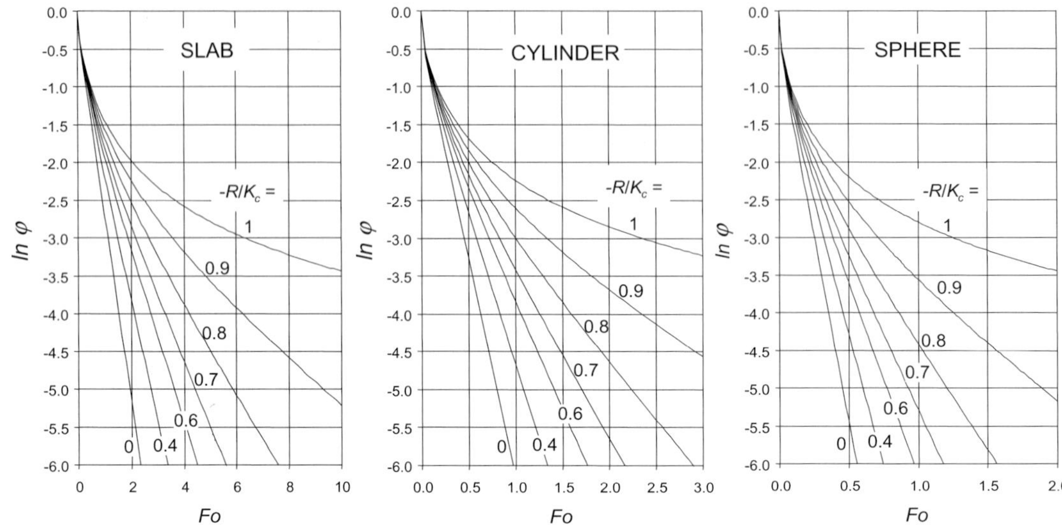

Figure 31.16 Fractional distance from equilibrium for counterflow leaching of simple shaped particles.

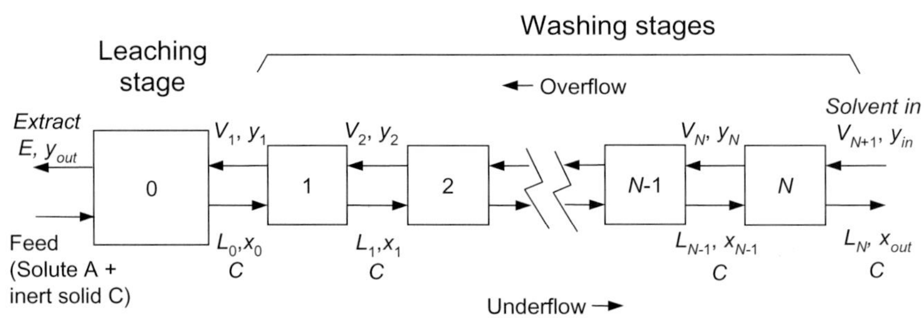

Figure 31.17 Countercurrent leaching and washing cascade. Note that L refers to the liquid part of the underflow only and does not include the carrier solid flow C.

Equilibrium Stage Countercurrent Leaching and Washing

In this section we consider a countercurrent cascade made up of equilibrium stages. In the stage receiving the fresh feed (the *leaching stage*, numbered 0 in Figure 31.17) all the solute is leached out of the particles before the mixture of depleted solids and extract is separated. Because mechanical separation is never complete, the solid leaves the stage as a sludge, which still carries a significant amount of solute in the liquid phase. Therefore, the sludge stream or *underflow* is washed in a countercurrent cascade of *washing stages* (numbered 1 to N) to remove the residual solute. The (solid-free) liquid solvent stream is termed the *overflow* and flows in the opposite direction.

The symbol V refers to the liquid overflow and L to the *liquid part* of the underflow, not including the solid.

We now make the following assumptions:

1. The carrier solid (solute-free solid), C, is completely inert, i.e. does not dissolves or disperse at all into the liquid.
2. No solid goes into the overflow (V) phase.
3. The liquid (solute + solvent)-to-solid ratio in the underflow is constant.

Assumptions 1 and 2 imply that the carrier solid flow is constant along the cascade. Assumption 3 then implies that the liquid flow in the underflow is also constant from the leaching stage onwards:

$$L_0 = L_1 = \ldots = L_N = L(\text{constant}) \quad (31.34)$$

From now on the carrier solid will be ignored, all flows and compositions being based on the liquid phase, i.e. solvent and solute, only. Material balance over each washing stage n gives:

$$L_{n-1} + V_{n+1} = L_n + V_n \quad (31.35)$$

and since $L_{n-1} = L_n$:

$$V_{n+1} = V_n = V(\text{constant}) \quad (31.36)$$

i.e. the overflow rate is also constant, except for the extract flow from the leaching stage. This flow, E, is given by material balance over the leaching stage:

$$E = V + \text{Entering solute} - L \quad (31.37)$$

Solute balance over stage 1 gives:

$$Lx_0 + Vy_2 = Lx_1 + Vy_1 \quad (31.38)$$

$$(y_1 - y_2) = \frac{L}{V}(x_0 - x_1) \quad (31.39)$$

where x and y denote the mass fraction of solute in the L and V streams respectively. Note that although the underflow also contains solid, the latter is ignored in calculating mass fractions, i.e. x, like y, is measured in kg solute/(kg solvent + kg solute).

The x and y values of the two streams leaving a given stage n will be equal, since they are essentially the same liquid solution:

$$y_n = x_n \quad (31.40)$$

and hence $x_0 = y_{out}$ and $x_1 = y_1$. Substituting this into the previous equation gives:

$$(y_1 - y_2) = \frac{L}{V}(y_{out} - y_1) \tag{31.41}$$

Similarly:

$$(y_2 - y_3) = \frac{L}{V}(y_1 - y_2) \tag{31.42}$$

$$\ldots$$

$$(y_{N-1} - y_N) = \frac{L}{V}(y_{N-2} - y_{N-1}) \tag{31.43}$$

$$(y_N - y_{in}) = \frac{L}{V}(y_{N-1} - y_N) \tag{31.44}$$

Multiplying together Equations 31.41–31.44 gives:

$$(y_N - y_{in}) = \left(\frac{L}{V}\right)^N (y_{out} - y_1) \tag{31.45}$$

Putting $y_N = x_{out}$ and rearranging gives the McCabe–Smith (1956) equation:

$$N = \frac{\ln\left(\frac{x_{out} - y_{in}}{y_{out} - y_1}\right)}{\ln(L/V)} \tag{31.46}$$

Example 5

Oil is leached from soybean in a countercurrent flow system consisting of a leaching stage and several washing stages. Inlet flow rates are 20000 kg·h^{-1} of soybeans and 30000 kg·h^{-1} of n-hexane. Raw soybean contains 20% oil by mass. 95% of the oil is recovered in the extract. The underflow from each stage contains 1.2 kg liquid for each kg of oil-free solid. Determine the number of equilibrium stages.

Solution

1. We apply material balances to calculate the solvent flows in each stream:
   Total oil entering system = 20000 × 0.20 = 4000 kg·h^{-1};
   Oil recovered in extract = 4000 × 0.95 = 800 kg·h^{-1};
   From material balance, the oil in the leaving underflow is 4000 – 3800 = 200 kg·h^{-1};

Soybean solids flow = 20000 – 4000 = 16000 kg·h^{-1}. This will remain constant throughout the cascade;

From the given liquid-to-solid ratio in the underflow, $L = 16000 \times 1.2 = 19200$ kg·h^{-1}. For the washing cascade (from solvent entry to just before the leaching stage) V is constant and equal to the entering solvent flow: $V = 30000$ kg·h^{-1}

Material balance over the leaching stage gives the flow rate of extract leaving the system, $E = 30000 + 4000 - 19200 = 14800$ kg·h^{-1}.

2. Next, we work out the compositions required for the McCabe–Smith method:
Oil mass fraction in extract:

$$y_{out} = 3800/14\,800 = 0.257$$

From equilibrium, $x_0 = y_{out} = 0.257$

Oil leaving in underflow from leaching stage = $Lx_0 = 19200 \times 0.257 = 4930$ kg·h^{-1}
From the oil balance over the leaching stage, mass of oil in V_1 = 4930 (in leaving underflow) + 3800 (in leaving extract) – 4000 (in entering soybean) = 4730 kg·h^{-1}

Mass fraction of oil in overflow V_1 entering leaching stage:

$$y_1 = 4730/V = 4730/30\,000 = 0.158$$

5% of the oil leaves with the exiting underflow, hence:

$$x_{out} = 4000 \times 0.05/19\,200 = 0.0104$$

The entering solvent is pure hexane, hence:

$$y_{in} = 0$$

3. Use the McCabe–Smith equation to calculate the number of washing stages, N:

$$N = \frac{\ln\left(\dfrac{0.0104 - 0}{0.257 - 0.158}\right)}{\ln(19\,200/30\,000)} = 5.0$$

Supercritical Fluid Extraction

What is Supercritical Fluid Extraction?

Fluids are classified as supercritical when they are maintained at conditions that exceed the critical temperature and pressure. Extraction with supercritical fluids (SCFs) is based on the experimental observation that many gases become good solvents for solids and liquids when compressed to conditions above the critical point.

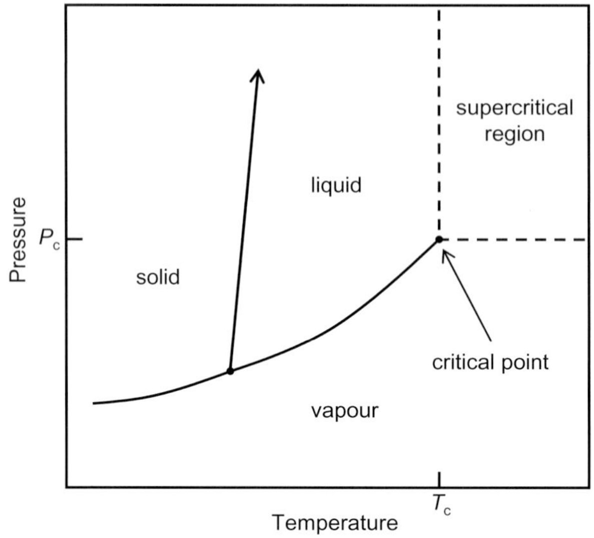

Figure 31.18 Phase diagram for a pure substance.

Table 31.3 Comparison of the physical properties of gases, liquids and supercritical fluids (SCFs) (Johnston, 1984).

Property	Gas	SCF	Liquid
Density ($g \cdot cm^{-3}$)	10^{-3}	0.3	1
Viscosity (cP)	10^{-2}	0.1	1
Diffusivity ($cm^2 \cdot s^{-1}$)	0.1	10^{-3}	5×10^{-6}

A supercritical region originating from the critical point (T_c, P_c) may be identified in the phase diagram for a pure substance, as shown in Figure 31.18.

The properties of SCFs vary over a wide range depending on the temperature and pressure, but are generally intermediate between those of gases and liquids. Selected physical properties of SCFs are shown in Table 31.3 along with typical values for gases and liquids. The density of a SCF lies closer to that for a liquid and it is this liquid-like density that accounts for most of the enhanced solvating power of SCFs. The diffusivity or diffusion coefficient and viscosity represent transport properties that affect rates of mass transfer. The transport properties of SCFs may be described as gas-like in that they are more favourable than those for liquids. SCFs therefore enable faster penetration of a solid matrix and a more efficient extraction of embedded solutes than would be obtained with liquid solvents.

The properties of SCFs are very sensitive to small changes in temperature and pressure in the vicinity of the critical point. This applies especially to density, as shown

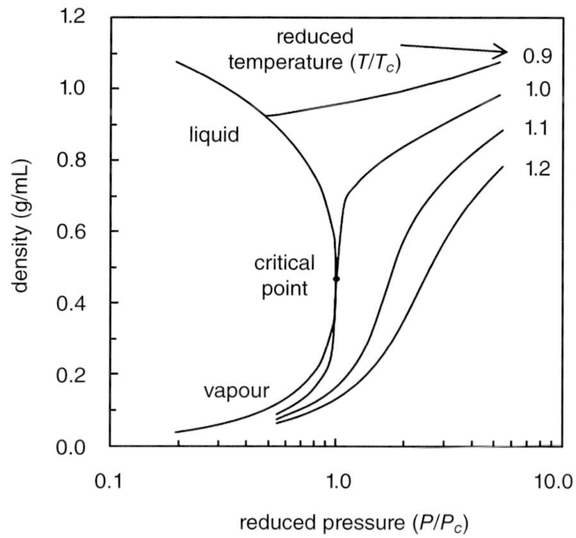

Figure 31.19 Variation of the density of CO_2 in the vicinity of the critical point.

Table 31.4 Critical properties of commonly used supercritical fluids (McHugh and Krukonis, 1994).

Solvent	Critical temperature (°C)	Critical pressure (bar)
CO_2	31.1	73.8
Ethane	32.2	48.8
Ethylene	9.3	50.4
Propane	96.7	42.5
Water	374.2	220.5

in Figure 31.19. Since density is a measure of the solvating power of a SCF, temperature and pressure can be used as variables to control the extraction and separation of a solute. The conditions of temperature and pressure of most interest in supercritical fluid extraction (SFE) are usually bounded by reduced temperatures between 1.0 and 1.2 and reduced pressures greater than 1.0.

SCFs have a number of distinct advantages over conventional liquid solvents. The adjustable solvent strength and favourable transport properties have already been mentioned and it is these features that really differentiate SCFs from liquid solvents. The more desirable SCFs in food applications are low molecular weight gases that have relatively low critical temperatures (Table 31.4). This permits extraction at moderate temperatures, which is desirable in the recovery of heat-sensitive food materials. Another important advantage is that after the release of pressure, the extract is left free of the residual supercritical solvent. Carbon dioxide is by far the most widely used SCF because it has the desirable properties of being non-toxic, non-flammable, readily available in high purity and inexpensive.

SFE has been investigated for the recovery of a wide range of food materials (Reverchon and De Marco, 2006). There are well-established industrial SFE processes for decaffeination of coffee and tea and for the production of hop extracts. Increasing government regulation on the use of organic solvents and growing consumer demand for natural products have also contributed to more widespread use of supercritical CO_2 in the processing of fats, oils, and other specialty lipids (Sahena *et al.*, 2009; Temelli, 2009; Catchpole *et al.*, 2009).

Equipment

SFE most often involves the removal of one or more solid/liquid components from a solid matrix. This type of process is preferably carried out in batch mode, in view of the difficulty of continuously feeding solid material into a pressurised vessel. Fractionation of liquid mixtures is more effectively carried out in a continuous countercurrent operation.

Batch Extraction

The basic equipment required for batch extraction of a solid with supercritical CO_2 is shown in Figure 31.20. The solid material is usually dried and ground into small particles to improve extraction kinetics. The solid is placed in the extraction vessel in the form of a fixed bed and in continuous contact with the supercritical solvent. Removable baskets can be employed for more rapid charging/discharging of the solid from the extraction vessel. Large-scale batch extractors are typically in the order of several cubic metres in volume. More recently, it has become common to employ smaller pressure vessels under 1 m^3 in volume (Brunner, 2005). Liquid CO_2 is fed to the pump and heated to the extraction temperature prior to entering the extraction

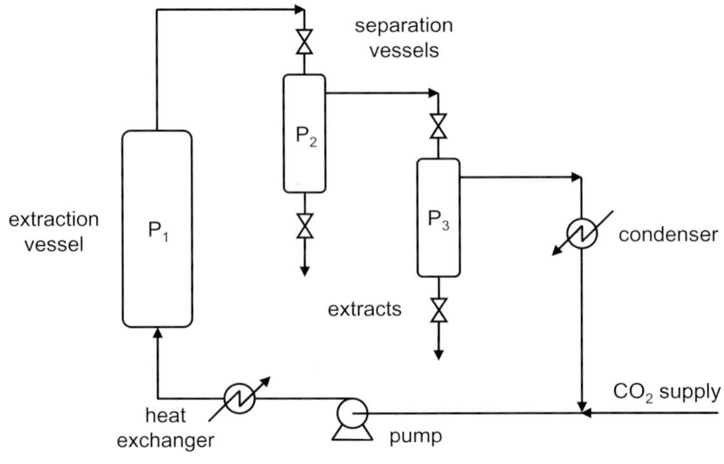

Figure 31.20 Schematic diagram of a batch extraction process with supercritical CO_2.

vessel. The loaded solvent then passes through a depressurisation valve into a separation vessel. Due to the lower pressure, the extract is separated from the CO_2 and recovered at the bottom of the vessel. Fractionation of the extract can be achieved by employing a cascade of separation vessels as shown in Figure 31.20. The second and subsequent separators are maintained at progressively decreasing pressures ($P_1 > P_2 > P_3$) so that the least soluble components are recovered in the first separator. Depending on the operating conditions of the last separator, a condenser may be employed for recycling CO_2 in the liquid state.

In a standard extraction at constant temperature and pressure, the more soluble components are preferentially extracted in the early stages of operation. However, the extraction may become inefficient and time consuming in the later stages because of the low solubilities of the components. An alternative mode of operation is to carry out extraction in several stages. Each stage is completed under a different set of operating conditions to enhance the loading of particular components. The composition of the supercritical solvent can also be adjusted through addition of small quantities of co-solvent (<10%). The co-solvent is typically a volatile liquid solvent such as ethanol and is beneficial for the extraction of polar or heavier components. Addition of co-solvent requires a separate dosing pump, usually located after the main CO_2 pump. Other standard designs for SFE and associated economics are given by Bertucco and Vetter (2001). The work also provides detailed information on additional equipment such as high pressure pumps, valves and fittings.

Countercurrent Extraction

Countercurrent extraction is more effective for reducing the consumption of the supercritical solvent and for increasing production rates. This mode of operation is performed with solids by employing several extraction vessels, each containing a fixed bed of material. The vessels are connected in series and operated on a semi-continuous basis that is similar to countercurrent extraction in leaching. The scheme shown in Figure 31.20 can also be adapted for continuous fractionation of liquid mixtures. The main difference is that the extraction vessel is replaced by a packed column. The packing consists of an inert material with high surface area to promote contact between the liquid and supercritical CO_2. Countercurrent operation is usually performed by feeding the supercritical CO_2 from the bottom to the top of the column while the liquid enters the column at the top. Other variations include: feeding the liquid at an intermediate location along the column; and refluxing part of the loaded solvent back into the column. Countercurrent extraction systems involving supercritical CO_2 and other pressurised fluids are described in more detail by Pronyk and Mazza (2009).

Process Design

Due to the high diffusivity of SCFs and resulting high mass transfer rates, SFE equipment usually operates at or close to equilibrium and the equilibrium stage methods described under LLE and solid leaching can be applied. This section will therefore concentrate on the estimation of solubilities in a supercritical solvent. To illustrate the principles involved in this type of phase equilibrium calculation we consider here the case of the solubility of a pure solid in supercritical CO_2. The defining equation for the mole fraction of solute i in the SCF phase is:

$$y_i = \frac{P_i^{sat}}{P} \cdot \frac{1}{\phi_i} \cdot \exp\left[\frac{v_i^s}{RT}(P - P_i^{sat})\right] \tag{31.47}$$

where P_i^{sat}, ϕ_i and v_i^s are the vapour pressure, fluid phase fugacity coefficient and molar volume of the solute respectively. Equation 31.47 is based on the equality of fugacities of component i in each phase as a condition of equilibrium. The main assumption involved is that CO_2 does not dissolve into the solid. The reader is referred elsewhere for details on the derivation of Equation 31.47 (Prausnitz *et al.*, 1986; Reid *et al.*, 1987).

Cubic equations of state are the most widely used method for estimating the fugacity coefficient of the solute in the SCF phase. The key advantages of this approach are: applicability over a wide range of temperature and pressure; easy adaptability to multicomponent systems; short computation time. Cubic equations of state generally conform to the following four-parameter expression (Schmidt and Wenzel, 1980):

$$P = \frac{RT}{v - b} - \frac{a}{(v^2 + ubv + wb^2)} \tag{31.48}$$

The most successful equations of this form have a temperature-dependent attractive parameter a, while the repulsive parameter b is usually made independent of temperature. These two parameters are substance dependent and calculated from the critical constants of the substance. Small integer values are assigned to u and w as shown in Table 31.5.

Cubic equations of state are not completely predictive for mixtures and require adjustable parameters that are fitted from experimental data. The adjustable parameters are normally incorporated into mixing rules for the attractive and repulsive parameters. The most commonly employed mixing rules for a and b are the van der Waals one-fluid mixing rules:

$$a_{mix} = \sum_i^N \sum_j^N y_i y_j \sqrt{a_i a_j}\,(1 - k_{ij}) \tag{31.49}$$

Table 31.5 Examples of cubic equations of state.

Equation of state	*u*	*w*
van der Waals	0	0
Soave Redlich–Kwong	1	0
Peng–Robinson	2	–1

$$b_{mix} = \sum_{i}^{N} y_i b_i \tag{31.50}$$

The binary interaction parameter k_{ij} is included to characterise the interactions between unlike molecular species. By convention, $k_{ij} = 0$ for identical molecules. The geometric mean term in Equation 31.49 usually causes an over-prediction of the mixture attractive parameter, in which case, k_{ij} assumes a small positive value between 0 and 1. This value is adjusted to make the equation of state fit the experimental data. Negative values for the interaction parameter are less frequently reported and indicate stronger than normal interactions between the molecular species (Saquing *et al.*, 1998).

Using the four-parameter cubic equation of state and conventional mixing rules, the general expression for the fugacity coefficient is:

$$\ln \phi_i = \frac{b_i}{b_{mix}}(Z-1) - \ln(Z - B^*) + \frac{A^*}{B^*\sqrt{u^2-4w}}\left(\frac{b_i}{b_{mix}} - \delta_i\right)\ln\frac{2Z + B^*\left(u+\sqrt{u^2-4w}\right)}{2Z + B^*\left(u-\sqrt{u^2-4w}\right)} \tag{31.51}$$

where:

$$A^* = \frac{a_{mix}P}{R^2T^2} \tag{31.52}$$

$$B^* = \frac{b_{mix}P}{RT} \tag{31.53}$$

$$\delta_i = \frac{2\sqrt{a_i}}{a_{mix}}\sum_j y_j\sqrt{a_j}\,(1-k_{ij}) \tag{31.54}$$

To facilitate the evaluation of the fugacity coefficient, Equation 31.48 can be rewritten as a cubic polynomial in terms of the compressibility factor Z. The mixture compressibility factor required in Equation 31.51 corresponds to the largest positive root of the following equation:

$$Z^3 - (1 + B^* - uB^*)Z^2 + (A^* + (w-u)B^{*2} - uB^*)Z - (A^*B^* + wB^{*2} + wB^{*3}) = 0 \tag{31.55}$$

In the case of the Peng–Robinson equation of state, u and w are assigned the respective values of 2 and –1:

$$P = \frac{RT}{v-b} - \frac{a}{v^2 + 2bv - b^2} \tag{31.56}$$

where:

$$a = \frac{0.45724R^2T_c^2}{P_c}\left[1+(0.37464+1.54226\omega-0.26992\omega^2)(1-\sqrt{T_r})\right]^2 \quad (31.57)$$

$$b = \frac{0.07780RT_c}{P_c} \quad (31.58)$$

and T_c, P_c, ω and T_r (= T/T_c) are the critical temperature, critical pressure, acentric factor and reduced temperature respectively.

In summary, the calculation of the solubility of a solid in supercritical CO_2 requires the following physical properties: P_c, T_c, and ω for CO_2; and P_c, T_c, ω, molar volume and vapour pressure for the solid. It should be noted that the calculation procedure for ϕ_i is an iterative process. The following procedure enables the solubility of a solid (2) in supercritical CO_2 (1) to be determined at a particular temperature and pressure:

1. Calculate the pure component parameters a_1, a_2, b_1 and b_2 using Equations 31.57 and 31.58.
2. Set an initial value for k_{12} (0 – 0.1) and the solubility y_2 (10^{-4} – 10^{-1}).
3. Calculate the mixture parameters a_{mix} and b_{mix} using Equations 31.49 and 31.50.
4. Solve Equation 31.55 for the fluid mixture compressibility factor Z.
5. Calculate ϕ_2 using Equation 31.51.
6. Substitute ϕ_2 into Equation 31.47 to obtain a new value for the solubility y_2.
7. If the new value of solubility is significantly different from the initial estimate, go to step 3 using the new value of solubility until the value for y_2 converges to a single value.

The above procedure also enables optimised values of k_{12} to be fitted from a set of experimental solubility data. For a given isotherm of data, the same value of k_{12} is used for data measured at different pressures. The optimised value of k_{12} is obtained by minimising an objective function that represents the difference between calculated and experimental solubilities, e.g. the sum of squared deviations.

The calculation procedure for the solubility of a liquid in a SCF is somewhat more complex and involves consideration of the vapour–liquid equilibria of the system. At elevated pressure, there is substantial dissolution of the SCF in the liquid phase. Fugacity coefficients therefore need to be calculated for both phases using iterative bubble point/dew point calculation procedures (McHugh and Krukonis, 1994; Smith *et al.*, 2001).

One limitation of the equation of state approach in food-related applications is the lack of physical property data for the pure components. This is particularly the case for high molecular weight materials. Many flavour and fragrance components, however, are well characterised and easily modelled with cubic equations of state. Critical constants and other physical properties for a wide range of pure components are available in the data compilation by Daubert and Danner (1989). An extensive compilation of high pressure phase equilibrium data is also given by Dohrn *et al.* (2010).

Hygienic Design Aspects

Food safety hazards can arise from microbiological growth, chemicals such as cleaning agents or lubricants, contamination by inappropriate materials of construction, ingress of foreign materials, and pests.

An assessment of food safety risks must be undertaken prior to the design or selection of food processing equipment (ISO, 2002). The main factors to be considered in this case are:

- the type of product to be processed;
- the degree of further processing;
- the specific application of the product;
- the mode of use of the machine – continuous or intermittent;
- the cleaning and inspection frequency.

Extraction usually does not involve products in the highest risk category, such as meat, dairy products and ready-to-eat food. Many extraction processes use non-aqueous solvents, concentrated salt or sugar solutions, acid solutions or caustic solutions, and the chances of microbial growth are low in these media. In many cases the products are subjected to further processing involving heating, such as distillation or drying, which reduces microbial risks still further.

SCFs may have beneficial hygienic effects. High pressure CO_2 has been investigated for some years as an alternative cold pasteurisation technique for foods (Spilimbergo and Bertucco, 2003; Garcia-Gonzalez *et al.*, 2007). This method permits processing at much lower temperature than thermal pasteurisation, although it has yet to be implemented on a large scale in the food industry. Typical operating conditions vary in the range of temperature from 25 to 45°C and pressures of 5–20 MPa. Under these conditions significant inactivation of bacteria, moulds and yeasts can be achieved in less than 1 hour of exposure. The technique has mainly been considered for liquid foods (e.g. fruit and vegetable juices). Processing of solids requires longer treatment times and introduces a greater likelihood of extraction of volatile food components from the solid matrix.

Microbial risks are greatest when extracting from solids with high nutritive values, when using dilute aqueous solvent, when the product does not undergo further thermal processing after extraction, and when the equipment is used intermittently. Extraction equipment comprises components such as mixing and settling vessels, piping and rotating parts (stirrers, pump shafts, screw conveyors, etc.) that may cause accumulation of material, contamination and/or microbial growth. Solid extraction involves particles, pastes or pulps that may be caught in piping and machinery, which must be designed to be easily accessed or dismantled to ensure adequate cleaning.

When food safety risks are present, the usual guidelines for hygienic design must be followed (ISO, 2002; Lelieveld *et al.*, 2000):

- *Materials*:
  - use suitable construction materials, such as 316 stainless steel, that are corrosion and abrasion resistant, non-porous and non-absorbent;
  - all materials potentially in contact with food must be non-toxic.
- *Surfaces*:
  - surfaces must be smooth and free of cracks;
  - joints such as welds must be smooth and continuous;
  - avoid sharp corners, projections, edges and recesses;
  - internal corners must be rounded. A minimum radius of about 6 mm or ¼ inch is generally recommended;
  - do not use screws and rivets in contact with food.
- *Doors and covers*:
  - all openings must be protected by doors and covers from unintended ingress of pests and contaminants;
  - doors and covers should be sloped to an outside edge to prevent accumulation of water.
- *Accessibility*:
  - equipment must be accessible for inspection and cleaning, with or without disassembly;
  - where there is a space between items of equipment, or between equipment and the enclosure (walls, ceiling and floor), there must be sufficient clearance to allow easy inspection and cleaning.
- *Drainage*:
  - ensure that fluids can be easily drained by providing sloping bottoms and ports at the lowest point of the equipment;
  - avoid dead spaces where material may accumulate;
  - piping not designed for routine disassembly must be sloped.
- *Moving parts:*
  - shaft seals must use only food-grade lubricants or be lubricated by the food itself;
  - bearings should be mounted outside the product area.
- *Piping, pumps and valves:*
  - pumps and valves must be of a sanitary design;
  - for valves, contact of moving mechanisms with food should be minimised or eliminated, since materials may be caught between moving and stationary parts;

- piping joints must be smooth and continuous. Protrusions and recesses will hinder the free flow of materials;
- dismountable joints must be easily disassembled and have true and hygienic fit.

Economics

When designing extraction equipment, contradictory requirements must be balanced in order to optimise performance and cost. For example, smaller droplet size (in LLE) or particle size (in leaching) leads to more efficient mixing and faster diffusion, but longer phase separation time. A smaller solvent-to-feed ratio reduces solvent or solvent recovery cost, but increases the necessary number of stages in countercurrent extraction. The use of mechanical agitation and centrifugation reduces residence time and equipment volume but increases capital cost and power consumption. The use of a SCF allows efficient extraction and easy separation of solvent and product, but capital costs are high due to the requirement for high pressure equipment and the cost of CO_2, or of re-compression if the CO_2 is recycled.

The cost of extraction equipment will depend on many factors:

- product throughput;
- process parameters: number of stages and/or extraction time, as determined by the design methods presented in this chapter;
- fluid mechanic requirements: mechanical components and power requirements for mixing, vessel volume for phase separation, conveying and pumping costs;
- solvent cost, or cost of solvent recovery;
- pre-processing (size reduction) and post-processing (solvent recovery, further purification of product);
- hygiene requirements. More stringent standards of hygiene will result in higher cost;
- high pressure equipment cost, if SFE is used.

These requirements must be carefully determined before equipment can be selected and sized, and costs can be determined according to the usual methods (e.g. Peters *et al.*, 2003).

Despite the relatively high capital investment required for SFE, the cost of production is competitive with traditional methods in many cases. Preparation of extracts via SFE may also be the only way to meet product specifications. Brunner (2005) cites the cost of production for extraction of solids in the range of US$3/kg feed for a capacity of around 1000 ton/year. Much higher production rates are expected to reduce the cost of production to below US$1/kg feed. Addition of co-solvent can also reduce production costs significantly. Pereira and Meireles (2010) have presented a simple method for estimating the cost of SFE of bioactive compounds from plant materials. Their analysis considers direct costs, fixed costs and general expenses. Estimated costs of manufacturing of SFE extracts vary in the range of US$1–10/kg dry feed and in the majority of cases these costs are lower than those obtained from steam distillation. Furthermore, the cost of raw material represents the largest component of the cost of manufacturing, rather than the capital cost as is often assumed. CO_2 recovery also impacts significantly on production costs. Weidner (2009) has discussed CO_2 recovery in the case of high pressure micronisation for food applications. Complete recovery of CO_2 leads to a reduction in the production cost by a factor of 2 (€0.15/kg product with 100% CO_2 recovery). Similar cost savings would be expected in large-scale SFE plants.

Summary and Future Needs

Extraction is one of the most important separation processes in the food industry, as it does not usually involve high temperature and therefore causes minimal degradation to the product. It is also a complex operation, possibly requiring significant effort in raw material preparation and product recovery. A wide variety of equipment is available, from simple mixer-settlers with their large volumes to more compact and expensive equipment involving centrifugal separation. Apart from the choice of equipment and solvent, the user must also optimise the balance between mass transfer and separation: better mass transfer requires more intimate mixing and smaller droplets or particles, which leads to more difficulty in separating out the two phases, requiring larger equipment or centrifugal acceleration.

A variety of design calculation methods is available for different situations and some have been presented in this paper. These mathematical methods often make implicit assumptions of ideal behaviour, such as attainment of equilibrium, constant physical properties, uniform particle size, etc., and the designer must keep these limitations in mind. Furthermore, data on distribution ratios, diffusivities and other physical properties are often unavailable as food material comes in many different varieties, and usually laboratory work is necessary to determine the required contact time and to optimise conditions. Nevertheless, the equations given when applied properly will enable equipment scaling up and prediction of the effects of varying operating conditions. A thorough understanding of the physical processes will point the way towards better design and operational practice.

Although extraction is an old process, recent innovations or potential commercial applications of laboratory techniques such as pressurised liquid extraction, microwave-assisted extraction (Kaufmann and Christen, 2002), ultrasonic-assisted extraction (Vilkhu *et al.*, 2008) and pulsed electric field-assisted extraction (Jaeger *et al.*, 2008) promise continuing improvements in efficiency and versatility.

While water and CO_2 are benign solvents, many organic solvents are potentially hazardous and subject to strict regulations (Wakelyn and Wan, 2003). With consumer health and environmental issues becoming increasingly important concerns, industry must ensure that safety regulations are strictly adhered to and that residue concentrations and emissions are minimised. At the same time new and safer solvents must be developed for both LLE and leaching.

Nomenclature

a	Half-thickness or radius of particle, m. Also, parameter in equation of state
A	Solute being extracted
A_i	Absorption factor for component i, $A_i \equiv L/(K_i V)$
b	Parameter in equation of state
B	Solvent
c	Solute concentration in solid phase, kg·m^{-3}
c^L	Solute concentration in liquid phase, kg·m^{-3}
C	Number of components in a mixture
C	Carrier fluid or solid
D_{AB}	Diffusivity, m^2·s^{-1}
e_i	Mass flow of component i in extract, kg·s^{-1}
E	Extract stream
E_i	Extraction factor for component i, $E_i \equiv K_i V/L$
f	Exponential decay factor
f_i	Mass flow of component i in feed, kg·s^{-1}
F	Feed stream
Fo	Fourier number $D_{AB}t/a^2$
H	Enthalpy, J·kg^{-1}
j	Pre-exponential factor
k_{ij}	Binary interaction parameter
K_c	Concentration equilibrium distribution ratio of solute, c^L/c
K_i	Mass fraction equilibrium distribution ratio of component i, y_i/x_i
K_i^{molar}	Mass fraction equilibrium distribution ratio of component i, ξ_i^V/ξ_i^L
l_i	Flow rate of component i in stream L, kg·s^{-1}
L	Flow rate of raffinate stream (LLE) or of underflow liquid (leaching), kg·s^{-1}
M	Molecular mass, kg·mol^{-1}
N	Number of stages in a countercurrent cascade
P_i^{sat}	Saturation vapour pressure, Pa
P	Total pressure, Pa
r_i	Mass flow of component i in raffinate, kg·s^{-1}
R	Raffinate stream
R	Ratio of solid volume to liquid volume; also, gas constant, J·K^{-1}·mol^{-1}
s_i	Mass flow of component i in solvent, kg·s^{-1}
S	Entering solvent stream.
T	Absolute temperature, K
u	Parameter in equation of state
v_i	Flow rate of component i in stream V, kg·s^{-1}

V	Flow rate of extract stream (LLE) or of overflow (leaching), $kg{\cdot}s^{-1}$
w	Parameter in equation of state
x	Mass fraction of solute in stream L
$\mathbf{x}$	Composition vector (x_A, x_B)
y	Mass fraction of solute in stream V (or mole fraction in SFE solubility calculations)
Z	Compressibility factor

Greek Symbols

ϕ	Fractional distance from equilibrium concentration (batch leaching)
ϕ_i	Fluid phase fugacity coefficient of component i
φ	Fractional distance from equilibrium concentration (countercurrent leaching)
ρ	Density, $kg{\cdot}m^{-3}$
ξ	Mole fraction
υ	Molar volume of vapour, $m^3{\cdot}mol^{-1}$
υ_i^s	Molar volume of the solid, $m^3{\cdot}mol^{-1}$
ω	Acentric factor

Subscripts

0	Initial value
1, 2 . . .	Stage number
A	Solute
B	Solvent
c	Critical
C	Carrier
eq	At equilibrium
i	i-th component
in	Stream entering system
L	First liquid phase
mix	Mixture
N	Stage N
m	Average value
n	Refers to a stage, or to the streams leaving that stage
out	Stream leaving system
r	Reduced (value of P or T divided by critical value)
V	Second liquid phase

Superscripts

L	In the (first) liquid phase
V	In the second liquid phase

References

Abrams, D.S. and Prausnitz, J.M. (1975) Statistical thermodynamics of liquid mixtures: a new expression for the excess Gibbs energy of partly or completely miscible systems. *AIChE Journal* 21: 116–128.

Arlt, W., Macedo, M.E.A., Rasmussen, P. and Sorensen, J.M. (1979–1987) *Chemical Data Series Volume V, Part 1: Binary Systems, Part 2+3: Ternary and Quaternary Systems, Part 4: Supplement 1*. DECHEMA, Frankfurt.

Batista, E., Monnerat, S., Stragevitch, L., Pina, C.G., Gonçalves, C.B. and Meirelles, A.J.A. (1999) Prediction of liquid–liquid equilibrium for systems of vegetable oils, fatty acids, and ethanol. *Journal of Chemical Engineering Data* 44: 1365–1369.

Batista, E.A.C., Meirelles, A.J.A., Rodrigues, C.E.C. and Gonçalves, C.B. (2009) Liquid–liquid rxtraction applied to the processing of vegetable oil. In: *Bioactive Compounds for Food Products, Theory and Applications* (ed M.A.A. Meireles). CRC Press, Boca Raton, Chapter 5, pp. 220–267.

Bertucco, A. and Vetter, G. (2001) *High Pressure Process Technology: Fundamentals and Applications*. Elsevier Science, Amsterdam.

Boston, J.F. and Sullivan, S.L. (1974) A new class of solution methods for multicomponent, multistage separation processes. *Canadian Journal of Chemical Engineering* 52: 52–63.

Brunner, G. (2005) Supercritical fluids: technology and application to food processing. *Journal of Food Engineering* 67: 21–33.

Catchpole, O.J., Tallon, S.J., Eltringham, W.E., Grey, J.B., Fenton, K.A., Vagi, E.M., Vyssotski, M.V., MacKenzie, A.N., Ryan, J. and Zhu, Y. (2009) The extraction and fractionation of specialty lipids using near critical fluids. *Journal of Supercritical Fluids* 47: 591–597.

Crank, J. (1975) *The Mathematics of Diffusion*. Oxford University Press, London.

Daubert, T.E. and Danner, R.P. (1989) *Physical and Thermodynamic Properties of Pure Chemicals: Data Compilation*. Hemisphere Publishing Corporation, New York.

DECHEMA (Society for Chemical Engineering and Biotechnology) (2009).

DETHERM, http://www.dechema.de/en/detherm.html (Accessed 17/12/2009).

Derr, E.L. and Deal, C.H. (1969) *Analytical Solution of Groups: Correlation of Activity Coefficients Through Structural Group Parameters. Institution of Chemical Engineers Symposium Series* 3(32): 40–51.

Dohrn, R., Peper, S. and Fonseca, J.M.S. (2010) High-pressure fluid-phase equilibria: Experimental methods and systems investigated (2000–2004). *Fluid Phase Equilibria* 288: 1–54.

Edmister, W.C. (1957) Absorption and stripping factor functions for distillation calculation by manual and digital computer methods. *AIChE Journal* 3: 165–171.

Fredenslund, A., Jones, R.L., Prausnitz, J.M. (1975) Group-contribution estimation of activity coefficients in nonideal liquid mixtures. *AIChE Journal* 21: 1086–1099.

Friday, J.R. and Smith, B.D. (1964) An analysis of the equilibrium stage separation problem formulation and convergence. *AIChE Journal* 10: 698–707.

Garcia-Gonzalez, L., Geeraerd, A.H., Spilimbergo, S., Elst, K., Van Ginneken, L., Debevere, J., Van Impe, J.F. and Devlieghere, F. (2007) High pressure carbon dioxide inactivation of microorganisms in foods: The past, the present and the future. *International Journal of Food Microbiology* 117: 1–28.

Gmehling, J., Lohmann, J., Jakob, A., Li, J. and Joh, R. (1998) A modified UNIFAC (Dortmund) model. 3. Revision and extension. *Industrial & Engineering Chemistry Research* 37: 4876–4882.

Hansen, H.K., Rasmussen, P., Fredenslund, A., Schiller, M. and Gmehling, J. (1991) Vapor–liquid equilibria by UNIFAC Group Contribution 5. Revision and extension. *Industrial & Engineering Chemistry Research* 30: 2352–2355.

ISO 14159 (2002) *Safety of Machinery – Hygiene Requirements for the Design of Machinery*. International Organization for Standardization.

Jaeger, H., Balasa, A. and Knorr, D. (2008) Food industry applications for pulsed electric fields. In: *Electrotechnologies for Extraction from Food Plants and Biomaterials* (eds E Vorobiev and N Lebovka). Springer, New York, pp. 181–216.

Johnston, K. (1984) Supercritical fluids. In: *Kirk-Othmer Encyclopedia of Chemical Technology*, 3rd ed. (eds M. Grayson and D. Eckroth), John Wiley & Sons, New York, supplement volume.

Kojima, K. and Tochigi, T. (1979) *Prediction of Vapor–Liquid Equilibrium by the ASOG Method*. Elsevier, Amsterdam.

Kaufmann, B. and Christen, P. (2002) Recent extraction techniques for natural products: microwave-assisted extraction and pressurised solvent extraction. *Phytochemical Analysis* 13: 105–113.

Kremser, A. (1930) Theoretical analysis of absorption process. *National Petroleum News* 22: 43–49.

Lelieveld, H.L.M., Mostert, M.A. and Curiel, G.J. (2000). Hygienic equipment design. In: *Hygiene in Food Processing* (eds H.L.M. Lelieveld, M.A. Mostert, J. Holah, B. White). Woodhead, Cambridge, pp. 122–166.

Lo, T.C., Baird, M.H.I., Hanson, C. (eds) (1983) *Handbook of Solvent Extraction*. Wiley Interscience, New York.

Magnussen, T., Rasmussen, P. and Fredenslund, A. (1981) UNIFAC parameter table for prediction of liquid-liquid equilibria. *Industrial and Engineering Chemistry Process Design and Development* 20: 331–339.

McCabe, W.L. and Smith, J.C. (1956) *Unit Operations of Chemical Engineering*. McGraw-Hill, New York.

McHugh, M.A. and Krukonis, V.J. (1994) *Supercritical Fluid Extraction*; 2nd ed. Butterworth-Heinemann, Boston.

Naphtali, L.M. and Sandholm, D.P. (1971) Multicomponent separation calculation by linearization. *AIChE Journal* 17: 148–153.

Pereira, C.G. and Meireles, M.A.A. (2010) Supercritical fluid extraction of bioactive compounds: Fundamentals, applications and economic perspectives. *Food and Bioprocess Technology* 3: 340–372.

Perry, R.H. and Green, D.W. (1997) *Perry's Chemical Engineers' Handbook*, 7th edn., McGraw-Hill, New York, pp. 15.23–15.47.

Peters, M.S., Timmerhaus, K. and West, R.E. (2003) *Plant Design and Economics for Chemical Engineers*, 5th ed. New York: McGraw-Hill.

Prausnitz, J.M., Lichtenthaler, R.N. and de Azevedo, E.G. (1986) *Molecular Thermodynamics of Fluid-Phase Equilibria*. Prentice-Hall Inc,: Englewood Cliffs, NJ.

Pronyk, C. and Mazza, G. (2009) Design and scale-up of pressurized fluid extractors for food and bioproducts. *Journal of Food Engineering* 95: 215–226.

Reid, R.C., Prausnitz, J.M. and Poling, B.E. (1987) *The Properties of Gases and Liquids*; McGraw-Hill Inc., New York, NY.

Reissinger, K.H. and Schroter, J. (1978) Selection criteria for liquid-liquid extraction. *Chemical Engineering* 85: 109–118.

Renon, H. and Prausnitz, J.M. (1968) Local composition in thermodynamic excess functions for liquid mixtures. *AIChE Journal* 14: 135–144.

Reverchon, E. and De Marco, I. (2006) Supercritical fluid extraction and fractionation of natural matter. *Journal of Supercritical Fluids* 38: 146–166.

Sahena, F., Zaidul, I.S.M., Jinap, S., Karim, A.A., Abbas, K.A., Norulaini, N.A.N. and Omar, A.K.M. (2009) Application of supercritical CO_2 in lipid extraction – a review. *Journal of Food Engineering* 95: 240–253.

Saquing, C.D., Lucien, F.P. and Foster, N.R. (1998) Steric effects and preferential interactions in supercritical carbon dioxide. *Industrial Engineering and Chemical Research* 37: 4190–4197.

Schmidt, G. and Wenzel, H. (1980) A modified van der Waals type equation of state. *Chemical Engineering Science* 35: 1503–1512.

Schwartzberg, H.G. (1980) Continuous counter-current extraction in the food industry. *Chemical Engineering Progress* 76: 67–85.

Schwartzberg, H.G. (1982) Solute diffusivities in leaching processes. *Food Technology* 36: 73–86.

Seader, J.D. and Henley, E.J. (2006) *Separation Processes Principles*. John Wiley, New York.

Smith, J.M., Van Ness, H.C. and Abbott, M.M. (2001) *Introduction to Chemical Engineering Thermodynamics*, 6th ed. McGraw-Hill, New York.

Spilimbergo, S. and Bertucco, A. (2003) Non-thermal bacteria inactivation with dense CO_2. *Biotechnology and Bioengineering* 84: 627–638.

Sujata, A.D. (1961) Absorber-stripper calculation made easier. Hydrocarbon *Processing* 40: 137–140.

Temelli, F. (2009) Perspectives on supercritical fluid processing of fats and oils, *Journal of Supercritical Fluids* 47: 583–590.

Tochigi, K., Tiegs, D., Gmehling, J. and Kojima, K. (1990) Determination of new ASOG parameters. *Journal of Chemical Engineering of Japan* 23: 453–463.

Treybal, R.E. (1963) *Liquid Extraction*, 2nd ed. McGraw-Hill, New York.

Vilkhu, K., Mawson, R., Simons, L. and Bates, D. (2008) Applications and opportunities for ultrasound assisted extraction in the food industry – A review. *Innovative Food Science and Emerging Technologies* 9: 161–169.

Wakelyn, P.J. and Wan, P.J. (2003) Solvent extraction: safety, health and environmental issues. In: *Extraction Optimization in Food Engineering* (C. Tzia and G. Liadakis, eds). Marcel Dekker, New York, pp. 391–427.

Weidlich, U. and Gmehling, J.A. (1987) Modified UNIFAC Model. 1. Prediction of VLE, h^E and γ^∞. *Industrial & Engineering Chemistry Research* 26: 1372–1381.

Weidner, E. (2009) High pressure micronization for food applications. *Journal of Supercritical Fluids* 47: 556–565.

32

Size Reduction Process Design

M. Reza Zareifard, Ali Esehaghbeygi and Amin Allah Masoumi

Introduction

Size reduction is an important unit operation in which the size of solid food materials is reduced by the application of grinding, compression, or impact forces. The production of powders and fine particles is known as comminuting. When applied to the reduction in size of globules of immiscible liquids (e.g. oil globules in water), size reduction is more frequently referred to as homogenization or emulsification. The purpose of size reduction as an auxiliary operation for other processes is crucial to the densification process. Particle size reduction increases the ratio of surface area to volume of the material. This is an important issue in different unit operations on solids, such as in drying, extracting, leaching, cooling, and heating, as well as for immiscible liquids and gases.

Many foods are marketed in cut form (including slices, cubes, strings, and bars): vegetables, fruit, meat, fish, bread, cheese, butter, animal feed, and tobacco. Other products are comminuted to puréed form, like mashed vegetables, tomato paste, apple sauce, fruit in jams and soft drinks, and meat in sausage and luncheon meat, and all kinds of emulsions. In these cases the objective is to obtain products that are homogeneous or quasi-homogeneous in appearance and/or taste. Size reduction may play a part in mechanical separations. A good example is the recovery of starch from

Handbook of Food Process Design, First Edition. Edited by Jasim Ahmed and Mohammad Shafiur Rahman.

potatoes. In the case of size reduction as an auxiliary action, the form in which the reduced product should be shaped is influenced by the principal operation concerned. Excellent examples are the cutting of sugar beets into strips and the crushing of oilseed to flakes; both shapes are very favorable for extraction of sugar and oil respectively. The finer the better rule does not always apply.

Size reduction has the following uses in food processing:

- There is an increase in the surface-to-volume ratio of the food, which increases the rate of drying, heating or cooling, as well as the extraction rate of the soluble component (e.g. juice extraction from cut fruit).
- When combined with screening, a predetermined range of particle sizes is produced, which is important for the correct functional properties of some products (e.g. icing sugar, spices, and cornstarch); a similar range of particle sizes also allows more complete mixing of ingredients (e.g. dried soups and cake mixes).
- Size reduction is frequently used in the food industry not only to obtain smaller particles but also to separate edible and inedible components of raw materials, e.g. hulling of grains and seeds, such as oats, barley, rice, and some oil-seeds; peeling of fruits, potatoes, and carrots; scraping of potatoes and carrots; stoning of cherries; tailing of berries and cherries; cutting of oats; stripping of tobacco; filleting of fish; and trimming of beans.

Size reduction and emulsification are sometimes used to improve the quality of food or to prepare it for further processing with little or no effect on the nutrition. In some foods, they may promote degradation by the release of naturally occurring enzymes from damaged tissues or by microbial activity treatments. Different methods of size reduction are classified according to the particle size range produced:

- Chopping, cutting, slicing, and dicing:
  - large to medium (stewing steak, cheese, and sliced fruit for canning);
  - medium to small (bacon, sliced beans, and diced carrot);
  - small to granular (minced or shredded meat, flaked fish or nuts, and shredded vegetables).
- Milling to powder or pastes of increasing fineness (spices, flours, fruit nectars, powdered sugar, starches, and smooth pastes).
- Emulsification and homogenization (mayonnaise, milk, essential oil, butter, ice cream, and margarine).

Edge mills, kneading machines, disintegrators, colloid mills, and homogenizers, which are used for this purpose, can be classified somewhere between mixers and mills. Materials with high liquid content might become a pulp or paste on extensive size reduction. In such a case, the term wet milling is used. Thus, ultra-fine milling is nearly always carried out in the presence of liquid, e.g. in disintegrators, colloid mills, roller mills, conches in the chocolate industry, etc. In dry milling, attention must be paid to the evaporation of water and other volatile components, and to thermal decom position and oxidation; these all result from the high temperatures that can occur. Milling is sometimes carried out at a low temperature or in an inert gas atmosphere, the latter particularly if there is the danger of dust explosions. The technical procedure for size reduction depends greatly on the properties of the material. If little water or oil is present, the material is relatively hard and brittle.

Texture of Materials

Cellular material of plant or animal origin is usually built from cells with highly specialized purposes and consequently very different mechanical properties. Knowledge of the plant structure is necessary to understand the plant material reaction to cutting forces and deformations. This makes it easier to find logical solutions to improved cutting device designs. Also, a thorough description of the mechanical properties of plant stems and cells, and the factors that influence these properties, are needed. Forage grasses, grain, and corn (maize) are the most common materials cut in agriculture, and knowledge of their strength properties is, consequently, of importance for the study of agricultural cutting processes. The study and characterization of plant material from a strength standpoint may be done at three different levels: entire stem section, and cell wall and individual layers in the cell wall magnification. A study of the entire stem section shows several areas with different characteristics when compared to each other, but each of which is fairly uniform within. Each area consists of a large number of similar cells. Dimensions may conveniently be expressed in millimeters (Persson, 1987).

Size Classifications

Size, shape, particle distribution, and uniformity of the material after size reduction depend upon the physical characteristics of the material, its previous history, and the method of reduction. Furthermore, it is extremely improbable that the shape of even a small percentage of, for example, grains would approximate any simple geometric figure. In theoretical studies, it is customary to represent an irregular particle by an equivalent sphere, cube or other geometric figure, and by the surface area or volume being used as the basis for comparison. Deviation from the performance of the idealized shape is recognized by the introduction of empirical factors. A factor is applied which indicates the degree of fit. Reduced materials may be placed in three groups or classes based upon size:

- *Dimension range*: particles or units that can be accurately measured and easily seen with minimum measurements of approximately 3.18 mm or more. Diced fruit and vegetables, and chopped forage are examples.

- *Sieve range*: particles with an approximate minimum dimension range of 3.18–0.074 mm. Granular materials such as ground feed and commercial fertilizers fall into this group.
- *Microscopic range*: particles with minimum dimensions of less than 0.074 mm. Materials such as chemical powders, dusts, and Portland cement are examples.

Grains are ground before they are fed to animals to improve their digestibility. The particle size distributions of feeds and forages influence their handling, storage, and utilization characteristics. Particle size distribution, f(x), can be described using the Weibull distribution, which is widely used in engineering practice for the description of fatigue and failure of components as follows:

$$f(x)=\left(\frac{n}{\bar{x}}\right)\left(\frac{x-x_0}{\bar{x}}\right)\exp\left(-\left(\frac{x-x_0}{\bar{x}}\right)^n\right) \quad x\geq 0 \qquad (32.1)$$

where $\bar{x}$ is the average value of x (a random variable), n is a constant shape factor characterizing the uniformity of x, and x_0 is the minimum value of x or minimum particle size.

When the log-normal and Weibull distributions are applied to particle size analysis, the number of particles of a given size must be determined. In the case of the log-normal distribution, the geometric mean diameter and geometric standard deviation may be estimated from the graph. Mean particle diameter is the diameter corresponding to a cumulative weight of 50% of the particles and the geometric mean is the ratio of the particle diameter corresponding to a cumulative weight of 84% to the mean diameter.

Commonly used techniques are microscopic analysis, electrolytic resistivity (the Coulter counter), air elutriation, centrifuging, sedimentation, and sieving. Only the microscopic analysis and electrolytic resistivity methods can determine discrete particle sizes and numbers of particles of each size. The microscopic analysis is slow and tedious, and involves examination and counting of particles under a microscope. In the Coulter counter, the particles are suspended in an electrically conductive liquid that is forced though a small aperture. The suspension is sufficiently dilute so that particles pass through the aperture one at a time and electrodes on opposite sides of the aperture measure the electrical resistance. The magnitude of the voltage pulse produced by the passage of a particle is assumed to be proportional to the volume of the particle. The remainder of the techniques separate the particles according to the number of particles in a given size range. In the air elutriation technique, the velocity of an airstream is adjusted so that particles less than a given diameter are suspended. After the particles within the size range are collected, the air velocity is increased. The centrifuging and sedimentation techniques are based on the fact that when particles of uniform density but differing sizes are suspended in a liquid, they will settle at different rates. Particles that settle in given time intervals are collected and weighed.

Sieve Analysis

The sieving technique is commonly used for larger particles as in feeds and powder or ground foods (ASABE Standards, 2006a, b; S319 and S424). The sieves are constructed of wires crossing each other in a regular grid and are numbered so that higher numbers correspond to smaller spacing between wires. Two standard sieve series are the United States Standard series and the Tyler series. For larger materials, the sieves can be stacked one upon the other and placed on an automatic shaker such as the Ro-Tap machine. This device shakes the sieves with a circular motion and has a metal arm that periodically strikes the top of the stack. When the sample contains very small particles, as is the case with a finely ground flour, the sieves are placed on the shaker one at a time and the sample is sieved using successively finer sieves. A dispersing agent may be added to improve sieving characteristics. A charge of 100 g should be used, although larger or smaller charges may be used if necessary. Extra care should be taken to recover all material from the sieves when small charges are used. The charge should be placed on one sieve or the top sieve of the nest of test sieves and shaken until the mass of material on that sieve reaches end-point. End-point is decided by determining the mass on each sieve at 1-min intervals after an initial sieving time of 10 min. If the mass on the smallest sieve containing a material changes by 0.1% or less of the charge mass during a 1-min period, the sieving is considered complete. For industrial applications, the end-point determination process can be omitted, and the end-point is set to a sieving time of 15 min.

A classification system fineness modulus or uniformity index is used to indicate the uniformity of grind or distribution of fines and coarses in the resultant product. The standard definition of fineness modulus is: "An empirical factor obtained by adding the total percentages of a sample of the aggregate retained on each of a specified series of sieves, and dividing the sum by 100" (CRD, 1980; CRD-C 104-80). Although the fineness modulus gives an average size, it does not indicate the distribution of the fines and coarses in any sample. These objections can be overcome by using the uniformity index, which is demonstrated on the basis of tabulated analysis. The general terms "fine, medium, and coarse" of uniformity index prevent meaningful comparisons of data. To overcome these limitations, the American Society of Agricultural Engineers developed methods to describe particle sizes more specifically. The average particle size is given as geometric mean diameter (GMD; Equation 32.2), expressed in millimeters or microns, and the range of variation is described by geometric standard deviation (GSD; Equation 32.3), with a larger GSD representing lower uniformity.

$$\mathrm{GMD} = (abc)^{1/3} \tag{32.2}$$

$$\mathrm{GSD} = \left[\frac{\sum_{i=1}^{n} \left(X - \bar{X}\right)^2}{n-1} \right]^{1/2} \tag{32.3}$$

where *a*, *b*, and *c* are the different character size, $\bar{X}$ is the sample geometric mean average, and *n* is the sample size.

GMD and GSD are accurate descriptors only when particle size distributions, expressed as log data, are distributed parametrically, i.e. log normally. Feed particle size is typically determined by dry sieving of a 100 g representative sample (Baker and Herrman, 2002). The feed sample is passed through a sieve stack on a shaker for 10 min. The amount of particles retained on each screen size is then determined, and the GMD and GSD of the sample calculated using standard formula or computer software. Particle size distribution may also be determined by wet sieving, but this method is used more often on digest and excreta samples. In the wet sieving method, feed samples are suspended in 50 mL of water and left to stand for 30 min prior to sieving to ensure adequate hydration (Lentle *et al.*, 2006). The sample is then washed through a set of sieves and the material retained on each sieve, plus a representative sample of elute, are subsequently filtered and dried for 24 h at 80 °C. The weight of the retained particles from each sieve is then expressed as percent of total dry matter recovered. Calculation of particle size, surface area, and number of particles by mass calculations is based on the assumption that particle sizes of all ground feeds and feed ingredients are log-normally distributed. The size of particles can be reported in terms of GMD (or median size) and GSD by mass. Calculation formulas, based on the derivations, are as follows:

$$d_{gw} = \log^{-1}\left[\frac{\sum_{i=1}^{n}\left(w_i \ \log \bar{d}_i\right)}{\sum_{i=1}^{n} w_i}\right] \tag{32.4}$$

$$S_{gw} \approx \frac{1}{2} d_{gw}\left[\log^{-1} S_{\log} - \left(\log^{-1} S_{\log}\right)^{-1}\right] \tag{32.5}$$

$$S_{\log} = \left[\frac{\sum_{i=1}^{n} w_i\left(\log \bar{d}_i - \log d_{gw}\right)^2}{\sum_{i=1}^{n} w_i}\right]^{1/2} = \frac{S_{\ln}}{2.3} \tag{32.6}$$

where d_i is the nominal sieve aperture size of the ith sieve (mm), d_{i+1} is the nominal sieve aperture size in the next larger than the ith sieve (just above in a set) (mm), d_{gw} is the GMD or median size of particles by mass (mm), S_{log} is the GSD of log-normal distribution by mass in 10-based logarithm (dimensionless), S_{ln} is the GSD of log-normal distribution by mass in natural logarithm (dimensionless), S_{gw} is the GSD of particle diameter by mass (mm), W_i is the mass on the $\underline{i}^{th}$ sieve (g), and *n* is the number of sieves +1 (pan).

The equation for estimating the total surface area of particles in a charge is:

$$A_{st} = \frac{\beta_s W_t}{\beta_v \rho} \exp\left(4.5\sigma_{\ln}^2 - \ln \mu_{gw}\right) \tag{32.7}$$

where A_{st} is the estimated total surface area of a charge (cm^2), β_s is the shape factor for calculating surface area of particles (for cubical, $\beta_s = 6$; for spherical, $\beta_s = \pi$), β_v is the shape factor for calculating volume of particles (for cubical, $\beta_v = 1$; for spherical, $\beta_v = \pi/6$), ρ is the particle density of the material (g·cm^{-3}), σ_{ln} is the log-normal GSD of parent population by mass in natural logarithm (use S_{ln} as an estimate), μ_{gw} is the geometric mean particle diameter of the parent population by mass (cm; use d_{gw} as an estimate) (note: μ_{gw} is expressed in cm and d_{gw} in mm), and W_t is the mass of a charge (g).

Similarly the number of particles in a charge is calculated as:

$$N_t = \frac{W_t}{\beta_v \rho} \exp\left(4.5\sigma_{\ln}^2 - 3\ln \mu_{gw}\right) \tag{32.8}$$

where N_t is the number of particles in a charge.

Size Reduction Procedures

The equipment used in size reduction is generally known as a crusher when it performs coarse reduction and as a mill when used for all other applications. According to Barbosa-Canovas *et al.* (2005), approximately 20 different designs exist in comminuting processes. The size of agricultural products is reduced by cutting, crushing, and shearing, either on their own or in combination.

Cutting

Cutting is separation or reduction, which is achieved by pushing or forcing a thin, sharp knife through the material to be reduced (Persson, 1987). This may result in minimum deformation and rupture of the reduced particles. The new surfaces that are produced by the sharp edge of the knife are relatively smooth for fruit and vegetables. Since the pores in the new surfaces are open because of minimum damage from the sharp edge, drying, leaching or any process requiring transfer of a liquid or vapor to or from the material proceeds at a maximum rate. The most satisfactory cutting device is an extremely sharp knife whose blade is as thin as structurally possible. The motion of the knife should be such that the edge has a sawing component in moving through the material. This provides a smoother cut and probably requires less energy to perform.

Investigations into parameters affecting cutting forces in foods were undertaken to identify basic trends, such as the relationship of cutting forces to cutting speeds and food temperatures. Literature searches reveal few published data on cutting forces, which could be used for design or optimization. As stated by McGorry *et al.* (2003), even for hand cutting operations, "The force exposure associated with meat cutting operations and the effect of knife sharpness on performance and productivity have not been well documented." To develop optimal cutting and slicing systems, data are required on cutting forces for different food types and how they vary with such factors as cutting temperature, speed, and type of cutting device, e.g. blade profile and edge angle. Brown *et al.* (2005) used a simple plain blade to cut three typical foodstuffs (cheese, bacon, and beef) at three feed speeds and three temperatures. Cutting forces for cheese decreased with increasing temperature and increased with cutting speed. The effect of cutting speed was not consistent for all forces, but higher speeds generally produced higher forces. For beef, there was a marked difference between frozen and unfrozen samples, but little difference between samples at different unfrozen temperatures. In unfrozen samples, cutting speed had little effect on forces, whereas faster cutting speeds produced higher forces in frozen samples. The proportion of total cutting forces made up by friction was found to be consistent over all temperatures and speeds for cheese and bacon, but markedly higher in the frozen beef samples compared to the unfrozen samples.

Crushing

Crushing is reduction by applying a force to the unit to be reduced in excess of its strength. The material ruptures in many directions and the resulting particles are irregular in shape and size. The characteristics of the new surfaces and particles are dependent upon the type of material and the method of force application. Limestone and other chemical fertilizers, ground feed for livestock, flour and meal, and fruit and vegetable purées are produced in part or whole by crushing. Crushing is used to extract juice from sugar and to break the structure of forage crops to speed drying. The force used in crushing can be applied statically, as is done when cracking a walnut in a vise, or dynamically as with a hammer. Crushing by means of a rigid roll or bed, such as the sorghum mill, is an example of static force application. The hammer mill exemplifies dynamic force application. Shearing is a combination of cutting and crushing. If the shearing edge is thin and sharp, performance approaches that of cutting. A thick, dull shearing edge performs more as a crusher.

Shearing

Shearing is usually used for reducing materials of a tough fibrous nature where some crushing may be advantageous and the resulting units are generally uniform in size. Cutting ensilage for livestock is an example. The shearing units consist of a sharp knife and a bar. The knife is usually thick to withstand the shock it receives when it hits the material. For best performance, the clearance between the bar and the knife should be as small as possible and the knife should be sharp. Esehaghbeygi *et al.* (2009b) showed that the shearing force of wheat stems decreased as the moisture content decreased and as the cutting height of the stalk increased, because of a reduction in stalk diameter. Shearing stress was reduced by using a smooth edge knife, because there is less friction than with a serrated one. A blade oblique angle of 30 degrees showed the least shearing stress in wheat.

Water-Jet Cutting Application

Traditional food slicing and cutting techniques that use blades, saws, knives, etc. suffer from drawbacks such as frequent cutting blades replacement, poor hygienic design, and difficult clean process; poor cutting properties can also result in excessive waste in the form of fine, tears, and miss-cuts. However, many relatively new slicing and cutting techniques that are used extensively in other nonfood industries or services may show benefits if applied to food products. There are several possibilities for slicing foods such as water-jet cutting; abrasive water-jet cutting; oil-jet cutting; cryogenic fluid spray cutting; supersonic air-jet cutting; laser cutting; maser cutting; ultrasonic cutting; electro-erosion cutting, and advanced mechanical and combination techniques.

Water-jet methods have great applicability in the food industry. A large volume of water is pumped through a small orifice in the nozzle (Figure 32.1). The constant volume of water traveling through a reduced cross-sectional area causes the particles to rapidly accelerate. This accelerated stream leaving the nozzle impacts on the material to be cut. The extreme pressure of the accelerated water particles contacts a small area of the work piece and small cracks develop in this small area. This type of cutting is limited to softer material or material with naturally occurring small cracks. An abrasive water-jet slurry system mixes abrasive with the water. Slurry systems accelerate the abrasive particles with the water throughout the system (Hashish, 1989). Tool steels and woodcutting are applications for abrasive water-jet cutting. For circuit boards, water-jet cutting is mostly used to cut out smaller boards from a large piece of stock. Wire stripping is another effective application of water-jet cutting. Certain foods such as bread can also be easily cut with water-jet cutting. Since the water jet exerts such a small force on the food, it does not crush it, and with a small "knife" width, very little is wasted. A wide range of water-jet settings can be used to cut potato slices. French fries cut with the water jet set to cause intermediate subsurface damage had increased color irregularities but no extreme taste or textural differences compared to conventionally cut controls (Becker and Gray, 2006). A successful approach to the problem of lamb chop fat trimming to a constant depth has been developed and a demonstrator experimental system for this has been constructed by Purnell and Brown (2004).

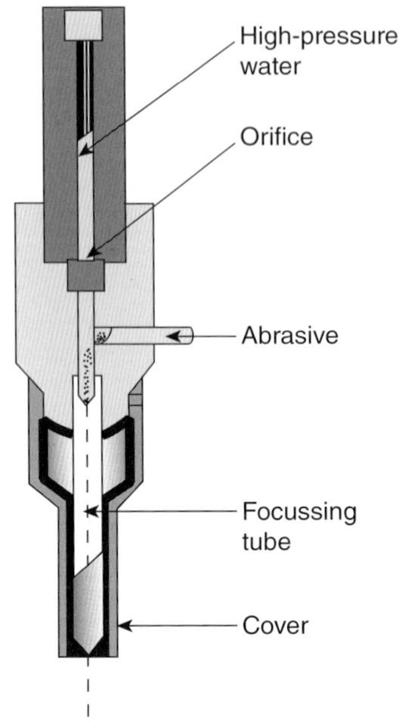

Figure 32.1 Schematic view of the water jet cutting head. (Courtesy of Omax Corporation, USA.)

Types of Stresses and Energy Requirements

There are three types of force used to reduce the size of foods: (i) compression forces, (ii) impact forces, and (iii) shearing (or attrition) forces. In most size reduction equipment, all three forces are present, but often one is more important than the others. When stress (force) is applied to a food, the resulting internal strain will exceed the elastic stress limit (E); when the stress is removed, the tissues will return to their original shape and release the stored energy as heat. As little as 1% of applied energy may be used for size reduction. However, when the strain within a localized area exceeds the elastic stress limit, the food is permanently deformed (Figure 32.2). If the stress is continued, the strain reaches a yield point (Y), above which the food begins to flow (known as the region ductility (Y–B in Figure 32.2)). Finally, the breaking stress is exceeded and the food fractures along a line of weakness. Part of the stored energy is then released as heat. As the size of the piece of food is reduced, there are fewer lines of weakness available, and the breaking stress that must be exceeded increases. When no lines of weakness remain, new fissures must be created to reduce the particle size further, and this requires an additional input of energy. There is therefore a substantial increase in energy requirement as the size of the particles is reduced. It is important to specify the required size distribution in the product to avoid the expenditure of time and energy in creating smaller particles than necessary.

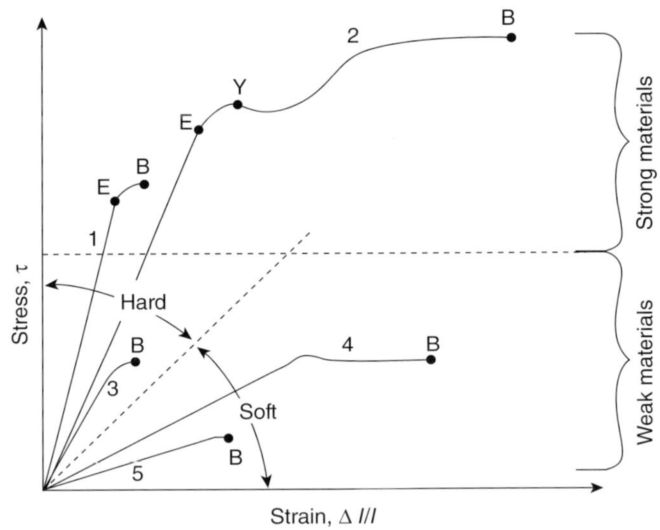

Figure 32.2 Stress–strain diagram for various foods. E, elastic limit; Y, yield point; B, breaking point; OE, elastic region; EY, inelastic deformation; YB, region of ductility. Material 1 (curve 1) is hard, strong, and brittle; material 2 (curve 2) is hard, strong, and ductile; material 3 (curve 3) is hard, weak, and brittle; material 4 (curve 4) is soft, weak, and ductile; and material 5 (curve 5) is soft, weak, and brittle. (From Loncin and Merson (1979), courtesy of Elsevier.)

The amount of energy that is absorbed by a food before it fractures is determined by its hardness and tendency to crack (friability), which in turn depend on the structure of the food. Harder foods absorb more energy and consequently require a greater energy input to create fractures. With increasing kinetic energy from an impact pendulum, the average breakage susceptibility index increased for rice grains (Esehaghbeygi *et al.*, 2009a). The more lines of weakness in a food, the lower is the energy input needed to cause fracturing. Compression forces are used to fracture friable or crystalline foods, combined impact and shearing forces are necessary for fibrous foods, and shearing forces are used for fine grinding of softer foods. It is thought that foods fracture at a lower stress level if force is applied for longer. The extent of size reduction, energy expended, and amount of heat generated in the food therefore depend on both the size of the forces and the time that the food is subjected to the force.

Other factors that influence the energy input are the moisture content and heat sensitivity of foods. The moisture content significantly affects both the degree of size reduction and mechanism of breakdown in some foods. Wheat, for example, is conditioned to optimum moisture content before milling. Maize is thoroughly soaked and wet milled in order to obtain complete disintegration of the starchy material. Further details are given be Kent (1983). However, excess moisture in a dry food can lead to agglomeration of particles, which then block the mill. Very dry foods create excessive dust, which is a health hazard, and is extremely inflammable and potentially explosive. Substantial amounts of heat are generated in high-speed mills. The heat sensitivity of the food determines the permissible temperature rise and the necessity to cool the mill. In cryogenic grinding, liquid nitrogen or solid carbon dioxide is mixed with foods (e.g. spices) before milling to cool the product and to retain volatiles or other heat-sensitive components. Solid carbon dioxide is used to cool meat during size reduction in the manufacture of sausage meat. A number of studies have been carried out to decrease stem shearing strength in different crops (Ince *et al.*, 2005; Skubisz *et al.*, 2007; Hoseinzadeh *et al.*, 2010). For canola stems, Esehaghbeygi *et al.* (2009c) showed that the specific shearing energy increased, and bending stress and Young's modulus decreased as the moisture content increased. Naimi *et al.* (2007) studied the modeling of biomass size reduction.

As far as the theory of size reduction is concerned, only a few comments will be made here. During size reduction a product is deformed and this results in strain. On further deformation the strain increases until the breaking strength is locally exceeded; the cohesion is then broken. High strain is concentrated occur at the edges of a small area where the cohesion is broken, and the cracks formed expand quickly. Work must be done to deform and break the material. This work is larger for higher breaking strengths and for greater deformation. For a purely elastic material (according to Hooke's law) the internal strain is proportional to the elastic modulus E and relative deformation Δ. Whether the breaking strength σ_{max} will be exceeded somewhere therefore depends on the ratio $E\Delta/\sigma_{max}$. The amount of work per unit volume required for the deformation is proportional to the integral:

$$\int_{\sigma}^{\Delta} E\Delta d\Delta = \frac{1}{2} E\Delta^2 \tag{32.9}$$

When deformation continues until breakage occurs, i.e. until $E\Delta/\sigma_{max}$ reaches a critical value, then $\frac{1}{2}E\Delta^2$ is proportional to σ_{max}^2/E. In this simple case, the work done to cause breaking is proportional to the square of the breaking strength and inversely proportional to the elastic modulus. Assuming that the above may be generalized, it may be concluded that a material is more difficult to reduce in size the greater its resistance to breakage and the lower its resistance to deformation. In simpler terms, this means that a slight deformation is sufficient to exceed the breaking strength of a brittle material and this is therefore easy to reduce in size; a large deformation is required to exceed the breaking strength of a tough material and such a material is difficult to reduce in size; a soft material has a high breaking strength and is easily reduced in size; and a hard material has a high breaking strength and is difficult to reduce in size. Deformation, which leads to breakage, may then be obtained by subjecting the material to a blow or to shear forces (friction). The energy required to reduce the size of solid foods is calculated using one of the following three equations.

Kick's Law

Kick's law states that the energy required to reduce the size of particles is proportional to the ratio of the initial size of a typical dimension (e.g. the diameter of the pieces) to the final size of that dimension:

$$E = K_K \ln\left(\frac{d_1}{d_2}\right) \tag{32.10}$$

where E (J) is the energy required per unit mass of feed, K_K is Kick's constant, d_1 (m) is the average initial size of pieces, and d_2 (m) is the average size of ground particles; d_1/d_2 is known as the size reduction ratio and is used to evaluate relative performance of different types of equipment.

Rittinger's Law

Rittinger's law states that the energy required for size reduction is proportional to change in surface area of the pieces of food (instead of a change in dimensions as described in Kick's law):

$$E = K_R\left(\frac{1}{d_2} - \frac{1}{d_1}\right) \tag{32.11}$$

where K_R is Rittinger's constant.

Bond's Law

Bond's law (Bond, 1952) is used to calculate the energy required for size reduction from:

$$E = K_B\left(\sqrt{1/d_{(80)2}} - \sqrt{1/d_{(80)1}}\right) \tag{32.12}$$

Note that the definition of particle size in Bond's law is different: d_{80} = particle size such that 80% by weight of the sample is smaller than it. Bond's law is often written in terms of the work index (W_i) as:

$$\frac{E}{W_i} = \sqrt{\left(\frac{100}{d_2}\right)} - \sqrt{\left(\frac{100}{d_1}\right)} \tag{32.13}$$

where W_i (40 000–80 000 $J{\cdot}kg^{-1}$ for hard foods) is the Bond work index, d_1 the diameter of the sieve aperture that allows 80% of the mass of the feed to pass, and d_2 the diameter of the sieve aperture that allows 80% of the mass of the ground material to pass (Loncin and Merson, 1979).

In practice it has been found that Kick's law gives reasonably good results with coarse grinding in which there is a relatively small increase in surface area per mass. Rittinger's law gives better results with fine grinding where there is a larger increase in surface area and Bond's law is intermediate between these. However, the Rittinger and Bond equations were developed from studies of hard materials (coal and limestone) and deviation from predicted results is likely with many foods.

Problem 1

Food is milled from 6 to 0.0012 mm using a 10-hp motor. Is this motor adequate to reduce the size of the particles to 0.0008 mm? Assume Rittinger's equation and that 1 hp = 745.7 W.

Solution

Using Rittinger's equation, first the constant value (K_R) can be calculated as follow:

$$7457 = K_R\left(\frac{1}{0.0012\times10^{-3}}\right)-\left(\frac{1}{6\times10^{-3}}\right)$$

Therefore:

$$K_R = \frac{7457}{1/1.2\times10^{-6} - 1/6\times10^{-3}} = 0.0089$$

To produce particles of 0.0008 mm:

$$E = 0.0089\frac{1}{0.0008\times10^{-3}} - \frac{1}{6\times10^{-3}} = 11123 \text{ kw} = 15 \text{ hp}$$

Therefore the motor is unsuitable and an increase in power of 50% is required.

Problem 2

If 5 hp-h are required to reduce a material from ¼ in size to 10 mesh, how much power would be required if the reduction were to 20 mesh?

Solution

According to Kick's law:

$$E = C\ln(L_1/L_2)$$

$$C = \frac{E}{\ln(L_1/L_2)} = \frac{5}{\ln(0.25/0.065)} = 8.56$$

$$E = 8.56 \ln(0.25/0.0328) = 7.55 \text{ hp-h}$$

According to Rittinger's law:

$$E = C\left(\frac{1}{L_2} - \frac{1}{L_1}\right)$$

$$C = \frac{E}{(1/L_2) - (1/L_1)} = \frac{5}{(1/0.065) - (1/0.25)} = 0.438$$

$$E = 0.438(1/0.0328) - (1/0.25) = 11.6 \text{ hp-h}$$

Kick's and Rittinger's laws were developed from studies of materials common to the chemical and mechanical engineering areas – talc, coal, limestone, etc., being examples. Such materials are different from agricultural materials such as forage, small grains, and fertilizer components. The particles resulting from the reduction could be any shape or a great number of shapes. The required energy must be related to the initial and reduced particle.

Mohsenin (1986) concluded that almost all of the energy in the grinding process is wasted as heat, and only about 0.06–1% of the input energy is consumed for disintegration of the material. Measuring the energy requirement for alfalfa size reduction could be very useful in developing strategies to reduce input energy in the process of converting biomass to bioenergy. Lopo (2002) reported that the energy consumption of grinding biomass depends on the ratio of particle size distribution of materials before and after milling, moisture content, bulk and particle densities, feed rate of the material, and machine variables. The amount of energy used for milling to obtain small particles is relatively high. Fang *et al.* (1997) reported that screen opening size was the most significant factor affecting mill performance. Samson *et al.* (2000) reported a specific energy consumption of 161.64 $kJ \cdot kg^{-1}$, when a hammer mill with a screen size of 5.6 mm was used to grind switch grass. The results of the specific energy used in particle size reduction can be applied to the selection of hammer mill operating factors to produce a particular size of switch grass, wheat straw, and corn stove grind, and will serve as a guide for relations between the energy and various analytic descriptors of biomass particle distributions (Bitra *et al.*, 2009a).

Different methods have been used to measure energy consumption of milling operation. Manlu *et al.* (2006) measured the specific energy requirement for grinding oak wood chips using torque and speed sensors connected to the engine shaft to measure the torque and engine rotational speed respectively. Tabil (1996) measured the specific energy consumption of the pellet mill for alfalfa at two hammer mill screen sizes of 2.4 and 3.2 mm with a W-h meter attached to a data logger with sampling time of 15 s. Ghorbani *et al.* (2010) measured the specific energy requirement for grinding alfalfa chops. They reported that, as size of screen on the hammer mill increased from 1.68 to 4.76 mm, the specific energy consumption decreased from 25.89 to 6.67 $kJ \cdot kg^{-1}$. In addition, they reported a simple linear model between the specific energy consumption and the ratio of initial to final size of screens used in size reduction. With an

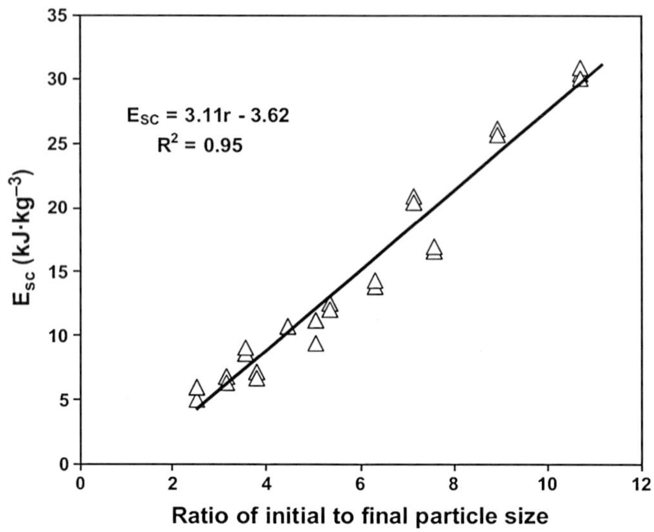

Figure 32.3 Relationship between specific energy consumption and ratio of initial to final particle size for grinding alfalfa. (From Ghorbani *et al.* (2010), courtesy of Elsevier.)

increasing ratio of initial to final size of screens, the specific energy consumption also increases (Figure 32.3). Mani *et al.* (2004) reported a simple linear relationship between the specific energy consumption and hammer mill screen size (from 3.175 to 0.794 mm) for four types of biomasses with a moisture content of 8% (w.b.).

Performance Characteristics

The performance of a machine for reducing the size of material is characterized by the capacity, power required per unit of material reduced, size and shape of the product before and after reduction, and range in size and shape of the resultant product. A size reducer operating ideally would have the following characteristics:

- product of uniform size;
- minimum temperature rise during reduction;
- minimum power requirement;
- trouble-free operation.

Considerable investigational work has been done regarding the performance of the various grinders used in agriculture, especially burr and hammer mills. The performance of these devices will be discussed in view of known applicable theory and test results.

Uniformity of Product

The burr mill is believed to produce a more uniform product than the hammer mill. If the hammer mill reduces by impact, two factors could contribute to this lower uniformity of product:

1. Each of the individual grains may be hit a number of times before it has an opportunity to pass through the screen. Since the path of travel of a grain through the grinder is random, the number of times a grain is hit varies and as a result the size of the product varies.
2. The energy dissipated upon contact between a grain and a hammer varies with the square of the velocity. Since the velocity or peripheral speed varies with the rotor radius, then the energy of impact varies according to the square of the radius, or the energy of impact at the end of a hammer is four times as great as at a point half-way between the end of the hammer and the center of the shaft. Consequently a grain that is hit near the end of a hammer is more finely divided than one that is hit closer to the shaft.

Power Requirements

The exact power required for a specific job is difficult to determine. The type of material, moisture content, fineness of grinding, rate of feed, type and condition of mill, etc., affect the power requirements. Moist grains are more difficult to grind than dry grain. The power required to operate the empty hammer mill increases quickly as the speed increases. The horsepower requirement for an empty mill without a fan is a straight-line function. Note that the useful power, the difference between horsepower consumed when running the mill empty and when full, is a small percentage of the input. From the standpoint of power consumption, operation below the rated speed is more advisable than operation above it. In order to maintain speeds above 3600 rpm, the feeding rate has to be reduced so that a larger portion of the available power can be used to maintain mill speed. Although it is generally believed that roller mills utilize energy more efficiently than hammer mills, Fang *et al.* (1997) compared the roller mill and hammer mill for energy consumption and efficiency.

Temperature Rise

The energy for grinding feed is dissipated as heat energy raises the temperature of the ground product, the mill, and the ambient air. Some heat energy is lost in moisture vaporization. The temperature may rise by 10 °C or more when grinding fibrous materials such as oats or ear corn in a burr mill, particularly if a relatively fine grind is being produced. The hammer mill produces a cooler product because of the large amount of air circulated with the ground grain. High temperatures contribute to decomposition of the ground material, especially if the moisture content is high.

Although grinding to fine particle size is thought to improve pellet quality, it will markedly increase energy consumption during milling. Systematic investigations on the relationships of feed particle size, temperature rise during the process, and diet uniformity with bird performance, gut health, and pellet quality is warranted if efficiency is to be optimized in respect of the energy expenditure of grinding (Amerah *et al.*, 2007).

Devices

Most meats, fruits, and vegetables fall into the general category of fibrous foods. Meats are usually frozen and tempered to just below their freezing point to improve the efficiency of cutting. Fruits and vegetables naturally have a specific fiber texture that is suitable for cutting at ambient or chill temperatures. There are at least eight main types of size reduction equipment, classified according to the size to which they decrease food materials.

Slicing Equipment

Slicing equipment consists of rotating or reciprocating blades which cut the food as it passes through. In some designs food is held against the blades by centrifugal force (Figure 32.4). For slicing meat products the food is held on a carriage as it travels across the blade. Harder fruits such as apples are simultaneously sliced and de-cored as they are forced over stationary knives fitted inside a tube. In a similar design (the hydrocutter), foods are conveyed by water at high speed over fixed blades. A comprehensive evaluation of the effects of moisture content, size, and surface area on the efficiency of mechanical slicing of Nigerian ginger was conducted by Onu and Okafor (2002). Size, surface area, and moisture content of the ginger rhizome affect the quality and efficiency of its slicing according to their study.

Dicing or Cutting Equipment

Dicing or cutting equipment is for vegetables, fruits, and meats. The food is first sliced and then cut into strips by rotating blades. The strips are fed to a second set of rotating knives which operate at right angles to the first set and cut the strips into cubes (Figure 32.4). For many other products such as meat, bread, butter, and cheese, special cutting machines are used. Bakeries also make frequent use of cutting machines; biscuits are cut from a slab of dough for instance. The Translicer, designed for food industry demands, has wide a variety of interchangeable cutting wheels available to produce a full range of slices, shreds, and julienne cuts (Figure 32.5).

A universal piece of equipment is the meat grinder; this is a lightly constructed screw press or extruder, with a cutting plate or rotating knives at its outlet (Figure 32.6). It is suitable for all kinds of sift, liquid-rich materials: vegetables, meat, peanuts,

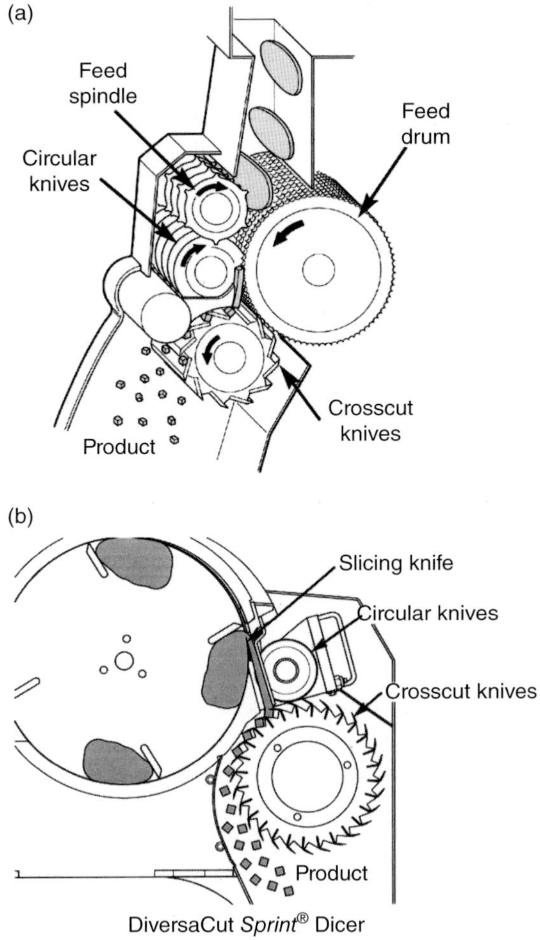

Figure 32.4 (a) Slicing equipment and (b) dicing equipment. (Courtesy of Urschel Ltd.)

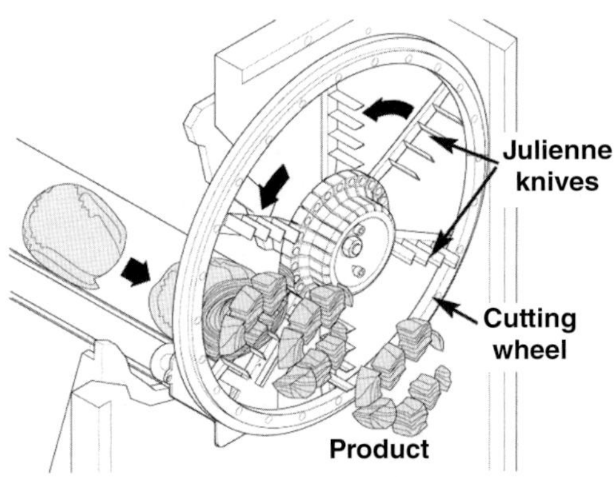

Figure 32.5 The translicer cutter. (Courtesy of Urschel Ltd.)

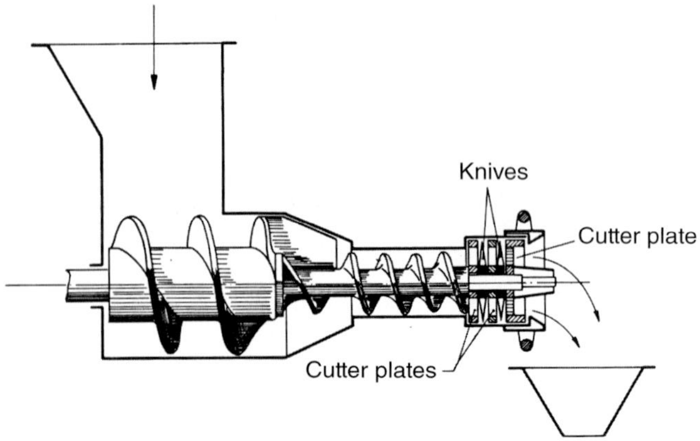

Figure 32.6 Meat grinder. (From Leniger and Beverloo (1975), courtesy of Springer.)

etc. A normal meat grinder has a closed housing and thus the product passes quantitatively through the cutting plate. The housing, however, may also be perforated; the soft part of the product then flows through the perforations, while larger and harder parts are removed axially.

Another group of cutting machines makes products of a particular shape. Some examples of the many possible variations are:

- Grass, straw, and other green feeds are chopped; an apparatus for this operation has thick rotating knives mounted on a spindle or drum.
- Tobacco is cut with a notching machine in which a compressed layer of tobacco is slowly moved forwards while a fast reciprocating knife cuts along it (guillotine system); the same mechanism is used for many other materials.
- Another frequently used apparatus consists of rotating rolls equipped with knives and which pass along a stationary knife; the knives are arranged in such a way that a scissor-like action results.

Vegetables and fruits are cut into many different shapes. Machines must not only produce the required shape of the cut product, but also accommodate the shape of the raw material. For this reason a very large number of models exist.

Flaking Equipment

Flaking equipment for fish, nuts or meat is similar to slicing equipment. Adjustment of the blade type and spacing is used to produce the flakes. Sowbhagya *et al.* (2007) observed that flaking of celery helped in overcoming the problem of clogging and choking, which is associated with conventional grinding. In addition, flaking had the advantage of higher yields of the volatile oil with better flavor quality.

Shredding Equipment

Shredding equipment typically is a modified hammer mill in which knives are used instead of hammers to produce a flailing or cutting action. A second type of shredder is known as the squirrel cage disintegrator. Here two concentric cylindrical cages inside a casing are fitted with knife blades along their length. The two cages rotate in opposite directions and food is subjected to powerful shearing and cutting forces as it passes between them.

Anonymous (2007) compared four different commercially available technologies for their ability to reduce the size of commonly processed feedstock mixtures: a rotary shear shredder, chipper shredder, macerator, and large bucket and spade. These feedstock mixtures included bread, meat, fruit, and vegetables. Results showed that for the size reduction of a large quantity of bread as a dry product, the rotary shear is a suitable technology, and has similar processing efficiency to the chipper shredder. It is important to note, however, that the bucket and spade technology is only suitable for processing small quantities (up to 50 kg at a time) of fruit and vegetables and that the structure of the end-product is poor. Larger quantities are more satisfactorily treated with the rotary shear, which also produces a better structured end-product with lower amounts of free liquid, which assists in handling.

Pulping Equipment

Pulping equipment is used for juice extraction from fruits or vegetables, and for seeds and meats. A combination of compression and shearing forces is used with each type of equipment. A rotary grape crusher consists of a metal cylinder fitted internally with high-speed rotating brushes or paddles (Nelson and Tressler, 1980). Grapes are heated if necessary to soften the tissues, and pulp is fed through the perforations of the screen by the brushes. The size of the perforations determines the fineness of the pulp. Skins, stalks, and pips are carded from the end of the screen. Other types of pulper, including roller presses and screw presses, are used for juice extraction.

Bowl Chopper

A bowl chopper is used to chop meat and harder fruits and vegetables with a coarse pulp (e.g. for sausage meat or mincemeat preserve). A horizontal, slowly rotating bowl moves the ingredients beneath a set of high-speed cutting blades. Food may be passed several times beneath the knives until the desired degree of size reduction and mixing has been achieved.

Crushers

Crushers reduce the material with a pressing or squeezing operation. A variety of machines is in use. Agricultural application of crushers is important but not extensive.

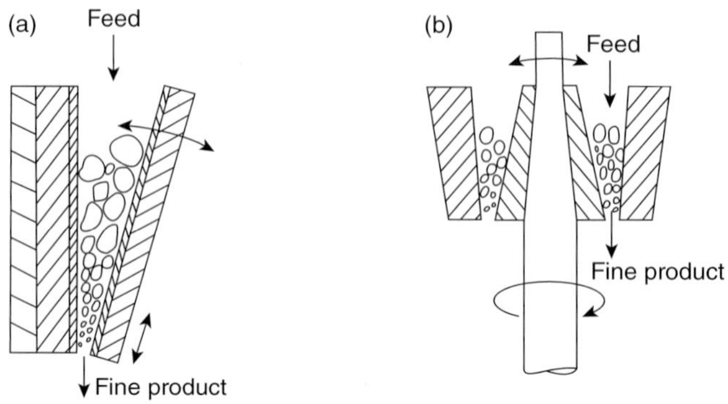

Figure 32.7 Crushers: (a) jaw and (b) gyratory. (From Earle (1983), courtesy of Pearson Education.)

Lime and other stones are given an initial reduction with a jaw or gyratory crusher (Figure 32.7). The jaw crusher is the cheaper and the slower of the two and is used for smaller operations. The resultant motion of the cone crushes the material in a manner similar to the jaw crusher, but operation is smoother and the relative capacity is higher. For primary reduction, roll crushers in various forms are used by themselves or, more frequently, in combination with burr or hammer mills. Jaw and gyratory crushers are heavy equipment and are not used extensively in the food industry. In a jaw crusher, the material is fed between two heavy jaws, one fixed and the other reciprocating, so as to work the material down into a narrower and narrower space, crushing it as it goes. The gyrator crusher consists of a truncated conical casing, inside which a crushing head rotates eccentrically. The crushing head is shaped into an inverted cone and the material being crushed is trapped between the outer fixed and the inner gyrating cones, and it is again forced into a narrower and narrower space.

Mills

The most common material characteristics that present challenges to milling are as follows:

- *Fibrous materials:* The major problem with reducing fibrous materials is that they are typically nonfriable, which means they require more energy for milling. This often creates excessive heat that can damage the particles and cause greater wear to the mill components. Examples of fibrous materials include corn, oat hulls, celery, carrots, various spices, and cellulose-based materials. Solutions for reducing fibrous materials can be as simple as installing the appropriate blade or classifying screen, or using an appropriate method, such as vacuum conveying, for assisting material flow through the mill and its screen to eliminate material build-up inside the grinding chamber.
- *Nonfriable materials:* Nonfriable materials such as polymers, resins, waxes, and rubber cannot be shattered or fractured using regular impact or direct-pressure

milling. Knife milling often cannot reduce a nonfriable material to a very fine particle size range. Solutions for reducing nonfriable materials typically require turning the nonfriable material into a friable material by freezing it. This includes injecting a cryogen such as liquid carbon dioxide or liquid nitrogen into the grinding chamber. In certain cases, preconditioning or exposing the particles to a cryogen may be necessary. For low-temperature milling with cryogens, the mill must be constructed of materials suitable for low-temperature operation. Over time, cryogenic processing at low temperatures can cause components to become brittle and certain lubricating greases to lose their viscosity and freeze.

- *Heat-sensitive materials:* If heat is created during milling, heat-sensitive materials can melt or soften in the grinding chamber or on the classifying screen, reducing the mill's overall efficiency. Examples of heat-sensitive materials include thermal plastics, polyester and epoxy resins, nutraceuticals, waxes, gelatin, spices, and sugar-based materials. Solutions to this melting or softening problem typically involve cooling the mill, cooling the surfaces inside the grinding chamber, or cooling the material prior to milling. Examples include water-jacketing the mill's critical components, directly injecting a cryogen into the grinding chamber, or exposing the material to a cryogen prior to milling.
- *Wet, fatty, or sticky materials:* Generally, low-moisture materials (containing $< 5\%$ moisture) and high-moisture materials (containing $> 50\%$ moisture) do not present major size reduction problems. However, some semi-solid materials or fatty materials can be extremely difficult to reduce because the material can stick to or become trapped in the mill's classifying screen, plugging the screen's holes. The material can also stick to the grinding chamber's sides. Examples of such wet, fatty, or sticky materials include cheeses, waxes, pepper sauces, and baby foods. Solutions to these sticky problems include installing a custom-designed screen that allows the appropriate sized particles to pass through without sticking or plugging and internal air or water injectors that clean the mill's internal surfaces, removing any stuck or trapped particles.
- *Hygroscopic and deliquescent materials:* These have a tendency to absorb and retain moisture from the air. When exposed to air, hygroscopic materials such as some chemical powders become sticky, while deliquescent materials such as calcium gluconate dissolve into a liquid. This can become a problem because both actions can impair the mill's effectiveness and efficiency and reduce the final product quality. Solutions involve limiting the materials exposure to air or humidity. This can be done by reducing the material in an inert environment created by purging the mill's grinding chamber with nitrogen or argon gas or by installing a dehumidifier to remove moisture from the grinding chamber air.
- *Dense or hard materials:* Reducing extremely dense or hard materials such as calcium carbonate, minerals, and silica-based materials can cause excessive damage and wear to the mill's contact surfaces, hammers or blades, and classifying screens. One way to avoid these problems is to mill in multiple passes rather than trying to achieve a particle size goal in one pass. Multiple passes allow for greater material

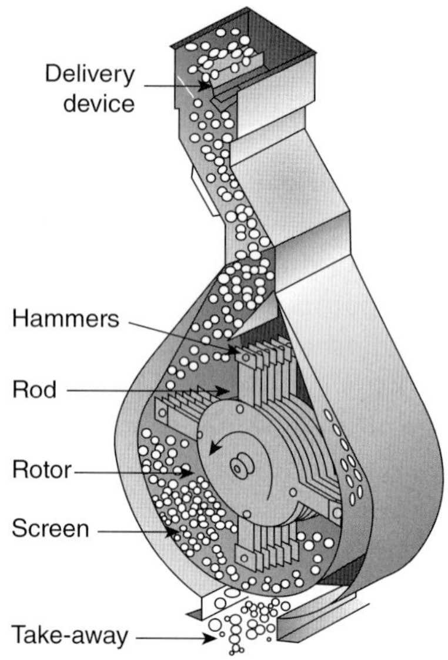

Figure 32.8 Phantom view of a hammer mill. (From www.feedmachinery.com.)

capacities, require less horsepower, and reduce part replacement costs. Wear-resistant components can also be installed in the mill or wear-resistant coatings (tungsten carbide and stellite) can be applied to the mill's contact surfaces.

Hammer Mills

A typical hammer mill mechanism with a cylindrical chamber, a high-speed rotor inside the chamber, and the hammers is shown in Figure 32.8. In operation, food is disintegrated mainly by impact forces as the hammers drive it against the breaker plate. In some designs the exit from the mill is restricted by a screen and food remains in the mill until the particles are sufficiently small to pass through the screen apertures. Under these choke conditions; shearing forces play a larger part in the size reduction. The hammer mill is used for a variety of size reduction or grinding jobs. Besides feed preparation, it is used for pulverizing limestone and ingredients for commercial fertilizers. It has many industrial applications. The material is introduced into the housing, and the beater, which consists of a series of hammers turning at 1500–4000 rpm, beats and pounds the material until it is small enough to pass through the screen at the bottom. Fineness of division is controlled mainly by the size of holes in the screen, although the rotor revolutions per minute and the rate of feed are additional control factors.

The hammers are rigidly fixed to the shaft or swing. There is less danger of the swinging hammer causing damage if a large metallic object gets into the mill by accident. There are many hammer striking edge designs, indicating that there is no one best design. The swinging hammers are usually reversible so that two or perhaps four edges are available for use per hammer. The hammer mill is assumed to reduce size by impact. The terrific speed of the hammer produces kinetic energy that is dissipated on the material, causing it to disintegrate. Although most of the size reduction is probably by impact between the material and the hammers, no doubt some shear between the screen or other parts of the mill and the particles takes place. After passing through the screen, the material is removed by shovel, auger, and chain elevator or by a fan.

Lengthy straw/stalk biomass cannot be directly fed into grinders such as hammer mills and disk refiners. Hence, biomass needs to be preprocessed using coarse grinders like a knife mill to allow for efficient feeding into refiner mills without bridging and choking (Bitra *et al.*, 2009b). The advantages of the hammer mill are: simplicity, versatility, freedom from significant damage due to foreign objects, freedom from damage when operating empty, and efficiency not materially reduced by hammer wear. The inability to produce a uniform grind and high power requirements are disadvantages. Hammer milling technology also has the advantage of low maintenance over other methods, including crushing, shearing, and roller milling. Yu *et al.*'s (2003) review of equipment design parameters showed that hammer mill tip speed, power requirements, grinding rate, screen size, and clearance affect performance to varying degrees. A slightly more robust apparatus for such a purpose is the Rietz disintegrator of which many models exist. The principle is illustrated in Figure 32.9; hammers beat and cut

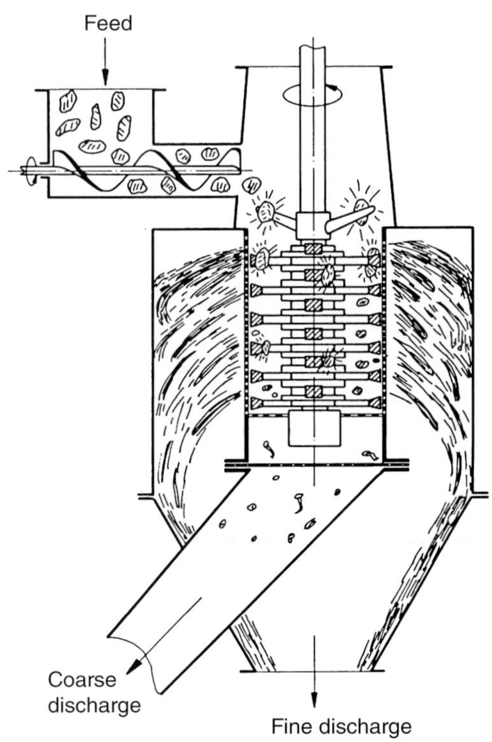

Figure 32.9 Hammers beat and cut. (From Leniger and Beverloo (1975), courtesy of Springer.)

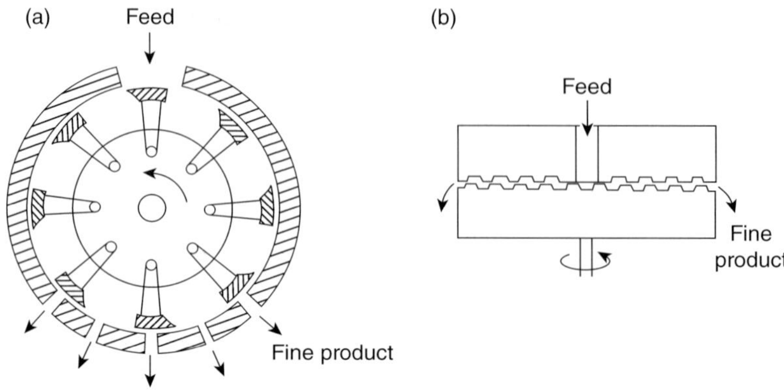

Figure 32.10 Grinders: (a) hammer mill and (b) plate mill. (From Earle (1983), courtesy of Pearson Education.)

the product and force it through the screen. It is normally used for materials with high liquid content. Another type of impact mill is the jet mill, which uses compressed gas to accelerate the particles, causing them to impact against each other in the process chamber. Impact mills can reduce both fine powders and large chunks of friable material such as pharmaceutical powders, sugar, salt, spices, plastics, coal, limestone, chemical powders, and fertilizer down to <10 (jet mills) to 50 μm (mechanical impact mills). Mechanical impact mill types include hammer mills, pin mills, cage mills, universal mills, and turbo mills.

Burr Mills

Burr mills or plate mills consist essentially of two roughened plates, one stationary and the other rotating (Figure 32.10). The material is fed between the plates and is reduced by crushing and shear. Burr mills are frequently combined with a single roller crusher so that large materials such as ear corn will be crushed to a suitable size for feeding between the plates. Grinders of this type are moderately versatile and widely used on farms. If the material is fed slowly so that the flutes are not filled, reduction is probably mainly by shear. With a faster feed and the flutes filled, both shear and crushing no doubt exist. Overfeeding reduces the effectiveness of the grinder and excessive heating results. The plates are designed for a variety of jobs and usually made of chilled cast iron, although alloy steel may be advisable in certain cases. Operating speeds are usually less than 1200 rpm.

The fineness of reduction is controlled by the type of plates and by their spacing. The spacing screw is spring loaded so that the space will increase in case of an overload or if a foreign object gets into the mill. Small rocks and metallic objects may not cause damage, but breakage can be expected if large objects are fed into the mill. The attrition mill is a heavy-duty precision burr mill used in the commercial preparation of feed and food products. Each burr rotates and is driven independently, speeds are much higher, and design and construction are more precise. The advantages of the burr mill are: low initial cost, uniformity of product, and low power requirements; while the breakage due to the foreign objects, excessive burr wear while operating empty, and poor yield are disadvantages.

Ball Mills

This type of mill consists of a slowly rotating, horizontal steel cylinder that is half filled with steel balls of 2.5–15 cm in diameter. At low speeds or when small balls are used, shearing forces predominate. With larger balls or at higher speeds, impact forces become more important. A modification of the ball mill is the rod mill; rods instead of balls are used to overcome problems associated with the balls sticking to foods. Empirical equations were developed to express the grinding rate constant of each material as a function of feed size and ball diameter (Kotake *et al.*, 2002).

Disk Mills

There are a large number of designs of disk mill. Each type employs shearing forces and may be improve the effectiveness of milling by creating additional impact and shearing forces. For all types of impact mill the degree of fineness depends mainly on the peripheral speed of the rotors and the supply rate of the materials. Figure 32.11 shows a disk mill in which food passes through an adjustable gap between a stationary casing and a grooved disk that rotates at high speed. The material enters the mill axially and then passes radially between two disks. In order to produce greater shearing forces, double-disk mills, in which two disks rotate at high speed in opposite directions, can be used. The distance between the rotors can be varied and equipped for internal pneumatic classification. Edge mills work batch-wise and have heavy rolls that roll and slip over a table (Figure 32.12). In general the rolls are driven, although sometimes the table is driven instead. Edge mills are often used for very fine milling, where the table and the rolls are made of smooth granite or steel. Pin-and-disk mills have intermeshing pins fixed either to the single disk and casing or to double disks.

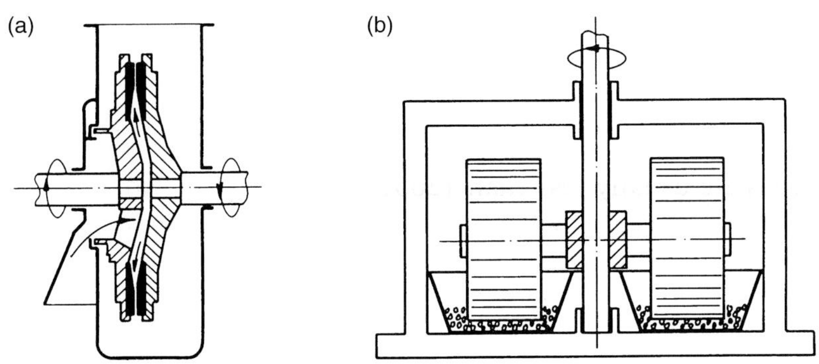

Figure 32.11 (a) Disk mill and (b) edge mill. (From Leniger and Beverloo, (1975), courtesy of Springer.)

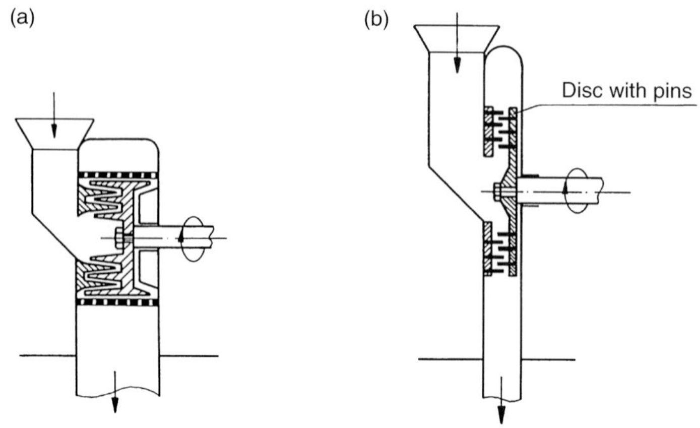

Figure 32.12 (a) Fixed and rotating disks with pegs and (b) disk mills with pins. (From Leniger and Beverloo (1975), courtesy of Springer.)

The mill shown in (Figure 32.12a) consists of a stationary disk with a rotating disk a short distance away; both disks are set with pegs, pins or teeth. These mills sometimes have a screen over the entire periphery. The material is supplied axially; it passes through the mill in the radial direction and is repeatedly reduced in size by stoke or impact. Figure 32.12b shows a pin-and-disk mill where the rotors have a peripheral speed of $160\,m{\cdot}s^{-1}$. These mills are screenless and the material is passed through the mill by an airstream. They are suitable for very fine grinding.

Roller Mills

Two or more steel rollers revolve towards each other and pull particles of food through the nip (the space between the rollers). The main force is compression but, if the rollers are rotated at different speeds, or if the rollers are fluted (shallow ridges along the length of the roller), there is an additional shearing force exerted on the food. The size of the nip is adjustable for different foods and overload springs protect against accidental damage from metal or stones. Double rollers, with or without serrated surfaces, produce a more uniform product than most other reducers. They are used extensively in the industrial preparation of cereals for human consumption. Figure 32.13 shows a roller crusher for farm preparation of animal feed. When only pressure is important, the mills are called roll crushers. Roller mills can easily be cooled or heated. The roller crusher for grinding grain on the farm was used to some extent years ago. The cost of roller mills is higher than burr mills but lower than hammer mills. The mills are durable and require little attention when operating. The product from the rolls in grinding barley, wheat, and corn is not dissimilar to that from the burr mill except that the percentage of fine material obtained is lower. The power required for rolling is somewhat greater than that used with burr mills. The continued use of roller mills will depend largely upon whether the coarser grinding meets farm requirements, but

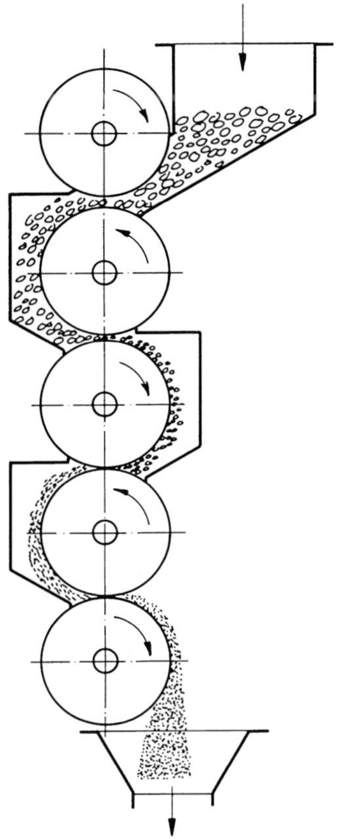

Figure 32.13 Roller crusher. (From Leniger and Beverloo (1975), courtesy of Springer.)

the possibilities of farm use of the roller crusher should be resurveyed. Roller mills are more efficient and require less energy for grinding than the hammer mill. The roller mill produces a more uniform particle size distribution with lower a proportion of fines (GMD < 500 μm) than the hammer mill (Nir and Ptichi, 2001). Commercially the hammer mill is more commonly used for grinding grains because it is easier to use and maintain. It must be noted that even within a grain type, grinding in the same mill type under similar conditions may result in different particle sizes due to variations in endosperm hardness. Lentle *et al.* (2006) reported that grinding grains from three cultivars of wheat in a hammer mill through the same sieve resulted in different particle size distributions.

Wet Milling

Wet milling has many advantages over dry size reduction and thus liquid is sometimes added to semi-dried materials or dried products to allow wet milling. The advantages are: easier to obtain a high degree of fineness, high temperatures can be prevented, and less danger of oxidation, which means less chance of the extreme case of oxidation, a dust explosion. Wet milling is used mainly because of the extremely fine product obtained. For wet size reduction of foods, a mill derived from the ball mill is used, also called an attritor mill or agitated ball mill. It consists of a stationary vertical cylinder that is filled with balls and fitted with a stirrer. The attritor mill is used for producing fine and homogeneous dispersions quickly under controlled conditions. This type of mill has been used for chocolates, chemicals, and pharmaceuticals. The material to be ground is charged into a vertical tank filled with grinding media. Both the material and grinding media are then agitated by a shaft with arms, causing the media to exert shearing and impact forces on the material. This action produces an extremely fine material measured in microns. No premixing is required. The attritor can be used batch-wise with the product circulating if necessary, or by continuous flow-through of the product (Figure 32.14). The choice of mill is determined in particular by the consistency of the solid–liquid mixture.

Apparatuses with frictional action, the so-called strainers, are used for very soft materials like soft fruit and cooked apples. They consist of a cylinder of screen gauze or a perforated plate (with holes of 1–3 mm) through which the soft product is forced by a rotating beater. Some separation takes place in this type of machine. The fruit pulp passes through the screen while harder and coarser material stays behind. Size reduction of soft materials is often called pulping, even when done with other types of equipment. A completely different type of mill for dry materials is the jet mill in which jets of air or superheated steam are used at high speed. When the material enters this mill, the particles collide and rub against each other as a result of the turbulence, and thus size reduction takes place. The advantages of this system are that there are no moving parts and very fine products can be obtained. Application of these mills has been limited up until now mainly because of their high power consumption.

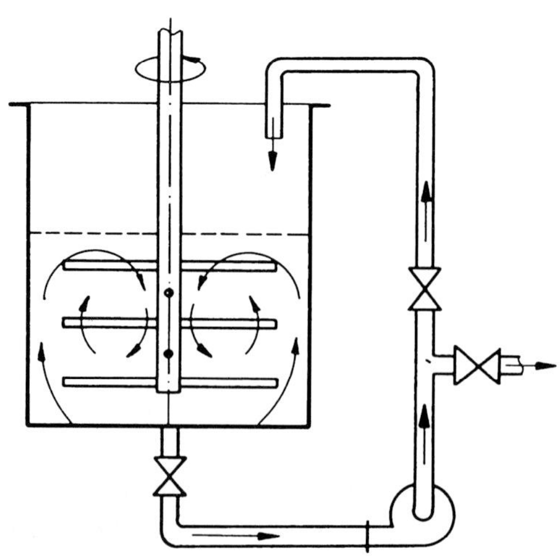

Figure 32.14 Attritor ball mill. (From Leniger and Beverloo (1975), courtesy of Springer.)

Knife Milling

In knife milling, a sharp blade applies high, head-on shear force to a large particle, cutting it to a predetermined size to create smaller particles and to minimize fines. This milling action is produced by a rotating assembly that uses sharp knives or blades to cut the particles. Knife mills can reduce 2-inch or larger chunks or slabs of material, including elastic or heat-sensitive materials such as various foods, rubber, and wax, down to 250–1200 μm. Mill types include knife cutters, dicing mills, and guillotine mills.

Direct-Pressure Milling

Direct-pressure milling occurs when a particle is crushed or pinched between two hardened surfaces. Two rotating bars or one rotating bar and a stationary plate generally produce this milling action. Direct-pressure mills typically reduce 1-inch or larger chunks of friable materials, such as minerals, chemicals, and some food products, down to 800–1000 μm. Types include roll mills, cracking mills, and oscillator mills. Most mills use a combination of these principles to apply more than one type of force to a particle. This allows mill manufacturers to custom-design their equipment based on a material's characteristics.

Solid Foods Size Reduction

Size reduction is a processing aid that permits control over the properties of foods, to improve the efficiency of mixing and heat transfer. The texture of solid processed foods, including different types of bakery products and pastry, processed meats, and vegetables, is controlled by the conditions used during size reduction. There is an indirect effect on the aroma and flavor of some foods while going through the size reduction process. The disruption of cells and resulting increase in surface area promotes oxidative deterioration and higher rates of microbiological and enzyme activity. Size reduction therefore has little or no preservative effect. Dry foods such as grains and nuts have a sufficiently low a_w to permit storage for several months after milling without substantial changes in nutritional value or eating quality. However, moist foods deteriorate rapidly if additional preservative operations (e.g. chilling, freezing, and heat processing) are not undertaken.

Changes During Size Reduction

There are small but largely unreported changes in the color, flavor, and aroma of dry foods during size reduction. There is a loss of volatile constituents from spices and some nuts, which is accelerated if the temperature is allowed to rise during milling. In moist foods the disruption of cells allows enzymes and substrates to become more intimately mixed, which causes accelerated deterioration of flavor, aroma, and color. Additionally the release of cellular materials provides a suitable substrate for microbiological growth and this can result in the development of off-flavors and aromas. The texture of foods is substantially altered by size reduction, both by the physical reduction in the size of tissues and by the release of hydrolytic enzymes (Fellows, 1998).

Nutritional value may change due to oxidation of fatty acids and vitamins as the surface area of foods increases. Losses of vitamin C and thiamin in chopped or sliced fruits and vegetables are substantial. For example, Erdman and Erdman (1982) reviewed changes in nutritional values and reported 78% reductions in vitamin C during slicing of cucumber. In dry foods the main loss in nutritional value results from screening operations performed on the product after size reduction. Results from Raji and Famurewa (2008) on the effects of hull on the physicochemical properties of flour obtained from processed soybean seed do not indicate any considerable reduction in the nutritional content of the fibrous food, and the different acceptabilies were all above the threshold. Hoque *et al.* (2007) analyzed the performance and productivity of size reduction equipment for fibrous materials.

Ultrasonic Cutting

Due to significant technical advances in the last years, high power ultrasonic cutting has become an alternative to many conventional food processing steps, such as homogenization, milling, high shear mixing, pasteurization, and solid/liquid separation. Also, it has been shown to improve the efficiency of traditional processes such as filtration/screening, extraction, crystallization, and fermentation. The use of ultrasonic cutting is often driven by economic benefits, yet in some cases unique product functionality can be achieved (Patist and Bates, 2008). Ultrasonic cutting is used in the food industry, especially for the separation of products consisting of layers with different rheological properties to achieve minimally damaged cut surfaces. This may be useful for selecting operational variables (cutting velocity, excitation amplitude) during cutting to achieve a controlled reduction of the cutting work.

When using cutting assemblies with ultrasonic support, its impact on cutting efficiency depends on factors such as blade geometry, direction of vibration relative to the cutting direction, and frequency and amplitude at the cutting edge (Rawson, 1998). Depending on the product to be separated, ultrasonic-induced mechanical and thermal effects occur in the separation zone at the cutting edge and/or along the flanks of the blade, and lead to a significant reduction of the cutting force which, when measured as a function of the cutting path, allows a better understanding of the cutting process (Brown *et al.*, 2005).

Target food materials for ultrasonic cutting are, predominantly, cheese, candy bars, bakery and confectionery products, and convenience food. Other target foods for ultrasonic cutting are relatively soft products containing solid fats and/or hydrocolloid gels, where dynamic friction rather than separation forces determine the total cutting work (Schneider *et al.*, 2002). In the time domain, a discrete particle at the edge of the cutting blade shows a harmonic oscillatory movement, y as a function of time, t as

($y = y_0 \sin \omega t$), where y_0 and ω ($rad \cdot s^{-1}$) correspond to the displacement amplitude and angular velocity respectively. Introducing frequency f (Hz) and using y_0, the maximum vibration speed at the cutting edge, v_0 ($m \cdot s^{-1}$), corresponds to the strain rate amplitude and can be calculated as described previously (Mason and Lorimer, 2002) by ($v_0 = \omega y_0 = 2\pi f y_0$). The effects of vertical cutting velocity and the magnitude of ultrasonic excitation on the reduction of the work necessary to separate various food materials were investigated by Zahn *et al.* (2006). Their results showed that cutting work increased with increasing cutting velocity but, at each particular cutting velocity, decreased with increasing magnitude of the ultrasonic excitation of the cutting tool. Interactions between cutting velocity and the maximum vibration speed at the cutting edge, which is determined by both excitation amplitude and excitation frequency, are significant. Depending on the food under action, the relative amount of cutting work reduction is either affected by the maximum vibration speed or, additionally, by vertical cutting velocity. No distinct effects of the excitation frequency (20 or 40 kHz) were observed (Zahn *et al.*, 2006).

Liquid Foods Size Reduction

The terms emulsifier and homogenizer are often used interchangeably for equipment used to produce emulsions. Emulsions are stable suspensions of one liquid in another, the liquids being immiscible. Stability of the emulsion is obtained by dispersion of very fine droplets of one liquid, called the disperse phase, through the other liquid, which is called the continuous phase. This can be achieved by imposing very high shearing forces upon the liquid to be dispersed and these break the material into a multitude of fine particles. Homogenization is the reduction in size (to 0.5–3 μm) and increase in number of solid or liquid particles of the dispersed phase by the application of intense shearing forces, to increase the intimacy and stability of the two immiscible liquids. Homogenization results in smaller and more uniform droplet sizes; practical examples are the homogenization of milk and ice cream mixes (dispersions of fat and air in sugar solutions). Homogenization is therefore a more severe operation than emulsification. Both operations are used to change the functional properties or eating quality of foods. They have little or no effect on nutritional value or shelf-life.

Size Reduction Mechanisms

Mechanical forces during homogenization cause droplet or particle size reduction by shear, turbulence, impact, and cavitation. Shear is caused by elongation and subsequent break-up of droplets due to acceleration of a liquid. Centrifugal orifices may be used to obtain the shearing action. In passing through the nozzle, very large shear forces are exerted on the liquid, dispersing it in small particles. Disks that spin at high velocities give rise to high shearing forces in liquids flowing over them. Cavitations are caused by an intense pressure drop, leading to formation of vapor bubbles in the liquid, which implode and cause shock waves in the fluid. This leads to the disruption of droplets, particles, and cell membranes. Shearing is generally achieved by passing the liquid through a high-pressure pump to bring it up to pressures of the order of 0.007 kPa, and then discharging this pressure suddenly by the expansion of the liquid through a small gap or nozzle; the equipment is often called a homogenizer. The higher the interfacial tension between the continuous and dispersed phases, the more difficult it is to form and maintain a stable emulsion. Emulsifying agents therefore lower the energy input needed to form an emulsion.

The four main types of homogenizer are high-speed mixers, pressure homogenizers, colloid mills, and ultrasonic homogenizers, which are described below and in more detail by Rees (1967).

High-Speed Mixers

Mixers are classified into two major types that are suitable for low- or high-viscosity liquids and dry powders or particulate solids. Paddle agitators with wide flat blades rotate at 20–150 rpm and are often pitched to promote longitudinal flow in unbaffled tanks. Impeller agitators with two or more blades operate at 30–500 rpm and are used for pre-mixing emulsions. Turbine- or propeller-type high-speed mixers are used for low-viscosity liquids. They operate by a shearing action on the food at the edges and tips of the blades. Propeller agitators operate at 400–1500 rpm and are used for blending miscible liquids, diluting concentrated solutions, preparing syrups or brines, and dissolving ingredients.

Pressure Homogenizers

These consist of a high-pressure pump, operating at 10 000–70 000 kPa, which is fitted with a homogenizing valve on the discharge side (Figure 32.15). Pumping liquid through a small adjustable gap between a valve and the valve seat results in high pressure and high liquid velocity (8400 $m \cdot s^{-1}$). There is then an almost instantaneous drop in velocity as the liquid emerges from the valve. These extreme conditions of turbulence produce powerful shearing forces. The collapse of air bubbles (cavitations) and impact forces created in some valves by placing a hard surface (a breaker ring) in the path of the liquid further reduce the globule size. In some foods (e.g. milk products) there may be inadequate distribution of the emulsifying agent over the newly formed surfaces, which causes fat globules to clump together. A second similar valve is then use to break up the clusters of globules. Pressure homogenizers are widely used before pasteurization and ultrahigh-temperature sterilization of milk, and in the production of salad creams, ice cream, and some sauces.

Colloid Mills

These homogenizers are essentially disk mills. Small clearance (0.05–1.3 mm) between a vertical disk rotating at 3000–15 000 rpm and a similar sized stationary disk creates high shearing forces. They are more effective than pressure homogenizers for high-

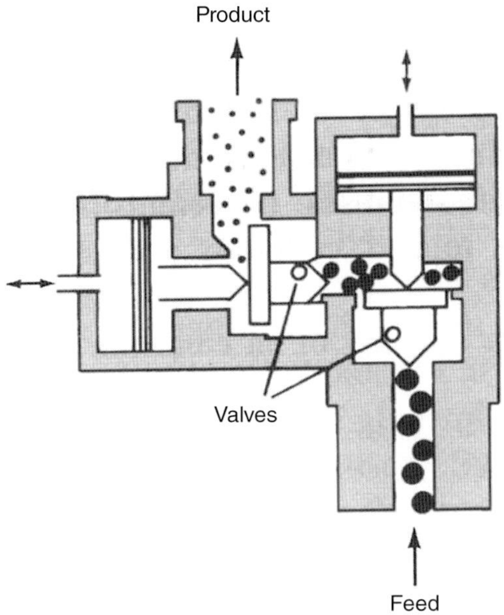

Figure 32.15 Hydraulic two-stage pressure homogenizing valve. (From Fellow, P., ed. (2005) *Food Processing Technology: Principle and Practice*, with permission of Woodhead Publishing Limited, CRC Press LLC.)

viscosity liquids, but with intermediate-viscosity liquids they tend to produce larger droplet sizes than pressure homogenizers do. Numerous designs of disk, including flat, corrugated, and conical shapes, are available for different applications. Modifications of this design include the use of two contrarotating disks or intermeshing pegs on the surface of the disks to increase the shearing action. For highly viscous foods (e.g. peanut butter, meat or fish pastes) the disks may be mounted horizontally (the paste mill). The greater friction created in viscous foods may require these mills to be cooled by the recirculation of water.

Ultrasonic Homogenizers

High-frequency sound waves (18–30 kHz) cause alternate cycles of compression and tension in low-viscosity liquids and create cavitations of air bubbles, to form an emulsion with droplet sizes of 1–2 μm. In operation, the dispersed phase of an emulsion is added to the continuous phase and both are pumped through the homogenizer at pressures of 340–1400 kPa. The ultrasonic energy is produced by a metal blade, which vibrates at its resonant frequency. Vibration is produced either electrically or by the liquid movement (Figure 32.16). The frequency is controlled by adjusting the clamping position of the blade. This type of homogenizer is used for the production of salad creams, ice cream, synthetic creams, and essential oil emulsions. It is also used for dispersing powders in liquids, and the ultrasonic-enhanced liquefaction and saccharification of corn in biofuel production (Khanal *et al.*, 2007).

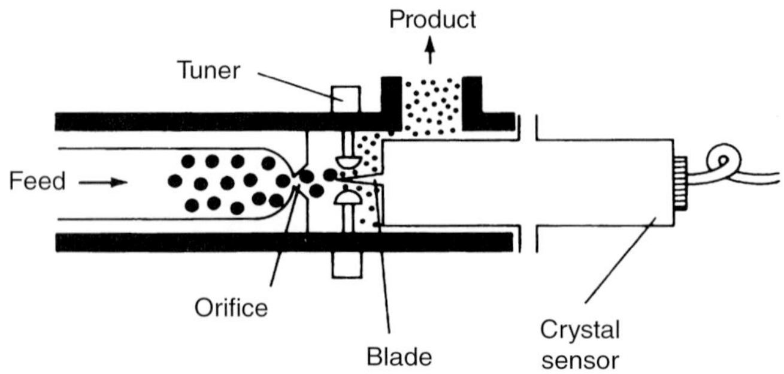

Figure 32.16 Ultrasonic homogenizer. (After Loncin and Merson (1979), courtesy of Elsevier.)

The results of Yamaguchi *et al.* (2009) indicate that a small number of cavitation events with strong physical disturbance of liposomes can reduce the size of the liposomes more efficiently than a large number of cavitations events with weaker disturbance. High-power ultrasound has become an efficient tool for large-scale commercial applications, such as emulsification, homogenization, extraction, crystallization, dewatering, low temperature pasteurization, degassing, de-foaming, activation and inactivation of enzymes, particle size reduction, and viscosity alteration. This can be attributed to improved equipment design and more efficient large-scale continuous flow-through systems. Like most innovative food processing technologies, high-power ultrasound is not an off-the-shelf technology and therefore needs to be developed and scaled up for each application (Patist and Bates, 2008).

Effect of Homogenization on Food Properties

Homogenization stabilizes a product, gives the emulsified food a uniform appearance, and prevents separation. In many countries, special regulations are in force to control hygienic standards during preparation of food emulsions due to the risk of dispersing pathogenic bacteria throughout the food. The stability of the continuous phase is determined in part by the water-holding capacity (WHC) and fat-holding capacity (FHC) of the food system. The factors that affect WHC and FHC are described by Laurie (1985). The quality of the emulsion is influenced by the ratio of water to fat, use of polyphosphates to bind water, and the time, temperature, and speed of homogenization. The emulsion is set by heat during cooking.

Viscosity or Texture

In many liquid and semi-liquid foods, the desired mouth feel is achieved by careful selection of the type of emulsifying agent and stabilizer, and control over homogenization reduces the average size of fat globules from 4 mm to less than 1 mm. The increase in viscosity is due to the higher number of globules and adsorption of casein onto the globule surface (Harper, 1979). Optimizing the particle size distribution in suspensions can produce an up to 50-fold reduction in shear viscosity (Servais *et al.*, 2002). In solid food emulsions the texture is determined by the composition of the food, homogenization conditions, and post-processing operations (e.g. heating or freezing).

Color, Aroma, and Nutritional Value

Homogenization has an effect on the color of some foods. For example, in milk the larger number of globules causes greater reflectance and light scattering, and an increase in whiteness. Flavor and aroma are improved in many emulsified foods. Volatile components are dispersed finely throughout the food and hence have greater contact with taste buds when eaten. The nutritional value of emulsified foods is changed if components are separated (e.g. in butter making), and there is improved digestibility of fats and proteins owing to the reduction in particle size (e.g. baby foods). Conventional grinding of cumin is associated with problems of temperature rise, caking, clogging of sieves, and deterioration of quality due to loss of volatiles. Shortcomings of the conventional size reduction have been overcome by flaking. Sowbhagya *et al.* (2008) showed that the flavor profiles were improved by flaking as this retained more of the volatile terpene compounds; cumin-aldehyde were higher in oil obtained from flakes as compared to powder.

Nanoparticles in the Food Industry

Particle size, morphology, and composition can be manipulated to produce materials with different properties. A wide range of properties change as the particle size of a material is reduced to below a critical size. The actual size at which these changes occur depends on the material. Particles with one or more dimensions of the order of 100 nm or less are called nanoparticles (Borm *et al.*, 2006). The novel properties that differentiate nanoparticles from the bulk material typically develop at a critical length of under 100 nm and stem entirely from the different physics of the nanoparticles. For example, nanoparticles can increase the strength and hardness of metals and ceramics and they can make protective coatings transparent. Nanoparticles are in effect a bridge between atomic or molecular structures and bulk materials (Henglein, 1993).

The field of nanoparticle research covers a wide range of interests in the fields of chemistry, physics, and materials science. Some nanoparticle applications include colloidal dispersions, metallic nanoparticles, biopolymers, and nanostructure materials. Nanoparticulated metallic films, in particular, have been the focus of intensive research, primarily due to the interest in their optical and electronic properties, thereby offering great potential for the development of novel electrical and optical sensors, and catalysts (Krasteva *et al.*, 2002).

Control of the nanoscale morphology enables precise control of the properties of the end-product. In order to elucidate the size-dependent physical and chemical properties of nanoparticles, it is critically important to control their size and shape. High-pressure homogenization and supercritical fluid processing are the technologies that are used currently or are being developed for nanoparticle generation (Chow, 1998). The patented technology enables us to measure nanoparticles down to less than 1 nm, i.e. 10^{-9} m.

The different physicochemical properties that nanoparticles exhibit as compared to their respective bulk materials include changes in optical properties, which can cause changes in color (e.g. gold colloids appear as deep red), thermal behavior, material strength, solubility, conductivity, and (photo-) catalytic activity (Hochella, 2002; Kamat, 2002; Burleson *et al.*, 2004). Probably the most significant influence on the behavior of nanoparticles is the change in surface-to-volume ratio (Banfield and Zhang, 2001). Volume decreases with size but the proportion of atoms at the particle surface increases, and the surface properties can, therefore, dominate the properties of the bulk material (Waychunas, 2001). Furthermore, the structure and properties of the surfaces of nanoparticles are substantially modified compared to those of the same materials in bulk form, e.g. proportionally high curvature of the nanoparticle surfaces, more surface defects and edges, and presence of catalytically highly active sites (Madden and Hochella, 2005).

The toxicity of engineered nanoparticles and their effects on human health, as well as their environmental fate and impact on water and soil, is still widely unknown (Burleson et al., 2004). Therefore, it is crucial that we begin to understand the behavior of engineered nanoparticles in food materials, consumer products, and environmental matrices, as well as their toxicity to humans and the environment. The potential toxicity and behavior of nanoparticles will be affected by a wide range of factors, including particle number and mass concentration, surface area, charge, chemistry and reactivity, size and size distribution, state of aggregation, elemental composition, as well as structure and shape (Borm *et al.*, 2006; Chau *et al.*, 2007).

Nanoparticle Sizing

Instrumentation techniques that have been used to evaluate the size of the nanoparticles include:

- scanning electron microscopy (SEM);
- optical microscopy;
- atomic force microscopy (AFM);
- single particle optical sensing;
- dynamic light scattering (DLS);
- nuclear magnetic resonance (NMR) microscopy;
- differential scanning calorimetry (DSC);
- X-ray diffraction;
- turbidimetry;
- filtration;
- field flow fractionation;
- hydrophobic interaction chromatography;
- electrophoresis;

- isopygnic centrifugation;
- zeta potential.

The Zetasizer Nano series measures the size of particles in a fluid down to less than 1 nm by observing the thermal movement, or Brownian motion, of the particle. The effect of pH, concentration of an additive or the ionic strength of the medium on the zeta potential and rheological properties can give information that allows a product to be formulated with maximum stability. The effect of these parameters on the stability of particle dispersion can be automatically determined using an auto-titrator. The Malvern Multi Purpose Titrator is a device capable of performing such titrations in conjunction with the Zetasizer Nano series. In addition, any one of the Malvern rheometer range can be used to provide complementary information.

Size distribution can be checked using the Coulter counter or other instruments like the Malvern particle analyzer. These submicron particle size techniques are usually expensive in terms of the capital cost of equipment, and the time and expertise required to produce a useful image. They also suffer from the statistical problem that only a few particles are observed, unless many experiments are done. Submicron particle sizes are measured by observing the scattering of laser light from these particles, determining the diffusion speed and deriving the size from this, using the Stokes–Einstein relationship. This method is called dynamic light scattering (DLS) (and sometimes photon correlation spectroscopy (PCS) or quasi-elastic light scattering (QELS)) and is used, for example, to size separate lipid nanocapsules (Yegin and Lamprecht, 2006). DLS is the only technique that is able to measure particles in a solution or dispersion in a fast, routine manner with little or no sample preparation. A number of related techniques such as scanning tunneling microscopy (STM) all require a well-trained technician to prepare the sample in a suitable manner for scanning with a probe (Wigginton *et al.*, 2007). The technique is only suitable for hard materials, i.e. those not affected by the preparation technique, and are poor from a statistical point of view as only tens or hundreds of particles are measured. It therefore suffers from the same disadvantages as the atomic force microscopy techniques (Balnois *et al.*, 2007); however, both techniques can give more information about the shape and surface structure of the particle than an ensemble technique like DLS. Transmission electron microscopy (TEM) has been traditionally used to divulge the particle size profile of nanometer-sized particles (Ngo and Pileni, 2000). Magnetic resonance imaging of submicron-sized MnAs ferromagnets grown on GaAs substrate has been performed with magnetic force microscopy equipment (Ando *et al.*, 1998).

Techniques based on or related to chromatography can be used for the separation of nanoparticles. These techniques are rapid, sensitive (detector dependent) and non-destructive, so samples are available for further analysis. Although some chromatographic tools allow a range of solvents to be used, samples usually cannot be run in their original media, which can cause sample alteration and sample–solvent interaction (Krueger *et al.*, 2005; Ziegler *et al.*, 2005; Huang *et al.*, 2005). Centrifugation and filtration techniques are well-established tools for the preparative size fractionation of samples. These are low-cost, high-speed, and high-volume techniques. Ultracentrifugation, for example, is capable of very high spinning speeds for accelerations up to 1 000 000 g. Preparative ultracentrifugation has been used for pelleting of fine particulate fractions, gradient separations (Bootz *et al.*, 2004), and harvesting aquatic colloids and nanoparticles from specific substrates (Mavrocordatos *et al.*, 2007; Balnois *et al.*, 2007). More techniques are described in Tiede *et al.* (2008).

It has been reported that the average size and size distribution of nanoparticles can significantly vary when comparing results for different techniques, such as electron microscopy, DLS, centrifugation and filtration, or ultracentrifugation (Bootz *et al.*, 2004). The lack of consistent reference materials and standards further exacerbates this problem (Lead and Wilkinson, 2006). Nanoparticle sizing standards, as well as standardized methods for sampling and measurement, are, therefore, urgently required to overcome the problem of inconsistent data (Borm *et al.*, 2006). To the best of our knowledge, standardized nanoparticles are not yet available and researchers have to rely on commercially available, often not well characterized, nanoparticles.

Properties of Nanoparticles

Many of these nonmaterials are made directly as dry powders, and it is generally believed that these powders stay in the same state when stored. However, they will rapidly aggregate (in a few seconds) through a solid bridging mechanism. Whether these aggregates are detrimental will depend entirely on the application of the nanomaterial. If the nanoparticles need to be kept separate, then they must be prepared and stored in a liquid medium designed to facilitate sufficient interparticle repulsion forces to prevent aggregation.

The effect of relative concentrations of surfactant and water on the size and surface roughness of ceria nanoparticles was examined using TEM and atomic force microscopy (AFM) respectively by Gupta *et al.* (2005). The investigation confirmed a relationship between the size and roughness of the nanoceria as a function of the surfactant-to-water ratio. With increasing dilution of the surfactant, the size distribution became narrower such that average particle size decreased linearly as the ratio increased, without affecting lower size threshold of nanoparticles. The surface roughness, on the other hand, increased with increasing water-to-surfactant ratio, implying that a dilute surfactant would provide ceria nanoparticles with rougher surfaces. This information can be used to tailor the adhesion properties of nanoceria by optimizing the size distribution as well as surface roughness as a function of water-to-surfactant ratio. (2005). AFM, TEM, and ferromagnetic resonance (FMR) were used to clarify the nanoparticle sizes in a ferrofluid sample (Lacava *et al.*, 2001). Compared to TEM, the AFM method showed a nanoparticle diameter reduction of 20% and standard deviation increase of 15%. The differences were associated with the AFM tip and the nanoparticle concentration of the substrate.

The great challenge faced by researchers in composite materials for structural applications is the fabrication of nanocomposite materials with monodispersion of nanoparticles in thermoset or thermoplastic polymers. Good dispersion of nanoparticles in polymer composite materials is extremely difficult to achieve, since nanoparticles tend to aggregate during synthesis. The degree to which the nanoparticles can be homogeneously dispersed in the polymer matrix would significantly influence the thermal, mechanical, and optoelectronic properties of the material (Rangari *et al.*, 2009).

Agglomeration of metal powders, fragmentation of brittle solids, and enhancement of the reactivity of metals, rates of intercalation in layered materials, and rates of catalytic reactions are some of the observed mechanochemical effects of ultrasound size reduction practice on the nanoparticles being processed. The poor dispersion of nanoparticles in the polymer matrix leads to inferior mechanical, thermal and opto-electrical properties (Barber and Putterman, 1992).

Particle Size Reduction

Size matching is important in carrying out any activity. Nanoparticles are best suited to reach subcellular levels. Cellular uptake of nanoparticles is greater than that of microparticles. However, it is more difficult to make nanoparticles from biological materials than from hard materials. Organic compounds are stickier in nature as compared to inorganic materials such as silica or metal oxides. In live systems, drug delivery is influenced by the biochemistry of the body.

Nanomaterials can be produced either by milling or by fast precipitation from solutions. For many particles, there is a limit to the grind ability in that beyond this limit no decrease in particle size is observed with further grinding. It is energy intensive to go down to a nanoparticle size range. Most pharmaceutical size reduction operations utilize high shear wet milling for production of nanoparticles. Typical operations for wet milling may take hours to days and the drug must be stable over that time period. One has to be aware also of contamination problems due to milling media.

Researchers have used several techniques for dispersing nanoparticles:(i) mechanical agitation, such as ball milling or magnetic stirring; (ii) ultrasonic vibration; (iii) shear mixing; (iv) noncontact mixing; and (v) using the dispersing agent (Rangari *et al.*, 2009). Polymer- or protein-based drug nanoparticles can be produced by the emulsification solvent evaporation process. With recent developments in homogenization, very fine emulsions can yield nanoparticles. The size, zeta potential, and hydrophilicity of the nanoparticles in the drug can be controlled by various parameters, including the amounts of emulsifier, drug and polymer intensity, duration of homogenization, and particle hardening profile.

Biazar *et al.* (2009) influenced the size reduction of acetaminophen particles by mechanical activation using a dry ball mill. The activated samples with an average particle size of 1 μm were then investigated over different time periods with infrared (IR), inductively coupled plasma (ICP), atomic force microscopy (AFM), and X-ray diffraction (XRD) methods. The results of the IR and XRD images showed no change in the drug structure after the mechanical activation of all samples. According to the peaks in the AFM images, the average size of the particles after 30 h of activation was 24 nm, with a normal particle distribution. The greatest reduction in size was after milling for 30 h.

The actual process of mechanical alloying starts with mixing the component powders in the correct proportion and loading the mixture into the grinding mill along with the grinding medium, generally steel balls. The mixture is then milled for the desired length of time until a steady-state is reached, i.e. the proportions of the components in the powder particles are the same as in the original mixture. The milled powder is then consolidated into a bulk shape and heat treated to obtain the desired microstructure and properties. Thus, the important components of the mechanical alloying process are the raw materials, the mill, and the process variables.

An increases in the contact surface area between particles is a direct consequence of the milling size reduction process. As the energy of the system increases there is an associated decrease in mullitization temperatures (Suryanarayana, 2001). The amount of energy required to affect fracture is dependent on the hardness and particle size of the material being milled and the type of stress applied. According to fracture theory, the hardness of a material increases as its particle size decreases; as a consequence the milling of finer materials requires more energy and therefore higher impact speeds.

Rangari *et al.* (2009) used the ultrasound irradiation technique in the presence of ethanol to produce a high-quality mono-dispersed nanocomposite of WO_3 epoxy, using the commercially available WO_3 (about 80 nm), and synthesized porous WO_3 nanoparticles (pore size 2–5 nm). Similar techniques have been applied by other researchers (Lee *et al.*, 2006) to reduce the kaolinite particle size.

Gold nanoparticles are commonly prepared by the chemical reduction of gold ions in a solution in the presence of a stabilizing reagent. Muto *et al.* (2008) found that the gold nanoparticles could be fragmented into smaller particles. In order to investigate the dynamics of fragmentation, the nanoparticles were irradiated with delayed double laser pulses, both of which had an identical wavelength and pulse energy (532 nm, 30 mJ). The average diameter of the product fragments was smallest when the two laser pulses simultaneously irradiated the nanoparticles. The diameter increased with an increase in the delay time from the first to the second laser pulse.

The size of the nanoparticles can be reduced in a controlled manner by laser irradiation of the nanoparticles, laser-induced size reduction (Mafune *et al.*, 2002a, b). The mechanism of the size reduction is still not well-known. Using picosecond photo absorption spectroscopy, Kamat *et al.* (1998) concluded that size reduction is a result of fragmentation caused by the Coulomb explosion of the photoionized metal nanoparticles. Plech *et al.* (2005) observed the structural changes of nanoparticles and water molecules in their vicinity by resolved X-ray scattering. They found that the particles undergo a melting transition within a time scale of 1 ns, and hence, they can be fragmented into small particles by the thermal process.

The above mentioned nano-related techniques have been used in size reduction of nonfood materials and to some extent in pharmaceutical industries in the last decade.

Food-related application of these approaches to size reduction of biological and eatable material is about to be investigated for the potential improvement of food quality and human health.

References

Amerah, A.M., Ravindran, V., Lentle, R.G. and Thomas, D.G. (2007) Feed particle size: Implications on the digestion and performance of poultry. *World Poultry Science Journal* 63: 439–455.

Ando, K., Chiba, A. and Tanoue, H. (1998) Uniaxial magnetic anisotropy of submicron MnAs ferromagnets in GaAs semiconductors. *Applied Physics Letters* 73: 387–389.

Anonymous (2007) *Evaluation of Size Reduction Technologies for On-site Organics Management Systems*, 2nd ed. Recycled Organics Unit, The University of New South Wales, Sydney, p. 1466.

ASABE Standards (2006a) *S319: Method of Determining and Expressing Fineness of Feed Materials by Sieving*. ASABE, St. Joseph, Michigan.

ASABE Standards (2006b) *S424: Method of Determining and Expressing Particle Size of Chopped Forage Materials by Screening*. ASABE, St. Joseph, Michigan.

Baker, S. and Herrman, T. (2002) *Evaluating Particle Size*. MF-2051 Feed Manufacturing, Department of Grain Science and Industry, Kansas State University, pp. 5.

Balnois, E., Papastavrou, G., and Wilkinson, K.J. (2007) Force microscopy and force measurements of environmental colloids. In: *Environmental Colloids and Particles: Behaviour, Structure and Characterization* (K.J. Wilkinson, J.R. Lead, eds). John Wiley & Sons, Chichester, pp. 405–468.

Banfield, J.F. and Zhang, H.Z. (2001) Nanoparticles in the environment. *Nanoparticles and the Environment* 44: 1–58.

Barber, B.P. and Putterman, S.J. (1992) Spectrum of synchronous picoseconds sonoluminescence. *Physics Review Letters* 69: 1182–1184.

Barbosa-Canovas, G.V., Ortega-Rivas, E., Juliano, P.and Yan, H. (2005) *Food Powders: Physical Properties, Processing, and Functionality*. Kluwer Academic, Plenum Publisher, New York, p. 163.

Becker, R. and Gray, G.M. (2006) Evaluation of a water jet cutting system for slicing potatoes. *Journal of Food Science* 57: 132–137.

Biazar, E., Beitollahi, A., Rezayat, S.M., Forati, T., Asefnejad, A., Rahimi, M., Zeinali, R., Ardeshir, M., Hatamjafari, F., Sahebalzamani, A. and Heidari, M. (2009) Effect of the mechanical activation on size reduction of crystalline acetaminophen drug particles. *International Journal of Nanomedications* 4: 283–287.

Bitra, V.S.P., Womac, A.R., Chevanan, N., Miu, P.I., Igathinathane, C. and Sokhansanj, S., Smith, D.R. (2009a) Direct mechanical energy measures of hammer mill comminution of switchgrass, wheat straw, and corn stover and analysis of their particle size distributions. *Powder Technology* 193: 32–45.

Bitra, V.S.P., Womac, A.R., Igathinathane, C., Miu, P.L., Yang, Y.T., Smith, D.R., Chevanan, N. and Sokhansanj, S. (2009b) Direct measures of mechanical energy for knife mill size reduction of switchgrass, wheat straw, and corn stover. *Bioresearch Technology* 100: 6578–6585.

Bond, F.C. (1952) The third theory of combination. *Transactions of the AIME* 193: 484–494.

Bootz, A., Vogel, V., Schubert, D. and Kreuter, J. (2004) Comparison of scanning electron microscopy, dynamic light scattering and analytical ultracentrifugation for the sizing of poly (butyl cyanoacrylate) nanoparticles. *European Journal of Pharmaceutics and Biopharmaceutics* 57: 369–375.

Borm, P.J.A., Robbins, D., Haubold, S., Kuhlbusch, T., Fissan, H., Donaldson, K., Schins, R., Stone, V., Kreyling, W. and Lademann, J. (2006) The potential risks of nanomaterials: a review carried out for ECETOC. *Particle Fibre Toxicology* 3.

Brown, T., James, S.J. and Purnell, G.L. (2005) Cutting forces in foods: experimental measurements. *Journal of Food Engineering* 70: 165–170.

Burleson, D.J., Driessen, M.D. and Penn, R.L. (2004) On the characterization of environmental nanoparticles. *Journal of Environmental Science and Health A* 39: 2707–2753.

Chau, C.F., Wu, S.H. and Yen, G.C. (2007) The development of regulations for food nanotechnology. *Trends Food Science Technology* 18: 269–280.

Chow, G.M. (1998) Nanostructured materials. In: *Nanostructured Materials Science and Technology* (G.M. Gan-Moog Chow and NI Noskova, eds). Heidelberg, Germany.

CRD (1980) *Method of Calculation of the Fineness Modulus of Aggregate*. CRD-C 104, pp. 1–2.

Earle, R.L. (1983) *Unit Operations in Food Processing*, 2nd ed. Pergamon Press, New York.

Erdman, J.W. and Erdman, E.A. (1982) Effect of home preparation practices on nutritive value of food. In: *Handbook of Nutritive Value of Processed Foods* (M. Redhcigl, ed.). CRC Press, Boca Raton, pp. 237–263.

Esehaghbeygi, A., Daeijavad, M. and Afkarisayyah, A.H. (2009a) Breakage susceptibility of rice grains by impact loading. *Applied Engineering and Agriculture* 25: 943–946.

Esehaghbeygi, A., Hoseinzadeh, B., Khazaei, M. and Masoumi, A.A. (2009b) Bending and shearing properties of wheat stem of Alvand variety. *World Applied Science Journal* 6: 1028–1032.

Esehaghbeygi, A., Hoseinzadeh, B. and Masoumi, A.A. (2009c) Effects of moisture content and urea fertilizer on bending and shearing properties of canola stem. *Applied Engineering and Agriculture* 25: 947–951.

Fang, Q., Boloni, I., Haque, E. and Spillman, C.K. (1997) Comparison of energy efficiency between a roller mill and a hammer mill. *Transactions of the ASAE* 13: 631–635.

Fellows, P. (1998) *Food Processing Technology (Principles and Practice)*. Ellis Horwood Ltd, Chichester.

Ghorbani, Z., Masoumi, A.A. and Hemmat, A. (2010) Specific energy consumption for reducing the size of alfalfa chops using a hammer mill. *Biosystems Engineering* 105: 34–40.

Gupta, S., Brouwer, P., Bandyopadhyay, S., Patil, S., Briggs, R., Jain, J. and Seal, S. (2005) TEM/AFM investigation of size and surface properties of nanocrystalline ceria. *Journal of Nanoscience and Nanotechnology* 5: 1101–1107.

Harper, W.J. (1979) *Process Induced Changes. Dairy Technology and Engineering.* AVI, Westport, Connecticut, pp. 561–568.

Hashish, M. (1989) A model for abrasive water jet machining. *Journal of Engineering Materials and Technology* 111: 154–162.

Henglein, A. (1993) Physiochemical properties of small metal particles in solution – microelectrode reactions, chemisorptions, composite metal particles, and the atom-to-metal transition. *Journal of Physical Chemistry* 97: 5457–5471.

Hochella, M.F. (2002) Nanoscience and technology the next revolution in the Earth sciences. *Earth Planet Science Letters* 203: 593–605.

Hoque, M., Sokhansanj, S., Naimi, L., Xiaotao, B., Lim, C.J. and Womac, A.R. (2007) Review and analysis of performance and productivity of size reduction equipment for fibrous materials. *ASAE Annual Meeting*, Paper No: 076164.

Hoseinzadeh, B., Esehaghbeygi, A. and Raghami, N. (2010) Effect of moisture content, bevel angle, and cutting speed on shearing energy of three wheat varieties. *World Applied Science Journal* 7: 1120–1123.

Huang, X.Y., Mclean, R.S. and Zheng, M. (2005) High-resolution length sorting and purification of DNA-wrapped carbon nanotubes by size-exclusion chromatography. *Analytical Chemistry* 77: 6225–6228.

Ince, A.S., Ugurluay, G.E. and Ozcan, M.T. (2005) Bending and shearing characteristics of sunflower stalk residue. *Biosystems Engineering* 92: 175–181.

Kamat, P.V. (2002) Photophysical, photochemical and photocatalytic aspects of metal nanoparticles. *Journal of Physical Chemistry B* 106: 7729–7744.

Kamat, P.V., Flumiani, M. and Hartland, G.V.J. (1998) Picosecond dynamices of silver nanoclusters; photoefection of electrons and fragmentation. *Journal of Physics and Chemistry B* 102: 3123–3128.

Kent, N.L. (1983) *Technology of Cereals*, 3rd ed. Pergamon Press, Oxford, pp. 73–85.

Khanal, S.K., Montalbo, M., Leeuwen, J., Srinivasan, G. and Grewell, D. (2007) Ultrasonic enhanced liquefaction and saccharification of corn for bio-fuel production. *ASAE Annual Meeting*, Paper No: 072710.

Kotake, N., Suzuki, K., Asahi, S. and Kanda, Y. (2002) Experimental study on the grinding rate constant of solid materials in a ball mill. *Powder Technology* 122: 101–108.

Krasteva, N., Besnard, I., Guse, B., Bauer, R.E., Muellen, K., Yasuda, A. and Vossmeyer, T. (2002) Self-assembled gold nanoparticles/dendrimer composite films for vapor setting applications. *Nano Letters* 2: 551.

Krueger, K.M., Al-Somali, A.M., Falkner, J.C., Colvin, V.L., (2005) Characterization of nanocrystalline CdSe by size exclusion chromatography. *Analytical Chemistry* 77: 3511–3515.

Lacava, L.M., Lacava, B.M., Azevedo, R.B., Lacava, Z.G.M., Buske, N., Tronconi, A.L. and Morais, P.C. (2001) Nanoparticle sizing: a comparative study using atomic force microscopy, transmission electron microscopy, and ferromagnetic resonance. *Journal of Magnetism and Magnetic Materials* 225: 79–83.

Laurie, R.A. (1985) *Meat Science*, 4th ed. Pergamon Press, Oxford, pp. 169–207.

Lead, J.R. and Wilkinson, K.J. (2006) Aquatic colloids and nanoparticles: Current knowledge and future trends. *Environmental Chemistry* 3: 159–171.

Lee, J., Ashokkumar, M., Kentish, S. and Grieser, F. (2006) Effect of alcohols on the initial growth of multibubble sonoluminescence. *Journal of Physical Chemistry B* 110: 17282–17285.

Leniger, H.A. and Beverloo, W.A. (1975) *Food Process Engineering*. D. Reidel Publishing Company, Dordrecht.

Lentle, R.G., Ravindran, V., Ravindran, G. and Thomas, D.V. (2006) Influence of feed particle size on the efficiency of broiler chickens fed wheat based diets. *Journal of Poultry Science* 43: 135–142.

Loncin, M. and Merson, R.L. (1979) *Food Engineering*. Academic Press, New York, pp. 246–264.

Lopo, P. (2002) The right grinding solution for you: roll, horizontal or vertical. *Feed Management* 53: 23–26.

Madden, A.S. and Hochella, M.F. (2005) A test of geochemical reactivity as a function of mineral size: Manganese oxidation promoted by hematite nanoparticles. *Geochimica Cosmochimica Acta* 69: 389–398.

Mafune, F., Kohno, J., Takeda, Y. and Kondow, T. (2002a) Full physical preparation of size-selected gold nanoparticles in solution: laser ablation and laser-induced size control. *Journal of Physical Chemistry B* 106: 7575–7577.

Mafune, F., Kohno, J., Takeda, Y. and Kondow, T. (2002b) Growth of gold clusters into nanoparticles in a solution following laser-induced fragmentation. *Journal of Physical Chemistry B* 106: 8555–8561.

Mani, S., Tabil, L.G. and Sokhansang, S. (2004) Grinding performance and physical properties of wheat and barley straws, corn stover and switch grass. *Bioma Bioenergy* 27: 339–352.

Manlu, Y.U., Womac, A.R., Miu, P.I., Igathinathance, C.I., Sokhansanj, S. and Narayan, S. (2006) Direct energy measurement system for rotary biomass grinder-hammer mill. *ASABE Annual International Meeting*, Convention Centre Portland, Oregon, Paper No: 66217.

Mason, T.J. and Lorimer, J.P. (2002) *Applied Sonochemistry*. Wiley-VCH, Weinheim.

Mavrocordatos, D., Perret, D. and Leppard, G.G. (2007) Strategies and advances in the characterization of environmental colloids by electron microscopy. In: *Environmental Colloids and Particles: Behaviour, Structure and Characterization* (K.J. Wilkinson and J.R. Lead, eds). John Wiley & Sons, Chichester, pp. 345–404.

McGorry, R.W., Dowd, P.C. and Dempsey, P.G. (2003) Cutting moments and grip forces in meat cutting operations and the effect of knife sharpness. *Applied Ergonomics* 34: 375–382.

Mohsenin, N. (1986) *Physical Properties of Plant and Animal Materials, Vol. I: Structure, Physical Characteristics and Mechanical Properties*. Gordon-Breach Science Publishers, New York.

Muto, H., Miyajima, K. and Mafune, F. (2008) Mechanism of laser-induced size reduction of gold nanoparticles as studied by single and double laser pulse excitation. *Journal of Physical Chemistry* C 112: 5810–5815.

Naimi, L.J., Sokhansanj, S., Xiaotao, B., Lim, C.J., Womac, A.R. and Mani, S. (2007) Modeling and characterization of biomass size reduction. *ASAE Annual Meeting*, Paper No: 076047.

Nelson, P.E. and Tressler, D.K. (1980) *Fruit and Vegetable Juice Processing Technology*, 3rd ed. AVI, Westport, Connecticut, pp. 268–309.

Ngo, A.T. and Pileni, M.P. (2000) Nanoparticles of cobalt ferrite: Influence of the applied field on the organization of the nanocrystals on a substrate and on their magnetic properties. *Advanced Materials* 12: 276.

Nir, I. and Ptichi, I. (2001) Feed particle size and hardness: Influence on performance, nutritional, behavioral and metabolic aspects. In: *Proceedings of the 1st World Feed Conference*, Utrecht, The Netherlands, pp. 157–186.

Onu, L.I. and Okafor, G.I. (2002) Effect of physical and chemical factor variations on the efficiency of mechanical slicing of Nigerian ginger (*Zingiber Officinale Rose*). *Journal of Food Engineering* 56: 43–47.

Patist, A. and Bates, D. (2008) Ultrasonic innovations in the food industry: From the laboratory to commercial production. *Innovations in Food Science and Emerging Technology* 9: 147–154.

Persson, S. (1987). *Mechanics of Cutting Plant Material*. ASAE Publications, Michigan.

Plech, A., Kotaidis, V., Lorenc, M. and Wulff, M. (2005) Thermal dynamics in laser excited metal nanoparticles. *Chemical Physics Letters* 401: 565.

Purnell, G. and Brown, T. (2004) Equipment for controlled fat trimming of lamb chops. *Computers and Electronics in Agriculture* 45: 109–124.

Raji, A.O. and Famurewa, J.A.V. (2008) Effect of hull on the physico-chemical properties of soy- flour. *Agricultural Engineering International: the CIGR E-J Manu FP 07018* 10: 1–14.

Rangari, V.K., Hassan, T.A., Mayo, Q. and Jeelani, S. (2009) Size reduction of WO_3 nanoparticles by ultrasound irradiation and its applications in structural nanocomposites. *Composite Science and Technology* 69: 2293–2300.

Rawson, F.F. (1998) An introduction to ultrasonic food cutting. In: *Ultrasound in Food Processing* (M.J.W. Povey and T.J. Mason, eds). Blackie Academic, London, pp. 254–269.

Rees, L.H. (1967) What to know about homogenizers. *Food Engineering* 39: 69–71.

Samson, P., Duxbury, P., Drisdelle, M. and Lapointe, C. (2000) Assessment of pelletized bio fuels. Available at: http://reap.ca/reports/pelletaug 2000.html

Schneider, Y., Zahn, S. and Linke, L. (2002) Qualitative process evaluation for ultrasonic cutting of food. *Engineering in Life Science* 2: 153–157.

Servais, C., Jones, R. and Roberts, I. (2002) The influence of particle size distribution on the processing of food. *Journal of Food Engineering* 51: 201–208.

Skubisz, G., Kravtsova, T.I. and Vielikanov, L. (2007) Analysis of the strength properties of pea stems. *International Agrophysics* 21: 189–197.

Sowbhagya, H.B., Sampathu, S.R. and Krishnamurthy, N. (2007) Evaluation of size reduction on the yield and quality of celery seed oil. *Journal of Food Engineering* 80: 1255–1260.

Sowbhagya, H.B., Sathyendra Rao, B.V. and Krishnamurthy, N. (2008) Evaluation of size reduction and expansion on yield and quality of cumin (*cuminum cyminum*) seed oil. *Journal of Food Engineering* 84: 595–600.

Suryanarayana, C. (2001) Mechanical alloying and milling. *Progress in Material Science* 46: 1–184.

Tabil, L.G. (1996) *Binding and pelleting characteristics of alfalfa*. PhD dissertation, Department of Agricultural Bioresource Engineering, University of Saskatchewan, Saskatchewan, Canada.

Tiede, K., Boxall, A.B.A., Tear, S.P., Lewis, J., David, H. and Hassellov, M (2008) Detection and characterization of engineered nanoparticles in food and the environment. *Food Additives and Contamination* 25: 795–821.

Waychunas, G.A. (2001) Structure, aggregation and characterization of nanoparticles. *Nanoparticles and the Environment* 44: 105–166.

Wigginton, N.S., Rosso, K.M., Lower, B.H., Shi, L. and Hochella, M.F. (2007) Electron tunneling properties of outer-membrane decaheme cytochromes from Shewanella oneidensis. *Geochimica Cosmochimia Acta* 71: 543–555.

Yamaguchi, T., Nomura, M., Matsuoka, T. and Koda, S. (2009) Effects of frequency and power of ultrasound on the size reduction of liposome. *Chemical and Physical Lipids* 160: 58–62.

Yegin, B.A., Lamprecht, A. (2006) Lipid nanocapsule size analysis by hydrodynamic chromatography and photon correlation spectroscopy. *International Journal of Pharmaceutics* 320: 165–170.

Yu, M., Womac, A.R. and Pordesimo, L.O. (2003) *Review of Biomass Size Reduction Technology*. ASAE, St Joseph, Paper No: 036077.

Zahn, S., Schneider, Y. and Rohm, H. (2006) Ultrasonic cutting of foods: Effects of excitation magnitude and cutting velocity on the reduction of cutting work. *Innovations in Food Science and Emerging Technology* 7: 288–293.

Ziegler, K.J., Schmidt, D.J., Rauwald, U., Shah, K.N. Flor, E.L. Hauge, R.H. and Smalley, R.E. (2005) Length-dependent extraction of single-walled carbon nanotubes. *Nano Letters* 5: 2355–2359.

33

Irradiation Process Design

Rod Chu

Introduction

Food irradiation is the exposure of food to a source of ionizing radiation to obtain a product with added desirable properties. Food is irradiated for preservation purposes such as the extension of shelf-life, microbial reduction or the elimination of pathogenic microorganisms, and for quarantine control purposes such as the disinfestation of insects or parasites. The safety and benefits of food irradiation have been well documented and demonstrated in studies over a number of decades (WHO, 1995; WHO, 1999). The commercial application of food irradiation has been slow, mainly because of the perceived problem of consumer acceptance (Frenzen *et al.*, 2001). An irradiation plant requires high capital investment, and many factors must be taken into account when choosing the optimum design for a specific application.

New interest in food irradiation has resulted from the increase in the number of outbreaks of food-borne diseases and the desire to replace chemical fumigants such as methyl bromide for phytosanitary treatments (Ross and Engeljohn, 2000). Food irradiation has emerged as a viable sanitary and phytosanitary treatment to help governments meet the requirements set out in the Agreement on the Application of Sanitary and Phytosanitary Measures (SPS) of the World Trade Organization (WTO, 1995).

Food irradiation is a highly regulated process, and approval is required from regulatory authorities before a food can be irradiated for sale. Classes of foods must be

Handbook of Food Process Design, First Edition. Edited by Jasim Ahmed and Mohammad Shafiur Rahman.

specifically approved for irradiation, and this process involves submission of a petition containing data on the safety and effectiveness of the process. The review of the petition is rigorous and may take years before clearances are granted. These clearances must be obtained for each country accepting the irradiated food.

The Codex Alimentarius Commission of the FAO/WHO Food Standard Program has issued a Codex General Standard for irradiated foods (FAO/WHO, 2003a) and an associated Code of Practice for the operation of irradiation facilities for the treatment of food (FAO/WHO, 2003b). The Codex General Standard contains provisions about irradiation facilities and process control that include, among other requirements, requirements for the keeping of adequate records, including quantitative dosimetry. The standard requires that an irradiation facility be licensed and registered for the purpose of food processing, and that the facility be staffed with trained and competent personnel.

At present, over 50 countries have implemented national regulations or issued specific approval for certain irradiated food items or classes of food to ensure their safety and quality, and to satisfy quarantine requirements related to trade. These food items include meat products, fresh fruits, spices and dried vegetable seasonings. A database showing the clearances that have been granted is maintained by the Food and Environmental Subprogram of the Joint FAO/IAEA Division of Nuclear Techniques in Food and Agriculture (FAO/IAEA, 2010a).

For food irradiation, the main quantity that must be measured and controlled is the absorbed dose. The irradiated food must receive an absorbed dose that is sufficient to achieve the required benefit, and excessive doses must be avoided to prevent deterioration of food quality. Dosimetry is performed to determine the operating parameters and for routine monitoring of the process.

Food irradiation is performed in an irradiator that is designed to economically process the food while ensuring that the absorbed-dose requirements are met by all of the processed product. The irradiator may need to process only a single product, or it may be necessary for the irradiator to be able to process a variety of products of different forms and with different environmental control requirements. There is no single design that is the best for all applications, and the potential user of a food irradiator must consider the type of radiation source and the irradiation geometry that are suitable for the specific application.

Since the construction of a new irradiation facility costs several million dollars and the staffing and operation of the irradiator lead to significant operating costs, it may be more economical initially to have the irradiation performed in a multipurpose facility providing a contract irradiation service. The Food and Environmental Subprogram of the Joint FAO/IAEA Division of Nuclear Techniques in Food and Agriculture maintains a database providing a list of food irradiation facilities worldwide. The current list shows 70 facilities (FAO/IAEA, 2010b). When a contract irradiation service is used, there are logistical challenges regarding transportation costs and facility availability, especially if the product is seasonal. If an irradiation facility is designed for a specific seasonal product, it may also be worthwhile to include a capability for the irradiation of other products during off-seasons.

Applications of Food Irradiation

The applications of food irradiation generally fall into one of three dose ranges. Low doses, up to 1.0 kGy, are used for sprout inhibition and insect disinfestation. Medium doses, from 1.0 kGy to 10 kGy, are used for the preservation of foods based on the inactivation of spoilage-causing and pathogenic microorganisms. High doses, above 10 kGy, are used to obtain sterile products. Typical absorbed-dose ranges for some approved applications of food irradiation are summarized in Table 33.1.

Table 33.1 Absorbed-dose ranges for some applications of food irradiation.

Range	Application	Typical absorbed-dose range (kGy)	Products	Description
Low	Sprout inhibition	0.02–0.15	Potatoes, onion, garlic, shallots, yams	Inhibition of sprouting of bulbs, roots and tubers during storage
	Delay of ripening	0.1–1.0	Fresh fruits	Delay of ripening or physiological growth
	Insect disinfestation	0.2–1.0	Grains, pulses, cereals, flour, coffee beans, spices, dried fruits, dried nuts, dried fishery products	Prevention of losses caused by insect pests
	Phytosanitary control	0.15–1.0	Insect pests such as fruit flies in fresh fruits and vegetables	Kills or inhibits further development of various life cycle stages of insect pests
Medium	Shelf-life extension	1–10	Fresh meat, seafood, vegetables and fruits	Enhancement of keeping quality through reduction in the number of spoilage microorganisms
	Microbial reduction or pasteurization	2–8	Ground beef (*E. coli*) and poultry (*Salmonella*)	Inactivation of pathogenic and most spoilage organisms
High	Microbial decontamination	10–30	Dried spices, herbs, and other dried vegetable seasonings	Alternative to the chemical fumigant ethylene oxide for reduction of microbial load
	Sterilization	25–75		Sterile foods for immunocompromised patients or astronauts

The clearances usually state the upper dose limits for the process and may also specify a minimum dose. However, the actual doses to be given to the food should be determined by the levels at which the sensory and functional properties of the food are not impaired. These values are based on research performed for each particular application of food irradiation and are specified by food technologists for each combination of process and product on the basis of the results obtained in experimental research preceding commercial-scale application. For applications where the acceptable dose range is small, the irradiator design must allow treatment of products with a low maximum-to-minimum dose ratio.

Foods irradiated for sale may require labeling to identify them as being irradiated. This may mean a brief description of the process and the use of the Radura radiation symbol (FAO/WHO, 2010). At present, the most common application of food irradiation is the irradiation of spices and dried vegetable seasonings, which represents about 50% of the total volume of irradiated food. Over 200 000 tonnes of spices are now being irradiated annually (Kume *et al.*, 2009).

The Food Irradiation Process

When a product is considered for irradiation, various factors must be taken into consideration to ensure that irradiation is a viable process. These include the following:

- annual throughput;
- seasonal variations in the throughput;
- the physical form of the product;
- handling and storage requirements;
- the acceptable dose range.

The optimum irradiator design will depend on the physical form of the product and the conditions under which the product will be transported, stored and processed. The product may be packed in standard-size packages or may be a powder or liquid that is stored in bulk and transported as a fluid. Other products may be packed into drums or packages of varying sizes and may be transported on pallets or as individual packages. Some foods are delicate and must be handled with care to minimize possible damage.

The size and shape of the product packages may have a significant impact on the economics of the process. If the product is to be irradiated in standard irradiation containers, the efficiency of the process will depend on how well the product packages fill the containers without requiring repacking. If the product is handled on pallets throughout the distribution system, there may be advantages in irradiating the product on the pallets on which it is distributed and stored. Irradiation on pallets may require a more complex irradiation geometry, resulting in increased capital costs. There may also be a reduction in the throughput per unit activity. The additional costs of irradiation of the product on pallets must be balanced against the cost of labor for handling the product or the cost of automated product handling, including palletizing and depalletizing.

For fruits and vegetables, the product may need to be irradiated within a short period of time after harvesting. If the product is seasonal, the availability of other products for irradiation during the off-season may help to improve the economics but may require changes to the irradiator design. Meat or seafood products may require refrigeration or may need to be processed while frozen.

The Food Irradiation Process Flow

Food irradiators consist of a shielded irradiation room, an irradiation source, and a system to transport the food from a storage area into the irradiation room, past the radiation source at a controlled rate, and out of the irradiation room to a separate processed-storage area. These facilities are capital-intensive and cost several millions of dollars. Because of the large capital cost, food irradiators are often multipurpose facilities operated as contract facilities. When different products are processed by a number of users at a contract facility, a technical agreement must be in place to separate the responsibilities of the operator of the irradiation facility and the food supplier.

For a typical product, the product flow is as follows:

- The food to be processed is delivered to the irradiation site on pallets.
- The pallets are moved to the nonirradiated storage area, where the environmental conditions required for the product are maintained; for example, the temperature and humidity are maintained at specified values for refrigerated or frozen food.
- The pallets are brought to the irradiator loading position, where the product is loaded into the irradiation containers (e.g., carriers or totes) used to transport the product in the specified loading configuration.
- The timer setting or conveyor speed is set to expose the product to the radiation source for the time required to receive the required dose, based on dosimetry data obtained during the performance qualification of the product.
- The product is transported past the radiation source at the specified speed to obtain the desired dose.
- The irradiation containers are moved to the unloading position, where the product is repacked onto pallets. The dosimeters used for routine monitoring of the dose are removed and read.
- When the dosimeter readings and the irradiation parameters have been reviewed and all of the process specifications have been shown to have been met, the product is released for shipment.

Product tracking from the receipt of the product at the irradiation facility until release of the product is often performed using product-tracking software. This

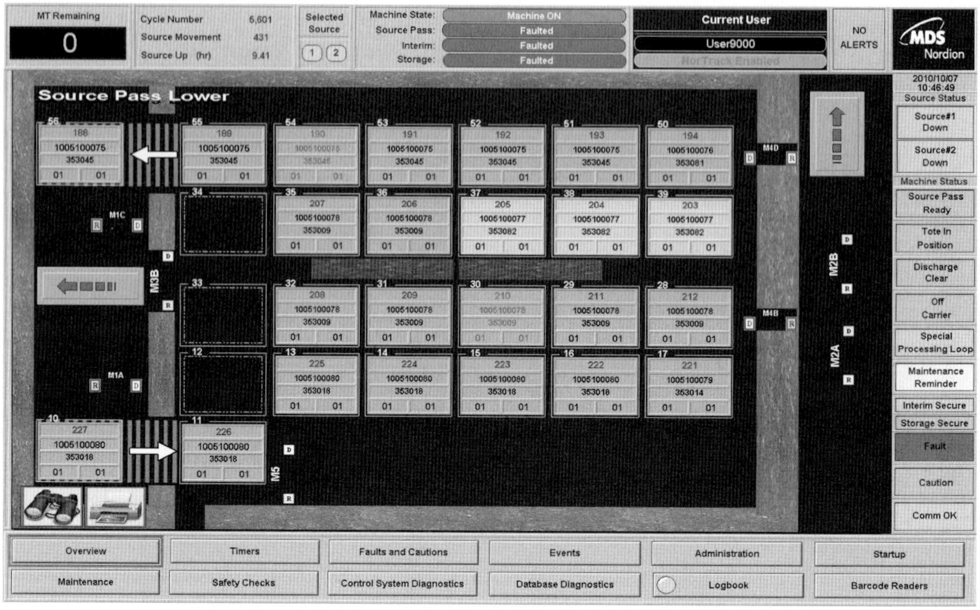

Figure 33.1 Product flow through irradiator controlled and monitored by NorTrack™ product-tracking software. (Courtesy of Nordion.)

product-tracking software displays the location of the product at each stage, including the unirradiated storage area, the irradiator and the irradiated storage area. An example of the information provided by such software during an irradiation process is shown in Figure 33.1. The software also tracks the product before and after irradiation, as shown in Figure 33.2.

Basic Theoretical Principles

The main quantity of interest in food irradiation is the absorbed dose, which is the mean energy imparted to matter by ionizing radiation. The design of food irradiators is based on exposing the product to ionizing radiation to obtain the desired absorbed dose. The desired effects occur as a result of ionization and excitation produced by the radiation.

"Ionizing radiation" is a broad term describing any radiation with sufficient energy to cause electrons to be released from atoms. The radiation may be photons from the decay of radioactive materials or X-rays, or high-energy electrons from an accelerator. For the irradiation of food, the Codex General Standard (FAO/WHO, 2003a) allows the use of the following types of ionizing radiation:

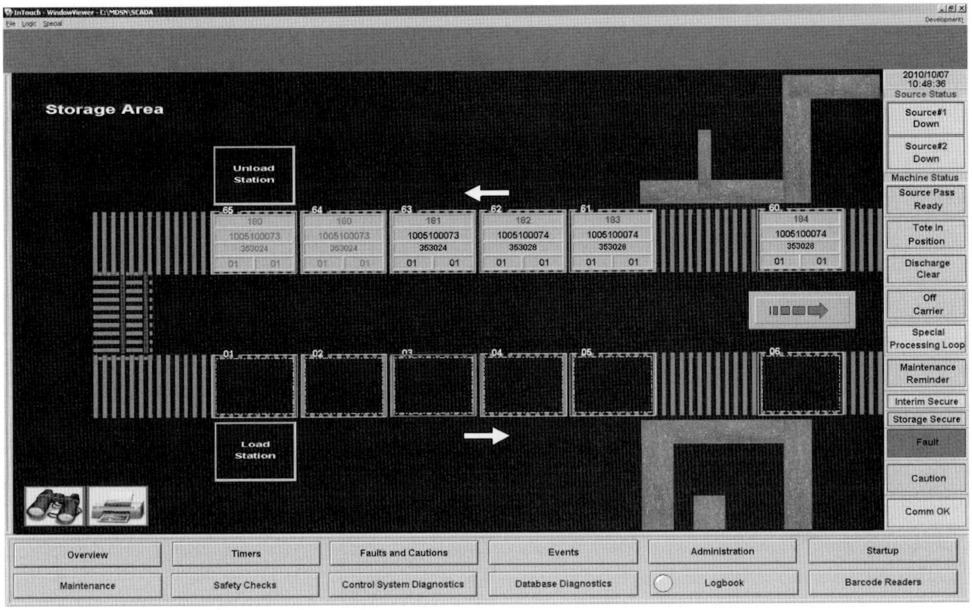

Figure 33.2 Product flow through storage area controlled and monitored by NorTrack™ product-tracking software. (Courtesy of Nordion.)

- gamma rays from the radionuclides ^{60}Co and ^{137}Cs;
- X-rays generated by machine sources operated at or below an energy level of 5 MeV;
- electrons generated by machine sources operated at or below an energy level of 10 MeV.

Because of concern regarding the solubility of the source material contained in ^{137}Cs sources, ^{60}Co is the only radionuclide presently used for food irradiation. The limits on the maximum energy for X-rays or electrons are based on concerns regarding the possibility of the formation of radioactive products in the food (IAEA, 2004). However, recent studies have indicated that X-ray energies up to 7.5 MeV will not produce significant amount of radioactive products (Grégoire *et al.*, 2003).

When food is irradiated, free radicals are formed. Microorganisms are destroyed by radicals formed in their cells reacting with their DNA, thereby breaking the chain bonds and blocking replication. Similar effects at much lower doses prevent insects from proliferating and delay the ripening of fruits, resulting in prolonged shelf-life.

Photon Interactions

Energetic photons transfer their energy to the irradiated material through interactions with electrons. For the photon energies used in food irradiation, this can occur through one of three processes – the photoelectric effect, Compton scattering and pair production. In the photoelectric effect, the entire energy of an incident photon is transferred to an orbital electron. The electron is ejected from its orbital, and the excess energy above that required to overcome the binding energy of the orbital electron appears as kinetic energy of the ejected electron. Photoelectric absorption is the predominant interaction for the absorption of low-energy photons in high-atomic-number materials. For water and other low-atomic-number materials, the photoelectric effect is small for the initial photons at the photon energies used for food irradiation.

For ^{60}Co sources, the average initial photon energy is 1.25 MeV. For gamma rays of this energy, the predominant effect when they pass through low-atomic-number materials is Compton scattering. In Compton scattering, the photon collides with an orbital electron, losing some of its energy and releasing the electron from its orbital binding. The photon is deflected and continues to travel in a new direction with reduced energy. The deflected photon makes additional collisions and has its energy further reduced until its energy is sufficiently low to be completely absorbed by the photoelectric effect.

For higher-energy photons, pair production is another method by which energy can be transferred to matter. Pair production occurs when an incident photon interacts with the field around a nucleus, resulting in the conversion of the energy of the photon into an electron–positron pair. The positron interacts with a nearby electron, releasing two X-ray photons of energy 0.511 MeV. This process can only occur when the photon energy is greater than 1.022 MeV, so the effect is small with ^{60}Co radiation and becomes significant only with higher-energy X-rays.

The total effect of the three principal types of interaction can be expressed by a mass attenuation coefficient, which describes the probability of the removal of a photon by absorption or scattering in a collision. Tables of mass attenuation coefficients for elements and compounds of general interest can be found on the NIST website (Hubbell and Seltzer, 1995). Mass attenuation coefficients for water and materials of dosimetric interest can be found in ICRU Report 80 (ICRU, 2008).

For a narrow beam of photons, such as a beam passing through a narrow slit, each interaction results in a photon being absorbed or scattered away from the target, and the dose rate is reduced exponentially with depth in the material:

$$D = D_0 \exp[-(\mu/\rho)x]$$

where D is the dose rate in the material, D_0 is the dose rate at the same location with no scattering or absorbing material, μ/ρ is the mass attenuation coefficient ($m^2 \cdot kg^{-1}$) and x is the distance in the material ($kg \cdot m^{-3}$).

For a broad beam of photons, not all of the photons are completely removed. Some of the scattered photons may be scattered back to the target volume, resulting in a larger dose than that calculated for exponential attenuation. The factor for the scattered photons is referred to as the buildup factor. Equations describing the passage of ionizing radiation through matter have been solved analytically to obtain buildup factors for simple geometries such as a point source in an infinite medium. Although the data for these simple geometries are not exact for the complex source and product geometries of actual food irradiators, point kernel calculations using exponential attenuation and buildup factors are very useful for estimating the performance of an irradiator design.

To calculate the performance of a food irradiator, published values of buildup factors for water with standard geometries, such as those calculated by Chilton *et al.* (1980), are often used. Approximate values for these buildup factors can be obtained using equations published by the American National Standards Institute and the American Nuclear Society (ANSI/ANS, 1991).

Electron Interactions

Electrons accelerated in an electron beam accelerator, and those released from atoms through Compton scattering, the photoelectric effect or pair production, interact with other electrons, resulting in a track of free electrons. As these energetic electrons pass through a material, their kinetic energy is transmitted to atomic electrons, mainly through Coulomb interactions with the fields of atoms in the material, resulting in excitation or ionization. Additional electrons are ejected from their orbits and travel at high speed, interacting with other electrons.

Because electrons have both mass and charge, the distance they travel is small and the energy loss from interactions is continuous. The average energy loss per unit path length is referred to as the linear stopping power or mass stopping power. Tables of mass stopping powers for water and materials of dosimetric interest can be found in ICRU Report 80 (ICRU, 2008). Accelerated electrons have a limited range, and measurements of the range are used to determine the effective energy of electrons. For electrons having an energy of 10 MeV, the maximum distance traveled in water is approximately 3 cm. When electrons interact with the fields of high-atomic-number atoms, another interaction occurs that involves the production of electromagnetic radiation (bremsstrahlung, or X-rays). In X-ray irradiators, bremsstrahlung is deliberately obtained by the use of a titanium or tungsten converter in an electron beam.

Design Considerations for Food Irradiators

All food irradiators must expose the product to the radiation source effectively; the radiation may be either from ^{60}Co or machine-generated. Many factors influence the design of a food irradiator and the choice of the radiation source. The major considerations include the uniformity of the distribution of the absorbed dose in the given product, efficient utilization of the radiation energy, and cost-effectiveness based on minimizing the combined capital and operating costs.

The acceptable dose range for the product has a significant effect on the design. For high-dose applications, the cost of the ^{60}Co or of the electricity for generation of electrons or X-rays from the machine source may be a major portion of the total cost. If the product must be processed within a very narrow dose range, the width of the product may need to be kept to a minimum, and doses within the acceptable limits may not be achievable with electrons. If the irradiation of a wide variety of products having different dose requirements is anticipated, the design may require additional features. The design should allow easy changing from one product to another without any product receiving doses outside the allowable limits. Limitations on the maximum mechanical speed of the conveying system may require the use of only part of the radionuclide source for low-dose applications.

Whatever type of irradiator is used, its operation must be reliable and must consistently provide a product that meets the design specifications. Controls and monitors must be provided that will stop the processing if the processing parameters are outside the design limits. Safety issues related to the product are a major concern when the process is designed to reduce the number of pathogenic organisms. All products must be processed within the required limits to ensure microbial safety. Safety systems to protect operating personnel from hazards caused by radiation must also be provided. These include appropriate shielding, radiation cell entry control and radiation-monitoring systems. The design of the irradiator should also minimize the risk from other industrial hazards.

Cobalt-60 Irradiators

The gamma rays used in radiation processing are usually obtained from large ^{60}Co radionuclide sources. In the decay of ^{60}Co, each disintegration results in the emission of two photons, one with an energy of 1.17 MeV and the other with an energy of 1.33 MeV. The SI unit of activity is the becquerel (Bq), which is one disintegration per second. However, because the becquerel is very small compared with the activities used for food irradiation, the activity of ^{60}Co is usually given in kilocuries (kCi) or megacuries (MCi). From the definition of the curie, a curie is the activity that yields 3.7×10^{10} disintegrations per second, or 1 kCi = 37 terabecquerels (TBq) and 1 MCi = 37 petabecquerels (PBq).

Cobalt-60 is obtained by irradiating cobalt metal in a nuclear reactor, often a power reactor used for the production of electricity. The total amount of ^{60}Co being used for radiation processing is several hundred megacuries. The typical activity of ^{60}Co in a single facility ranges from several hundred kilocuries to several megacuries.

Cobalt-60 decays by beta decay to stable ^{60}Ni. The emission of a beta particle is followed by the emission of two gamma rays, one at 1.17 MeV and one at 1.33 MeV. This energy is optimum for radiation-processing applications because it is sufficient to provide good penetration but low enough to avoid the induction of radioactivity. The half-life for the decay of cobalt-60 is approximately 5.27 years, resulting in a decrease in activity by about 1.1% per month. To compensate for this decrease in

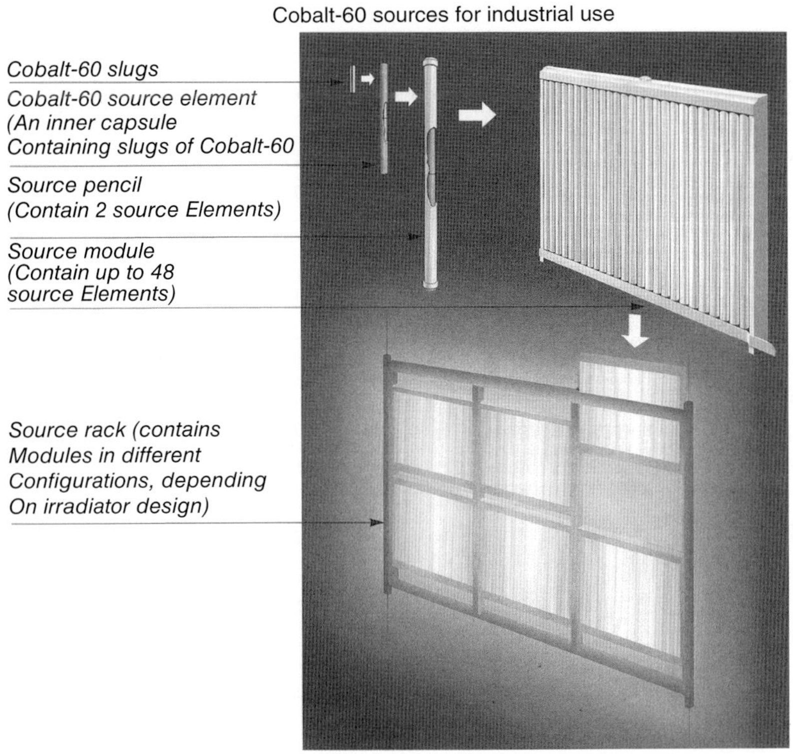

Figure 33.3 Schematic illustration of a typical source rack composed of source modules, source pencils, inner capsules and cobalt slugs. (Courtesy of Nordion.)

activity, the irradiator timer controls can be adjusted monthly. The decay rate is sufficiently low to allow replenishment of the source to be performed annually or at longer increments. During source replenishment, additional activity may also be added to take increases in product throughput into account.

The ^{60}Co source elements are usually in the form of cylindrical pencils. A number of source pencils may be installed into individual modules to simplify the handling of the source. The modules are then distributed in a rack to form the desired source array. For panoramic irradiators with sources stored in a water-filled pool, the source handling is usually performed underwater. Figure 33.3 shows source elements, source modules and a typical rectangular source rack. The size of the source rack should be sufficient to allow the addition of source pencils without requiring the return of source pencils until the end of their working life. Generally, a source rack is designed to allow the addition of new source pencils without removing sources for approximately 20 years.

Since the radiation from a ^{60}Co source is emitted isotropically, the product must surround the source for efficient utilization of the radiation. In many ^{60}Co irradiator designs, the product is moved in parallel rows on both sides of a vertical rectangular source rack. The product may move at a controlled speed or may spend specified time increments in different static locations (shuffle dwell). The drive mechanism for moving the product may be pneumatic, hydraulic or electric. Any components such as wiring, switches and motors within the radiation room will receive a high radiation dose and must be made of radiation-resistant materials.

The product can be loaded into individual metal containers (totes) and transported on conveyors, or it can be loaded into hanging containers (carriers) for transport past the source. The product can be transported past the source rack at only one level (in an overlapping-source arrangement) or past the source at two or more levels, with the source effectively buried in the product (in an overlapping-product arrangement). For high-dose applications, the cost of the source is a major portion of the processing cost, so the design must utilize the radiation source efficiently. For high radiation utilization efficiency, it may be necessary to have an overlapping product and have the product move in several passes (in a multipass arrangement) on each side of the source rack.

Generally, the narrower the dimensions of the product in the direction perpendicular to the source rack, the more uniform the dose is, and the more rows of product, the better the efficiency. However, the distance from the source to the product and the amount and location of metal in the containers, conveyors and supporting structure also affect the throughput and the dose uniformity. For larger product volumes, such as entire pallet loads, it may be difficult to obtain the required dose uniformity, especially for higher-density products. A different source and product geometry may be used to allow the product to be irradiated from four sides to improve dose uniformity.

An example of a product geometry where the product is moved in parallel rows is shown in Figure 33.4. In this product geometry, the product is moved in tall irradia-

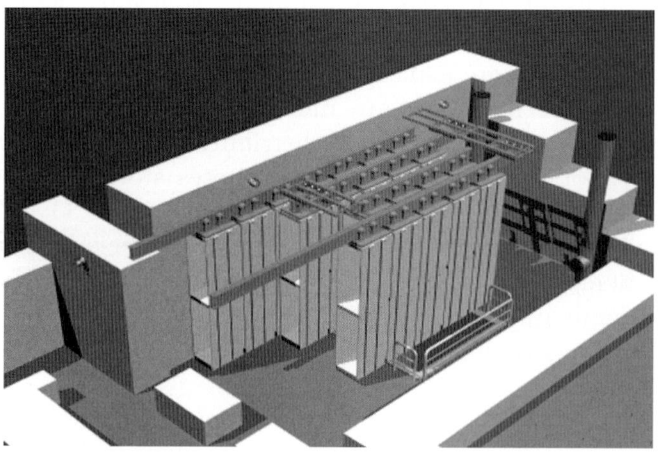

Figure 33.4 A product irradiation geometry in a ^{60}Co gamma irradiator. (Courtesy of Nordion.)

tion containers past a vertical rectangular source rack located near the center of the product volume. The product passes the source in four rows, two on each side of the source, and passes the source at two levels. The product extends past the ends of the source in both the vertical and the horizontal direction to minimize radiation loss and to obtain high radiation utilization.

Electron Beam Irradiators

Modern accelerators are available with power levels up to several hundred kilowatts. A beam power of 14.8 kW, equivalent to the power from a 1 MCi ^{60}Co source, is sufficient for many low- and medium-dose applications. The actual beam power required to process the same throughput as with a ^{60}Co source with the same power output depends on the size, shape and density of the product. For low-density, uniform products, the throughput for an electron beam irradiator can be greater than that for a ^{60}Co irradiator with the same power. However, for high-density, heterogeneous products, the penetration of the electrons may not be sufficient to allow direct use of electrons.

Electron beams suitable for food irradiation can be obtained from several different types of accelerators. In direct-current accelerators, electrons are accelerated by a constant voltage to give a continuous beam of electrons with energies up to 5 MeV (Cleland *et al.*, 1993; Salimov *et al.*, 2000; Galloway *et al.*, 2004b). In linear accelerators, a pulsed electron beam is obtained by acceleration of electrons using a high-frequency electromagnetic field (Keraron and Santos, 1990; Kerluke and McKeown, 1993). In the Rhodotron™ accelerator shown in Figure 33.5, electrons are accelerated in a complex pattern, resulting in a compact unit with high power capability (Jongen

Figure 33.5 Rhodotron accelerator. (Courtesy of IBA.)

et al., 1994; Defrise *et al.*, 1995; Abs *et al.*, 2004). The maximum energy of the electrons from linear accelerators and Rhodotrons™ is limited to 10 MeV to ensure that no significant radioactivity is produced in the food.

For electron beam irradiators, the product is usually transported at a controlled speed past a scanned beam of electrons. Because the electrons are stopped by a short depth of product, the irradiation zone is usually small and the dose is delivered in a very short time. Measurements of the variation of absorbed dose with depth in a homogeneous material are used to estimate electron energies and to determine the thickness of product that can be irradiated. At 10 MeV, the range of electrons in water is approximately 3 cm, and irradiation from two sides is required for thicker products. For two-sided irradiation, the maximum thickness of water that can be irradiated is approximately 8 cm.

X-ray Irradiators

The electron beam from an accelerator can be converted into X-rays (bremsstrahlung) by being made to strike a high-atomic-number absorber (Seltzer *et al.*, 1983). The electrons are stopped in a converter, and the kinetic energy of the decelerated electrons emerges in the form of bremsstrahlung. In contrast to radionuclide sources, which emit nearly monoenergetic photons, bremsstrahlung sources emit a broadband spectrum consisting of photons of numerous energies.

The design of an X-ray irradiator is similar to that of a ^{60}Co irradiator. However, when electrons are directed onto a converter target to obtain bremsstrahlung, most of the photons travel in the forward direction from the target, and the product needs to be located on only one side of the target. In the case of a product that is moved past the beam, the product must be turned and moved past the beam a second time to receive a uniform dose.

High-power accelerators with power levels up to 700 kW are now available (Abs *et al.*, 2004). If the conversion efficiency from electron power to X-ray power is 10% and the radiation utilization efficiency of an X-ray irradiator is approximately the same as that for a ^{60}Co irradiator, a 700 kW X-ray facility can provide a throughput equivalent to that of a gamma irradiator containing 4.7 MCi of ^{60}Co. Since most of the energy is deposited as heat in the converter target, care must be taken to ensure that sufficient cooling is provided to adequately remove the heat generated in the target.

Rules of Thumb

If the product throughput is known, an initial estimate of the strength of the source required can be obtained by making an assumption about the efficiency of the utilization of radiation. For most commercial systems in use today, the aim is to obtain an efficiency in the range of 20–50%. Higher-efficiency irradiators require less source activity or less electrical power consumption, but the product movement may be more complicated. The optimum design must take both capital costs and operating costs into consideration.

An initial estimate of the required source activity for a ^{60}Co gamma irradiator for a desired product throughput can be obtained by converting the source activity to the power emitted and assuming an approximate efficiency for the irradiator. For a ^{60}Co source, each disintegration results in the emission of two photons, one with an energy of 1.17 MeV and the other with an energy of 1.33 MeV. For a ^{60}Co industrial irradiation facility with a source of strength 1 MCi (3.7×10^{16} Bq), the total power emitted in the form of gamma rays is given by the following:

$$\text{Energy emitted per second from 1 MCi} = 2.5 \times 3.7 \times 10^{16}\ \text{MeV} \cdot \text{s}^{-1}$$

$$1\ \text{MeV} = 1.60206 \times 10^{-13}\ \text{J}$$

$$\text{Energy emitted per second from 1 MCi} = 14.8 \times 10^{3}\ \text{J} \cdot \text{s}^{-1}$$

$$1\ \text{W} = 1\ \text{J} \cdot \text{s}^{-1}$$

Therefore, the energy per second, or power, from 1 MCi of ^{60}Co is equal to 14.8 kW.

If all of this gamma power could be usefully absorbed in the product, the throughput would be given by the following relationship:

$$\text{Power from 1 MCi } ^{60}\text{Co source} = 1.48 \times 10^{4}\ \text{J} \cdot \text{s}^{-1} = 5.328 \times 10^{7}\ \text{J} \cdot \text{h}^{-1}$$

For an absorbed dose of 1 kGy, the throughput per hour can be calculated as follows:

$$\text{Throughput from 1 MCi (100\% efficiency)} = 5.328 \times 10^{7}\ \text{J} \cdot \text{h}^{-1} \div 1000\ \text{J} \cdot \text{kg}^{-1}$$

$$1\ \text{kGy} = 1000\ \text{J} \cdot \text{kg}^{-1}$$

$$\text{Throughput from 1 MCi (100\% efficiency)} = 5.328 \times 10^{4}\ \text{kg} \cdot \text{h}^{-1}$$

For a first approximation, assume an efficiency of 30% based on the minimum dose. (Some parts of the product will receive higher doses, so the percentage of the emitted radiation absorbed will be greater than 30%. The percentage of the emitted radiation absorbed will depend on the dose uniformity achievable for the irradiation geometry used.)

$$\text{Throughput from 1 MCi} = 1.60 \times 10^{4}\ \text{kg} \cdot \text{h}^{-1} \text{ or } 16 \text{ tonnes} \cdot \text{h}^{-1}$$
$$\text{for a minimum absorbed dose of 1 kGy}$$

In a ^{60}Co irradiator, the source emits radiation continuously, so the most economic method of operation is usually continuous operation 24 hours per day, 7 days per week.

If the product is provided at a uniform rate, the irradiator can be operated for approximately 8000 hours per year. The annual throughput for a 1 MCi ^{60}Co source is therefore approximately 1.28×10^8 kg per year, or 128 000 tonnes per year.

For an electron beam irradiator with an efficiency of 30%, the same throughput can be obtained from an accelerator having a beam power of 14.8 kW.

For an X-ray irradiator, the throughput will depend on the conversion efficiency from electrons to X-rays. If the conversion efficiency is 10% and the irradiator efficiency is 30%, the electron accelerator providing the electrons to the target would need to have a power of about 148 kW to provide the same throughput as a 1 MCi ^{60}Co source.

Generally, when a food producer needs to determine the feasibility of irradiating a product, the product throughput and the dose requirements are known. Simple calculations similar to those shown above can then be performed to determine the feasibility of building an irradiator facility for processing the product. These calculations should only be used to obtain an initial estimate of the economics, as many other factors must be taken into account before a full economic picture can be obtained.

Simple Equations

For accurate calculations of the performance of a food irradiator, detailed modeling must be performed to determine the effects of all components affecting the performance. However, calculations using simple equations can provide useful information about the performance of an irradiator and the effect of variations of the size and density of the product.

Simple calculations of the dose rate in products located at different locations in a ^{60}Co irradiator can be performed by approximating the source by a number of point sources. The intensity of the radiation from each point source depends on the distance from the source, and varies inversely with the square of the distance. The dose rate D_2 at a distance d_2 is related to the dose rate D_1 at a distance d_1 by the equation

$$D_2 = D_1 \times (d_1)^2 / (d_2)^2$$

For a radioactive source, the output of the source is related to the activity by a value called the air kerma-rate constant, which gives the absorbed dose rate in air at one meter from an unattenuated point source. The air kerma-rate constant for ^{60}Co is approximately $1.1 \times 10^2\,\mathrm{Gy \cdot h^{-1}}$ per curie. To calculate the performance of a food irradiator, we need to know the absorbed dose rate in water. To convert from absorbed dose rate in air to absorbed dose rate in water, a factor of approximately 1.1 must be applied. The dose rate in water at 1 m from a 1 Ci ^{60}Co source is approximately 1.2×10^2 Gy/h.

A first approximation to the unattenuated dose rate at a location in a food irradiator can be obtained by summing the contributions from each point source, using the absorbed dose rate at 1 m and applying an inverse square correction for distance. The reduction in the dose rate by attenuation in the product and other materials can then be calculated from the equation

$$D = D_0 B \exp[-(\mu/\rho) \cdot x]$$

where D is the dose rate in the material, D_0 is the dose rate at the same location with no scattering or absorbing material, B is the buildup factor, μ/ρ is the mass attenuation coefficient ($m^2 \cdot g^{-1}$ in SI units, but usually tabulated in units of cm^2/g) and x is the distance in the material ($kg \cdot m^{-3}$ in SI units, but usually expressed in g/cm^2).

Values of the mass attenuation coefficient for water can be found in ICRU Report 80 (ICRU, 2008). Published values for the buildup factor B for water can be found in the ANSI/ANS standard (ANSI/ANS, 1991). In calculations of the attenuation for a food product, the product can be assumed to have the same attenuation properties as water. For structural materials, a first approximation may be made by assuming an equivalent thickness of water.

This simple calculation is based on the point kernel method, which is used extensively by irradiator designers. More sophisticated methods for accounting for heterogeneous materials can be used for these calculations, but the results will still be only approximations. The published values for the buildup factor B that can be found in the literature are for ideal conditions, such as a point source in an infinite medium. The reduction in the buildup factor near the edges of a product volume will result in some error. Detailed calculations for heterogeneous materials can only be performed using Monte Carlo methods.

Tables of throughput and uniformity as a function of density are often provided in brochures produced by commercial irradiator suppliers. These published values can often be used to obtain preliminary estimates to determine the feasibility of an irradiator design. A typical table showing the throughput and dose uniformity for a ^{60}Co irradiator in which products are loaded into aluminum totes for irradiation is shown in Table 33.2.

Suppose that a mango producer desires to export mangoes to a country where the required minimum dose is 400 Gy and the maximum dose cannot exceed 1000 Gy.

Table 33.2 Performance of a ^{60}Co tote irradiator for fruits and vegetables.

Product packing density ($kg \cdot m^{-3}$)	Throughput per hour per 100 kCi ^{60}Co, 1 kGy minimum dose			Dose uniformity ratio
	Totes	m^3	kg	
200	28.1	6.33	1270	1.42
250	26.3	5.91	1480	1.46
300	24.4	5.49	1650	1.51
350	22.6	5.09	1780	1.57
400	20.9	4.70	1881	1.64

Table 33.3 Typical operating parameters for a ^{60}Co tote irradiator processing mangoes for phytosanitary treatment.

Operating parameter	Typical value
Initial source activity	125 kCi
Timer setting for 400 Gy	51 s
Timer setting used	55 s
Estimated minimum dose	430 Gy (required minimum dose = 400 Gy)
Estimated maximum dose	680 Gy (maximum allowable dose = 1000 Gy)
Hourly throughput	5200 kg
Days required for 10 000 tonnes	80

The mangoes are packed into cases that are 50 cm long by 25 cm wide and 15 cm high. Each case weighs 6.5 kg. Based on the throughputs shown in Table 43.2 and an anticipated throughput of 10 000 tonnes in 3 months (approximately 110 000 kg·day^{-1} or 4600 kg·h^{-1}), the producer has estimated that it can achieve the desired throughput with a source of 110 kCi. To account for source decay and variations in daily throughput, the source activity installed in the irradiator is 125 kCi. The operating parameters for this source activity are shown in Table 33.3.

Food irradiation is a process that imparts little energy to the food being processed. Although the process occurs through the interaction of high-energy photons or electrons, the total amount of energy imparted to the food is small. For a dose of 1 kGy, the energy imparted is 1000 J·kg^{-1}. Since the amount of heat needed to raise the temperature of 1 kg of water by 1 °C is 4184 J and most food products have a specific heat capacity similar to that of water, a dose of 1 kGy will raise the temperature of the food by about 0.24 °C.

The power from a 1 kCi ^{60}Co source is approximately 15.4 W (0.3 W from the beta decay followed by 14.8 W from the gamma photons), so the total power from a source containing 125 kCi of ^{60}Co is approximately 2000 W. When the source is stored in a water-filled pool, all of this heat is absorbed in the water. When the source is in the irradiation position, some of the energy is absorbed by the product and the remaining energy is absorbed by the source pencils, the source rack, the irradiation containers, the support structures and the shield. Most of the heat is removed from the radiation room by the ventilation system used to reduce the amount of ozone in the room arising from the radiolysis of air. For large quantities of ^{60}Co, a heat exchanger may be installed in the pool to remove the heat absorbed in the water.

If an electron beam accelerator were used for the irradiation and the same radiation utilization efficiency could be obtained, the same total amount of heat would be absorbed by the product and by materials in the area near the electron beam. However, the mains power required for the generation of the accelerated beam of electrons would be two to five times the beam power, requiring an electrical power input of 4–10 kW. If the irradiation were performed by an X-ray irradiator, since the conversion efficiency from electrons to X-rays is on the order of 10%, the mains power required would be approximately 40–100 kW. In addition, a significant amount of power would be required to cool the target.

The other power requirements for a food irradiator are small. The main requirements are for room ventilation and for the product transport mechanism.

Process Control

The control of a food irradiation process should ensure that processing is performed with full adherence to current quality requirements. The documentation should include evidence that all critical process parameters have been established and operational ranges defined. All critical steps of the irradiation process must be validated and the process must be reliably controlled. For ^{60}Co irradiators, the most critical processing parameter is the irradiation time. For machine sources, the irradiation time and the beam parameters must be controlled and monitored. Processing outside of the specified ranges of parameters may result in doses outside the acceptable range and have a detrimental effect on the quality of the food. The relationship between dosimetry data and process monitoring, including a justification of the acceptance criteria, must be documented. Clear, specified procedures describing the reporting and the actions to be taken in the case of rejection must be provided.

Dosimetry

The control of a food irradiation process is based on control of the process parameters and measurement of the absorbed dose. Dosimetry is performed at all stages of the process, beginning with operational-qualification measurements to confirm that the performance of the irradiator is in accordance with the design specifications. Dosimetry is then used in performance qualification measurements to determine the processing parameters for specific products. Dosimetry is also used for routine monitoring of the process. Dosimetry requirements for food irradiation facilities are given in ISO/ASTM Practices 51204 and 51431 (ISO/ASTM, 2004a; ISO/ASTM, 2005a). General dosimetry requirements for radiation-processing facilities are given in ISO/ASTM Practices 51702, 51608 and 51649 (ISO/ASTM, 2004b; ISO/ASTM, 2005b; ISO/ASTM, 2005c).

Calibration of Dosimetry System

The dosimetry systems used for the control of food irradiation are similar to those used in other applications, including external-beam radiation therapy. Many national standards laboratories maintain primary or secondary standards for the measurement of absorbed doses in the range required for radiation therapy. Some national standards laboratories also provide dosimetry standards for higher doses in the range required for use in food irradiation applications.

Dosimetry systems for measurement of absorbed doses in the range required for food irradiation are also used extensively in other radiation-processing applications, particularly radiation sterilization of medical products. Food irradiation and radiation sterilization are controlled processes where traceability of absorbed-dose measurements to national standards is required. Standards for the calibration and use of these dosimetry systems have been produced by ASTM International, and many of these standards are now published jointly by ISO and ASTM International as ISO/ASTM standards. ASTM E 2628 (ASTM, 2009a) provides the basis for radiation-processing dosimetry and provides references to other radiation dosimetry standards. Guidance on the calibration of dosimetry systems is given in ISO/ASTM 51261 (ISO/ASTM, 2002).

Installation Qualification Measurements

For electron beam and X-ray irradiators, measurements must be performed during the installation qualification of the irradiator to determine the beam characteristics. These may include measurement of the beam energy, beam current, scan width and uniformity of scan width. For ^{60}Co gamma irradiators, no specific installation qualification measurements need to be performed. The source activity and source geometry must be documented, taking into account the changes in the source activity due to source decay.

Operational-Qualification Dosimetry

Dosimetry to determine the radiation characteristics of the irradiator is performed during the operational qualification of the irradiator. Dose mapping is performed to determine the distribution of absorbed dose over the product volume in irradiation containers filled with a homogeneous product. For machine sources, operational-qualification measurements are also performed to determine the relationship of the beam parameters and conveyor speed to the absorbed dose.

Performance Qualification Dosimetry

Following operational-qualification dosimetry, performance qualification dosimetry is performed to determine the dose distribution in the actual products that will be processed. This dosimetry takes into account the heterogeneity introduced by the actual products and is used to determine the processing parameters to be used for the treatment of each product. During the operational-qualification and performance qualification exercises, the doses at reference monitoring locations suitable for use during routine processing are also determined.

Routine-Monitoring Dosimetry

During routine processing, dosimeters are located at one or more routine-monitoring locations. Confirmation that the measured doses are within the specified values is required before the product can be released. For food irradiation, it is important that the dosimetry systems are suitable for the conditions of use. Many of the dosimetry systems available for dose measurements in the range required for food irradiation may have responses that are affected by factors such as the temperature and dose rate (ASTM, 2009b). The calibration of the dosimetry systems must ensure that the measurements are valid for the conditions of use. For food that is processed under extreme conditions, for example processing while frozen, it may be difficult to find suitable dosimetry systems that provide accurate measurements under the processing conditions used. It may be necessary to perform dose mapping at room temperature and to isolate the dosimeter from the low temperature during the dosimetry performed for routine monitoring.

Operating Parameters

For ^{60}Co gamma irradiators, the operating parameters are determined during the operational-qualification and performance qualification measurements and adjusted for decay of the source. The operating parameters are determined for different products with different densities and different dose requirements. When the source is replenished, some of the original dose-mapping measurements are repeated to confirm the operation of the irradiator with the new source configuration.

For machine sources, the dose-mapping measurements performed during operational qualification and performance qualification are used to determine the relationship between the beam parameters (such as the beam current and the scan width), the conveyor speed and the absorbed dose. The beam and conveyor parameters can then be varied to obtain the desired dose. When critical components are replaced or when maintenance tasks are performed that may affect the output of the accelerator, the dose-mapping measurements may need to be repeated.

Software for Modeling Food Irradiators

Modeling

Simple calculations can be performed for gamma irradiators by approximating the source geometry by a number of point sources and following the gamma rays to a number of target points. The attenuation of the intervening material can be approximated using values of the mass attenuation coefficient and published buildup factors. However, for determining accurately the performance of a food irradiator design, the initial calculations done using simple equations and algorithms are usually not adequate.

Detailed modeling of both the source and the product geometries is required for accurate calculations. This can be performed using codes written by the irradiator designer or by using published software. ASTM Guide E 2232 (ASTM, 2010) describes many of the software packages for modeling the transport of radiation through product materials. A users' group, the Radiation Process Simulation and Modeling User Group (RPSMUG) (www.rpsmug.org) has been formed to assist users and to provide benchmark calculations to test the validity of models.

Although the concepts behind the calculational methods are straightforward, the modeling of the irradiator geometry, including the product containers, conveying mechanisms and structural components, is difficult and time-consuming. The use of computer-aided design software that can provide the product geometry to the calculational software is a big help in the calculations.

Many software packages are now available for following the attenuation and scattering of photons and electrons in products being transported through a radiation field. These are often general-purpose software packages designed for the calculation of the transport of charged or neutral particles and photons from various sources of ionizing radiation. The software is generally capable of performing other calculations, such as the scattering of neutrons, that are not of interest for the design of food irradiators.

Many of these software packages are useful tools for the determination of spatial dose distributions of photons emitted following the decay of ^{60}Co, distributions of energetic electrons from particle accelerators or distributions of bremsstrahlung generated by electron accelerators. Most of the commercial codes have front ends that provide a means for inputting information about the irradiator geometry.

Point Kernel Calculations

For calculations using the point kernel method, a complex source array is broken down into a number of point sources, and the effects of intervening material between the source and the target are considered using transmission data and buildup data. For a narrow beam of photons, collisions will result in the loss of photons, and the reduction in the number of photons can be calculated from mass attenuation coefficients. For a broad beam of photons, the scattered photons may undergo additional scattering and contribute to the dose. This additional dose contribution has been calculated for simple geometries and published as buildup data. Using this buildup data, dose contributions in the source geometries used in food irradiators can be estimated.

Point kernel calculations are limited in that the buildup factors are given for simple geometries, whereas information for complex geometries needs to be generated. A number of commercial point kernel codes are now available that simplify the input of data for complex geometries. However, there still remains the fundamental assumption that the effects related to complex geometries are small.

The point kernel method follows the paths of photons, taking into account objects between the photon source and the target. Some codes, such as the point kernel code QAD-CGGP, are available free of charge. Commercial software for calculating the performance of gamma irradiators using the point kernel method is also available. A list of available codes and information on how they can be obtained are given in ASTM

Guide E 2232 (ASTM, 2010). The point kernel software codes developed by companies designing irradiators are often considered to be proprietary information.

Monte Carlo Calculations

The Monte Carlo method is a simulation method that can be used for calculating the absorbed dose in a volume of interest using a statistical summary of the interactions of the radiation. A Monte Carlo calculation consists of running a large number of particle histories until some acceptable statistical uncertainty in the desired calculated quantity is reached. It uses random numbers to follow photons or electrons passing through materials. This method allows computation of dose distributions in irradiated product samples, even for complex source and product geometries. The uncertainties in Monte Carlo calculations are based on counting statistics.

The increase in the speed of computers now allows many histories to be followed in a relatively short period of time, even on a desktop computer. However, it may still be necessary to use variance reduction techniques such as biasing to reduce the estimated uncertainty or the computer run time. Simulations using the natural probabilities of physical events may require unacceptably long run times to obtain adequate statistics for rare events. The number of particles tracked and the weights given to particles may be adjusted to ensure a statistically valid sample from the probability distribution. Appropriate biasing requires a detailed knowledge of the model.

Solving particle transport problems with the Monte Carlo method is theoretically a simple task. All that is necessary is to simulate the behavior of various particles. However, in practice the modeling is not so simple. The theory is quite straightforward and is based on transport simulation, random number generation, random sampling, computational geometry, collision physics, tallies, statistics, eigenvalue calculations, variance reduction and parallel algorithms. These fundamental methods are used in all modern Monte Carlo particle transport codes, but some codes are more user-friendly than others for irradiator design applications.

A number of Monte Carlo codes, including ITS, MCNP, EGS and PENELOPE, are available from the Radiation Safety Information Computational Center (RSICC), Oak Ridge National Laboratory. Other codes are available from their developers. ASTM Guide E 2232 (ASTM, 2010) lists the available codes and provides information on how they can be obtained. Many codes are available free of charge, but it is recommended that users of the codes take courses from the code developers to determine their optimum use. The subjects covered in these courses include basic geometry and advanced geometry, source definitions, tallies, data, variance reduction, statistical analysis and plotting of geometry.

An example of a Monte Carlo code used in the design of food irradiators is the general-purpose *N*-particle code MCNP ("Monte Carlo *N*-Particle") (Breismeister, 2000), which can be used for neutron, photon, electron or coupled neutron/photon/electron transport. This code treats an arbitrary three-dimensional configuration of materials in bounded geometric cells. MCNP is very versatile and can be used for many purposes, including radiation protection and dosimetry, radiation shielding, radiography, medical physics, nuclear criticality safety, detector design and analysis, nuclear oil well logging, accelerator target design, fission and fusion reactor design, decontamination, and decommissioning. For the design of food irradiators, only the transport of photons and electrons is of interest.

The important standard features that make MCNP very versatile and easy to use include a powerful general source code, a rich collection of variance reduction techniques, a flexible tally structure, an extensive collection of cross section data and a powerful visual editor. The MCNP Visual Editor is a graphical user interface with powerful display capabilities, including the ability to display multiple cross-sectional views of the geometry, with optional displays of the geometry in 3D. Additional capabilities include plotting of the source and the display of particle tracks during random walks. The Visual Editor also includes geometry creation capabilities that allow the user to create MCNP geometries directly from the plot window using a mouse. These capabilities allow the MCNP programmer the tools to quickly create complex geometries and display important features of the transport process.

Cleaning and Sanitation Methods

There are no special requirements for cleaning and sanitation for food irradiation performed using an electron beam or X-ray irradiator. The normal cleaning procedures related to the handling of the specific foods being processed need to be followed.

For food irradiation performed using a ^{60}Co gamma irradiator, precautions must be taken to ensure that there is no spread of radioactive contamination in the event of an incident where capsules containing the source material may be damaged, resulting in the escape of radioactive material from the source. Procedures for monitoring for radioactive contamination are required as conditions for the licensing of a ^{60}Co gamma irradiator. These procedures may include continuous monitoring of the water in the source storage pool or routine wipe testing of the irradiation containers and areas near the source rack. Prior to cleaning of the irradiation containers or the floor of the irradiation room, there should be confirmation that there is no radioactive contamination. All workers at the irradiation facility should receive training in radiation safety and in the precautions to be taken to prevent the spread of radioactive contamination.

The cleaning and sanitation procedures for each food irradiation facility should be clearly stated in a Standard Operating Procedure (SOP). This SOP should include any special requirements related to the food irradiation process. For example, for the irradiation of meat, the cleaning procedure should ensure that bacteria from the nonirradiated product cannot be transferred to the processed product. The SOP should include the method and frequency of cleaning irradiation containers and the method of handling damaged packages. Each person working at the facility should be encouraged to maintain a high degree of personal cleanliness and should be given training in personal hygiene and facility sanitation. This training should be repeated at regular intervals.

Capital and Operating Costs for Different Sizes of Equipment

Food irradiation is a capital- and cost-intensive process. Not only do the costs include the initial cost of the source and the conveying mechanism, but the process must also take place within a shielded room with concrete walls approximately 1.5–2.0 m thick. Capital costs need to be balanced against operational costs. In some situations, it may be worthwhile to install a complex transport system with high source utilization efficiency and a high degree of automation. In other situations, a simple transport system with reduced source utilization efficiency may provide improved economics. The labor costs are highly dependent on the labor rates in the area. For products that are seasonal, there may be an advantage in designing the irradiator to allow the processing of other products during the off-season.

In estimating the cost of the irradiation of a food product, all components contributing to the cost must be considered. These include:

- estimated annual throughput – initial throughput and expected throughput at capacity;
- seasonal variation in throughput;
- dose requirements – the minimum dose required to achieve the specified benefit and the maximum acceptable dose;
- the source activity or machine power required to process the initial throughput and throughput at capacity;
- replenishment costs for ^{60}Co sources or maintenance and electricity costs for machine sources;
- land and construction costs for the irradiation facility, including costs of the radiation shield, offices, laboratories and warehouse;
- labor rates and salaries;
- the cost of the product transport system, including the load/unload system;
- the control system, and safety interlocks and monitors;
- auxiliary equipment, including ventilation systems and monitors;
- utilities, taxes and insurance;
- dosimetry and other quality assurance costs.

Treatment costs vary as a function of dose and facility usage. Many studies have been performed on the relative costs of ^{60}Co irradiators, electron beam irradiators and X-ray irradiators. These studies indicate general conditions where one particular type of source may provide lower processing costs. For example, electron beam irradiators have been found to have economic advantages over gamma irradiators when the product throughput is large, the thickness of the product being treated is small, and continuous treatment is possible by integrating the irradiator into the production line. However, before one decides on the type of irradiator to be used, each situation should be assessed individually.

In a study performed in 1990, the average costs per kilogram of irradiating selected foods were found to be similar for the electron accelerator and ^{60}Co irradiators analyzed in the study (Morrison, 1990). The initial investment costs generally varied by US$1 million. The costs of irradiation treatment ranged from 1 to 15 US cents per kilogram for the foods and annual volumes examined, with larger volumes having lower treatment costs. Cobalt-60 was less expensive than electrons when the annual volume was below 23 million kilograms. For radiation source requirements above the equivalent of about 1 MCi of ^{60}Co, electrons became more economical. The largest differences in costs occurred with papaya irradiators, where using X-rays to penetrate the fruit was more expensive than using ^{60}Co.

Summary and Future Needs

The irradiation of foods has the potential to be a major processing method to meet the various needs being faced by the food industry. However, the future demand for food irradiators will depend on what applications the food producers and consumers feel are needed. A sense of urgency caused by new concerns regarding pathogens in foods or new directives to replace chemical fumigants may result in an increased desire from the food industry and consumer groups for irradiated foods.

Safety of food supplies is a major concern, but consumers need to be confident that the elimination of one safety concern is not replaced by another. The safety of irradiated foods continues to be an issue that is difficult to overcome. Many authorities have reviewed the experimental evidence obtained by researchers and have concluded that irradiated foods do not pose significant health risks. However, there still remains some doubt regarding some of the products that are formed during irradiation, and continued research is required to address these concerns.

The optimum doses for different applications still need to be determined. When changes to the taste or smell limit the maximum dose that can be applied, the product must still be shown to meet sanitary and phytosanitary requirements. The dose required to meet the technical requirements may be close to the maximum acceptable dose, so that the product may need to receive doses within a very narrow range. This will affect the source type and the irradiator design that can be used.

With the large number of foods and different processing conditions required, it is difficult to obtain general data suitable for all foods. Clearances for each application are required from regulatory agencies, but regulatory agencies are reluctant to put a high priority on reviewing submitted data if there is no pressing need from the food processing industry. When there is no current market for an irradiated food, there is little incentive for food processors to spend money and effort on research.

If there is a situation where an immediate need arises for the irradiation of a significant amount of a product, the present irradiation facilities are limited in the amount of product that can be processed. The throughput of gamma facilities can be increased to the maximum capacity of the facility by the addition of ^{60}Co sources if adequate supplies are available. However, the producers of ^{60}Co are reluctant to produce more sources than the amount required from projected market demands.

Because of the requirements of regulatory authorities and local bodies regarding the licensing of new gamma irradiation facilities, the time from the planning to the commissioning of a new gamma facility is considerable. In the future, there must be good methods for projecting market demands to allow the timely production of adequate quantities of ^{60}Co and the planning of the construction of new facilities. For ^{60}Co sources, issues related to the transport and disposal of radioactive materials must also be addressed.

The construction of new electron beam irradiators has fewer barriers, but the penetration of the electrons may limit their application. Electron beam accelerators can be equipped with a converter target to generate X-rays, but the low conversion ratio for the generation of X-rays results in increased costs.

Examples of Food Irradiators

To help readers understand the factors to be considered during the design of food irradiators, some examples of patented designs claimed to offer advantages over standard designs are given in this section. The following irradiators are not necessarily designed exclusively for food irradiation, and may instead have been designed for general radiation-processing applications. The designs were patented to protect the claims of improved performance.

The Palletron™ Irradiator (Stichelbaut *et al.*, 2004; Jongen *et al.*, 2004) uses a combination of rotation and the movement of shielding material to optimize the dose distribution for a product irradiated by a bremsstrahlung source. The dose uniformity ratio in a Palletron™ is better than 1.5 in palletized products for a wide range of product densities. This uniform dose is obtained by the use of a turntable in front of the X-ray beam to allow irradiation of the product from all sides. A collimator is used to shape the beam and prevent overirradiation of portions of the product.

The Quadura™ (Beers, 2006) is a ^{60}Co irradiator that also provides uniform doses to full pallets of products. Extensive experience in the development of irradiator designs for actual applications and the MCNP modeling software were used to address dose delivery issues. With this design, the delivery of tight dose uniformity to a full pallet of food product is possible.

A different concept is used in the Brevion™ (Brinston and Levesque, 2004). In this design, the entire product-conveying system is attached to an end shielding wall. When the irradiation is complete, the shielding wall is moved away, exposing the irradiated product for transfer to the unloading area. New unirradiated product is then introduced.

Other irradiators with unique features have been described in papers by Tomita and Sugimoto (1977), Beers (1990), Curzio and Croci (1990), Keraron and Santos (1990), McKinnon and Beers (1993), Möhlmann (1993), Defrise *et al.* (1995), and Galloway *et al.* (2004a).

References

Abs, M., Jongen, Y., Poncelet, E. and Bol, J.L. (2004) The IBA Rhodotron TT1000: a very high power e-beam accelerator. *Radiation Physics and Chemistry* 71: 287–290.

ANSI/ANS (1991) *Gamma Ray Attenuation Coefficients and Buildup Factors for Engineering Materials*, ANSI/ANS-6.4.3-1991. American Nuclear Society, LaGrange Park, IL.

ASTM (2009a) *Practice for Dosimetry in Radiation Processing*, ASTM E 2628. ASTM International, West Conshohocken, PA.

ASTM (2009b) *Guide for Performance Characterization of Dosimeters and Dosimetry Systems for Use in Radiation Processing*, ASTM E 2701. ASTM International, West Conshohocken, PA.

ASTM (2010) *Guide for Selection and Use of Mathematical Methods for Calculating Absorbed Dose in Radiation Processing Applications*, ASTM E 2232. ASTM International, West Conshohocken, PA.

Beers, E.W. (1990) Innovations in irradiator design. *Radiation Physics and Chemistry* 35: 539–546.

Beers, E. (2006) *Innovation of Design and Supply in the Gamma Process Industry*. MDS Nordion, Ottawa. Available from International Irradiation Association, www.iiaglobal.org.

Breismeister, J.F. (ed.) (2000) *MCNP: A General Monte Carlo N-Particle Transport Code*, Version 4C, LA-13709-M. Los Alamos National Laboratory, Los Alamos, NM.

Brinston, R.M. and Levesque, D.G. (2004) Irradiation on a new scale: introducing Brevion. In: *Emerging Applications of Radiation Processing*, Proceedings of a technical meeting held in Vienna, April 28–30, 2003. International Atomic Energy Agency, Vienna, IAEA-TECDOC-1386, pp. 162–165.

Chilton, A.B., Eisenhauer, C.M. and Simmons, G.L. (1980) Photon point source factors for air, water, and iron. *Nuclear Science and Engineering* 73: 97–107.

Cleland, M.R., Thomson, C.C., Saito, H., Lisanti, T.F., Burgess, R.G., Malone, H.F., Loby, R.J. and Galloway, R.A. (1993) New high-current Dynamitron accelerators for electron-beam processing. *Nuclear Instruments and Methods in Physics Research* 79: 861–864.

Curzio, O.A. and Croci, C.A. (1990) A ^{60}Co multipurpose radiation processing facility at Bahia Blanca, Argentina. *Radiation Physics and Chemistry* 35: 590–594.

Defrise, D., Abs, M., Genin, F. and Jongen, Y. (1995) Technical status of the first industrial unit of the 10 MeV, 100 kW Rhodotron. *Radiation Physics and Chemistry* 46: 473–476.

FAO/IAEA (2010a) *Food Irradiation Clearances Database*. Food and Agriculture Organization/International Atomic Energy Agency. http://nucleus.iaea.org/FICDB/DatabaseHome.html. Accessed 2010.

FAO/IAEA (2010b) *Food Irradiation Facilities Database*. Food and Agriculture Organization/International Atomic Energy Agency. http://nucleus.iaea.org/FIFDB/DatabaseHome.html. Accessed 2010.

FAO/WHO (2003a) *Codex General Standard for Irradiated Foods*, CODEX STAN 106-1983, Rev.-2003, Codex Alimentarius Commission. Food and Agriculture Organization and World Health Organization, Rome.

FAO/WHO (2003b) *Recommended Code of Practice for the Operation of Radiation Facilities Used for the Treatment of Foods*, CAC/RCP 19-1979, Rev.-2003, Codex Alimentarius Commission. Food and Agriculture Organization and World Health Organization, Rome.

FAO/WHO (2010) *General Standard for the Labelling of Prepackaged Foods*, CODEX STAN 1-1985, Amended 2010, Codex Alimentarius Commission. Food and Agriculture Organization and World Health Organization, Rome.

Frenzen, P.D., DeBess, E.E., Hechemy K.E., Kassenborg, H., Kennedy, M., McCombs, K., McNees, A. and Foodnet Working Group (2001) Consumer acceptance of irradiated meat and poultry in the United States. *Journal of Food Protection* 64 (12): 2020–2026.

Galloway, R.A., DeNeuter, S., Lisanti, T.F. and Cleland, M.R. (2004a) The new IBA self-shielded Dynamitron accelerator for industrial applications. *Radiation Physics and Chemistry* 71: 283–285.

Galloway, R.A., Lisanti, T.F. and Cleland, M.R. (2004b) A new 5 MeV–300 kW Dynamitron for radiation processing. *Radiation Physics and Chemistry* 71: 551–553.

Grégoire, O., Cleland, M.R., Mittendorfer, J., Dababneh, S., Ehlermann, D.A.E., Fan, X., Käppeler, F., Logar, J., Meissner, J., Mullier, B., Stichelbaut, F. and Thayer, D.W. (2003) Radiological safety of food irradiation with high energy X-rays: theoretical expectations and experimental evidence. *Radiation Physics and Chemistry* 67: 169–183.

Hubbell, J.H. and Seltzer, S.M. (1995) *Tables of X-ray Mass Attenuation Coefficients and Mass Energy-Absorption Coefficients from 1 keV to 20 MeV for Elements Z = 1 to 92 and 48 Additional Substances of Dosimetric Interest*, Publication NISTIR 95-5632. National Institute of Standards and Technology, Gaithersburg, MD, http://www.nist.gov/pml/data/xraycoef/index.cfm.

IAEA (2002) *Natural and Induced Radioactivity in Food*, IAEA-TEC DOC-1287. International Atomic Energy Agency, Vienna.

ICRU (2008) *Dosimetry Systems for Use in Radiation Processing*, ICRU Report 80. International Commission on Radiation Units and Measurements, Bethesda, MD.

ISO/ASTM (2002) *Guide for Selection and Calibration of Dosimetry Systems for Radiation Processing*, ISO/ASTM 51261. ASTM International, West Conshohocken, PA.

ISO/ASTM (2004a) *Practice for Dosimetry in Gamma Irradiation Facilities for Food Processing*, ISO/ASTM 51204. ASTM International, West Conshohocken, PA.

ISO/ASTM (2004b) *Practice for Dosimetry in Gamma Irradiation Facilities for Radiation Processing*, ISO/ASTM 51702. ASTM International, West Conshohocken, PA.

ISO/ASTM (2005a) *Practice for Dosimetry in Electron Beam and X-ray (Bremsstrahlung) Irradiation Facilities for Food Processing*, ISO/ASTM 51431. ASTM International, West Conshohocken, PA.

ISO/ASTM (2005b) *Practice for Dosimetry in an X-ray (Bremsstrahlung) Facility for Radiation Processing*, ISO/ASTM 51608. ASTM International, West Conshohocken, PA.

ISO/ASTM (2005c) *Practice for Dosimetry in an Electron Beam Facility for Radiation Processing at Energies between 300 keV and 25 MeV*, ISO/ASTM 51649. ASTM International, West Conshohocken, PA.

Jongen, Y., Abs, M., Capdevila, J.M., Defrise, D., Genin, F. and Nguyen, A. (1994) The Rhodotron, a new high-energy, high power, CW electron accelerator. *Nuclear Instruments and Methods in Physics Research* 89: 60–64.

Jongen, Y., Abs, M., Bol, J.-.L, Mullier, B., Poncelet, E., Rose, G. and Stichelbaut, F. (2004) Advances in sterilization with X rays, using a very high power Rhodotron and a very low DUR pallet irradiator. In: *Emerging Applications of Radiation Processing*, Proceedings of a technical meeting held in Vienna, April 28–30, 2003. International Atomic Energy Agency, Vienna, IAEA-TECDOC-1386, pp. 44–54

Keraron, Y. and Santos, P.L. (1990) Amphytrion: example of a high capacity irradiator. *Radiation Physics and Chemistry* 35: 597–601.

Kerluke, D.R. and McKeown, J. (1993) The commercial launch of Impala™. *Radiation Physics and Chemistry* 42: 511–514.

Kume, T., Furuta, M., Todoriki, S., Uenoyama, N. and Kobayashi, Y. (2009) Status of food irradiation in the world. *Radiation Physics and Chemistry* 78: 222–226.

McKinnon, R.G. and Beers, E.W. (1993) Pallet irradiators: circumstances under which this type of system is appropriate. *Radiation Physics and Chemistry* 42: 539–542.

Möhlmann, J.H.F. (1993) New pallet irradiator for Gammaster International. *Radiation Physics and Chemistry* 42: 443–446.

Morrison, R.M. (1990) Economics of food irradiation: comparison between electron accelerators and cobalt-60. *Radiation Physics and Chemistry* 35: 673–679.

Ross, R.T. and Engeljohn, D. (2000) Food irradiation in the United States: irradiation as a physanitary treatment for fresh fruits and vegetables and for control of microorganisms in meat and poultry. *Radiation Physics and Chemistry* 57: 211–214.

Salimov, R.A., Cherepkov, V.G., Golubenko, J.I., Krianov, G.S., Korabelnikov, B.M., Kuznetsov, S.A., Kuksanov, N.K., Malinin, A.B., Nemytov, P.I., Petrov, S.E., Prudnikov, V.V., Fadeev, S.N. and Veis, M.E. (2000) D.C. high power electron accelerators of ELV-series: status, development, applications. *Radiation Physics and Chemistry* 57: 661–665.

Seltzer, S.M., Farrell, J.P. and Silvermann, J. (1983) Bremsstrahlung beams from high-power electron-accelerators for use in radiation processing. *IEEE Transactions on Nuclear Science* 30: 1629–1633.

Stichelbaut, F., Bol, J.-L., Cleland, M.R., Grégoire, O., Herer, A.S., Jongen, Y. and Mullier, B. (2004) The Palletron™: a high-dose uniformity pallet irradiator with X-rays. *Radiation Physics and Chemistry* 71: 291–295.

Tomita, L. and Sugimoto, S. (1977) A commercial gamma ray irradiation plant in Japan. *Radiation Physics and Chemistry* 9: 567–573.

WHO (1995) *Review of Data on High Dose (10–70 kGy) Irradiation of Food*, International Consultative Group on Food Irradiation, Report of Consultation in Karlsruhe, August/September 1994. World Health Organization, Geneva.

WHO (1999) *High-Dose Irradiation: Wholesomeness of Food Irradiated with Doses above 10 kGy*, WHO Technical Report Series, No. 890. World Health Organization, Geneva.

WTO (1995) *Agreement on the Application of Sanitary and Phytosanitary Measures (SPS)*. World Trade Organization, Geneva.

34

Design for High-Pressure Processing

Tatiana Koutchma

Introduction

During the last decade, consumers have come to prefer minimally processed, additive-free foods with fresh-like characteristics. A number of novel processing technologies, including high-hydrostatic-pressure (HHP) processing, have been investigated as complements or alternatives to conventional thermal technologies. HHP processing is a method of food processing where foods are subjected to elevated pressures, between 100 MPa (~990 atm) to 900 MPa (~8900 atm), to achieve microbial inactivation or alterations of food attributes such as shelf-life extension while retaining the desired qualities of fresh food. HHP technology is known for its potential to manufacture novel value-added foods with retained heat-labile nutrients, flavors and aromas packed in individual or institutional-size packages (e.g., 6 kg packages of sliced turkey breast), and as an alternative to traditional thermal pasteurization and sterilization. In addition to the added value of the product, however, when novel techniques are introduced to replace or enhance conventional processes, their economic efficiency has to be proven in terms of production costs, sustainability, energy efficiency, and new waste products or by-products. This chapter will review the potential of HHP processing as a unit operation in the food industry, and critical factors affecting the design and development of high-pressure–low-temperature processing cycles and processing cycles using high pressure combined with high temperature, depending on the application.

Handbook of Food Process Design, First Edition. Edited by Jasim Ahmed and Mohammad Shafiur Rahman.

The Commercial Market for HHP-Processed Products

The use of high pressure as a method of food processing dates back to 1899, when Hite (1899) observed that milk, treated at a pressure of 600 MPa for 1 h, maintained its sweet taste and exhibited a longer shelf-life than that of its untreated counterpart. However, it was nearly a century later before a resurgence in using high pressure as an alternative method of food processing occurred (Patterson, 2005). In the early 1990s, the first commercial application of high-pressure technology to food was observed, with the application of HHP treatment to a high-acid jam produced by the Japanese company Medi-Ya (Mertens,1995). More recently, the use of high pressure on foods has extended to meat-based products (31% of all industrial applications), vegetable products (35%), juices and beverages (12%), seafood and fish (14%), and other products (8%). The fast development of HHP technology has demonstrated that HHP treatment can be used to process both liquid foods and high-moisture-content solid foods. Currently, the commercial applications are limited to pasteurized food products, including fruit juices, ready-to-eat meals, meat and salsas (San Martin *et al.*, 2002). However, the use of high pressures combined with high temperatures to prolong the shelf-life of foods at room temperature to months or even years by inactivating spores is within reach, owing to ongoing research and development and recent regulatory approval of the pressure-assisted thermal sterilization (PATS) process by the US FDA (Matser *et al.*, 2004).

The Potential of HHP Technology as a Unit Operation

As of today, there are only a few novel unit operations in the processing of meat, poultry, seafood, dairy, fruit and vegetables that are based on HHP technology. Microbial inactivation and preservation are, apparently, the HHP applications that have made the technology known and accepted in the food industry. However, HHP equipment has the potential for other unit operations such as modification of the structure of protein-based products, meat separation, freezing and thawing. The reported examples of commercial applications that are currently available or under development are summarized in Table 34.1.

Meat Processing

Microbial intervention, primarily against pathogens, is the application of HHP that has been studied, reported and commercialized the most intensively. The Spanish meat processor Espuña is using HHP equipment for decontamination of cured, cooked ham with the aim of prolonging the shelf-life to 60 days in an undisturbed cooling chain. In the USA, Hormel Foods is using HHP processing mainly for the elimination of *Listeria monocytogenes*, since the United States Department of Agriculture (USDA) regulatory authority maintains a policy of zero tolerance. HHP processing offers

Table 34.1 Unit operations based on application of HHP technology in the food industry.

Industry	Unit operation	Product	Company
Meat	Raw-material preparation	Ham products	Nestlé
	Curing	Cured and cooked ham	Fujichiku Company, Japan
	Drying and aging	Uncooked ready-to-eat meats	Espuña, Spain
	Decontamination and preservation	Ham	Hormel Foods, USA
	Accelerated thawing	Meat-containing entrees	Maple Leaf Consumer Foods, Canada
	Tenderization	Meat-containing salads	Under development
	Marinating	Frozen pork loins	Meat Research Institute, Denmark
			Avure, USA
Dairy	Pasteurization	Milk	Fonterra, New Zealand
	Sterilization	Milk	Under development
	Shelf-life extension	Yoghurt	Fonterra, New Zealand; Hyperbaric, Spain
	Rennet or acid coagulation	Milk	Danon, France
	Fruit preparation	Yoghurt	Under development
	Modification of whey protein	Ice cream	
	Ripening	Cheddar, goat cheese	
Sea products	Meat separation	Oysters	Motivatit Seafoods, LA, USA
	Decontamination	Raw and frozen lobsters	Joey Oysters, Amite, LA, USA
			Nisbet Oyster Co., WA, USA
			Maine Lobster, Richmond, ME
			Clearwater Seafoods, Nova Scotia, Canada
Fruit and vegetables	Shelf-life extension	Guacamole, salsa	Fresherized Foods, USA
	Pasteurization	Fruit smoothies	Syros NV, Belgium
	Sterilization	Apple sauce/fruit blends	Leahy Orchard Inc.; Franklin Centre, QC, Canada
		UltiFruit®; fresh orange juices	Pernod Ricard Company, France
		Ready-to-eat meals	US Army, Natick Soldier Center
		(mashed potatoes)	
Poultry	Shelf-life extension	Poultry deli products	Maple Lodge Farms; Pillars; Santa Marie Foods Ltd., Ontario, Canada
		Oven-roasted whole bird turkey	Perdue Farms; Foster Farms; Oscar Mayer; Hormel; Tyson, USA

additional nonmicrobiological advantages and opportunities for the development of unique processes and meat products because of features of the behavior of raw proteins and other food components under pressure. For example, HHP-accelerated freezing and thawing is one seldom-mentioned area of application. The basic concept in HHP-assisted thawing is that high pressure causes phase changes, lowering the melting point of ice. In other words, HHP can "convert" ice to water without one having to increase the temperature. Limited research has been carried out on HHP-assisted curing and aging. Nestlé has patented a method of improving the water-holding capacity of ham muscles, which are HHP-treated prior to multineedle injection. A raw ham product, which was matured considerably faster by using HHP than was traditionally possible, has been marketed by the Fujichiku Company, Japan. The DMRI has studied the acceleration of brine diffusion by HHP. Relatively low pressures can be used to tenderize and marinate meats.

Milk and Dairy Products

In the dairy industry, the potential for the use of HHP technology exists not only from the standpoint of microbial and enzyme inactivation in raw milk but also for the purpose of improving the quality and yield of dairy products such as yoghurt and exploring new functional properties of traditional dairy ingredients such as casein proteins. Trujillo (2002) reported that HHP processing improved the coagulation of milk by rennet or acid without detrimental effects on important quality characteristics, such as taste, flavor, vitamin content and nutrient content. An increase in the flavor-binding affinity of whey protein concentrate after it was treated with HHP was reported by researchers at Washington State University (WSU, 2008; Lim *et al.*, 2008). The selective barosensibility of spoilage-causing and pathogenic microorganisms has been used in the postpackaging HHP processing of cultured foods, as was reported by Hyperbaric Company, Spain. Inactivation of yeasts and molds resulted in up to 3 months' conservation of a reduction in the *Lactobacillus* count and retention of bioactive components such as lactoferrin and inmunoglobulins, without alteration of their physiological properties. In addition, HHP processing is being used for the selective control of yoghurt starter cultures.

Fruit, Vegetables and Derived Products

HHP processing was introduced to consumers with the taste of all-natural refrigerated guacamole (Fresherized Foods, USA), soon followed by ripe avocado halves with a shelf-life of up to 30 days. Later, the use of HHP technology in fruit and vegetable processing expanded to salsa, prechopped onions, organic juices, fruit smoothies and apple sauce (Leahy Orchard Inc., Canada). HHP processing inactivates *Salmonella*, *E. coli* and *Listeria monocytogenes* in fruit and vegetable products. The color of fruit and vegetable products such as jams, fruit juices and purees is generally preserved if thresholds of temperature and/or pH are observed. Excellent retention of fresh-like flavors for far greater time periods than those obtained with conventional thermal treatment has also been observed under optimal storage conditions (Ludikhuyze and Hendrickx, 2001). Processors use HHP in their HACCP program to achieve the US FDA requirement of a 5-log reduction in pathogens in fresh juices. European companies presently employing this technology include UltiFruit® (for orange juice), the Pernod Ricard Company, France, and Solofruita, Italy (for fruit jams) (Norton and Da-Wen Sun, 2008).

Seafood

Clean, virtually 100% separation of meat from lobsters, oysters, clams and other fresh products can be achieved by pressure-denaturing of the specific protein that holds the meat to the shell. HHP processing allows the maximum product yield without causing any mechanical damage to the product, regardless of size. By adjusting the processing conditions, beneficial texture changes can also be created by improving the moisture retention ability of proteins, thus resulting in less water loss during storage or cooking. Additionally, HHP processing is being used as a means to accomplish nonthermal treatment for reducing the bacterial load of raw or fresh seafood products.

Poultry

Purdue Farms, Oscar Mayer, Tyson Foods and Hormel are big-name companies who are utilizing high pressure to add value to poultry offerings. Maple Lodge Farms (Ontario, Canada) was the first poultry processor in Canada to incorporate HHP technology into its operations. Maple Lodge has implemented a new method of ensuring that precooked sliced luncheon meat that has already been packaged is safer for consumers. In addition to the added safety benefits, the process also reduces the need for the company to add preservatives and other chemicals to its products, making it possible to make healthier and more natural food.

The HHP Processing Cycle

In the high-pressure vessel used for HHP processing, food items can be subjected to pressures between 100 and 900 MPa. Pressure reduces the volume of water by 10% at 300 MPa and by 15% at 600 MPa. Pressurization of compressible substances such as food products results in a temperature increase due to the work of compression related to the pressure increase. This is known as adiabatic heating and should be given consideration in the case of preservation processes. The extent of the temperature rise depends on the pressure P, the rate of pressurization, the initial temperature T_0, and the thermal and physical properties of the food sample and the packaging. The pressure, the temperature T and the time t are critical process parameters for establishing process specifications. The process temperature range can be specified from 0 °C to temperatures higher than 100 °C. In food sterilization, adiabatic heating can be used advantageously to provide heating without the presence of large thermal gradients.

Commercial processing times can range from a pulse of few seconds to over 20 min. The come-up time (CUT) to the process pressure ranges from 1 to 5 min and depends on the design of the vessel and the power of the intensifier pump used. Immediately after release of the pressure, the temperature of the product returns to its initial value. The come-down time (CDT) to atmospheric pressure is almost instantaneous. The fast cooling capacity of HHP processing is of most interest in the production of high-quality foods. Obviously, long CUTs will add appreciably to the total process time and affect the process throughput, but these periods will also affect the inactivation kinetics of microorganisms. Therefore, consistency and control of these times are important in the development of HHP techniques.

Preheating or precooling of products and temperature equilibration are important steps in HHP processing to achieve the required process temperature. For example, in order to achieve a final process temperature of 121 °C in a high-pressure sterilization process, a preheating step up to a temperature of 90 °C is necessary and can take up to 20 min, depending on the size and physical properties of the product, packaging and heating medium and the design of the preheating tank. A uniform initial target temperature T_i of the food sample is desirable in order to achieve a uniform temperature increase in a homogeneous food system during compression. If cold spots are present within the food or the food system is not homogeneous, parts of the product will not achieve the target process temperature T_f during pressurization. Still-water baths (at T_i or greater), steam, steam injection in water and dielectric heating have been suggested as alternative methods of preheating (Juliano *et al.*, 2008a). Faster preheating methods provide less uniformity, and thus require a longer time for equilibration to achieve temperature homogeneity. In practice, the sample may undergo either equilibration in a thermostat or water bath or equilibration inside the chamber after the product has been inserted and the chamber has been closed.

In some cases, keeping the initial temperature low (~4 °C) can assist in avoiding a temperature increase during pressurization, and a precooling step may be required. Figure 34.1 shows schematically the temperature and pressure cycles during high-pressure–low-temperature (HP-LT) and high-pressure–high-temperature (HP-HT) processing utilized for pasteurization and sterilization, respectively. The CUT, the CDT and the duration of the processing cycle shown in the figure are specific to the design of HHP vessel used. Typically, commercial HHP vessels are equipped with pressure control by means of pressure transducers with measurements that differ by not more than 1%. The temperature control includes control of the temperature of the pressurization water from 5 to 30 °C. However, there is no control of the water temperature inside the vessel. The high-pressure vessel does not have any temperature control, and it is at the temperature of the room where it is located.

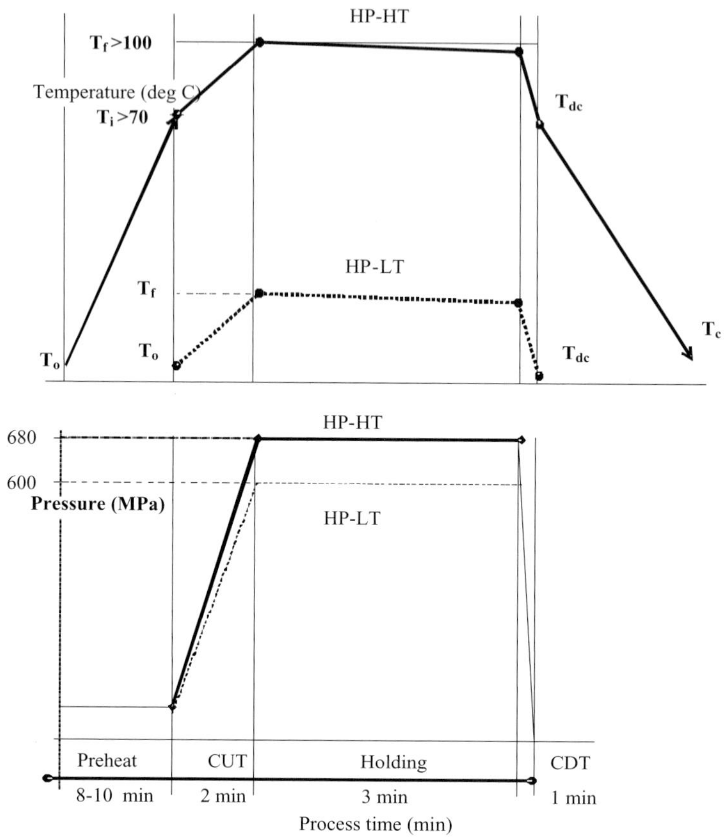

Figure 34.1 Temperature and pressure histories during high-pressure sterilization (HP-HT, solid line) and pasteurization (HP-LT, dotted line) processes. The symbols represent the product temperature at the start of the process (T_0), after preheating (T_i), during the application of high pressure (T_f), after decompression (T_{dc}), and after cooling (T_c). CUT, come-up time; CDT, come-down time.

HHP Pasteurization

Commercial HP-LT pasteurization processes achieve inactivation of vegetative microorganisms at hydrostatic pressures of 600 MPa (or less) and process temperatures lower than 45 °C for 1 to 15 min, depending upon the safety requirements of the product and process (Juliano, 2006). The use of lower temperatures has allowed better retention of the sensory attributes characteristic of "fresh" or "just prepared" foods, as well as nutritional components (Cano and de Ancos, 2005). As a result, HP-LT processing has become a postpackaging technology that is convenient for foods whose quality would otherwise be altered by heat pasteurization.

The ability of HHP processing to eliminate microorganisms has been demonstrated in many studies describing the response of important pathogens (*E. coli*,

L. monocytogenes, *Salmonella* spp., etc.) to high pressure (Patterson and Kilpatrick, 1998). For vegetative cells, microbial inactivation via HHP processing is variable and is influenced by many factors. In general terms, Gram-positive bacteria tend to be more pressure-resistant than Gram-negative bacteria. It is hypothesized that this is due to the complexity of the envelope of Gram-negative cells, making the bacteria more susceptible to changes caused by pressure (Shigehisa *et al.*, 1991). Exceptions exist, however, as certain strains of *Escherichia coli* O157:H7 can be very pressure-resistant (Benito *et al.*, 1999). While generally not associated with food-borne disease, yeasts and molds often play an important role in the spoilage of food. These multicellular microorganisms are relatively sensitive to pressure, presenting a useful application of HHP in the treatment of fruit products to extend the shelf-life of these foods.

HHP Sterilization

As defined by the US Food and Drugs Administration (FDA), sterilization is a process to remove or destroy all viable forms of microbial life, including bacterial spores. HP-HT processing is an emerging preservation technique for the development of shelf-stable low-acid food (LAF) products, as a promising alternative to the conventional thermal sterilization method (Barbosa-Cánovas and Juliano, 2008). This technique provides relatively fast heating and fast cooling due to hydrostatic compression and decompression, respectively, of a preheated food package. In HP-HT sterilization, a temperature increase to above 100 °C throughout the volume when a pressure above 600 MPa is used for a holding times of 3–5 min, compared with the 20–40 min used in retort processing, has been shown to achieve inactivation of microbial spores (Koutchma *et al.*, 2005). Therefore, the reduced exposure of the product to high temperature can produce shelf-stable products with sensory characteristics not achievable by conventional retort processing (Juliano *et al.*, 2007; Leadley *et al.*, 2008; Matser *et al.*, 2004). For instance, pressurization temperatures of 90–116 °C combined with pressures of 500–700 MPa for 3–5 min have been used to inactivate a number of strains of *C. botulinum* spores (Farkas and Hoover, 2000; Margosch *et al.*, 2004). The critical parameters to establish the requirements for a pasteurization or sterilization process are summarized in Table 34.2, where the product and HHP process parameters, target microorganisms, packaging, and conditions of storage required in each case are compared.

Mode of Operation

High pressures are generated either by direct compression, indirect compression or thermal expansion. With direct compression, the volume of the vessel is reduced by the action of a hydraulic pressure applied to a piston. In indirect compression systems, an intensifier, or high-pressure pump, is used to pump a pressure-transmission fluid

Table 34.2 Critical product and process conditions for establishment of requirements for HHP preservation.

	HP-LT	HP-HT
Product parameters		
pH, a_w	3.5 < pH < 4.6; pH < 3.5	pH > 4.6; a_w > 0.86
Process parameters		
Temperature, °C	≤45	>100
Pressure, MPa	≤600	>700
Target microorganisms		
Pathogenic	*Escherichia coli*; *Listeria*; *Salmonella*	*Clostridium botulinum* spores
Spoilage	Lactic bacteria Yeasts Molds	*Geobacillus* spp. *Bacillus cereus*
Storage	Refrigerated conditions	Ambient temperature
Packaging	Hermetically sealed flexible containers	Hermetically sealed flexible containers

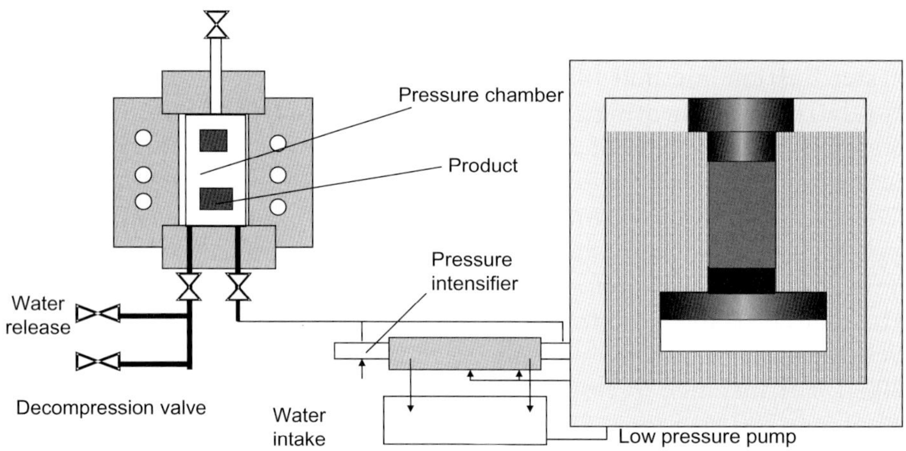

Figure 34.2 Schematic diagrams of direct (right) and indirect (left) pressurization systems.

directly into the vessel to achieve a given pressure. Schematic diagrams of these two pressure generation systems are shown in Figure 34.2. The indirect method of pressurization is currently used for the application of HHP in food processes. The third pressurization method involves the heating of the pressure-transmission fluid inside the vessel to cause expansion by the increase in temperature (San Martin *et al.*, 2002). A typical HHP process uses food products packaged in high-barrier, flexible pouches or in plastic containers. The packages are loaded into the high-pressure chamber. The vessel is sealed and then filled with a pressure-transmitting fluid (normally water) and pressurized by the use of a high-pressure pump (an intensifier), which injects additional quantities of fluid. The packages of food, surrounded by the pressure-transmitting fluid, are subjected to the same pressure as exists in the vessel itself. After the product has been held for the desired time at the target pressure, the vessel is decompressed by releasing the pressure-transmitting fluid. For most applications, products are held for 3–5 min at 600 MPa. Approximately 5–6 cycles per hour are possible, allowing time for compression, holding, decompression, loading and unloading. Slightly higher cycle rates may be possible using fully automated loading and unloading systems. After pressure treatment, the processed product is removed from the vessel and stored or distributed in the conventional manner.

Mode of Operation of the HP-HT Process

The HP-HT process (Figure 34.3) consists of (i) preheating food packages in a carrier outside the vessel, (ii) transferring the preheated carrier into the vessel and

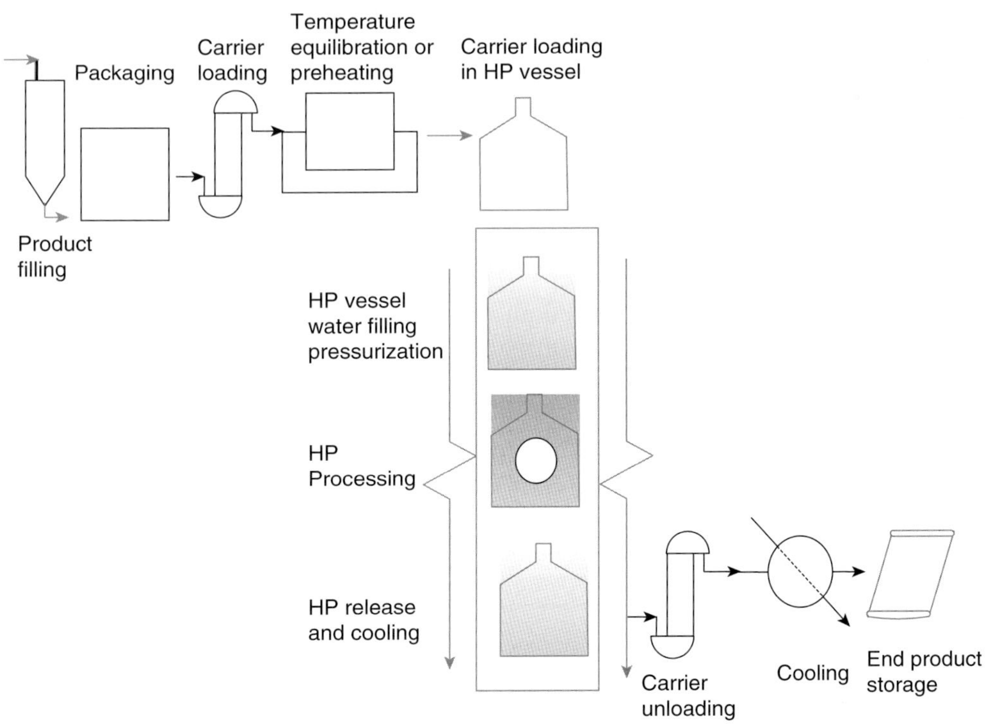

Figure 34.3 Flow diagram of HP process.

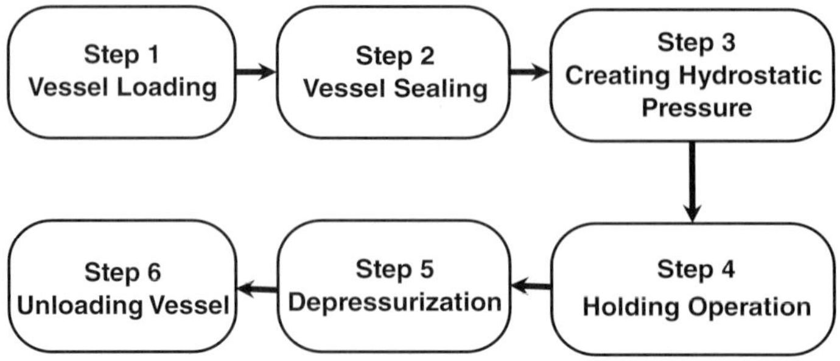

Figure 34.4 Examples of processing steps in the Quintus 35-L high-pressure unit.

equilibrating up to the initial temperature, (iii) pressurizing and holding the carrier at the target pressure, (iv) releasing the pressure, (v) removing the carrier from the vessel, and (vi) cooling down the products in the carrier and removing the products. Therefore, the temperature history inside an HP-HT-processed food is determined by six main time intervals in the process (Juliano, 2006; Barbosa-Cánovas and Juliano, 2008) (Figure 34.1): (i) preheating of the product to the target temperature T_i, (ii) equilibration of the product to an initial temperature equal to T_i, (iii) an increase in the temperature of the product to T_f due to compression heating, (iv) cooling down of the product to T_{f2} due to heat removal in the chamber, (v) a decrease in the temperature of the product to T_{dc} during decompression, and (vi) cooling of the product to T_c.

The commercial Quintus 35-L Food Press manufactured by Avure Technologies (Kent, WA) operates at pressures up to 700 MPa; the process temperature can vary between 4 and 40 °C and the processing cycle time from 3 to 15 min, excluding the loading and unloading times. The typical high-pressure processing cycle shown in Figure 34.4 includes the following steps: (1) the packaged food items are placed in the pressure vessel; (2) the vessel is sealed and filled with water; (3) a pump forces more water into the vessel to create a hydrostatic pressure; the pressure is isostatically transmitted by the fluid medium; (4) the pressure in the vessel is maintained for a predetermined holding time; (5) when the cycle is completed, the vessel is quickly depressurized, and the temperature returns to the starting temperature; and (6) the high-pressure vessel is opened and the product is removed.

The Design of High-Pressure Vessels

The typical HHP system consists of a high-pressure vessel, a means to close the vessel off, a system for pressure generation, a system for temperature and pressure control, and a material-handling system. The primary components of an HHP system include a pressure vessel, a closure or closures for sealing the vessel, a device for holding the closure(s) in place while the vessel is under pressure (e.g., a yoke), low-pressure fill systems to fill the pressure chamber with a temperature-regulated pressure medium (water) and remove air when the system is closed, a high-pressure intensifier pump or pumps to build up the pressure using a food-grade hydraulic oil, a system for controlling and monitoring the pressure and (optionally) temperature, and a product-handling system for transferring product to and from the pressure vessel. The machinery required is complex, and requires extremely high precision in its construction, use and maintenance. For HP-HT treatment at pressures over 400 MPa, pressure vessels can be built from two or more concentric cylinders made of high-tensile-strength steel. An outer cylinder confines the inner cylinders such that the wall of the pressure chamber is always under some residual compression at the design operating pressure. In some cases, the cylinders and the vessel frame are prestressed by winding wire under tension layer upon layer. The tension of the wire compresses the cylinders, reducing the diameter of the cylinders (Hjelmqwist, 2005). This special arrangement allows an equipment lifetime of over 100 000 cycles at pressures of 680 MPa or higher.

HP-HT sterilization systems are equipped with a temperature-controlled preheating tank to achieve the minimum initial temperature of the product. For example, in the Quintus Press QFP 35L S unit designed for operation to 690 MPa and temperatures of up to 130 °C from Avure, the preheating tank utilizes immersion in circulating water to maintain temperature uniformity prior to processing. As the containers of product are filled, they are immediately placed in the carrier and semicontinuously lowered into the heating bath. Since the temperature of the tank is above the minimum process temperature, heat loss is prevented and the product is always in excess of the measured minimum initial temperature. The set-point for the preheating tank is at a minimum of 3 °C above the target minimum initial product temperature. This ensures that the coldest container never loses heat and that each and every package going into the carrier is above the minimum initial temperature. The cooling tank has a refrigeration system for cooling the water to the desired temperature. Rapid cooling maintains product quality and minimizes detrimental chemical reactions.

Over the past few years, manufacturers have made significant progress in simplifying the installation and maintenance of HHP equipment. Normally, perforated baskets are used to insert and remove prepackaged food products from the pressure vessel. The systems also have provisions for filtering and reusing the compression fluid (usually water or a food-grade solution). Stainless steel construction is used in key areas of the vessel for reliability and easy washing down in the most demanding environments. The corrosion protection meets regulatory requirements for food processing plants. As pressure vessels of all types utilize potentially hazardous energy, the relevant regulations seek to identify good design, good manufacturing practices and detailed safety assessments for the safe operation and maintenance of the vessels and auxiliary parts.

Commercial HHP Vessels and Process Economics

Batch and semicontinuous systems are currently commercially available as options for HHP processing equipment. Semicontinuous HHP systems are used in only a few cases, to directly process pumpable products that then need to be aseptically packaged. Batch processing that requires in-container prepackaged products is the most conventional of the two options and was relatively easy to implement in the food industry. In batch HHP systems, the product is generally treated in its final primary package; commonly, the food and its package are treated together, and so the entire pack remains a "secure unit" until the consumer opens it. In-container processing requires packages in the form of pouches, large bulk bags or container–lid combinations. According to Lambert *et al.* (2000), 90% of HHP-processed foods are expected to be processed in flexible or partially rigid packages.

Commercial batch vessels have internal volumes ranging from 30 to more than 600 L. Avure Technologies (USA), NC Hyperbaric (Spain) and Uhde (Germany) are major suppliers of commercial-scale pressure equipment. Both horizontal and vertical pressure-vessel configurations are available. The horizontal-cylinder design is intended to serve in installations where vertical space is limited and minimal plant alteration is desired. The horizontal orientation allows single-direction in-line product flow. In addition, there are differences in capacities, cycle times, achievable pressures, capital costs, etc.

The Wave 6000/300 Tandem system (maximum working pressure of 600 MPa) from NC Hyperbaric is the biggest HHP system for industrial production. In this system, two vessels and their peripherals work together, sharing the same intensifier pumping group. The operation is dephased: when one vessel is in the phase of increasing pressure, the other one is in the holding-time phase. Owing to its volume and vessel diameter (two vessels, each vessel of capacity 300 L and 300 mm in diameter), it is the most productive machine in the range.

The 687L system from Avure, with a horizontal design, was built to serve the seafood industry, where continuous flow is of the essence. With its large 687 L capacity and 310 MPa pressure capability, this system can reliably process up to 5000 kg of raw product per hour.

Commercial-scale high-pressure processing systems cost approximately $300 000 to $2.5 million, depending on the equipment capacity and the extent of automation. Currently, HHP treatment costs are quoted as ranging from 4 to 10 cents/lb. The critical factors that impact process cost and throughput are the vessel size, the pumping power, application and processing conditions such as the cycle time, process pressure and packaging, and the level of automation. Calculations of processing costs have been done taking account of amortization of the equipment cost over five years, assuming 3000 working days per year and 16 hours per day; operating costs; wear and tear of parts; and utilities. Amortization of equipment is responsible for about 60% of the processing cost, wear and tear of parts represents 36%, energy represents less than 4% and the cost of water consumption is negligible. These processing costs should be

Table 34.3 Economic model for the Avure 215 L HHP machine.

	250 per cycle	300 per cycle
Vessel price (each)	$1 450 000	$1 450 000
Employees per shift	2	2
Labor cost per hour ($13.00/h burdened)	$26.00	$26.00
Energy cost/hour ($0.045/kWh × % pressure-up time)	$7.83	$7.83
Pounds per cycle	250	300
Cycles per hour	8	8
Percentage operating time	94%	94%
Average spare parts per cycle	$3.50	$3.50
Annual cost for HHP equipment	$150 000	$150 000
Annual depreciation cost for conveyors and automation	$35 000	$35 000
Annual depreciation cost for building	$4 000	$4 000
Annual spare parts cost	$157 920	$157 920
Annual labor cost	$156 000	$156 000
Annual electricity cost	$46 953	$46 953
Total annual cost	$549 873	$549 873
Cost per pound – HHP equipment, conveyors and automation	$0.0164	$0.0137
Depreciation cost per pound – building	$0.0004	$0.0003
Spare parts cost per pound	$0.0140	$0.0117
Labor cost per pound	$0.0138	$0.0115
Electricity cost per pound	$0.0042	$0.0035
Total cost per pound	$0.0487	$0.0406

increased by 10–40% to include labor costs, depending on the level of automation; this includes loading products into baskets and drying after processing. An economic model of the 215 L Avure machine with various product loads is given in Table 34.3.

As the demand for HHP equipment grows, innovation is expected to reduce capital and operating costs further. In general, during the last few years the cost of treatment per kilogram of HHP-processed products has improved. Capacities have increased owing to design improvements and optimized loading and unloading systems. The production cost of a process must be lower than the value added to the product. The value added by HHP processing can be measured in terms of higher product quality, increased product safety and a longer product shelf-life. These issues can translate further into reduced transportation, storage, insurance and labor costs, and into consumer convenience and enhanced safety. Food companies must be able to make a realistic cost–benefit analysis of the potential rewards of investment in HHP processing. The value of HHP processing in terms of increasing food safety assurance may alone be sufficient to justify an investment in some cases.

The Regulatory Status of HHP Processing

The FDA and USDA have approved HHP processing as a postpackage pasteurization technology for the manufacture of shelf-stable high-acid foods and pasteurized low-acid food products, and developed guidelines and regulations for those products (21CFR §114 and 21CFR §113). In 2009, the FDA approved a petition for the commercial use of a pressure-assisted thermal sterilization process (PATS) in the production of low-acid foods, which was filed initially for processing in a 35 L high-pressure sterilization vessel. Earlier, the European Commission adapted existing legislation (EC258/97) on novel foods to products processed by HHP (European Commission, 2002). Later, Health Canada (*Novel Food Decisions*, available on Health Canada's website at http://www.novelfoods.gc.ca) issued no-objection decisions about ready-to-eat meats and poultry that have been treated by HHP for the control of *Listeria monocytogenes* in 2006; about meat-containing entrees, meat-containing salads and meat products in 2006; and about apple sauce and apple sauce/fruit blends in 2004.

Basic Theoretical Principles

The governing principles of HHP processing are based on the assumption that foods under high pressure in a vessel follow the isostatic rule regardless of their size or shape. The isostatic rule states that pressure is instantaneously and uniformly transmitted throughout a sample, whether the sample is in direct contact with the pressure medium or hermetically sealed in a flexible package. Therefore, in contrast to thermal processing, the time necessary for HHP processing should be independent of the sample size. The effect of HHP processing on food chemistry and microbiology is governed by Le Chatelier's principle. This principle states that when a system in equilibrium is disturbed, the system responds in a way that tends to minimize the disturbance (Pauling, 1964). In other words, high pressure stimulates some phenomena (e.g., phase transitions, chemical reactions and changes in molecular configuration) that are accompanied by a decrease in volume, but opposes reactions that involve an increase in volume. The effects of pressure on protein stabilization are also governed by this principle, i.e., the negative changes in volume that occur with an increase in pressure cause an equilibrium shift towards bond formation. Alongside this, the breaking of ionic bonds is also enhanced by HHP, as this leads to a volume decrease due to the electrostriction of water. Moreover, hydrogen bonds are stabilized by high pressure, as their formation involves a volume decrease. Pressure does not generally affect covalent bonds. Consequently, HHP can disrupt large molecules and microbial cell structures, such as enzymes, proteins, lipids and cell membranes, and leave small molecules such as vitamins and flavor components unaffected (Linton and Patterson, 2000).

Adiabatic-Compression Heating

During pressurization, the packaged product undergoes isostatic compression by the pressure-transmitting fluid, causing a reduction in the package volume (compressibility) of up to 19%, depending on the final pressure and temperature reached (Caner *et al.*, 2004; Mertens, 1995). Fluids such as air, water and food materials undergo adiabatic heating during compression, and cooling during decompression. The following equation expresses this adiabatic heating and has been utilized by several authors (Juliano *et al.*, 2008b):

$$\frac{dT}{dP} = \frac{T\alpha_p}{\rho C_p} \tag{34.1}$$

As indicated by Equation 34.1, the temperature increase dT depends on the volumetric expansion coefficient α_p (measured in units of K^{-1}), the density ρ ($kg{\cdot}m^{-3}$) and the isobaric heat capacity C_p ($J{\cdot}kg^{-1}{\cdot}K^{-1}$) of the material, as well as on its initial temperature T (K). In general, the temperature of both the product and the compression fluid may rise by 20–40 °C during HHP treatment. Solid metallic materials do not experience significant compression heating. Therefore, the temperature increase may vary in foods with relatively complex compositions (Patazca *et al.*, 2007). The quasi-adiabatic temperature increase δ_S upon compression of food samples can be evaluated as the ratio of the temperature increase to the pressure increase using the following equation:

$$\delta_S = 100 \times \frac{\Delta T}{\Delta P} = 100 \times \frac{T_f - T_i}{P - P_{atm}} \cong 100 \times \frac{T_f - T_i}{P} \left[\frac{°C}{100\ \text{MPa}}\right] \tag{34.2}$$

where T_f is the temperature of the sample at the applied pressure, T_i is the initial temperature and P is the applied pressure. Water, carbohydrates, fats and proteins are the main components of the complex food matrix that respond uniquely under compression. It has been reported that water has the lowest rate of temperature increase under compression, about 3 °C per 100 MPa at 25 °C, whereas fats have the highest value, up to 6.7–8 °C per 100 MPa (Barbosa-Cánovas and Juliano, 2008; Rasanayagam *et al.*, 2003; Ting *et al.*, 2002). Nevertheless, only limited information is available on the temperature rise with compression of real foods with complex fat compositions. Patazca *et al.* (2007) studied the temperature increase δ_S under adiabatic compression heating of selected food substances. Significant differences in the compression heating behavior were observed in foods with a high oil/fat content such as vegetable oil, cheese and mayonnaise and foods with a high water content such as milk. Foods with a high water content such as milk experienced a temperature increase with pressure change similar to that of water (the polar component), in the range of 3 °C per 100 MPa (Figure 34.5). Another pattern of behavior under HHP was found for the group of foods with significant proportions of nonpolar components such as cheese, mayonnaise and vegetable oil. The magnitude of the temperature increase decreased with increasing pressure for oil and mayonnaise and did not change significantly for cheese. Similarly, the temperature increase for foods with a high content of proteins such as meat (beef and chicken breast) did not deviate significantly from that of water-like products. Patazca *et al.* (2007) calculated δ_S values for food samples assuming that the δ_S values

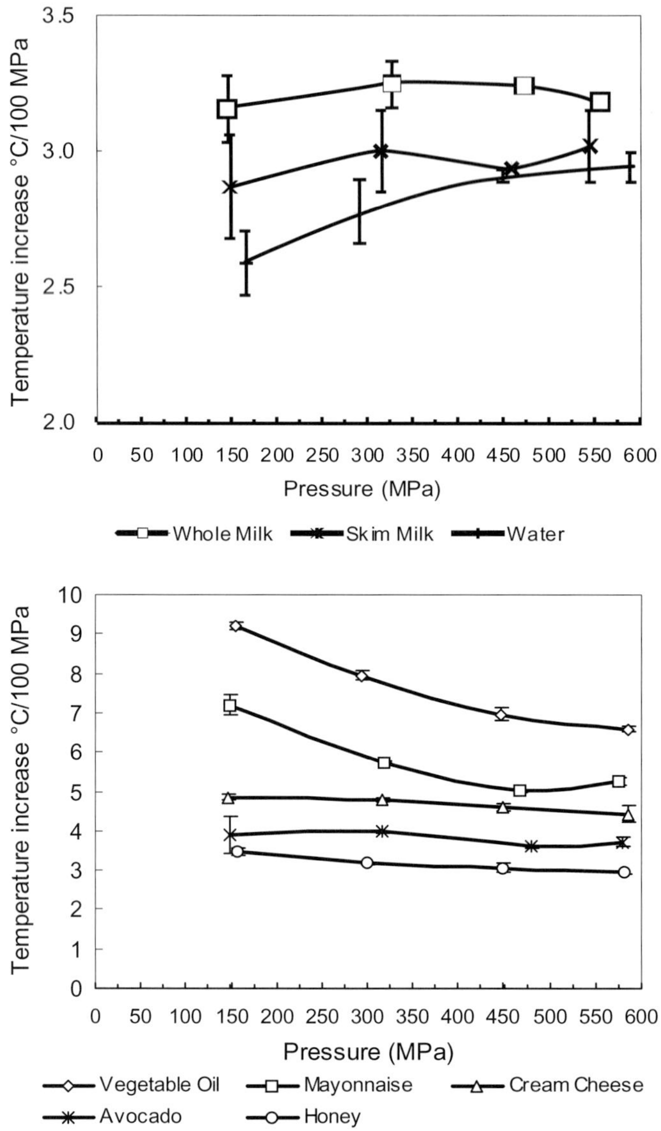

Figure 34.5 Adiabatic heats of compression of selected foods.

for protein, carbohydrate and ash in solution in water were similar to that of water. Experimental δ_S values for oil/fat and water obtained in this study were used as reference values. The calculated δ_S values (Table 34.4) showed a correlation with the experimentally measured data for all products tested except for mayonnaise dressing. The δ_S values appear to be primarily proportional to the water and fat/oil contents of the foods. The homogeneous-mixture equation proposed by Rasanayagam *et al.* (2003) can be used to estimate δ_S values for foods based on their composition.

Table 34.4 Experimental and calculated δ_S values [$100\Delta T/\Delta P$ (°C/100 MPa)] obtained for an initial temperature of 25 °C and pressures from 150 to 600 MPa. The calculated values were evaluated using $\Delta T/(100\Delta P)$ ratios for vegetable oil and water at different pressures as reference values and assuming that fat and water are the main components of the food (adapted from Patazca *et al.*, 2007).

Food	Composition (%)					ΔT/100 MPa	
	Fat	Water	Protein	Carbohydrate	Ash	Calculated	Experimental
Mayonnaise dressing	33.40	39.90	0.90	23.90	1.90	4.8–4.2	7.2–5.3
Cream cheese	28.60	58.50	7.10	3.50	2.30	4.5–4.0	4.9–4.7
Egg yolk	26.54	52.31	15.86	3.58	1.71	4.3–3.9	4.5–4.3
Hass avocado	15.41	72.33	1.96	8.64	1.66	3.6–3.5	4.1–3.7
Beef, ground	10.70	68.50	19.80	0.00	1.00	3.3	3.2
Egg, whole	9.94	75.84	12.58	0.77	0.87	3.3	3.3
Whole milk	3.25	88.32	3.22	4.52	0.69	3.0	3.2
Gravy, beef	2.36	87.48	3.75	4.81	1.60	2.9	3.0
Egg white	0.17	87.57	10.90	0.73	0.63	2.8	2.8
Skim milk	0.08	90.84	3.37	4.96	0.75	2.8	3.0

The adiabatic temperature change of an isotropically compressed or decompressed material follows the differential equation given in Equation 34.1, which can be rearranged into the following equation as described by Knoerzer *et al.* (2010):

$$\frac{dT}{dP} = \frac{\alpha_P}{\rho \cdot C_P} \cdot T = k_C \cdot T \tag{34.3}$$

where $k_C = \alpha_P/\rho C_P$ is the compression heating coefficient (Pa^{-1}). As shown in Equation 34.3, the temperature change depends on a complex interaction of the thermal expansion coefficient, the density and the specific heat capacity of the material. It is challenging to determine these properties separately under high-pressure conditions. Therefore, to predict the extent of compression heating during a high-pressure process, the pressure-and-temperature-dependent properties can be combined into one pressure-and-temperature-dependent parameter referred to as the compression heating coefficient, k_C. For modeling HHP processes, a knowledge of these properties is imperative for all materials involved in the modeled scenario. The variation of the compression heating of water based on thermophysical properties taken from the NIST database for water and steam (Harvey *et al.*, 1996) has been summarized in the literature. Limited information is available on the compression heating of food materials. The compression-heating coefficients of food materials can be derived from temperature–pressure profiles obtained using an adiabatic high-pressure system. There are certain types of thermocouples (the K and T types) that can reliably measure temperature under high-pressure conditions. A detailed description of this method can be found elsewhere (Knoerzer *et al.*, 2010).

Packaging materials also undergo compression heating. Knoerzer *et al.* (2010) have shown that polypropylene (PP) and polyethylene (PE) undergo compression heating greater than that of water under both HP-LT (10 °C and 50 °C) and HP-HT (90 °C) conditions up to 750 MPa. In particular, the temperature increase with pressure was not linear and, therefore, the relative increase with respect to water depended on the pressure range selected as well as on the initial temperature. For instance, PE showed higher temperatures than water under both HP-LT and HP-HT conditions throughout the whole pressure range. In contrast, although the temperature of PP remained higher than that of water throughout the pressure range under HP-LT conditions, the compression-heating curve intersected with that of water at 500 MPa under HP-HT conditions. Other authors have observed a temperature increase with compression of 4.5 °C per 100 MPa in PP (Schauwecker *et al.*, 2002).

Energy Balance

Barbosa-Cánovas and Rodríguez (2005) applied a typical energy balance to a pressure system, obtaining the following equation:

$$E_{st} = E_{in} - E_{out} + E_g \tag{34.4}$$

where E_{in} is the energy entering the system, or the work of compression; E_{out} is the energy leaving the system (i.e., the heat loss through the vessel walls during compression, the holding time and decompression); E_g is the energy of compression inside the system; and E_{st} is the energy accumulated or stored inside the system. It can be assumed that no energy is produced from any chemical reaction (i.e., $E_g = 0$). If the system is adiabatic (i.e., $E_{out} = 0$), then Equation 34.3 can be reduced to a single-term equation,

$$E_{in} = E_{st} \tag{34.5}$$

In this case, energy is generated owing to compression heating during the come-up time. Assuming the HHP system is adiabatic, Equation 34.1 can be rearranged into Equation 34.5 and differentiated as a function of time to express the conversion of mechanical energy into thermal (internal) energy due to compression during a reversible adiabatic change:

$$\rho C_p \frac{\partial T}{\partial t} = T\alpha_p \frac{\partial P}{\partial t} \tag{34.6}$$

Equation 34.6 indicates that the rate of energy accumulation due to the temperature increase is equivalent to the heat produced due to compression during pressurization at a given temperature in an adiabatic (isolated) system. When the system is adiabatic, Equation 34.5 is valid for the holding (i.e., $\partial P/\partial t = 0$) and decompression steps.

By comparing specific energy inputs of a thermal and a combined thermal and HHP sterilization process, Toepfl *et al.* (2006) estimated that the specific energy input required for the sterilization of cans can be reduced from 300 to 270 $kJ \cdot kg^{-1}$ when a high pressure is applied. In the case of high-pressure processing, a compression energy recovery rate of 50% can be estimated when a two-vessel system or a pressure storage system is used. Making use of energy recovery, a specific energy input of 242 $kJ \cdot kg^{-1}$ will be required for sterilization, corresponding to a reduction of 20%.

Design and Calculations for HHP Preservation Processes

In the design of a "preservation specification" for a thermal pasteurization or sterilization process, the processing time F_p is traditionally defined by the initial load of resistant organisms N_0, the endpoint of the process N_F, the specific logarithmic reduction *SLR* and the logarithmic resistance *D* of the target bacteria under defined conditions, and it is calculated using the following equation:

$$F_p = D \times (Log\, N_0 - Log\, N_F) = SLR \times D \tag{34.7}$$

The establishment of process criteria for HHP pasteurization and sterilization processes requires optimization of the process temperature and pressure to inactivate the target pathogenic and spoilage-causing bacteria, based on knowledge of the behavior of the food under pressure and of the rate constants for microbial inactivation.

Microbial-Destruction Kinetics

A kinetic analysis of microbial destruction can be carried out under isobaric and isothermal conditions after the come-up time portion of the pressure cycle has been completed and the maximum process pressure attained. As has been extensively reported, various modeling approaches have been used to describe microbial inactivation during HHP processing, and these provide a basis for predicting reduction levels in HHP-treated foods.

Linear Models

Patazca *et al.* (2006) used first-order kinetics (as expressed in the following equation) to analyze the effect of temperature and pressure on inactivation during the hold time:

$$\frac{dN}{dt} = -kN \tag{34.8}$$

where *N* is the spore count after exposure to a lethal treatment for a specific time *t*. In calculations for thermal processes, the first-order kinetics is often described by a decimal reduction time, or thermal D-value. This D-value can be obtained as the negative reciprocal slope of log (N/N_0) vs. time and is reciprocally related to k, where N_0 is the initial spore count measured immediately after the come-up time of the process (thermal or pressure).

The thermal D-value at constant pressure, $D_{T,P}$, can be calculated from survival curves plotted at constant pressure, as expressed by the following equation:

$$D_{T,P} = \frac{t}{\log\left(\frac{N}{N_0}\right)} \tag{34.9}$$

The D-value at constant temperature, $D_{P,T}$, can be calculated from survival curves plotted at constant temperature:

$$D_{P,T} = \frac{t}{\log\left(\frac{N}{N_0}\right)} \tag{34.10}$$

The thermal resistance z_T (the temperature required to reduce the value of $D_{T,P}$ by 90% at constant pressure) and the pressure resistance z_P (the pressure required to reduce the value of $D_{P,T}$ by 90% at constant temperature) are determined from the following equations:

$$z_T = -\frac{1}{[\text{slope}]} = \frac{T_2 - T_1}{\log\left(\frac{D_{T_1,P}}{D_{T_2,P}}\right)} \tag{34.11}$$

$$z_P = -\frac{1}{[\text{slope}]} = \frac{(P_2 - P_1)}{\log\left(\frac{D_{T,P_1}}{D_{T,P_2}}\right)} \tag{34.12}$$

Biphasic kinetics for the destruction of vegetative cells has been applied to evaluate HHP inactivation (Ramaswamy and Marcotte, 2006). The first phase is a kill due to the pressure pulse, and a first-order kinetic model is applied to the microbial kill during the holding time. However, a phenomenon that is common to HHP processing is the persistence of pressure-resistant microorganisms during treatment. When microorganisms are subjected to high pressure, initially there is a linear decrease in the microbial load; however, as time progresses, the rate of microbial inactivation decreases, ultimately resulting in a survival curve with a "pressure-resistant tail." The existence of this subpopulation of pressure-resistant microorganisms is poorly understood. The "tailing effect" may be due to inherent phenotype variation in pressure resistance amongst the target microorganisms; however, the substrate and growth conditions may also have an influence. The logarithmic pattern of microbial destruction is not exclusive to HHP processing, but is common to many lethal treatments (Cerf, 1977).

Nonlinear models

During HHP processing, the resistance of the microbial population to high pressure may not be the same for all of the population, and this may result in survival curves showing shoulders or tailing, and which deviate from a linear curve. This type of distribution is described by a power-law or Weibull distribution model (Peleg and Cole, 1998), given by the following equation:

$$\mathrm{Log}\frac{N_t}{N_0} = -bt^n \tag{34.13}$$

where b is a scale factor and n is a shape parameter: $n < 1$ for curves that are concave upwards, $n > 1$ for curves that are concave downwards and $n = 1$ for linear curves. The scale factor is considered to be a nonlinear rate paramcter, which primarily reflects the overall steepness of the survival curves when the power n is fixed.

Feeherry *et al.* (2005) used a quasi-chemical differential equation model to evaluate the linear and nonlinear ("shoulders" and "tailing") inactivation kinetics of *Escherichia coli* and *Listeria monocytogenes*. The log-logistic equation was used by Chen and Hoover (2003) to describe the nonlinear thermal inactivation of microorganisms. Chen (2007) tested linear, Weibull and log-logistic functions to model the pressure inactivation of seven food-borne pathogens in milk and concluded that the Weibull distribution model fitted better than the linear model for all seven pathogens.

Critical Factors for Microbial Inactivation

Inactivation of vegetative pathogenic microorganisms of primary public health significance and inactivation of spoilage microorganisms in high-acid ($\mathrm{pH} < 3.7$) and medium-acid ($3.7 < \mathrm{pH} < 4.6$) foods is a requirement for HHP pasteurization that is conducted at pressures above 200 MPa and at chilled or process temperatures less than 45 °C. Pasteurization demands a logarithmic reduction of 5 or 6 in pathogens. *Listeria monocytogenes* is considered as a target microorganism of public health concern in dairy and meat products, *Salmonella* is a target microorganism in eggs, and *Escherichia coli* is a target microorganism in fruit and vegetable juices and other fruit-based products. The rate of inactivation is strongly influenced by the peak pressure (Patterson, 2005; Lau and Turek, 2007). Commercially, higher pressures are preferred as a means of accelerating the inactivation process, and current practice is to operate at 600 MPa, except for those products where protein denaturation needs to be avoided. The pressure resistance of vegetative microorganisms often reaches a maximum at ambient temperatures, so the initial temperature of the food prior to HHP processing can be reduced or elevated to improve inactivation at the processing temperature (i.e., the temperature at pressure). The extent of inactivation also depends on the type of microorganism and on the composition, pH and water activity of the food. Gram-positive bacteria are more resistant than Gram-negative. Most yeasts are inactivated by exposure to 300–400 MPa at 25 °C within a few minutes. However, yeast ascospores may require treatment at higher pressures. The pressure inactivation of molds appears to be similar to yeast inactivation. However, additional studies on molds of interest in food preservation are needed.

Reduced pH is generally synergistic with pressure in eliminating microorganisms. Reduced water activity, however, tends to inhibit pressure inactivation, with noticeable retardation as the water activity falls below ~0.95. Studying the effect of water activity on the HHP inactivation of *L. innocua*, Mariappagoudar (2007) found that reducing the water activity reduces the inactivation rate and results in an increase in D-values. These results appear to be consistent with previous research done by Palou *et al.* (1997) and Setikaite *et al.* (2009), who studied the combined effect of the water activity a_w and HHP processing on the inactivation of *Zygosaccharomyces bailli* and *Escherichia coli*. Comparison of D-values for *L. innocua* at $a_w = 0.95$ showed that the values for various model solutes decreased in the order sorbitol > fructose > NaCl > glycerol. However, at $a_w = 0.99$, NaCl had the most baroprotective effect and the D-values at $a_w = 0.99$ decreased in the order NaCl > glycerol > sorbitol > fructose. The inactivation curves of *L. innocua* were nonlinear, with shoulders and tailing. As a result, both a linear and a nonlinear (Weibull) model were used for evaluation of the inactivation kinetics. The scale factor b of the Weibull model was found to be proportional to the rate of inactivation k. The differences in the D-values for *L. innocua* at 350 MPa and 23 °C in solutions of all four solutes at $a_w = 0.99$ and 0.95 are shown in Figure 34.6. The D-value for sorbitol at $a_w = 0.95$ was 46.3 min, whereas the D-value for glycerol was 9.35 min, which is nearly equal to the D-value for salt, 8.66 min. The D-value decreased from 46.3 min to 2.9 min when the a_w value of the sorbitol-based

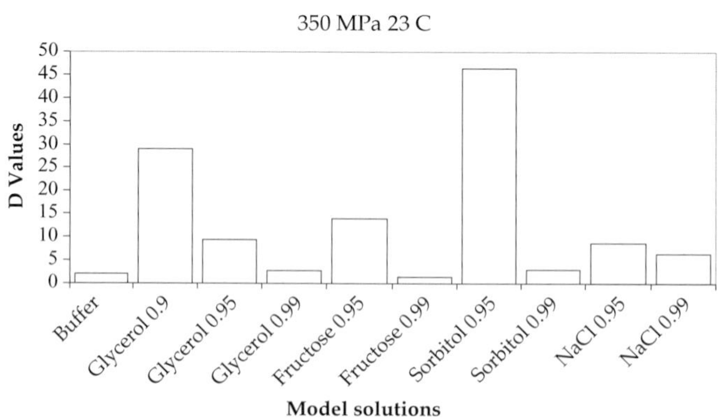

Figure 34.6 D-values for *L. innocua* 227 at a_w values of 0.95 and 0.99 in sorbitol, glycerol, NaCl and fructose solute models.

Table 34.5 Effect of water activity and type of solute on D-values for *L. innocua* under HHP inactivation in the range of 350 to 450 MPa at 23 °C.

Water activity	Pressure (MPa)	*k* (min^{-1})	D-value (min)	*B*	*n*
Glycerol					
0.9		0.035	28.99	0.03	1.03
0.95	350	0.107	9.35	0.05	1.25
0.99		0.354	2.82	0.09	1.65
0.9		0.118	8.50	0.03	1.46
0.95	400	0.172	5.83	0.29	1.07
0.99		0.577	1.73	0.45	1.09
0.9		0.270	3.71	0.14	1.34
0.95	450	0.437	2.29	0.39	1.06
0.99		1.347	0.74	1.68	0.83
orbitol					
0.95	350	0.022	46.30	0.07	0.61
0.99		0.337	2.97	0.29	1.06
0.95	400	0.023	43.67	0.10	0.46
0.99		0.718	1.39	0.48	1.23
0.95	450	0.083	12.12	0.18	0.66
0.99		1.096	0.91	0.93	1.12
NaCl					
0.95	350	0.116	8.66	0.21	0.78
0.99		0.154	6.51	0.39	0.65
0.95	400	0.215	4.65	0.18	1.10
0.99		0.413	2.42	0.43	0.98
0.95	450	0.290	3.45	0.44	0.78
0.99		0.617	1.62	0.78	0.83
Fructose					
0.95	350	0.072	13.83	0.07	1.01
0.99		0.771	1.30	2.23	0.61
0.95	400	0.166	6.02	0.09	1.21
0.99		2.210	0.45	2.40	0.82
0.95	450	0.232	4.31	0.21	1.04
0.99		4.734	0.21	4.70	0.97

model was increased from 0.95 to 0.99. It can be concluded that reducing a_w makes *L innocua* 227 more resistant to HHP processing. In contrast, salt, which has the least protective effect against inactivation of *L. innocua* 227 at $a_w = 0.95$, has more protective effect at $a_w = 0.99$, with a D-value of 6.5 min. The model fructose solution at $a_w = 0.99$ resulted in the least resistance of *L. innocua* 227 to HHP processing, with a D-value of 1.3 min. Since the effect of processing temperature on HHP inactivation of *L. innocua* in the range of 15 °C to 34 °C was found not to be statistically significant, a comparison of the effects of solute type and a_w on the D-values and the parameters *b* and *n* in the Weibull model in the range of 350 to 450 MPa at 23 °C is given in Table 34.5.

The F_0-value for HP-HT Sterilization

The establishment of process criteria for HP-HT sterilization processes requires optimization of the process temperature and pressure to inactivate the target pathogenic and spoilage-causing spore-forming bacteria. Nonpathogenic *Clostridium sporogenes* spores, which have a heat resistance similar to *Clostridium botulinum* spores, are often used for this purpose (Koutchma *et al.*, 2005). *C. sporogenes* PA 3679 has been the primary choice of researchers in most studies of thermal processing of inoculated packages. The classical heat-resistant, nonpathogenic, thermophilic spores of *Geobacillus stearothermophilus* have been sufficiently characterized to design and validate thermal sterilization processes. However, it was reported that the pressure resistance of bacterial spores did not correlate with their heat resistance. Thus, the species and strains whose spores are used as resistant targets in the case of thermal processing need to be reidentified in the case of pressure processing.

Two paths towards addressing the issues of spoilage and pathogens of public health significance, as relating to commercial sterility, are used. First, the establishment of the process temperature, pressure and time for HP-HT sterilization processes requires a knowledge of the inactivation kinetic parameters of the target pathogenic and spoilage-causing spore-forming bacteria. Secondly, the outcome of the process, or a performance criterion, must be defined as a prerequisite to further validation of an established sterilization process. Historically, the endpoint of sterilization processes has not been clearly defined. The 12D concept established by Stumbo (1973) does not establish an endpoint, but merely defines that the process should accomplish a 12-decade reduction of *C. botulinum*. The concept of defining the specification of a thermal process by describing the outcome or "end point of a preservation process" as the probability of a nonsterile unit (PNSU) was first introduced by Pflug and Zeghman (1985) and can be adapted to determine an F_0-value for the HP-HT sterilization of foods. A spoilage failure rate of 10^{-6} was selected for mesophilic spores, represented by *C. sporogenes* PA 3679, and 10^{-3} for the thermophilic spores of *G. stearothermophilus*.

Patazca *et al.* (2006) and Koutchma *et al.* (2005) demonstrated that log–linear models are suitable for predicting the inactivation of both of these classical surrogates used in the case of thermal processing. The F_0-values were calculated for a 7-log reduction for PA 3679 spores and a 5-log reduction for *G. stearothermophilus* spores based on the assumptions of Pflug and Zeghman (1985). The calculations showed that an HP-HT process that delivers 3.2 min at 121 °C and a pressure of 600 or 700 MPa, or 1.5 min at 800 MPa and 121 °C will be adequate to prevent spoilage by mesophilic and thermophilic spore-forming organisms. When the HP-HT sterilization process is compared with the 27 min of heat sterilization that would be required in order to achieve commercial sterility in terms of spoilage spore formers, it is apparent that the length of the HP-HT process is drastically shorter. According to the differences in D-values between HP-HT and classic thermal sterilization, combining temperature with pressure reduces the process time. Nevertheless, the design of

HP-HT processes to make foods safe from a public health point of view requires a knowledge of the D-values for *C. botulinum* spores under similar high-pressure conditions.

Computer-Aided Design of HHP Processes

Juliano *et al.* (2008a) have summarized the extensive research that has been done to develop modeling approaches that can be utilized to predict temperatures in a high-pressure system. The modeling approaches tested include analytical, numerical, macroscopic and artificial-neural-network (ANN) models to predict transient temperature distributions, uniformity, and the loss of compression heating through the walls of the high-pressure vessel during all steps of high-pressure processing. The analytical modeling of high-pressure processing involves the fitting of experimental temperature data from a heating and cooling curve obtained during pressurization and depressurization.

Discrete numerical modeling and computational fluid dynamics (CFD) have been used to predict the distribution of the temperature and flow inside the high-pressure vessel (Denys *et al.*, 2000; Hartmann and Delgado, 2002; Hartmann and Delgado, 2003; Hartmann and Delgado, 2004). Some recently reported models have included solid materials (Knoerzer *et al.*, 2007; Otero *et al.*, 2007a,b; Juliano *et al.*, 2008b; Juliano *et al.*, 2009), whereas other models were used to predict temperature distributions in three dimensions (Ghani and Farid, 2006) and under high-pressure sterilization conditions. CFD software packages can not only assist in calculating the temperature evolution, but also allow coupling the governing equations of fluid dynamics with kinetic equations for inactivation to predict the distribution of enzyme and microbial inactivation from the temperature and flow distribution. Special attention has been given to the distribution of enzyme and microbial inactivation throughout the chamber and the packages, particularly for *Clostridium botulinum* in the case of HPHT sterilization. Juliano *et al.* (2008a) described how CFD models were developed and applied to evaluate the effects of varying inlet velocity, the size of the vessel, the viscosity of the fluid, the presence of packages, the presence of solid foods, and the sample carriers on the distribution of the temperature, flow and/or inactivation.

CFD modeling can also assist in studies of scaling-up, as shown by Hartmann and Delgado (2002) and Hartmann and Delgado (2003), some of which have shown there is higher heat retention in machines of industrial size. Commercial-size vessels require short come-up times and therefore high inflow velocities, which provide turbulent flow conditions in the inlet area. Macroscopic models have been developed to integrate the effects of different portions of the high-pressure system that contribute to heat transfer or heat retention to predict the final temperature inside the vessel (Otero *et al.*, 2007a).

Furthermore, a few publications have described the application of ANNs to predict temperatures in a high-pressure vessel (Torrecilla *et al.*, 2004; Torrecilla *et al.*, 2005). As indicated by Juliano *et al.* (2008a), the use of ANNs is the most limited method, since it is restricted to providing a single value (e.g., temperature or time). Numerical models have shown excellent adaptability to system modifications, and allow the optimization of processing conditions and system structure. The development of 3D models will be of further aid in modeling horizontal high-pressure vessels and the effect of packages inside them on the distribution of temperature, flow and inactivation.

Summary and Future Needs

HHP-processed products such as deli meats, shellfish, fruits, salsa and guacamole are commercially available in the US, Canada, Europe, Australia and Japan. For most commercial products, HHP processing has been applied as a final preservation step after packaging, to extend the shelf-life of the product. HHP processing is applied to packaged meats for additional microbial safety assurance owing to the risk of post-processing contamination from pathogens such as *L. monocytogenes*. However, products must still be stored at refrigeration temperatures. Pressure treatment of vegetable products is currently limited, owing to their relatively high pH and the possible survival of pathogenic spore-forming microorganisms. In addition to its use as a preservation step, HHP technology is being applied for raw-meat separation in seafood products, pressure-assisted freezing and thawing, and texture modification. HHP thermally assisted sterilization processes for prepacked low-acid foods are not available commercially, although they have been approved by the US FDA. The HHP vessels commercially available, both vertical and horizontal, have internal volumes ranging from 30 to more than 600 L. The biggest obstacle for HHP systems is the initial capital investment required, which currently limits their application to high-value products. The actual cost of operating an HHP plant depends on many factors, ranging from the operating pressure, cycle time and product geometry to the labor skills available and energy costs.

The phenomenon of a temperature increase in foods plays a critical role in the establishment of the conditions for an HHP process. The magnitude of the temperature increase due to compression per 100 MPa pressure increase needs to be taken into account, and is dependent on the product composition and the initial temperature. The establishment of process criteria for HHP pasteurization and sterilization processes requires optimization of the process pressure, temperature and time to inactivate target pathogenic and spoilage-causing bacteria based on knowledge of the behavior of the food under pressure. Even though a first-order destruction model can be applied to microorganisms subjected to HHP, it is observed that as the HHP process time progresses, the rate of microbial inactivation decreases and results in a survival curve with a "pressure-resistant tail." Owing to the complicated intrinsic and extrinsic factors, the HHP conditions need to be verified case by case to ensure the required microbial reduction. CFD software packages can assist in modeling the HHP process, in optimization of the processing conditions and system structure, and in performance

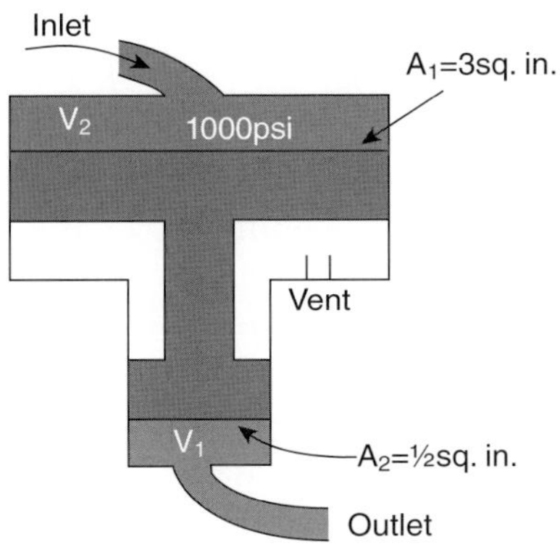

Figure 34.7 Hydraulic device considered in Problem 2.

validation. However, a knowledge of the physical properties of all materials involved in the modeled scenario is imperative.

Problems

1. What will be the final temperature of a water sample at a pressure of 600 MPa if the initial temperature of the water is 2 °C or 50 °C? What will be the final temperature of oil processed under the same conditions? Explain the difference.
2. The actuating cylinder shown in Figure 34.7 is supplied with hydraulic fluid at a pressure of p_1 = 1000 psi, and this pressure acts on a piston of surface area A_1 = 3 sq. in. Calculate the pressure developed on the outlet side of the smaller cylinder if A_2 = 0.5 sq. in.
3. The spoilage failure rates for mesophilic spores of *C. sporogenes* PA 3679 and thermophilic spores of *G. stearothermophilus* are 10^{-6} and 10^{-3}, respectively. Assuming that the initial population of mesophilic spores is equal to 1000 CFU·ml^{-1} and that of thermophilic spores is 10 000 CFU·ml^{-1}, calculate F_0 for a high-pressure sterilization process using pressures of 600 MPa and 700 MPa at 121 °C, and compare the HHP process with thermal sterilization at atmospheric pressure. The decimal reduction times are as follows. For *C. sporogenes* PA 3679, $D_{600,\,121}$ = 21 s, $D_{700,\,121}$ = 19 s and $D_{0,\,121}$ = 48 s. For *G. stearothermophilus*, $D_{600,\,121}$ = 6 s, $D_{700,\,121}$ = 5 s and $D_{0,\,121}$ = 330 s.

References

Barbosa-Cánovas, G.V. and Juliano, P. (2008) Food sterilization by combining high pressure and heat. In: *Food Engineering: Integrated Approaches* (eds G.F. Gutierrez-López, G. Barbosa-Cánovas, J. Welti-Chanes and E. Paradas-Arias). Springer, New York, pp. 9–46.

Barbosa-Cánovas, G.V. and Rodríguez, J.J. (2005) Thermodynamic aspects of high hydrostaticpressure food processing. In: *Novel Food Processing Technologies* (eds G.V. Barbosa-Cánovas, M.S. Tapia and M.P. Cano). CRC, New York, pp. 183–206.

Benito, A., Ventoura, G., Casadei, M., Robinson, T. and Mackey, B. (1999) Variation in resistance of natural isolates of *Escherichia coli* O157 to high hydrostatic pressure, mild heat, and other stresses. *Applied Environmental Microbiology* 65: 1564–1569.

Caner, C., Hernandez, R.J., Pascall, M., Balasubramaniam, V.M. and Harte, B.R. (2004) The effect of high-pressure food processing on the sorption behaviour of selected packaging materials. *Packaging Technology and Science* 17: 139–153.

Cano, M.P. and de Ancos, B. (2005) Advances in use of high pressure to processing and preservation of plant foods. In: *Novel Food Processing Technologies* (eds G.V. Barbosa-Cánovas, M.S. Tapia and M.P. Cano). CRC, Boca Raton, FL, pp. 361–373.

Cerf, O. (1977) Tailing of survival curves of bacterial spores. *Journal of Applied Biotechnology* 42: 1–10.

Chen, H. (2007) Use of linear, Weibull and log-logistic functions to model pressure inactivation of seven foodborne pathogens in milk. *Journal of Food Microbiology* 24(3): 197–204

Chen, H. and Hoover, D.G. (2003) Modeling the combined effect of high hydrostatic pressure and mild heat on the inactivation kinetics of *Listeria monocytogenes* Scott A in whole milk. *Innovative Food Science and Emerging Technologies* 4: 25–34.

Denys, S., Van Loey, A.M. and Hendrickx, M.E. (2000) Modelling conductive heat transfer during high-pressure thawing processes: determination of latent heat as a function of pressure. *Biotechnology Progress* 16(3): 447–455.

European Commission (2002) *Novel Foods and Novel Food Ingredients: Introduction and Legislation*. http://ec.europa.eu/food/food/biotechnology/novelfood/index_en.htm.

Farkas, D. and Hoover, D.G. (2000) High pressure processing. *Journal of Food Science Supplement* 65: 47–64.

Feeherry, F.E., Doona, C. and Ross, E.W. (2005) The quasi-chemical kinetics model for the inactivation of microbial pathogens using high pressure processing. In: *3rd International Symposium on Applications of Modelling as an Innovative Technology in the Agri-Food Chain, MODEL-IT* (eds M.L.A.T.M. Hertog and B.M. Nicolaï), May 29, 2005.

Ghani, A.G.A. and Farid, M.M. (2006) Numerical simulation of solid–liquid food mixture in a high pressure processing unit using computational fluid dynamics. *Journal of Food Engineering* 80(4): 1031–1042.

Hartmann, C. and Delgado, A. (2002) Numerical simulation of convective and diffusive transport effects on a high-pressure induced inactivation process. *Biotechnology and Bioengineering* 79(1): 94–104.

Hartmann, C. and Delgado, A. (2003) The influence of transport phenomena during high-pressure processing of packed food on the uniformity of enzyme inactivation. *Biotechnology and Bioengineering* 82(6): 725–735.

Hartmann, C. and Delgado, A. (2004) Numerical simulation of the mechanics of a yeast cell under high hydrostatic pressure. *Journal of Food Engineering* 37(7): 977–987.

Harvey, A.H., Peskin, A.P. and Sanford, A.K. (1996) *NIST/ASTME-IAPSW Standard Reference Database 10*, version 2.2.

Hite, B.H. (1899) The effect of pressure in the preservation of milk. *Washington, Va. University, Agriculture Experiment Station, Bulletin* 58: 15–35.

Hjelmqwist, J. (2005) Commercial high-pressure equipment. In: *Novel Food Processing Technologies* (eds G.V. Barbosa-Cánovas, M.S. Tapia and M.P. Cano). CRC, New York, pp. 361–373.

Juliano, P. (2006) *High Pressure Thermal Sterilization of Egg Products*. PhD thesis, Washington State University, Pullman, WA.

Juliano, P., Clark, S., Koutchma, T., Ouattara, M., Mathews, J.W., Dunne, C.P. and Barbosa-Cánovas, G.V. (2007) Consumer and trained panel evaluation of high pressure thermally treated scrambled egg patties. *Journal of Food Quality* 30: 57–80.

Juliano, P., Knoerzer, K. and Barbosa-Cánovas, G.V. (2008a) High pressure processes: thermal and fluid dynamic modeling applications. In: *Engineering Aspects of Thermal Processing* (ed. R. Simpson). CRC, Boca Raton, FL, pp. 159–207.

Juliano, P., Knoerzer, K. and Barbosa-Cánovas, G.V. (2008b) High pressure thermal processes: thermal and fluid dynamic modeling principles. In: *Engineering Aspects of Thermal Processing* (ed. R. Simpson). CRC, Boca Raton, FL, pp. 91–158.

Juliano, P., Knoerzer, K., Fryer, P. and Versteeg, C. (2009) *C. botulinum* inactivation kinetics implemented in a computational model of a high pressure sterilization process. *Biotechnology Progress* 25(1): 163–175.

Knoerzer, K., Juliano, P., Gladman, S., Versteeg, C. and Fryer, P. (2007) A computational model for temperature and sterility distributions in a pilot-scale high-pressure high temperature process. *AIChE Journal* 53: 2996–3010.

Knoerzer, K., Buckow, R., Sanguansri, P. and Versteeg, C. (2010). Adiabatic compression heating coefficients for high-pressure processing of water, propylene-glycol and mixtures – a combined experimental and numerical approach. *Journal of Food Engineering* 96: 229–238.

Koutchma, T., Biao, G., Patazca, E. and Parisi, B. (2005) High pressure-high temperature sterilization: from kinetic analysis to process verification. *Journal of Food Process Engineering* 28(6): 610–629.

Lambert, Y., Demazeau, G., Largeteau, A., Bouvier, J.M., Laborde-Croubit, S. and Cabannes, M. (2000) Packaging for high-pressure treatments in the food industry. *Packaging Technology and Science* 13: 63–71.

Lau, M.H. and Turek, E.J. (2007) Determination of quality difference in low-acid foods sterilized by high pressure versus retorting. In: *High Pressure Processing of Foods* (eds C.J. Doona and F.E. Feeherry). Blackwell Publishing/IFT Press, Ames, IA, pp. 195–217.

Leadley, C., Tucker, G.S. and Fryer, P.J. (2008) A comparative study of high pressure sterilisation and conventional thermal sterilisation: quality effects in green beans. *Innovative Food Science & Emerging Technologies* 9: 70–79.

Lim, S.-Y., Swanson, B.G., Ross, C.F. and Clark, S. (2008) High hydrostatic pressure modification of whey protein concentrate for improved body and texture of low fat ice cream. *Journal of Dairy Science* 91(4): 1308–1316.

Linton, M. and Patterson, M.F. (2000) High pressure processing of foods for microbiological safety and quality. *Acta Microbiologica et Immunologica Hungarica* 47(2–3): 175–182.

Ludikhuyze, L. and Hendrickx, M.E. (2001) Effects of high pressure on chemical reactions related to food quality. In: *Ultra High Pressure Treatment of Food* (eds M.E. Hendrickx and D. Knorr). Kluwer, London, pp. 17–185.

Margosch, D., Ehrmann, M.A., Ganzle, M.G. and Vogel, R.F. (2004) Comparison of pressure and heat resistance of *Clostridium botulinum* and other endospores in mashed carrots. *Journal of Food Protection* 67: 2530–2537.

Mariappagoudar, P. (2007) *Effect of Water Activity, Solute Type and Temperature on Inactivation of L. innocua during High Pressure Processing*. MS thesis, IIT, Chicago.

Matser, A.M., Krebbers, B., van den Berg, R.W. and Bartels, P.V. (2004). Advantages of high pressure sterilisation on quality of food products. *Trends in Food Science and Technology* 15: 79–85.

Mertens, B. (1995) Hydrostatic pressure treatment of foods: equipment and processing. In: *New Methods of Food Preservation* (ed. E.W. Gould). Blackie Academic & Professional, London, pp. 135–158.

Norton, T. and Da-Wen Sun (2008) Recent advances in the use of high pressure as an effective processing technique in the food industry. *Food and Bioprocess Technology* 1: 2–34.

Otero, L., Ramos, A.M., de Elvira, C. and Sanz, P.D. (2007a) A model to design high-pressure processes towards an uniform temperature distribution. *Journal of Food Engineering* 78(4): 1463–1470.

Otero, L., Ousegui, A., Urrutia-Benet, G., de Elvira, C., Havet, M., LeBail, A. (2007b) Modelling industrial scale highpressure- low-temperature processes. *Journal of Food Engineering* 83(2): 136–141.

Palou, E., Pez-Malo, A., Barbosa-Cánovas, G.V., Welti-Chanes, J. and Swanson, B.G. (1997) Effect of water activity on high hydrostatic pressure inhibition of *Zigosacharomyces bailii*. *Letters in Applied Microbiology* 24: 417–420.

Patazca, E., Koutchma, T. and Ramaswamy, H.S. (2006) Inactivation kinetics of *Geobacillus stearothermophilus* spores in water using high-pressure processing at elevated temperatures. *Journal of Food Science* 71(3): M110–M116.

Patazca, E., Koutchma, T. and Balasubramaniam, V.M. (2007) Quasi-adiabatic temperature increase during high pressure processing of selected foods. *Journal of Food Engineering* 80: 199–205.

Patterson, M.F. (2005) Microbiology of pressure-treated foods – a review. *Journal of Applied Microbiology* 98(6): 1400–1409.

Patterson, M. and Kilpatrick, D. (1998) The combined effect of high hydrostatic pressure and mild heat on inactivation of pathogens in milk and poultry. *Journal of Food Protection* 61(4) 432–436.

Pauling, L. (1964) *College Chemistry: An Introductory Textbook of General Chemistry*. Freeman and Company, San Francisco.

Peleg, M. and Cole, M.B. (1998) Reinterpretation of microbial survival curves. *Critical Reviews in Food Science and Nutrition* 38: 353–380.

Pflug, I.J. and Zeghman, L.G. (1985). Microbial death kinetics in the heat processing of food: determining an F-value. In: *Proceedings of Aseptic Processing and Packaging of Foods*, September 9–12, 1985, Tylosand, Sweden, pp. 211–220.

Ramaswamy, H. and Marcotte, M. (2006) Thermal processing. In: *Food Processing: Principles and Applications* (ed. H. Ramaswamy). CRC, New York, pp. 67–168.

Rasanayagam, V., Balasubramaniam, V.M., Ting, E., Sizer, C.E., Anderson, C. and Bush, C. (2003) Compression heating of selected fatty food substances during high pressure processing. *Journal of Food Science* 68(1): 254–259

San Martin, M.F., Barbosa-Cánovas, G.V. and Swanson, B.G. (2002) Food processing by high hydrostatic pressure. *Critical Reviews in Food Science and Nutrition* 42: 627–645.

Setikaite, I., Koutchma, T., Patazca, E. and Parisi, B. (2009) Effects of water activity in model systems on high-pressure inactivation of *Escherichia coli*. *Food and Bioprocess Technology* V2(2): 213–221.

Schauwecker, A., Balasubramaniam, V.M., Sadler, G., Pascall, M.A. and Adhikari, C. (2002) Influence of high-pressure processing on selected polymeric materials and on the migration of a pressure-transmitting fluid. *Packaging Technology and Science* 15: 255–262.

Shigehisa, T., Ohmori, T., Saito, A., Taji, S. and Hayashi, R. (1991). Effects of high pressure on the characteristics of pork surries and inactivation of microorganisms associated with meat and meat products. *International Journal of Food Microbiology* 12: 207–216.

Stumbo, C.R. (1973) *Thermobacteriology in Food Processing*. Academic Press, New York.

Ting, E., Balasubramaniam, V.M. and Raghubeer, E. (2002) Determining thermal effects in high-pressure processing. *Food Technology* 56: 31–35.

Toepfl, S., Mathys, A., Heinz, V. and Knorr, D. (2006) Review: potential of high hydrostatic pressure and pulsed electric fields for energy efficient and environmentally friendly food processing. *Food Reviews International* 22(4) 405–423.

Torrecilla, J.S., Otero, L. and Sanz, P.D. (2004) A neural network approach for thermal/pressure food processing. *Journal of Food Engineering* 62: 89–95.

Torrecilla, J.S., Otero, L., and Sanz, P.D. (2005) Artificial neural networks: a promising tool to design and optimize high-pressure food processes. *Journal of Food Engineering* 69: 299–306.

Trujillo, J. (2002) Applications of high-hydrostatic pressure on milk and dairy products. *High Pressure Research* 22(3): 619–626.

WSU (2008) *Pressure-Modified Whey Protein Concentrate May Have Little Impact on Ice Cream Flavor*, Emerging Food R&D Report. Washington State University, Pullman, WA.

35

Microwave and Radio-Frequency Heating Processes for Food

Francesco Marra

Introduction

Microwave (MW) and radio-frequency (RF) heating are forms of indirect electroheating, where the electrical energy is first converted to electromagnetic radiation, which subsequently generates heat within a lossy product. The purpose of both MW and RF heating processes is to provide heat to products in such a way as to overcome the limits set by conventional heat transfer processes, where internal heat conduction and external heat convection are the controlling steps, often resulting in slow and uneven heating. Both MW and RF heating processes are applied in the food industry, but they differ from each other in terms of principles, design, and industrial and domestic applications.

The scientific literature indicates a growing interest in the application of MW and RF heating (Marra *et al.*, 2009) and, interestingly, a number of contributions have appeared in books (Metaxas and Meredith, 1983; Datta and Anantheswaran, 2001; Rowley, 2001). The application of RF and MW technology is not limited solely to the food industry. This chapter, however, is intended to provide the reader with essential guidelines for designing food processes and equipment based on MW and RF heating: suggestions for references to be used as deeper sources of information are given at various steps, together with relevant online resources provided by government agencies and institutions.

Handbook of Food Process Design, First Edition. Edited by Jasim Ahmed and Mohammad Shafiur Rahman.

In fact, during MW and RF heating, the product to be heated receives a density of power internally, thanks to interactions between the electromagnetic field and the dielectric nature of the product: under these conditions, conduction may not limit the heating, since the density of power can be rather uniform and can cover the entire product, and the time needed to reach a specified temperature is shortened. In simple terms, heat transfer processes assisted by MW or RF heating can reach an efficiency (defined as the ratio of the energy absorbed by the product to be heated to the energy provided to the process) of up to 70%, compared with an optimal efficiency 5–10% for traditional heat transfer processes. MW and RF heating can even be more effective when the purpose of heating is a change in the physical state of the product (i.e., thawing or tempering) or a change in its composition as a result of a change in the physical state of one component (as in drying, when bound and free water are removed from the product by vaporization): in such cases, in fact, the energy provided during heating has to balance the energy losses due to changes in physical state. Furthermore, in traditional processes such as convective drying, where the driving force is not due just to a difference in temperature but also to differences in composition, when the driving force is small, the rate of moisture evaporation tends to zero. In the case of MW-assisted drying, however, the internal heat generation provides the energy needed to remove all of the residual moisture (Marra *et al.*, 2010).

Although MW and RF heating have similar principles and may be used with the same goals, these two processes differ very much in their technology and uses (not to mention the fact that everybody has seen a domestic MW oven at home, whereas few have had a chance to look at an RF-based oven). So, the following sections will be devoted to describing the basics of MW and RF heating, starting from the principles and theoretical fundamentals and proceeding to potential products and processes, schematic flow sheets and diagrams, the equipment available, and design procedures. Hazards and safety will be discussed too. Finally, some tips on economics and costs will be given.

Indirect Electroheating: Basic Information about MW and RF Heating

In an MW oven, food is cooked by exposing it to microwave radiation. Most household microwave ovens operate at a frequency of 2450 MHz in a continuous-wave (CW) mode. The source of the radiation in an MW oven is an MW generator, usually a magnetron tube. The magnetron, basically, converts 50 Hz or 60 Hz power-line electric current to electromagnetic radiation of frequency 2450 MHz. The high voltage (typically 3000–4000 V) that powers the magnetron tube is produced by a step-up transformer, a rectifier and a filter, which convert the alternating current to 4 kV direct current. A waveguide section transfers the MW energy from the magnetron to the oven cavity. Often, the oven cavity is equipped with a mode stirrer that helps to distribute the MW energy more or less evenly throughout the oven.

When an alternating electric field is applied to a food, one phenomenon that occurs is a movement of positive ions in the material towards negative regions of the electric field and a movement of negative ions towards positive regions of the field (Buffler, 1993, pp. 46–69). This movement of ions in this fashion is often referred to as ionic depolarization and is essentially resistance heating, like that found in ohmic heating, which is another electroassisted process. Heating occurs because the field is not static; its polarity changes continually at a high frequency (e.g., 27.12 MHz for RF or 2450 MHz for MW), is in contrast to ohmic heating, where the polarity changes at a much lower frequency (i.e., 50 Hz in Europe and 60 Hz in the USA). However, regardless of the frequency, the continued reversal of polarity in the electric field leads to the oscillation of ions forwards and backwards in the product, the net effect of this being the internal generation of heat within the product by friction (Buffler, 1993) (thereby avoiding a temperature lag between the surface and the center of a solid product). In addition to the movement of ions, dipolar molecules such as water in a material will also attempt to align themselves appropriately with the changing polarity of an electric field (a phenomenon known as dipole rotation). The movement of these dipoles can also cause friction between molecules, which can also lead to heat generation. Although MW and RF heating are both classed as dielectric heating methods (i.e., heating due to energy absorption by a lossy dielectric when it is placed in a high-frequency electrical field, as in Rowley (2001)), it is generally accepted that ionic depolarization tends to be the dominant heating mechanism at the lower frequencies encountered in the RF range, whereas both ionic depolarization and dipole rotation can be dominant loss mechanisms at the frequencies relevant to MW heating (i.e., 400–3000 MHz), depending upon the moisture and salt contents of the product (Tang, 2005). Therefore, in RF range, dissolved ions are more important for heat generation than the dipoles of the water in which they are dissolved (Ohlsson, 1983).

Theoretical Principles of Indirect Electroheating

According to the United States frequency allocation chart, the frequencies in the radio spectrum for industrial, scientific and medical (ISM) use are 6.78 ± 0.015 MHz, 13.560 ± 0.007 MHz, 27.12 ± 0.163 MHz, 40.68 ± 0.02 MHz, 915.0 ± 13 MHz, 2450.0 ± 50 MHz, 5.8 ± 0.075 GHz, 24.125 ± 0.125 GHz, 61.25 ± 0.250 GHz and 122.5 ± 0.500 GHz. At present, the industrial equipment available to be purchased operates only in frequency bands centered on 13.56 MHz, 27.12 MHz and 896 MHz (in the United Kingdom), 915 MHz (in the United States) and 2450 MHz (in the rest of the world).

As indirect electroheating involves the conversion of electromagnetic energy into heat, it is important to have a mathematical description of the power absorbed per unit volume of a product placed in the electromagnetic field developed in an MW cavity or an RF applicator. The product to be heated has different dielectric properties from the surrounding environment, and the effects of the electromagnetic field are attenuated.

The power density Q_v is proportional to the frequency f of the applied electric field and the dielectric loss factor ε″, and is also proportional to the square of the local electric field, E^2 (which plays a key role in determining how the material will absorb energy in the AC electric field), as in the following equation:

$$Q_v = \omega\varepsilon_0\varepsilon_r''E^2 = 2\pi f\varepsilon''E^2 \tag{35.1}$$

The electromagnetic field that is established in an MW cavity or an RF apparatus can be described in terms of the electric field intensity $\underline{E}$, the magnetic field intensity $\underline{H}$ and the electric flux density $\underline{D}$, by Maxwell's equations. When the medium involved is isotropic, linear and homogeneous, Maxwell's equations assume the following form:

$$\underline{\nabla}\times\underline{E} = -\mu\frac{\partial}{\partial t}(\underline{H}) \tag{35.2}$$

$$\underline{\nabla}\times\underline{H} = \left(\varepsilon_0\varepsilon_r\frac{\partial}{\partial t} + \sigma_c\right)\underline{E} \tag{35.3}$$

$$\underline{\nabla}\cdot\underline{D} = \rho_c \tag{35.4}$$

$$\underline{\nabla}\cdot\underline{H} = 0 \tag{35.5}$$

where σ_c is the effective electrical conductivity and ρ_c is the charge density.

The reader may recognize that the equations above are also known as Faraday's law (Equation 35.2), Ampère's law (Equation 35.3), Gauss's law for the electric field (Equation 35.4) and Gauss's law for the magnetic field (Equation 35.5).

Maxwell's equations need some assumptions about the dependence of the fields on time. A common assumption is to consider the fields to be time-harmonic:

$$\underline{E} = \underline{E}_0 e^{j\omega t} \tag{35.6}$$

$$\underline{H} = \underline{H}_0 e^{j\omega t} \tag{35.7}$$

When the time-harmonic assumption is made, Equations 35.2 and 35.3 become

$$\underline{\nabla}\times\underline{E} = -j\mu\omega\underline{H} \tag{35.8}$$

$$\underline{\nabla}\times\underline{H} = -j\varepsilon_0\varepsilon_r\omega\underline{E} \tag{35.9}$$

Combining the last two equations, one obtains the following wave equation, which is used for calculations in the frequency domain:

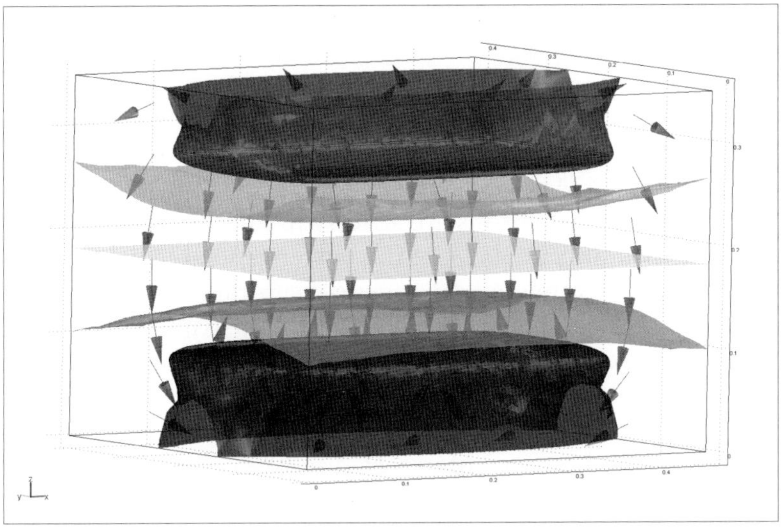

Figure 35.1 Distribution of $\underline{E}$ field (arrows) and positions of isopotential surfaces for a parallel capacitor placed in an RF cavity.

$$\underline{\nabla} \times \frac{1}{\mu_r} \underline{\nabla} \times \underline{E} - \omega^2 \mu_0 \varepsilon_0 \varepsilon_r \underline{E} = 0 \tag{35.10}$$

Boundary conditions are also needed in order to obtain a unique, correct solution. The external boundaries of the cavity can be considered as perfect conductors, and the following boundary condition is then used:

$$\underline{n} \times \underline{E} = 0 \tag{35.11}$$

where $\underline{n}$ is the unit vector normal to the boundary surface.

The solution of the set of electromagnetic equations presented above allows one to reproduce the $\underline{E}$-field distribution, for example that in an RF cavity (Figure 35.1).

Under a number of assumptions (the food product is regularly shaped, in the form of a slab; it is placed in an electromagnetically shielded cavity; only the upper surface of the product is exposed to MW radiation, assumed to be of uniform intensity; the product is homogeneous and isotropic; its dielectric properties are constant; and neither shrinkage nor deformation of the sample is taken into account), Q_v can be represented by an analytical expression (Fleischman, 1996; Fleischman, 1999), such as

$$\frac{Q_v(z)}{Q_0} = f(z) = \frac{2\varepsilon' |\tau|^2 (\cosh 2\alpha z + \cos 2\beta z)}{|r|^2 \exp(-2\alpha H_s) + \exp(2\alpha H_s) + 2|r| \cos(2\beta H_s + \arg r)} \tag{35.12}$$

where ε'' is the dielectric loss factor, α is the wave number, β is the attenuation constant, r is the argument of the reflection coefficient, or phase angle, z is the distance from the slab surface exposed to MW perturbation, $|\tau|$ is the magnitude of the transmission coefficient, and H is the slab thickness. Alternatively, a semiempirical expression, such as Lambert's equation (Romano *et al.*, 2005), can be used. For a general geometry, the solution of Maxwell's equations needs numerical methods (Ayappa *et al.*, 1991), and Maxwell's equations are strictly linked to the heat equation and vice versa. However, a preliminary analysis of the power generated per unit volume allows one to foresee where the heating will be most concentrated. As shown by Marra *et al.* (2010), even for a simple geometry (i.e., a slab), the distribution of power generated per unit volume varies very much with the nature of the food substrate to be treated and the size of the sample (Figure 35.2).

But there are also a number of processes, such as RF heating processes, where a quasi-static approach can be used (Marra *et al.*, 2007; Marra *et al.*, 2009). In these cases, the mathematical analysis of the process is simplified, since Maxwell's equations collapse to the following equation:

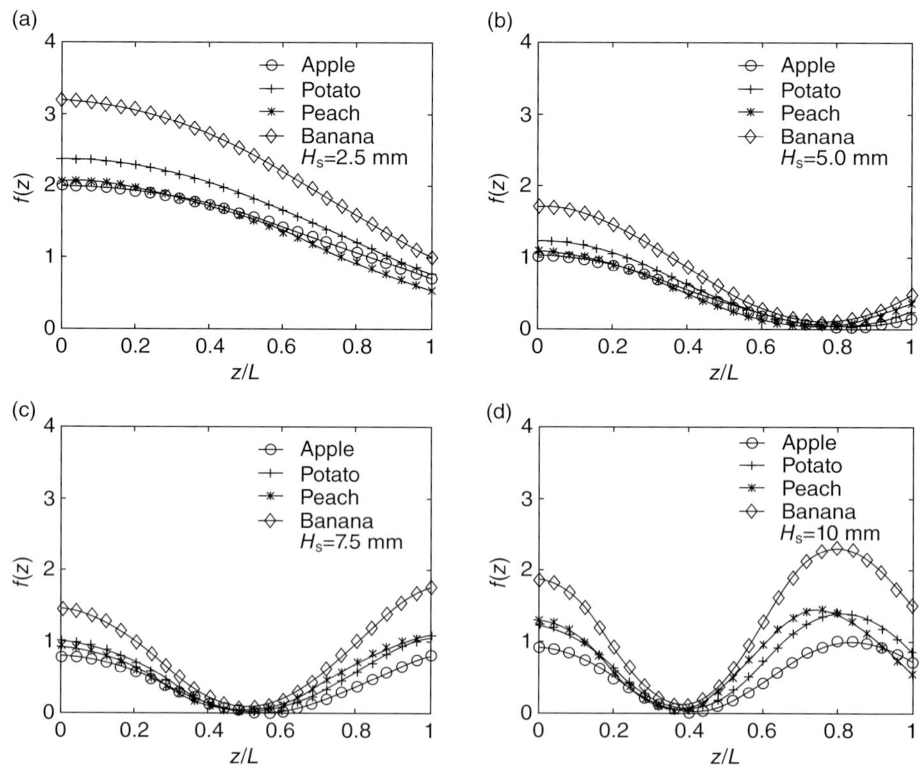

Figure 35.2 Examples of the distribution of power generated per unit volume in a slab undergoing MW heating. (From Marra *et al.*, (2010), courtesy of Elsevier).

$$\underline{\nabla} \cdot [(\sigma + j\omega\varepsilon)\underline{\nabla} V] = 0 \tag{35.13}$$

where V is the electrical potential, related to the electric field by

$$\underline{E} = -\underline{\nabla} V \tag{35.14}$$

In this case, the boundary conditions are that a source electrical potential V_0 is applied to the upper electrode of the capacitor and the bottom is electrically grounded ($V = 0$). The shells of RF applicators are electrically insulated, so the boundary condition on the shell is

$$\underline{\nabla} \cdot \underline{E} = 0 \tag{35.15}$$

A mathematical description of the heat transfer within the food product placed between the electrodes is given by the unsteady heat-conduction equation (assuming that a solid-like foodstuff is being processed in the RF applicator), with a generation term represented by the power density already given in Equation 35.1:

$$\rho_{\mathrm{f}} c_{p\mathrm{f}} \frac{\partial T_{\mathrm{f}}}{\partial t} = \nabla \cdot (\lambda_{\mathrm{f}} \nabla T_{\mathrm{f}}) + Q_{\mathrm{v}} \tag{35.16}$$

where T is the temperature in the sample, t is the process time, α is the thermal diffusivity, ρ is the density, C_p is the heat capacity and the subscript "f" denotes the food.

The heat transport equation to be solved needs initial and boundary conditions. For the initial condition, a uniform temperature T_0 can often be assumed within the food sample. On the boundary, as a general condition, convective heat transfer from the external surface in accordance with Newton's law can be assumed:

$$-k\nabla T = U(T - T_{\mathrm{air}}) \tag{35.17}$$

where U is the overall convective heat transfer coefficient and T_{air} is the temperature inside the oven. But, at the boundary, other important phenomena have to be taken into account, such as heat losses related to moisture transfer and radiative heat flux (as in the case of some combined apparatuses).

More recently, researchers have discussed the need for a conjugate approach to better couple external and internal heating effects in combined MW and convective heating processes (Marra *et al.*, 2010). The latter approach is based on a deeper knowledge of transport phenomena and may require dedicated software and hardware. Readers are referred to the more recent literature (Salagnac *et al.*, 2004; Kowalski *et al.*, 2010; Marra *et al.*, 2010) on this topic.

Empirical Data and Properties Needed for Designing MW and RF Processes

The conceptual design of both RF and MW heating processes needs an accurate knowledge of the physical properties of the food material, above all the dielectric properties, which determine the calculation of the electric field in the product to be heated and, thus, the power needed.

Dielectric Properties

Permeability and Permittivity

The dielectric properties of food materials can be divided into two parts, known as the permeability and permittivity (ε). The permeability values for foodstuffs are generally similar to that of free space and, as a result, are not believed to contribute to heating (Zhang and Datta, 2001). However, the permittivity, which determines the dielectric constant ε' and the loss factor ε'', influences RF heating. The quantities ε' and ε'', which are the real and imaginary parts, respectively, of ε, are given by

$$\varepsilon = \varepsilon' - j\varepsilon'' \qquad (35.18)$$

The quantity ε' is a characteristic property of a material and is a measure of the capacity of the material to absorb, transmit and reflect energy from the electric portion of the electromagnetic field (Engelder and Buffler, 1991); it is a constant for any material at a given frequency under constant conditions. The value of ε' is a measure of the polarizing effect of the applied electric field (i.e., how easily the medium is polarized), whereas ε'' measures the amount of energy that is lost from the electric field, which is related to how the energy from the field is absorbed and converted to heat by the material (Engelder and Buffler, 1991). A material with a low value of ε'' will absorb less energy and could be expected to heat poorly in an electric field owing to its greater transparency to electromagnetic energy (Decareau, 1985). However, it is important to emphasize that the dielectric properties are but one of a range of properties that influence the temperature rise, and other properties (e.g., the specific heat capacity) will also have an influence on the magnitude of the temperature rise obtained.

Electrical Conductivity

The electrical conductivity σ indicates the ability of a material to conduct an electric current. In a dielectric food system, σ is related to ionic rotation. It contributes to ε'' and, in the RF range, it can be calculated from the following equation (Piyasena et al., 2003a):

$$\sigma = 2\pi f\varepsilon'' \qquad (35.19)$$

It is also worth noting that at RF frequencies, ε″ can be estimated with reasonable accuracy for liquid or semiliquid foods by measuring the σ value of the material using an electrical-conductivity meter (Guan *et al.*, 2004).

Loss Tangent

The tangent of the dielectric loss angle (tan δ) is often called the loss tangent or the dissipation (power) factor of the material. For a given material, this is equivalent to the ratio of ε″ to ε′:

$$\tan\delta = \frac{\varepsilon''}{\varepsilon'} \quad (35.20)$$

Penetration Depth

Bengtsson and Risman (1971) defined the penetration depth d_p as the depth in a material where the energy of a plane wave propagating perpendicular to the surface has decreased to $1/e$ (1/2.72) of the surface value. It can be expressed mathematically as

$$d_{\mathrm{p}} = \frac{c}{2\pi f\sqrt{2\varepsilon'}\sqrt{\sqrt{1+(\tan\delta)^2}-1}} \quad (35.21)$$

where c is the speed of propagation of waves in a vacuum ($3 \times 10^8\,\mathrm{m{\cdot}s^{-1}}$) and d_{p} is expressed in meters. When $\tan\delta$ is low (i.e., far less than 1), Equation 35.20 can be simplified to

$$d_{\mathrm{p}} = \frac{c}{2\pi f\sqrt{\varepsilon'}\tan\delta} = \frac{4.47\times10^7}{f\sqrt{\varepsilon'}\tan\delta} \quad (35.22)$$

Equations 35.8 and 35.9 illustrate the effect of f, ε′ and ε″ on d_p. Bengtsson and Risman (1971) found that the greatest values of d_{p} occurred when both ε′ and ε″ were low.

Factors Influencing the Dielectric Properties of Foods

The dielectric properties of foods play an important role in both RF and MW heating (Piyasena *et al.*, 2003b; Ahmed *et al.*, 2007a,b). The dielectric properties of most materials are influenced by a variety of factors. The moisture content is generally a critical factor (Tang, 2005), but the frequency of the applied alternating field, the temperature of the material and also the density, chemical composition (i.e., the amounts of fat, protein, carbohydrate and salt) and structure of the material all have an influence (Piyasena *et al.*, 2003b; Ahmed *et al.*, 2007a,b). In terms of bulk density, samples of an air–particle mixture with higher densities generally have higher ε′ and ε″ values because less air is incorporated into the samples (Nelson and Datta, 2001). In relation to composition, Nelson and Datta (2001) stated that the dielectric properties of materials are dependent on chemical composition and, especially, on the presence of mobile ions and the permanent dipole moments associated with water. While a considerable amount of work on dielectric properties at MW frequencies has been published, a more limited number of studies have examined the dielectric properties of food and agricultural products at RF frequencies. The work completed to date was summarized by Piyasena *et al.* (2003a) and by Marra *et al.* (2009).

Conceptual Design of Electroheating Processes

In both RF and MW heating processes, the development of the process flow sheet begins by identifying the purpose of the process itself and the material to be processed (and, thus, its thermophysical and dielectric properties and characteristics as related to electroassisted heating), and by choosing between a batch and a continuous process.

The required power is usually calculated based on an initial assessment of the proposed process and verified by actual testing once an initial oven concept, layout and size have been established. The key parameters to be verified include the heating-rate sensitivity, temperature uniformity and process efficiency. The heating-rate sensitivity, which can be a problem in some drying applications, may force the use of a longer cavity to increase the process time at the expense of process efficiency.

Because the MW power that will be provided in the process will mainly contribute to the moisture losses, the nominal power of the plant, in an initial shortcut procedure, can be calculated as

$$P_{\mathrm{nMW}} = \frac{P_{\mathrm{req}}}{\eta_{\mathrm{MW}}} \tag{35.23}$$

where P_{nMW} is the nominal power of the MW generator, P_{req} is the power requested to accomplish the process and η_{MW} is the overall efficiency of the MW process, which can be assumed to be 50%. In this way, a shortcut estimation of the MW generator power can be done. Because the power of a magnetron is limited, the nominal power of the MW generator is divided by the power of a single generator (in many industrial cases, 10–25 kW) in order to obtain the number of MW generators needed for the process.

Once the MW power generation system has been chosen, the flow sheet for the design must take into account any other secondary heat sources that may be needed for the purposes of the process. If the purpose is drying, convection heating using warm air can help. This can be achieved by a fan blowing warm air over the product being dried or by a series of heaters or gas burners placed along the path.

An example of a custom-designed MW process is shown in Figure 35.3. Here, a train of open bottles containing minimally processed baby food arrives at an MW pasteurizer, where the products receive the power needed to raise the internal temperature

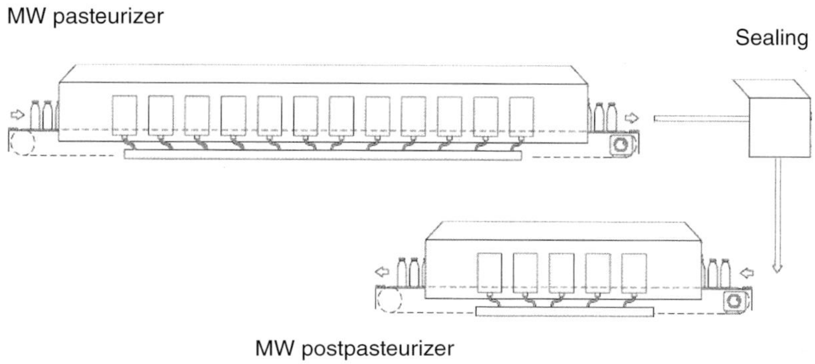

Figure 35.3 Custom-designed MW process.

of the product to some desired level. The bottles are then sealed and finally sent to an MW postpasteurizer, where the product undergoes another heating step.

Equipment

Characteristics of MW Systems

The main characteristics of an MW system can be summarized by considering three basic parts: the MW source, the waveguide and the applicator. The first type of MW generator widely used was the klystron, soon replaced in most applications by the magnetron. At the customary microwave frequencies, the magnetron is certainly the most economical choice; it is capable of functioning properly in both household ovens and large industrial applications. Because magnetrons are produced in millions, their cost is rather low; also, magnetrons have advantages such as small size, low weight and high efficiency. The magnetron is categorized as a crossed-field MW tube, so called because the basic interaction depends upon electron motion in electric and magnetic fields that are perpendicular to one another. The magnetron consists of a vacuum tube with a central electron-emitting cathode at a highly negative potential, surrounded by a structured anode. In its most familiar embodiment, a cylindrical electron emitter, or cathode, is surrounded by a cylindrical structure, or anode, which is at a high potential and capable of supporting microwave fields. Magnets are arranged around it to supply a magnetic field parallel to the axis and hence perpendicular to the electric field between the anode and cathode (NMAB, 1994). The interaction between the electrons traveling in this crossed field and the microwave field supplied by the anode causes a net energy transfer from the applied DC voltage to the MW field. The interaction occurs continuously as the electrons traverse the cathode–anode region. The anode contains cavities, which are coupled to the fringing fields and have a resonant frequency equal to the intended MW frequency. The stored electromagnetic energy can be coupled out by a circular loop antenna in one of the cavities into a waveguide or a coaxial line.

The power output of a magnetron can be controlled by the tube current or the magnetic field strength. In order to prevent melting of the anode, its temperature has to be limited and, thus, the maximum power of the magnetron has to be limited too. The practical limits at 2.45 GHz are approximately 1.5 kW and 25 kW for air- and water-cooled anodes, respectively. The 915 MHz magnetrons have larger cavities (a lower resonant frequency means a longer wavelength) and thus can achieve a higher power per unit. The efficiencies of modern 2.45 GHz magnetrons are around 70–80%, most being limited by the magnetic flux of the ferrite magnets used for reasons of economy, but the total efficiency of MW heating applications is often lower owing to unmatched loads.

Simply stated, MW applicators are devices that are designed to heat a material by exposing it to an MW field in a controlled environment. The objective of their use is to cause a controlled interaction between the microwave energy and the material to occur under safe, reliable, repeatable and economical operating conditions. Applicators may be conveyor-operated or of batch type, or a combination of both.

Several industrial applications also require other energy sources, depending on the particular details of the process: in these cases, MW energy may be combined inside the applicator with hot air, with or without forced convection, or with infrared radiation or steam, as discussed elsewhere in this chapter. Other special requirements (both for the process and for the product to be processed) have led to the design of MW applicators capable of controlled interaction under a variety of ambient conditions, ranging from vacuum to high pressure and humidity.

MW applicators can be divided into two main classes: multimode and single-mode cavities. The multimode cavity is, by far, the most widely used MW applicator, since it is used in domestic MW ovens, but it is also used in a large number of industrial power units. It is a simple configuration, where a wide range of loads can be processed. It can be a simple closed metal box, of course with some means of coupling in power from an MW generator. The product dimensions often determine its dimensions, and in any case the cavity has to be several wavelengths long in at least two dimensions. This type of applicator has the disadvantage of a spatially nonuniform distribution of heating. To improve the uniformity of heating, movement of the product to be heated (using rotating turntables or straight-line motion along a travel axis inclined with respect to the principal axes of the applicator) can be used; also, metal stirrers, i.e., moving devices capable of reflecting energy and changing the field distribution continuously, can be employed.

The most efficient type of applicator is the single-mode resonant cavity (Figure 35.4), but it has some limitations. A single-mode cavity may be cylindrical or rectangular, and within it only one mode of propagation (the simplest being TM_{010}) is permitted: thus, it is possible to design an applicator that has a single resonance near to the operating frequency. In this way, the electromagnetic field is well defined in space (Dibben, 2001). Unfortunately, because of the typical values of the dielectric properties of food products, this kind of applicator is suitable only for small samples, and this limits its use for food processing.

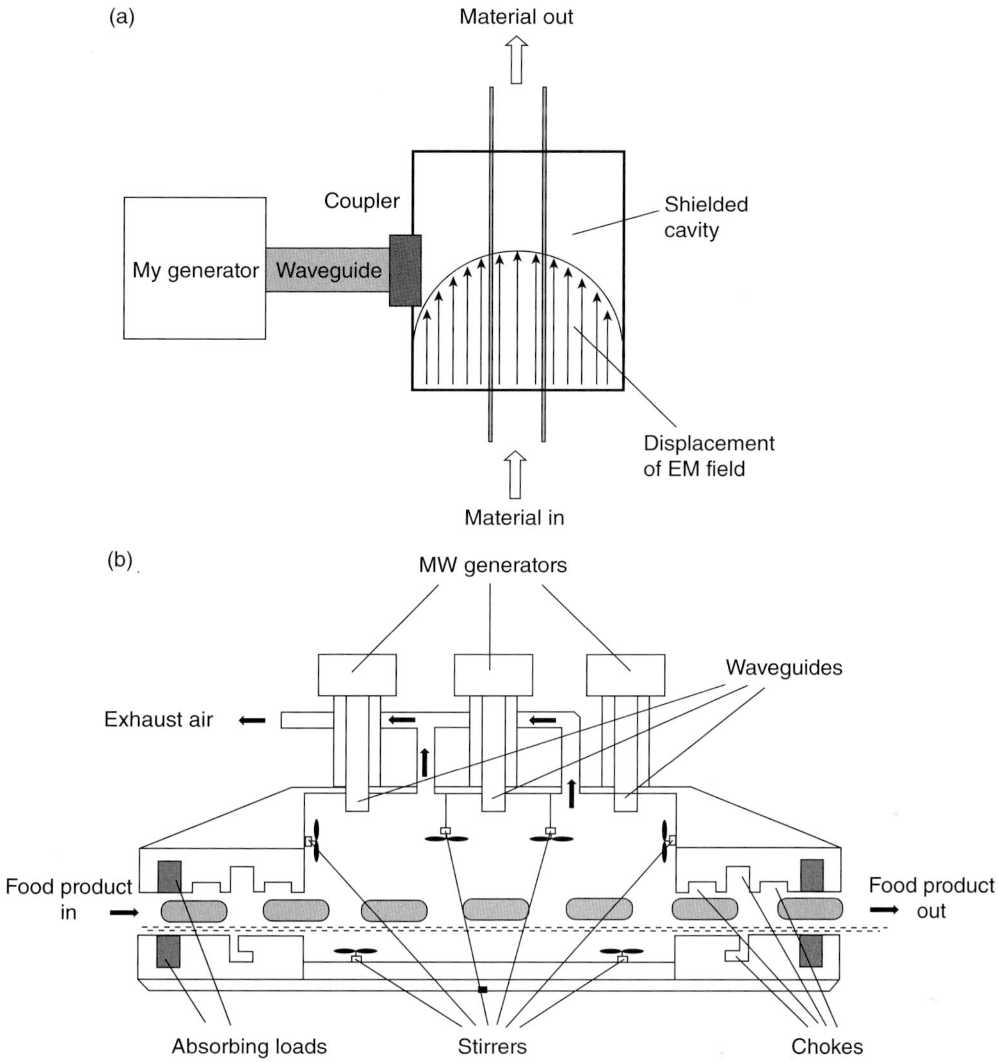

Figure 35.4 (a) MW single-mode applicator; (b) MW multi-mode applicator

Characteristics of RF Systems

In contrast to MW cavities, RF energy is generated by a triode valve and is applied to the material via a pair of electrodes (Rowley, 2001). In the the type of applicator that is simplest to assemble, the parallel-plate RF system depicted in Figure 35.1, one of these electrodes is grounded, which creates a capacitor that stores electrical energy. The target material to be heated is placed between, but not touching, the parallel electrodes. It should be noted that while parallel-plate electrodes (or "through-field" applicators, as in Figure 35.5) are the most commonly used electrode configuration for heating thicker materials, two other types of configuration are available: "fringe-field" applicators, which consist of a series of bar, rod or narrow plate electrodes and are

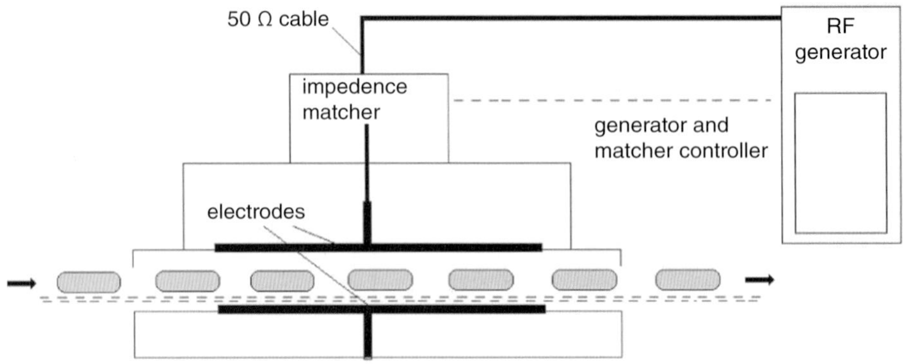

Figure 35.5 Through-field 50 Ω RF applicator.

most suited for heating or drying thin layers (<10 mm thick) and "staggered through-field" applicators, which consist of rod- or tube-shaped electrodes in a staggered arrangement on either side of a belt, which are used for heating products of intermediate thickness (Marra *et al.*, 2009).

Processes, Products and Potential Products

This section deals with the products that are currently being processed by MW and RF heating and also discusses some potential products relevant to the food industry. Electroassisted heating can offer high-temperature and short-time processing, resulting in quality advantages (Zhao *et al.*, 2000).

MW Cooking

Cooking can cause partial or total loss of valuable nutrients, depending on the process parameters. Research has demonstrated that many vitamins in foods are thermolabile and leach out during thermal processing. Over the years, microwave processing has been proposed as an advantageous method over traditional processing, with reduced nutrient losses (Orzaez Villanueva *et al.*, 2000; Begum and Brewer, 2001; Brewer and Begum, 2003). Ohlsson and Bengtsson (2001) discussed the industrial application of microwave cooking of bacon: this process yields rapid output and the collection of high-quality rendered fat for reuse or resale, and it also yields very good product appearance and convenience and diminishes the risk of nitrosamine formation. Combination ovens are now being marketed for commercial use (in restaurants and other food services) to offer a competitive edge using an innovative combination of convection, radiant and microwave technology.

MW Pasteurization/Sterilization

Industrial microwave pasteurization of packaged foods is another potential application of electroassisted heating. In a study by Lau and Tang (2002), pickled asparagus in glass jars was pasteurized using microwaves at 915 MHz. The process produced uniform heating while reducing the process time by at least one-half compared with water-bath heating, and significantly reduced the thermal degradation of asparagus. When MW heating is used for pasteurization, care must be taken to ensure homogeneity of the thermal process within the product and that the target lethal temperature is maintained for a sufficient period of time to provide a safe product.

Tang *et al.* (2008) investigated the feasibility of developing a short-time sterilization protocol for a highly inhomogeneous food (consisting of sliced beef and gravy) prepackaged in polymeric trays using 915 MHz MW energy. These authors concluded that 915 MHz single-mode MW sterilization technology was effective for processing the nonhomogeneous food investigated, but would also be effective for other packaged nonhomogeneous foods, such as chicken meat in gravy in trays, and salmon in sauce in pouches.

MW Tempering

Frozen food often needs to be thawed or tempered ahead of processing to allow further mechanical handling or processing (slicing, cutting, molding, etc.). Industrially, this has been achieved in the past by immersing the product in water for an extended period. Nowadays, this inefficient method can be replaced by MW and RF processes (Farag *et al.*, 2010). The aim of MW tempering of frozen food products is to raise the temperature of the product as uniformly as possible, while maintaining the quality of the product at refrigeration temperature. With rapid MW tempering, there is no temperature abuse of the product and there are reduced drip losses and reduced space and inventory requirements. Various suppliers (Microdry Incorporated, Sairem, Defreeze Corporation, Amtek Inc. and Bielle) are providing conveyor tunnels equipped for MW tempering for industrial purposes.

MW Drying

It is known that, during drying, changes in the physical and biological structure of products are inevitable. Traditionally, drying is performed by convection, using dry air to remove surface water from the product to be dried, therefore creating a pressure gradient between the surface and the inside of the product, which causes water migration from the inside to the surface. In this process, an increased temperature will enhance the ability of dry air to remove water from the surface and increase the water migration rate within the commodity. However, during the time the process takes to reach equilibrium, there is potentially a serious overheating of the product surface. In air-drying systems, hot air removes water in the free state from the surface of the product, whereas MW energy removes the water inside the product, which is again in the free state (Schiffmann, 1995). The mechanism of this process pushes water out of the product with great efficiency as the moisture content of the product decreases.

An advantage of using MW energy is the possibility of combining multiple drying methods. Drying systems in which air and MW drying are combined not only increase the drying rate of the product, but also increase the quality of the dry product obtained

(Alibas, 2007). In addition to hot air, MW energy can also be combined with freeze drying or vacuum drying.

A significant industrial application was described in a paper by Osepchuk (2002), showing how the moisture content of pasta, crackers and chips can be leveled off by combined MW/air tunnel drying. The use of MW energy for drying offers some advantages, as it complements conventional drying well in the later stages by targeting specifically the internal moisture of the product. However, when the heat generation within the product is too rapid, it can cause the generation of a high internal pressure of steam, resulting in expansion, which can lead to collapse of the product or explosion of material (puffing). Microwave drying techniques have proved to be effective for a number of agricultural products (Vadivambal and Jayas, 2007).

Numerous studies have been carried out with microwave–air combined dryers, and many agricultural products have been dried successfully. Crops such as carrots, apples, mushrooms, American ginseng roots and potatoes (Khraisheh *et al.*, 2004) can be dried successfully using a combined drying technique. Combined MW/hot-air tunnel dryers are commercially available (from Bielle, Italy, and Sairem, France).

MW Baking

Microwave baking has been the focus of much research and development since the 1950s, with variable success. All results point to the general rule that to achieve success, considerable product reformulation must be considered (Decareau, 1985). According to Schiffmann (1992), the use of MW baking for cookies results in a more uniform moisture distribution than does forced convection. The effects of post-baking by MW were studied by Bernussi *et al.* (1998), who analyzed the moisture gradient and overall quality of cookies. After forced-convection heating at 240 °C for 4 min, the cookies were baked further in an MW oven: MW baking significantly reduced the moisture gradient and total moisture content of the cookies, with certain undesired effects such as a darker surface color and a shrinkage effect attributed to the microwave treatment.

RF Meat Processing

Meat and meat products are the area in which most of the information on RF heating treatments has been published to date, both in terms of processing and in terms of the effects of RF treatment on the quality of foods. RF processing was found to reduce cooking times to up to 1/25 of the time for conventional cooking in a water bath (Laycock *et al.*, 2003). These authors compared the heating rate, time–temperature profiles and quality of three meat products (ground, comminuted and noncomminuted muscle) cooked in a water bath or by a 1.5 kW RF heater (a cylindrical chamber between two electrodes wrapped around the RF applicator). Instrumental quality measurements indicated that the RF-cooked samples had lower juice losses and were also acceptable in terms of color and water-holding capacity. However, the same workers found that the texture of ground beef was too chewy and elastic, and the color of whole beef was considered inferior. These authors also postulated that well-mixed comminuted and ground meat products appeared to be the most promising for RF cooking.

Orsat *et al.* (2004) pasteurized vacuum-packaged ham slices in an RF applicator (600 W at 27.12 MHz) and subsequently examined samples for moisture loss, color changes, total bacterial surface counts and sensory quality attributes, such as off-odors and sliminess, concluding that RF heating, coupled with appropriate packaging, could improve the storability of repacked hams by decreasing the bacterial load and reducing the moisture loss, while maintaining overall a greater sensory quality of the product. Also in the area of microbial quality, Guo *et al.* (2006) compared the effectiveness of RF versus hot-water-bath cooking for the inactivation of *E. coli* K12 in ground beef. These workers, who used an RF apparatus similar to the system used by Laycock *et al.* (2003), reduced the cooking time to 1/30 of the time for immersion cooking and found lower temperature variations. They noted that both methods significantly reduced *E. coli* and extended the product shelf-life, and suggested that RF cooking had great potential as an alternative method to immersion cooking.

Extensive investigations of the processing and quality of RF-cooked meats were published between 2004 and 2007 (Brunton *et al.*, 2005; Lyng *et al.*, 2007; McKenna *et al.*, 2006; Tang *et al.*, 2005, 2006; Zhang *et al.*, 2004; Zhang *et al.*, 2006; Zhang *et al.*, 2007) by a team from UCD Dublin, Ireland, who heated products using a 27.12 MHz RF applicator with a maximum power output of 600 W (Lyng, 2007). These workers developed systems for RF cooking of cased meat samples that avoided arcing on the casings, one of most undesirable side effects that occurred during RF cooking in air. These systems involved submerging the cased products in circulating water at 80 °C during cooking in purpose-built polyethylene cells. The quality attributes of the RF-cooked products that were examined included yield, texture, color and flavor. In the case of comminuted products, no yield differences were noted between conventional and RF-cooked samples. However, for noncomminuted products (ham and beef), in all cases the RF-cooked products had higher yields than those of steam-cooked products, the difference being about 1–1.5% in the case of ham and 4–6% for beef. When products cooked by conventional or RF cooking methods were compared in terms of instrumental texture, no definite trend emerged; the differences, where they occurred, were generally small (Lyng, 2007). The rates of oxidation in beef and turkey were slightly less in RF-cooked samples than in steam-cooked samples, but the magnitude of the difference was so low that it was unlikely to be detected by sensory analysis. In addition, no major differences were noted in sulfur volatiles in RF-cooked versus steam-cooked samples. Zhang *et al.* (2004) developed an optimized cooking protocol for pasteurizing meat emulsion samples. These authors reported reductions of up to 79% in pasteurization times for meat products compared with equivalent steam-cooked samples, although they claimed that the use of a higher-power RF source could reduce cooking times further.

RF Postharvest Treatment and Disinfestation of Fruit

Nowadays, RF radiation is often considered as an alternative method for the disinfestation of fruit and nuts. This is largely due to the fact that processors are seeking alternatives to some of the more traditional fumigation agents such as methyl bromide, which may be discontinued because of environmental concerns and also issues regarding health and safety. Researchers from Washington State University in the USA have performed the most extensive investigations on this topic (Birla *et al.*, 2004; Birla *et al.*, 2005; Hansen *et al.*, 2006a,b; Ikediala *et al.*, 2002; Mitcham *et al.*, 2004; Monzon *et al.*, 2006; Tang *et al.*, 2000; Wang *et al.*, 2002; Wang *et al.*, 2003; Wang *et al.*, 2006a), including discussion of the scaling-up of RF plants and the design of commercial RF processes (Wang *et al.*, 2006b; Wang *et al.*, 2007a,b). A kinetic-model study describing the intrinsic thermal mortality of insect pests (Tang *et al.*, 2000) demonstrated the potential to develop high-temperature–short-time thermal treatments for controlling codling moth in fruits. Wang *et al.* (2003) examined the dielectric properties of fruits and insect pests in the context of RF and MW treatments. When a selection of fruits and insect larvae were examined, these authors found that the loss factors of common pest insects at RF frequencies were clearly larger than that of nuts, suggesting possible differential, faster heating of insects compared with nuts when the two were treated simultaneously in an RF applicator.

An interesting improvement in the design of RF-based treatments for fruits was proposed by Birla *et al.* (2004), who developed a laboratory-scale fruit-mover that was capable of continuously rotating a fresh fruit (an orange and an apple were the test samples) using a series of water jets, with the purpose of achieving uniform RF heating. The published results showed that rotation and movement of the fruit minimized the adverse effects of the nonuniformity of the RF field and the irregular geometry of the fruit and led to an overall improvement in the heating uniformity. It is important to emphasize that these authors found that the geometry of the fruit influenced the heating pattern.

Wang *et al.* (2006b, 2007a,b) considered the design and scaling-up of plants and processes for industrial-scale postharvest RF treatment of walnuts, working with continuous apparatus characterized by high power (up to 25 kW) and a high product load (up to 187 kg of product moving on a belt). The design and running of industrial-scale RF processes is one of the main challenges in this field, since Wang *et al.* (2007b) stated that many factors (e.g., the orientation of the walnuts, differential heating between walnuts with open and closed shells, and the actual poststorage and energy costs) that influence commercial RF processes must be taken into account. The energy costs may be comparable to the costs of traditional fumigation processes, but at present the use of RF installations would incur considerably higher capital costs (Marra *et al.*, 2009).

MW and RF Safety Guidelines

In an earlier section, it was stated that that according to the United States frequency allocation chart, the allowed frequencies for ISM use are 6.78 ± 0.015 MHz, 13.560 ± 0.007 MHz, 27.12 ± 0.163 MHz, 40.68 ± 0.02 MHz, 915.0 ± 13 MHz, 2450.0 ± 50 MHz, 5.8 ± 0.075 GHz, 24.125 ± 0.125 GHz, 61.25 ± 0.250 GHz and 122.5 ± 0.500 GHz. The nature of their use is classified as unconditional: this means that there is no interference-based radiation limit within these frequency bands and that radio communication services operating within these bands must accept any harmful interference that may be caused by ISM applications (Metaxas and Meredith, 1983). Independently of the frequency used, all MW- and RF-based equipment must comply with specific regulations designed to keep radiation leakage below given levels. The currently accepted standard is the guidelines developed by the American National Standards Institute of a power density of 10 mW/cm^2 for exposure (ANSI C95.1-1991). The power density guideline is based on a maximum permissible exposure of 0.4 W/kg specific absorption rate, which is a factor of 10 less severe than the threshold absorption level that has been determined (Redhead, 1992). Standards based on the Institute's guidelines include the Food and Drug Administration's emission standard of 5 mW/cm^2 at 5 cm for microwave ovens (HHS, 1991) and the Occupational Safety and Health Administration's exposure standard of 10 mW/cm^2 (HHS, 1991). These exposure limits are averaged over a 6 min (0.1 h) period.

As the MW power levels for industrial processing systems increase, the potential hazards associated with exposure to radiation become more important. Extensive work, as summarized in a review article (Michaelson and Lin, 1987), has indicated that the effects on biological tissue from exposure at MW frequencies are purely thermal in nature. Unlike the higher-energy, ionizing region of the electromagnetic spectrum, including X-rays and gamma rays, the nonionizing bands from DC to visible light do not carry enough energy to break chemical bonds (Redhead, 1992). At present, the only confirmed effect is warming, from the conversion of electromagnetic energy to heat. Thus, MW exposure standards are based on the thermal effects of exposure, which has to be minimized. In order to minimize exposure, MW systems need to be designed with effective leakage suppression, with viewing and ventilation screens, and with an interlock system on doors and access apertures to shut off power when the doors are opened (Osepchuk, 2002).

In order to prevent leakage of MW radiation, a number of practical solutions (for both batch systems and online systems) are available. For batch systems, the only source of MW radiation leakage is represented by the door seals. The door seals should retain their integrity throughout the life of the MW apparatus even when slightly damaged or contaminated with dirt. Internally, they have to provide a conducting surface identical to the body of the apparatus. A fail-safe interlock system is required too, to prevent MW power being supplied unless the doors are closed and fully clamped.

Whatever the industrial application, designers, installers and operators should also pay attention to the general recommendations presented in Table 35.1.

Table 35.1 Safety tips for the operation of MW ovens and apparatuses.

Operations	Operators	Maintenance
Do not operate MW ovens and apparatuses when empty	Read operating manuals carefully and pay attention to labels placed in the most hazardous locations of the equipment	Ensure that the microwave oven is unplugged or disconnected from electrical power before reaching into any accessible openings or attempting any repairs
The power generated must never be allowed to radiate freely into occupied areas	When required by circumstances, remain outside the Faraday cage or behind any additional screen	
Reduce stray radiation by applying adequate reactive and absorbent loads at entry and exit ports	Always dress appropriately	The services of a qualified operator should be sought when any malfunction is suspected
Do not test an MW power-generating component without an appropriate load connected to its output	Remember that MWs can interfere with pacemakers. Persons with pacemaker implants should not be near MW facilities unless they are sure that it is in good operating condition and there is no leakage of microwave radiation	Ensure that the adjustment of voltages, replacement of MW power-generating components, dismantling of components of the apparatus and refitting of waveguides are undertaken only by persons who have been specially trained for such tasks
Check that the door seals and the inside surfaces of the door and the oven cavity are clean after each use	Warning lights must be placed around the zone of the MW apparatus to avoid personnel other than operators approaching hazardous areas	
Take special care to ensure that no damage occurs to the part of the oven making contact with the door or door seals		
Never put hands or foreign objects in entry ports while the system is operating		

The Economics of MW and RF Processing

Establishing what are the key parameters that determine the costs associated with MW or RF processing and how these costs compare with the costs of conventional processing is not trivial. The costs have to be examined based on the cost of capital equipment, operating costs (energy, replacement and maintenance), the energy requirements of each part (energy efficiency) and the cost of energy, deletions and addition of steps compared with conventional processing schemes, savings in time and space, and changes in yield compared with conventional processing.

Considering just the operating costs, the economic feasibility of such processes is a function of local variations in energy costs and environmental laws. MW and RF processing can improve the yield and the productivity, and, in the case of large production volumes, when the tasks of supervisors can be important, labor costs play a significant role. On the other hand, further development of MW and RF processing is expected soon, because of the enormous advantages that these technologies give in terms of yield and product improvement, so saving on unit operating costs.

What is definitely not trivial is to provide capital costs for turnkey plants and processes: MW technology is well established in some sectors (domestic ovens) but is still developing in the field of industrial applications, the range of such applications being quite large. RF technology is in its infancy. One must also consider the fact that the cost of MW equipment depends on the size, the power rating, the frequency, the applicator design, other services such as gas control systems, peripherals, the manufacturer, and the size of the market for that particular type of equipment. Because of these dependencies, capital costs vary widely depending on the application. Also, the equipment used for industrial MW processing is generally custom-designed and optimized based on specific application needs.

On the basis of the available literature (Sheppard, 1988; Edgar and Osepchuk, 2001) and the author's experience, the capital cost of a complete system, for both MW and RF industrial processes, can vary between 2500 and 6000 euros per installed kilowatt (i.e., an industrial MW plant with 50 kW installed may cost between €125 000 and €300 000, depending the particular process that has to be carried out). The components of the costs are summarized in Table 35.2, where the costs related to installation are also included.

Owing to the differences in the configuration and the processing approach between MW and conventional systems, it is very difficult to perform a general comparison of capital costs in a meaningful way. However, it is worth mentioning that MW processing equipment is almost always more expensive than conventional systems and the cost/benefit ratio will always be product-specific. Edgar and Osepchuk (2001) reported a series of practical case studies (on tempering, bacon cooking and potato chip processing). MW and RF processing are unlikely to be economically competitive with processing using natural gas in the foreseeable future because of the difference in costs between natural gas and electrical power.

The intrinsic performance characteristics of MW and RF heating make these processes attractive when the improvements in properties obtainable are significant, when plant space is limited, when electricity is available at a competitive price and when minimization of handling is advantageous. Also, the need to maintain a very clean, controlled processing environment may lead one to choose an MW or RF heating process.

Table 35.2 Capital costs of MW and RF industrial technology.

Power generator	Applicator	Power transmission	Ancillary instrumentation	Installation and startup
20–30%	>35%	5–10%	5–30%	5–15%

References

Ahmed, J., Ramaswamy, H.S. and Raghavan, G.S.V. (2007a) Dielectric properties of Indian Basmati rice flour slurry. *Journal of Food Engineering* 80: 1125–1133.

Ahmed, J., Ramaswamy, H.S. and Raghavan, G.S.V. (2007b) Dielectric properties of butter in the MW frequency range as affected by salt and temperature. *Journal of Food Engineering* 82: 351–358.

Alibas I. (2007) Microwave, air and combined microwave–air-drying parameters of pumpkin slices. *LWT: Food Science and Technology* 40: 1445–1451.

Ayappa, K.G., Davis, H.T., Davis, E.A. and Gordon, J. (1991) Microwave heating: an evaluation of power formulation. *Chemical Engineering Science* 46: 1005–1016.

Begum, S. and Brewer, M.S. (2001) Physical, chemical and sensory quality of microwave blanched snow peas. *Journal of Food Quality* 24: 479–493.

Bengtsson, N.E. and Risman, P.O. (1971) Dielectric properties of foods at 3 GHz as determined by cavity perturbation technique. *Journal of Microwave Power* 6: 107–123.

Bernussi, A.L.M., Chang, Y.K. and Marinez-Bustos, F. (1998) Effects of production by microwave heating after conventional baking on moisture gradient and product quality of biscuits (cookies). *Cereal Chemistry* 75: 606–611.

Birla, S.L., Wang, S., Tang, J. and Hallman, G. (2004) Improving heating uniformity of fresh fruit in radio frequency treatments for pest control. *Postharvest Biology and Technology* 33(2): 205–217.

Birla, S.L., Wang, S., Tang, J., Fellman, J.K., Mattinson, D.S. and Lurie, S. (2005) Quality of oranges as influenced by potential radio frequency heat treatments against Mediterranean fruit flies. *Postharvest Biology and Technology* 38(1): 66–79.

Brewer, M.S. and Begum, S. (2003) Effect of microwave power level and time on ascorbic acid content, peroxidase activity and colour of selected vegetables. *Journal of Food Processing and Preservation* 27: 411–426.

Brunton, N.P., Lyng, J.G., Li, W., Cronin, D.A., Morgan, D. and McKenna, B. (2005) Effect of radio frequency (RF) heating on the texture, colour and sensory properties of a comminuted pork meat product. *Food Research International* 38(3): 337–344.

Buffler, C.R. (1993) *Microwave Cooking and Processing*. Van Nostrand Reinhold, New York.

Datta, A.K. and Anantheswaran, R.C. (2001) *Handbook of Microwave Technology for Food Applications*. Marcel Dekker, New York.

Decareau, R.V. (1985) *Food Industry and Trade: Microwave Heating: Industrial Applications*. Academic Press, Orlando, FL.

Dibben, D. (2001) Electromagnetics: fundamental aspects and numerical modeling. In: *Handbook of Microwave Technology For Food Applications* (eds A.K. Datta and R.C. Anantheswaran). Marcel Dekker, New York, pp. 1–28.

Edgar, R.H. and Osepchuk, J. (2001) Consumer, commercial, and industrial microwave ovens and heating systems. In: *Handbook of Microwave Technology for Food*

Applications (eds A.K. Datta and R.C. Anantheswaran). Marcel Dekker, New York, pp. 215–277.

Engelder, D.S. and Buffler, C.R. (1991) Measuring dielectric properties of food products at microwave frequencies. *Microwave World* 12(2): 2–11.

Farag, K.W., Marra, F., Lyng, J.G., Morgan, D.J. and Cronin, D.A. (2010) Temperature changes and power consumption during radio frequency tempering of beef lean/fat formulations. *Food and Bioprocess Technology* 3(5): 732–740.

Fleischman, G. (1996) Predicting temperature range in food slabs undergoing long term/low power microwave heating. *Journal of Food Engineering* 27: 337–351.

Fleischman, G. (1999) Predicting temperature range in food slabs undergoing short term/high power microwave heating. *Journal of Food Engineering* 40: 81–88.

Guan, D., Cheng, M., Wang, Y. and Tang, J. (2004) Dielectric properties of mashed potatoes relevant to microwave and radio-frequency pasteurization and sterilization processes. *Journal of Food Science* 69(1): 30–37.

Guo, Q., Piyasena, P., Mittal, G.S., Si, W. and Gong, J. (2006) Efficacy of radio frequency cooling in the reduction of *Escherichia coli* and shelf stability of ground beef. *Food Microbiology* 23(2): 112–118.

Hansen, J.D., Drake, S.R., Heidit, M.L., Watkins, M.A., Tang, J. and Wang, S. (2006a) Radio frequency–hot water dips for postharvest codling moth control in apples. *Journal of Food Processing and Preservation* 30: 631–642.

Hansen, J.D., Drake, S.R., Watkins, M.A., Heidit, M.L., Anderson, P.A. and Tang, J. (2006b) Radio frequency pulse application for heating uniformity in postharvest codling moth (Lepidoptera: tortricidae) control of fresh apples (*Malus domestica* borkh.). *Journal of Food Quality* 29: 492–504.

HHS (1991) *Food and Drug Administration 21 CFR 1030.10*. U.S. Department of Health and Human Services, Public Health Service, Washington, DC.

Ikediala, J.N., Hansen, J.D., Tang, J., Drake, S.R. and Wang, S. (2002) Development of a saline water immersion technique with RF energy as a postharvest treatment against codling moth in cherries. *Postharvest Biology and Technology* 24(1): 25–37.

Khraisheh, M.A.M., McMinn, W.A.M. and Magee, T.R.A. (2004) Quality and structural changes in starchy foods during microwave and convective drying. *Food Research International* 37(5): 497–503.

Kowalski, S.J., Musielak, G. and Banaszak, J. (2010) Heat and mass transfer during microwave-convective drying. *AIChE Journal* 56(1): 24–35.

Lau, M.H. and Tang, J. (2002) Pasteurization of pickled asparagus using 915 MHz microwaves. *Journal of Food Engineering* 51: 283–290.

Laycock, L., Piyasena, P. and Mittal, G.S. (2003) Radio frequency cooking of ground, comminuted and muscle meat products. *Meat Science* 65(3): 959–965.

Lyng, J.G. (2007) Rapid pasteurisation of meats using radio frequency or ohmic heating. *New Food* 3: 58–62.

Lyng, J.G., Cronin, D.A., Brunton, N.P., Li, W. and Gu, X. (2007) An examination of factors affecting radio frequency heating of an encased meat emulsion. *Meat Science* 75(3): 470–479.

Marra, F., Lyng, J., Romano, V. and McKenna, B. (2007) Radio-frequency heating of foodstuff: solution and validation of a mathematical model. *Journal of Food Engineering* 79(3): 998–1006.

Marra, F., Zhang, L. and Lyng, J.G. (2009) Radio frequency treatment of foods: review of recent advances. *Journal of Food Engineering* 91(4): 497–508.

Marra, F., De Bonis, M.V. and Ruocco, G. (2010) Combined microwaves and convection heating: a conjugate approach. *Journal of Food Engineering* 97: 31–39.

McKenna, B.M., Lyng, J., Brunton, N. and Shirsat, N. (2006) Advances in radio frequency and ohmic heating of meats. *Journal of Food Engineering* 77(2): 215–229.

Metaxas, A.C. and Meredith, R.J. (1983) *Industrial Microwave Heating*. Peregrinus, London.

Michaelson, S.M. and Lin, J.C. (1987) *Biological Effects and Health Implications of Radiofrequency Radiation*. Plenum, New York.

Mitcham, E.J., Veltman, R.H., Feng, X., de Castro, E., Johnson, J.A., Simpson, T.L., Biasi, W.V., Wang, S. and Tang, J. (2004) Application of radio frequency treatments to control insects in in-shell walnuts. *Postharvest Biology and Technology* 33: 93–100.

Monzon, M.E., Biasi, B., Simpson, T.L., Jhonson, J., Feng, X., Slaughter, D.C. and Mitcham, E.J. (2006) Effect of radio frequency heating as a potential quarantine treatment on the quality of 'Bing' sweet cherry fruit and mortality of codling moth larvae. *Postharvest Biology and Technology* 40: 197–203.

Nelson, S.O. and Datta, A.K. (2001) Dielectric properties of food materials and electric field interactions. In: *Handbook of Microwave Technology for Food Applications* (eds A.K. Datta and R.C. Anantheswaran). Marcel Dekker, New York, pp. 69–114.

NMAB (1994) *Microwave Processing of Materials*. National Materials Advisory Board, Washington, DC.

Ohlsson, T. (1983) Fundamentals of microwave cooking. *Microwave World* 4: 4–9.

Ohlsson, T. and Bengtsson, N. (2001) Microwave technology and foods. In: *Advances in Food and Nutrition Research*, Vol. 43 (ed. S.L. Taylor). Academic Press, San Diego, pp. 65–140.

Orsat, V., Bai, L., Raghavan, G.S.V. and Smith, J.P. (2004) Radio-frequency heating of ham to enhance shelf-life in vacuum packaging. *Journal of Food Process Engineering* 27, 267–283.

Orzaez Villanueva, M.T., Diaz Marquina, A., Franco Vargas, E. and Blazquez Abellan, G. (2000) Modification of vitamins B1 and B2 by culinary processes: traditional systems and microwaves. *Food Chemistry* 71: 417–421.

Osepchuk, J.M. (2002) Microwave power applications. *IEEE Transactions on Microwave Theory and Techniques* 50(3): 975–985.

Piyasena, P., Dussault, C., Koutchma, T., Ramaswamy, H.S. and Awuah, G.B. (2003a) Radio frequency heating of foods: principles, applications and related properties. A review. *Critical Reviews in Food Science and Nutrition* 43(6): 587–606.

Piyasena, P., Ramaswamy, H.S., Awuah, G.B. and Defelice, C. (2003b) Dielectric properties of starch solutions as influenced by temperature, concentration, frequency and salt. *Journal of Food Process Engineering* 26: 93–119.

Redhead, C.S. (1992) *Police Traffic Radar Safety*. Congressional Research Service, Report 92-618 SPR.

Romano, V., Marra, F. and Tammaro, U. (2005) Modelling of microwave heating of foodstuff: study on the influence of sample dimensions with a FEM approach. *Journal of Food Engineering* 71: 233–241.

Rowley, A.T. (2001) Radio frequency heating. In: *Thermal Technologies in Food Processing* (ed. P. Richardson). Woodhead, Cambridge, pp. 162–177.

Salagnac, P., Glouannec, P. and Lecharpentier, D. (2004) Numerical modelling of heat and mass transfer in porous medium during combined hot air, infrared and microwave drying. *International Journal of Heat and Mass Transfer* 47: 4479–4489.

Schiffmann, R.F. (1992) Microwave processing in the U.S. food industry. *Food Technology* 48: 50–52, 56.

Schiffmann, R.F. (1995) Microwave and dielectric drying. In: *Handbook of Industrial Drying* (ed. A.S. Mujumdar), 2nd ed.). Marcel Dekker, New York, pp. 345–371.

Sheppard, L.M. (1988) Manufacturing ceramics with microwaves: the potential for economic production. *Ceramic Bulletin* 67(10): 1556–1515.

Tang, J. (2005) Dielectric properties of foods. In: *The Microwave Processing of Foods* (eds H. Schubert and M. Regier). Woodhead, Cambridge, pp. 22–40.

Tang, J., Ikediala, J.N., Wang, S., Hansen, J.D. and Cavalieri, R.P. (2000) High-temperature–short-time thermal quarantine methods. *Postharvest Biology and Technology* 21: 129–145.

Tang, X., Cronin, D.A. and Brunton, N.P. (2005) The effect of radio frequency heating on chemical, physical and sensory aspects of quality in turkey breast rolls. *Food Chemistry* 93: 1–7.

Tang, X., Lyng, J.G., Cronin, D.A. and Durand, C. (2006) Radio frequency heating of beef rolls from *biceps femoris* muscle. *Meat Science* 72(3): 467–474.

Tang, Z., Mikhaylenko, G., Liu, F., Mah, J., Pandit, R., Younce, F. and Tang, J. (2008) Microwave sterilization of sliced beef in gravy in 7-oz trays. *Journal of Food Engineering* 89: 375–383.

Vadivambal, R. and Jayas, D.S. (2007) Changes in quality of microwave-treated agricultural products – a review. *Biosystems Engineering* 98: 1–16.

Wang, S., Tang, J., Johnson, J.A., Mitcham, E., Hansen, J.D., Cavalieri, R.P., Bower, J. and Biasi, B. (2002) Process protocols based on radio frequency energy to control field and storage pests in in-shell walnuts. *Postharvest Biology and Technology* 26: 265–273.

Wang, S., Tang, J., Johnson, J.A., Mitcham, E., Hansen, J.D., Hallman, G., Drake, S.R. and Wang, Y. (2003) Dielectric properties of fruits and insect pests as related to radio frequency and microwave treatments. *Biosystems Engineering* 85: 201–212.

Wang, S., Birla, S.L., Tang, J. and Hansen, J.D. (2006a) Postharvest treatment to control codling moth in fresh apples using water assisted radio frequency heating. *Postharvest Biology and Technology* 40(1): 89–96.

Wang, S., Tang, J., Sun, T., Mitcham, E.J., Koral, T. and Birla, S.L. (2006b) Consideration in design of commercial radio frequency treatments for postharvest pest control in in-shell walnuts. *Journal of Food Engineering* 77: 304–312.

Wang, S., Monzon, M., Johnson, J.A., Mitcham, E.J. and Tang, J. (2007a) Industrial-scale radio frequency treatments for insect control in walnuts I: heating uniformity and energy efficiency. *Postharvest Biology and Technology* 45, 240–246.

Wang, S., Monzon, M., Johnson, J.A., Mitcham, E.J. and Tang, J. (2007b) Industrial-scale radio frequency treatments for insect control in walnuts II: insect mortality and product quality. *Postharvest Biology and Technology* 45: 247–253.

Zhang, H. and Datta, A.K. (2001) Electromagnetics of microwave heating: magnitude and uniformity of energy absorption in an oven. In: *Handbook of Microwave Technology for Food Applications* (eds A.K. Datta and R.C. Anantheswaran). Marcel Dekker, New York, pp. 1–28.

Zhang, L., Lyng, J.G. and Brunton, N.P. (2004) Effect of radio frequency cooking on texture, colour and sensory properties of a large diameter comminuted meat product. *Meat Science* 68(2): 257–268.

Zhang, L., Lyng, J.G. and Brunton, N.P. (2006) Quality of radio frequency heated pork leg and shoulder ham. *Journal of Food Engineering* 75(2): 275–287.

Zhang, L., Lyng, J.G. and Brunton, N.P. (2007) The effect of fat, water and salt on the thermal and dielectric properties of meat batter and its temperature following microwave or radio frequency heating. *Journal of Food Engineering* 80(1): 142–151.

Zhao, Y., Flugstad, B., Kolbe, E., Park, J.E. and Wells, J.H. (2000) Using capacitive (radio frequency) dielectric heating in food processing and preservation – a review. *Journal of Food Process Engineering* 23: 25–55.

36

Design of Ohmic Heating Processes

Ilkay Sensoy

Introduction

Ohmic heating is a thermal processing technology where an alternating electrical current passes through an electrically conducting food. The electrical energy supplied is released within the food in the form of heat. The heating rate is determined by the electrical conductivities of the components rather than by heat transfer. Ohmic heating is also known as Joule heating, electrical resistance heating, direct electrical resistance heating, electroheating and electroconductive heating.

Ohmic heating has a high energy efficiency, as almost all of the electrical power supplied is transformed into heat (Fellows, 2000; Ohlsson *et al.*, 2002; Jun and Sastry, 2009). The advantage of ohmic heating over conventional methods of heating is the volumetric nature of the heating effect. Heat is generated very fast, owing to the electrical current, within the food material rather than being transferred via conduction or convection. Therefore, ohmic heating reduces the time required for heating as compared with conventional heat transfer mechanisms. The high rate of heat transfer makes the achievement of the desired level of microbial safety possible without causing product degradation. This is especially important for particulate foods, where ohmic heating can be used to sterilize them under high-temperature–short-time (HTST) conditions without causing thermal damage to the liquid carrier

Handbook of Food Process Design, First Edition. Edited by Jasim Ahmed and Mohammad Shafiur Rahman.

or overcooking of the outside of the particles, which are shortcomings of conventional sterilization processes.

The use of ohmic heating in food processing has been studied since the early 1900s. It was used as a commercial process in the early twentieth century for the pasteurization of milk. However, between the late 1930s and the 1960s the process was discontinued owing to the cost of electricity and the lack of suitable electrode materials (Bengston *et al.*, 2006). Interest in ohmic heating reemerged in the 1980s when researchers were seeking suitable methods to sterilize mixtures of large particles and liquids effectively (Sastry *et al.*, 2002; Sarang *et al.*, 2008; Sastry, 2008; Ghnimi *et al.*, 2009). It is possible to heat products containing large particles (up to 2.5 cm in size), which would be damaged in conventional equipment, to sterilization temperatures of up to 140 °C in less than 90 s (Ruan *et al.*, 2004). The products are cooled back down to ambient temperature within 15 min (Ruan *et al.*, 2004). Cooling may be the process that limits the particle size (Ruan *et al.*, 2004). These processing times are significantly shorter than the typical two-hour process cycles for various in-can sterilization techniques (Ruan *et al.*, 2004). The flow of a liquid–particle mixture approaches plug flow at high concentrations (20–70%) (Kim *et al.*, 1996; Zitoun, 1996; Bengston *et al.*, 2006). Ohmic heating is currently being utilized to process food products in Europe, Asia and the USA (Bengston *et al.*, 2006; Ghnimi *et al.*, 2009).

The electricity for ohmic heating is taken from the mains supply (50 Hz in Europe and 60 Hz in the USA) and stepped up. Normally, voltages of up to 5000 V are applied (Ohlsson *et al.*, 2002). The effects of electric fields on microorganisms has been known for some time. They have been used in biotechnology for electroporation and electrofusion. In the field of food processing, knowledge of the effects of electric fields on microorganisms led to the emergence of the nonthermal processing method known as pulsed-electric-field (PEF) processing. As Sastry (2008) and Sastry *et al.* (2002) mentioned, even though the voltages applied during ohmic heating are low compared with those used in PEF technology, the nonthermal effects seen in ohmic heating deserve attention. This resulted in a process termed moderate-electric-field (MEF) processing (Sastry *et al.*, 2002; Sastry, 2008). There is no formal definition of the term "MEF." However, the term can be used for processes where the electric field strength is less than or equal to 1000 $V \cdot cm^{-1}$ (Sastry, 2008). MEF processing is different from PEF processing, where a very high electric field strength is applied in the form of pulses while the temperature of the product is kept low. The nonthermal effects seen in MEF processing have resulted in the emergence of many new applications, such as extraction, drying and fermentation, in the last ten years (Sastry, 2008).

The advantages of ohmic heating can be summarized as follows: it has a high energy efficiency; there are no hot heat transfer surfaces, resulting in reduced fouling; it provides rapid, uniform treatment of liquid and solid phases with minimal heat damage; it causes less nutrient loss; it is good for shear-sensitive products, owing to the low velocity of the product; it has a high solid-loading capacity; it allows simple process control, with instant switch-on and shutdown; it results in reduced maintenance costs; and it allows environmentally friendly systems to be built (Ruan *et al.*, 2004; Vicente and Castro, 2007).

Applications of Ohmic Heating and Moderate-Electric-Field Processing

Ohmic Heating for Pasteurization and Sterilization

Ohmic heating is applicable to a wide variety of foods, encompassing liquids, solids and solid–liquid mixtures. It is in commercial use in Europe, Asia and the USA for the aseptic processing of high-value-added meals, pasteurization of particulate foods for hot filling, preheating products before canning, the production of liquid egg and the processing of delicate whole fruits such as strawberries (Ohlsson *et al.*, 2002). Ohmic heating is especially advantageous where the heating speed is important. Several researchers have reported that ohmic heating was an effective means of minimizing textural degradation of fish muscle caused by endogenous heat-stable proteases (Yongsawatdigul *et al.*, 1995; Yongsawatdigul and Park, 1996; Park *et al.*, 1998; AbuDagga and Kolbe, 2000). The rapid increase in temperature causes proteases to be inactivated before they can significantly hydrolyze myofibrilar proteins, giving a better-quality product (Park *et al.*, 1998). Better product quality may also be achieved when ohmic heating is used for heating high-acid food products such as tomato-based sauces prior to hot filling (Ruan *et al.*, 2004). Furthermore, ohmic heating has the advantage of heating particulate foods more uniformly than conventional heating, owing to the volumetric heating that it provides. In conventional aseptic processing, the time required to correctly process the centers of particles causes overprocessing of the surrounding volume (Vicente and Castro, 2007). During ohmic heating, a particle–liquid mixture can be heated fast and uniformly to a high temperature irrespective of particle size when the electrical conductivities of the liquid and the particles are similar or when the solids concentration, viscosity, conductivity and flow rate are appropriate (Fryer *et al.*, 1993; Khalaf and Sastry, 1996).

The lack of hot heat transfer surfaces in ohmic heaters reduces the risk of burning or overprocessing of the surface (Ghnimi *et al.*, 2009). The electrode temperatures may be 2–3 °C cooler than the bulk liquid (Fryer *et al.*, 1993). Although the electrodes are carrying the same current density as the liquid, their lower resistivity produces much less resistance heating than in the liquid. This results in having no hot surface, which reduces the fouling problem. In commercial ohmic units, foods with a solids fraction of up to 80% are heated while in flow (Fryer and Li, 1993; Yang, 2000). A high solids content is desirable in ohmic heating because it leads to faster heating of particles with a conductivity that is the same as or lower than that of the liquid (Sastry, 1992; Sastry and Palaniappan, 1992a,b; Orangi *et al.*, 1998; Benabderrahmane and Pain, 2000). In addition, when the solids content is high, the flow of a liquid–particle mixture approaches plug flow (Fryer and Li, 1993; Kim *et al.*, 1996; Zitoun, 1996).

The identification, control and validation of the critical control points required to demonstrate the achievement of commercial sterility is more difficult with ohmic processing than with conventional heating processes (Vicente and Castro, 2007). The contribution of the electric field, the physical properties of the components, and the existence of different particle concentration/orientation and flow conditions make ohmic heating of particulate food rather complex. Therefore, research is needed to identify, measure and test any cold spots that may occur during ohmic heating of particulate food (Sastry, 2008). Several research studies have demonstrated the effect of conductivity, concentration, particle orientation and the flow regime around a particle (de Alwis *et al.*, 1989; Fryer and de Alwis, 1989; de Alwis and Fryer, 1990; Sastry, 1992; Sastry and Palaniappan, 1992a,b; Fryer and Li, 1993; Orangi *et al.*, 1998; Zhao *et al.*, 1999). The worst-case heating scenarios typically happen when a particle has an electrical conductivity significantly different from its surroundings (Salengke and Sastry, 2007b,c; Sastry, 2008).

The heating rate is affected by the electrical conductivities of the constituents of the food. For most solid foods, the electrical conductivity increases considerably after the cellular structure of the food is disrupted (Sastry and Palaniappan, 1992b). The electrical conductivity of a food sample can be a strong function of frequency, it can be anisotropic and the conductivity–temperature profile can differ significantly between ohmic and conventional heating (Fryer and de Alwis, 1989; Halden *et al.*, 1990; Sastry and Palaniappan, 1992a,b; Fryer and Li, 1993). The electrical conductivity is affected by many things that occur during conventional heating, such as starch transitions, melting of fats and cell structure changes (Wang and Sastry, 1997).

Alternating current at low frequencies such as 50 and 60 Hz has an electrolytic effect (Reznick, 1996). High frequencies above 100 kHz can be used to overcome the problem of electrolytic effects (Reznick, 1996). Alternatively, insoluble, especially treated pure carbon electrodes can be used safely with electrical power at a frequency of 50 or 60 Hz (Reznick, 1996). In addition, with the development of solid-state power supply technology, it is possible to economically control electrolytic effects by using ohmic heating in a pulse mode (Sastry, 2008).

Novel Applications of Moderate-Electric-Field Processing

Several researchers have shown that molecular movement or diffusion is increased when an electric field is applied to foods with a cellular structure (Schreier *et al.*, 1993; Stapley *et al.*, 1995; Mizrahi, 1996). In addition, it has been shown that for foods with a cellular structure, the heating rate can be increased at lower frequencies (such as 50 Hz) as a result of permeability and molecular movement (Imai *et al.*, 1995). At high frequencies (such as 10 kHz) the heating rate increases owing to dielectric loss, where the contribution of dipole rotation of water becomes significant as in microwave and radio frequency heating (Imai *et al.*, 1995; Park *et al.*, 1995; Imai *et al.*, 1998). Recent studies have shown that a mild electroporation-type mechanism might have been seen during ohmic heating (Sastry *et al.*, 2002). These have become the basis for the surfacing of several novel applications such as extraction, dehydration, blanching and fermentation.

Extraction

One application of the increased permeability induced by an electric field is in juice extraction; for example, the juice yield of pressed apple cossettes was increased by electroplasmolysis pretreatment (Lima and Sastry, 1999; Bazhal and Vorobiev, 2000; Wang and Sastry, 2000). Lima and Sastry (1999) also showed that a higher yield was obtained at lower frequencies. Electroplasmolysis has been shown to increase the extraction speed for beet cossettes (Praporscic *et al.*, 2005). Sensoy and Sastry (2004b) showed that MEF processing increased the extraction yield for fresh mint leaves, where a higher yield was achieved at low frequency (50 Hz). The work of Kulshrestha and Sastry (2003) and Lima *et al.* (2001) showed that treatment of beet tissue by ohmic heating resulted in leaching of dyes from cells. Thus, MEF treatment can be an alternative method for the extraction of specific cell metabolites (macromolecules, sugars and pigments), where heat and the electric field work together to increase the permeability.

Dehydration

Several researchers (Lima and Sastry, 1999; Wang and Sastry, 2000; Zhong and Lima 2003) have observed a higher drying rate with samples pretreated by ohmic heating. Stapley *et al.* (1995) studied the self-diffusion of water in samples of carrot, either raw or cooked by either ohmic or conventional heating. They used pulsed-field-gradient NMR to characterize the self-diffusion of water, and observed the disappearance of structure with cooking. The difference between samples cooked electrically and conventionally was much less than the difference between cooked and uncooked material, but was statistically significant. Thus, MEF processing shows promise as a pretreatment for drying.

Blanching

Blanching is a pretreatment for vegetables to achieve adequate quality in dehydrated, canned and frozen products. Mizrahi (1996) reported that owing to the rapid, uniform heating during MEF treatment, no dicing was required for vegetables. Blanching uses large quantities of water. Sensoy and Sastry (2004a) found that mushrooms could be preshrunk at significantly lower temperatures, and with considerably lower water use, compared with conventional blanching. These studies show that MEF blanching offers an economical alternative.

Fermentation

Cho *et al.* (1996) showed that MEF treatment decreased the lag time of *Lactobacillus acidophilus* but found that the productivity of fermentation was reduced during later stages. Unal *et al.* (1998) obtained similar results for *Lactococcus lactis*. Other researchers have shown that cell membrane permeabilization occurred for *Lactobacillus acidophilus* (Loghavi *et al.*, 2009) and bacteriocin activity increased in the early stages of fermentation during MEF treatment (Loghavi *et al.*, 2007). These results suggest that MEF technology can be used in the earlier stages of fermentation.

Summary and Other Applications

Moderate-electric-field processing has been shown to increase mass transfer and therefore has potential for several applications, such as blanching, drying, extraction and fermentation. In addition, Salengke and Sastry (2007a) have shown that this type of processing may have potential as a pretreatment to reduce oil uptake during frying.

The Ohmic Heating Process

In an ohmic heating process, the food product is simply pumped between electrodes, either parallel or perpendicular to flow, which are in direct contact with the food (Ohlsson *et al.*, 2002). A schematic diagram of a continuous-flow ohmic heating process is shown in Figure 36.1, in which the electrodes are perpendicular to the flow. After heating, the product passes through a holding tube for a fixed time to achieve commercial sterility (Skudder and Biss, 1987). The product is then cooled by a series of tubular heat exchangers, by a scraped-surface heat exchanger or by flash cooling,

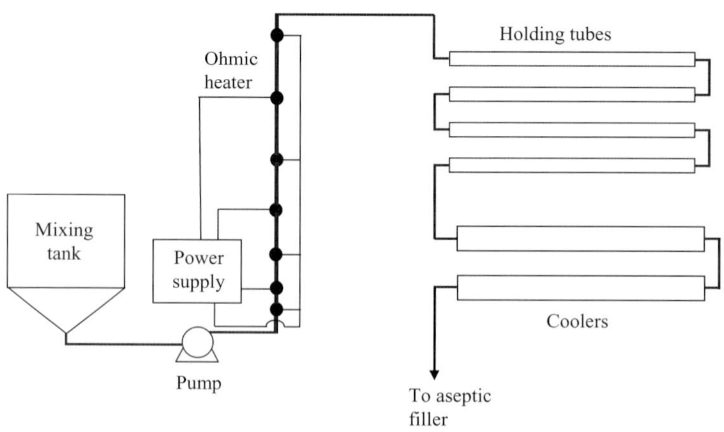

Figure 36.1 Schematic diagram of a continuous-flow ohmic heating process (adapted from Parrot, 1992). Reproduced with permission of the Institute of Food Technologists.

depending on the nature of the product. After cooling, the product enters a storage tank prior to aseptic filling.

An ohmic heating system consists of three parts, namely a power supply, a heater assembly and a control panel (Ruan *et al.*, 2004). Commercial ohmic heaters may be either of the co-field type, where the electrodes are positioned along the length of the product flow, or of the cross-field type, where the electrodes are located perpendicular to the flow (Ruan *et al.*, 2001; Ruan *et al.*, 2004; Vicente *et al.*, 2006). Differences in electrical conductivity between phases cause a heterogeneous temperature increase during ohmic heating. Therefore, foods may need pretreatment to increase the electrical conductivity of the solid phase (Wang and Sastry, 1993). The adverse electrolytic effects that were neglected in the early designs have been eliminated by using new technologies that use food-compatible electrode materials and, in some cases, high frequencies (Ruan *et al.*, 2004).

Fundamentals of Ohmic Heating

The principal mode of conduction in electrolytes is ionic conduction. Thus, the presence of ionic constituents is necessary for heating to occur during ohmic heating. Most foods are electrically conductive owing to their acid and salt content (Tulsiyan *et al.*, 2008). At higher frequencies such as those used in microwave and radio frequency heating, the contribution of dipolar rotation of water molecules to the heating effect becomes significant. In ohmic heating, a voltage gradient must be applied between the electrodes to generate heat within the food (Figure 36.2). The electric field is governed by the current continuity equation (Equation 36.1 below), and the flow and thermal fields are governed by equations of continuity, motion and energy change (Orangi *et al.*, 1998). The current continuity equation is

$$\nabla \cdot \mathbf{J} = -\frac{\partial \rho}{\partial t} \; (\mathrm{A \cdot m^{-3}}) \tag{36.1}$$

where $\mathbf{J}$ is the current density (measured in $\mathrm{A{\cdot}m^{-2}}$), ρ is the charge density ($\mathrm{C{\cdot}m^{-3}}$) and t is the time (s). For steady electric currents, the current continuity equation becomes

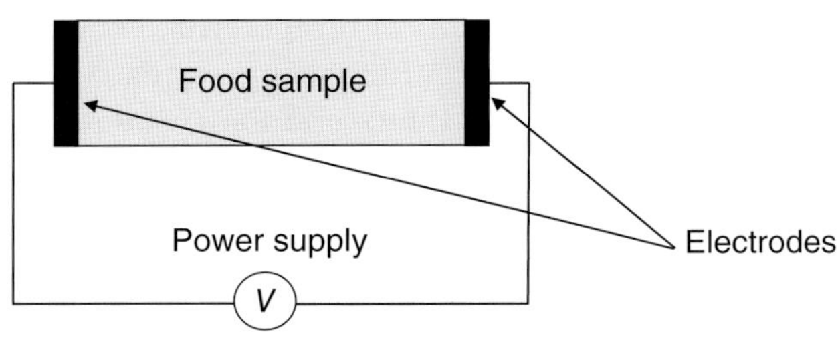

Figure 36.2 Principle of ohmic heating.

$$\nabla \cdot \mathbf{J} = 0 \; (\mathrm{A/m^3}) \tag{36.2}$$

The point form of Ohm's law, which holds at all points in space, is

$$\mathbf{J} = \sigma \mathbf{E} \; (\mathrm{A/m^2}) \tag{36.3}$$

where σ, the electrical conductivity ($\mathrm{S \cdot m^{-1}}$), can be a function of the spatial coordinates. The electric field intensity **E** is equal to the negative of the voltage gradient:

$$\mathbf{E} = -\nabla V \tag{36.4}$$

The steady-state form of the continuity equation for the current (Equation 36.2) can be combined with the point form of Ohm's law to solve for the voltage distribution (Equation 36.5 below). The electric field intensity **E** is replaced by the negative of the voltage gradient (Equation 36.4) to obtain

$$\nabla \cdot (\sigma \nabla V) = 0 \tag{36.5}$$

The local temperature inside the food can be determined by using the energy equation for a constant-density fluid,

$$\rho_f C_p \frac{DT}{Dt} = k \nabla^2 T + \dot{q} \tag{36.6}$$

The local electric power converted to heat is, via Joule's law, $\mathbf{E \cdot J}$, which is equal to $\sigma |\nabla V|^2$. Therefore, the heat generation rate $\dot{q}$ ($\mathrm{W \cdot m^{-3}}$) in the ohmic heater is directly proportional to the square of the local electric field strength and the electrical conductivity:

$$\dot{q} = \sigma |\nabla V|^2 \tag{36.7}$$

Laplace's equation (Equation 36.5) and the energy equation (Equation 36.6) are solved with the appropriate boundary and initial conditions to obtain the voltage and temperature distribution within the food.

The physical properties can be functions of temperature. The electrical conductivity of food products usually increases with increasing water content and temperature (Benabderrahmane and Pain, 2000; Goullieux and Pain, 2005). The electrical conductivity of both solid foods and liquids is often considered to be a linear function of temperature:

$$\sigma_T = \sigma_{\mathrm{ref}} [1 + m(T - T_{\mathrm{ref}})] \tag{36.8}$$

Table 36.1 shows some electrical properties of selected foods.

Table 36.1 Electrical properties of selected foods.

	Potato	Carrot	Turnip	Reference
σ_{70} (S·m^{-1})	1.11	1.05	1.21	Chen *et al.* (2010)
m (S·m^{-1}·°C^{-1})	0.0104	0.0098	0.0106	Chen *et al.* (2010)

The rate of ohmic heating depends on the electric field distribution, and the uniformity of heating depends on the electrical conductivity and the geometry of the system. In a static ohmic heater, when a particle is in series with a liquid, a less conductive particle tends to heat at a higher rate (de Alwis and Fryer, 1992). On the other hand, if the particle is more conductive than the liquid, it heats more slowly than the liquid. In the case of continuous flow of a liquid with particles suspended in it, the situation is more complicated. The complex electric field distribution, the residence time distribution of the particles and interphase heat transfer between the particles and the fluid make even numerical solution very complicated. Assumptions, simplifications and suggestions for models have been made for the solution of several cases in order to understand the interactions of these physical parameters with each other (Sastry and Palaniappan, 1992a,b; Fryer *et al.*, 1993; Zhang and Fryer, 1993; Orangi *et al.*, 1998; Benabderrahmane and Pain, 2000; Salengke and Sastry, 2007b,c). Further developments in modeling will make the design and optimization of ohmic heating processes for industrial application easier.

Computer Modeling

Computer modeling is a necessary tool for the aseptic processing of particulate foods by ohmic heating, where internal temperatures cannot be measured without interfering with the flow (Chen *et al.*, 2010). One computer model that predicts temperatures in the particles and the liquid in a continuous-flow ohmic heater (Figure 36.3) is based on a model developed by Sastry (1992). The model assumes plug flow through the heater due to the high solids concentrations normally used in ohmic heaters. The model involves separate energy balances for the fluid and the particles (Sastry, 1992). For the liquid in an incremental section of thickness Δx (Figure 36.4) of the pipe, between locations n and $n + 1$, the following energy balance is used:

$$\dot{v}_f \rho_f C_{pf}(T_f^{n+1} - T_f^n) = \dot{q}_f v_f + N_p h_{fp} A_p (T_{sm} - T_{fm}) + UA_{pi}(T_a - T_{fm}) \quad (36.9)$$

where

$$T_{fm} = \frac{T_f^{n+1} + T_f^n}{2}$$

$$T_{sm} = \frac{T_s^{n+1} + T_s^n}{2}$$

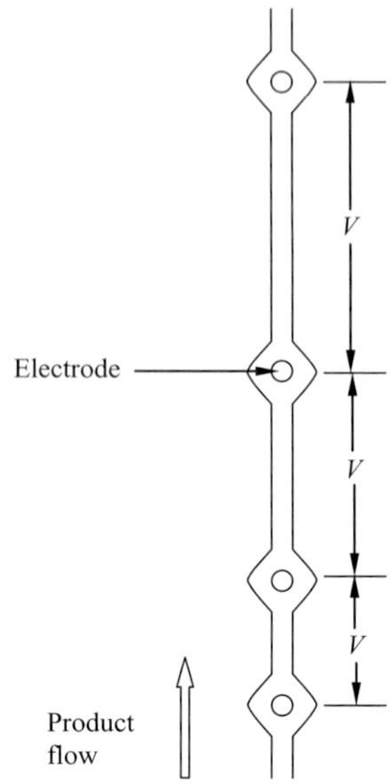

Figure 36.3 Schematic diagram of ohmic heater. (Adapted from Sastry (1992), courtesy of John Wiley & Sons).

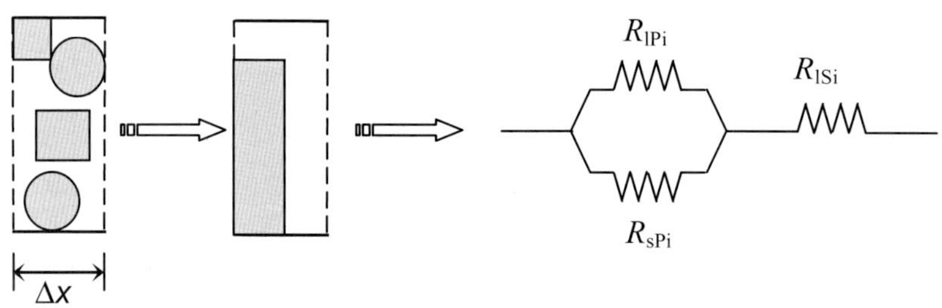

Figure 36.4 Equivalent circuit for an incremental section.

Here, $\dot{v}_f$ is the volumetric flow rate, ρ_f is the density, C_{pf} is the specific heat capacity and v_f is the volume of the fluid in the incremental section. N_p is the number of particles. The subscripts "p," "pi" and "a" represent a particle, the pipe and the ambient air, respectively. The temperature T_f of the fluid at each incremental location $(n + 1)$ can be determined from Equation 36.9 if the voltage field V and the mean particle surface temperature T_{sm} are known. The heat transfer problem for the particles is the conduction heat transfer equation with temperature-dependent internal heat generation:

$$\rho_p C_{pp} \frac{\partial T_p}{\partial t} = \nabla \cdot (k_p \nabla T_p) + \dot{q}_p(T_p) \tag{36.10}$$

where the subscript "p" again represents a particle, and k is the thermal conductivity. The time-dependent boundary condition is

$$k_p \nabla T_p \cdot \vec{n} = h_{fp}[T_{sp} - T_f(t)] \quad (36.11)$$

where T_{sp} is the particle surface temperature.

The electric field distribution arises from the continuity equation (Equation 36.5) for the current, which is difficult to solve for a multiparticle mixture. Doing this would require knowledge of the location and properties of every particle at all points in time. The electric field distribution can instead be solved using circuit theory, where the ohmic-heater column is considered to be a set of equivalent resistances in series (Figure 36.4). This is a useful and realistic approximation because voltage gradients are likely to occur primarily along the length of the heater with high solids concentrations (Sastry, 1992). The voltage drop in each incremental section must be determined separately, because the temperature and conductivity of the product increase as the product passes through the heater. For each incremental section i of thickness Δx lying between locations n and $n + 1$, the resistance is calculated from the values for the continuous phase (the carrier fluid) and the discontinuous phase (the solid) as

$$R_i = R_{lSi} + \frac{R_{lPi} R_{sPi}}{P_{lPi} + R_{sPi}} \quad (36.12)$$

where

$$R_{lSi} = \frac{\Delta x_{lSi}}{A_{lSi}\sigma_{li}} \quad (36.13)$$

$$R_{sPi} = \frac{\Delta x_{sPi}}{A_{sPi}\sigma_{si}} \quad (36.14)$$

$$R_{lPi} = \frac{\Delta x_{lPi}}{A_{lPi}\sigma_{li}} \quad (36.15)$$

$$A_{lSi} = \frac{\pi d^2}{4} = A_{sPi} + A_{lPi} \quad (36.16)$$

The total resistance of the ohmic heater column is then

$$R = \sum_{i=1}^{N} R_i \quad (36.17)$$

The total current flowing through the system is

$$I = \frac{\Delta V}{R} \tag{36.18}$$

The voltage distribution is calculated assuming that all equipotential surfaces are planar and perpendicular to the tube walls. Thus, the voltage drop across the incremental section i is calculated as

$$\Delta V_i = IR_i \tag{36.19}$$

and the results are used to calculate the voltage gradient and the energy generation in each incremental section.

Systems Available

The fundamental requirements for ohmic heating equipment are electrodes, a non-conducting enclosure for the food material and a power supply (Figure 36.2). The ohmic heater can be a batch or a continuous system. In a batch system, the electrodes are placed on opposite side of the container. The electric field lines are distributed according to the shape of the container and the electrodes. For industrial applications, however, continuous systems are preferred. In a continuous-flow system, either a co-field or a cross-field arrangement can be used.

Batch processes are usually used for cooking meat products (Vicente *et al.*, 2006). Continuous ohmic heaters are used for viscous fluids and fluids with particles (Vicente *et al.*, 2006). There are several manufacturers of ohmic heating equipment, including APV Baker, Ltd. (Crawley, UK), Raztek Corp. (Sunnyvale, CA, USA) and Asepsystems S.r.l. (Parma, Italy). Several APV Baker ohmic heaters have been installed for processing a wide variety of high-quality, value-added pumpable particulate food products such as meals and fruit preparations (Ruan *et al.*, 2004). Raztek's Electroheating™ technology has been implemented in New Jersey, USA, to extend the shelf-life of liquid egg products and minimize the production downtime (Ruan *et al.*, 2004). The Emmepiemme high-frequency ohmic heater from Asepsystems is being used for diced fruit and vegetable products (http://www.asepsystems.com).

Ohmic heaters are usually used in steps to provide better control over heat-channeling effects that can cause hot spots. In one APV Baker ohmic heater, the food flows upward in an insulating tube, with cylindrical electrodes placed along the tube (Ruan *et al.*, 2004). This is a co-field ohmic heater. During heating, as the product approaches the outlet, its electrical conductivity increases as a result of the increasing temperature. A higher electrical conductivity causes a higher heating rate, and this may cause runaway heating. To minimize or provide better control of this effect, multiple electrodes are used in series. The distance between the electrodes can be increased as the flow approaches the outlet to take account of the increase in electrical conductivity with temperature.

Design

The design of an ohmic heater for a particular application is somewhat more specific to the application than designing or choosing a heat exchanger (Ruan *et al.*, 2004). In order to benefit from the advantages of ohmic heating over conventional heating and to optimize the process, the ohmic heater needs to be tailored to the specifications of the application (Ruan *et al.*, 2004). The critical processing conditions and the procedures used must be defined. The definitions of these conditions and procedures should include electrical conductivities, premix formulation procedures, initial temperatures, flow rates or particle residence times, exit temperatures and solid loading rates, and the control of these. The process parameters should ensure product safety while simultaneously maximizing product quality. These parameters are specific to individual systems and formulations.

An essential requirement in the design of commercial continuous-flow sterilization processes for viscous and particulate foods is to calculate the temperature history of the center of the largest particle, and to estimate the resulting inactivation of microorganisms (Maroulis and Saravacos, 2003). However, for some types of product, the coldest point in the ohmic heater is not as obvious. The complexity of the ohmic heating process is due to the complex coupling between the temperature field, the electric field and process parameters such as the size and shape of the particles and their orientation relative to the electric field (Jun and Sastry 2009). Understanding the roles of these factors require rigorous testing and reevaluation.

The design considerations for ohmic heaters include the electrode configuration, the heater geometry, the current density, the applied voltage and the velocity profile of the product. Some other parameters that are important are properties of the product such as the electrical conductivity, ratio of solids, viscosity, and specific heat capacity (Bengston *et al.*, 2006). Having particles with a lower electrical conductivity than the fluid may yield a higher heating rate for the particles (Orangi *et al.*, 1998), which is an advantage in ohmic heating. Products may need pretreatment before the ohmic heater, because the electrical conductivities may need adjustment prior to the ohmic heating process. Having particles with different electrical conductivities in the system causes strong temperature inhomogeneities, especially in the case of unmixed batch systems (Davies *et al.*, 1999; Goullieux and Pain, 2005).

A knowledge of the spread of residence times for the fluid and particles is essential for determining the thermal treatment that any product has received. The residence time distribution of the particles depends on the flow rate and viscosity of the carrier medium, and also on the size, density and concentration of the particles (Sandeep and Puri, 2009). For continuous ohmic heaters, the electrical conductivity and the residence time distribution of the particles, and the fluid–particle heat transfer coefficient are the crucial parameters when one is designing an ohmic heating system. In summary, the design considerations for ohmic heating systems should include the following:

- *Product characteristics*. The resistance of an ohmic heating device is a function of the electrical resistance of the product and the geometry of the equipment (Ruan *et al.*, 2004). The electrode configuration and the power requirements can be determined from the electrical conductivity of the product and its variation over the range of temperature used. For particulate food products, the electrical conductivity, specific heat capacity and density of each phase, the solids concentration, the viscosity of the liquid, and the heat transfer coefficient between the phases are important parameters in determining the heating rate.
- *Processing parameters*. The heating is very fast in an ohmic heating system, and therefore even small delays in the flow may lead to high temperature differences. Therefore, it is important to minimize even small differences in flow velocity in the cross section (Ruan *et al.*, 2004). Plug flow can be achieved with high solids concentrations. The concentration of particles (for two-phase systems) and the initial temperature of the product are important parameters.

Process Control

The inlet temperature, flow rate and specific heat capacity of the product are the parameters that affect the outlet temperature. Ohmic heaters usually contain multiple sets of electrodes to provide better control of the process. Problems may arise if a single pair of electrodes is used to heat the food so as to cause a large temperature change (Ruan *et al.*, 2004). Significant changes in electrical conductivity along the length of the system may cause substantial changes in the heating rate, which is hard to control. Raztek's full-scale industrial system (Ruan *et al.*, 2004) uses a feedback system to control the outlet temperature.

Cleaning and Sanitation

An important aspect of the presterilization of an ohmic heater assembly is to use a cleaning solution that has approximately the same electrical conductivity as the food material to be processed. The adjustment of the electrical power during the change to the product is minimized by doing this. This ensures a smooth transition with little temperature fluctuation (Yang, 2000). Presterilization of the ohmic heater, holding tube and coolers is achieved by recirculating a solution such as a sodium sulfate solution at a concentration that approximates the electrical conductivity of the food material to be processed (Skudder and Biss, 1987; Yang, 2000). After processing, the power is shut down and cleaning is done with a 2% (w/v) solution of caustic soda recirculated at 60–70 °C for 30 min (Yang, 2000). The cleaning solution is heated by conventional methods, owing to its high electrical conductivity (Yang, 2000).

Capital and Operating Costs

The capital cost of ohmic heaters is low compared with other new technologies and is becoming less expensive with improvements in the technology (Sastry *et al.*, 2009). According to Ghnimi *et al.* (2009), the cost of ohmic sterilizers dropped by a factor of 10 between 1992 and 2003. The operational costs of ohmic heating were found to be comparable to those for freezing and retorting of low-acid products (Allen *et al.*, 1996; Zoltai and Swearingen, 1996). Studies such as that of Allen *et al.* (1996) have shown that the process is a viable alternative to conventional processing technologies for low-acid foods and it may be a useful technology for high-quality high-acid foods. Some studies with meat products in Canada showed 70% energy savings when ohmic heating was used instead of traditional smokehouse cooking (Vicente *et al.*, 2006). Savings in time would be an additional benefit (Vicente *et al.*, 2006).

Summary and Future Needs

Ohmic heating offers unique advantages, such as achieving higher temperatures in particles than in liquids, which is impossible with current sterilization techniques; reduced fouling; reduced burning of food; minimal mechanical damage; better nutrient and vitamin retention; easy process control, with instant switch-on and shutdown; a high solids loading capacity; and quiet and environmentally friendly systems. However, enhancements are always needed, which can be summarized as follows. There is a need to improve noninvasive temperature-monitoring techniques for profiling the heat distribution and locating potential cold or hot spots during the ohmic heating process. Other techniques may interfere with the electric field. There is a need to develop reliable mathematical models that correlate the physical and electrical properties of products with process parameters so that one can calculate values for sterility and the extent of cooking. There is also a need to quantify the effects of the electric field on mass transfer for the applications of ohmic heating in drying, extraction, blanching and fermentation. Finally, better electrode materials or new types of power supply equipment (e.g., high-frequency or pulse mode) are needed to minimize electrolysis.

Nomenclature

A Surface area (m^2)
C_p Specific heat capacity ($kJ \cdot kg^{-1} \cdot °C^{-1}$)
d Diameter of heater tube (m)
$\mathbf{E}$ Electric field intensity ($V \cdot m^{-1}$)
h Heat transfer coefficient between carrier fluid and particle surface ($W \cdot m^{-2} \cdot °C^{-1}$)
I Current (A)
$\mathbf{J}$ Current density ($A \cdot m^{-2}$)

k Thermal conductivity ($W \cdot m^{-1} \cdot °C^{-1}$)
m Temperature coefficient ($°C^{-1}$)
N Number
$\vec{n}$ Unit normal vector
$\dot{q}$ Heat generation rate ($W \cdot m^{-3}$)
R Resistance (Ω)
T Temperature (°C)
t Time (s)
U Overall heat transfer coefficient ($W \cdot m^{-2} \cdot °C^{-1}$)
V Electric potential (V)
v Volume (m^3)
$\dot{v}$ Volumetric flow rate ($m^3 \cdot s^{-1}$)
x Coordinate dimension (m)
Δ Increment
ρ Charge density ($C \cdot m^{-3}$) when used in current continuity equation, density ($kg \cdot m^{-3}$) when used in energy balance equation
σ Electrical conductivity ($S \cdot m^{-1}$)
∇ Gradient

Subscripts and Superscripts

a Ambient air
f Fluid
i Index of incremental section
l Liquid
m Mean value
n Time step index
P Parallel
p Particle
pi Pipe
ref Reference temperature
S Series
s Surface (when used with temperature), solid (when used with resistance)
T Temperature

References

AbuDagga, Y. and Kolbe, E. (2000) Analysis of heat transfer in surimi paste heated by conventional and ohmic means. *Journal of Aquatic Food Product Technology* 9(2): 43–54.

Allen, K., Eidman, V. and Kinsey, J. (1996) An economic-engineering study of ohmic food processing. *Food Technology* 50(5): 269–273.

Bazhal, M. and Vorobiev, E. (2000) Electrical treatment of apple cossettes for intensifying juice pressing. *Journal of the Science of Food and Agriculture* 80(11): 1668–1674.

Benabderrahmane, Y. and Pain, J. (2000) Thermal behaviour of a solid/liquid mixture in an ohmic heating sterilizer – slip phase model. *Chemical Engineering Science* 55(8): 1371–1384.

Bengston, R., Birdsall, E., Feilden, S., Bhattiprolu, S., Bhale, S. and Lima, M. (2006) Ohmic and inductive heating. In: *Handbook of Food Science, Technology and Engineering*, Vol. 3 (ed. Y.H. Hui). CRC, Boca Raton, FL, pp. 120:1–8

Chen, C., Abdelrahim, K. and Beckerich, I. (2010) Sensitivity analysis of continuous ohmic heating process for multiphase foods. *Journal of Food Engineering* 98(2): 257–265.

Cho, H., Yousef, A. and Sastry, S. (1996) Growth kinetics of *Lactobacillus acidophilus* under ohmic heating. *Biotechnology and Bioengineering* 49(3): 334–340.

Davies, L., Kemp, M. and Fryer, P. (1999) The geometry of shadows: effects of inhomogeneities in electrical field processing. *Journal of Food Engineering* 40(4): 245–258.

de Alwis, A. and Fryer, P. (1990) A finite-element analysis of heat-generation and transfer during ohmic heating of food. *Chemical Engineering Science* 45(6): 1547–1559.

de Alwis, A. and Fryer, P (1992) Operability of the ohmic heating process – electrical conductivity effects. *Journal of Food Engineering* 15(1): 21–48.

de Alwis, A., Halden, K. and Fryer, P. (1989) Shape and conductivity effect in the ohmic heating of foods. *Chemical Engineering Research and Design* 67(2): 159–168.

Fellows, P.J. (2000). *Food Processing Technology: Principles and Practice*, 2nd edn. CRC, Boca Raton, FL.

Fryer, P. and de Alwis, A. (1989) Validation of the APV ohmic heating process. *Chemistry and Industry* 19: 630–634.

Fryer, P. and Li, Z. (1993) Electrical resistance heating of foods. *Trends in Food Science and Technology* 4: 364–369.

Fryer, P., de Alwis, A., Koury, E., Stapley, A. and Zhang, L. (1993) Ohmic processing of solid liquid mixtures heat generation and convection effects. *Journal of Food Engineering* 18(2): 101–125.

Ghnimi, S., Zaid, I., Maingonnat, J. and Delaplace, G. (2009) Axial temperature profile of ohmically heated fluid jet: analytical model and experimental validation. *Chemical Engineering Science* 64(13): 3188–3196.

Goullieux, A. and Pain, J.-P. (2005) Ohmic heating. In: *Emerging Technologies for Food Processing* (ed. D.-W. Sun). Elsevier, London, pp. 469–505.

Halden, K., de Alwis, A. and Fryer, P. (1990) Changes in the electrical conductivity of foods during ohmic heating. *International Journal of Food Science and Technology* 25(1): 9–25.

Imai, T., Uemura, K., Ishida, N., Yoshizaki, S. and Noguchi, A. (1995) Ohmic heating of Japanese white radish *Rhaphanus sativus L. International Journal of Food Science and Technology* 30(4): 461–472.

Imai, T., Uemura, K. and Noguchi, A. (1998) Heating rate of egg albumin solution and its change during ohmic heating. *Process-Induced Chemical Changes in Food* 434: 101–108.

Jun, S. and Sastry, S. (2009) Modeling of ohmic heating of foods. In: *Food Processing Operations Modeling: Design and Analysis*, 2nd edn (eds S. Jun and J.M. Irudayaraj). CRC, Boca Raton, FL, pp. 143–171.

Khalaf, W. and Sastry, S. (1996) Effect of fluid viscosity on the ohmic heating rate of solid–liquid mixtures. *Journal of Food Engineering* 27(2): 145–158.

Kim, H., Choi, Y., Yang, T., Taub, I., Tempest, P., Skudder, P., Tucker, G. and Parrott, D.L. (1996) Validation of ohmic heating for quality enhancement of food products. *Food Technology* 50(5): 253–261.

Kulshrestha, S. and Sastry, S. (2003) Frequency and voltage effects on enhanced diffusion during moderate electric field (MEF) treatment. *Innovative Food Science and Emerging Technologies* 4: 189–194.

Lima, M. and Sastry, S. (1999) The effects of ohmic heating frequency on hot-air drying rate and juice yield. *Journal of Food Engineering* 41(2): 115–119.

Lima, M., Heskitt, B. and Sastry, S. (2001) Diffusion of beet dye during electrical and conventional heating at steady-state temperature. *Journal of Food Process Engineering* 24(5): 331–340.

Loghavi, L., Sastry, S.K. and Yousef, A.E. (2007) Effect of moderate electric field on the metabolic activity and growth kinetics of *Lactobacillus acidophilus*. *Biotechnology and Bioengineering* 98(4): 872–881.

Loghavi, L., Sastry, S.K. and Yousef, A.E. (2009) Effect of moderate electric field frequency and growth stage on the cell membrane permeability of *Lactobacillus acidophilus*. *Biotechnology Progress* 25(1): 85–94.

Maroulis, Z.B. and Saravacos, G.D. (2003) *Food Process Design*. Marcel Dekker, New York.

Mizrahi, S. (1996) Leaching of soluble solids during blanching of vegetables by ohmic heating. *Journal of Food Engineering* 29(2): 153–166.

Ohlsson, T., Bengtsson, G. and Bengtsson, N. (2002) *Minimal Processing Technologies in the Food Industry*. Woodhead, Cambridge.

Orangi, S., Sastry, S. and Li, Q. (1998) A numerical investigation of electroconductive heating in solid–liquid mixtures. *International Journal of Heat and Mass Transfer* 41(14): 2211–2220.

Park, S.J., Kim, D., Uemura, K. and Noguchi, A. (1995) Influence of frequency on ohmic heating of fish-protein gel. *Journal of the Japanese Society for Food Science and Technology – Nippon Shokuhin Kagaku Kogaku Kaishi* 42(8): 569–574.

Park, J.W., Yongsawatdigul, J. and Kolbe, E. (1998) Proteolysis and gelation of fish proteins under ohmic heating. *Process-Induced Chemical Changes in Food* 434: 25–34.

Parrot, D. (1992) Use of ohmic heating for aseptic processing of particulates. *Food Technology* 46(12): 68–72.

Praporscic, I., Ghnimi, S. and Vorobiev, E. (2005) Enhancement of pressing of sugar beet cuts by combined ohmic heating and pulsed electric field treatment. *Journal of Food Processing and Preservation* 29(5–6): 378–389.

Reznick, D. (1996). Ohmic heating of fluid foods. *Food Technology* 50(5): 250–251.

Ruan, R., Ye, X., Chen, P., Doona, C.J. and Taub, I. (2001) Ohmic heating. In: *Thermal Technologies in Food Processing* (ed. P. Richardson). Woodhead, Cambridge, pp. 241–265.

Ruan, R., Ye, X., Chen, P., Doona, C. and Yang, T. (2004) Developments in ohmic heating. In: *Improving the Thermal Processing of Foods* (ed. P. Richardson). Woodhead, Cambridge, pp. 224–252.

Salengke, S. and Sastry, S.K. (2007a) Effects of ohmic pretreatment on oil uptake of potato slices during frying and subsequent cooling. *Journal of Food Process Engineering* 30(1): 1–12.

Salengke, S. and Sastry, S.K. (2007b) Experimental investigation of ohmic heating of solid–liquid mixtures under worst-case heating scenarios. *Journal of Food Engineering* 83(3): 324–336.

Salengke, S. and Sastry, S.K. (2007c) Models for ohmic heating of solid–liquid mixtures under worst-case heating scenarios. *Journal of Food Engineering* 83(3): 337–355.

Sandeep, K.P. and Puri, V.M. (2009) Aseptic processing of liquid and particulate foods. In: *Food Processing Operations Modeling: Design and Analysis*, 2nd edn (eds S. Jun and J.M. Irudayaraj). CRC, Boca Raton, FL, pp.13–52.

Sarang, S., Sastry, S.K. and Knipe, L. (2008) Electrical conductivity of fruits and meats during ohmic heating. *Journal of Food Engineering* 87(3), 351–356.

Sastry, S.K. (1992) A model for heating of liquid–particle mixtures in a continuous flow ohmic heater. *Journal of Food Process Engineering* 15, 263–278.

Sastry, S.K. (2008) Ohmic heating and moderate electric field processing. *Food Science and Technology International* 14(5): 419–422.

Sastry, S.K. and Palaniappan, S. (1992a) Ohmic heating of liquid particle mixtures. *Food Technology* 46(12): 64–67.

Sastry, S.K. and Palaniappan, S. (1992b) Influence of particle orientation on the effective electrical resistance and ohmic heating rate of a liquid–particle mixture. *Journal of Process Engineering* 15: 213–227.

Sastry, S.K., Yousef, A., Cho, H.-Y., Unal, R., Salengke, S., Wang, W.-C., Lima, M., Kulshrestha, S., Wongsa-Ngasri, P. and Sensoy, I. (2002) Ohmic heating and moderate electric field (MEF) processing. In: *Engineering and Food for the 21st Century* (eds J. Welti-Chanes, G.V. Barbosa-Cánovas and J.M. Aguilera). CRC, Boca Raton, FL, pp. 785–794.

Sastry, S.K., Jun, S., Somavat, R., Samaranayake, C., Yousef, A. and Pandit, R. (2009) Heating and sterilization technology for long-duration space missions transport processes in a reusable package. *Interdisciplinary Transport Phenomena: Fluid, Thermal, Biological, Materials, and Space Sciences* 1161: 562–569.

Schreier, P.J.R., Reid, D.G. and Fryer, P.J. (1993) Enhanced diffusion during the electrical heating of foods. *International Journal of Food Science and Technology* 28(3): 249–260.

Sensoy, I. and Sastry, S.K. (2004a) Ohmic blanching of mushrooms. *Journal of Food Process Engineering* 27(1): 1–15.

Sensoy, I. and Sastry, S.K. (2004b) Extraction using moderate electric fields. *Journal of Food Science* 69(1): E7–E13.

Skudder, P. and Biss, C. (1987) Aseptic processing of food products using ohmic heating. *The Chemical Engineer* (433): 26–28.

Stapley, A., Goncalves, J., Hollewand, M., Gladden, L. and Fryer, P. (1995) An NMR pulsed field gradient study of the electrical and conventional heating of carrot. *International Journal of Food Science and Technology* 30(5): 639–654.

Tulsiyan, P., Sarang, S. and Sastry, S. (2008) Electrical conductivity of multicomponent systems during ohmic heating. *International Journal of Food Properties* 11(1): 233–241.

Unal, R., Yousef, A.E. and Sastry, S.K. (1998) Growth kinetics of Lactococcus lactis subsp. Lactis under ohmic heating. Abstract No. 59C-5. In: *IFT Annual Meeting*, Atlanta, GA, June 20–24 1998.

Vicente, A. and Castro, I.A. (2007) Novel thermal processing technologies. In: *Advances in Thermal and Non-Thermal Food Preservation* (eds G. Tewari and V.K. Juneja). Blackwell Publishing, Ames, IA, pp. 99–130.

Vicente, A., de Castro, I. and Tiexeira, J.A. (2006) Ohmic heating for food processing. In: *Thermal Food Processing: New Technologies and Quality Issues* (ed. D.-W. Sun). CRC, Boca Raton, FL, pp. 425–468.

Wang, W. and Sastry, S.K. (1993) Salt diffusion into vegetable tissue as a pretreatment for ohmic heating – electrical conductivity profiles and vacuum infusion studies. *Journal of Food Engineering* 20(4): 299–309.

Wang, W.C. and Sastry, S.K. (1997) Changes in electrical conductivity of selected vegetables during multiple thermal treatments. *Journal of Food Process Engineering* 20(6): 499–516.

Wang, W.C. and Sastry, S.K. (2000) Effects of thermal and electrothermal pretreatments on hot air drying rate of vegetable tissue. *Journal of Food Process Engineering* 23(4): 299–319.

Yang, T.C.S. (2000) Aseptic processing: ohmic heating. In: *Wiley Encyclopedia of Food Science and Technology*, 2nd edn (ed. F.J. Francis). John Wiley & Sons, Inc., New York, pp. 128–132.

Yongsawatdigul, J. and Park, J.W. (1996) Linear heating rate affects gelation of Alaska pollack and Pacific white surimi. *Journal of Food Science* 61(1): 149–153.

Yongsawatdigul, J., Park, J., Kolbe, E., AbuDagga, Y. and Morrissey, M. (1995) Ohmic heating maximizes gel functionality of Pacific whiting surimi. *Journal of Food Science* 60(1): 10–14.

Zhang, L. and Fryer, P. (1993) Models for the electrical heating of solid liquid food mixtures. *Chemical Engineering Science* 48(4): 633–642.

Zhao, Y., Kolbe, E. and Flugstad, B. (1999) A method to characterize electrode corrosion during ohmic heating. *Journal of Food Process Engineering* 22(1): 81–89.

Zhong, T. and Lima, M. (2003) The effect of ohmic heating on vacuum drying rate of sweet potato tissue. *Bioresource Technology* 87(3): 215–220.

Zitoun, K.B. (1996) *Continuous Flow of Solid–Liquid Food Mixtures during Ohmic Heating: Fluid Interstitial Velocities, Solid Area Fraction, Orientation and Rotation*. PhD thesis, The Ohio State University, Columbus, OH.

Zoltai, P. and Swearingen, P. (1996) Product development considerations for ohmic processing. *Food Technology* 50(5): 263–266.

37

Design of Equipment for Pulsed Electric Field Processing

Federico Gómez Galindo and Pär Henriksson

Introduction

In recent years, there has been increasing interest in the use of pulsed electric field (PEF) technology owing to its potential to induce nonthermal permeabilization of cell membranes. The external application of an electric field ($\leq$1 kV·cm^{-1}) induces a potential difference of ~0.3–1 V across the cytoplasmic membrane for a period long enough (microseconds to milliseconds) to induce pore formation in the plasma membrane (Weaver, 2000). Depending on the properties (i.e., size, conductivity, shape and orientation) of the cells and the electropulsation parameters (i.e., field strength, duration and number of pulses), the application of a PEF may cause lethal damage to the cells due to irreversible loss of cell membrane permeability properties, leakage of cytoplasmic contents and lysis (Aronsson *et al.* 2001). Based on this phenomenon, most of the applications in food science that have been studied have concentrated on two main aspects. (1) The effects of high-field-strength PEF treatments on the inactivation of microorganisms for pasteurization of liquid foods without a substantial increase in temperature (Pothakamuri *et al.* 1997; Qin *et al.* 1998). Fruit juices (Walkling-Ribeiro *et al.* 2009; Aguiló-Aguayo *et al.*, 2009; Akin and Evrendilek, 2009), milk (Fernandez-Molina *et al.*, 2005) and waste water (Gusbeth *et al.* 2009) are a few examples of products that have been treated with PEFs for bacterial reduction. (2) The irreversible permeabilization of cell membranes in plant tissues can, at a lower field strength, be

Handbook of Food Process Design, First Edition. Edited by Jasim Ahmed and Mohammad Shafiur Rahman.

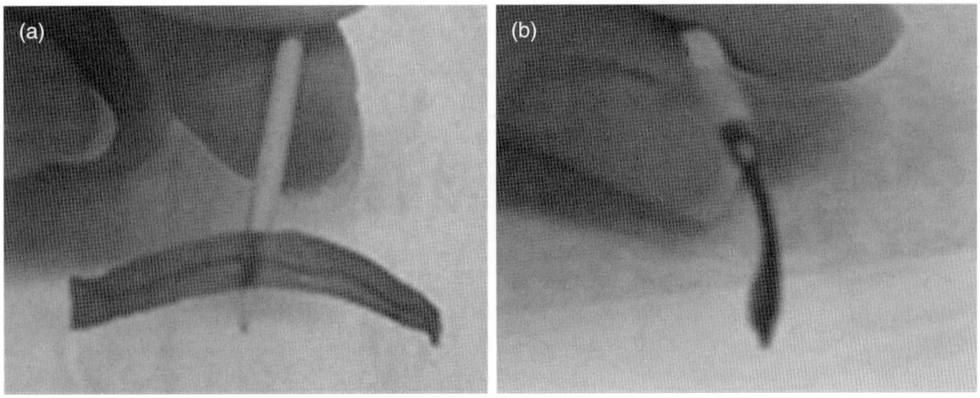

Figure 37.1 Spinach leaves treated with a freeze–thaw cycle. (a) The leaf cells were infused with the cryoprotectant trehalose before the freezing–thawing cycle. PEF treatment was used for the delivery of the cryoprotectant into the cytoplasm of the cells. (b) Control sample that did not undergo PEF-assisted infusion of the cryoprotectant before the freezing–thawing cycle.

used for the improvement of mass transfer processes such as obtaining an increased extraction yield of fruit juices (Fincan *et al.* 2004; Chalermchat and Dejmek, 2005; Turk *et al.*, 2009), extraction of biomolecules (e.g., colorants and antioxidants) (Chalermchat *et al.*, 2004, Jaeger *et al.*, 2008) and increasing the rate of drying (Ade-Omowaye *et al.*, 2003; Lebovka *et al.*, 2007a).

Permeabilization may be done without affecting cell viability through strict control of the electropulsation parameters. It is routinely used for gene transfer with walled and wall-less systems (Ganeva *et al.*, 1995). It is now being proposed as an efficient method for delivery of drugs, antibodies and plasmids *in vivo* in clinical biotechnological applications. The delivery of a cryoprotectant was recently proposed in the context of the food industry as a method for improving the freezing tolerance of leafy vegetables (Phoon *et al.*, 2008); remarkably, cell viability is preserved after a freezing–thawing cycle (Figure 37.1).

It should be emphasized that very little is known about what actually occurs in the cell and its membranes at the molecular level upon reversible electroporation, and the physiological responses to PEF-induced stress are still largely unknown. It has been reported that PEF treatment affects metabolism, including an oxidative burst with consequent generation of reactive oxygen species (Gabriel and Teissié, 1994; Sabri *et al.*, 1996), and stimulation of the production of secondary metabolites, such as increased yields of a cytostatic compound in a cell culture of *Taxus chinensis* (Ye *et al.*, 2004) and of antioxidants and phytoesterols from oil seeds and fruits (Guderjan *et al.*, 2005; Gunderjan *et al.*, 2007).

Recent developments using more sophisticated technologies (nanosecond pulses, nanoPEFs) have shown that cytosolic targeting of the effect of the field is possible. A nanosecond pulse is shorter in duration and higher in electric field by orders of

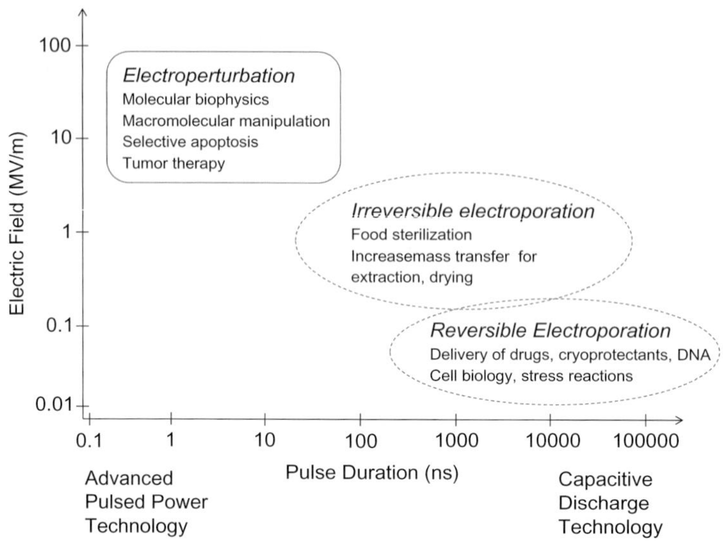

Figure 37.2 Schematic representation of the various areas of application of PEF technology (adapted from Martin Gundersen's webpage, Pulsed Power Group, University of Southern California, http://www.usc.edu/dept/ee/Gundersen/, with permission).

magnitude. Therefore, the effect of the pulse is completely different, as the pulse duration is shorter than the charging time of the plasma membrane, and the cell response changes significantly. The fast-rising electric pulse instead facilitates penetration of the electric field into the cell interior, charging intracellular structures and directly inducing intracellular responses. As a result, nanoPEF treatment could be referred to as a method of "electroperturbation" rather than "electroporation" (Beebe and Schoenbach, 2005), in which membrane charging is the main mechanism that yields effects such as stress responses (Gundersen *et al.*, 2004). To the best of our knowledge, this technology has not been tested on food systems.

Figure 37.2 is a schematic diagram summarizing the various areas of application of pulsed electric fields as a function of pulse duration and electric field strength.

Principles and Technology

In this chapter, we illustrate the principles of the application of PEFs and the use of the necessary equipment.

Pulse Shapes and Protocols

Square wave pulses are normally delivered by PEF systems, either monopolar or bipolar, and maintain a constant peak voltage through the pulse duration, with very

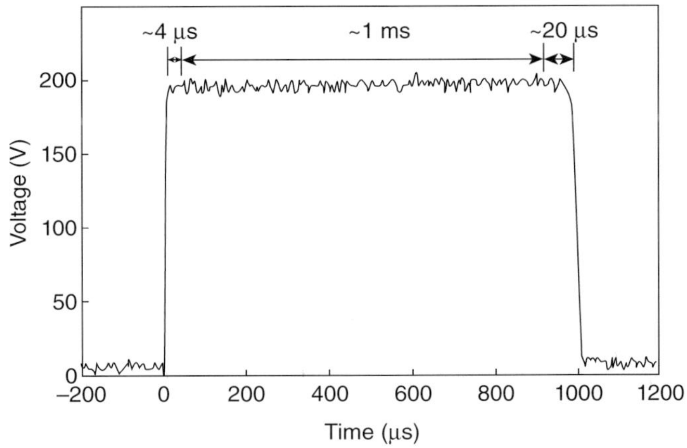

Figure 37.3 Characteristic square wave pulse as a function of time. The pulse width is 1 ms; the rise and fall times are shown. (From Figure 1B of Gómez Galindo *et al.* (2008)).

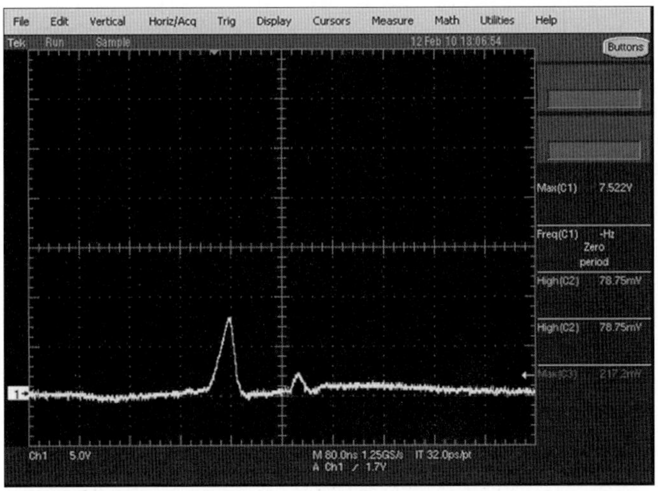

Figure 37.4 Typical short-duration pulse (nanoseconds) as a function of time, as shown on a conventional oscilloscope measurement device. At high voltages, short-time reflections may accompany the pulse. In the picture, each vertical division represents 80 ns and each horizontal division represents 2500 V.

short rise and fall times in comparison with the total duration of the pulse. An example is given in Figure 37.3 of a 1 ms pulse. When pulses of very short duration (e.g., on the order of nanoseconds) are required, square pulses are not longer possible, owing to the very fast rise and fall times required. Reactive properties such as cable capacitance and inductance will affect the pulse. The peak electric field is maintained for only a short time before the field decreases to its final value, as shown in Figure 37.4.

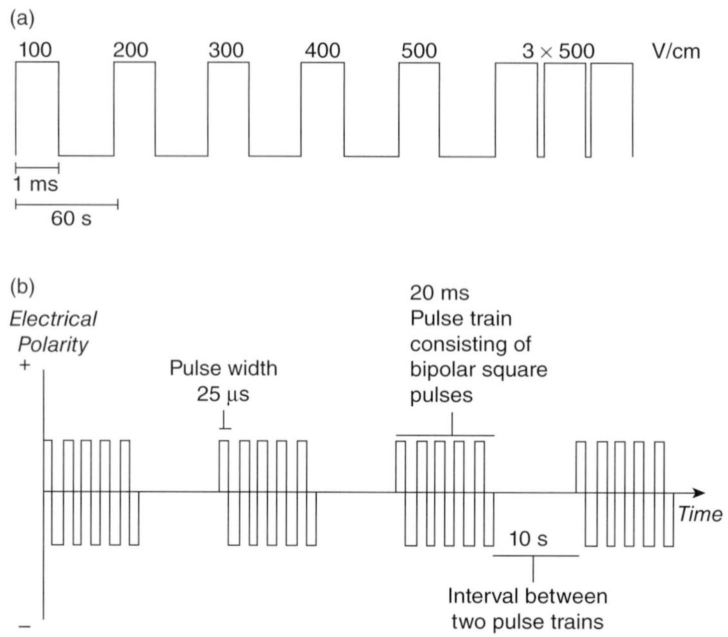

Figure 37.5 Schematic illustration of PEF delivery protocols in which the pulse width and strength, the interval between pulses and the interval between trains of pulses are considered. (a) Monopolar pulses, (b) bipolar pulses. (From Figure 1B of Phoon *et al.* (2008)).

The parameters to be considered when one is treating samples with electric pulses are the pulse width and strength, the number of pulses, the space between the pulses, and the space between trains of pulses. Figure 37.5 gives examples of some protocols for the use of electric pulses.

Equipment

Several different kinds of pulsed power technology can be used to generate fast electrical discharges. The main components are a power source, a charging device, a charge storage device, a switch, and an applicator or treatment chamber in which the product to be treated is situated (Figure 37.6).

PEF Delivery Equipment: The "Spark Gap" Design

The most traditional approach in the design of PEF delivery equipment is to use a "spark gap" (Figure 37.7). Two conducting electrodes are separated by an air gap or a gas-filled gap. When a voltage with sufficient amplitude is applied to the gap, a spark is formed, which ionizes the gas and drastically reduces the electrical resistance between the two electrodes. The point where the spark appears depends on the dis-

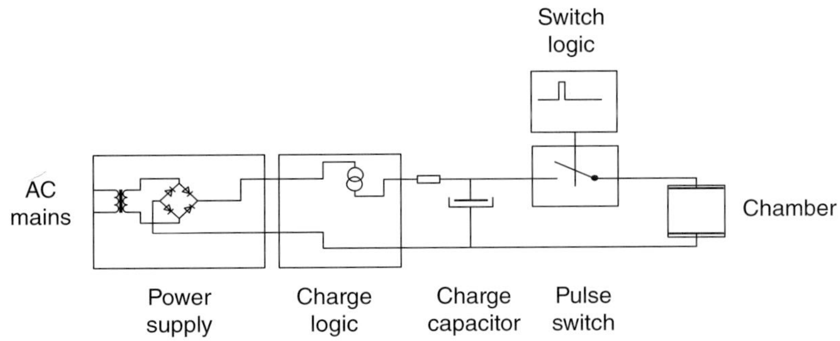

Figure 37.6 Block diagram with typical components.

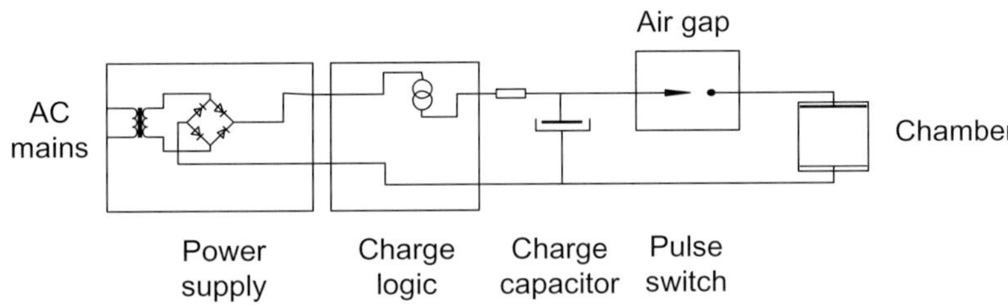

Figure 37.7 Typical air gap system.

tance between the two electrodes, the gap and the gas used. When the gas is ionized, the current continues to flow until the voltage source has discharged and the current is below a certain threshold level. If the gap is connected to a high-voltage capacitor that is charged using a constant-current source or similar, a train of pulses can be achieved. The pulse amplitude can be controlled by an adjustment device connected to the gap that is used to adjust the distance between the electrodes. The frequency of the pulses can be set by adjustment of the current from the constant-current source. For various reasons, however, this is not a suitable way of creating high-voltage pulses, as more accurate control of the amplitude and also of the frequency and pulse width is needed. The pulse width in this configuration is influenced by the impedance of the product to be treated.

Modern Designs of Pulse Generators

A number of different switching devices are available on the market and are constantly being developed. These devices are designed to handle high voltages, although the current capacity is low. In older designs, various forms of vacuum tubes are used; however, vacuum tubes have the drawback that their efficiency is normally low and their lifetime is limited. Modern designs use semiconductors. There are various types of semiconductors, and a complete survey is beyond the scope of this document, although a few can be mentioned. A thyristor-based circuit has similar properties to an air gap system; the main difference is that the thyristor can be controlled to a certain extent. The point where the device "fires" can be controlled, but the thyristor remains active or conducting until the voltage across the capacitor is below a threshold level. For some applications, thyristors can be a cost-effective solution.

In general, conventional transistors can handle high currents but voltage handling is more limited. Bipolar transistors are current amplifiers which can handle very high currents, and voltages up to 1 kV or slightly higher. A field effect transistor (FET) or the more powerful metal–oxide–semiconductor FET (MOSFET) is basically a voltage amplifier and works in a manner similar to a vacuum tube. A more recent design is the insulated-gate bipolar transistor (IGBT), a simplified transistor where the output stage is designed as a bipolar transistor and the input stage is designed as an FET. This device combines the best of both worlds and is currently under development.

If the generator is designed to be connected to a mains power supply that has an alternating voltage supply in the range of 90–230 V AC (or 400 V AC, which is commonly available for industrial applications), it is obvious that this voltage has to be converted up to a level of several kilovolts. Most commonly, an arrangement using transformers is used for this purpose, as semiconductors can traditionally handle high currents but not withstand high voltages; the most popular approach is to use a pulse transformer at the output stage, that is, to switch at a relatively low voltage and use a transformer to convert a very high-current, low-voltage pulse to a high-voltage, low-current pulse. In other words, the current is traded off against the voltage. This arrangement is similar to that found in most internal combustion engines to create the ignition pulse. In a car, a charge at a voltage of 12 V is stored in a relatively small capacitor – a few microfarads – and this capacitor is then discharged through a pulse transformer that has a ratio of, for instance, 1 : 15 000. This means that an alternating voltage on the primary side (input) with an amplitude of 1 V will result in an alternating voltage with an amplitude of 15 000 V on the secondary side. This increased voltage is traded off against the current; approximately, the current on the primary side is 15 000 times higher than that available on the secondary side. A resulting high-voltage pulse with an amplitude of up to 20 kV or more can thus be achieved, although very high currents are required on the switching side.

The transformer approach has limitations that are most apparent when high-power pulses are needed. The capacity of the charge storage device has to be very high. If the primary switching voltage is just over 320 V DC (which is achieved if the 230 V AC mains from a common outlet is rectified into a direct current) and a pulse with an amplitude of 20 kV is desired, then a transformer with a ratio of at least 1 : 60 is needed. To be exact, the input pulse of 320 V has to be amplified 60 times and the amplitude has to be 19 200 V. If a current in the range of 100 A is desired, then a primary current of 6000 A is required.

A capacitor is normally used to store the charge, and this device has a capacitance that is measured in farads (F) (A·s). If the duration of the pulse is 10 μs, then a 300 V capacitor capable of storing $10\,\mu s \times 6000\,A = 60\,000\,\mu F$ is needed. A device with this capacity does not exist, and has to be constructed using a large number of devices connected in parallel (to increase the capacitance) or in series (to increase the voltage-handling capacity). In other words, this device has to be formed using a relatively large number of capacitors. Furthermore, to create the pulse on the primary side of the transformer, a switch capable of handling a very high current, in this case 6000 A, is necessary, and an arrangement using a number of switching devices in parallel would be required.

Another disadvantage of using a pulse transformer is the fact that a transformer is a reactive device and the shape of the pulse cannot be squared, owing to resonances, ringing and other phenomena.

Another approach, therefore, is to use a transformer before the switching device and then feed thc treatment applicator directly from the switching device. In this method, the voltage is increased to the required level, and then a switch is assembled using a number of semiconductor devices connected in series in order to be able to withstand the high voltage. Also, charge storage capacitors have to be connected in series to withstand the high voltage. To further stabilize the design, resistors and diodes are needed to ensure that the voltage is evenly spread across the switches and the capacitors. By using such a design, the disadvantages of the pulse transformer are avoided and a "perfect" square wave pulse with a very high rise time can be created (see Figure 37.8). However, the drawback of this design is that very high voltages are difficult to handle. They are always present in the generator, and thus the design has to be carefully considered and the use of the system needs precautions. Insulation of wires, distance between high-voltage terminals and various levels of insulation are needed to render the equipment safe.

A novel approach is used in the CEPT® device (Figure 37.9), which differs from the traditional designs (Henriksson, 2008; Henriksson, 2009) in that instead of creating a pulse with a very high amplitude, a pulse with an amplitude in the range of 1–2 kV is produced. Depending on the amplitude required for the application, a number of pulse generators are used, and these are connected in series and synchronized from an external controller. The resulting high-voltage pulse will only appear when all the

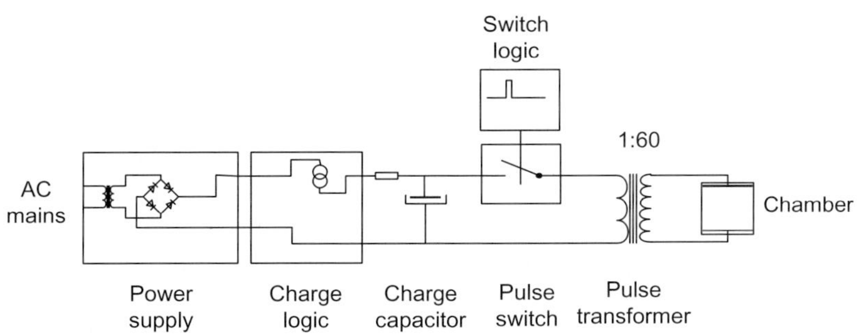

Figure 37.8 Typical pulse transformer design.

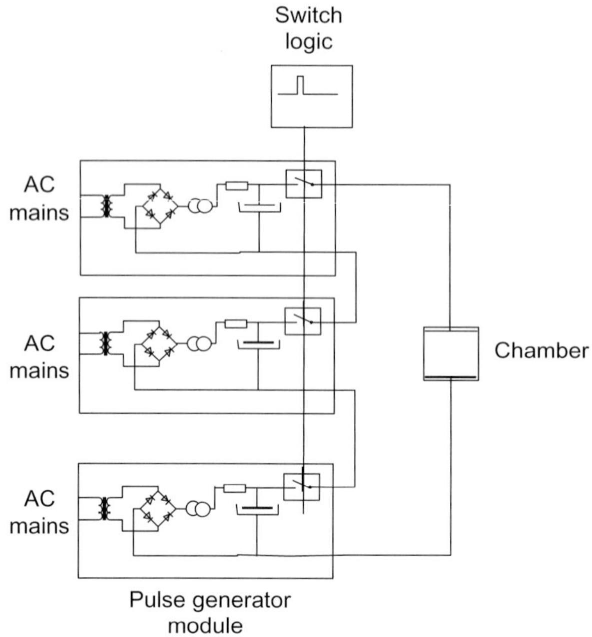

Figure 37.9 Simplified CEPT® generator.

generators are synchronized, using an external controller device to simultaneously and for a very short time apply a voltage to the output of each module. As the modules are connected in series, the resulting voltage is the sum of the voltages from all the modules.

The advantage of using this approach is that each generator is relatively simple and uses standard components designed for voltages in the range of up to 2 kV. These components are widely available and thus very cost-effective, the drawbacks of using transformers are avoided, and the fast rise and fall timed obtained from simple designs are maintained. This approach makes it possible to design a modular low-cost generator that can be adapted to particular needs by adding more modules and increased capacity. The pulse width, amplitude and frequency can be controlled and fine-tuned.

Treatment Chambers for Continuous Processing

When PEF treatment is used in an industrial situation, a continuous or semicontinuous flow is required. Many designs and PEF systems for this purpose have been reported (for a review, see Huang and Wang, 2009). In this section, we describe the basic principles of design and some new state-of-the-art systems. The most straightforward design includes the electrodes in the tube used for the transportation of the product.

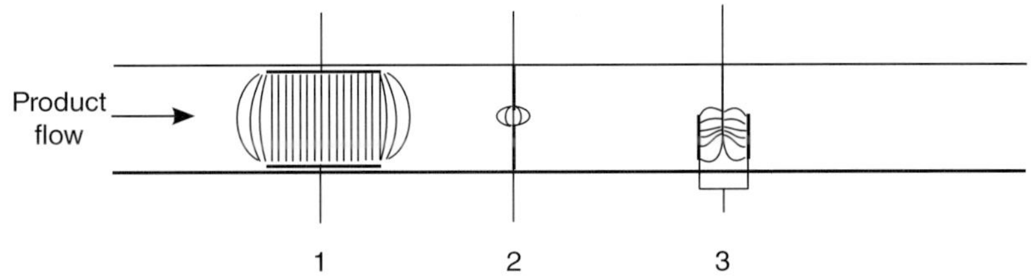

Figure 37.10 Three different arrangements of electrodes in a transport tube.

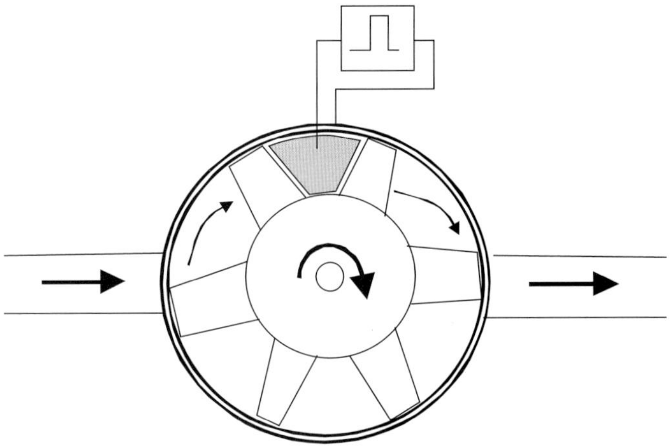

Figure 37.11 PEF treatment chamber integrated into a pump.

In Figure 37.10, a few possible arrangements of electrodes are shown. The simplest (1) has two plates integrated into the tube, and the product passes between these two plates. This configuration is similar to the cuvette used in the laboratory, but the product is in motion while being treated. The flow inside a tube is not linear, owing to various effects, and therefore the pulse frequency has to be adapted to the flow to ensure that all parts of the product are exposed to the number of pulses required for treatment.

The electrode arrangements 2 and 3 (Figure 37.10) are examples of alternative designs where an aperture is used to force the product to pass between more narrowly spaced electrodes. With this design, the electric field is more concentrated and has a higher strength. The disadvantage is that this configuration presents resistance to the flow and also creates turbulence within the product during treatment.

In a few more recent designs, the PEF treatment chamber is integrated into a conventional rotating pump, similar to those often used in the food industry (Lindgren, 2005) (Figure 37.11). The pumpable product is transported from the inlet to the outlet and, during transportation, a volume is enclosed between the pump blades, which

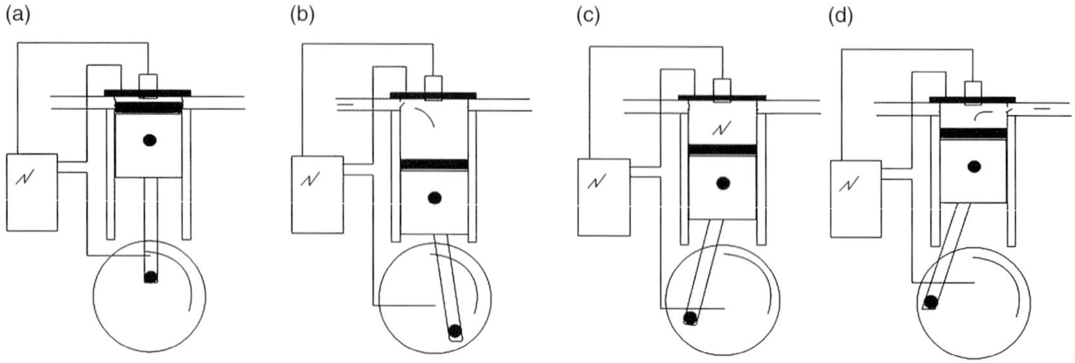

Figure 37.12 CEPT® PEF treatment chamber.

form a small compartment. This product volume passes over an electrode arrangement and, while passing, the product is exposed to pulses. The pump produces a continuous flow and at the same time acts as a PEF treatment chamber. The negative aspect of this design is that the pump may produce an underpressure and air bubbles might form in the product, affecting the discharges. Another disadvantage is that the product cannot be fed back to the supply line in case of failures during the treatment process.

In the CEPT® (Closed Environment PEF Treatment) system, a conventional piston pump is used; Figure 37.12 illustrates how this system works. In Figure 37.12a, the pump is empty, and when the piston is moving downwards (Figure 37.12b), the compartment is filled. When the piston moves upwards (Figure 37.12c), the pressure rises and, at a predefined overpressure, the high-voltage pulses pass though the product and the product is fed forward to the production line (Figure 37.12d). The advantage of this design is that the product is more or less stationary during exposure to the pulses. The product is pressurized to avoid formation of air bubbles, and there is the possibility to monitor each volume so that it is properly treated and also to decide whether to open the valve to feed the product forward as a clean product or, in case of a failure, feed it to a drain and return it to the untreated-product reservoir.

Preservation of Liquid Foods by Pulsed Electric Fields

Nonthermal processes have gained importance as a potential technology to replace or complement the traditional thermal processing of food. Compared with thermal processing, nonthermal processes offer the advantages of a low processing temperature, low energy utilization, and the retention of flavors, nutrients and a fresh taste, while inactivating spoilage-causing microorganisms.

During the last decade, an unprecedented quantity of research and development activities has been carried out regarding nonthermal processing. Most of the work has concentrated on high hydrostatic pressure and pulsed electric fields, with microbial inactivation, quality improvement and shelf life extension forming the initial primary goals of those efforts (Bendicho *et al.*, 2002; Palou *et al.*, 1999; Vega-Mercado *et al.*, 1997).

Important industrial developments of this technique have been achieved through evaluation of the efficiency of PEF treatment for microbial inactivation. The level of microbial inactivation achieved with PEF treatment depends mainly on the field strength and the number of pulses applied during the process (Martín *et al.*, 1997; Pothakamuri *et al.*, 1997; Qin *et al.*, 1995).

As examples, a few patents such as US Patents 4695472 (Dunn and Pearlman, 1987), 5034235 (Dunn *et al.*, 1991) and 5235905 (Bushnell et al., 1993) describe the preservation of fluid foods such as dairy products, fruit juices and liquid eggs by treatment with an electric field strength in the range of 12–25 $kV \cdot cm^{-1}$ for 1–100 µs. The lethal effects observed during PEF treatments were remarkably higher than those achieved in a parallel test using only thermal pasteurization. The PEF process is safe because no dangerous chemical reactions have been detected; furthermore, it is reliable because the same results can be obtained repeatedly. US Patent 4695472 reported a reduction by five log cycles in the microbial count of naturally occurring microorganisms in orange juice after 35 pulses of length 100 µs at a voltage intensity of 33.6–35.7 $kV \cdot cm^{-1}$ and a process temperature of 42–65 °C. The effect of PEF treatment on raw milk has also been studied. Raw milk was subjected to several treatment conditions and, if stored under refrigeration, the PEF-processed milk was found to have a microbial shelf life of 2 weeks (Qin *et al.*, 1995). These reports of microbial inactivation were accompanied by evidence showing that the taste of the treated liquids did not change following the electric-field treatments, which is an extremely important finding as today's consumers prefer "fresh" flavors. Other products with reported successful PEF treatments include raw peach juice, skim milk, beaten eggs, pea soup, apple juice and reconstituted apple juice (Zhang *et al.*, 1995).

More detailed scientific studies of the mechanisms of microbial inactivation by PEF have concluded that morphological and environmental factors are also crucial for the inactivation levels that can be achieved. For example, it has been reported that yeast cells are more sensitive to PEF treatment than bacterial cells, and Gram-negative bacterial cells are more sensitive than Gram-positive cells (Wan *et al.*, 2009). Also, PEF is widely reported as being ineffective for the inactivation of bacterial spores.

The information contained in the literature about the influence of the conductivity and pH of the treatment medium on the efficacy of microbial inactivation by PEF treatment is not conclusive. Some studies have shown that lowering the conductivity of the medium increases the degree of microbial inactivation (Vega-Mercado *et al.*, 1996; Wouters *et al.*, 1999). This factor has been shown not to be dependent on the type of target microorganism (Alvarez *et al.*, 2003; Sepúlveda *et al.*, 2006). Early studies reported that the pH of the medium did not affect microbial inactivation by PEF treatment (Sale and Hamilton, 1967; Hülsheger *et al.*, 1981), whereas later studies reported that the inactivation was enhanced by a decrease in the pH of the medium (Geveke and Kozempel, 2003). The influence of the pH on the sensitivity of microorganisms to PEF treatment appears to be dependent on the target microflora (Pagan *et al.*, 2005).

Other environmental factors that have been studied in the literature for their influence on the effects of PEF treatment are the presence of divalent cations such as Ca^{2+} and Mg^{2+}, which increase the resistance of *E. coli* to PEF treatment (Hülsheger *et al.*, 1981), and the fat content (of milk), with results showing an effect dependent on the target microflora (Wan *et al.*, 2009).

Improvement of Mass Transfer Processes in Plant Materials

Juice Extraction

It has been demonstrated in several different plant tissues that electroporation enhances mass transfer during pressure extraction (increases in the hydraulic permeability and pressure conductivity of the tissue are observed) and aqueous extraction (an increase in solute diffusivity is observed) (Vorobiev, 2009). However, the data in the literature with respect to juice yield obtained using PEF treatment are contradictory. Considerable increases in yield have been reported for carrots (Tedjo *et al.*, 2002) and grapes (Grimi *et al.*, 2009) (Figure 37.13), whereas an improved yield was not shown for PEF-treated apple mash (McLellan *et al.*, 1991; Schilling *et al.*, 2008). Nevertheless, there is general agreement in the literature that a combination of PEF treatment with other treatments such as heating at moderate temperatures (~40 °C) (Jemai and Vorobiev, 2002; Lebovka *et al.*, 2007b) or the use of pressure (Bazhal *et al.*, 2001) is effective in increasing the yield of fruit juices. The use of PEF technology in sugar extraction from sugar beet has been described as a more energy-efficient extraction method, used in combination with the method of alkaline extraction (Sack *et al.*, 2009).

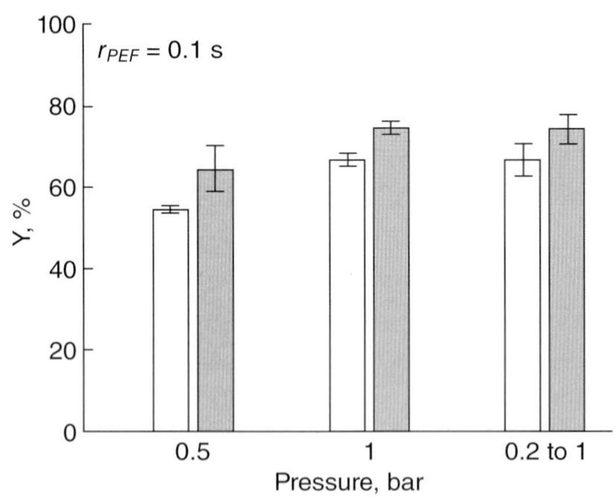

Figure 37.13 Final juice yield *Y* for untreated and PEF-treated grapes under different pressing conditions without PEF treatment (left bars) and with PEF treatment (right bars). The pulse width was 0.1 s. (From Figure 7 of Grimi *et al.* (2009), with kind permission from Springer Science+Business Media.)

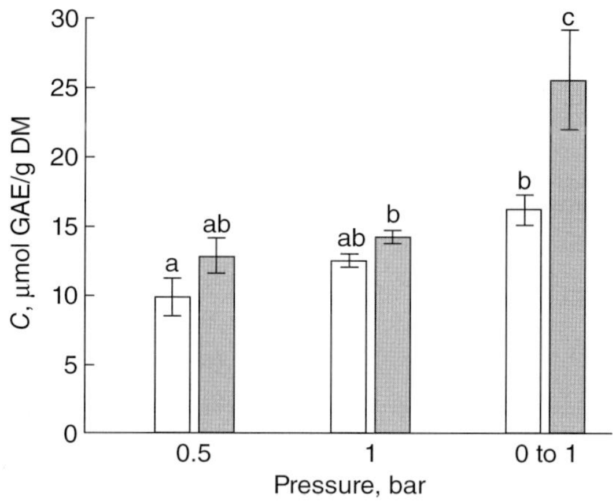

Figure 37.14 Content of total polyphenols, *C*, for 1 h of pressing at constant pressure ($P = 0.5$ bar and 1.0 bar) and with a progressive increase from 0 to 1 bar. Left bars, no PEF treatment; right bars, PEF-treated samples. The yields marked "a" were not significantly different (LSD test, 5% level), and similarly for the yields marked "b" and "c." (From Figure 9 of Grimi *et al.* (2009), with kind permission from Springer Science+Business Media.)

Higher extraction of nutritionally important compounds such as various macromolecules, flavors, pigments and other cell metabolites when PEF treatment is applied has been consistently reported in the literature. The electroporation of mashed carrots and grapes resulted in juices with a higher β-carotene content and higher release of polyphenols and pigment than from samples processed in a conventional way (Jaeger *et al.*, 2008; Sack *et al.*, 2009; Grimi *et al.*, 2009) (Figure 37.14). When PEF treatment is employed instead of enzymatic mash maceration, it is possible to recover antioxidant substances from plant residual material such as grape pomace (Balasa *et al.*, 2006; Boussetta *et al.*, 2009). The energetically efficient enhanced extraction of betalains, amaranthin and anthraquinones from beetroot, *Chenopodium rubrum* and *Morinda citrifolia*, respectively, has also been reported (Dörnenburg and Knorr, 1993; Fincan *et al.*, 2004).

Dehydration

Drying is one of the oldest and most traditional methods of food preservation. The commonly used freeze-drying and conventional drying techniques are limited by high energy consumption. Moreover, elevated temperatures affect the physical and biochemical status of foods, including undesirable changes in pigments, vitamins, and flavor and texture.

In general, drying consumes a considerable part of the total amount of energy used in the food industry and, therefore, various pretreatments that allow plant tissues to be dehydrated faster and/or at lower temperatures (e.g., microwave heating, ohmic

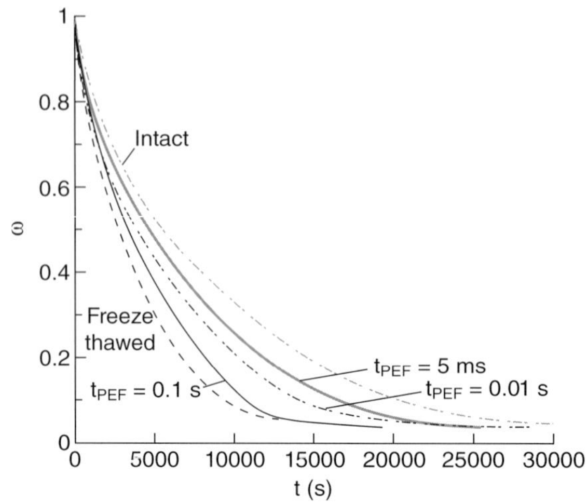

Figure 37.15 Moisture ratio ω versus drying time t for intact, PEF-treated and freeze–thaw-treated (dashed line) tissues at a drying temperature of 50 °C. PEF treatment was performed at 400 V·cm^{-1}, a pulse duration of 1 ms, a pulse repetition time of 0.001 s and different treatment times t_{PEF}, as shown in the figure. (Adapted from Figure 4 of Lebovka *et al.* (2007), with kind permission from Elsevier.)

heating, and osmotic pretreatments) have been the subject of extensive research. PEF treatment is a promising nonthermal method with great advantages for enhancement of the drying of thermally sensitive food materials. PEF-induced cell damage facilitates moisture diffusion, which enhances drying (Lebovka *et al.*, 2007a) (Figure 37.15). As an example, PEF-enhanced permeability of red beetroot allowed a reduction of the drying temperature by 20–25 °C, with higher preservation of the colorants in the beetroot (Shynkaryk *et al.*, 2008). Shorter drying times were demonstrated for PEF-pretreated potato tissue (Angerbach and Knorr, 1997). However, the greater cell damage caused by PEF treatment increases rehydration time, as reported for PEF-treated beetroot (Shynkaryk *et al.*, 2008) and carrots (Gachovska *et al.*, 2009).

Diffusion coefficients and drying rates in osmotic dehydration are also accelerated by PEF pretreatments. As reported by Rastogi *et al.* (1999), PEF treatment with an applied energy in the range of 0.04–2.25 kJ·kg^{-1} was enough to enhance diffusion coefficients without an increase in temperature.

PEF Treatment of Meat Products

PEF treatment of meat products has been used in processes such as drying, marination and curing, where improved mass transport is of interest (Toepfl *et al.*, 2006). Changes in tissue structure upon PEF treatment of ham at field strength from 1 to 3 kV·cm^{-1} lead to a weight increase after brine injection, as well as greater water-holding capacity and less loss during cooking. PEF causes porous, swamp-like structures that hold the injected brine through capillary forces better than the untreated ham, resulting in a softer and more tender product (Toepfl *et al.*, 2005). A more porous structure has also been observed when fish such as salmon, pollock and whelk were treated with PEFs (Gudmundsson and Hafsteinsson, 2001; Klonowski *et al.*, 2006).

Calculations, Monitoring and Optimization of Treatment Parameters

Electric Field Intensity

Upon application of an electric field, cell membrane breakdown occurs when the potential difference induced by the pulse, ΔV, exceeds the critical breakdown voltage at a critical field strength E_c.

If the membrane is regarded as a perfect insulating spherical layer surrounded by a conductive medium, and the membrane potential difference is taken as the potential outside the cell minus the potential inside the cell, ΔV can be expressed as (Teissié and Rols, 1993; Kotnik *et al.*, 1997)

$$\Delta V(E, M, t) = -fgrE_c \cos\theta(M)(1 - e^{-t/\tau}) \tag{37.1}$$

where E is the intensity of the electric field (in units of $V{\cdot}m^{-1}$), M is a point on the cell surface (Figure 37.16), t is the time after the field is delivered, f is a factor correlated with the shape of the cell, g is a factor correlated with the electrical conductivities of the extracellular space, the membrane and the cytoplasm (see Equation 37.2 below), r is the size of the pulsed cell (assumed to be a sphere), $\theta(M)$ is the angle between the direction of the field and the normal to the cell surface at M pointing out of the cell, and τ is a characteristic time constant (in the microsecond range). It is generally assumed that the charging time τ is much shorter than the pulse duration t.

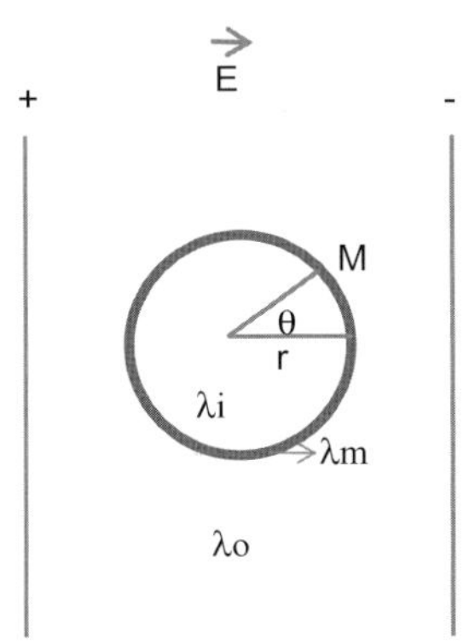

Figure 37.16 Graphical representation of the external electric field E, which induces a position-dependent modulation of the membrane potential difference, superimposed on the critical breakdown voltage.

The factor g is defined as

$$g = \frac{2\lambda_o \lambda_i}{(2\lambda_o + \lambda_m)(2\lambda_m + \lambda_i) + (r/d)\lambda_m(2\lambda_o + \lambda_i)} \quad (37.2)$$

where λ_o, λ_m and λ_i are the conductivities of the extracellular medium, the membrane and the cytoplasm, respectively, and d is the thickness of the membrane.

As permeabilization is observed when $\theta = \pi$, the critical threshold in terms of potential difference is given by

$$\Delta V = fgrE_c \quad (37.3)$$

The value of the factor fg has been estimated as 0.8 for Chinese hamster ovary cells and 0.3 for plant protoplasts (Teissié and Rols, 1993).

Energy Input and Temperature Increase

The electrical energy E (in units of joules) dissipated during the discharge of a capacitor with capacitance C, charged to a voltage V, is given by

$$E = 0.5CV^2 \quad (37.4)$$

The energy density Q' ($J \cdot m^{-3}$) dissipated in a food sample per applied pulse is defined by (Picart *et al.*, 2003)

$$Q' = \frac{1}{u}\int V(t) \cdot i(t) \cdot dt = \frac{1}{u}\int (V^2(t) \cdot R_s^{-1}) \cdot dt \quad (37.5)$$

where $V(t)$ and $i(t)$ are the instantaneous voltage across the electrodes and the instantaneous current crossing the sample, respectively; u is the volume of the treatment chamber (m^3); and R_s (Ω) is the resistance of the sample. The total accumulated treatment time t (s) is defined as

$$t = n \cdot \tau \quad (37.6)$$

where τ is the time width of a single pulse. The total energy density dissipated in the sample, Q_t ($J \cdot m^{-3}$), is defined by

$$Q_t = n \cdot Q' \quad (37.7)$$

In an industrial operation in which a liquid is being pumped into a PEF treatment cell under adiabatic conditions, the temperature increase ΔT after one pulse can be predicted from

$$\Delta T = \frac{Q'}{\rho C_p} \tag{37.8}$$

where ρ and C_p are the density and the heat capacity, respectively, of the liquid being PEF-treated. If the volume flow rate q of the liquid is taken into consideration, Equation 37.8 becomes

$$\Delta T = \frac{Q'}{q\rho C_p} \tag{37.9}$$

The energy input Q' per unit volume of treated material can also be defined as (Dejmek and Trägårdh, 1994)

$$Q' = E \cdot \lambda \cdot t \tag{37.10}$$

where E ($kV \cdot m^{-1}$) is the electric field strength, λ ($S \cdot m^{-1}$) is the conductivity of the liquid and t is the treatment time.

Models for Microbial Inactivation

Several models have been published to describe the relation between the survival fraction of microorganisms and the electric field strength and/or the pulse treatment time. Hülsheger *et al.* (1981) established that there are critical threshold values of the electric field intensity and treatment time at which a PEF treatment starts to effectively kill microorganisms. Based on their experimental observations, they published the following equation:

$$\frac{N}{N_0} = \left(\frac{t}{t_c}\right)^{-(E-E_c)/k} \tag{37.11}$$

where N_0 and N are the microbial counts ($CFU \cdot ml^{-1}$) before and after treatment, respectively; t is the pulse treatment time; t_c is the critical or threshold pulse treatment time; E_c is the critical or threshold electric field; and k is an empirical constant. The constant k is defined by

$$k = \frac{E_i - E_c}{b_{t(E_i)}} = \frac{\ln(t_i/t_c)}{b_{E(t_i)}} \tag{37.12}$$

where b_E and b_t are the regression coefficients of plots of E vs. survival rate and t vs. survival rate, respectively. The parameters E_c, t_c and k are solely determined by the chosen microorganism. A small value of E_c or t_c indicates a high sensitivity of the microorganism to PEF treatment.

For milk pasteurization, Barkström and Tortensson (1999) developed the following equations for inactivation of *L. innocua* and *P. fluorescens*, respectively:

$$Y = 2.7801 + 0.0279 \cdot n - 0.1442 \cdot E + 0.0001 \cdot n^2 + 0.0016E^2 - 0.0012 \cdot n \cdot E \quad (37.13)$$

$$Y = -1.184 + 0.064 \cdot E - 0.066 \cdot n - 0.0001 \cdot E^2 - 0.003 \cdot n \cdot E \quad (37.14)$$

where Y is the response of the survivors (log c/c_0), n is the number of pulses, E is the electric field intensity, and c and c_0 are the initial and final microorganism concentrations.

Models for Extraction

Cell Damage

Evaluation of the effect of cell damage due to PEF application can be performed by measuring the change in electrical conductivity of the tissue between its values before and after the treatment (Chalermchat *et al.*, 2004). The conductivity disintegration index Z is calculated as follows (Vorobiev and Lebovka, 2009):

$$Z = \frac{\sigma - \sigma_i}{\sigma_d - \sigma_i} \quad (37.15)$$

where σ is the electrical conductivity measured at low frequency (normally 1–5 kHz), and the indices "i" and "d" refer to the conductivities of the intact and the totally destroyed cellular system, respectively. The determination of σ_d can be carried out by measuring the conductivity of the tissue after a freezing/thawing cycle.

Diffusion Models

Solid–liquid extraction from cellular tissue can be described by Fick's diffusion equation, which, for an infinite slab, is solved as follows:

$$y = \frac{Y_t - Y_\infty}{Y_0 - Y_\infty} = \sum_{n=1}^{\infty} C_n e^{\lambda_n t} \quad (37.16)$$

where y is the concentration ratio, and the subscripts 0, t and ∞ represent the relevant concentrations at $t = 0$, t and ∞ (equilibrium), respectively. λ_n and C_n are defined as follows:

$$\lambda_n = -\frac{Dq_n^2}{l^2} \quad (37.17)$$

$$C_n = \frac{2\alpha(1+\alpha)}{1+\alpha+\alpha^2+\alpha^2 q_n^2} \tag{37.18}$$

where $2l$ is the solid thickness, D is the diffusion coefficient, the q_n are the nonzero positive roots of the equation $\tan q_n = -\alpha q_n$, and $\alpha = n(Y/X)^{\infty}$, where Y and X are the solute concentrations in the extract and the solid (at the end of extraction), respectively. The liquid/solid ratio n is equal to L/S, where L is the extract volume and S is the solid volume.

As described by El-Belghiti *et al.* (2005), Equation 37.15 fails to predict the kinetics of mass transfer during the initial period of extraction. As an alternative, the mathematical model proposed by So and Macdonald (1986) describes two simultaneous processes with different kinetic coefficients: (i) an initial process of fast solute transfer from the solid surface and outer broken cells to the solvent and (ii) a second process of slower solute transfer from the inside of the solid. These processes are described by

$$c^* = c_w^*(1-e^{-k_w t}) + c_d^*(1-e^{-k_d t}) \tag{37.19}$$

where $c^* = c/c_\infty$, c is the solute concentration in the solution at any time during the extraction, $c_\infty \approx c_0/(n+1)$ is the theoretical equilibrium concentration, c_0 is the concentration of the substance to be extracted in the fresh tissue, $c_w^* = c_w/c_\infty$, c_w is the hypothetical final solute concentration due to the first process alone, $c_d^* = c_d/c_\infty$, c_d is the hypothetical final solute concentration in the solution due to the second process alone, k_w is the kinetic coefficient of the first process, k_d is the diffusion coefficient of the second process and t is the time.

The bimodal diffusion model of Equation 37.18 has been reported to better describe low levels of PEF permeabilization, whereas at high degrees of permeabilization, a single (fast) diffusion coefficient is sufficient to explain the extraction.

Capital and Operating Costs

The amount of energy used for PEF treatment is very low owing to the short duration of the pulses. As an example, the operating cost of an electric-pulse process for the pasteurization of fluid foods was evaluated by Barsotti and Cheftel (1999) as approximately 0.17 US cents·L^{-1} for electricity and approximately 0.22 cents·L^{-1} for maintenance. These authors assumed that the amount of energy dissipated in the food would be 100 J·mL^{-1} and calculated an energy cost of 0.17 cents·L^{-1}, on the basis of 6 cents per kWh.

The cost of consumable parts is difficult to predict. It is too early to indicate the reliability and the lifetime of the electronic components used for PEF equipment. The lifetime is affected by parameters such as the number of pulses, the energy in a pulse

Table 37.1 Estimated costs of a PEF system for the preservation of liquid food in comparison with conventional heat treatment.

Cost (US$)	PEF	Heat treatment
Investment cost	300000	450000
Cost per tonne	0.8	3
Annual running costs	16000	60000
Yearly maintenance and consumables	30000	20000
Total yearly cost	76000	125000

and the temperature of the generator. It is, however, certain that safety and measurement devices constitute a sizable part of the capital investment. The need for pumps and aseptic packaging systems should also be considered.

For the preservation of liquid food, we estimate that the investment required for a system with a capacity of $5\,m^3 \cdot h^{-1}$ running 80 hours per week with a depreciation time of 10 years would be in the range of US$300000. This estimate does not include additional benefits such as the preservation of product quality by PEF treatment and reduction of processing times. If PEF treatment at ambient temperature is desired, the electrical energy input required for microbial decontamination needs to be removed by a cooling system, adding costs for operation and investment.

Table 37.1 reports estimated costs for a system that could be used in the production example described above, in comparison with the costs of heat treatment. In these estimates, we have not taken account of the fact that the excess heat from the heat treatment system may be reused in a factory nor the fact that the PEF system is much smaller and needs less room for installation.

Summary and Future Needs

We have presented here an overview of PEF treatment as a unit operation. The basic theoretical principles, modes of operation and various levels of design have been covered in this chapter. We have also given an overview of various applications and advantages of the technology. However, none or very few commercial-scale PEF treatment systems have emerged on the market. A number of critical factors have to be addressed before PEF technology experiences a full market breakthrough. This section gives an overview of some of these factors.

There is much to be discussed in regard to equipment and processing design. Some important aspects are:

- Emerging technologies always met with skepticism. In the case of PEF treatment, it is important to find clear methods to monitor the process, otherwise it will be difficult to obtain the certifications needed for some products. Processing-line

design with efficient monitoring of the treatment should allow one, in the case of any failure, to repeat the treatment or feed the product back to the product supply line.
- A capacity of several cubic meters of product per hour is needed for industrial applications. Large treated volumes call for high-power pulses that are more expensive and more difficult to control. A possible solution is to use parallel systems with a number of treatment chambers rather than trying to build larger systems. If ten parallel systems are working together in a production line and one of these systems fails, then nine will still remain operational. This means that capacity decreases but production can continue.
- When a high voltage is applied to a product, it is important that no air bubbles exist in the product, as air is not conductive and high voltages may arise across the bubbles. High voltages might result in discharges and minor explosions that will affect the treatment of the product, and may also result in damage to the product. This effect can be reduced by applying pressure to the product during the treatment.
- PEF systems handle very high voltages and have high power capacity, and are thus very dangerous devices if not designed and handled properly. The design must include several levels of interlocks to protect the users and the environment from hazards. One issue that we have not discussed in this chapter is electromagnetic compliance (EMC) requirements. These requirements are very demanding today. A device with a switching power in the range of many kilowatts will have an EMC impact and might interfere with radio communication systems and other sensitive electronic devices.
- The development of fast-switching IGBTs that have high capacity with regard to both current and voltage handling will make the design of PEF systems simpler. These devices are developing quickly owing to market demand.

Microbial inactivation through the application of PEFs has clear advantages for treating heat-sensitive liquid foods. However, opportunities for research and improvement remain, such as:

- The effects of food composition (e.g., milk constituents and juice fibers) on microbial resistance to PEF treatment are complex and not well understood.
- The uniformity of the treatment with regard to field variation in the treated volume and irregularities of flow during processing needs to be studied.
- Optimization of process energy consumption needs to be done.

Cell damage in plant tissue, aimed at accelerating mass transfer in operations such as extraction and drying, is a potential industrial application of PEF treatment. A few aspects of this application need to be better understood for optimization of the treatment parameters:

- Asymmetric permeabilization needs to be understood better. The permeabilization of plant cells in a tissue is not uniform. Neither the factors inducing asymmetric permeabilization nor its mechanism are understood; this is clearly a subject worthy of further investigation.
- The influence of the conductivity of the treatment medium on permeabilization is an issue, where contradictory results have been reported in the literature. The conductivity of the treatment medium is very likely to influence membrane permeabilization and the interactions between the tissue and the electrodes.
- The PEF treatment protocol plays a critical role in tissue permeabilization and, obviously, the pulse strength and width are important. The role of the interpulse delay is a topic of current research. Longer interpulse delays seem to have a stronger effect on cells, as each pulse may be experienced like a new event. Clearly, as the interval between pulses approaches zero, the effect will be equivalent to one continuous pulse. So the question is: How long does it take for the system (cell membrane and associated structures) to "recover" enough so that a subsequent pulse looks like a new event?

Although we have concentrated on the industrial applications of irreversible electroporation, interest in reversible electroporation, which retains cell viability, has increased in the last few years. Potential applications such as the impregnation of plant tissues may be revolutionary for the food industry (see Figure 37.1). Induction of metabolic stress by the application of PEFs has been suggested to increase the concentration of metabolites of nutritional interest and to influence the development of growing tissues such as germinating seeds.

The questions that we have outlined here concerning the industrial applications of irreversible electroporation and new potential applications of reversible electroporation call for a multidisciplinary approach aimed at a deeper understanding of the effects of electric fields at the cellular level. This forms the basis for a fascinating field of research and technological innovation.

References

Ade-Omowaye, B.I.O., Rastogi, N.K., Angersbach, A. and Knorr, D. (2003) Combined effect of pulsed electric field pre-treatment and partial osmotic dehydration on air drying behaviour of red bell peppers. *Journal of Food Engineering* 60: 89–98.

Alvarez, I., Pagán, R., Condón, S. and Raso, J. (2003) The influence of process parameters for the inactivation of Listeria monocytogenes by pulsed electric fields. *Journal of Food Microbiology* 87: 87–95.

Angerbach, A. and Knorr, D. (1997) Anwendung elektrischer hochspannungsimpulse als vorbehandlungsverfahren zur beeinflussung der trocknungscharakteristika und rehydration-seigenschaften von kartoffelwürfels. *Nahrung* 41: 194–200.

Aguiló-Aguayo, R., Soliva-Fortuny, O. and Martín-Belloso, O. (2009) Effects of high-intensity pulsed electric fields on Lypoxigenase and Hydroperoxide Lyase activities in tomato juice. *Journal of Food Science* 8: 595–601.

Akin, E. and Evrendilek, G.A. (2009) Effects of pulsed electric fields on physical, chemical and microbiological properties of formulated carrot juice. *Food Science and Technology International* 15: 275–282.

Aronsson, K., Lindgren, M., Johansson, B.R. and Rönner, U. (2001) Inactivation of microorganisms using pulsed electric fields: the influence of process parameters on *Escherichia coli, Listeria innocua, Leuconostoc mesenteroides* and *Saccharomyces cerevisiae. Innovative Food Science & Emerging Technologies* 2: 41–54.

Balasa, A., Toepfl, S. and Knorr, D. (2006) Influence of pulsed electric field treatment on total polyphenolic content of grape products. In: *Food Factory of the Future*, Gothenburg, Sweden.

Barkström, E. and Tortensson, P. (1999) *Inactivation of Microorganisms in Milk Using Pulsed Electric Fields (PEF)*. MSc thesis, Lund University, Sweden.

Barsotti, L. and Cheftel, J.C. (1999) Food processing by pulsed electric fields. II. Biological aspects. *Food Reviews International* 15: 181–213.

Bazhal, M., Lebovka, N. and Vorobiev, E. (2001) Pulsed electric field treatment of apple tissue during compression for juice extraction. *Journal of Food Engineering* 50: 129–139.

Beebe, S. and Schoenbach, K.H. (2005) Nanosecond pulsed electric fields: a new stimulus to activate intracellular signaling. *Journal of Biomedicine and Biotechnology* 4: 297–300.

Bendicho, S., Barbosa-Cánovas, G.V. and Martín, O. (2002) Milk processing by high intensity pulsed electric fields. *Trends in Food Science and Technology* 13: 195–204.

Boussetta, N., Reess, T., Pecastaing, L., De Ferron, A., Falcimaigne-Cordin, A., Lanoisellé, J.L. and Vorobiev, E. (2009) Application of high voltage electrical discharges at semi-pilot scale: extraction of polyphenols from grape pomace. In: *Proceedings of the International Conference on Bio and Food Electrotechnologies*, Compiegne, France, pp. 108–113.

Bushnell, A.H., Dunn, J.E., Clark, R.W. and Pearlman, J.S., inventors; Foodco Corporation, assignee (1993) High pulsed voltage systems for extending the shelf-life of pumpable food products. US Patent 5235905.

Chalermchat, Y. and Dejmek, P. (2005) Effect of pulsed electric field pretreatment on solid–liquid expression from potato tissue. *Journal of Food Engineering* 71: 164–169.

Chalermchat, Y., Fincan, M. and Dejmek, P. (2004) Pulsed electric field treatment for solid–liquid extraction of red beetroot pigment: mathematical modeling of mass transfer. *Journal of Food Engineering* 64: 229–236.

Dejmek, P. and Trägårdh, C. (1994) Energy cost of high electric field pulse treatment. *Trends in Food Science and Technology* 5: 265.

Dörnenburg, H. and Knorr, D. (1993) Cellular permeabilization of cultured plant tissues by high electric pulses or ultra high pressure for the recovery of secondary metabolites. *Food Biotechnology* 7: 35–48.

Dunn, J.E. and Pearlman, J.S., inventors; Maxwell Laboratories Inc., assignee (1987) Methods and apparatus for extending the shelf-life of fluid food products. US Patent 4695472.

Dunn, J.E., Clark, W.R., Asmus, J.F., Pearlman, J.S., Boyer, K., Painchaud, F. and Hofmann, G.A., inventors; Maxwell Laboratories Inc., assignee (1991) Methods of preservation of foodstuffs. US Patent 5034235.

El-Belghiti, K., Rabhi, Z. and Vorobiev, E. (2005) Kinetic model of sugar diffusion from sugar beet tissue treated by pulsed electric field. *Journal of the Science of Food and Agriculture* 85: 213–218.

Fernandez-Molina, J.J., Barbosa-Canovas, G.V. and Swanson, B.G. (2005) Skim milk processing by combining pulsed electric fields and thermal treatments. *Journal of Food Processing and Preservation* 29: 291–306.

Fincan, M., DeVito, F. and Dejmek, P. (2004) Pulsed electric field treatment for solid–liquid extraction of red beetroot pigment. *Journal of Food Engineering* 64: 381–388.

Gabriel, B. and Teissié, J. (1994) Generation of reactive oxygen species induced by electropermeabilization of Chinese hamster ovary cells and their consequence on cell viability. *European Journal of Biochemistry* 223: 25–33.

Gachovska, T.K., Simpson, M.V., Ngadi, M.O. and Raghavan, G.S.V. (2009) Pulsed electric field treatment of carrots before drying and rehydration. *Journal of the Science of Food and Agriculture* 89: 2372–2376.

Ganeva, V., Galutzov, B. and Teissie, J. (1995) Electric field mediated loading of macromolecules in intact yeast cells is critically controlled at the wall level. *Biochimica et Biophysica Acta* 1240: 229–236.

Geveke, D.J. and Kozempel, M.F. (2003) Pulsed electric field effects on bacteria and yeast cells. *Journal of Food Processing and Preservation* 27: 65–72.

Gómez Galindo, F., Vernier, P.T., Dejmek, P., Vicente, A. and Gundersen, M.A. (2008) Pulsed electric field reduces the permeability of potato cell wall. *Bioelectromagnetics* 29: 296–301.

Grimi, N., Rinville, X., Vorobiev, E., David, P. and Vaxelaire, J. (2009) Expression yield and qualitative characteristics of juice obtained from electrically treated red grapes. In: *Proceedings of the International Conference on Bio and Food Electrotechnologies*, Compiegne, France, pp. 200–205.

Guderjan, M., Töpfl, S., Angersbach, A. and Knorr, D. (2005) Impact of pulsed electric field treatment on the recovery and quality of plant cells. *Journal of Food Engineering* 67: 281–287.

Gudmundsson, M. and Hafsteinsson, H. (2001) Effect of electric field pulses on microstructure of muscle foods and roes. *Trends in Food Science and Technology* 12: 122–128.

Gunderjan, M., Elez-Martínez, P. and Knorr, D. (2007) Application of pulsed electric fields at oil yield and content of functional food ingredients at the production of rapeseed oil. *Innovative Food Science and Emerging Technologies* 8: 55–62.

Gundersen, M.A., Berhrend, M.R., Sun, Y.H., Vernier, P.T. and Kuthi, A. (2004) Four-channel pulse generator for real-time biological investigations. In: *Proceedings of the 26th International Power Modulator Symposium and 2004 High-Voltage Workshop*, pp. 210–215.

Gusbeth, C., Frey, W., Volkmann, H., Schwartz, T. and Bluhm, H. (2009) Pulsed electric field treatment for bacteria reduction and its impact on hospital wastewater. *Chemosphere* 75: 228, 233.

Henriksson, P., inventor (2008) Arrangemang och förfarande för oskadliggörande av mikroorganismer med ett elektriskt fält samt användningar av arrangemanget. Swedish Patent SE 531 797 C2.

Henriksson, P., inventor (2009) Arrangement for neutralisation of microorganisms. International patent application published under the Patent Cooperation Treaty (PCT), WO2009/126084 A1.

Huang, K. and Wang, J. (2009) Designs of pulsed electric fields treatment chambers for liquid foods pasteurization process: a review. *Journal of Food Engineering* 95: 227–239.

Hülsheger, H., Potel, J. and Niemann, E.G. (1981) Killing of bacteria with electric pulses of high field strength. *Radiation and Environmental Biophysics* 20: 53–65.

Jaeger, H., Balasa, A. and Knorr, D. (2008) Food industry applications for pulsed electric field. In: *Electrotechnologies for Extraction from Food Plants and Biomaterials* (eds E. Vorobiev and N. Lebovka). Springer, Berlin, pp. 181–216.

Jemai, A.B. and Vorobiev, E. (2002) Effect of moderate electric field pulses on the diffusion coefficient of soluble substances from apple slices. *International Journal of Food Science and Technology* 37: 73–86.

Klonowski, I., Heinz, V., Toepfl, S., Gunnarsson, G. and Þorkelsson, G. (2006) *Applications of Pulsed Electric Field Technology for the Food Industry*. R&D Report Summary 06-06, Icelandic Fisheries Laboratories.

Kotnik, T., Bobanovic, F. and Miklavcic, D. (1997) Sensitivity of transmembrane voltage induced by applied electric fields – a theoretical analysis. *Bioelectrochemistry and Bioenergetics* 43: 285–291.

Lebovka, N.I., Shynkaryk, N.V. and Vorobiev, E. (2007a) Pulsed electric field enhanced drying of potato tissue. *Journal of Food Engineering* 78: 606–613.

Lebovka, N.I., Shynkaryk, M. and Vorobiev, E. (2007b) Moderate electric field treatment of sugarbeet tissues. *Biosystems Engineering* 96: 47–56.

Lindgren, M., inventor (2005) Arrangement and method for treatment of pumpable substances. US Patent 2005/0008740 A1.

Martín, O., Qin, B.L., Chang, F.J., Barbosa-Cánovas, G.V. and Swanson, B.G. (1997). Inactivation of *Escherichia coli* in skim milk by high intensity pulsed electric fields. *Journal of Food Processing Engineering* 20: 317–336.

McLellan, M.R., Kime, R.L. and Lind, L.R. (1991) Electroplasmolysis and other treatments to improve apple juice yield. *Journal of the Science of Food and Agriculture* 57: 303–306.

Pagan, R., Condón, S. and Raso, J. (2005) Microbial inactivation by pulsed electric fields. In: *Novel Food Processing Technologies* (eds G.V. Barbosa-Canovas, M.S. Tapia and M.P. Cano). CRC Press, Boca Raton, pp. 45–68.

Palou, E., Lopez-Malo, A., Barbosa-Canovas, G.V. and Swanson, B.G. (1999) High pressure treatment in food preservation. In: *Handbook of Food Preservation* (ed. M.S. Rahman). Marcel Dekker, New York, pp. 533–576.

Phoon, P.Y., Gómez Galindo, F., Vicente, A. and Dejmek, P. (2008). Pulsed electric field in combination with vacuum impregnation with trehalose improves the freezing tolerance of spinach leaves. *Journal of Food Engineering* 88: 144–148.

Picart, L. and Cheftel, J.C. (2003) Pulsed electric fields. In: *Food Preservation Techniques* (eds P. Zeuthen and L. Bögh-Sorensen). Woodhead, Cambridge, pp. 360–427.

Pothakamuri, U.R., Barbosa-Cánovas, G.V., Swanson, B.G. and Spence, K.D. (1997) Ultrastructural changes in *Staphylococcus aureus* treated with pulsed electric fields. *Food Science and Technology International* 3: 113–121.

Qin, B., Pothakamury, U.R., Vega, H., Martin, O., Barbosa-Cánovas, G.V. and Swanson, B.G. (1995) Food pasteurization using high-intensity pulsed electric fields. *Food Technology* 49: 53–60.

Qin, B.L., Barbosa-Cánovas, G.B., Swanson, B.G., Pedrow, P.D. and Olsen, R.G. (1998) Inactivating microorganisms using a pulsed electric field continuous treatment system. *IEEE Transactions on Industry Applications* 34: 43–50.

Rastogi, N.K., Eshtiaghi, M.N. and Knorr, D. (1999) Accelerated mass transfer during osmotic dehydration of high intensity electrical field pulse pretreated carrots. *Journal of Food Science* 64: 1020–1023.

Sabri, N., Pulissier, B. and Teissie, J. (1996) Electropermeabilization of intact maize cells induces an oxidative stress. *European Journal of Biochemistry* 238: 737–743.

Sack, M., Eing, C., Frey, W., Schultheiss, C., Bluhm, H., Attmann, F., Stängle, R., Wolf, A., Muller, G., Sigler, J., Stukenbrock, L., Frenzel, S., Arnold, J. and Michelberger, T. (2009) Research on industrial-scale electroporation devices at Forschungszentrum Karlsruhe and cooperation partners. In: *Proceedings of the International Conference on Bio and Food Electrotechnologies*, Compiegne, France, pp. 265–270.

Sale, A.J.H. and Hamilton, W.A. (1967) Effects of high electric fields on microorganisms. I. Killing of bacteria and yeasts. *Biochimica et Biophysica Acta* 148: 781–788.

Schilling, S., Toepfl, S., Ludwig, M., Dietrich, H., Knorr, D., Neidhart, S., Schieber, A. and Carle, R. (2008) Comparative study of juice production by pulsed electric field treatment and enzymatic maceration of apple mash. *European Food Research and Technology* 226: 1389–1398.

Sepúlveda, D.R., Guerrero, J.A. and Barbosa-Cánovas, G.V. (2006) Influence of electric current density on the bactericidal effectiveness of pulsed electric field treatments. *Journal of Food Engineering* 76: 656–663.

Shynkaryk, M.V., Lebovka, N.I. and Vorobiev, E. (2008) Pulsed electric fields and temperature effects on drying and rehydration of red beetroots. *Drying Technology* 26: 695–704.

So, G.C. and Macdonald, D.G. (1986). Kinetic of oil extraction from canola (rapeseed). *Canadian Journal of Chemical Engineering* 64: 80–86.

Tedjo, W., Eshtiaghi, M.N. and Knorr, D. (2002) Einsatz, nicht-thermischer verfahren zur zell-permeabilisierung von weintrauben ung gewinnung von inhaltsstoffen. *Fluessiges Obst* 9: 578–583.

Teissié, J. and Rols, M.P. (1993) An experimental evaluation of the critical potential difference inducing cell membrane electropermeabilization. *Biophysical Journal* 65: 409–413.

Toepfl, S., Ruehle, C., Heinz, V. and Knorr, D. (2005) PEF fish and meat processing. In: *Workshop of IFT Nonthermal Processing Division*, Philadelphia, PA.

Toepfl, S., Heinz, V. and Knorr, D. (2006) Pulsed electric fields (PEF) processing of meat. In: *Proceedings of the International Conference IUFOST*, Shanghai, China.

Turk, M., Vorobiev, E. and Baron, A. (2009) Pulsed electric field assisted pressing of apple mash on a continuous pilot scale plant: extraction yield and qualitative characteristics of cider juice. In: *Proceedings of the International Conference on Bio and Food Electrotechnologies*, Compiegne, France, pp. 114–119.

Vega-Mercado, H., Pothakamuri, U.R., Chang, F.J., Barbosa-Cánovas, G. and Swanson, B.G. (1996) Inactivation of Escherichia coli by combining pH, ionic strength and pulsed electric fields hurdles. *Food Research International* 29: 117–121.

Vega-Mercado, H., Martín-Belloso, O., Qin, B.L., Chang, F.J., Góngora-Nieto, M., Barbosa-Cánovas, G. and Swanson, B.G. (1997) Non-thermal food preservation: pulsed electric fields. *Trends in Food Science and Technology* 8: 151–157.

Vorobiev, E. (2009) Electric field enhanced extraction and separation process. In: *Proceedings of the International Conference on Bio and Food Biotechnologies*, Compiegne, France.

Vorobiev, E. and Lebovka, N. (2009) Pulse-electric-fields-induced effects in plant tissues: fundamental aspects and perspectives of applications. In: *Electrotechnologies for Extraction from Food Plants and Biomaterials* (eds E. Vorobiev and N. Lebovka). Springer, New York, pp. 1–43.

Walkling-Ribeiro, M., Noci, F., Cronin, D.A., Lyng, J.G. and Morgan, D.J. (2009) Shelf life and sensorial evaluation of orange juice after exposure to thermosonication and pulsed electric fields. *Food and Bioproducts Processing* 87: 102–107.

Wan, J., Coventry, J., Swiergon, P., Sanguansri, P. and Versteeg, C. (2009) Advances in innovative processing technologies for microbial inactivation and enhancement of food safety – pulsed electric field and low-temperature plasma. *Trends in Food Science and Technology* 20: 414–424.

Weaver, J. (2000) Electroporation of cells and tissues. *IEEE Transactions on Plasma Science* 28: 24–33.

Wouters, P.C., Dutreux, N., Smelt, J.P.P.M. and Lelieveld, H.L.M. (1999) Effects of pulsed electric fields on inactivation kinetics of *Listeria innocua*. *Applied and Environmental Microbiology* 65: 5364–5371.

Ye, H., Huang, L.L., Chen, S.D. and Zhong, J.J. (2004) Pulsed electric field stimulates plant secondary metabolism in suspension cultures of *Taxus chinensis*. *Biotechnology and Bioengineering* 88: 788–795.

Zhang, Q., Barbosa-Canovas, G. and Swanson, B.G. (1995) Engineering aspects of pulsed electric field pasteurization. *Journal of Food Engineering* 25: 261–281.

38

Process Design Involving Ultrasound

Jordi Salazar, Antoni Turó, Juan A. Chávez and Miguel J. García-Hernández

Introduction

Ultrasound refers to sound waves, i.e., mechanical vibrations, which propagate through solids, liquids or gases with a frequency greater than the upper limit of human hearing. Although this limit varies from person to person, the ultrasonic frequency range is considered to be that of frequencies over 20 kHz. The upper limit of the frequency range of ultrasound is limited mainly by the ability to generate ultrasonic signals.

The food industry offers manifold possibilities for the use of ultrasound. Although ultrasound has recently attracted considerable interest in the food industry for both analysis and processing of foods, ultrasonic techniques have been available in the industry since World War II (Povey and McClements, 1988). Typically, the applications of ultrasound are divided into two broad categories: low- and high-intensity applications.

Low-intensity applications are characterized mainly by their use of frequencies above 100 kHz and energies typically below $1\,W{\cdot}cm^{-2}$. In addition, the ultrasonic wave does not have any significant effect on the material under test, in contrast to high-intensity applications. In the food industry, low-intensity ultrasound is used as an analytical technique either to measure and/or control process variables, such as liquid level or flow velocity, or to obtain information about various physicochemical properties of foods: for instance, it can provide information about air bubbles in aerated foods,

Note: Parts of this chapter have been adapted from Salazar *et al.* (2010).

Handbook of Food Process Design, First Edition. Edited by Jasim Ahmed and Mohammad Shafiur Rahman.

the ratio of fat in meats, various characteristics of vegetables and fruit, the quality of eggs, cracks in cheese, the texture of biscuits, and milk coagulation, and it can be used for wine fermentation control and dough characterization (Povey, 1989; Contreras *et al.*, 1992; McClements, 1997; Benedito *et al.*, 2002; Simal *et al.*, 2003; Salazar *et al.*, 2004; Resa *et al.*, 2007; Álava *et al.*, 2007; Mizrach, 2008).

On the other hand, high-intensity, or power, ultrasound applications tend to use frequencies below 100 kHz and energies above $10\,W{\cdot}cm^{-2}$. These applications are intended to have an effect on, or cause a change in, the material, usually by generation of intense cavitation (Mason and Lorimer, 1988). Cavitation is an important physical phenomenon of high-intensity ultrasound and involves the formation, growth and implosive collapse of bubbles in a liquid. Some typical examples of effects include disruption of biological cells, emulsification, drying, mixing of materials and microbial inactivation (Shoh, 1975; Raso *et al.*, 1998; Villamiel and Jong, 2000; Gallego-Juárez *et al.*, 2007). In this type of application, the ultrasonic frequency, intensity and exposure time are the process parameters that must be controlled. In addition, many of the applications of power ultrasound are not exclusively ultrasonic processes but ultrasonically assisted processes such as filtration, drying and extraction. In other applications, such as cutting and slicing, power ultrasound is used as a tool without the aim of altering food properties.

Fundamentals of Ultrasound

An ultrasonic wave that propagates through a material can be described by its amplitude A and frequency f, which are design parameters chosen by the user, and its wavelength λ, velocity of propagation v and attenuation coefficient α, which are parameters dependent on the material properties. Figure 38.1 depicts some of these wave parameters.

The amplitude is the maximum displacement of the medium from its equilibrium, and the frequency is the number of oscillations per second. Conversely, the period T is the time required for one complete cycle to occur, so that $T = 1/f$. The wavelength

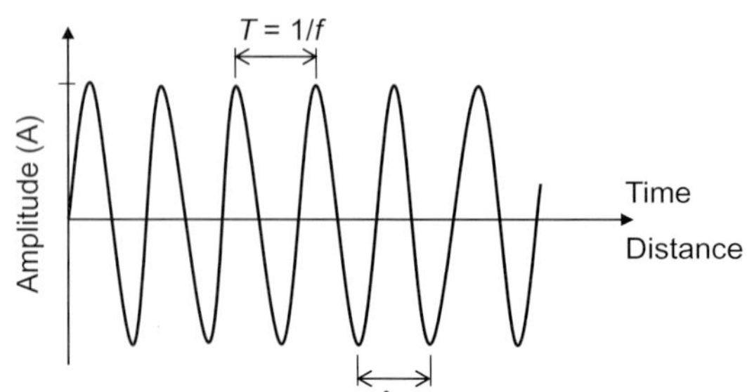

Figure 38.1 Wave parameters.

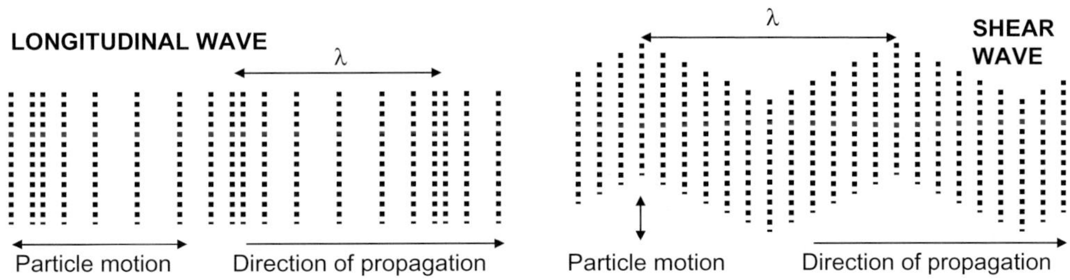

Figure 38.2 Types of wave.

is the spatial period of the wave and is related to the velocity and frequency by the following equation:

$$\lambda = \frac{V}{f} \tag{38.1}$$

It is important to note that the frequency of a wave does not depend on the material characteristics. Therefore, when an ultrasonic wave passes from one medium to another, its frequency remains the same. However, if the velocities of propagation in the two media are different, then the wavelengths will also be different.

The velocity of propagation and the attenuation coefficient will be discussed in more detail later in this chapter.

Most applications of ultrasonics in the food industry use either longitudinal or shear waves. In a longitudinal wave, also called a compression wave, the particle displacement is in the same direction as the propagation of the wave. However, in a shear wave, the particle displacement is perpendicular to the direction of propagation. These waves are also called transverse waves. In addition, the velocity of a shear wave is a little less than half that of a longitudinal wave. Figure 38.2 illustrates the particle displacement versus the direction of propagation for longitudinal and shear waves. There are also other types of waves such as surface (Rayleigh) and plate (Lamb) waves, which are used in some specialized cases.

Unlike longitudinal waves, shear waves can hardly propagate in most liquids and gases, because these materials do not support shear stresses. Thus, the less viscous a material is, the more resistance to shear wave propagation there is. Because of the high attenuation of shear waves in low-viscosity materials, the measurement distances must be extremely short.

Low-Intensity Ultrasound

In the case of low-intensity ultrasound, the ultrasonic wave parameters generally measured include the velocity of propagation, the attenuation coefficient of the acoustic wave traveling through the sample and the acoustic impedance of the material. These parameters can be related to some of the physicochemical properties of foods.

In this section, the basic theoretical principles underlying the ultrasonic measurement techniques used to study food materials are given, with a special emphasis on the most relevant applications in the food industry and their results.

Parameters Measured in Ultrasonic Techniques

Ultrasonic waves are normally generated and received by piezoelectric transducers. These convert electrical energy into mechanical energy and vice versa. When they propagate in a material, ultrasonic waves are affected by the physical properties of the material, such as the hardness, density, elastic modulus and grain structure. Thus, it is possible to characterize the fundamental physical properties of a material by measuring the velocity of propagation of the wave through the material, and the attenuation coefficient and the acoustic impedance of the material. The ultrasonic velocity and attenuation coefficient are, generally, the two parameters most widely used to evaluate the properties of materials. More in-depth studies of sound propagation and material properties can be found in Mason (1958) and Auld (1990).

Ultrasonic Velocity

The velocity of propagation v in a solid material depends mainly on the density ρ and the elastic modulus E of the material. The relationship between these three variables is given by the following equation:

$$v = \sqrt{\frac{E}{\rho}} \tag{38.2}$$

where v is the velocity of propagation, measured in $m \cdot s^{-1}$, and E can be expressed, in the case of a solid, as $K + (4/3)G$, where K is the bulk modulus and G is the shear modulus.

In the case of a fluid medium, the bulk modulus is much larger than the shear modulus and Equation 38.2 can be simplified to

$$v = \sqrt{\frac{K}{\rho}} \tag{38.3}$$

The reciprocal of the bulk modulus is called the adiabatic compressibility. So, Equation 38.3 can be rewritten alternatively as

$$v = \sqrt{\frac{1}{\rho \cdot \beta}} \tag{38.4}$$

For viscoelastic materials such as dough, the elastic modulus E of the material in Equation 38.2 is complex and frequency-dependent: $E^*(\omega) = E'(\omega) + jE''(\omega) = K^*(\omega) + (4/3)G^*(\omega)$, where $E'(\omega)$ is the storage modulus and $E''(\omega)$ is the loss modulus.

$E'(\omega)$ and $E''(\omega)$ provide an indication of the elastic and the viscous properties, respectively, of the material. $E^*(\omega)$ is, however, a combination of the bulk and shear moduli. Nevertheless, when pure shear waves are used, $E^*(\omega) = G^*(\omega)$. This is of great interest for evaluating the rheological properties of viscoelastic materials. Like the use of traditional rheometers that measure the dynamic shear viscosity, the ultrasonic determination of G', the shear storage modulus, and G'', the shear loss modulus, allows the characterization of the viscoelastic properties of foods.

The relationship of G' and G'' to the velocity and attenuation of an ultrasonic shear wave at a specific frequency ω is given by the following equations (Kono, 1960):

$$E' = G' = \frac{\rho v^2 \left(1 - \frac{\alpha^2 v^2}{\omega^2}\right)}{\left(1 + \frac{\alpha^2 v^2}{\omega^2}\right)^2} \text{ and } E'' = G'' = \frac{2\rho v^2 \frac{\alpha v}{\omega}}{\left(1 + \frac{\alpha^2 v^2}{\omega^2}\right)^2} \tag{38.5}$$

provided that the condition $\alpha v/\omega < 1$ is satisfied, i.e., a low-attenuation material. Otherwise, G' would take an invalid negative value. Also, it should be noted that the frequencies used in ultrasonic experiments (usually over 100 kHz) are some orders of magnitude higher than those used in traditional rheological tests (usually under 1 kHz). In addition, viscosity measurements are frequency-dependent and, consequently, the ultrasonic G' and G'' values will differ from those encountered through traditional rheological tests and therefore cannot be compared directly.

Generally, the ultrasonic velocity is determined by measuring the time t taken for an ultrasonic wave to travel a known distance d ($v = d/t$). Often, this time is referred as the time of flight (TOF). Although the ultrasonic velocity depends on both the elastic modulus and the density, the elastic properties of a material have a greater influence on the ultrasonic velocity than does the density. The reason for this is that the differences or changes in the elastic moduli of materials are usually greater than those that occur in the density. So, for instance, ultrasonic-velocity measurements can provide valuable information about the elastic properties of foods. For instance, monitoring the ultrasonic velocity can be used to indicate phase changes in foods.

Attenuation Coefficient

Attenuation is the energy loss associated with the decrease in the amplitude of a wave when traveling through a material and is generally due to scattering and absorption mechanisms.

Scattering results from the fact that the material is not strictly homogeneous, and this is especially important in heterogeneous materials. When an ultrasonic wave traveling in a material meets an inhomogeneity, it is forced to deviate from its original trajectory and is scattered in all directions. Such inhomogeneities tend to behave like diffuse reflectors and normally include particles, grains, bubbles, defects and variations in density. Scattering usually occurs when the size of the inhomogeneity is smaller than the wavelength; otherwise, specular reflection occurs instead.

Absorption is the process by which part of the energy of an ultrasonic wave is converted into heat, and it occurs in both homogeneous and heterogeneous materials. Absorption is caused mainly by three mechanisms: viscosity, thermal conduction and molecular relaxation processes.

The attenuation coefficient of a material, α, is defined by

$$A = A_0 e^{-\alpha x} \tag{38.6}$$

where A_0 is the initial amplitude at some point in the material and A is the reduced amplitude after the wave has traveled a distance x from the initial point. The ultrasonic attenuation exhibits a power dependence on frequency, where the power coefficient is between 1 and 2, depending on the propagation medium, which makes the use of high frequencies unsuitable in the case of highly attenuating materials. The units of the attenuation coefficient are nepers per meter ($Np{\cdot}m^{-1}$), although units of decibels per meter ($dB{\cdot}m^{-1}$) are usually used. The relationship between the two units is that $1\,N{\cdot}m^{-1}$ is equal to $8.68\,dB{\cdot}m^{-1}$.

Acoustic Impedance

The acoustic impedance Z of a material is defined as the product of the density and the velocity of propagation. Thus,

$$Z = \rho \cdot v \tag{38.7}$$

The acoustic impedance is usually measured in rayls, or, equivalently, in $kg{\cdot}m^{-2}{\cdot}s^{-1}$. This parameter is used in the determination of the fractions of an ultrasonic wave that are transmitted and reflected at a boundary between two materials having different acoustic impedances. For normal incidence of a plane wave (see Figure 38.3), the reflection and transmission coefficients R and T, respectively, are given by

$$R = \frac{A_r}{A_i} = \frac{Z_2 - Z_1}{Z_2 + Z_1} \tag{38.8}$$

$$T = \frac{A_t}{A_i} = 1 + R = \frac{2Z_2}{Z_2 + Z_1} \tag{38.9}$$

where A_r, A_i and A_t are the amplitudes of the reflected, incident and transmitted waves, respectively, and Z_1 and Z_2 are the acoustic impedance of the material the acoustic wave is traveling in and the material the wave is reflected by, respectively.

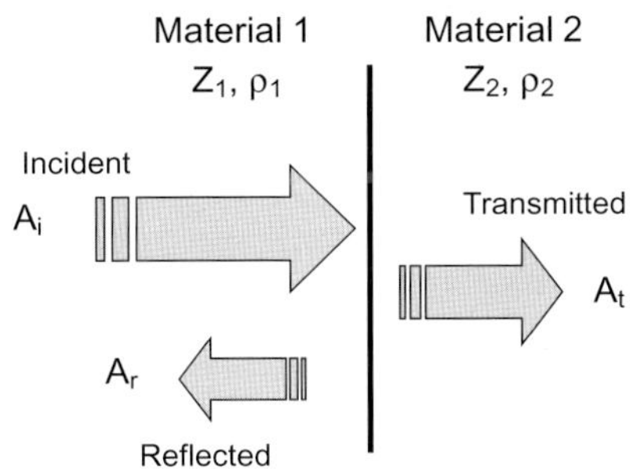

Figure 38.3 Reflection and transmission of acoustic waves at an interface between materials of different acoustic impedances.

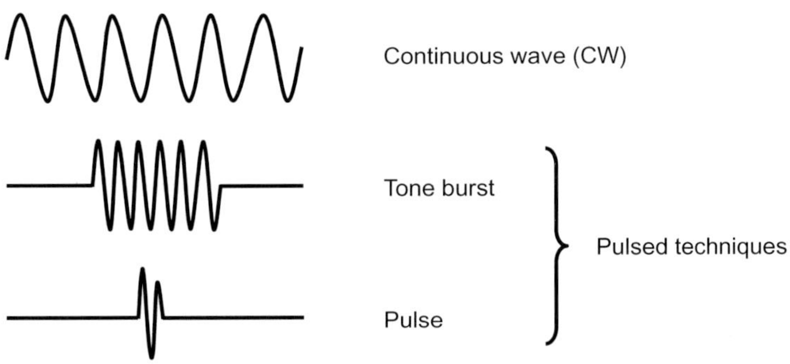

Figure 38.4 Different types of signal used in ultrasonic measurements.

The reflection coefficient can be determined by measuring the amplitudes of the incident and reflected waves. Then, if the acoustic impedance of the material the incident wave is traveling in is known, the acoustic impedance of the transmitted medium can be determined from the reflection coefficient, according to Equation 38.8, as

$$Z_2 = Z_1 \frac{1+R}{1-R} \tag{38.10}$$

Measurement Techniques

Ultrasonic measurement techniques can be separated into two broad categories, namely pulsed and continuous techniques, according to the type of excitation signal applied to the transducer, which may be a continuous wave, a tone burst or a pulse. Figure 38.4 shows these three types of wave. A continuous wave has a constant amplitude and frequency and an infinite duration. A tone burst is a wave formed by one or

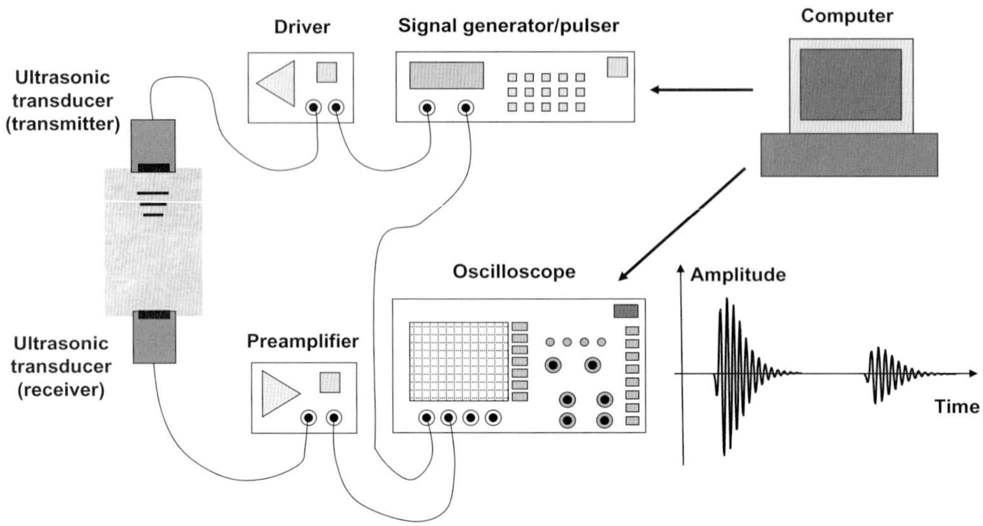

Figure 38.5 Typical measurement setup for an ultrasonic measurement system.

more cycles of a continuous wave at a certain frequency. Finally, a pulse is a short signal, sometimes a spike of very high voltage. Both tone burst and pulse waveforms are normally repeated in time at a given pulse repetition frequency (PRF).

Regardless of the measurement technique used, the hardware of a typical ultrasonic measurement system is composed of a signal generator or pulser, a driver or amplifier, ultrasonic transducers, a measurement cell, a preamplifier, a signal receiver or oscilloscope, and a computer, as depicted in Figure 38.5. The signal generator produces the desired electrical waveform, which is amplified by the driver and then applied to the ultrasonic transducer. This latter converts the electrical signal into a sound signal, which propagates through the sample under test until it reaches the receiver transducer, which converts the sound signal into an electrical signal again. Generally, this electrical signal is very weak and needs first to be amplified before it can be displayed properly on the oscilloscope and/or sent to the computer.

Pulsed-Wave Techniques

Pulse methods are fairly straightforward measuring methods. They basically consist of propagating pulses through the material under test and measuring the time taken to travel a given distance. Also, the attenuation coefficient of the material can be determined from the relative amplitudes of pulses that have traveled different distances. Pulse techniques are applied using one of three different configurations: pulse-echo, through-transmission, and pitch-and-catch.

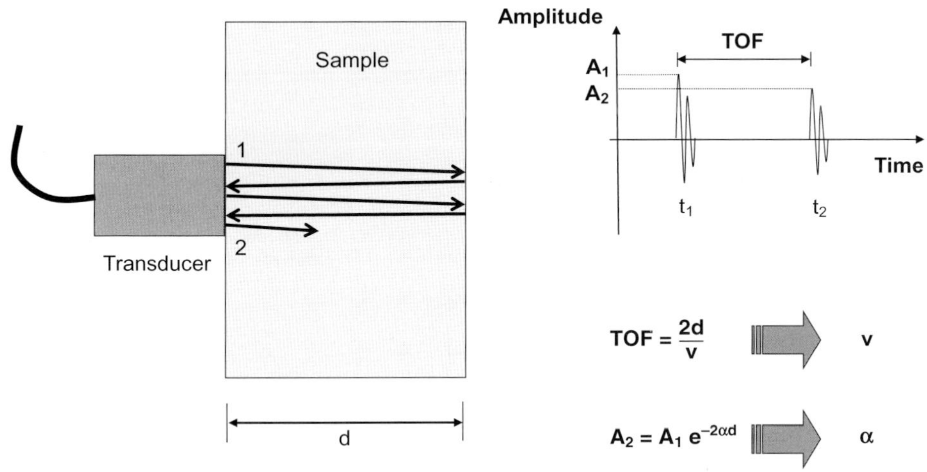

Figure 38.6 Pulse-echo technique.

Pulse-Echo

The pulse-echo technique is probably by far the most widely used ultrasonic measurement technique. The main advantage of this configuration is that only one side of the material under test needs to be accessible. Figure 38.6 demonstrates the basic principle of operation of the pulse-echo technique. Only one transducer is used in this configuration, acting first as an emitter and then as a receiver. The transducer converts the electric signal into an ultrasonic wave that travels through the sample until it reaches the far boundary of the sample, which reflects the wave. The reflected ultrasonic wave travels back through the sample to the receiver, which converts the ultrasonic wave into an electric signal (echo 1). This echo will be displayed on the oscilloscope with an amplitude A_1 and time t_1. It is important to note that the ultrasonic wave is partially transmitted and partially reflected at sample boundaries. Therefore, successive echoes, A_2 at time t_2 and so on, will also be displayed on the oscilloscope. The distance traveled by each echo is twice the sample thickness, i.e., $2d$.

From the information provided by the received echoes, both the velocity and the attenuation coefficient can be determined (see Figure 38.6). Thus, for instance, the ultrasonic velocity is obtained by measuring the TOF between two consecutive echoes and, using Equation 38.6, the attenuation coefficient is given by the decrease in amplitude between two consecutive echoes.

In addition, by placing a buffer rod of well-known acoustic impedance between the transducer and the sample, the acoustic impedance of the sample can be determined by measuring the amplitudes of the echoes received from the buffer rod/sample interface and the sample/air interface, using Equations 38.6 and 38.10, as described by Lynnworth (1989).

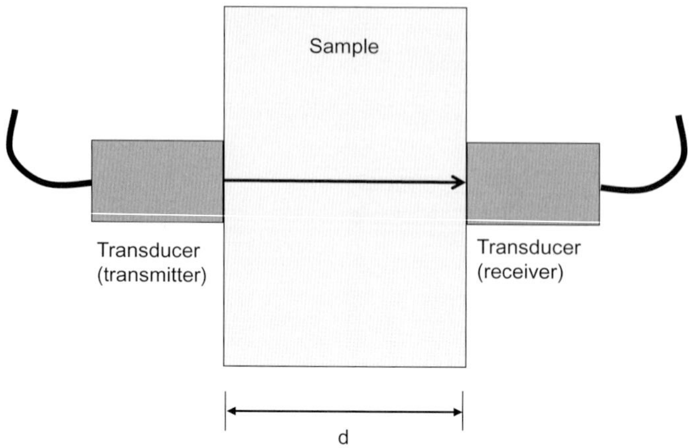

Figure 38.7 Through-transmission technique.

Through-Transmission

In the through-transmission configuration, two transducers are used, one as a transmitter and the other as a receiver. The two transducers are located on opposite sides of the material under test, as depicted in Figure 38.7. This configuration gives very good sensitivity and is suitable for inspecting highly attenuating materials but is limited by the need to access both sides of the material under test, and, depending on the application (such as defect inspection), to coordinate the movements of the two transducers. The ultrasonic velocity can be determined directly from measurements of the TOF and the sample thickness d.

In order to counteract the effects of the attenuation introduced at every interface between the sample and the transducers, measurements at two different distances (d_1 and d_2) can be carried out. Assuming that d_1 is greater than d_2, the attenuation coefficient of the material can be determined from the following expression:

$$\alpha = \frac{1}{d_1 - d_2} \ln\left(\frac{A_2}{A_1}\right) \qquad (38.11)$$

where A_1 and A_2 are the amplitudes at distances d_1 and d_2, respectively.

Pitch-and-Catch

This technique is a variation of the pulse-echo configuration. The pitch-and-catch technique involves a setup with separated transmitter and receiver transducers, both located at the same side of the material under test and positioned close together (Figure 38.8). Sometimes the two transducers are constructed as a single assembly. This configuration is intended to overcome the difficulties that sometimes arise in recording

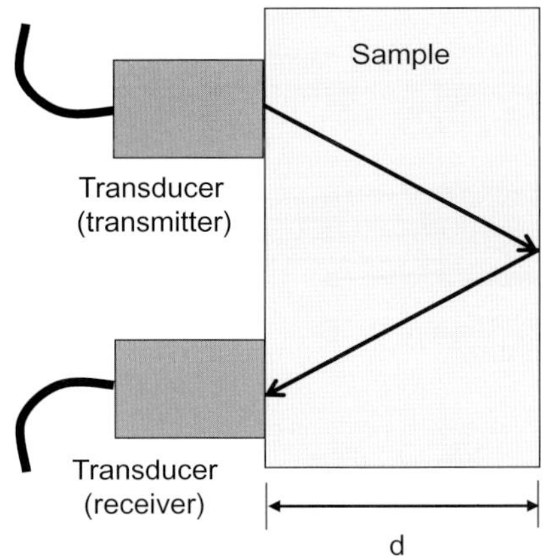

Figure 38.8 Pitch-and-catch technique.

the echoes if the same transducer is used for transmitting and receiving, as occurs with the pulse-echo configuration.

Continuous-Wave Techniques

Continuous-wave measuring techniques are mostly restricted to research laboratories (Papadakis, 1975). They are also referred as resonance methods. The main features of these techniques are they provide highly accurate measurements and are able to detect very small changes in attenuation. The principle of operation is based on the establishment of standing waves within the material under test. Two different configurations can be used, with a fixed and a variable length of the measuring path. The former can be used for solids and fluids, and the latter only for fluids.

Fixed-Length Path

This configuration involves frequency tuning with a constant measuring length of the material and is particularly suitable for measurements of the properties of solids. The experimental setup includes a signal generator, an oscilloscope and two transducers aligned as in the arrangement for the through-transmission method (Figure 38.9). A sinusoidal continuous wave of appropriate frequency and amplitude is applied to the transmitter transducer and propagates through the material under test. The ultrasonic wave undergoes multiple reflections between the two transducers, which results in the formation of a standing wave.

The amplitude of the received wave exhibits a series of maxima and minima when the frequency of the ultrasonic wave is varied. The ultrasonic velocity within the sample is given by

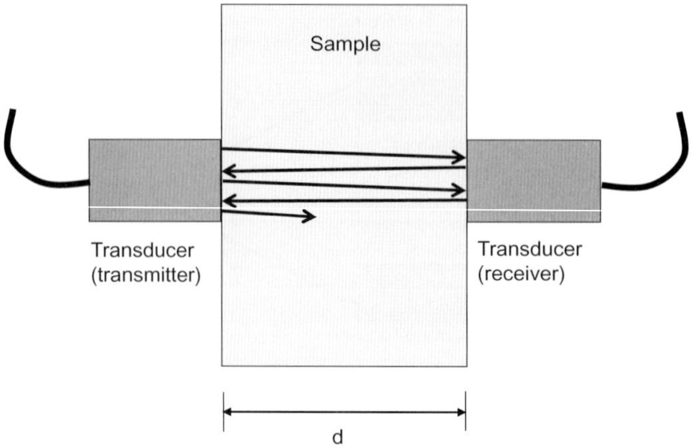

Figure 38.9 Fixed-length-path continuous-wave technique.

$$c = 2d\,\Delta f \tag{38.12}$$

where d is the distance between the transducers and Δf is the frequency difference between two consecutive maxima or minima at the receiver transducer.

The attenuation coefficient is somewhat more complicated to determine. There is, however, a first approximation for the attenuation per wavelength, $\alpha\lambda$, which is given by the following expression:

$$\alpha\lambda = \pi \frac{\Delta f_n}{f_n} \tag{38.13}$$

where f_n is the frequency of a maximum and Δf_n is the –3 dB bandwidth of the resonance.

Variable-Length Path

This configuration is more commonly used with gases and liquids. In this configuration, the receiver transducer is replaced by a movable reflector. The ultrasonic wave generated by the transmitter transducer is reflected back by the reflector and, after multiple reflections, a standing wave is formed (Figure 38.10). The received signal goes through maxima and minima as the reflector is moved. Two consecutive maxima or minima are separated by one-half of a wavelength ($\Delta d = \lambda/2$). So, using Equation 38.1, the ultrasonic velocity can be determined directly from

$$v = f\lambda = 2f\,\Delta d \tag{38.14}$$

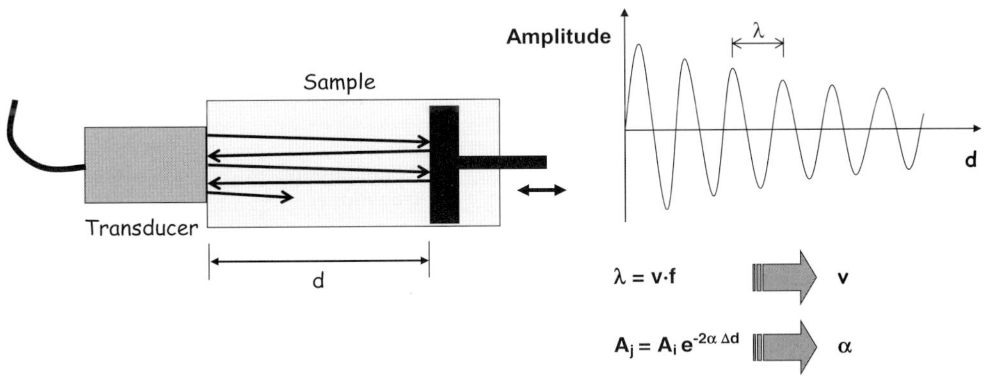

Figure 38.10 Variable-length-path continuous-wave technique.

Similarly, if A_i and A_j are the amplitudes of two consecutive maxima, using Equation 38.6, the attenuation coefficient can be determined from

$$\alpha = \frac{1}{2\Delta d}\ln\left(\frac{A_i}{A_j}\right) \tag{38.15}$$

Low-Intensity Applications

In this section, the most relevant applications of low-intensity ultrasound in the food industry are reviewed. The scope of applications is vast and includes detection of foreign bodies, detection of fill levels, determination of the composition in liquid mixtures and characterization of food properties. In terms of the kind of information provided by ultrasonic measurement, applications can be grouped into two different groups: those intended for process control and those for characterizing the physicochemical properties of food.

Ultrasonic Control of Processes

An excellent review of this type of application has been given by Lynnworth (1989). Among such applications, the measurement of flow rate, density, level, temperature, pressure, position, thickness and other things are treated in detail. Although the parameters of the ultrasonic wave are modified by factors such as temperature and pressure, measurement of these parameters is done by conventional methods (Hauptmann *et al.*, 2001). In this section, only flowmetry, level measurement and foreign-body detection will be considered.

Ultrasonic devices for flowmetry and level detection are by far the most widely commercially available devices of this kind. However, ultrasound is also regarded as a promising and cost-effective technology for the detection of foreign bodies in foods.

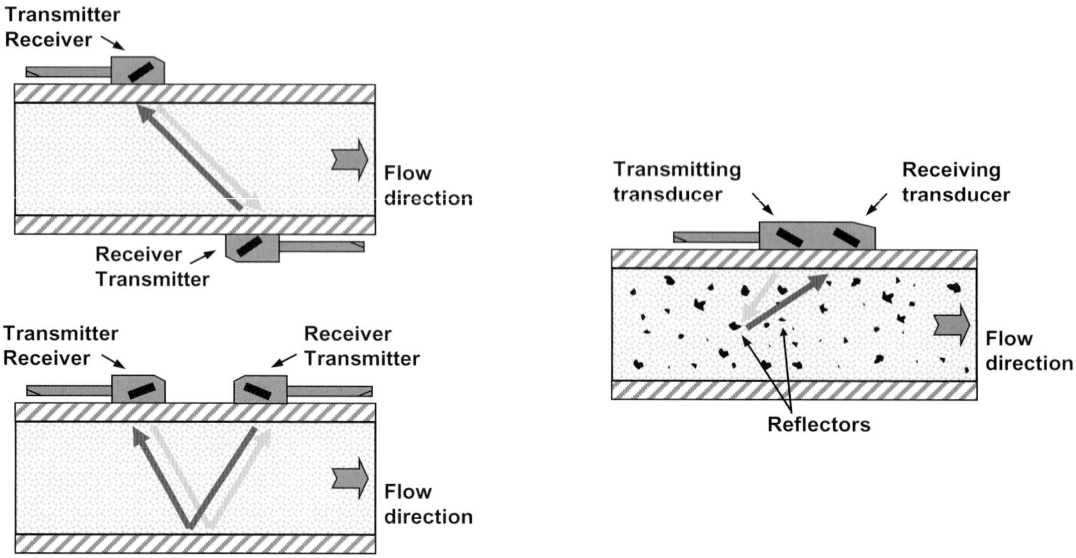

Figure 38.11 Ultrasonic flowmeters. Left: transit-time flowmeter configurations, with transducers diagonally mounted on opposite sides (top) and on the same side (bottom). Right: Doppler flowmeter.

Flowmetry

An ultrasonic flowmeter measures the velocity at which a food fluid travels through a pipe. Several different types of flowmeters, with different principles of measurement, exist and are available commercially. The most popular are transit-time and Doppler types.

Transit-time flowmeters measure the flow rate by means of the difference between the transit times of downstream and upstream ultrasonic pulses. In this case, the two transducers are mounted diagonally on opposite sides of the pipe, and the ultrasonic pulse crosses the pipe flow once. Alternatively, the two transducers can be mounted on the same side of the pipe, and the ultrasonic pulse crosses the pipe flow twice. This is the most commonly used installation method, and can be applied to pipe sizes from 1 inch up to 12 inches. Both configurations are depicted in Figure 38.11 (left).

Doppler flowmeters use reflected ultrasonic echoes to measure the flow rate. For proper operation, the fluid flowing through the pipe must contain sonically a reflective material such as solid particles or entrained air bubbles (see Figure 38.11 (right)). The movement of these materials alters the frequency of the beam reflected onto the receiver transducer. The frequency shift is linearly proportional to the rate of flow of the material in the pipe.

Both types of flowmeter usually incorporate the latest electronics and digital signal-processing technologies, thus providing high performance and easy operation. The

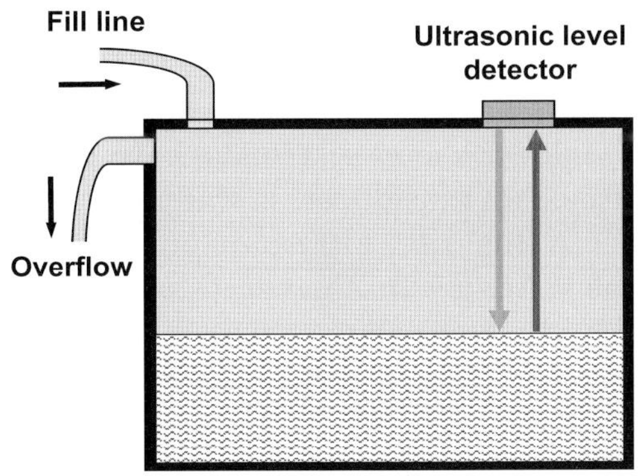

Figure 38.12 Liquid level measurement.

measurement accuracy can be in the range of 1% of the flow rate, and the speed of response can be as fast as 1 s.

Level Measurement

This type of sensor is often used as a level detector in a tank, as shown in Figure 38.12. It is normally mounted at the top of the tank and measures the time delay between an emitted ultrasonic pulse and its reflection from the substance in the tank. An ultrasonic level sensor may contain a temperature sensor to compensate for changes in operating temperature that would alter the speed of sound and hence the distance calculation that is necessary for making an accurate level measurement.

Detection of Foreign Bodies

The term *foreign body* refers to any undesirable piece of solid matter present in a food product. When food products are manufactured or packaged, small foreign bodies may end up in the product. The presence of foreign bodies is one of the major concerns of the food industry and the biggest single source of customer complaints, which are expensive in terms of both money and goodwill for the manufacturer.

X-ray and optical inspection systems, and metal detection systems, are well established across the food industry. However, these systems fail with organic materials such as animal bones, fruit stones and fragments of glass that may be found in canned food.

Ultrasound is regarded as a potential technology, but the development of such systems is still in progress (Graves *et al.*, 1998). The major drawback of this technology is that the transducer and the product to be inspected need to be acoustically coupled. To overcome this problem, the product must be immersed in an acoustic medium such as water. Hæggström and Luukkala (2001) have detected and identified foreign bodies in pieces of cheese and in marmalade by a water immersion method, by using a pulse-echo transducer and checking the amplitude pattern of the reflected echoes. Discrimination of foreign bodies was more difficult in inhomogeneous samples than in homogeneous samples. Zhao *et al.* (2003a) have developed a laboratory prototype composed of a pulse-echo transducer, a water jet nozzle and an *x–y* table to inspect bottled beverages. In this case, ultrasonic pulses are coupled to the bottom of the bottle by means of the water jet. In Zhao *et al.*'s experiments, foreign bodies such as pieces of glass, metal and plastic were successfully detected by scanning the flat bottom of the bottles. Effective detection was accomplished for sample sizes from 10 to 2.5 mm^2.

Ultrasonic Food Characterization

Despite extensive experiments with different types of foods, ultrasonic food characterization has still not been applied commercially, and almost all applications are restricted to the laboratory level. Ultrasonic measurements on food products have special requirements. The reason is that the range of foods is very wide, and foods are generally more complex than other materials to inspect ultrasonically. Specific ultrasound instrumentation needs to be developed or adjusted for every different application, and even for the same application when different foods are tested. However, ultrasonic inspection has great advantages, such as the fact that it has a low cost, is nondestructive and makes it feasible to inspect materials that are packaged or opaque to light.

A short overview of the state of the art in the use of ultrasonics in food characterization applications and several of the current trends is given below. The applications have been grouped and reviewed according to common food market sectors.

Bakery Products

One of the first experiments was carried out on biscuits by Povey and Harden (1981). These authors measured the crispness of biscuits and found that the ultrasonic velocity correlated with crispness better than did either the ultrasonically derived Young's modulus or the modulus derived from measurements with an Instron universal testing machine.

In recent years, ultrasound has been widely used to study flour products. These studies have dealt mainly with the analysis of the properties of bread dough, at specific stages of preparation, with variation or addition of some ingredients or with changes in the production procedure. Dough is a highly attenuating material and therefore its inspection by means of ultrasound represents a big challenge. One of the first studies was done by Lee *et al.* (1992). In these experiments, good qualitative agreement between ultrasonic measurements and traditional rheometry was found, but the results obtained with ultrasonic shear waves were somewhat unrealistic; the values obtained were more typical of solid materials than of flour–water products. A decade later, Létang *et al.* (2001), Salazar *et al.* (2002) and Ross *et al.* (2004) described the effects on the ultrasound parameters of changes in some properties of bread dough or in its processing, such as the water content, the mixing time or the type of flour. Also, Kidmose *et al.* (2001), Scanlon *et al.* (2002), Elmehdi *et al.* (2003) and Lee *et al.* (2004a) studied the changes in the ultrasonic characteristics of doughs during fermentation. The influence of gas bubbles on the acoustic properties of bread dough was studied by Elmehdi *et al.* (2004).

Recent work has studied the variation of the acoustic parameters of dough with the operation conditions of extrusion processes (Owolabi *et al.*, 2008). García-Álvarez *et al.* (2005) studied the time dependence of the mechanical properties of dough for the purpose of flour strength evaluation. The ultrasonic attenuation of the dough was monitored as a function of time during resting. The capability of ultrasound to discriminate flours for different purposes was demonstrated by García-Álvarez *et al.* (2006) and Álava *et al.* (2007). Ultrasound measurements were able to separate types of flour with specific properties of velocity and attenuation. In Figure 38.13, individual values of velocity for doughs made from a total of 36 different flours are plotted against the attenuation, and the plot shows that there is a difference in ultrasound properties depending on the flour used. Strong flours tend to be situated at the top left corner, with high velocity and low attenuation values, whereas weak flours tend to be situated at the opposite corner. According to García-Álvarez *et al.* (2006) and Álava *et al.* (2007), the attenuation/velocity ratio could be used as an indicator of dough quality.

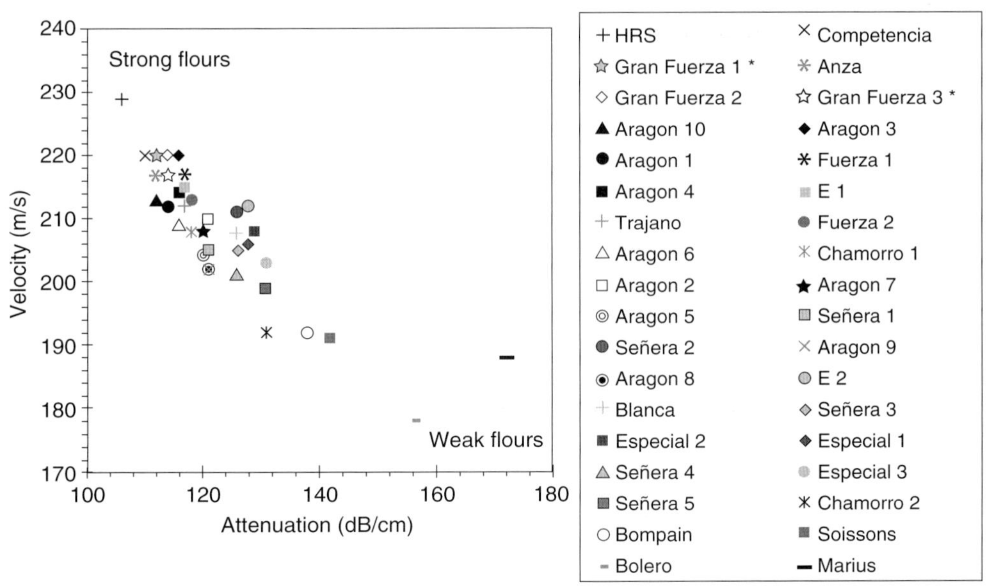

Figure 38.13 Ultrasound measurements of dough quality using a range of flours mixed using the Alveograph method.

Recently, ultrasound studies have been carried out to assess the rheological properties of gluten-free rice flour dough and the potential of ultrasound to distinguish changes induced by protein ingredients such as soybean protein isolate and processing aids such as transglutaminase (Rosell *et al.*, 2010). The experimental results have demonstrated that ultrasound is sensitive to the effects on the properties of rice flour dough of the addition of both soybean protein and transglutaminase at different temperatures, although the changes at high temperatures are less marked.

In contrast, there are comparatively few published studies of ultrasound characterization of flour batters. Batters are complex macroemulsion systems whose physical properties play an important role in determining the characteristics of the resulting cakes. The most relevant of these properties are density and rheology. Some ultrasound methods such as the through-transmission technique, which is commonly used in bread dough analysis, are hard to apply to batters owing to their usually more highly aerated nature. This high gas content can make batters very attenuating and dispersive media, which often means that ultrasound measurements are hard to perform accurately. In addition, some ultrasound signals such the shear waves used in the ultrasound rheology of bread dough are difficult to apply to the study of batters, because batters usually have more liquid-like characteristics.

Consequently, special sensors based on reflection techniques and longitudinal waves have been used instead. With such sensors, the specific gravity (Fox *et al.*, 2002; Fox *et al.*, 2004) and changes in the acoustic impedance (Salazar *et al.*, 2004) of batters have been studied. In order to determine the feasibility of using online ultrasonic sensors in the control of cake-manufacturing processes, Gómez *et al.* (2008) characterized the physical properties of a set of 27 batters by using an ultrasonic sensor based on acoustic-impedance measurements described by Salazar *et al.* (2004). It was found that the acoustic impedance of the batter correlated, even better than conventional measurements, with the quality of the resulting baked cake. Table 38.1 shows the significant correlation found between the acoustic impedance and the quality param-

Table 38.1 Correlation coefficients between physical properties of cakes and batters, and the acoustic impedance of the batter.

	Cake volume	Cake density	Volume index	Cake symmetry	Central height
Batter density	0.49*	–0.46*	0.25	0.34	0.31
Consistency	0.14	–0.08	–0.1	–0.2	–0.17
Flow index	–0.51**	0.51**	–0.51**	–0.55**	–0.56**
G'	0.27	–0.22	0.06	0.05	0.06
G''	–0.04	0.11	–0.32	–0.39*	–0.38*
G^*	0.12	–0.06	–0.13	–0.17	–0.16
Acoustic impedance	–0.45*	0.53**	–0.77***	–0.74***	–0.79***

*$p < 0.05$;
**$p < 0.01$;
***$p < 0.001$.

eters of the cakes. The correlation is especially high for the central height, volume index and symmetry, which are parameters related to the expansion produced and the shape of the final product. The observed correlations may be due mainly to the capability of acoustic-impedance measurements to detect very small changes in the amount of air added to the batter, very small changes in the stability of the bubbles and their distribution, very small changes in the viscosity of the batter (both the initial viscosity and the changes in viscosity during the baking process), and very small changes in the gelatinization temperature of the starch; the links between these changes may also be important.

Fruits and Vegetables

There are many properties of fruits and vegetables that are of importance and determine their quality, among them freshness, texture, color and taste. Sensory characteristics must be continually evaluated to monitor quality standards in order to, for instance, establish the optimum time to pick the best crop and produce a consistently good finished product.

Fruits and vegetables are perhaps the sector of the food market where the application of ultrasonic technology is finding the most difficulties. Fruits and vegetables are delicate, and damage to their tissues needs to be avoided. In addition, the presence of intercellular air spaces makes them very highly attenuating materials. Another problem that is often encountered with fruits and vegetables is the natural variability of their physical properties, which makes a large number of measurements necessary in order to obtain statistically significant results (McClements, 1997).

Mizrach (2008) has provided a complete review summarizing the last two decades of studies, adaptations and modifications of ultrasonic technology and devices, and innovations in this field, for the determination of material properties of fresh fruit and vegetable tissues in both preharvest and postharvest applications. Experiments have been carried out on tissues and on whole fruits and vegetables. The ultrasonic properties of avocados, mangoes, apples, plums, melons, potatoes, tomatoes and other fruits and vegetables have been related to specific quality indices of products such as firmness, dry weight, sugar content and ripeness. For instance, nondestructive ultrasonic monitoring of tomato quality revealed a decreasing trend in the attenuation coefficient over the course of time (Mizrach, 2007), and a parabolic expression yielded the best-fitting curve for the plot, with $R^2 = 0.965$ (see Figure 38.14, left). In addition, a very good correlation factor between the firmness and the attenuation coefficient was found in the experiments (see Figure 38.14, right). In this case, a linear regression with $R^2 = 0.916$ was selected. According to this result, the firmness of tomato fruits could be predicted by nondestructive attenuation measurements on sample fruits.

Despite the difficulties and limitations in applying ultrasonic technology, it should be noted that there has been considerable progress since the early studies that were undertaken several years ago.

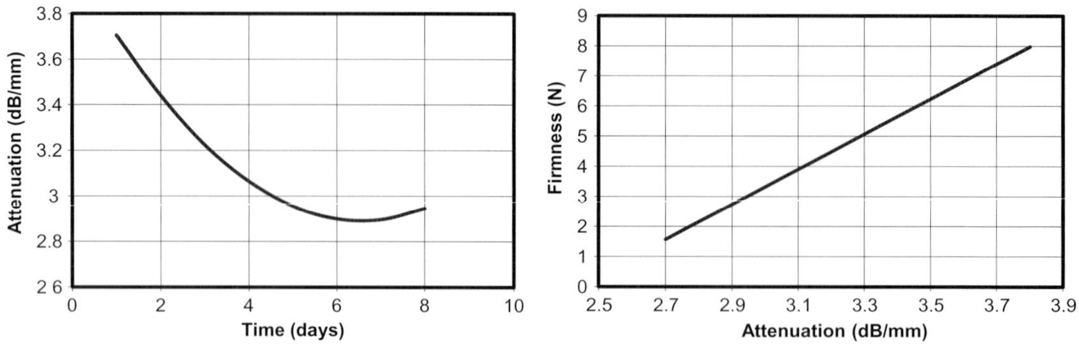

Figure 38.14 Left: variation of the attenuation coefficient of ultrasonic waves in tomatoes as a function of storage time. Right: relationship between firmness and attenuation coefficient.

Meat and Fish

Ultrasonic methods have been widely used for measuring meat composition (fat content, nonfat solids content and moisture content) and the back-fat thickness of meat in live animals, carcasses and meat-based products (Whittaker *et al.* 1992; Cross and Belk, 1994; Benedito *et al.* 2001b; Llull *et al.* 2002; Simal *et al.* 2003). Lean meat and fat have quite different values of ultrasonic velocity, of around $1580\,m{\cdot}s^{-1}$ and $1450\,m{\cdot}s^{-1}$, respectively; the ultrasonic velocity in fat is about 10% less than in lean meat. Thus, it is possible to deduce meat composition from a velocity measurement once a suitable calibration of the system has been carried out. The back-fat thickness is determined from the time of flight of the echoes reflected at the interface between the fat and lean meat.

Whittaker *et al.* (1992) and Park *et al.* (1994) used ultrasonic-velocity values to predict the intramuscular fat content, or marbling, in beef. These results enable researchers to map intramuscular fat deposition, which could subsequently allow beef producers to evaluate the genetic potential for marbling, and then feeders could finish an individual animal to a predetermined marbling end point more accurately. A more sophisticated approach was used by Ozutsumi *et al.* (1996), who used a 2 MHz color scanning scope to evaluate marbling in the *M. longissimus thoracis* of live Japanese Black steers. Reflected waves received from the interfaces between different tissues such as fat layers, muscle and bone were converted into digital signals with 12 bits and stored in order to obtain a 15 color picture of the cross-sectional area of the back of the animal according to the signal strength. The estimates of the subcutaneous fat thickness and the cross-sectional area of the *M. longissimus thoracis* obtained from the scans were in good agreement with conventional carcass measurements carried out one week after slaughter. It was concluded that the subcutaneous fat thickness, and the area and marbling of the *M. longissimus thoracis* could be estimated with useful precision using this color scanning scope.

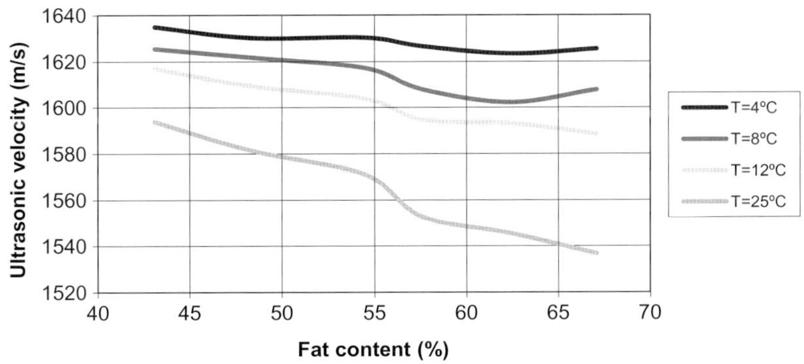

Figure 38.15 Ultrasonic velocity in a meat-based product as a function of moisture content at different temperatures.

More recent studies have shown that the ultrasonic velocity can also be related to the composition of chicken meat (Chanamai and McClements, 1999). By measuring the ultrasonic velocity in chicken meat over a range of temperatures, it was possible to determine the fat, nonfat solids and solid-fat contents. This measurement technique has been successfully extended to meat-based products such as dry fermented sausages (Benedito *et al.*, 2001b; Llull *et al.* 2002; Simal *et al.* 2003). The ultrasonic velocity decreases with increasing temperature. As an example, Figure 38.15 shows the ultrasonic velocity in a meat-based product as a function of moisture content at different temperatures. The results obtained by these authors demonstrate that ultrasonic velocity can be used to accurately estimate the chemical composition of any meat-based product.

The ultrasonic velocity can also be related to the moisture content and textural parameters of a meat-based product during an initial period of ripening. Llull *et al.* (2002) found that the ultrasonic velocity increased linearly ($R^2 = 0.9362$) with ripening time, while the moisture content decreased progressively throughout ripening. This decrease was most noticeable during the first days of ripening, as depicted in Figure 38.16.

In contrast to most other applications of ultrasound in the food industry, which have mainly been confined to the laboratory, there are many commercially available instruments for measuring meat composition (McClements, 1997). One example is the Autofom system (from Carometec A/S, formerly known as SFK Technology A/S) for pork grading. Autofom is an automatic classification system that measures fat and muscle depth in pig carcasses by means of ultrasound. The scanning device consists of 16 2.0 MHz transducers embedded in a U-shaped frame with a spacing of 25 mm between transducers. The Autofom system may process up to 1200 carcasses per hour, providing predictions of the lean meat percentage in the carcass, the fat thickness, the commercial value of each carcass and the weight of saleable meat in specific commercial cuts.

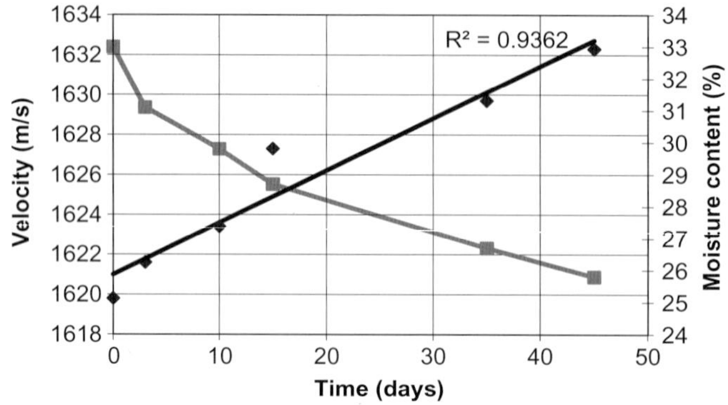

Figure 38.16 Ultrasonic velocity (diamonds) and moisture content (squares) during the ripening of dry fermented sausages.

Information about fish composition is also important to have, since it often determines the quality, value and utilization of fish. Ghaedian *et al.* (1998) examined the influence of fat, moisture and nonfat solids content on the ultrasonic velocity in fish tissue. They developed empirical equations to relate the ultrasonic velocity in fish tissue to its composition at any temperature. These equations were used to predict the composition of a variety of fish species such as catfish, cod, flounder, mackerel and salmon (Suvanich *et al.*, 1998).

Dairy Products

A huge variety of different studies have been carried out to investigate the potential of ultrasound for characterizing dairy products. The most popular application of ultrasound in the dairy industry has been to the determination of the concentrations of various components in milk and its many products (e.g., butter, cheese, yoghurt and cream) (McClements, 1997).

Milk is a complex liquid that, roughly, consists of emulsified fat globules in an aqueous phase containing dissolved proteins and lactose. Ultrasonic techniques can provide valuable information about the fat content, the droplet size distribution and the kinetics of product variation as a function of time. Physical properties of oil/water emulsions, such as droplet size and volume fraction, have been studied in depth in the past by Povey (1988) and McClements and Povey (1989). By measuring the ultrasonic velocity of an emulsion whose particle size distribution is known, it is possible to determine the volume fraction of the disperse phase. Alternatively, if the disperse-phase volume fraction is known, then the particle size can be determined.

In the last few years, ultrasound has been used by several researchers to investigate the formation of gels in milk. Coagulation plays a decisive role in determining the quality of most dairy products. In addition, the coagulation time is often used as a

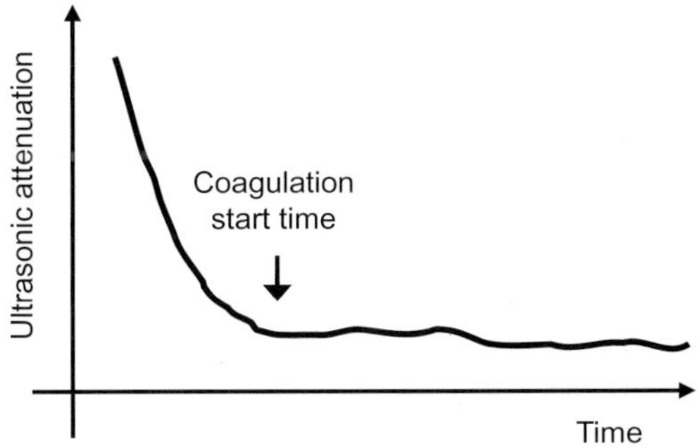

Figure 38.17 Ultrasonic attenuation during milk coagulation.

reference for determining the time at which curd is cut to drain the whey (Weber, 1987). Milk coagulation time has been determined by measurements of the apparent viscosity and ultrasonic attenuation (Gunasekaran and Ay, 1994). Statistical analysis indicated that there was no significant difference between the coagulation times determined by the two methods. It was found that the ultrasonic attenuation showed an abrupt decrease during the first stage of coagulation, and then decreased slowly in the second stage (see Figure 38.17). Based on the rate of change of the ultrasonic attenuation during coagulation, the coagulation start time was defined as the time at which the rate of change of the attenuation was $-0.1\,\text{neper}\cdot\text{m}^{-1}\cdot\text{min}^{-1}$. However, ultrasonic-velocity measurements were not considered, since the ultrasonic velocity showed fluctuations while coagulation progressed. These fluctuations were similar to those of the density of the milk. This is in accordance with the strong relation between the ultrasonic velocity and the density of a medium. Nassar *et al.* (2001) used variations in the time of flight to monitor structural changes during milk gelation. Another ultrasonic technique was reported by Bakkali *et al.* (2001), who measured milk coagulation time using ultrasonic velocity and attenuation and a pulse-echo immersion technique. Wang *et al.* (2007) applied ultrasonic and rheological methods to assess the renneting properties of casein solutions after UHT pretreatment. A linear correlation was found between the coagulation times determined by rheological and ultrasonic measurements on samples heated at 120 °C. Dukhin *et al.* (2005) used acoustic spectroscopy to monitor droplet size distributions in whole, homogenized and unhomogenized milk, and in light butter. Others (Ogasawara *et al.*, 2006) have monitored the end of yoghurt production by measuring variations in the velocity of ultrasonic waves propagating through the solution.

Another application of ultrasound to dairy products has been to the characterization of various properties of cheeses, for example the rheological properties (Lee *et al.*, 1992) and the maturity of cheese samples (Benedito *et al.*, 2001a). In addition, Orlandini and Annibaldi (1983) suggested that ultrasonic methods could be used to detect structural defects during the early stages of the maturation of Parmesan cheese. Gunasekaran

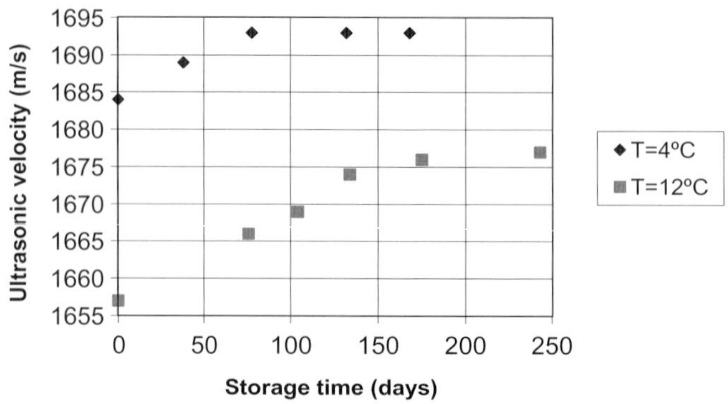

Figure 38.18 Relationship between ultrasonic velocity and storage time for Cheddar cheese.

and Ay (1996) applied ultrasonic techniques to evaluate the firmness of curd in order to determine the optimum cut-time for cheesemaking. Benedito *et al.* (2001a) determined the degree of maturity of cheeses from ultrasonic-velocity measurements performed in the through-transmission mode, and detected cracks within the cheeses using the pulse-echo technique. Figure 38.18 shows the evolution of the ultrasonic velocity during maturation of Cheddar cheese at two different storage temperatures. In both cases, the ultrasonic velocity increases quickly during the first few days of storage, and afterwards increases more slowly. In addition, ultrasound has been used to replace destructive tests to assess the texture of Manchego cheese (Benedito *et al.*, 2006). The changes in texture during maturation are considered as good predictors of the maturity of the cheese.

Not only quality but also safety is important for food producers and consumers. Elvira *et al.* (2005, 2006) have developed an eight-channel ultrasonic detecting device for evaluation of the microbiological quality of packed liquid foods, based on noninvasive, nondestructive ultrasonic methods. This device makes possible the noninvasive detection of microbial growth in liquid foods, with no need to open the carton-based packages in which they are contained. In the experiments, three different strains were inoculated at different concentrations into UHT milk packs. The changes in the liquid medium produced by the growth of microorganisms induce variations in the ultrasonic propagation parameters, giving noninvasive evidence of the developing contamination. These variations may affect the elastic and viscous properties of the milk, changing the propagation velocity and the attenuation of an ultrasonic wave. Different patterns appeared depending on the microorganism tested, which indicates that their metabolisms induce different changes in milk. Growth detection was achieved between 7 and 48 h, depending on the number and type of bacteria inoculated.

Fruit Juices and Drinks

Fruit juices and drinks are basically solutions or liquid mixtures, often with particles in suspension. Ultrasonic propagation through solutions and liquid mixtures has been

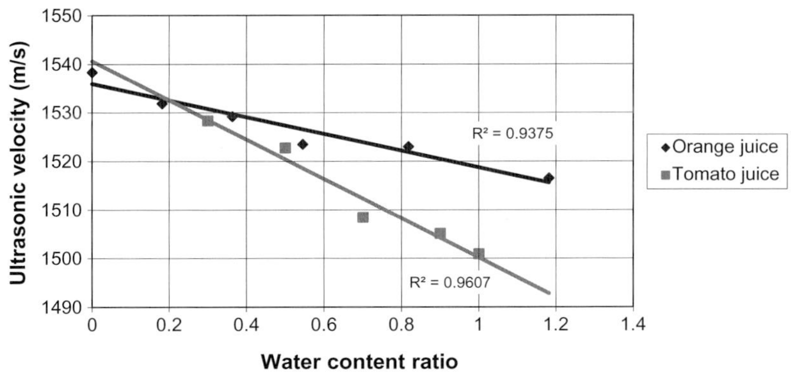

Figure 38.19 Measured ultrasound velocity in diluted fruit juices.

the subject of several studies in the past (Urick, 1947; Nomoto, 1958; Harker and Temple, 1988; Pavlovskaya *et al.*, 1992; McClements, 1995; Povey, 1997). Ultrasonic-velocity measurements can be used to determine the sugar content of beverages and the composition of liquid mixtures. Contreras *et al.* (1992) studied the relation between the speed of sound in sugar solutions and the concentration and temperature, and summarized the results with an empirical equation. Zhao *et al.* (2003b) conducted experiments to estimate the apparent viscosity of orange and tomato juices using the pulse-echo method. The results showed that an increase in the number of solid particles in the juice increased both the apparent viscosity and the ultrasound velocity, which can be attributed to the effects of friction between the solid and liquid, and stiffening of the bulk modulus. One year later, the same authors (Zhao *et al.*, 2004) presented a conceptual design and evaluation of an online ultrasonic system for quality measurement and control of bottled beverages. By using pulse-echo and through-transmission techniques simultaneously, the ultrasonic velocities in orange and tomato juices diluted with water were measured in glass bottles without the need to know the diameter or wall thickness of the bottle, as seen in Figure 38.19. The ultrasound velocity was then correlated with the viscosity of the beverage. The experiment demonstrated that ultrasound velocity is a good index for correlating with the apparent viscosity of beverages. More recently, Kuo *et al.* (2007, 2008) measured the sugar content and viscosity of watermelon and orange juice. They found that the ultrasonic velocity tends to respond linearly to both sugar content and viscosity. However, attenuation measurements were found to be ineffective for such a purpose.

Ultrasonic measurements can also be used for nondestructive evaluation of the percentage of water that is frozen in frozen foods. Monitoring of the freezing of foods is important for quality purposes because the freezing process may damage the structure of foods and consequently affect their quality; this is due mainly to the increase in size of the ice crystals after nucleation. Lee *et al.* (2004b) measured the velocities of longitudinal and shear waves and the attenuation of longitudinal waves in freezing orange juice; these quantities showed a strong correlation with the amount of unfrozen water in the juice. Specifically, for orange juice, the experimental data showed a strong velocity increase below −10 °C and an attenuation peak at −20 °C, corresponding to a proportion of unfrozen water in the juice of nearly 20%. Carcione et al. (2007) developed a poroelastic model of ultrasonic waves propagating through orange juice in order to predict the degree of freezing of orange juice. The model showed excellent agreement with the experimental data, particularly the ultrasonic velocities.

Alcoholic fermentation processes have become a topic of interest in the last few years. In early work, Winder *et al.* (1970) proposed an ultrasonic method to determine the alcohol and extract contents of wines; although they did not study the alcoholic fermentation process itself, an empirical relation between ultrasonic parameters and the concentrations of alcohol and soluble solids was proposed, indicating that these concentrations are the main factors that determine the speed of the sound in fermenting media. Later, Becker *et al.* (2001, 2002) used an ultrasonic technique to determine the density of beer during fermentation. They discarded the possibility of obtaining a theoretical relation between density and ultrasonic propagation velocity, being aware of the lack of a representative model for liquid mixtures and solutions. Resa *et al.* (2004) showed that the changes occurring during the course of the alcoholic fermentation of several aqueous mixtures can be monitored online by measuring the velocity of an elastic wave propagating through the fermenting medium. In addition, Resa *et al.* (2005) proposed a semiempirical approach to improving theoretical models describing the density and velocity of liquid mixtures of low alcohol concentration (under 20%). More recently, Van Sint Jan *et al.* (2008) have developed an ultrasound-based method for measuring sugar and alcohol content simultaneously in hydroalcoholic solutions mimicking fermenting musts in semi-industrial environment.

Several different approaches have been proposed for measuring density, and a large number of possibilities offered by ultrasound. Greenwood *et al.* (1999) and Greenwood and Bamberger (2002) developed an online ultrasonic density sensor composed of six transducers mounted on a plastic wedge with its base in contact with the liquid. The amount of reflection at the plastic–liquid interface depends on the density of the liquid, the ultrasonic velocity in the liquid and the parameters of the wedge. Sugar–water solutions and oils were successfully characterized. The main advantage of this technique is that highly attenuating liquids can be characterized because the operation of the sensor depends on reflection at the interface rather than transmission of ultrasound through the liquid. Püttmer *et al.* (2000) investigated a novel ultrasonic density sensor for online application to liquid systems. The sensor is composed of a piezoceramic disk mounted between two reference rods of quartz glass of different lengths. Experiments under laboratory conditions provided a relative error of about ±0.1% full scale. In order to simplify the calibration process of these sensors, Mathieu and Schweitzer (2004) proposed a new method of measurement of liquid density from analysis of echoes backscattered by wires immersed in the liquid. According to Mathieu and Schweitzer (2004), this approach avoids some of the difficulties of calibration, in the sense that the calibration can be performed with any liquid, although the precision in the experiments was only about 1%. Recently, Adamowski *et al.*

(2010) performed measurements of the density of several liquids with an accuracy better than 0.2%, with a measurement cell using a conventional NDE (nondestructive-evaluation) transducer as an emitter and a 70 mm diameter, 52 µm thick P(VDF–TrFE) (poly(vinylidene fluoride–trifluoroethylene) copolymer) membrane as a receiver. In addition, the large-aperture PVDF receiver used in the sensor allowed the determination of the elastic constants of anisotropic materials without diffraction effects and showed good agreement with the results of tensile tests.

Devices and Instrumentation

The typical setup for ultrasonic measurements was described earlier in this chapter and is depicted in Figure 38.5. Depending on the type of application, all or only some of the instruments will be necessary to make up an inspection system. In this section, a brief description of the basic elements of the system will be given.

Transducers

Ultrasonic transducers convert electrical energy into mechanical energy, and vice versa. They are built around an active element, generally a piezoelectric ceramic, that vibrates at a given frequency when a voltage is applied and generates a voltage when it vibrates. The ultrasonic transducer is the most important and critical part of the whole inspection system. They come in different types, which must be properly chosen according to the specific application. Thus, for instance, general-purpose, direct-contact, delay-line, dual-element, immersion, angle-beam, normal-incidence shear wave and more types of transducer are commercially available.

Some important parameters for the selection of a transducer are its frequency, axial resolution, sensitivity and lateral resolution. Given a food material with a fixed ultrasonic velocity, the frequency of the transducer determines the wavelength of the ultrasonic wave, which is related to the desired size of discontinuities or particles to be detected, i.e., the axial resolution. A general rule of thumb is that the discontinuity or particle size must be larger than one-half of the wavelength. Normally, a broadband or highly damped transducer helps to shorten the reflected pulse, allowing the transducer to resolve closely spaced discontinuities or particles. The axial resolution also generally increases as the frequency increases. The sensitivity is the ability to detect small discontinuities or particles, and the lateral resolution is the ability of a transducer to separate signals produced by two reflectors close together. The sensitivity depends greatly on the piezoelectric material used. The most popular one used in measurement applications is lead zirconate titanate of the PZT-5 type. The lateral resolution depends on the diameter of the transducer. The larger the diameter of the transducer, the broader the ultrasonic beam is. A good alternative method to increase lateral resolution and sensitivity is to focus the ultrasonic beam at the desired distance by means of either an acoustic lens or a curved piezoelectric element.

Pulser/Receivers

A pulser/receiver is an instrument that provides both a pulser unit to drive the ultrasonic transducer with the required high-voltage pulse and a receiver unit that preamplifies and filters the returning ultrasonic echo. A pulser/receiver, along with appropriate transducers and an analog or digital oscilloscope, forms the basic ultrasonic inspection system. The pulser/receiver can operate in either pulse-echo or through-transmission mode.

The main controllable parameters of a pulser unit are the energy, duration and damping of the electrical pulses, and the pulse repetition rate. Commercial pulsers can provide high-voltage pulses from 100 to 900 V. The receiver unit controls the preamplifier gain, normally user-adjustable, and the cutoff frequency of the high- and low-pass filters included in the unit. Bandwidths from 10 to 150 MHz are commercially available, which enables the use of broadband transducers.

Signal Acquisition

The very weak electrical signal provided by the ultrasonic transducer needs to be filtered and amplified before being acquired. Filtering and amplification can be done either by the receiver unit of the pulser/receiver or by a specific low-noise amplifier module. The amplified signal is then converted to digital format by a high-speed analog-to-digital converter (ADC). This digitized data is then transferred to a personal computer, which permits one to store, process, analyze and display the signal. The sampling rate must, in accordance with Nyquist's sampling theorem, be at least twice the highest frequency of the ultrasonic signal. Alternatively, digitization can be carried out directly by a digital oscilloscope, which can also store and display the acquired signal, and transfer it to a personal computer for processing purposes.

High-Intensity Ultrasound

The use of power ultrasound in the food industry has increased during the last few years. An increasing number of studies have been carried out in this direction. Ultrasonic energy represents a promising tool for producing foods or enhancing a series of food processes without affecting the quality of the foodstuff.

Although extensive research has been carried out over many years into the application of power ultrasound, a great deal of future research is still required into the development of ultrasonic technology and also into the effects of power ultrasound on the properties of foods. On the one hand, power ultrasound technology needs to be specifically developed and scaled up for specific applications. On the other hand, a better understanding is required of how and to what extent ultrasonic effects vary under different experimental conditions in order to know and quantify the contribution of each variable parameter to the effects of power ultrasound. However, currently there is quite a large amount of commercially available instrumentation for the most popular applications.

Cavitation

Several mechanisms can be activated by power ultrasound, such as heating, turbulence, agitation, friction and surface instability. Although not all of the mechanisms involved in power ultrasound are known or well understood, most of the mechanisms important in the food industry can be attributed to a very complex nonlinear phenomenon known as cavitation. The phenomenon of cavitation was first reported by Thornycroft and Barnaby (1895), who postulated that the reason for the poor performance of a new torpedo was inefficiency and loss of power caused by the formation of cavitics in the water.

Power ultrasound enhances chemical and physical changes in a liquid medium through the generation and subsequent destruction of cavitation bubbles. When a liquid is subjected to ultrasound of sufficient power, the liquid is alternately compressed and expanded, forming small bubbles, or cavities. These bubbles react to the compression and expansion cycles of the ultrasonic wave, causing them to expand and, when attain a critical size, eventually collapse. This process is depicted in Figure 38.20. The collapse of bubbles can be so violent that it can cause considerable damage to the surrounding medium. The gas in the bubbles has been estimated to reach temperatures of around 5000 °C and pressures of more than 1000 atmospheres on a nanosecond timescale (Suslick and Price, 1999). This intense local heating, or the generation of hot spots, can drive significant gas phase chemical reactions, which are important in a variety of applications (Gould, 2001). However, this heat is transient and very localized, so the overall temperature increase in the matrix is relatively low, around 5 °C. When power ultrasound is used in combination with conventional heating, the effect of ultrasound treatment increases.

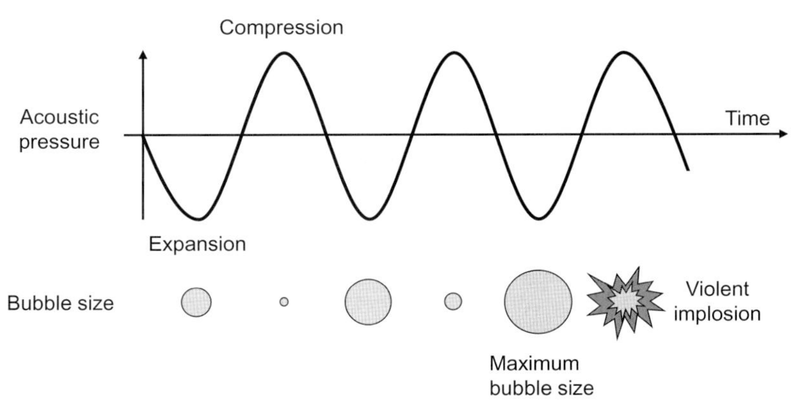

Figure 38.20 Cavitation bubble collapse.

Process Parameters

Generally, the variable process parameters are the intensity and frequency of the power ultrasound, the exposure time, the temperature, the pressure, and the volume of food to be processed. These parameters affect the phenomenon of cavitation and consequently the results of the process, such as the extraction rate, the particle size or the destruction of microorganisms. Other factors that also affect cavitation significantly are the type of treatment, choice of solvent and sample preparation.

The results of a process for a given process parameter configuration depend on the energy per unit volume (measured in $kWh \cdot L^{-1}$). It is also important to note that since power ultrasound applications differ a lot from each other, the optimal process parameter configuration must be adjusted individually for each application. In addition, one of the key parameters that strongly influences the results is the type of food (Salazar *et al.*, 2010).

The existence of many process parameters affecting the output of the process necessarily means that it takes time and effort to scale and fine-tune the whole process, with the goal being to achieve the maximum result with the minimum amount of energy.

Energy, Intensity and Exposure Time

The output power is obtained by multiplying the intensity by the effective radiating area of the tip of the ultrasonic probe. The amount of energy transferred to the process is then calculated as the product of the output power and the exposure time. Consequently, when the exposure time is increased, the amount of energy supplied to the process increases.

Normally, an increase in the intensity causes the effects of cavitation to increase to a maximum and then decrease; this is probably due to the formation of a dense concentration of bubbles near the probe tip, which acts as a barrier to the transfer of energy to the process. In addition, an excessively high intensity could have negative effects on the properties of the product. Therefore, in this case, a longer exposure time and a lower intensity could be a better option for obtaining a higher level of energy transmitted to the process.

Frequency

Low frequencies produce more violent cavitation and higher localized temperatures and pressures. A frequency range of 20–100 kHz is commonly used in power ultrasound, although most studies have been restricted to the range of 20–40 kHz. The reason why low frequencies produce more violent cavitation is that at very high frequencies, the formation of cavitation bubbles is reduced because either the rarefaction cycle of the sound wave produces a negative pressure that is insufficient in its duration and/or intensity to initiate cavitation, or the compression cycle occurs faster than the time required for a microbubble to collapse (Thompson and Doraiswamy, 1999). However, in applications such as reduction and oxidation reactions, a higher frequency is desired, since the number of free radicals increases and higher reaction rates are achieved.

Temperature and Pressure

Increasing the temperature reduces cavitation effects. Although an increase in temperature increases the formation of cavitation bubbles, these contain more vapor and consequently lead to a less violent collapse. In contrast, increasing the pressure increases the pressure in the bubbles and thus leads to a more rapid but more violent bubble collapse. Consequently, cavitation effects are enhanced.

In general, the largest sonochemical effects are observed at lower temperatures and higher pressures, where the majority of the content of the bubbles is gas.

High-Intensity Applications

In this section, the most important applications of high-intensity ultrasound in the food industry are reviewed. The scope of applications is vast, including microbial and enzyme inactivation, filtration, emulsification, drying, and meat tenderization. In terms of the kind of effect on the food material, the applications have been grouped into two different groups: those which do not change the physicochemical properties of the food, and those intended for food processing, which modify the properties or characteristics of the food.

Ultrasound-Assisted Processing Tools

Cutting and Slicing

Ultrasonic cutting is a well-established technology that is widely commercially available, and has been available to the industry for more than 50 years. In addition, in recent years, this technology has been introduced into the food industry in such diverse applications as the cutting of frozen fish, cheese, ice cream bars and bakery products. Ultrasonic food cutting, compared with other conventional method of cutting, provides many significant benefits. Thus, for instance, the cut face is especially clean in visual appearance, with more consistency, and the cutting operation is faster and more sanitary, which is of significant importance in the food preparation industry.

An ultrasonic cutter uses a blade horn, which is caused to vibrate at a frequency of 20–40 kHz with a power output of up to 2 kW at 20 kHz and 700 W at 40 kHz. The cutting blade is made of titanium and moves rapidly back and forth, reducing the friction resistance at the cutting surface and making the blade cut smoothly.

Atomization

Atomization is a process in which a liquid is broken into small droplets. Conventional atomizers force a liquid to pass through a small orifice at high velocity, which restricts their usage to very low-viscosity liquids. In contrast, ultrasonic atomizers require much less energy and overspray is practically eliminated, resulting in reduced consumption with better performance.

Ultrasonic atomization is based on an atomizing probe or nozzle that vibrates at a very high frequency. The liquid passes through the probe and reaches the probe tip, where the probe vibration is focused and the atomization takes place. The vibration of the tip causes cavitation and the formation of micron-scale droplets of high sphericity and uniform size distribution. The drop size is a function of frequency. Very small droplets can be produced by increasing the frequency. The ultrasonic frequency used for the atomization of liquids typically ranges from 10 kHz to 3 MHz.

Ultrasonic atomizers are used in a variety of industrial and medical applications. In the food industry, the applications of this technology include the coating of foodstuffs and food packaging materials with various liquid-state chemicals and ingredients. For instance, some crackers are coated with a thin layer of oil, and commercially manufactured tortilla chips are typically sprayed with one or more chemical preservatives to extend their shelf-life.

Food-Processing Applications

The use of power ultrasound in the food processing industry is a promising area. Ultrasound technology has a wide range of current and future applications. Table 38.2

Table 38.2 List of applications of power ultrasound in the food industry.

Application	Benefits
Microbial and enzyme inactivation	Destruction of microorganisms or inactivation of enzymes at lower temperatures
Crystallization	Formation of very small and uniform crystals with enhanced rate of seeding
Filtration	Higher filtration rates, cleaner filters and prevention of fouling
Drying	Increased product throughput at lower temperatures, avoiding degradation of food
Extraction	Increased yield of extracted components, higher extraction efficiency and higher processing throughput
Emulsification	Online use for flow processes, providing very effective, stable and homogeneous mixing of two or more immiscible liquids without or with very little additives
Meat tenderization	Increased tenderness by disruption of myofibrillar components in shorter aging periods
Degassing	Rapid removal of unwanted air or gas from liquids
Defoaming	Enhanced degree of defoaming
Freezing	Reduced freezing time with smaller ice crystal size distribution, leading to a product of better quality
Oxidation processes	Early maturation of alcoholic beverages

shows a list summarizing the current and potential applications of power ultrasound and their benefits in different fields of the food industry. However, despite the large number of potential ultrasonic processes, most of them are still restricted to a research environment, and only a few of them have been introduced into the industry. The main reason is that power ultrasound is not an off-the-shelf technology. Each application needs to be specifically developed and scaled up.

This section provides an overview of the major applications of power ultrasound in several different fields.

Microbial and Enzyme Inactivation

This application is a promising field that relies on the ability of power ultrasound to disrupt biological cell walls through intense cavitation. However, the effectiveness of ultrasound is dependent on the type of microorganism or enzyme being treated, the amplitude and frequency of the ultrasonic signal, the exposure time, the volume of food to be processed, and the type of food. In addition, power ultrasound alone is not very effective in the destruction of microorganisms or inactivation of enzymes unless very high intensities are used, and, precisely for this reason, ultrasound is generally used in conjunction with another technique such as pressure (manosonication), heat (thermosonication) or both (manothermosonication) to achieve efficient destruction of microorganisms or inactivation of enzymes. Thus, for instance, using pressure during power ultrasound treatment increases the rate of microbial inactivation for a variety of microorganisms, even at temperatures well below the boiling point of the medium. Figure 38.21 shows the effectiveness of this process at a very low temperature when the pressure is varied from 0 to 400 kPa. Such processes have been studied

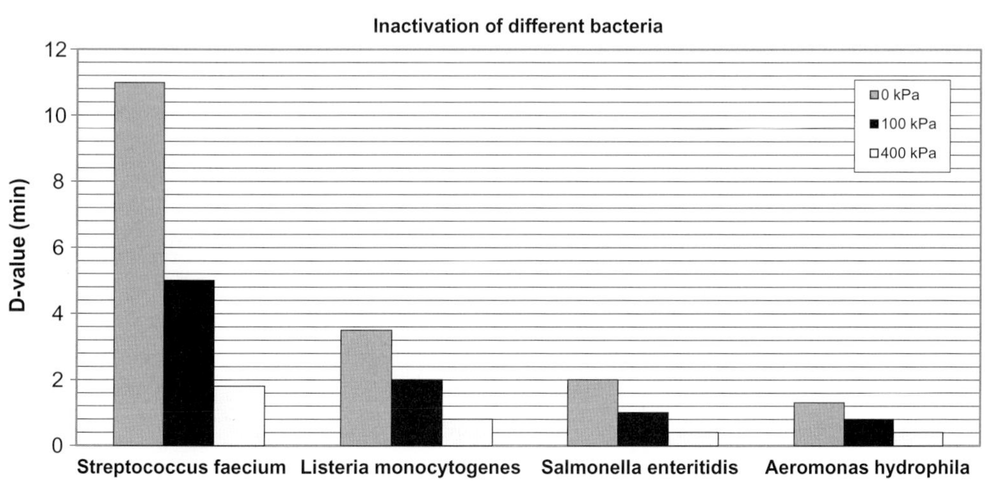

Figure 38.21 D-values of various bacteria treated with ultrasound (20 kHz, 117 μm amplitude) at 40 °C when the pressure is varied from 0 to 400 kPa.

and described in depth by several researchers (Ordoñez *et al.*, 1984; Raso *et al.*, 1998; Mañas *et al.*, 2000; Mañas and Pagán, 2005).

So far, research has been successfully done with a great number of bacteria, such as *Escherichia coli*, *Listeria monocytogenes*, *Salmonella*, *Staphylococcus aureus* and *Saccharomyces cerevisiae* (Piyasena *et al.* 2003). The exposure time, the temperature, and the intensity and frequency of the ultrasonic wave are key factors for successful destruction of microorganisms. Also, new areas, such as the winemaking industry, where power ultrasound has not yet been applied, are offering potential for its use. Jiranek *et al.* (2008) have explored the application of power ultrasound to modulating microbial activity and load at various stages of winemaking process in juices, musts and wines and in the barrels. Thus, for instance, rather than washing fermentation tanks and barrels with high-pressure hot water, the use of power ultrasound might achieve more effect in a reduced period of time with a lower temperature of the water.

Research into the inactivation of various enzymes under various conditions has also been carried out. Thakur and Nelson (1997) showed that the pH of the medium was the most important parameter for the inactivation of lipoxygenase. Raviyan *et al.* (2005) studied the effect of cavitation intensity and temperature on the inactivation of tomato pectinmethylesterase (PME). A comparison of thermal treatment, sonication and thermosonication under various treatment conditions was conducted. Increasing the temperature increases the inactivation rate; this has been demonstrated in many tests of thermal inactivation of tomato enzymes (López *et al.*, 1997; Crelier *et al.*, 2001). From Figure 38.22, it is clear that the use of thermosonication increases the inactivation rate. In addition, taking into account the fact that the inactivation caused exclusively by thermal treatment at 50 °C is negligible, the inactivation by

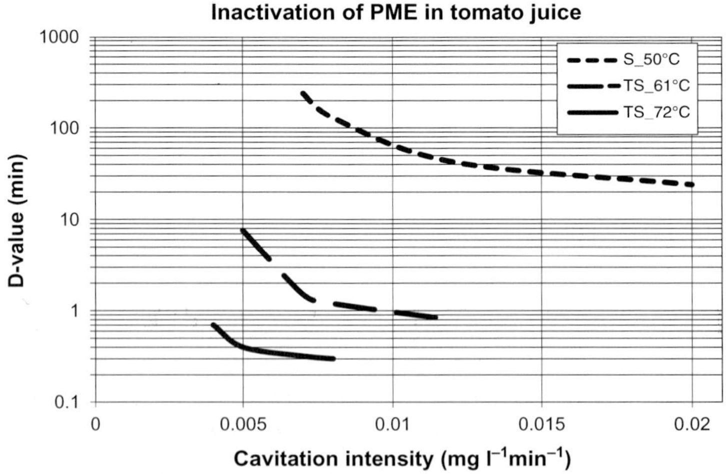

Figure 38.22 Inactivation of PME obtained from tomato juice using sonication and thermosonication at different cavitation intensities and temperatures. S, sonication; TS, thermosonication.

ultrasonic treatment at 50 °C must be due to sonication itself. With reference to the different values of the reduction in the residual activity at various temperatures reported by different researchers, the discrepancies might be caused by differences in tomato variety, degree of ripening, heating and temperature measurement techniques, and enzyme preparation and assay methods.

Experiments under pressure at different temperatures were carried out by Mañas *et al.* (2006) in order to study the inactivation of lysozyme. Ultrasonic waves at room temperature and pressure inactivate the enzyme very little. However, the application of an external pressure of 200 kPa and temperatures between 60 and 80 °C increased the inactivating effect of ultrasound.

It seems clear that the effect of ultrasound on enzymes is dependent on the type of the enzyme and is also influenced by many parameters. Kadkhodaee and Povey (2008) studied how and to what extent ultrasonic effects vary under different experimental conditions in order to develop a mathematical model that quantified the contribution of variable parameters to the overall effect of ultrasound and, consequently, allowed one to predict changes in the efficiency of a process. The results showed that ultrasound effectively inactivated α-amylase, with the minimum overall inactivation rate at 50 °C. Also, it was shown that the contribution of the various parameters to the inactivation of the enzyme seems to be dominated by the temperature and the acoustic intensity, although the amino acid composition and the conformational structure of the enzyme also have a great influence.

Crystallization

The application of power ultrasound to the crystallization process was first reported more than 50 years ago, but it has only been in the last decade that it has received much attention. The most important effect of ultrasound on crystallization is the induction of nucleation, which is again based mainly on cavitation phenomena. During cavitation, it is possible to break down crystals and nuclei into a large number of uniform small crystals that can act as nucleation centers. This effect is often referred to as sonocrystallization. A schematic illustration of sonocrystallization induced by ultrasonic cavitation is shown in Figure 38.23. The crystallization process

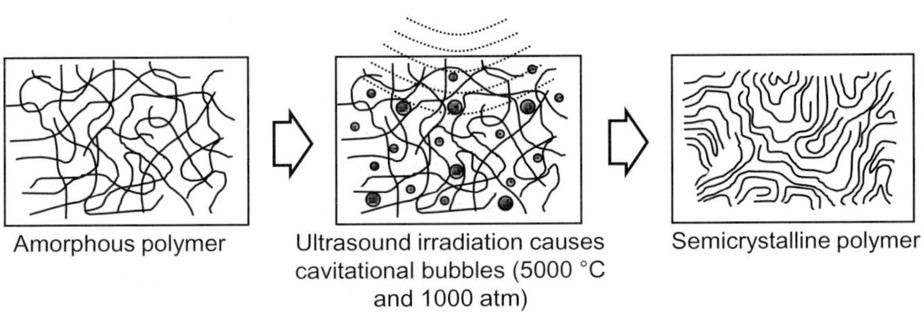

Figure 38.23 Schematic illustration of sonocrystallization induced by ultrasonic cavitation.

Table 38.3 Ultrasonic crystallization of sugar solutions.

Solute	Quantity dissolved in 10 ml of water at 50 °C	Temperature at which solid appears	
		Control	Ultrasound treatment
D-xylose	25 g	36 °C	46 °C
D-sucrose	18 g	<40 °C	47 °C
D-lactose	5.5 g	41 °C	43 °C
D-maltose	13 g	No crystals appeared at 20 °C	40 °C
D-cellobiose	2.0 g	No crystals appeared at 20 °C	42 °C

can be controlled by means of the amplitude and frequency of the ultrasonic signal and the exposure time, thus allowing manipulation of the crystal size distribution and the point at which crystallization occurs (Povey and Mason, 1998).

The application of power ultrasound to crystallization has been found to be particularly appropriate for pharmaceuticals and fine chemicals, which are amongst the hardest materials to crystallize well because they tend to be high-molecular weight organic compounds (Ruecroft *et al.*, 2005). There are also some interesting applications in the food industry based on the modification and control of the crystallization process in many food products. One of these applications deals with the control of sugar crystallite size. This application is of particular interest since the texture of food products can be affected by the size of undissolved sugar crystals dispersed in the material (Mason *et al.*, 1996). Table 38.3 gives some measured data on the reduction in the width of the metastable zone for a range of solutions cooled from 50 to 20 °C (Ruecroft *et al.*, 2005). In all cases except D-lactose, the width of the zone was significantly reduced.

The application of power ultrasound to honey has long been studied (Kaloyereas, 1955; Liebl, 1978; Thrasyvoulou *et al.* 1994). Honey is a supersaturated sugar solution, whose natural tendency is to crystallize. This is an undesirable process that must be avoided during the production process of liquid honey. Traditionally, crystallization of honey is delayed by applying a heat treatment. Heat helps to dissolve D-glucose monohydrate crystals but also negatively affects the delicate flavors of honey. The application of power ultrasound as an alternative eliminates existing crystals in honey and also retards the crystallization process, resulting in a cost-effective technology. Analysis of the crystallization process shows that honey treated by power ultrasound remains in the liquid state for a much longer period than does heat-treated honey. In addition, no significant effects on the quality parameters of the honey, such as moisture content, electrical conductivity and pH, are observed.

Crystallization of fats is another area that is gaining considerable interest in the chocolate, butter and ice cream industries. Thus, for instance, the effects of power ultrasound (20 kHz and 100–300 W) on tripalmitoylglycerol and cocoa butter have been studied (Higaki *et al.*, 2001; Ueno *et al.*, 2003). The results indicate that ultrasound irradiation is an efficient tool for controlling the polymorphic crystallization of fats and that the induction time is shortened.

Several theories have been proposed in order to explain the relationship between cavitation and nucleation, but the contribution of ultrasound to crystallization is still not fully understood. Therefore, further development of models is still required.

Filtration

Solid–liquid filtration is a common process in many industries today. Filtration can be applied either to clarify a liquid that contains solids or to remove solids from a liquid. Ultrasound-assisted filtration has been widely studied in the past. In addition to increasing substantially the rate of flow of the liquid through the filter, the application of ultrasound also keeps the filter cleaner and prevents fouling (Fairbanks, 1973; Semmelink, 1973). The enhancement of filtration relies on two effects: first, sonication causes agglomeration of fine particles and, second, sonication supplies enough vibratory energy to the system to keep the particles partly suspended and therefore leaves more free channels for solvent elution (Mason *et al.*, 1996).

Ultrasound-assisted filtration has been applied to several sectors of the food and beverage industry. The technique has been applied to fruit extracts and drinks to increase the amount of juice extracted from pulp (Mason *et al.*, 1996). Other potential areas of application include sugar, beer, wine and edible oils (Brennan *et al.*, 2006). Thus, for instance, the juice produced by extraction from sugar beet is treated with lime and filtered to produce a clear juice for further processing. Beer is clarified by filtration to remove deposits of yeast and trub formed on the bottom of the maturation tank during the maturation stage. In addition, wine and edible oils are filtered at various stages of production.

In an extensive review, Kyllönen *et al.* (2005) studied several parameters that influence the effectiveness of ultrasound filtration, such as frequency, power intensity, feed properties, cross-flow velocity and temperature. In general, higher filtration efficiencies are achieved with lower ultrasound frequencies and high power intensities using intermittent ultrasonic sonication. However, ultrasound is less effective for highly concentrated suspensions such as slurries. One of the current requirements in many industrial applications is the processing of highly concentrated suspensions of fine particles. Conventional filtration techniques are not satisfactory, because fouling often occurs, resulting in slow filtration rates. Gallego-Juárez *et al.* (2003) have been working on the development and application on a pilot scale of power ultrasonic technology to assist in cake deliquoring in the filtration of concentrated fine-particle slurries in rotary vacuum disk filters. One or several power transducers are placed parallel to the filter surface and very close to it. An increase in the filtration efficiency was obtained with two extremely fine, highly concentrated suspensions of particulate materials. The experimental ultrasonic system is schematically depicted in Figure 38.24.

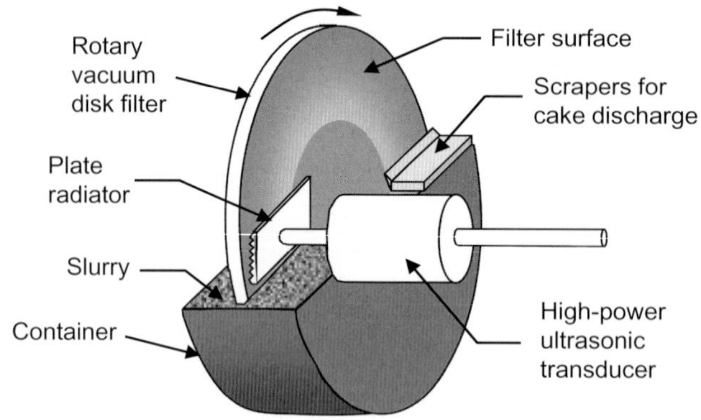

Figure 38.24 Schematic diagram of an ultrasonic-assisted filtration system.

Drying

Conventional systems for food dehydration are mainly based on either hot-air drying or freeze drying. Hot-air drying is a widely used method, but the food product can suffer deterioration. In freeze-drying, product deterioration is negligible but the process is expensive. In contrast to conventional dehydration methods, power ultrasound reduces the cost of the process and the probability of food degradation since the drying process can be carried out at lower temperatures.

Ultrasonic dehydration is still under development; the low drying rate of the ultrasonic process is one of its major drawbacks. However, there has been a lot of interest in recent years in the development of this potential application.

When power ultrasound is directly coupled to a food material to be dried, a rapid series of alternating compressions and expansions produce a kind of sponge effect and a quick migration of moisture from the product (Floros and Liang, 1994; Gallego-Juárez *et al.*, 2007). In addition, ultrasound produces cavitation, which may be beneficial for the removal of moisture that is strongly attached (Tarleton and Wakeman, 1998; Fuente-Blanco *et al.*, 2006).

Power ultrasound has been extensively applied as a pretreatment for drying fruits and vegetables such as apples, melons, bananas, papayas, cauliflowers and carrots. In all cases, the drying time after ultrasound treatment is shortened as compared with untreated products. The effect of pretreatments and drying conditions on the structure of products such as mushrooms, Brussels sprouts and cauliflowers has been investigated in order to see how later rehydration is influenced (Jambrak *et al.*, 2007). When compared with the freeze-drying method (taken as the reference method), the results indicated that the rehydration properties (measured by percentage weight gain) were better for freeze-dried samples. However, the rehydration properties for ultrasound-treated samples were higher than for untreated samples, so the rehydration behavior of plant-based foods can also be enhanced by the use of power ultrasound.

Fernandes and Rodrigues (2007) and Fernandes *et al.* (2007, 2008) have investigated the effect of ultrasonic pretreatment prior to air-drying on the dehydration of bananas, papayas and melons. In all cases, the integrated process (power ultrasound and air-drying) was optimized, searching for the operating condition that minimized total processing time. Time reductions of between 10% and 16% were achieved when ultrasound pretreatment was applied for approximately 20 min.

With the aim of scaling up the drying process, Gallego-Juárez *et al.* (1999) have focused their efforts on developing a new technology for vegetable dehydration based on airborne power ultrasound that applies ultrasonic vibration directly to the product, avoiding the usual ultrasonic bath. Two experimental procedures have been developed: forced-air drying assisted by airborne ultrasound, and ultrasonic dehydration by applying ultrasound in direct contact with the material. In addition, a parametric study of the relative influence of the main physical parameters involved in the process was carried out using a stepped-plate transducer working at 20 kHz and a power capacity of about 100 W (García-Pérez *et al.*, 2006), in order to establish starting points for the development of the system at a preindustrial stage. Experimental results for carrots, apples and potatoes and an extensive review of this new technology can be found in Fuente-Blanco *et al.* (2006) and Gallego-Juárez *et al.* (2007). The results obtained using direct contact show that the new technology provides a very effective method for food dehydration. However, the improvement obtained when high-intensity airborne ultrasound is applied is less than for direct-contact application. The main reason for this is the low penetration of ultrasonic energy into the vegetable material, caused by the mismatch between the acoustic impedances of the transducer, air and food.

Extraction

Ultrasonic extraction is based on cavitation phenomena and has become an efficient alternative to traditional methods for solvent extraction. Cavitation breaks down the walls of biological cells and releases the cell contents into the solvent (Mason *et al.*, 1996). Thus, the extraction of sugar from sugar beet has been found to be improved with the use of power ultrasound (Chendke and Fogler, 1975). Wang (1975) studied protein extraction from defatted soybeans using a 550 W probe operating at a frequency of 20 kHz, which resulted in a more efficient method of extraction than any other previous technology. Later, the experiment was scaled up to pilot plant level for the extraction of soybean protein (Moulton and Wang, 1982).

More recently, an ultrasonic extraction procedure for Ca, K and Mg from *in vitro* embryogenic and nonembryogenic *Citrus* sp. cultures was proposed by Arruda *et al.* (2003). In experiments, several different extracting media and sonication times of 5, 10, 15 and 30 min were used to optimize the procedure, which resulted in an excellent alternative to the traditional procedure, with an important reduction in sample handling and in operational costs. Albu *et al.* (2004) used power ultrasound to increase the efficiency of extraction of carnosic acid from the herb *Rosmarinus officinalis* using butanone, ethyl acetate and ethanol as solvents. Power ultrasound reduced the dependence on the extraction solvent and greatly enhanced the performance of ethanol, a poor solvent for conventional extraction. In addition, sonication appears to have great potential as a method for the extraction of antioxidant materials: comparable levels of extracted carnosic acid were obtained by employing an ultrasonic bath and by employing a probe system, thus indicating the potential for scaling-up of the extraction process. Another interesting application is to the extraction of phenolic compounds from coconut shell powder (Rodrigues and Pinto, 2007). The effects of toasting time, toasting temperature and extraction time were evaluated. The results indicated that high amounts of phenolics can be extracted from coconut shells by ultrasound-assisted extraction technology, the extraction time being the most significant parameter for the process. The best condition for obtaining a high phenolic content (406.93 $mg \cdot L^{-1}$) in the extracts was found to be a toasting time of 60 min at 100 °C and an ultrasound extraction time of 60 min.

Table 38.4 provides a list of applications of ultrasound-assisted extraction. A more extensive review of applications of power-ultrasound-assisted extraction has been given by Vilkhu *et al.* (2008), including the extraction of herbal, oil, protein and bioactive materials.

Emulsification

One of the first applications of power ultrasound was to emulsification, and was first reported by Wood and Loomis (1927). This application, which is also based on cavitation, provides a very effective, stable and homogenous mixing of two or more immiscible liquids without or with very few additives (Shoh, 1975; Mason *et al.*, 1996). The great advantage of emulsification is that it can be used online in flow processes. Thus, volumes up to 12 000 $L \cdot h^{-1}$ can be processed, as is the case in the manufacture of fruit juices, tomato ketchup and mayonnaise (Mason *et al.*, 1996).

The effects of power ultrasound on milk homogenization have been studied by various authors (Wu *et al.*, 2001; Villamiel and Jong, 2000; Ertugay *et al.*, 2004). It was found that sonication of fresh cow milk at 20 kHz resulted in a reduction in the size of the fat globules; power ultrasound is very effective in reducing the size of fat globules. An average size of the fat globules of less than 1 μm was achieved. Also, the application of power ultrasound results in more uniformly distributed fat particles than does the conventional homogenization method. Thus, milk homogenized by a conventional homogenizer has smaller fat globules than nonhomogenized milk, and milk homogenized by power ultrasound has even smaller fat globules than milk homogenized by a conventional homogenizer. Figure 38.25 compares the average fat globule diameter of nonhomogenized milk with the diameter achieved after conventional homogenization and under different ultrasonic treatment conditions. The fat globule diameter obtained in conventional homogenization could be achieved by means of ultrasonic emulsification at a power level of 180 W for 10 min (Villamiel and

Table 38.4 Applications of ultrasound-assisted extraction, and experimental results.

Product	Ultrasound system	Solvent	Results	Reference
Phenolic compounds from coconut shell powder	Batch, 25 kHz	Water and ethanol	The extraction of phenolic content was affected mainly by toasting time and extraction time	Rodrigues and Pinto (2007)
Hesperidin from penggan peel	Batch, 20 kHz, 60 kHz and 100 kHz	Methanol, ethanol and isopropanol	Shorter extraction time (by up to 8 times) with lower temperatures than in conventional method	Ma *et al.* (2008)
Pungent compounds from ginger	Batch, 20 kHz	Supercritical CO_2	Up to 30% increased yield	Balachandran *et al.* (2006)
Almond oils	Batch, 20 kHz	Supercritical CO_2	Up to 20% increased yield and similar yields in about 30% shorter time	Riera *et al.* (2004)
Carnosic acid from rosemary	Batch, 20 kHz and 40 kHz	Butanone and ethyl acetate	Reduction in extraction time	Albu *et al.* (2004)
Soy isoflavones	Batch, 24 kHz	Ethanol, methyl cyanide and methanol	Up to 15% increase	Rostagno *et al.* (2003)
Ca, K and Mg from *in vitro* citrus culture	Batch, 47 kHz	Water	Remarkable efficiency, handless manipulation and reduced operational costs	Arruda *et al.* (2003)
Rutin from flower buds of *Sophora japonica*	Batch, 20 kHz	Water and methanol	Up to 20% increase in 30 min when using methanol compared with conventional method	Paniwynk *et al.* (2001)
Herbal extracts	Stirred batch, 20 kHz to 1000 kHz	Water and ethanol	Up to 34% increased yield over stirred-batch method	Vinatoru (2001)
Soy protein	Continuous, 20 kHz	Water and alkali	53% and 23% yield increases over equivalent ultrasonic-batch conditions	Moulton and Wang (1982)

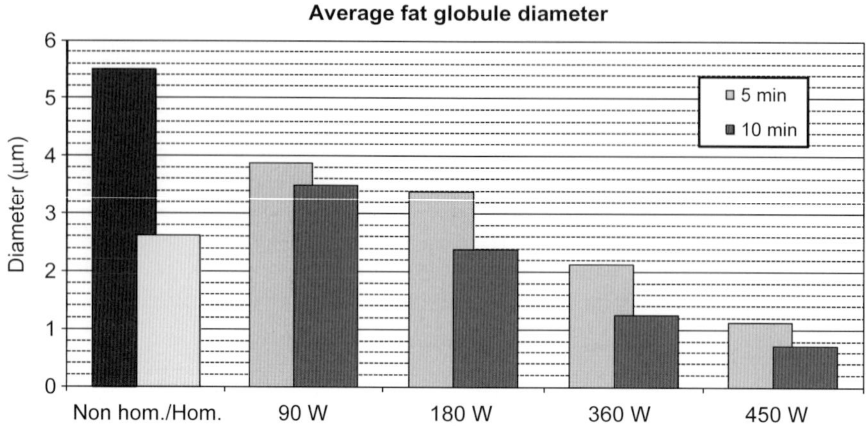

Figure 38.25 Effect of ultrasonic homogenization on fat globule diameter in milk (20 kHz ultrasonic probe) as a function of power level and exposure time. The first two bars refer to nonhomogenized milk ("Nonhom.") and milk homogenized by a conventional homogenizer ("Hom.").

Jong, 2000). In general, a substantial reduction in fat globule size, by up to 81.5%, was achieved after all treatments.

Wu *et al.* (2001) also studied the effect of power ultrasound on yoghurt fermentation. Ultrasound treatment before inoculation of the yoghurt starter resulted in an increase in water-holding capacity. In addition, when ultrasonic treatment was applied after inoculation, the total fermentation time was reduced by 0.5 h, which implies clear industrial benefits.

In more recent work, the effects of irradiation time, irradiation power and the physicochemical properties of the oil on the volume and droplet size of the dispersed phase have been studied in typical emulsions consisting of oil and water (Gaikwad and Pandit, 2008). Oils with different viscosities were used in the experiments. In particular, two edible oils and two mineral oils were used. The results indicated that with an increase in the irradiation time, the dispersed phase volume increased and the droplet size of the dispersed phase decreased. Likewise, as the irradiation power was increased, there was an increase in the hold-up of the dispersed phase and the droplet size of the dispersed phase decreased. A higher viscosity of the oil was related to a larger droplet size in the emulsion.

Meat Tenderization

Tenderness is one of the most important features of meat quality perceived by the consumer. Meat tenderness varies considerably among species, among animals within a species and among different muscles, but is highly affected by the way the meat is cooked as well. In general, meat is tender just after slaughter; however, tenderness is negatively affected by changes that occur during rigor mortis and post-rigor aging. Traditionally, aging has been regarded as a fail-safe method of producing tender meat. Carcasses are hung for several days at a temperature just above freezing to allow the meat to tenderize, but this process adds considerable cost to the production process because of the requirements for space and controlled temperatures.

When power ultrasound is used, the muscle and cell structure of tissues can be disrupted by means of cavitation effects, leading to tenderization of the meat. However, the existing studies are somewhat contradictory, since the ultrasound treatment conditions vary widely and, consequently, the effects on meat tenderization vary as well. In some cases, power ultrasound proved effective, although in a number of other studies the treatment conditions were found to have no effect on tenderness or even to lead to decreased tenderness. Table 38.5 summarizes experimental results from some studies of meat tenderization.

Smith *et al.* (1991) suggested that the tenderization of meat samples (*semitendinosus* muscle from cattle) induced by the application of power ultrasound is of sufficient magnitude to be detected by the average consumer. They found a significant increase in tenderness after 2–4 min of sonication. Nevertheless, opposite effects were observed when meat was sonicated for more than 8 min, but the reasons for this behavior were not explained by these authors. In two consecutive studies published by Lyng *et al.* (1997, 1998a) using the *longissimus thoracis et lumborum* and *semimembranosus* muscles, no significant effects on tenderness due to power ultrasound treatment were found compared with untreated samples. According to Lyng *et al.*, the reason could be associated with the method of treatment, as well as the ultrasound intensity levels and exposure times used. These factors, as well as the use of different frequencies and the type of muscles used, might have caused such inconsistent results.

The influence of high-frequency power ultrasound has also been investigated (Got *et al.*, 1999). These authors applied a 2.6 MHz transducer to *semimembranosus* muscles but no conclusive effects on meat tenderization were observed. One of the reasons could be that in this study, the benefits of cavitation were not achieved, since cavitation is unlikely at such high frequencies.

Recently, Jayasooriya *et al.* (2004) reviewed the effects of high-power ultrasound on the physical, biochemical and microbial properties of meat, and the potential benefits. The specific effects on tenderness were discussed. Power ultrasound appears to be a very attractive alternative technique for modifying the properties of meat and meat products, although further research is needed. To this end, Jayasooriya *et al.* (2007) studied the effect of power ultrasound and aging on the physical properties of bovine *semitendinosus* and *longissimus* muscles in order to determine the exposure time needed to tenderize different types of muscles, and studied the impact of power ultrasound on aging as well, to determine whether power ultrasound negatively affects other characteristics of meat such as color and pH. After 60 s of power ultrasound treatment (24 kHz and 12 $W \cdot cm^{-2}$), the reduction in the Warner–Bratzler shear force and hardness was comparable to that for meat aged for three to five days. The benefits of ultrasound were reduced as the aging time increased. In addition, the improvement

Table 38.5 Experimental results from some studies of meat tenderization by power ultrasound.

Muscle	Ultrasound conditions	Results	Reference
Beef *semitendinosus*	25.9 kHz, 1000 W and 0, 2, 4, 8, 16 min treatment	Significant decrease in shear force at 2 and 4 min of exposure, but shear force increased after 8 min of exposure.	Smith *et al.* (1991)
Chicken muscle	40 kHz, 2400 W and 15 min treatment	Significant reduction in shear force.	Dickens *et al.* (1991)
Lamb *longissimus thoracis et lumborum*	20 kHz, 56–62 $W{\cdot}cm^{-2}$ and 10–180 s treatment	Ultrasonic treatment of fibers enhanced proteolytic activity. Cell damage was not very extensive, although it depended upon ultrasonication and fiber conditions.	Roncalés *et al.* (1993)
Beef *longissimus thoracis et lumborum, semimembranosus* and *biceps femoris*	30–47 kHz, 0.62–0.29 $W{\cdot}cm^{-2}$ and 30 min per side	Peak force values were not affected by ultrasound treatment.	Lyng *et al.* (1997)
Beef *longissimus thoracis et lumborum* and *semimembranosus*	20 kHz, 62 $W{\cdot}cm^{-2}$ and 15 s treatment	No significant effects on tenderness were found.	Lyng *et al.* (1998a)
Lamb *longissimus thoracis et lumborum*	20 kHz, 63 $W{\cdot}cm^{-2}$ and 15 s treatment	Bite-force tenderometry, sensory analysis and collagen solubility all showed no effect of ultrasound treatment.	Lyng *et al.* (1998b)
Beef *semimembranosus*	2.6 MHz, 10 $W{\cdot}cm^{-2}$ and 2 consecutive periods of 15 s	No conclusive effects on the post mortem rate of tenderization of meat.	Got *et al.* (1999)
Beef *longissimus thoracis et lumborum and semimembranosus*	24 kHz, 12 $W{\cdot}cm^{-2}$ and 0, 30, 60, 120, 240 s treatment	Improvement in tenderness without detrimental effects on drip, cooking or total losses, or on color. The benefits obtained decreased with increasing aging time.	Jayasooriya *et al.* (2007)
Beef *semimembranosus*	45 kHz, 2 $W{\cdot}cm^{-2}$ and 120 s treatment	The high water-holding capacity of sonicated samples suggests an acceleration of rigor mortis followed by fragmentation of the protein structures of cells.	Stadnik *et al.* (2008)
Pork *longissimus dorsi*	20 kHz, 2–4 $W{\cdot}cm^{-2}$ and 30–180 min treatment	Water-holding capacity and textural properties were improved by ultrasonic treatment. However, higher intensities and/or longer treatment times caused denaturation of proteins.	Siró *et al.* (2009)

in tenderness was achieved without detrimental effects on the drip, cooking or total losses, or on the color.

Similarly, Stadnik *et al.* (2008) studied whether treatment with low-frequency and relatively low-intensity ultrasound (45 kHz, $2\,W{\cdot}cm^{-1}$) at 24 h post mortem affected the pH of meat and its water-holding capacity during 96 h of aging. Ultrasound treatment did not have a statistically significant effect on the pH of the meat. In addition, the high water-holding capacity of sonicated samples, typical of meat in advanced post mortem stages, suggests that because of the ultrasound treatment, acceleration of rigor mortis followed by fragmentation of the protein structures of cells occurred.

Power Ultrasound Instrumentation

In this section, several different configurations of ultrasonic equipment are discussed. Although considerable effort has been put into developing standard commercial ultrasonic equipment, scaling-up involves much more than the construction of a larger version of the equipment; a specific development process is required for every individual application.

As the applications of high-intensity ultrasound involve only coupling acoustic energy to the food material, no detection of the ultrasonic waves is required. Therefore, the setup shown schematically in Figure 38.5 can be greatly simplified, and the instrumentation is reduced to an ultrasonic transducer, a signal generator and a power amplifier.

Transducers

Most high-intensity ultrasonic transducers are of the piezoelectric type and have structures based on the classical prestressed sandwich transducer arrangement (Neppiras, 1973; Gallego-Juárez, 1990). A sandwich transducer is composed of one or several piezoelectric disks or rings between two metallic blocks. This assembly is mechanically prestressed by means of a bolt and operates as a half-wave resonator. The substitution of part of the piezoelectric material by metallic blocks permits one to overcome the low tensile strength of the piezoceramic, which is very undesirable in high-intensity applications, and this reduces the dimensions of the transducer. The most suitable piezoelectric material for these applications is lead zirconate titanate (PZT); the PZT-4 and PZT-8 compositions are the most popular, owing to their low internal losses. The metallic parts are usually made of duralumin, stainless steel or titanium.

Sandwich transducers are used directly in applications to liquids by bonding them to the liquid recipient. In contrast, transducers for applications of high-intensity ultrasound to solids normally require a large vibration amplitude at the face in contact with the material. Vibration amplification is achieved by adding to the front face of the sandwich transducer another metallic part of special shape, called a sonotrode or horn. The shape and dimensions of the horn are usually custom designed for each application.

Ultrasonic Equipment

There are several different types of ultrasonic equipment commercially available. The most frequently used are the ultrasonic bath, the ultrasonic horn (or probe), the cup-horn reactor, the liquid whistle reactor and the flow cell. The basic characteristics of each of them are described below.

The Ultrasonic Bath

The ultrasonic bath is perhaps the most widely used type of equipment because it is cheap, easy to operate and readily available. It consists mainly of a stainless steel tank with a number of transducers bonded to the base of the tank, as depicted in Figure 38.26. The tank is filled either with a coupling liquid, in which the reaction vessel is immersed, or directly with the mixture to be processed. In order to achieve a uniform distribution of the ultrasonic irradiation, it may be necessary to stir the mixture mechanically. However, the disadvantages of this type of equipment are that temperature control of the reaction mixture is difficult and that it is difficult to obtain reproducible results. In addition, the equipment has a relatively low power density because undetermined losses due to transfer through the vessel walls are introduced and also because the power of the transducers has to be limited in order to avoid cavitation damage to the tank walls.

The Ultrasonic Horn

The ultrasonic horn, also known as the ultrasonic probe system, consists of a metal horn coupled to an ultrasonic transducer. The length of the horn is a multiple of half a wavelength and its purpose is to amplify the vibration produced by the transducer.

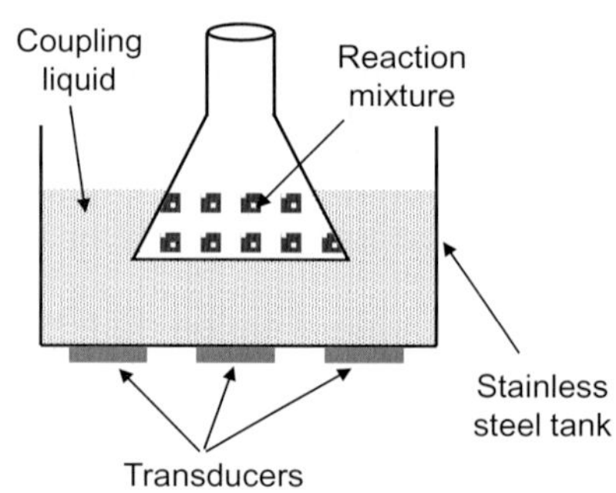

Figure 38.26 Ultrasonic bath.

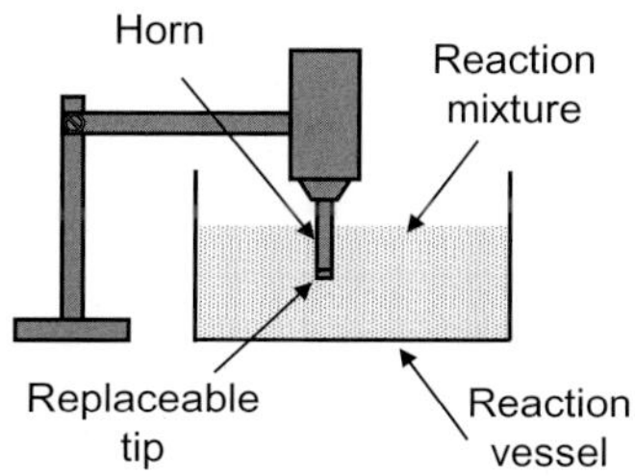

Figure 38.27 Ultrasonic horn.

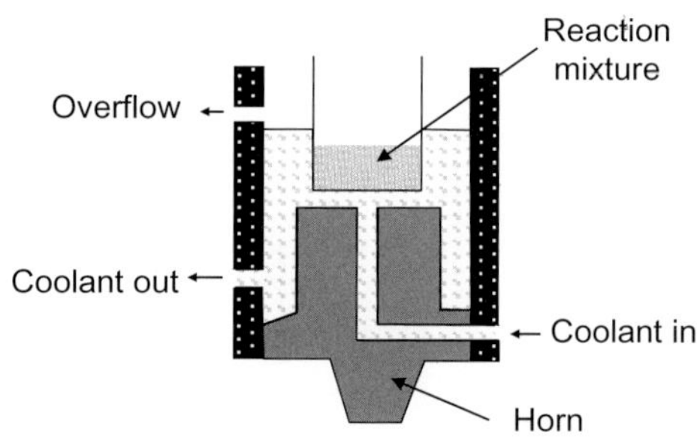

Figure 38.28 Cup–horn reactor.

The horn is immersed in the reaction mixture of interest and then intense ultrasound is generated directly at the tip of the horn (Figure 38.27). In contrast to the ultrasonic bath, the ultrasonic horn system is able to deliver large amounts of power directly to the reaction mixture, although it is not distributed uniformly. One restriction of this equipment is that erosion of the tip of the horn by cavitation requires frequent replacement of the tip, which might otherwise cause contamination of the reaction mixture.

The Cup–Horn Reactor

The cup–horn reactor (Figure 38.28) is composed of a small ultrasonic bath, a probe horn in contact with the reaction mixture and a cooling system for better temperature control than that provided by the ultrasonic bath. As it has a reduced power compared with the ultrasonic horn, there is no contamination by erosion of the horn tip material. The reaction vessel has a limited volume but provides more reproducibility of sonochemical effects.

The Liquid Whistle Reactor

This type of equipment (Figure 38.29) is very useful and provides an inexpensive method that has been used successfully in applications such as homogenization and

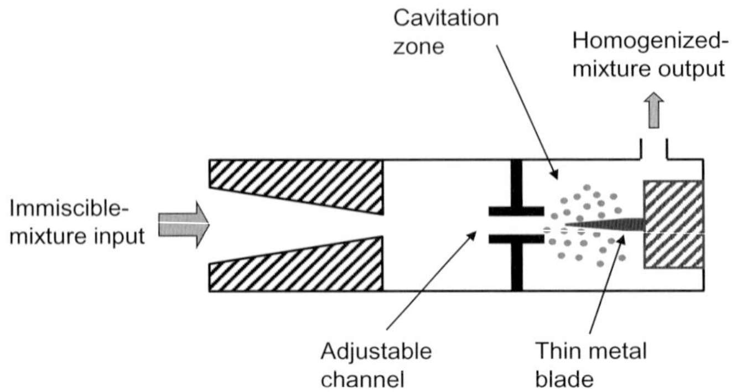

Figure 38.29 Liquid whistle reactor.

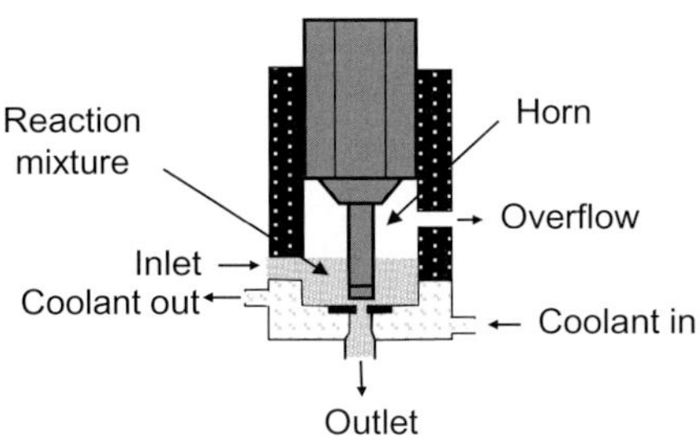

Figure 38.30 Flow cell.

emulsification. The whistle creates a cavitation zone without the need for a transducer. The liquid mixture is forced to flow rapidly through an orifice, from which it emerges as a jet and impacts on a thin metal blade, which is caused to vibrate at a frequency dependent on the flow rate. When a mixture of immiscible liquids is forced across the blade of the liquid whistle, the resulting cavitational mixing produces extremely efficient homogenization (Mason *et al.*, 1996). Because of its principle of operation, an obvious benefit of the liquid whistle is that it can be used on a large scale for flow processing and can be installed online. However, the intensity of the ultrasound generated is very low.

The Flow Cell

The flow cell (Figure 38.30) can be used with a continuous flow in large-scale processing. The ultrasonic horn is in direct contact with the reaction mixture that is being pumped through the cell, and consequently it is recommended for low-viscosity mixtures when the required sonication time is relatively low. Generally, it is used to emulsify, disperse and homogenize. The degree of processing of the reaction mixture can be controlled by varying the flow rate and the intensity of the ultrasonic horn. The flow cell is provided with a cooling jacket, through which a cooling liquid can be circulated to enable cooling of the reaction mixture while it is being processed.

Conclusions

In a wide variety of different applications in the food industry, low- and high-intensity ultrasound technology has been successfully proven in the laboratory, in pilot plants and, in less numerous cases, in large-scale applications. In recent years, ultrasound technology has attracted considerable interest in the food industry for both the analysis and the processing of foods. The benefits of using this technology are significant. Low-intensity ultrasound has major advantages over many other analytical methods because it is nondestructive, rapid, precise, relatively inexpensive and highly repeatable. In addition, it can be applied to concentrated and optically opaque foods, and can easily be adapted for online applications and thus allows automation and better control of product quality. In the case of high-intensity ultrasound, most of the studies reported in the literature provide evidence of benefits in various applications, either enhancing product quality or process efficiency, and no detrimental effects on food quality attributes have been observed, or at least they seem not to be especially affected.

Characterization of fundamental physical properties of foods such as elastic modulus, grain size, hardness and density is accomplished by measuring the ultrasonic velocity, the attenuation coefficient and the acoustic impedance of the food product, although the ultrasonic velocity and the attenuation coefficient are generally the two most widely used parameters. Ultrasonic food processing is based on the very complex nonlinear phenomenon known as cavitation. Generally, the intensity and frequency of the power ultrasound, the exposure time, the temperature, the pressure, and the volume of food to be processed affect the phenomenon of cavitation and consequently the results of processes, such as the extraction rate, particle size or destruction of microorganisms. Other factors that also affect cavitation significantly are the type of treatment, the choice of solvent and sample preparation. It is usually rather difficult to make comparisons between experiments carried out by different authors, since one of the key parameters that strongly influences the results is the type of food.

Both low- and high-intensity ultrasound place special demands on the equipment used. The reason is that the range of foods is very wide, and foods are generally more complex than other materials to inspect or process ultrasonically. Specific ultrasound instrumentation needs to be developed or adjusted for every different application, and even for the same application when different foods are to be tested or processed. Consequently, great efforts are being made by manufacturers and scientists to adapt equipment and measurement or processing techniques from a laboratory scale to large-scale application.

References

Adamowski, J.C., Buiochi, F. and Higuti, R.T. (2010) Ultrasonic material characterization using large-aperture PVDF receivers. *Ultrasonics* 50: 110–115.

Álava, J.M., Sahi, S.S., García-Álvarez, J., Turó, A., Chávez, J.A., García, M.J. and Salazar, J. (2007) Use of ultrasound for the determination of flour quality. *Ultrasonics* 46: 270–276.

Albu, S., Joyce, E., Paniwnyk, L., Lorimer, J.P. and Mason, T.J. (2004) Potential for the use of ultrasound in the extraction of antioxidants from *Rosmarinus officinalis* for the food and pharmaceutical industry. *Ultrasonics Sonochemistry* 11: 261–265.

Arruda, S.C.C., Rodriguez, A.P.M. and Arruda, M.A.Z. (2003) Ultrasound-assisted extraction of Ca, K and Mg from *in vitro* citrus culture. *Journal of the Brazilian Chemical Society* 14: 470–474.

Auld, B.A. (1990) *Acoustic Fields and Waves in Solids*, Vol. 1. Krieger, Malabar, FL.

Bakkali, F., Moudden, A., Faiz, B., Amghar, A., Maze, G., Montero de Espinosa, F. and Akhnak, M. (2001) Ultrasonic measurement of milk coagulation time. *Measurement Science and Technology* 12: 2154–2159.

Balachandran, S., Kentish, S.E., Mawson, R. and Ashokkumar, M. (2006) Ultrasonic enhancement of the supercritical extraction from ginger. *Ultrasonics Sonochemistry* 13: 471–479.

Becker, T., Mitzscherling, M. and Delgado, A. (2001) Ultrasonic velocity – a non-invasive method for the determination of density during beer fermentation. *Engineering Life Science* 1: 61–67.

Becker, T., Mitzscherling, M. and Delgado, A. (2002) Hybrid data model for the improvement of an ultrasonic-based gravity measurement system. *Food Control* 13: 223–233.

Benedito, J., Carcel, J., Gisbert, M. and Mulet, A. (2001a) Quality control of cheese maturation and defects using ultrasonics. *Journal of Food Science* 66: 100–104.

Benedito, J., Carcel, J., Rosselló, C. and Mulet, A. (2001b) Composition assessment of raw meat mixtures using ultrasonics. *Meat Science* 57, 365–370.

Benedito, J., Carcel, J.A., González, R. and Mulet, A. (2002) Application of low intensity ultrasonics to cheese manufacturing processes. *Ultrasonics* 40: 19–23.

Benedito, J., Simal, S., Clemente, G. and Mulet, A. (2006) Manchego cheese texture evaluation by ultrasonics and surface probes. *International Dairy Journal* 16: 431–438.

Brennan, J.G., Grandison, A.S. and Lewis, M. (2006) Separations in food processing. In: *Food Processing Handbook* (ed. J.G. Brennan), Wiley-VCH, Weinheim.

Carcione, J.M., Campanella, O.H. and Santos, J.E. (2007) A poroelastic model for wave propagation in partially frozen orange juice. *Journal of Food Engineering* 80: 11–17.

Chanamai, R. and McClements, D.J. (1999) Ultrasonic determination of chicken composition. *Journal of Agricultural and Food Chemistry* 47: 4686–4692.

Chendke, P.K. and Fogler, H.S. (1975) Macrosonics in industry. Part 4: chemical processing. *Ultrasonics* 13: 31–37.

Contreras, N.I., Fairley, P., McClements, D.J. and Povey, M.J.W. (1992) Analysis of the sugar content of fruit juices and drinks using ultrasound velocity measurements. *International Journal of Food Science and Technology* 27: 515–529.

Crelier, S., Robert, M.-C., Claude, J. and Juillerat, M.-A. (2001) Tomato (*Lycopersicon esculentum*) pectinmethylesterase and polygalacturonase behaviors regarding heat- and pressure-induced inactivation. *Journal of Agricultural and Food Chemistry* 49: 5566–5575.

Cross, H.R. and Belk, K.E. (1994) Objective measurements of carcass and meat quality. *Meat Science* 36: 191–202.

Dickens, J.A., Lyon, C.E. and Wilson, R.L. (1991) Effect of ultrasonic radiation on some physical characteristics of broiler breast muscle and cooked meat. *Poultry Science* 70: 389–396.

Dukhin, A.S., Goetz, P.J. and Travers, B. (2005) Use of ultrasound to characterizing dairy products. *Journal of Dairy Science* 88: 1320–1334.

Elmehdi, H.M., Page, J.H. and Scanlon M.G. (2003) Monitoring dough fermentation using acoustic waves. *Food and Bioproducts Processing* 81: 217–223.

Elmehdi, H.M., Page, J.H. and Scanlon, M.G. (2004) Ultrasonic investigation of the effect of mixing under reduced pressure on the mechanical properties of bread dough. *Cereal Chemistry* 81: 504–510.

Elvira, L., Sampedro, L., Matesanz, J., Gómez-Ullate, Y., Resa, P., Iglesias, J.R., Echevarría, F.J. and Montero de Espinosa, F. (2005) Non-invasive and non-destructive ultrasonic technique for the detection of microbial contamination in packed UHT milk. *Food Research International* 38: 631–638.

Elvira, L., Sampedro, L., Montero de Espinosa, F., Matesanz, J., Gómez-Ullate, Y., Resa, P., Echevarría, F.J. and Iglesias, J.R. (2006) Eight-channel ultrasonic device for non-invasive quality evaluation in packed milk. *Ultrasonics* 45: 92–99.

Ertugay, M.F., Şengül, M. and Şengül, M. (2004) Effect of ultrasound treatment on milk homogenisation and particle size distribution of fat. *Turkish Journal of Veterinary and Animal Sciences* 28: 303–308.

Fairbanks, H.V. (1973) Use of ultrasound to increase filtration rate. In: *Proceedings of the Ultrasonics International Conference*, pp. 11–15.

Fernandes, F.A.N. and Rodrigues, S. (2007) Ultrasound as pre-treatment for drying of fruits: dehydration of banana. *Journal of Food Engineering* 82: 261–267.

Fernandes, F.A.N., Oliveira, F.I.P. and Rodrigues, S. (2007) Use of ultrasound for dehydration of papayas. *Food Bioprocess Technology* 1: 339–345.

Fernandes, F.A.N., Gallão, M.I. and Rodrigues, S. (2008). Effect of osmotic dehydration and ultrasound pre-treatment on cell structure: melon dehydration. *LWT – Food Science and Technology* 41: 604–610.

Floros, J.D. and Liang, H. (1994) Acoustically assisted diffusion through membranes and biomaterials. *Food Technology* 48: 79–84.

Fox, P., Smith, P.P. and Sahi, S.S. (2002) Buffer rod design for measurement of specific gravity in the processing of industrial food batters. *IEEE Ultrasonics Symposium* 1: 679–682.

Fox, P., Smith, P.P. and Sahi, S.S. (2004) Ultrasound measurements to monitor the specific gravity of food batters. *Journal of Food Engineering* 65: 317–324.

Fuente-Blanco, S., Riera-Franco, E., Acosta-Aparicio, V.M., Blanco-Blanco, A. and Gallego-Juárez, J.A. (2006) Food drying process by power ultrasound. *Ultrasonics* 44: e523–e527.

Gaikwad, S.G. and Pandit, A.B. (2008) Ultrasound emulsification: effect of ultrasonic and physicochemical properties on dispersed phase volume and droplet size. *Ultrasonics Sonochemistry* 15: 554–563.

Gallego-Juárez, J.A. (1990) Transducer needs for macrosonics. In: *Power Transducers for Sonics and Ultrasonics* (eds B. Hamonic and J.N. Decarpigny). Springer, Berlin, pp. 35–47.

Gallego-Juárez, J.A., Rodríguez-Corral, G., Gálvez-Moraleda, J.C. and Yang, T.S. (1999) A new high intensity ultrasonic technology for food dehydration. *Drying Technology* 17: 597–608.

Gallego-Juárez, J.A., Elvira-Segura, L. and Rodríguez-Corral, G. (2003) A power ultrasonic technology for deliquoring. *Ultrasonics* 41: 255–259.

Gallego-Juárez, J.A., Riera, E., Fuente-Blanco, S., Rodríguez-Corral, G., Acosta-Aparicio, V.M. and Blanco, A. (2007) Application of high-power ultrasound for dehydration of vegetables: processes and devices. *Drying Technology* 25: 1893–1901.

García-Álvarez, J., Rodriguez, J.M., Yanez, Y., Turo, A., Chavez, J.A., Garcia, M.J. and Salazar J. (2005) Study of the time-dependence of the mechanical properties of doughs for flour strength evaluation. In: *Ultrasonics Symposium* (ed. J.M. Rodriguez). IEEE, Piscataway, NJ, pp. 1480–1483.

García-Álvarez, J., Alava, J.M., Chavez, J.A., Turo, A., Garcia, M.J. and Salazar, J. (2006) Ultrasonic characterisation of flour–water systems: a new approach to investigate dough properties. *Ultrasonics* 44: e1051–e1055.

García-Pérez, J.V., Cárcel, J.A., Fuente-Blanco, S. and Riera-Franco, E. (2006) Ultrasonic drying of foodstuff in a fluidized bed: parametric study. *Ultrasonics* 44: e539–e543.

Ghaedian, R., Coupland, J.N., Decker, E.A. and McClements, D.J. (1998) Ultrasonic determination of fish composition. *Journal of Food Engineering* 35: 323–337.

Gómez, M., Oliete, B., García-Álvarez, J., Ronda, F. and Salazar, J. (2008) Characterization of cake batters by ultrasound measurements. *Journal of Food Engineering* 89: 408–413.

Got, F., Culioli, J., Berge, P., Vignon, X., Astruc, T., Quideau, J.M. and Lethiecq, M. (1999) Effects of high intensity high frequency ultrasound on ageing rate, ultrastructure and some physico-chemical properties of beef. *Meat Science* 51: 35–42.

Gould, G.W. (2001) New processing technologies: an overview. *Proceedings of the Nutritional Society* 60: 463–474.

Graves, M., Smith, A. and Batchelor, B. (1998) Approaches to foreign body detection in foods. *Trends in Food science & Technology* 9: 21–27.

Greenwood, M.S. and Bamberger, J.A. (2002) Measurement of viscosity and shear wave velocity of a liquid or slurry for on-line process control. *Ultrasonics* 39: 623–630.

Greenwood, M.S., Skorpik, J.R. and Bamberger, J.A. (1999) On-line sensor for density and viscosity measurement of a liquid or slurry for process control in the food industry. In: *Proceedings of the AIChE Sixth Conference on Food Engineering*, Dallas, TX, pp. 691–696.

Gunasekaran, S. and Ay, C. (1994) Evaluating milk coagulation with ultrasonics. *Food Technology* 48: 74–78.

Gunasekaran, S. and Ay, C. (1996). Milk coagulation cut-time determination using ultrasonics. *Journal of Food Process Engineering* 19: 63–73.

Hæggström, E. and Luukkala, M. (2001) Ultrasound detection and identification of foreign bodies in food products. *Food Control* 12: 37–45.

Harker, A.H. and Temple, J.A.G. (1988) Velocity and attenuation of ultrasound in suspensions of particles in fluids. *Journal of Physics D: Applied Physics* 21: 1576–1588.

Hauptmann, P., Hoppe, N. and Puettmer, A. (2001) Ultrasonic sensors for process industry. In: *Proceedings of the IEEE Ultrasonics Symposium*, pp. 369–378.

Higaki, K., Ueno, S., Koyano, T. and Sato, K. (2001) Effects of ultrasonic irradiation on crystallization behaviour of tripalmitoylglycerol and cocoa butter. *Journal of the American Oil Chemists' Society* 78: 513–518.

Jambrak, A.R., Mason, T.J., Paniwnyk, L. and Lelas, V. (2007) Accelerated drying of button mushrooms, Brussels sprouts and cauliflower by applying power ultrasound and its rehydration properties. *Journal of Food Engineering* 81: 88–97.

Jayasooriya, S.D., Bhandari, B.R., Torley, P. and D'Arcy, B.R. (2004) Effect of high power ultrasound waves on properties of meat: a review. *International Journal of Food Properties* 7: 301–319.

Jayasooriya, B.R., Torley, P.J., D'Arcy, B.R. and Bhandari, S.D. (2007) Effect of high power ultrasound and ageing on the physical properties of bovine *Semitendinosus* and *Longissimus* muscles. *Meat Science* 75: 628–639.

Jiranek, V., Grbin, P., Yap, A., Barnes, M. and Bates, D. (2008) High power ultrasonics as a novel tool offering new opportunities for managing wine microbiology. *Biotechnology Letters* 30: 1–6.

Kadkhodaee, R. and Povey, M.J.W (2008) Ultrasonic inactivation of Bacillus α-amylase. I. Effect of gas content and emitting face of probe. *Ultrasonics Sonochemistry* 15: 133–142.

Kaloyereas, S.A. (1955) Preliminary report on the effect of ultrasonic waves on the crystallisation of honey. *Science* 121: 339–340.

Kidmose, U., Pedersen, L. and Nielsen, M. (2001) Ultrasonics in evaluating rheological properties of dough from different wheat varieties and during ageing. *Journal of Texture Studies* 32: 321–324.

Kono, R. (1960) The dynamic bulk viscosity of polystyrene and polymethyl methacrylate. *Journal of the Physical Society of Japan* 15: 718–725.

Kuo, F.J., Sheng, C.T. and Ting, C.H. (2007) Velocity of ultrasound as an effective indicator of the sugar content and viscosity of watermelon juice. *Agricultural Engineering Journal* 16: 169–178.

Kuo, F.J., Sheng, C.T. and Ting, C.H. (2008) Evaluation of ultrasonic propagation to measure sugar content and viscosity of reconstituted orange juice. *Journal of Food Engineering* 86: 84–90.

Kyllönen, H.M., Pirkonen, P. and Nyström M. (2005) Membrane filtration enhanced by ultrasound: a review. *Desalination* 181: 319–335.

Lee, H.O., Luan, H. and Daut, D.G. (1992) Use of an ultrasonic technique to evaluate the rheological properties of cheese and dough. *Journal of Food Engineering* 16: 127–150.

Lee, S., Pyrak-Nolte, L.J. and Campanella, O. (2004a) Determination of ultrasonic-based rheological properties of dough during fermentation. *Journal of Texture Studies* 35: 33–52.

Lee, S., Pyrak-Nolte, L.J., Cornillon, P. and Campanella, O. (2004b) Characterisation of frozen orange juice by ultrasound and wavelet analysis. *Journal of the Science of Food and Agriculture* 84: 405–410.

Létang, C., Piau, M., Verdier, C. and Lefebvre, L. (2001) Characterization of wheat-flour–water doughs: a new method using ultrasound. *Ultrasonics* 39: 133–141.

Liebl, D.E. (1978) Ultrasound and granulation of honey. *American Bee Journal* 2: 107.

Llull, P., Simal, S., Femenia, A., Benedito, J. and Rosselló, C. (2002) The use of ultrasound velocity measurement to evaluate the textural properties of sobrassada from Mallorca. *Journal of Food Engineering* 52: 323–330.

López, P., Sánchez, A.C., Vercet, A. and Burgos, J. (1997) Thermal resistance of tomato polygalacturonase and pectinmethylesterase at physiological pH. *Zeitschrift für Lebensmittel-Untersuchung und -Forschung* 204: 146–150.

Lyng, J.G. and Allen, P. (1997) The influence of high intensity ultrasound bath on aspects of beef tenderness. *Journal of Muscle Foods* 8: 237–249.

Lyng, J.G., Allen, P. and McKenna, B.M. (1998a) The effect on aspects of beef tenderness of pre- and post-rigor exposure to a high intensity ultrasound probe. *Journal of the Science of Food and Agriculture* 78: 308–314.

Lyng, J.G., Allen, P. and McKenna, B.M. (1998b) The effects of *Pre-* and *Post-rigor* high-intensity ultrasound treatment on aspects of lamb tenderness. *Lebensmittel-Wissenschaft und-Technologie* 31: 334–338.

Lynnworth, L.C. (1989) *Ultrasonic Measurement for Process Control: Theory, Techniques and Applications*. Academic Press, New York.

Ma, Y., Ye, X., Hao, Y., Xu, G., Xu, G. and Liu, D. (2008) Ultrasound-assisted extraction of heperidin from penggan (*Citrus reticulata*) peel. *Ultrasonics Sonochemistry* 15: 227–232.

Mañas, P. and Pagán, R. (2005) Microbial inactivation by new technologies of food preservation. *Journal of Applied Microbiology* 98: 1387–1399.

Mañas, P., Pagan, R., Raso, J., Sala, F.J. and Condón, S. (2000) Inactivation of *Salmonella typhimurium* and *Salmonella senftenberg* by ultrasonic waves under pressure. *Journal of Food Protection* 63: 451–456.

Mañas, P., Muñoz, B., Sanz, D., Condón, S. (2006) Inactivation of lysozyme by ultrasonic waves under pressure at different temperatures. *Enzyme and Microbial Technology* 39: 1177–1182.

Mason, W. P. (1958) *Physical Acoustics and the Properties of Solids.* Van Nostrand, Princeton, NJ.

Mason, T.J. and Lorimer, J.P. (1988) *Sonochemistry: Theory, Applications and Uses of Ultrasound in Chemistry.* John Wiley & Sons, New York.

Mason, T.J., Paniwnyk, L. and Lorimer, J.P. (1996) The uses of ultrasound in food technology. *Ultrasonics Sonochemistry* 3: S253–S260.

Mathieu, J. and Schweitzer, P. (2004) Measurement of liquid density by ultrasound backscattering analysis. *Measurement Science and Technology* 15: 869–876.

McClements, D.J. (1995) Advances in the application of ultrasound in flood analysis and processing. *Trends in Food Science and Technology* 6: 293–299.

McClements, D.J. (1997) Ultrasonic characterization of foods and drinks: principles, methods, and applications. *Critical Reviews in Food Science and Nutrition* 37: 1–46.

McClements, D.J. and Povey, M.J.W. (1989) Scattering of ultrasound by emulsions. *Journal of Physics D: Applied Physics* 22: 38–47.

Mizrach, A. (2007) Nondestructive ultrasonic monitoring of tomato quality during shelf-life storage. *Postharvest Biology and Technology* 46: 271–274.

Mizrach, A. (2008) Ultrasonic technology for quality evaluation of fresh fruit and vegetables in pre- and postharvest processes. *Postharvest Biology and Technology* 48: 315–330.

Moulton, K.J. and Wang, L.C. (1982) A pilot-plant study of continuous ultrasonic extraction of soybean protein. *Journal of Food Science* 47: 1127–1129.

Nassar, G., Nongaillard, B. and Noel, Y. (2001) Monitoring of milk gelation using a low-frequency ultrasonic technique. *Journal of Food Engineering* 48: 351–359.

Neppiras, E.A. (1973) The prestressed piezoelectric sandwich transducer. In: *Proceedings of the Ultrasonics International Conference*, pp. 295–302.

Nomoto, O. (1958) Empirical formula for sound velocity in liquid mixtures. *Journal of the Physical Society of Japan* 13: 1528–1532.

Ogasawara, H., Mizutani, K., Ohbuchi, T. and Nakamura, T. (2006) Acoustical experiment of yogurt fermentation process. *Ultrasonics* 44: e727–e730.

Ordoñez, J.A., Sanz, B., Hernández, P.E. and López-Lorenzo, P. (1984) A note on the effect of combined ultrasonic and heat treatments on the survival of thermoduric streptococci. *Journal of Applied Bacteriology* 54: 175–177.

Orlandini, I. and S. Annibaldi (1983) New techniques in evaluation of the structure of Parmesan cheese: ultrasonic and X-rays. *Scienzia e Technica Latiero-Caseria* 34: 20–30.

Owolabi, G.M., Bassim, M.N., Page, J.H. and Scanlon, M.G. (2008) The influence of specific mechanical energy on the ultrasonic characteristics of extruded dough. *Journal of Food Engineering* 86: 202–206.

Ozutsumi, K., Nade, T., Watanabe, H., Tsujimoto, K., Aoki, Y. and Aso, H. (1996) Non-destructive, ultrasonic evaluation of meat quality in live Japanese black steers from coloured images produced by a new ultrasound scanner. *Meat Science* 43: 61–69.

Paniwynk, L., Beaufoy, E., Lorimer, J.P. and Mason, T.J. (2001) The extraction of rutin from flower buds of *Sophora japonica*. *Ultrasonics Sonochemistry* 8: 299–301.

Papadakis, E.P. (1975) *Ultrasonic Velocity and Attenuation: Measurement Methods with Scientific and Industrial Applications*, Vol. XI. Academic Press, New York.

Park, B., Whittaker, A.D., Miller, R.K. and Hale, D.S. (1994) Predicting intramuscular fat in beef longuissimus muscle from speed of sound. *Journal of Animal Science* 72: 109–116.

Pavlovskaya, G., McClements, D.J. and Povey, M.J.W. (1992) Ultrasonic investigation of aqueous solutions of a globular protein. *Food Hydrocolloids* 6: 253–262.

Piyasena, P., Mohareb, E. and McKellar, R.C. (2003) Inactivation of microbes using ultrasound: a review. *International Journal of Food Microbiology* 87: 207–216.

Povey, M.J.W. (1988) Ultrasonics as a probe of food emulsions and dispersions. In: *Advances in Food Emulsions and Foams* (eds E. Dickinson and G. Stainsby). Elsevier, London, pp. 285–327.

Povey, M.J.W. (1989) Ultrasonics in food engineering. Part II: applications. *Journal of Food Engineering* 9: 1–20.

Povey, M.J.W. (1997) *Ultrasonic Techniques for Fluids Characterization*. Academic Press, San Diego, CA.

Povey, M.J.W. and Harden, C.A. (1981) An application of the ultrasonic pulse echo technique to the measurement of crispness of biscuits. *Journal of Food Technology* 16: 167–175.

Povey, M.J.W. and Mason, T.J. (1998) *Ultrasound in Food Processing*. Blackie Academic and Professional, London.

Povey, M.J.W. and McClements, D.J. (1988) Ultrasonics in food engineering. Part I: introduction and experimental methods. *Journal of Food Engineering* 8: 217–245.

Püttmer, A., Hauptmann, P. and Henning, B. (2000) Ultrasonic density sensor for liquids. *IEEE Transactions on Ultrasonics, Ferroelectrics, and Frequency Control* 47: 85–92.

Raso, J., Palo, A., Pagan, R. and Condón, S. (1998) Inactivation of *Bacillus subtilis* spores by combining ultrasonic waves under pressure and mild heat treatment. *Journal of Applied Microbiology* 85: 849–854.

Raviyan, P., Zhang, Z. and Feng, H. (2005) Ultrasonication for tomato pectinmethylesterase inactivation: effect of cavitation intensity and temperature on inactivation. *Journal of Food Engineering* 70: 189–196.

Resa, P., Elvira, L. and Montero de Espinosa, F. (2004) Concentration control in alcoholic fermentation processes from ultrasonic velocity measurements. *Food Research International* 37: 587–594.

Resa, P., Elvira, L., Montero de Espinosa, F. and Gómez-Ullate, Y. (2005) Ultrasonic velocity in water–ethanol–sucrose mixtures during alcoholic fermentation. *Ultrasonics* 43: 247–252.

Resa, P., Bolumar, T., Elvira, L., Pérez, G. and Montero de Espinosa, F. (2007) Monitoring lactic acid fermentation in culture broth using ultrasonic velocity. *Journal of Food Engineering* 78: 1083–1091.

Riera, E., Golás, Y., Blanco, A., Gallego, J.A., Blasco, M. and Mulet, A. (2004) Mass transfer enhancement in supercritical fluids extraction by means of power ultrasound. *Ultrasonics Sonochemistry* 11: 241–244.

Rodrigues, S. and Pinto, G.A.S. (2007) Ultrasound extraction of phenolic compounds from coconut (*Cocos nucifera*) shell powder. *Journal of Food Engineering* 80: 869–872.

Roncalés, P., Ceña, P., Beltrán, J.A. and Jaime, L. (1993) Ultrasonication of Lamb skeletal muscle fibres enhances *post mortem* proteolysis. *Zeitschrift für Lebensmittel-Untersuchung und -Forschung* 196: 339–342.

Ross, K.A., Pyrak-Nolte, L.J. and Campanella, O.H. (2004) The use of ultrasound and shear oscillatory tests to characterize the effect of mixing time on the rheological properties of dough. *Food Research International* 37: 567–577.

Rosell, C.M., Marco, C.J., García-Alvárez, J. and Salazar, J. (2010) Rheological properties of rice–soybean protein composite flours assessed by Mixolab® and ultrasound. *Journal of Food Process Engineering* (available online).

Rostagno, M.A., Palma, M. and Barroso, C.G. (2003) Ultrasound-assisted extraction of soy isoflavones. *Journal of Chromatography A* 1012: 119–128.

Ruecroft, G., Hipkiss, D., Ly, T., Maxted, N. and Cains, P.W. (2005) Sonocrystallization: the use of ultrasound for improved industrial crystallization. *Organic Process Research & Development* 9: 923–932.

Salazar, J., Alava, J.M., Sahi, S.S., Turo, A., Chavez, J.A. and Garcia, M.J. (2002) Ultrasound measurements for determining rheological properties of flour–water systems. *Proceedings of the Ultrasonics Symposium* 1: 877–880.

Salazar, J., Turó, A., Chávez, J.A. and García, M.J. (2004) Ultrasonic inspection of batters for on-line process monitoring. *Ultrasonics* 42: 155–159.

Salazar, J., Chávez, J.A., Turó, A. and García-Hernández, M.J. (2010) Effect of ultrasound on food processing. In: *Novel Food Processing: Effects on Rheological and Functional Properties* (eds J. Ahmed, H.S. Ramaswamy, S. Kasapis and J.I. Boye). Taylor & Francis, New York, pp. 65–84.

Scanlon, M.G., Elmehdi, H.M. and Page, J.H. (2002) Probing gluten interactions with low-intensity ultrasound. In: *Wheat Quality Elucidation: The Bushuk Legacy* (eds P.K.W. Ng and C.W. Wrigley). AACC Press, St. Paul, MN, pp. 170–182.

Semmelink, A. (1973) Ultrasonically enhanced liquid filtering. In: *Proceedings of the Ultrasonics International Conference*, pp. 7–10.

Shoh, A. (1975) Industrial applications of ultrasound – a review. I. High-power ultrasound. *IEEE Transactions on Sonics and Ultrasonics* SU-22: 60–71.

Simal, S., Benedito, J., Clemente, G., Femenia, A. and Rosselló, C. (2003) Ultrasonic determination of the composition of a meat-based product. *Journal of Food Engineering* 58: 253–257.

Siró, I., Vén, C., Balla, C., Jónás, G., Zeke, I. and Friedrich, L. (2009) Application of an ultrasonic assisted curing technique for improving the diffusion of sodium chloride in porcine meat. *Journal of Food Engineering* 91: 353–362.

Smith, N.B., Cannon, J.E., Novakofski, J.E., McKeith, F.K. and O'Brien, W.D. (1991) Tenderization of semitendinosus muscle using high intensity ultrasound. In: *Ultrasonics Symposium*, pp. 1371–1373.

Stadnik, J., Dolatowski, Z.J. and Baranowska, H.M. (2008) Effect of ultrasound treatment on water holding properties and microstructure of beef (*m. semimembranosus*) during ageing. *LWT – Food Science and Technology* 41: 2151–2158.

Suslick, K.S. and Price, G.J. (1999) Applications of ultrasound to materials chemistry. *Annual Review of Materials Science* 29: 295–326.

Suvanich, V., Ghaedian, R., Decker, E.A. and McClements, D.J. (1998) Prediction of proximate fish composition from ultrasonic properties: catfish, cod, flounder, mackerel and salmon. *Journal of Food Science* 63: 966–968.

Tarleton, E.S. and Wakeman, R.J. (1998) Ultrasonically assisted separation process. In: *Ultrasound in Food Processing* (eds M.J.W. Povey and T.M. Mason). Blackie Academic and Professional, London, pp. 193–218.

Thakur, B.R. and Nelson, P.E. (1997) Inactivation of lipoxygenase in whole soy flour suspension by ultrasonic cavitation. *Die Nahrung* 41: 299–301.

Thompson, L.H. and Doraiswamy, L.K. (1999) Sonochemistry: science and engineering. *Industrial & Engineering Chemistry Research* 38: 1215–1249.

Thornycroft, J. and Barnaby S.W. (1895) Torpedo boat destroyers. *Proceedings of the Institution of Civil Engineers* 122: 51–103.

Thrasyvoulou, A., Manikis, J. and Tselios, D. (1994) Liquefying crystallized honey with ultrasonic waves. *Apidologie* 25: 297–302.

Ueno, S., Ristic, R.I., Higaki, K. and Sato, K. (2003) *In situ* studies of ultrasound-stimulated fat crystallization using synchrotron radiation. *Journal of Physical Chemistry B* 107: 4927–4935.

Urick, J.R. (1947) A sound velocity method for determining the compressibility of finely divided substances, *Journal of Applied Physics*, 18: 983–987.

Van Sint Jan, M., Guarini, M., Guesalaga, A., Pérez-Correa, J.R. and Vargas, Y. (2008) Ultrasound based measurements of sugar and ethanol concentrations in hydroalcoholic solutions. *Food Control* 19: 31–35.

Vilkhu, K., Mawson, R., Simons, L. and Bates, D. (2008) Applications and opportunities for ultrasound assisted extraction in the food industry – a review. *Innovative Food Science and Emerging Technologies* 9: 161–169.

Villamiel, M. and Jong, P. de (2000) Influence of high-intensity ultrasound and heat treatment in continuous flow on fat, proteins, and native enzymes of milk. *Journal of Agriculture and Food Chemistry* 48: 472–478.

Vinatoru, M. (2001) An overview of the ultrasonically assisted extraction of bioactive principles from herbs. *Ultrasonics Sonochemistry* 8: 303–313.

Wang, L.C. (1975) Ultrasonic extraction of proteins from autoclaved soybean flakes. *Journal of Food Science* 40: 549–551.

Wang, Q., Bulca, S. and Kulozik, U. (2007) A comparison of low-intensity ultrasound and oscillating rheology to assess the renneting properties of casein solutions after UHT heat pre-treatment. *International Dairy Journal* 17: 50–58.

Weber, F. (1987) Curd drainage. In: *Cheesemaking: Science and Technology* (ed. A. Eck). Lavoisier, New York, pp. 22–25.

Whittaker, A.D., Park, B., Thane, B.R., Miller, R.K. and Savell, J.W. (1992) Principles of ultrasound and measurement of intramuscular fat. *Journal of Animal Science* 70: 942–952.

Winder, W.C., Aulik, D.J. and Rice, A.C. (1970) An ultrasonic method for direct and simultaneous determination of alcohol and extract content of wines. *American Journal of Enology and Viticulture* 21: 1–11.

Wood, R.W. and Loomis, A.L. (1927) The physical and biological effects of high frequency sound waves of great intensity. *Philosophical Magazine* 4: 417–436.

Wu, H., Hulbert, G.J. and Mount, J.R. (2001) Effects of ultrasound on milk homogenization and fermentation with yogurt starter. *Innovative Food Science & Emerging Technologies* 1: 211–218.

Zhao, B., Basir, O.A. and Mittal, G.S. (2003a) Detection of metal, glass and plastic pieces in bottled beverages using ultrasound. *Food Research International* 36: 513–521.

Zhao, B., Basir, O.A. and Mittal, G.S. (2003b) Correlation analysis between beverage apparent viscosity and ultrasound velocity. *International Journal of Food Properties* 6: 443–448.

Zhao, B., Basir, O.A. and Mittal, G.S. (2004) Conceptual design and evaluation of an ultrasonic system for bottled beverages' quality measurement and control. *Transactions of the Institute of Measurement and Control* 26: 41–54.

39

Process Design Involving Pulsed UV Light

Ali Demirci and Nene M. Keklik

Introduction

Every year, millions of people suffer from infectious diseases due to the consumption of contaminated food and water. Food-borne illnesses caused by pathogenic microorganisms lead not only to significant health issues and deaths, but also to huge economic losses for society. Moreover, the safety of food materials, which is the most important indicator of their quality, is becoming more challenging as the human population grows. The responsibility for ensuring food safety is shared mainly between governments and the food industry. While governments make the necessary regulations for food manufacturers and organize educational programs for consumers, the food industry follows procedures to ensure the safety of their products. These procedures involve decontamination techniques, which are used to eliminate pathogenic microorganisms in foods while retaining the quality of the foods.

Many new and innovative techniques to decontaminate food products have been developed and evaluated by researchers; some of these have been integrated into food processing plants and are currently in operation. Most of the decontamination techniques used in food processing plants include thermal and chemical treatment of the foods. Thermal treatments are highly effective against microorganisms in foods; however, they cannot be used for minimally processed or raw foods, as excessive

Handbook of Food Process Design, First Edition. Edited by Jasim Ahmed and Mohammad Shafiur Rahman.

heating causes cooking and loss of sensory attributes. On the other hand, chemical methods of treating foods may leave toxic residues.

Because of these issues, researchers and the food industry are continuing to seek alternative nonthermal and nonchemical techniques, by which a variety of foods, including minimally processed and raw products, could be decontaminated without the use of potentially toxic chemicals. Irradiation, high hydrostatic pressure, pulsed electric fields, ultrasound, ultraviolet (UV) light and pulsed UV light are among the alternative techniques that have been developed and have been evaluated for their efficacies on various foods. UV light technology is a nonchemical, nonthermal, simple, inexpensive approach to disinfection, which requires very low maintenance. UV light is defined as the electromagnetic radiation in the spectral region from 100 to 400 nm and is classified into four wavelength ranges: UV-A (315–400 nm), UV-B (280–315 nm), UV-C (200–280 nm) and vacuum UV (100–200 nm) (Krishnamurthy *et al.*, 2008b). UV-C light damages the DNA of microorganisms by causing the formation of lethal photoproducts and pyrimidine dimers in the microbial DNA (Miller *et al.*, 1999; Rupp, 2006; Xenon, 2006a). The formation of such bonds prevents the microbial cell from reproducing by blocking DNA replication.

Pulsed UV light technology, on the other hand, is a novel technology utilizing UV light, and is more efficient and effective than conventional UV light (Bialka, 2006). Pulsed UV light is also nonchemical, and is considered to be nonthermal for short treatment times. Microbial inactivation by intense UV light pulses using inert-gas flashlamps was first used in Japan during the late 1970s (Oms-Oliu *et al.*, 2010). In the USA, the FDA approved pulsed-light treatment of foods in 1999 (Federal Register, 1999). In a pulsed UV light system, an energized xenon gas lamp emits intense light pulses with a continuum broadband spectrum from the deep UV to the infrared; the spectrum is especially rich and efficient in the UV range (Demirci and Panico, 2008). The duration of a pulse is as short as a few hundred microseconds. In this chapter, we focus on the use of pulsed UV light as a potential technology to decontaminate food products, including raw and minimally processed foods. Various scientific and technical aspects of this promising technology are discussed to help the reader to better understand the technology.

End Products of the Process

The inactivation of microorganisms by UV light is caused mainly by photochemical reactions that take place when photons in the UV range are absorbed by microbial constituents. UV photons, the elementary particles of UV light, carry a high amount of energy, and have the ability to break or form chemical bonds. In the case of microbial cells, the chemical reactions induced by the photons occur mostly in the structure of proteins and DNA (Bank *et al.*, 1990; Blatchley and Peel, 2001; Jay et al., 2005; Miller *et al.*, 1999). UV light induces breakage of DNA chain, crosslinking of DNA strands, hydration of pyrimidines and formation of pyrimidine dimers between adjacent residues. The specific products of these photochemical reactions include cyclobutyl pyrimidine dimers, pyrimidine pyrimidinone-[6,4] photoproducts, Dewar pyrimidinone, adenine–thymine heterodimers, cytosine photohydrate, thymine photohydrate, single-strand breaks and DNA–protein crosslinks (Krishnamurthy *et al.*, 2008b).

Furthermore, the inactivation of microorganisms by pulsed UV light, which involves polychromatic radiation in a spectral region from the UV to the infrared, is not limited to photochemical damage to microbial DNA. Besides this photochemical effect, pulsed UV light is suggested to have photothermal and photophysical effects on microbial cells (Fine and Gervais, 2004; Krishnamurthy *et al.*, 2007; Wekhof, 2003). Wekhof (2003) and Fine and Gervais (2004) suggested that pulsed UV light may have a thermal effect on microorganisms. The thermal stress due to the difference between the absorption of pulsed UV light by a microorganism and by the surrounding medium causes vaporization of water in the bacterial cell, leading to rupture of the cell; this is induced mainly by the UV components of the light pulse (a photothermal effect).

Krishnamurthy *et al.* (2010) studied the damage caused by pulsed UV light treatment to *Staphylococcus aureus* using microspectrometry and transmission electron microscopy. After 5 s treatment with pulsed UV light, severe damage to *S. aureus* cells was observed by transmission electron microscopy. Cell wall damage and leakage of the cell contents occurred at several locations (Figure 39.1).

Furthermore, shrinkage of the cytoplasmic membrane and collapse of the internal cellular structure was observed. The semipermeability of the membrane was lost, and thus the osmotic equilibrium of the cell was damaged. As a result, cellular contents leaked from the cytoplasm and cell death occurred (Figure 39.2). It was also suggested that the intermittent high-intensity pulses may have caused disturbance to the cell structure, which eventually damaged the cells (a photophysical effect). This study also

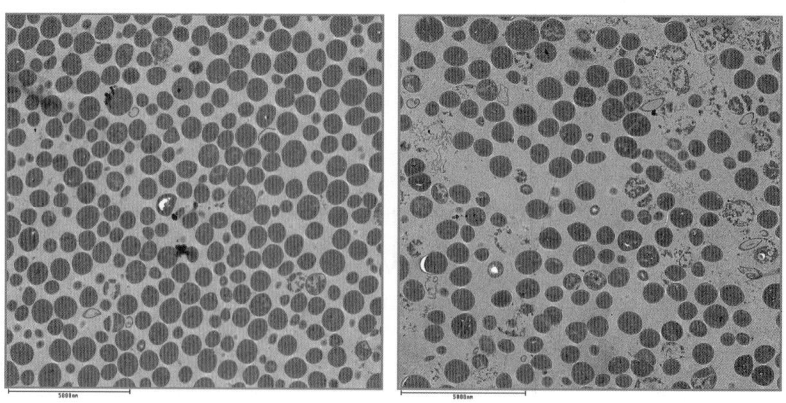

Figure 39.1 Comparison of cellular damage observed by transmission electron microscopy: control sample (left) and pulsed-UV-light-treated sample (right). (From Krishnamurthy et al., (2010), courtesy of Springer.)

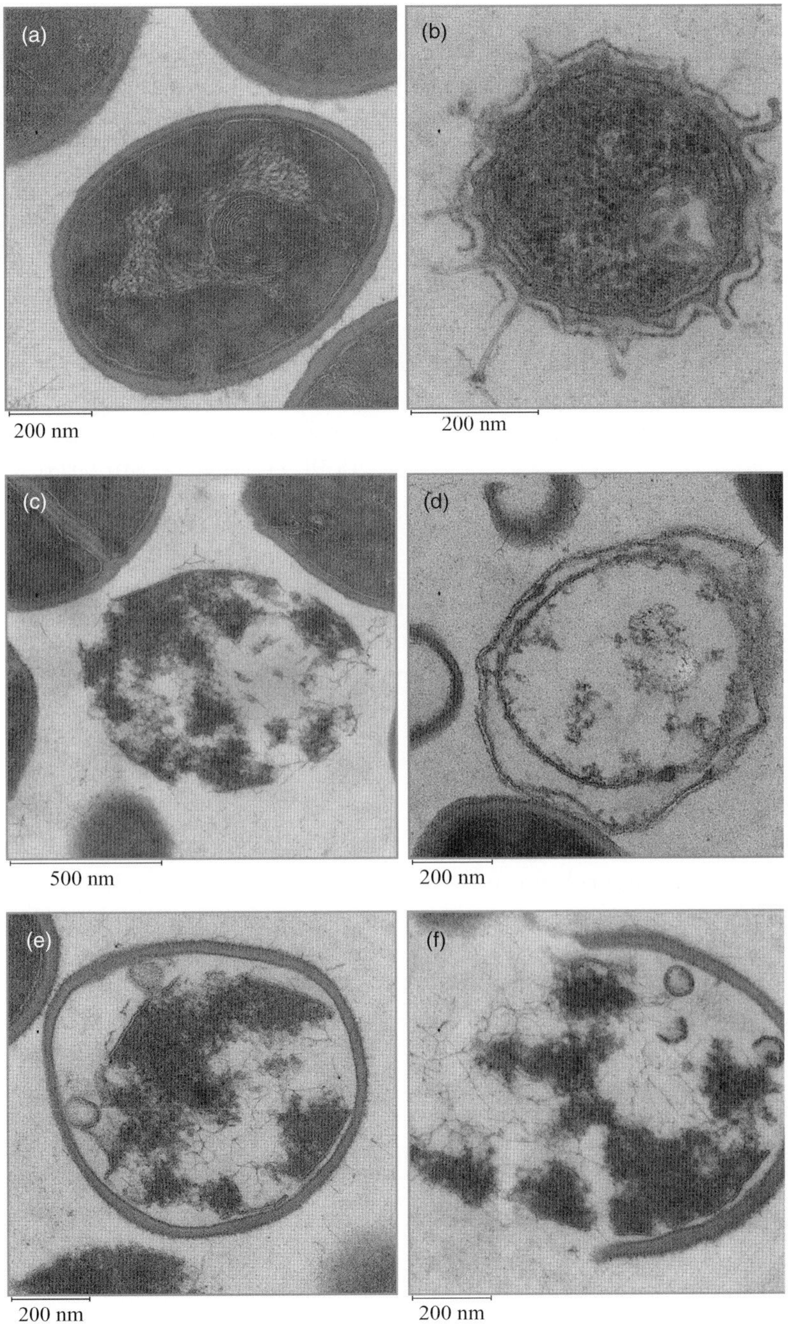

Figure 39.2 Effects of pulsed UV light treatment on *S. aureus* after 5 s. (From Krishnamurthy et al., (2010), courtesy of Springer.)

confirmed that some of the cells were inactivated without any structural damage, indicating the formation of lethal thymine dimers.

Pulsed UV light may also induce other biochemical changes in the target object in addition to microbial inactivation. In photochemical reactions, chemical bonds are dissociated when a photon's wavelength-dependent quantum energy is at least equal to the energy of the molecular bond upon which the light is incident (Xenon, 2006a). Having high photon energy, UV light is able to dissociate a number of chemical bonds, giving rise to new chemical species (Krishnamurthy *et al.*, 2008b). Having high intensity and a polychromatic nature, pulsed UV light may promote the formation of secondary radical reactions, causing the formation of germicidal chemicals such as hydroxyl radicals, ozone or hydrogen peroxide (Malley, 2002).

It has been reported that pulsed UV light at high intensities may induce lipid oxidation and color changes in foods (Gould, 2001; Keklik *et al.*, 2009), as well as alterations in plastic packaging materials (Keklik *et al.*, 2009). However, compared with continuous UV light, the oxidation reactions are limited in pulsed light systems because of the short pulse duration (e.g., 100 μs), the half-life of π-bonds (10^{-9} to 10^{-4} s), and the low number of pulses (1 to 3) (Fine and Gervais, 2004). In another study (Elmnasser *et al.*, 2008), treatment of milk proteins and lipids with pulsed UV light (2.2 J·cm^{-2}, 1–10 pulses) did not cause significant oxidation products, changes in the amino acid composition or changes in the protein conformation.

Process Components

Basically, in a pulsed light system, high-power electrical pulses are generated and converted into high-power light pulses (Figure 39.3). Firstly, line low-voltage AC power collected from a primary energy source is converted into high-voltage DC power by a power supply unit. Electrical energy is then accumulated and temporarily stored in capacitors (Palmieri and Cacace, 2005). The capacitor, together with an inductor, defines the energy and pulse profile of the system. An inert-gas arc lamp is used, where the gas is essentially insulating. A trigger electrode is used to build up an electrical

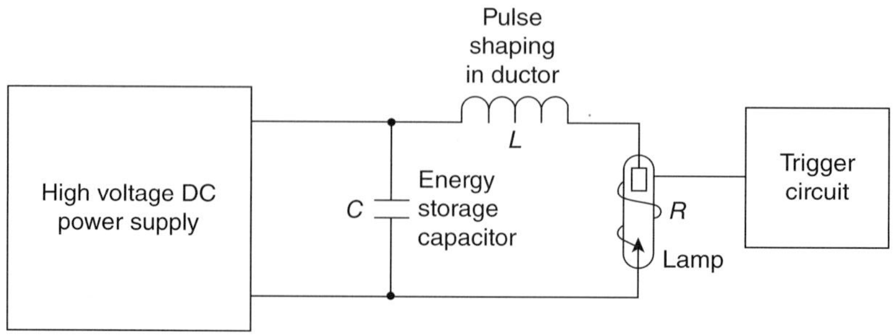

Figure 39.3 Flow diagram of a typical pulsed light system. (Courtesy of Xenon, 2008.)

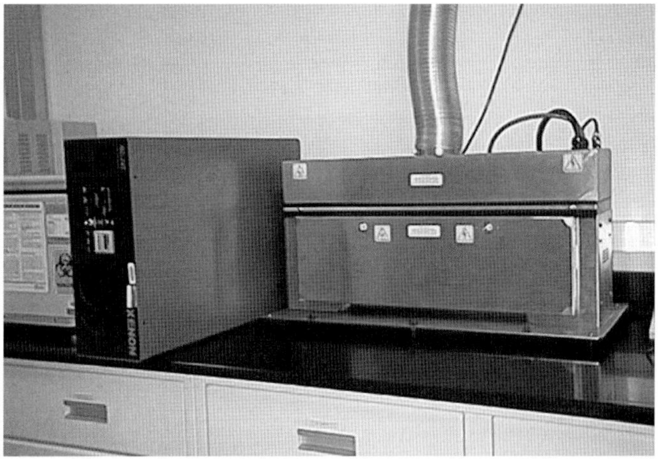

Figure 39.4 Laboratory-scale pulsed UV light system. (Xenon Corporation®, Wilmington, MA.)

potential inside the lamp. As this potential is increased, the gas inside the lamp starts to break down and becomes conducting. As energy flows from the capacitor though the lamp in this state, high-energy light is generated in the form of a glow discharge.

Pulsed light sources use power-compression technology to transfer stored electrical energy into short pulses with high peak power. For instance, a pulsed power source operating at 1 kV dissipates 20 J of energy stored in a 40 μF capacitor within 30 μs, creating a peak power of 1 MW (Wang *et al.*, 2005). Figure 39.4 shows a laboratory-scale pulsed UV light system, where the food item is treated in a chamber by pulsed UV light produced by a xenon gas lamp. Figure 39.5 shows a schematic diagram of this laboratory-scale system with its components.

A pulsed UV lamp delivers electromagnetic radiation in the range from 100 to 1100 nm (~54% of the light is in the UV range) (Krishnamurthy *et al.*, 2010), whereas continuous UV lamps deliver UV light in the range from 200 to 300 nm (Linden *et al.*, 2007). Pulsed UV light dissipates a much higher peak energy in a short time than does continuous UV light, which emits essentially constant energy over a longer time period. The inactivation efficiency of pulsed UV light is four times higher than that of continuous UV light (Bialka, 2006).

Basic Theoretical Principles and Mode of Operation

The light pulses emitted by a pulsed UV lamp consist of electromagnetic radiation in the wavelength range from 100 to 1100 nm, which includes the UV, visible and infrared spectra (Oms-Oliu *et al*, 2010). The UV region is responsible for most of the inactivation caused by pulsed light. UV-C light induces chemical reactions in DNA.

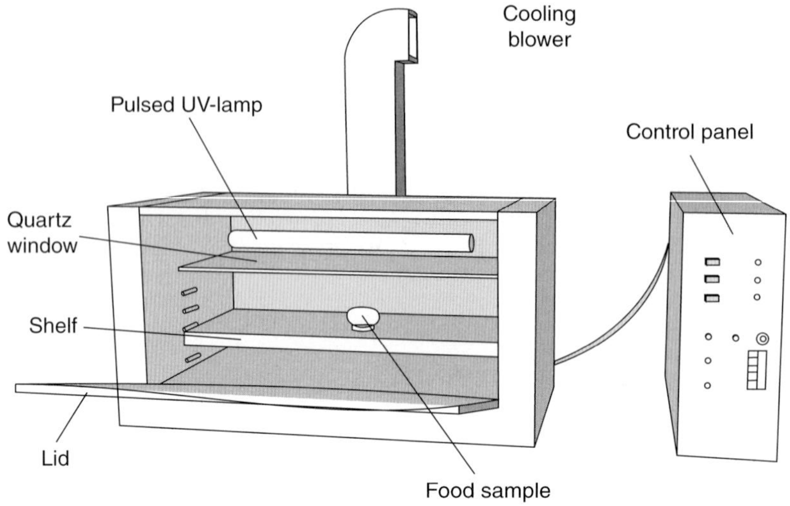

Figure 39.5 Schematic diagram of the laboratory-scale pulsed UV light system shown in Figure 39.4 (Keklik, 2009).

On the other hand, the entire UV light incident on the target is thought to be responsible for photothermal effects (Wekhof, 2000), while the photophysical effect may be due to the entire spectrum, including the visible and infrared portions (Krishnamurthy *et al.*, 2010). The inactivation efficacy of pulsed light is dependent mainly on the light intensity (measured in $J{\cdot}cm^{-2}$) and the number of pulses or treatment time (Elmnasser *et al.*, 2007). The pulse duration varies between 100 ns and 2 ms (Demirci and Panico, 2008), and the lamp pulses at a rate of about 1–20 times per second (Dunn, 1996). Pulsed light can be used to disinfect various surfaces, including packaging materials, packaging and processing equipment, foods, and medical devices, as well as bulk materials that allow penetrations of the light, such as water and air (Dunn, 1996).

The operation of a pulsed UV light system can be either batch-type or continuous-flow, and may be designed for disinfecting either food surfaces or liquid foods. A batch-type pulsed UV light system consists of a power unit, and a chamber in which one food item is treated at a time by one or more xenon gas lamps. An example of a batch-type pulsed UV light system is the lab-scale system that was shown in Figure 39.4. A continuous pulsed UV light system (in a real-world case) would involve continuous-flow exposure of food items to pulsed UV light on a moving platform; either the food items or the lamps would be moved by conveyors at a preset conveyor velocity, to achieve a preset treatment time. A pulsed UV light system designed for the treatment of liquid foods would be different from a system for solid foods; the liquid food, flowing in a transparent pipe, would be exposed to the pulsed light (Krishnamurthy *et al.*, 2007). The treatment time would be dependent on the flow rate of the liquid. In any pulsed UV light system, the distance between the food and the lamp surface needs to be adjusted so that the desired level of the light intensity at the food surface is attained, to maximize the microbial inactivation while retaining the quality of the food.

Equipment (Advantages and Limitations) and Parameters

A basic pulsed light system consists of three main parts: a lamp, a power supply and a pulse configuration device (controller) (Elmnasser *et al.*, 2007). However, the specifications of equipment can vary widely to meet the needs of particular manufacturing operations. There are two types of pulsed lamps: the flashlamp and the surface discharge lamp (Bohrerova *et al.*, 2008). In a flashlamp, the discharge is confined in a small envelope, whereas a surface discharge lamp uses a plasma discharge on the outside of a dielectric material placed inside a wide tube. Surface discharge lamps are able to produce higher UV efficiency and tolerate stress levels above the explosion limit of flashlamps (Schaefer *et al.*, 2007). The choice of a lamp with a specific spectral cutoff that produces the desired wavelength properties helps in controlling issues such as ozone production (Xenon, 2007b). Also, power supplies offering different energy levels are available to optimize the performance and the cost efficiency. Configuration of the pulse energy and rate, which is critical in providing the optimal amount of light received by the target, is managed by the controller. An interweave system and timer control are among the options available for a controller. In an interweave system, two lamps flashing at a high repetition rate are operated from a single power supply and controller to increase the delivery of UV light to the target, which is ideal for continuous processes. The controller can be incorporated together with the high-voltage power supply into a cabinet, which provides easy setup and practical operation from a front-panel control (Xenon, 2007b).

Pulsed UV light is more advantageous than continuous-wave UV light in terms of the lamps used to generate the light, the wavelength of the light delivered by the lamps, the light intensity, the inactivation mechanisms, the processing time and the penetration ability (Bialka, 2006). Pulsed UV light is typically produced by a xenon gas lamp, whereas continuous UV light is commonly produced by a mercury lamp (Xenon, 2006b). Mercury can be hazardous to humans in case of leakage due to breakage of the lamp, particularly in food applications. The inactivation mechanisms of pulsed UV light involve photochemical, photothermal and photophysical effects (Krishnamurthy *et al.*, 2010), whereas continuous UV light inactivates microorganisms by photochemical effects only (Bialka, 2006). Pulsed UV light requires shorter processing times compared with continuous UV light, which contributes to the efficiency by increasing the process speed. In addition, pulsed UV light works instantly, i.e., it does not require a long warm-up period as continuous UV light does. Owing to the high peak power delivered, pulsed UV light can penetrate deeper into food than can continuous UV light (Xenon, 2006b). However, there are two main concerns about pulsed UV light technology: heat generation and limited penetration of the light. Because the temperature rise is negligible (i.e., less than 1 °C) for short processing times (a few seconds), this process is considered to be a nonthermal process. However, for long processing times, heat is generated owing to the infrared portion of the pulsed UV light, which results in a temperature increase in the food product. The other concern is the limited penetration of the UV light. Although pulsed UV light penetrates deeper than continuous UV light (Xenon, 2007a), the penetration of pulsed UV light into a food may still be limited. An important factor affecting light penetration is the color of the food. If the food is transparent, the amount of UV light absorbed by the food becomes much more than that absorbed by an opaque food. In a study by Bialka *et al.* (2008a), the effects of various parameters, including the depth of the material, on the inactivation of *E. coli* K12 were investigated. Two different materials, agar and denatured whey protein isolate (DWPI), were used as model systems. It was demonstrated that the energy transmitted through the gel decreased by 30% when the thickness of agar was increased from 2 to 10 mm, whereas the decrease in the energy transmitted was 45% for the same thicknesses of DWPI; this indicates that pulsed UV light penetrates better in a transparent food than in an opaque food. It was also found that pulsed UV light was able to penetrate several millimeters into an opaque solid medium (10% DWPI).

The spectral distribution and the total fluence (intensity) of the light pulses are the most important "source" parameters in a pulsed-light treatment; both of these need to be adjusted according to the type of food and the desired level of microbial inactivation (Palmieri and Cacace, 2005). The pulse duration and frequency are also significant parameters in determining the best treatment time. The most important "target" parameter is the transparency of the food item, which needs to be as high as possible to increase the efficiency of microbial inactivation in deeper layers. For instance, water is highly transparent in the wavelength range of visible and UV light, whereas sugar solutions and wines display less transparency (Palmieri and Cacace, 2005). Also, it must be noted that the intricate shape of a food item may have a shadowing effect, preventing the pulsed UV light from reaching microorganisms present in cavities in the food.

Empirical Data and Rules of Thumb

To manage the microbial safety of foods by using pulsed UV light, it is essential to understand the behavior of microbial cells toward their environment during pulsed UV light processing. In other words, without knowing how the environmental factors affect the trends of microbial inactivation by pulsed UV light, one cannot perform reliable design calculations for a pulsed UV light unit. Thus, several researchers have investigated the trends of microbial survival during pulsed UV light treatment of foods, and developed empirical models to accurately predict the consequences of pulsed UV light treatment for the inactivation of pathogens.

In a study by Sharma and Demirci (2003), the effects of treatment time, distance from the UV strobe and the thickness of the product on the inactivation of *E. coli* O157:H7 on alfalfa seeds were investigated. Empirical models were developed from the collected data to predict the reduction in the bacterial population. Response surface regression analysis with uncoded units was performed to determine the regression coefficients using the general equation

$$\log_{10} \text{reduction} = \alpha_0 + \alpha_1 x_1 + \alpha_2 x_2 + \alpha_3 x_1^2 + \alpha_4 x_2^2 + \alpha_5 x_1 x_2 \tag{39.1}$$

where α_0 is a constant term, $\alpha_{1,\ldots5}$ are coefficients, x_1 is the time of treatment and x_2 is the thickness of the seed layer or the distance from the UV strobe. Based on Equation 39.1, two models, named the "thickness model" and the "distance model," were developed and validated for the treatment of alfalfa seeds with pulsed UV light.

The "thickness model" was given by

$$\log_{10} \text{reduction} = 3.498 + 0.049t - 1.883H + 0.216H^2 - 0.008H \cdot t \tag{39.2}$$

where H is the thickness of the seed layer (mm) and t is the treatment time (s). Analysis of variance of the $\log_{10}$ reduction given by Equation 39.2 demonstrated that both the thickness and the time had a significant effect ($p < 0.05$) on the prediction ability of the model, although the interaction between H and t was unclear.

The "distance model," which was the second model developed and validated in the study, was given by

$$\log_{10} \text{reduction} = -0.5485 + 0.0472t + 0.0826D + 0.0006t^2 + 0.0012D^2 - 0.0065Dt \tag{39.3}$$

where D is the distance from the UV strobe (cm) and t is the treatment time (s). All of the terms in the model in Equation 39.3 were significant ($p < 0.05$). This study demonstrated that shorter distances, longer treatment times (higher number of pulses) and smaller thicknesses resulted in higher inactivation of *E. coli* O157:H7 on alfalfa seeds. Under ideal conditions (such as a higher initial bacterial population and no thermal effects of the pulsed UV light), a $\log_{10}$ reduction in *E. coli* O157:H7 by up to 8 in terms of $\text{CFU}\cdot\text{g}^{-1}$ could be achieved.

In another study (Bialka *et al.*, 2008b), two inactivation models, namely a log–linear model and a Weibull model, were constructed to describe the inactivation *of E. coli* O157:H7 and *S. enterica* on raspberries and strawberries treated with pulsed UV light. First-order kinetics is traditionally used to describe microbial inactivation, based on the fact that enzyme inactivation is mostly governed by first-order kinetics (Van Boekel, 2002). The log–linear model represents first-order kinetics, and follows Equation 39.4 (Bialka *et al.*, 2008b):

$$\log\left(\frac{N}{N_0}\right) = -kt = -\frac{t}{D} \tag{39.4}$$

where N is the number of microorganisms at time t (CFU·g^{-1}), N_0 is the initial number of microorganisms (CFU·g^{-1}), t is the treatment time (s or min), k is the first-order extinction coefficient (s^{-1} or min^{-1}), and D is the decimal reduction time (s or min).

The second model used in the study by Bialka *et al.* (2008b) was the Weibull model, which is a distribution function generally utilized in reliability engineering for the description of the time-to-failure of electronic and mechanical systems (Van Boekel, 2002). The Weibull model is represented in terms of a survival curve by the following equation:

$$\log\left(\frac{N}{N_0}\right) = -\frac{1}{2.303}\left(\frac{t}{\alpha}\right)^{\beta} \tag{39.5}$$

where N is the number of microorganisms after a pulsed UV dose (CFU·g^{-1}), N_0 is the initial number of microorganisms (CFU·g^{-1}), t is the treatment time (s or min), α is a characteristic time (s or min), and β is a dimensionless shape parameter. The study of Bialka *et al.* (2008b) demonstrated that the Weibull model estimated the inactivation more accurately than did the log–linear model; this is due to the presence of concavity in the survival curves. Although the use of first-order kinetics is a common approach to the prediction of microbial inactivation, many survival curves do not follow first-order kinetics. These curves often have upward or downward concavity, a lag time or shoulders (Van Boekel, 2002).

The Weibull model, which is a statistical model that provides an alternative to the log–linear model, takes into account the nonlinearity of semilogarithmic survival curves. The parameter β describes the concavity, and the survival curve exhibits upward concavity when $\beta < 1$, and downward concavity when $\beta > 1$ (Van Boekel, 2002). The reliable life t_R, which is the 90th percentile of the failure time distribution, is analogous to the D-value of the log–linear equation:

$$t_R = \alpha * (2.303)^{1/\beta} \tag{39.6}$$

In another study by Bialka *et al.* (2008a), the Weibull model was used to describe the inactivation of *E. coli* K12 as a function of the energy dose, treatment time, and material depth. Two different materials, agar and denatured whey protein isolate (DWPI), were used as model systems. A model to estimate the microbial inactivation was obtained, given by

$$\log\left(\frac{N}{N_0}\right) = -\frac{1}{2.303}\left(\frac{I_0 e^{-kx} * t}{\alpha}\right)^{\beta} \tag{39.7}$$

where k is the extinction coefficient (mm^{-1}) and x is the material depth (mm). The R^2 and RMSE values of this model were 0.84 and 0.48 for agar and 0.81 and 0.13 for DWPI, respectively.

Estimation of the Design Parameters

The lethal action of pulsed UV light on microbial cells increases with increasing amounts of energy absorbed by the food. Therefore, it is a priority to adjust the energy dose delivered to the food to optimize the inactivation process. If the energy of one pulse is E_p ($J \cdot cm^{-2}$), the total dose D of light energy ($J \cdot cm^{-2}$) can be calculated as follows (Luksiene *et al.*, 2007):

$$D = E_p * t * f \tag{39.8}$$

where t is the treatment time (s) and f is the pulse frequency (Hz). In another study (Bialka *et al.*, 2008a), the treatment time t was replaced with the energy dose D in the Weibull model (Equation 39.9), and was given as follows:

$$\log\left(\frac{N}{N_0}\right) = -\frac{1}{2.303}\left(\frac{D}{\alpha}\right)^{\beta} \tag{39.9}$$

where the energy dose D was defined in an additional model as

$$D = I_0 e^{-kx} * t \tag{39.10}$$

Here, D is the energy dose ($J \cdot cm^{-2}$), I_0 is the initial intensity at the top of the sample ($J \cdot cm^{-2} \cdot s^{-1}$), k is the extinction coefficient (mm^{-1}), x is the material depth (mm), and t is the treatment time (s).

A useful approach to optimizing the energy dose was summarized by Wekhof (1991), where calculations of power ratios were offered to aid in optimization of the destruction of toxic organics in water, air and soil by UV flashlamps. Here, the peak power P_p is defined as the ratio between the energy of a single pulse E_p and the pulse duration τ:

$$P_p = \frac{E_p}{\tau} \tag{39.11}$$

The average power P_a gives the combined energy of all pulses delivered per second at the pulse frequency f:

$$P_a = E_p * f \tag{39.12}$$

The characteristic ratio of peak power to average power is in the range of 1000:1 to 10000:1 for a pulsed lamp. The most effective power ratio, i.e., the power ratio at which the peak wavelength of the UV generated is in the region of the targeted absorption bands, needs to be determined experimentally for each particular toxic organic that is to be destroyed (Wekhof, 1991).

Process Control, Operations and Maintenance

In pulsed UV light processing, the parameters that need to be monitored if the system is to work properly include the magnitude of the voltage supplied to the system, the pulse duration, the number of pulses per second (the pulse frequency), the intensity of light from the lamp (which depends on the condition of the lamp) and the treatment time (or the speed of the conveyor that carries the food in front of the pulsed UV lamp in the case of a continuous pulsed UV light system).

The magnitude of the voltage supplied to the system is important in terms of delivering the desired light intensity to the food. If the voltage supply is insufficient, then a lower light intensity will be delivered to the food, which may result in lower inactivation in that period of treatment time. The pulse duration and pulse frequency play a role in obtaining the best power ratio P_p/P_a, i.e., the value that maximizes the efficiency of microbial destruction. This is not surprising, since the total treatment time required to sufficiently inactivate microorganisms depends on the total number of pulses. If the pulse frequency changes accidentally during the treatment, the food will be either undertreated or overtreated in the prespecified treatment time. The intensity of light absorbed at the distance from the lamp at which the food is placed is another factor that needs to be monitored to ensure the required inactivation of microorganisms while keeping good quality of the food. Although sufficient voltage may be supplied to the system, the light intensity may still be lower or higher than required because of the condition of the lamp (e.g., because of dirt or age or because it is broken). Furthermore, the positions of the lamps in a pulsed UV light system need to be adjusted so that they are at a distance and angle with respect to the food item that maximizes the exposure of the food to the light. The treatment time or the speed of the conveyor also needs to be monitored to avoid either insufficient treatment (too short a time or too high a speed) or excessive treatment (too long a time or too low a speed), which would adversely affect the microbiological safety or the overall quality of the food, respectively. Finally, routine laboratory tests, including microbiological analyses and chemical and physical quality measurements, need to be performed to ensure that the food being processed by pulsed UV light is safe and wholesome.

Advanced Levels of Process Design for Complicated Systems

Although many researchers have evaluated the effectiveness of pulsed UV light for various foods, more information about process optimization needs to be collected before an advanced pulsed UV light system is designed. As the system gets bigger and more complex, additional environmental and geometric factors need to be controlled in order to optimize the efficiency of the system. Future research regarding the inactivation kinetics, optimization of processes, and the development of pilot-scale pulsed UV light systems would facilitate the successful integration of a pulsed UV light system into a food processing plant.

Previous studies have indicated that one of the most prominent issues to be confronted in relation to the treatment of foods with pulsed UV light is the temperature increase, which is caused mainly by the infrared portion of the light (Bialka, 2006). The temperature increase may become significant during long treatments owing to the accumulation of heat over time. A temperature increase may cause undesirable sensory changes in foods and also complicate the task of achieving the desired effect of the pulsed UV light on microbial cells by limiting the treatment time. Therefore, it is highly important to eliminate or minimize the temperature increase during pulsed UV light treatments. The possible ways to control the temperature during pulsed UV light treatments include filtration of the infrared portion of the light from the lamp via a specially designed filter, and installation of a thermostatic cooler to control the temperature increase inside the treatment chamber.

The industrial application of pulsed UV light would require all surfaces of the food products to be exposed evenly to the pulsed UV light, which is a challenge because of the irregular and nonsmooth surfaces of foods. To overcome this, the product needs to be randomly exposed to multidirectional light beams for a sufficient period of time; the randomness of the exposure can be enhanced by using dispersing reflectors (Lagunas-Solar *et al.*, 2006). Better randomization could be obtained by making slight changes to the current conveyorized operations, such as by providing rotation and movement of the food product in multiple planes. This would also minimize the effect of shadowing during the pulsed UV treatment.

Cleaning and Sanitation Techniques

Consisting of complex electronic devices and sensitive lamps, pulsed UV light systems need to be cleaned with extra care. In particular, since smoothness and transparency of the quartz windows of lamp housings are indispensable for transmission of the light from the source to the target, the mechanical and chemical cleaning equipment needs to be chosen carefully to prevent any streaking, etching or accumulation of residues on the windows. Periodic cleaning of pulsed UV light systems is required to prevent any contamination risk and also to address problems of lamp fouling. In order to minimize the risk of contamination of food products with undesirable microorganisms in the chamber or system, appropriate cleaning methods need to be used to prevent any damage to the electronic equipment. Periodic application of intense pulsed UV light produced by the system itself could be a way to decontaminate the chamber from microbial contaminants. The cleaning could be either manual or automated. In the manual cleaning of conventional UV systems, the system is taken offline, the quartz sleeves are removed and then the surface is cleaned with hydrochloric acid (Malley, 2002). However, manual cleaning may be time-consuming, expensive, and inappropriate for the cleaning of equipment designed for continuous operation, in which a cleaning system involving automatic acid wash cycles could be utilized (Malley, 2002). During this washing cycle, the pulsed UV system can be turned on to ensure complete inactivation of microorganisms. This approach was successfully used by Demirci and Krishnamurthy (2006).

Capital and Operating Costs

Pulsed UV light can be more cost-efficient than continuous UV light. Conventional continuous UV light systems produce continuous UV light with a power output in the range of 100 to 1000 W (Demirci and Panico, 2008). Generating these high power levels of high-intensity conventional UV light can be costly to the user. For this reason, systems are designed to efficiently maximize the conversion and collection of UV radiation. However, pulsed UV systems can generate many megawatts of electrical power (MacGregor *et al.*, 1997). Therefore, a modest energy input can yield high peak power levels (~35 $MW \cdot cm^{-2}$). For example, the capital and operating costs of sterilizing one million cups of yoghurt with pulsed UV light add up to about €100, which implies the cost-effectiveness of pulsed UV light (Wekhof, 2003).

Owing to the lack of sufficient data on either pilot or full-scale pulsed UV light processing, it is hard to determine the cost-efficiency of pulsed UV light processing for foods. In a pulsed UV light system, the electronic components and the lamps would have a significant cost. For instance, if the pulse frequency is 50 s^{-1}, then a xenon arc lamp would have a life on the order of 500 to 600 h (Malley, 2002). Pulsed UV light processing could, however, be considered economical in terms of waste management, as the system does not produce significant solid or liquid wastes (Palmieri and Cacace, 2005).

Examples of Studies

Many studies have demonstrated the effectiveness of pulsed UV light on foods. Rowan *et al.* (1999) reported a $\log_{10}$ reduction of about 6 in *Listeria monocytogenes*, *Escherichia coli* O157:H7, *Salmonella* Enteritidis, *Pseudomonas aeruginosa*, *Bacillus cereus* and *Staphylococcus aureus*, which were seeded on agar plates, when 200 pulses of UV light pulses (pulse duration ~100 ns) were used. Jun *et al.* (2003) investigated the use of pulsed UV light to inactivate *Aspergillus niger* spores in cornmeal. For a treatment time of 100 s, a $\log_{10}$ reduction of 4.95 $CFU \cdot g^{-1}$ was achieved when the distance between the sample and the strobe was 8 cm. Sharma and Demirci (2003) exposed alfalfa seeds inoculated with *E. coli* O157:H7 to pulsed UV light. They found that when a seed layer thickness of 1.02 mm was treated for 30 s, a $\log_{10}$ reduction of 4.80 in terms of $CFU \cdot g^{-1}$ was achieved. Ozer and Demirci (2006) studied the efficacy of pulsed UV light treatment for inactivating *Escherichia coli* O157:H7 and *Listeria monocytogenes* in raw salmon. According to their study, a $\log_{10}$ reduction of 1 (~90%) for both *E. coli* O157:H7 and *L. monocytogenes* Scott A was obtained for 60 s treatment (with three pulses per second) at 8 cm from the UV strobe (5.6 $J \cdot cm^{-2}$ per pulse on the strobe surface) without any change in the quality of the sample. Krishnamurthy *et al.* (2004) investigated the use of pulsed UV light to inactivate *Staphylococcus aureus* in buffer solution and on agar-seeded plates. They found a $\log_{10}$ reduction of 7 to 8 in terms of $CFU \cdot ml^{-1}$ for *S. aureus* on seeded agar plates and in buffer solution for treatment times of less than 5 s. Demirci and Krishnamurthy (2006) studied the disinfection of water by a flow-through pulsed UV light system. The pulsed UV light treatment resulted in complete inactivation of *Bacillus subtilis* spores in water ($\log_{10}$ reduction ≥ 5.5) for all flow rates used in the study (2–14 $L \cdot min^{-1}$). In the study by Bialka and Demirci (2008b), treatment of raspberries with pulsed UV light (72 $J \cdot cm^{-2}$) resulted in a maximum $\log_{10}$ reduction of 3.9 $CFU \cdot g^{-1}$ in *E. coli* O157:H7. Similarly, treatment of strawberries with pulsed UV light (34.2 $J \cdot cm^{-2}$) yielded a maximum $\log_{10}$ reduction of 2.8 $CFU \cdot g^{-1}$ in *Salmonella*. Uesugi and Moraru (2009) investigated the combined effect of pulsed light and nisin on *Listeria innocua* populations in ready-to-eat sausages. Individual treatments with pulsed light (9.4 $J \cdot cm^{-2}$) and nisin (5000 $IU \cdot ml^{-1}$) yielded $\log_{10}$ reductions of 1.37 and 2.35, respectively, in *L. innocua* counts. However, when pulsed light was combined with nisin, a $\log_{10}$ reduction of 4.03 was achieved after 48 h storage at 4 °C, implying an additive effect of the combined treatment.

Dunn (1996) studied the effect of pulsed light on *S.* Enteritidis on shell eggs. No colonies of *S.* Enteritidis were recovered from shell eggs after treatment with pulsed light at 4 $J \cdot cm^{-2}$. Based on the initial population of *S.* Enteritidis, a $\log_{10}$ reduction of as much as 8 was obtained. In a recent study conducted by Keklik (2009), a $\log_{10}$ reduction of 7.7 (in terms of $CFU \cdot cm^{-2}$) was obtained for *S.* Enteritidis on shell eggs after 20 s treatment at 9.5 cm from the pulsed UV lamp (with a total dose of 23.6 $J \cdot cm^{-2}$) without any visual damage to the eggs. Hierro *et al.* (2009) obtained a maximum $\log_{10}$ reduction of 3.6 in terms of CFU/egg in *Salmonella enterica* serovar Enteritidis on 24–80% of unwashed eggs treated with pulsed light of different fluences (up to 12 $J \cdot cm^{-2}$), while treatment of washed eggs (lacking a cuticle) yielded a maximum $\log_{10}$ reduction of 1.8 with a fluence of 12 $J \cdot cm^{-2}$. The decreased effectiveness of the treatment on the washed eggs was probably due to the lack of a cuticle, which allowed the bacteria to migrate into inner layers of the egg where the UV light could not penetrate. Takeshita *et al.* (2003) obtained a $\log_{10}$ reduction of about 5.8 in *Saccharomyces cerevisiae* cells after treating the yeast suspension with five flashes of pulsed UV light (total fluence 3.5 $J \cdot cm^{-2}$). The effectiveness of pulsed UV light has also been shown on viruses. Lamont *et al.* (2007) eliminated infectious poliovirus by treating a viral suspension with 25 flashes of pulsed UV light, while a $\log_{10}$ reduction of more than 4 in adenovirus was obtained after 200 pulses.

Keklik (2009) designed and evaluated a pilot-scale pulsed UV light system for the decontamination of chicken carcasses, where each carcass was moved in the pulsed-UV-light chamber on a linear rail and treated by four lamps, two on each side of the carcass. The optimum treatment time was identified as 45 s (actuator velocity 52 $cm \cdot min^{-1}$), which resulted in a $\log_{10}$ reduction of about 1. Pulsed UV light has also been studied for the decontamination of liquid foods. Hillegas and Demirci (2003) investigated the inactivation of *Clostridium sporogenes* in clover honey by pulsed UV light treatment. Inoculated honey samples of different depths were treated with pulsed UV light (5.6 J·cm^{-2} per pulse on the strobe surface) at different shelf heights and with different numbers of pulses. A spore reduction of 87.6% was observed for a shelf height of 8 cm, 135 pulses and a 2 mm depth of honey. Hillegas and Demirci concluded that raising the shelf (toward the UV strobe) and increasing the number of pulses resulted in an increase in the spore reduction. Krishnamurthy *et al.* (2008a) studied the pasteurization of milk by pulsed UV light treatment. According to an analysis of the surface response of pulsed-UV-light-treated milk samples, complete inactivation of *S. aureus* was achieved after treatment of 30 ml of milk at 8 cm for 180 s and after treatment of 12 ml of milk at 10.5 cm for 180 s. Krishnamurthy *et al.* (2007) also studied the inactivation of *S. aureus* using pulsed UV light for continuous treatment of milk. The effects of the distance of the sample from the UV strobe, the number of passes and the flow rate were investigated. Complete inactivation of *S. aureus* was observed after a single pass (28 cm in length) at a flow rate of 20 ml·min^{-1} and a distance of 8 cm from the UV strobe (5.6 J·cm^{-2} per pulse), and also after two passes at a flow rate of 20 ml·min^{-1} at a distance of 11 cm.

Worked Examples

Problem 1

Using Equations 39.2 and 39.3, i.e., the equations for the thickness and distance models developed for the inactivation of *E. coli* O157:H7 on alfalfa seeds with pulsed UV light (Sharma and Demirci, 2003), calculate:

(a) the estimated $\log_{10}$ reduction in *E. coli* O157:H7 after treatment of alfalfa seeds 1.2 mm thick with pulsed UV light for 90 s;
(b) the distance from the pulsed UV strobe.

Solution

(a) Substituting the known values of the time t and seed thickness H in Equation 39.2,

$$\log_{10} \text{reduction} = 3.498 + 0.049 \times 90 - 1.883 \times 1.2 + 0.216 \times (1.2)^2 - 0.008 \times 1.2 \times 90 \quad (39.13)$$

and solving Equation 39.13 yields a $\log_{10}$ reduction of ~5.1.
(b) Since the values of $\log_{10}$ reduction and time are known, substituting these values into Equation 39.3 gives

$$5.1 = -0.5485 + 0.0472 \times 90 + 0.0826D + 0.0006 \times (90)^2 + 0.0012D^2 - 0.0065D \times 90 \quad (39.14)$$

and solving Equation 39.14 for D gives ~7 cm as the answer.

Summary and Future Needs

In this chapter, we described the principles of the use of pulsed UV light, and the advantages and issues related to this emerging technology. Pulsed UV light offers microbial disinfection of food surfaces in very short times, and thus has the potential to be utilized in the food industry as a decontamination technique for foods. A pulsed UV light system consists of light pulses delivered by an inert-gas lamp (e.g., a xenon lamp). The light pulses include electromagnetic radiation in the wavelength range from the UV to the infrared, and are germicidal to microbial cells by several inactivation mechanisms leading to photochemical, photothermal, and photophysical changes in the cells. In a typical pulsed light system, the main components are a lamp, a power supply and a pulse configuration device, or controller. The pulsed UV lamp converts the pulsed electrical energy into pulsed high-power light energy, which is delivered to the target food surface. The parameters that need to be monitored during pulsed UV light processing include the magnitude of the voltage supplied to the system, the pulse duration, the pulse frequency, the light intensity, and the treatment time or the conveyor speed.

Various studies of the use of pulsed UV light have revealed that the efficiency of microbial inactivation depends on many factors, including the "source," "time", and "target" parameters. The light intensity received at the food surface, a source parameter, decreases with increasing distance between the lamp and the food, thus reducing the efficiency of microbial inactivation for foods at longer distances from the lamp. On the other hand, pulsed UV treatment for longer treatment times or at shorter distances increases microbial inactivation, but it may cause some degradation of the quality of the food and degradation of the packaging material owing to the accumulation of heat during a lengthy treatment. Therefore, the temperature needs to be controlled during treatment in order to evaluate the efficacy of pulsed UV light systems more precisely. A thermostatic cooling system could be used to maintain the desired temperature level in the chamber, and filtering to obtain a light spectrum having a high UV/infrared ratio could limit the temperature increase during treatment.

It should also be noted that the transparency of the food and the characteristics of the food surface play an important role in the absorption of the pulsed light by the food. Although it is able to penetrate into food surfaces better than conventional UV light, pulsed UV light still has limited penetration ability compared with gamma radiation. Therefore, it is necessary to evaluate the effectiveness of pulsed UV light in deeper layers of foods as well as on the surface. Further studies of how well pulsed UV light penetrates into food products and how the penetration can be enhanced would be a useful tool for improving the technology.

One of the most important areas of study that needs to be expanded and concentrated on in the future is the mathematical modeling of microbial inactivation on foods by pulsed UV light. Prediction of microbial behavior during inactivation using mathematical expressions would be an important tool for the enhancement of the technology, and would help in assessing microbiological risks during processing and storage in order to better control and prevent microbiological hazards. Furthermore, studies of the shelf-life of food products treated with pulsed UV light would provide a real-world evaluation of the effects of pulsed UV light on the microbiological quality and other quality parameters of food products in long-term storage.

Finally, more research is needed into finding ways of integrating pulsed UV light technology into the food industry. More studies involving pilot-scale design of pulsed UV light systems would help in optimizing the system parameters for larger processing units. For instance, the design of pulsed UV light systems in which all surfaces of the food product could be equally exposed to the light would provide important information about the feasibility of pulsed UV light technology for industrial application. Overall, pulsed UV light technology is a novel and promising technology that offers benefits for food safety.

References

Bank, H.L., John, J., Schmehl, M.K. and Dratch, R.J. (1990) Bacteriocidal effectiveness of modulated UV-light. *Applied and Environmental Microbiology* 56(12): 3888–3889.

Bialka, K.L. (2006) *Decontamination of Berries with Ozone and Pulsed UV-Light*. PhD thesis, Pennsylvania State University, University Park, PA.

Bialka, K.L. and Demirci, A. (2008) Efficacy of pulsed UV-light for the decontamination of *Escherichia coli* O157:H7 and *Salmonella* spp. on raspberries and strawberries. *Journal of Food Science* 73(5): M201–M207.

Bialka, K.L., Walker, P.N., Puri, V.M. and Demirci, A. (2008a) Pulsed UV-light penetration of characterization and the inactivation of *Escherichia coli* K12 in solid model systems. *Transactions of the ASABE* 51(1): 195–204.

Bialka, K.L., Demirci, A. and Puri, V.M. (2008b) Modeling the inactivation of *Escherichia coli* O157:H7 and *Salmonella enterica* on raspberries and strawberries resulting from exposure to ozone or pulsed UV-light. *Journal of Food Engineering* 85: 444–449.

Blatchley, E.R., III and Peel, M.M. (2001) Disinfection by ultraviolet irradiation. In: *Disinfection, Sterilization, and Preservation*, 5th edn (ed. S.S. Block). Lippincott Williams & Wilkins, Philadelphia, PA, pp. 823–851.

Bohrerova, Z., Shemer, H., Lantis, R., Impellitteri, C.A. and Linden, K.G. (2008) Comparative disinfection efficiency of pulsed and continuous-wave UV irradiation technologies. *Water Research* 42: 2975–2982.

Demirci, A. and Krishnamurthy, K. (2006) Disinfection of water by flow-through pulsed ultraviolet light sterilization system. *Ultrapure Water Journal* 24(1): 35–40.

Demirci, A. and Panico, L. (2008) Pulsed ultraviolet light. *Food Science and Technology International* 14(5): 443–446.

Dunn, J. (1996) Pulsed light and pulsed electric field for foods and eggs. *Poultry Science* 75: 1133–1136.

Elmnasser, N., Guillou, S., Leroi, F., Orange, N., Bakhrouf, A. and Federighi, M. (2007) Pulsed-light system as a novel food decontamination technology: a review. *Canadian Journal of Microbiology* 53: 813–821.

Elmnasser, N., Dalgalarrondo, M., Orange, N., Bakhrouf, A., Haertle, T., Federighi, M. and Chobert, J.M. (2008) Effect of pulsed-light treatment on milk proteins and lipids. *Journal of Agricultural and Food Chemistry* 56: 1984–1991.

Federal Register (1999) Pulsed light treatment of food. *Federal Register* 66, 338829–338830.

Fine, F. and Gervais, P. (2004) Efficiency of pulsed UV light for microbial decontamination of food powders. *Journal of Food Protection* 67(4): 787–792.

Gould, G.W. (2001) New processing technologies: an overview. *Proceedings of the Nutrition Society* 60: 463–474.

Hierro, E., Manzano, S., Ordonez, J.A., de la Hoz, L. and Fernandez, M. (2009) Inactivation of *Salmonella enterica* serovar Enteritidis on shell eggs by pulsed light technology. *International Journal of Food Microbiology* 135: 125–130.

Hillegas, S.L. and Demirci, A. (2003) Inactivation of *Clostridium sporogenes* in clover honey by pulsed UV-light treatment. *Agricultural Engineering International: The CIGR Journal of Scientific Research and Development* V:7, Manuscript FP 03-009.

Jay, J.M., Loessner, M.J. and Golden, D.A. (2005) Radiation protection of foods, and nature of microbial radiation resistance. In: *Modern Food Microbiology*, 7th edn. Springer Science and Business Media, New York, p. 372.

Jun, S., Irudayaraj, J.M., Demirci, A. and Geiser, D. (2003) Pulsed UV-light treatment of corn meal for inactivation of *Aspergillus niger* spores. *International Journal of Food Science & Technology* 38(8): 883–888.

Keklik, N.M. (2009) *Decontamination of Poultry Products by Pulsed UV-Light*. PhD thesis, Pennsylvania State University, University Park, PA.

Keklik, N.M., Demirci, A. and Puri, V.M. (2009) Inactivation of *Listeria monocytogenes* on unpackaged and vacuum-packaged chicken frankfurters using pulsed UV-light. *Journal of Food Science* 74(8): M431–M439.

Krishnamurthy, K., Demirci, A. and Irudayaraj, J.M. (2004) Inactivation of *Staphylococcus aureus* by pulsed UV-light sterilization. *Journal of Food Protection* 67(5): 1027–1030.

Krishnamurthy, K., Demirci, A. and Irudayaraj, J. (2007) Inactivation of *Staphylococcus aureus* in milk using flow-through pulsed UV-light treatment system. *Journal of Food Science* 72(7): M233–M239.

Krishnamurthy, K., Demirci, A. and Irudayaraj, J. (2008a). Inactivation of *Staphylococcus aureus* in milk and milk foam by pulsed UV-light treatment and surface response modeling. *Transactions of the ASABE* 51(6): 2083–2090.

Krishnamurthy, K., Irudayaraj, J., Demirci, A. and Yang, W. (2008b) UV pasteurization of food materials. Pp. 281-299. In: *Food Processing Operations Modeling: Design and Analysis*, 2nd edn (eds S. Jun and J.M. Irudayaraj). CRC Press, Boca Raton, FL, pp. 281–299.

Krishnamurthy, K., Tewari, J.C., Irudayaraj, J. and Demirci, A. (2010) Microscopic and spectroscopic evaluation of inactivation of *Staphylococcus aureus* by pulsed UV light and infrared heating. *Food and Bioprocess Technology* 3(1): 93–104.

Lagunas-Solar, M.C., Pina, C., MacDonald, J.D. and Bolkan, L. (2006) Development of pulsed UV light processes for surface fungal disinfection of fresh fruits. *Journal of Food Protection* 69(2): 376–384.

Lamont, Y., Rzezutka, A., Anderson, J.G., MacGregor, S.J., Given, M.J., Deppe, C. and Cook, N. (2007) Pulsed UV-light inactivation of poliovirus and adenovirus. *Letters in Applied Microbiology* 45: 564–567.

Linden, K.G., Thurston, J., Schaefer, R. and Malley, J.P., Jr. (2007) Enhanced UV inactivation of adenoviruses under polychromatic UV lamps. *Applied and Environmental Microbiology* 73(23): 7571–7574.

Luksiene, Z., Gudelis, V., Buchovec, I. and Raudeliuniene, J. (2007) Advanced high-power pulsed light device to decontaminate food from pathogens: effects on *Salmonella* typhimurium viability in vitro. *Journal of Applied Microbiology* 103: 1545–1552.

MacGregor, S.J., Turnbull, S.M., Tuema, F.A. and Farish, O. (1997) Factors affecting and methods of improving the pulse repetition frequency of pulse-charged and de-charged high pressure gas switches. *IEEE Transactions on Plasma Science* 25: 110–117.

Malley, J.P. (2002). Ultraviolet disinfection. Pp. 213-235. In: *Control of Microorganisms in Drinking Water* (ed. S. Lingireddy). American Society of Civil Engineers, Reston, VA, pp. 213–235.

Miller, R.V., Mitchell, J.D. and Elasri, M. (1999) Bacterial responses to ultraviolet light. *ASM News* 65: 535–541.

Oms-Oliu, G., Martín-Belloso, O. and Soliva-Fortuny, R. (2010) Pulsed light treatments for food preservation. A review. *Food and Bioprocess Technology* 3: 13–23.

Ozer, N. and Demirci, A. (2006) Inactivation of *Escherichia coli* O157:H7 and *Listeria monocytogenes* inoculated on raw salmon fillets by pulsed UV-light treatment. *International Journal of Food Science & Technology* 41: 354–360.

Palmieri, L. and Cacace, D. (2005) High intensity pulsed light technology. In: *Emerging Technologies for Food Processing* (ed. D.W. Sun). London, UK: Elsevier Academic Press, London, pp. 279–304.

Rowan, N.J., MacGregor S.J., Anderson, G., Fouracre, R.A., Mcllvaney, L. and Farish, O. (1999) Pulsed-light inactivation of food-related microorganisms. *Applied and Environmental Microbiology* 65: 1312–1315.

Rupp, W.D. (2006) *Molecular Mechanisms of DNA Damage*. Yale University, New Haven, CT. Available at: http://radonc.yale.edu/training/pdf/molecular_mechanisms.pdf. Accessed on February 23 2010.

Schaefer, R., Grapperhaus, M., Schaefer, I. and Linden, K. (2007) Pulsed UV lamp performance and comparison with UV mercury lamps. *Journal of Environmental Engineering and Science* 6: 303–310.

Sharma, R.R. and Demirci, A. (2003) Inactivation of *Escherichia coli* O157: H7 on inoculated alfalfa seeds with pulsed ultraviolet light and response surface modeling. *Journal of Food Science* 68(4): 1448–1453.

Takeshita, K., Shibato, J., Sameshima, T., Fukunaga, S., Isobe, S., Arihara, K. and Itoh, M. (2003) Damage of yeast cells induced by pulsed light irradiation. *International Journal of Food Microbiology* 85: 151–158.

Uesugi, A.R. and Moraru, C.I. (2009) Reduction of *Listeria* on ready-to-eat sausages after exposure to a combination of pulsed light and nisin. *Journal of Food Protection* 72(2): 347–353.

Van Boekel, M.A.J.S. (2002) On the use of the Weibull model to describe thermal inactivation of microbial vegetative cells. *International Journal of Food Microbiology* 74: 139–159.

Wang, T., MacGregor, S.J., Anderson, J.G. and Woolsey, G.A. (2005) Pulsed ultra-violet inactivation spectrum of *Escherichia coli*. *Water Research* 39: 2921–2925.

Wekhof, A. (1991) Treatment of contaminated water, air and soil with UV flashlamps. *Environmental Progress* 10(4): 241–247.

Wekhof, A. (2000) Disinfection with flash lamps. *PDA Journal of Pharmaceutical Science & Technology* 54(3): 264–276.

Wekhof, A. (2003) Ultra-fast sterilization by disintegration of microorganisms with intense pulsed UV light. *Business Briefing: Global Surgery. Technology & Services*, pp. 1–5.

Xenon (2006a) *Photochemical Sterilization*. Application Note AN 111. Xenon Corporation, Wilmington, MA. Available at: http://www.xenoncorp.com/Literature/PDF/AN-111.pdf. Accessed on February 22 2010.

Xenon (2006b) *Pulsed UV Treatment for Sanitation and Sterilization. Pulsed UV Technology*. Xenon Corporation, Wilmington, MA. Available at: http://www.xenoncorp.com/Literature/PDF/BrochureSteri.pdf. Accessed on June 10 2009.

Xenon (2007a) *Pulsed UV-Light Benefits*. Xenon Corporation, Wilmington, MA. Available at: http://www.xenoncorp.com/perf_cure.html. Accessed on February 23 2010.

Xenon (2007b) *RC-800 Series. UV Curing Systems*. Xenon Corporation, Wilmington, MA. Available at: http://uv-curing.xenoncorp.com/Literature/PDF/RC-800%20Series%20Brochure.pdf. Accessed on February 22, 2010.

Xenon (2008) *RC-801, RC-802 Pulsed UV Light Curing Systems. Installation and Maintenance Manual*. Part No. 810-0092. Xenon Corporation, Wilmington, MA, p. 5.

40

High-Voltage Food Processing Technology: Theory, Processing Design and Applications

Paul Takhistov

Introduction

The main purpose of food processing is food preservation, which means maintaining the highest quality and microbiological safety of a food product for a desired time frame. In the search for alternatives to traditional thermal food processing, currently used for both pasteurization and sterilization, various nonthermal food-processing methods have been explored during the past few years. One group of alternative technologies, based on electric-field treatment of a food product, has attracted special attention from the academic and industrial communities because of the high durability of these technologies, their simplicity of technical realization, and their ability to minimize deterioration of food quality during the treatment (Jeyamkondan *et al.*, 1999). These technologies include ohmic heating, pulsed electric field (PEF) treatment, microwave (MW) processing and radio frequency (RF) processing. The author of this chapter has limited his consideration to processes operating at frequencies below 1 GHz, i.e., gamma and UV irradiation food treatment technologies are not included in the model presented. The technological processes considered are briefly described below.

Ohmic and mild electric-field heating. Ohmic heating is a method of food processing involving heating of a food product by passing electric current through it. The heating is caused by dissipation of electric-field energy (i.e., Joule heating) within

Handbook of Food Process Design, First Edition. Edited by Jasim Ahmed and Mohammad Shafiur Rahman.

the food product. The applications of ohmic heating include blanching, evaporation, dehydration, fermentation and extraction. The principal advantage of ohmic heating lies in its ability to heat food products rapidly and uniformly, including products containing particulates. Ohmic heating is a thermally based process; therefore the electric field strength, the residence time, and the electrical conductivity of the food product are the major critical process factors. These factors strongly influence the temperature reached and the efficacy of the ohmic heating.

Pulsed electric field processing. PEF food processing involves applying short pulses of a high electric field to a bulk food. The duration of the pulses ranges from microseconds to milliseconds, and their intensity can be between 20 and 50 $kV \cdot cm^{-1}$ (EPRI, 1998). PEF technology is considered to be superior to conventional heat treatment, because it avoids or reduces harmful changes in the sensory and physical properties of foods (Quass, 1997). PEF processing has been tested on various foods, including dairy products, juices, soups and sauces (Ade-Omowaye *et al.*, 2001; Bazhal *et al.*, 2001; Lebovka *et al.*, 2003). However, its application is restricted to food products that can hold high electric fields, have low electrical conductivity and do not contain or form bubbles.

Microwave and radio frequency processing. Microwave and radio frequency food treatment processes utilize electromagnetic waves of certain frequencies to produce heat in the treated product (Ponne and Bartels, 1995). Microwave and radio frequency heating need less time to achieve the processing temperatures for pasteurization and sterilization, so they are preferred to conventional heating methods. Reviews of various food and chemical processes utilizing radio frequency can be found in Thostenson and Chou (1999) and Venkatesh and Raghavan (2004).

Unfortunately, the current use of electric-field-based food processing technologies is limited to academic research laboratories and sporadic industrial applications. The two major obstacles that prevent spreading of these promising technologies are (1) the lack of a common validation methodology that would allow various treatment processes based on different physical principles to be compared, and (2) the difficulty of scaling up these food treatment technologies to adapt them to industrial needs. Additionally, the design of a technological chain of processes with a synergistic action (e.g., a combination of thermal and nonthermal processes (Leistner and Gould, 2002; Ross *et al.*, 2003)) requires an ability to represent all processes in similar terms. All these issues can be addressed through the development of unified process models, based on the dimensionless analysis of process variables.

Process developers and researchers are concerned with the industrial implementation of food treatment processes in which chemical or microbiological conversion of food products takes place in conjunction with the transfer of mass, heat and momentum. These processes are scale-dependent, i.e., they have different characteristics in small (lab-scale) and large (industrial) applications. To compare processes for validation purposes, one needs to describe them in unified terms with comparable process parameters and food product properties. A unified model for various processes is also needed in situations where a process already exists but is not functioning properly, and therefore appropriate tests have to be carried out to discover the cause of the difficulties and provide a solution. In general, a unified process model should be able to answer the question of how to validate, assess or compare selected technologies.

Unified Analysis of Electric-Field-Based Food Processing Technologies

To compare electric-field-based food processing technologies, it is necessary to understand which factors distinguish one process from another. The electric field intensity, which might be a natural choice for the distinguishing parameter for these processes, cannot be used as a basis for classification. The same electric field intensity can be utilized in different processes, and there is no specific electric field strength that can serve as the unique parameter of a given technology, the only exception being PEF treatment. An analysis of the post-treatment biological effects and technological implementation of different technologies and the physical principles behind them shows that the frequency of the electric field is the only unique characteristic of any particular process. Electric-field-based food treatment processes can be classified based on the specific regions of the electromagnetic spectrum that they occupy, as depicted in Figure 40.1.

Set of Variables

Any electric-field-based food processing technology can be characterized by a number of parameters that can be divided into three groups: material (food product) properties; process variables, characterizing the operation regime and performance; and geometric (scale-up) factors, determined by the equipment design.

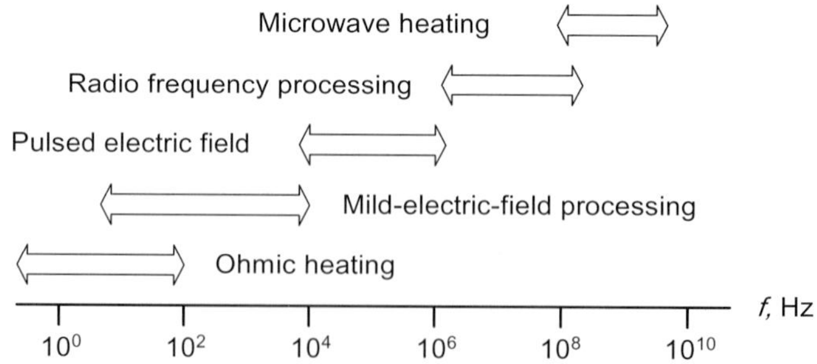

Figure 40.1 Distribution of electric-field-based food treatment processes over the electromagnetic spectrum.

Process Variables

Electric Field Strength (E)

The electric field intensity is one of the main factors that affect microbial inactivation (Hülsheger *et al.*, 1981; Dunne *et al.*, 1996). The microbial inactivation efficacy increases with increasing electric field intensity, which has to be above the critical transmembrane potential (Qin *et al.*, 1998). However, an increase in electric field strength may also result in an undesirable increase in the temperature of the food and degradation of food components.

Processing (Treatment) Time (t)

The treatment time is defined as the residence time of the treated product in the treatment chamber, or the overall duration of the treatment. An increase in this parameter increases microbial inactivation (Sale and Hamilton, 1967b); however, it may also result in an undesirable increase in the temperature of the food. The treatment time depends on the intensity of the applied electric field. Above the critical value of the electric field, the critical treatment time decreases with increasing electric field strength (Barbosa-Canovas *et al.*, 1999).

Process Temperature (T)

Experimental results obtained in challenge tests have demonstrated that both the treatment temperature and the process temperature impact microbial survival and recovery. At a constant electric field strength, the inactivation efficacy increases with an increase in process temperature. However, application of a high-intensity electric field causes an increase in the temperature of the food, and hence proper cooling is necessary to maintain the product temperature below that of the thermal pasteurization process.

Food Product Velocity (U)

The food product velocity is the main quantitative parameter of momentum transfer, and is common to all technologies of interest.

Geometric Factors

Several different laboratory- and pilot-scale treatment chambers have been designed and used for nonthermal food processing. The length L of the treatment chamber (for liquid foods) or the linear size of a solid food product can be used as the *characteristic length* in the process model.

Food Product Properties

The interaction of a food product with an electric field is a very complex process that includes dielectric phenomena, ionic transport, and mass and heat transfer. The set of food product properties includes the following parameters, representing the mechanical, transport and electrical characteristics of the foodstuff.

The mechanical-energy dissipation and inertial properties of foods are represented by the *kinematic viscosity* (ν) of the product. The ability of a food product to transfer heat and accumulate thermal energy is characterized by the *thermal diffusivity* (α) and the *specific heat capacity* (C_p), respectively. Clearly, all transport coefficients (diffusivity, thermal conductivity and viscosity) have the same dimensions, and they will be treated as one parameter in the proposed method of analysis.

Electrical Properties: Energy Transfer and Dissipation

The existing electric-field-based technologies cover a frequency range of over ten orders of magnitude (see Figure 40.1), and therefore the electrical properties of the food product to be treated require in-depth consideration. These properties of the food are the single most important factor that influences the performance of any given technology. They determine how effectively a food product can be polarized, heated by the field or conduct an electric current. Food composition has much greater influence on electric-field-based food processing than on conventional processing, because it impacts the conductive and dielectric properties of a foodstuff. High salt and moisture contents increase the efficacy of energy absorption by influencing the penetration depth and energy dissipation in the food product. An increase in electrical conductivity causes an increase in the overall energy dissipation and changes in the product temperature during electric-field-based treatment (Zhang *et al.*, 1995b). However, highly conductive foods are less suitable for electric-field-based treatment (Barbosa-Canovas *et al.*, 1999; Gongora-Nieto *et al.*, 2003) owing to the Joule heating effect. Dielectric phenomena in food products with a high salt content are influenced by their ionic conductivity. Thus, the polarization of a food material is affected by additional polarization due to the induced charge displacement. Therefore, the complex dielectric constant of a product must be used when one is examining electric-field-based processing of real food products.

Polarization phenomena in imperfect dielectrics can be described with equations formally similar to those for the polarization of perfect dielectrics, using the complex dielectric constant ε^*. The need for a generalized dielectric constant arises from the need to examine various food processing technologies influenced by the dielectric properties of food products in an alternating electric field:

$$\varepsilon^* = \varepsilon' + \varepsilon'' = \varepsilon' + j\frac{\sigma'}{2\pi\omega} \quad (40.1)$$

Here ε' is the dielectric constant characterizing the capacity of the food product to store electrical energy, and ε'' is the dielectric loss factor, which reflects the energy loss due to relaxation processes that occur in the system in an AC field.

The polarization process is characterized by a finite relaxation time, so that at sufficiently high frequencies, when the period of oscillation of the field is comparable to the relaxation time, the phase shift between the polarization of the food product and the field becomes perceptible and manifests itself in dielectric losses. It is clear, in this connection, that the capacitance and, correspondingly, the dielectric constant of a food product are not only complex but also frequency-dependent quantities. The frequency dependence of $\varepsilon^\star$ in the presence of a frequency-independent conductivity is reflected in Equation 40.1.

The real part $\mathrm{Re}(\varepsilon^\star)$ of the complex dielectric constant is also frequency-dependent at sufficiently high frequencies, since the polarization starts to lag behind the electric field. As follows from physical considerations, the main contribution to the food-product/electric-field interaction in the high-frequency region comes from the dielectric properties of the food, whereas in the low-frequency region, the principal role is played by the conductivity. Hence, taking Equation 40.1 into account, it is clear that the complex dielectric permittivity $\varepsilon^\star$ has two limits:

$$\begin{aligned} \varepsilon^\star &\to \varepsilon', \quad \omega \to \infty \\ \varepsilon^\star &\to \sigma', \quad \omega \to 0 \end{aligned} \tag{40.2}$$

where ω is the characteristic frequency of the foodstuff. The question of whether the conductive or the dielectric behavior of a product is dominant is determined by the characteristic (critical) frequency

$$\omega_{\mathrm{cr}} = \frac{1}{\tau} = \frac{\varepsilon'}{\sigma'} \tag{40.3}$$

where τ is the dielectric relaxation time of the food product, determined by the displacement and orientation polarizations. The dielectric relaxation time is the time needed to reach a new state of equilibrium after the excitation is changed by an external electric field (Frubing, 2001).

Therefore, the behavior of the food product should be considered as a function of the electric-field frequency. The frequency range investigated can be divided into three separate regions with distinct properties. In the low-frequency region ($\omega \ll \omega_{\mathrm{cr}}$), the conductivity of the food product dominates over its dielectric permittivity, and the corresponding technological processes can be modeled as an electric current flowing in a conductive medium with an active (ohmic) load. In the second region, of high frequencies ($\omega \gg \omega_{\mathrm{cr}}$), there is a strong dominance of dielectric processes in the system. The corresponding food treatment processes can be modeled as an electrostatic problem with a reactive (capacitive) load. In the most complicated (third) transition region, where $\omega \approx \omega_{\mathrm{cr}}$, both components of the complex permittivity play a significant role and must be taken into account.

Despite the diversity of electric-field-based food treatment technologies, all of them can be described using their common components and the similarity between the interactions of an electric field with food products.

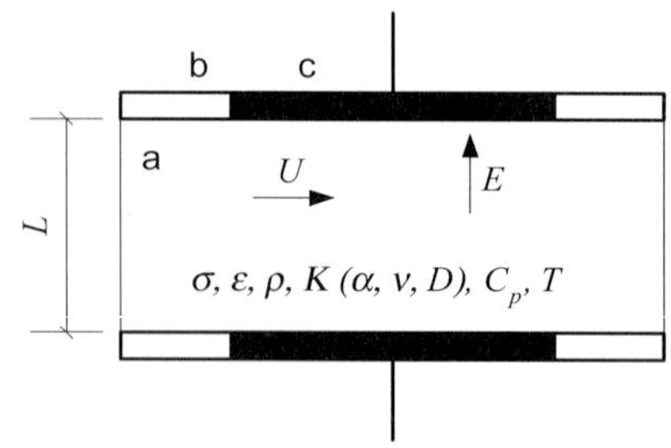

Figure 40.2 General scheme of an electric-field-based food treatment process.

Table 40.1 A set of process variables and physical properties for a food product.

Variables/Dimensions	*K*	*ρ*	*σ*	C_p	*ε*	*t*	*E*	*U*	*T*	*L*
Length, L	2	−3	−3	2	−3	0	1	1	0	1
Time, t	−1	0	3	−2	4	1	−3	−1	0	0
Mass, M	0	1	−1	0	−1	0	1	0	0	0
Temperature, ° (degrees)	0	0	0	−1	0	0	0	0	1	0
Current, I	0	0	2	0	2	0	−1	0	0	0

A simplified model representation of an electric-field-based food treatment process is depicted in Figure 40.2. All such processes use reaction chambers for the treatment. The chamber holds a food product to be processed (a), and consists of a main body (b) with integrated electrodes (c). Depending on the process involved, the electrodes can be in direct contact (PEF treatment, ohmic heating or microwave processing) or indirect contact (RF and microwave processing) with the product.

Once in the reaction chamber, the food product is transported between the electrodes, placed at a distance *L* apart, which is a design factor and a characteristic length scale of the system at the same time. The process variables include the applied electric field strength $\vec{E}$ (defined as the ratio between the applied voltage and the characteristic length *L*), the food velocity $\vec{U}$ and the process temperature *T*. The properties of the food product can be divided into four groups: transport, mechanical, thermal and electrical properties.

In general, the mass, momentum and heat transfer properties of a food product can be represented by a single transport coefficient *K* with dimensions of $[L^2t^{-1}]$, which combines thermal (*α*, the thermal diffusivity), mechanical/momentum (*ν*, the kinematic viscosity) and mass transfer (*D*, the diffusivity) coefficients. Additionally, a food product is characterized by mechanical properties (*ρ*, the density) and thermal parameters (*T*, the temperature, and C_p, the specific heat capacity). A complete set of parameters characterizing electric-field-based food treatment processes is presented in Table 40.1.

Dimensionless Analysis of Electric-Field-Based Food Processing

Dimensionless analysis is a powerful tool that is used to replace a system of equations relating n variables with a smaller system relating $(n - r)$ dimensionless parameters. Each dimensionless parameter (Π-term) is a product of independent variables, each of them raised to a power. Reducing the number of variables in a model minimizes the number of experiments necessary to establish the relationships between these variables. The behavior of a food product during a treatment can be predicted from the behavior of a model provided that the dimensionless parameters of the product are equivalent to those of the model.

Low-Frequency Region ($\omega \ll \omega_{cr}$)

In this region, the process is practically not affected by the dielectric properties of the foodstuff. An electric current passes through the food material by an ionic transport mechanism. A nonthermal food treatment process can be described by the following set of process- and material-dependent variables:

$$\Pi = \mathrm{K}^{e1}\mathrm{L}^{e2}\rho^{e3}\sigma^{e4}\mathrm{C}_{\mathrm{p}}^{e5}(t, U, E, T)^{e6} \tag{40.4}$$

This is an underdetermined system with five equations and eight independent variables. Of these, the time, temperature, electric field strength and food product velocity are the most important ones, and should appear in only one dimensionless product each. Using the rule of exponents, a set of four dimensionless products can be obtained:

$$\Pi_{\sigma 1} = \frac{Kt}{L^2}; \quad \Pi_{\sigma 2} = \frac{K}{LU}; \quad \Pi_{\sigma 3} = \frac{L^3E^2\sigma}{K^3\rho}; \quad \Pi_{\sigma 4} = \frac{C_pL^2T}{K^2} \tag{40.5}$$

The first Π-term is a very important parameter, characterizing the timescales of various physical processes occurring in a food product during treatment. By substituting the corresponding values of the transport coefficient K into $\Pi_{\sigma 1}$ and rearranging, one obtains three well-known characteristic times of the system:

$$t_{\alpha,D,v} = \frac{L^2}{\alpha}; \quad \frac{L^2}{D}; \quad \frac{L^2}{v} \tag{40.6}$$

The Fourier time t_α determines the speed of heating of the product. The second (Einstein) time, t_D, represents the rate of mass transfer in the system. The time t_v characterizes mechanical perturbations (the momentum transfer rate) in the system. From the values of these times, one can determine the dominant transport phenomenon for a given system. The Lewis number (defined as the ratio of the Fourier and Einstein characteristic times) is useful when choosing between the dominance of heat and mass transport in the system:

$$\mathrm{Le} = \frac{\alpha}{D} \tag{40.7}$$

The ratio of $\Pi_{\sigma 1}$ and $\Pi_{\sigma 3}$ when $K = \alpha$ is a well-known expression for the heating temperature, which characterizes the temperature rise in a food product depending on the specific heat capacity of the food, the thermal conductivity and the processing time:

$$\Pi_{\sigma 1} : \Pi_{\sigma 3} = \frac{C_p T t}{\alpha} - \text{heating temperature} \tag{40.8}$$

By substituting $K = v$ in the second Π-term, one obtains the Reynolds number, characterizing momentum transfer in the treatment chamber:

$$\Pi_{\sigma 2}^{-1} = \frac{LU}{v} = \mathrm{Re} \tag{40.9}$$

Finally, by dividing the third Π-term by $\Pi_{\sigma 14}$, one can obtain the ratio between the total energy of the electric field applied to the system and the total amount of heat generated as a result of dissipation of electrical energy (Joule heating):

$$\Pi_{\sigma 4} = \frac{LE^2\sigma}{K\rho C_p T} = \frac{\text{Total electric field energy}}{\text{Heat generated}} \tag{40.10}$$

This ratio can be used to evaluate the amount of energy dissipated during a food treatment.

Middle-Frequency Region ($\omega \sim \omega_{cr}$)

In this case, the capacitive and ohmic resistances of the food product are of the same order of magnitude, and hence neither of them can be neglected. The complete set of variables for this frequency region is

$$\Pi = \mathrm{K}^{e1} \rho^{e2} \varepsilon^{e3} \sigma^{e4} C_p^{e5} (L, T, E, U, t)^{e6} \tag{40.11}$$

A new set of Π-terms can be obtained by solving the system in Equation 40.11:

$$\Pi_1 = \frac{\sigma t}{\varepsilon}; \quad \Pi_2 = \frac{K\rho\sigma}{\varepsilon^2 E^2}; \quad \Pi_3 = \frac{\varepsilon U^2}{K\sigma}; \quad \Pi_4 = \frac{C_p \varepsilon T}{K\sigma}; \quad \Pi_5 = \frac{\varepsilon K}{\sigma L^2} \tag{40.12}$$

By rearranging Π_1, one obtains the characteristic relaxation frequency of the food product,

$$\omega_{cr} = \frac{\sigma}{\varepsilon} \tag{40.13}$$

and

$$\Pi_2 : \Pi_1 = \frac{K\rho}{\varepsilon E^2 t} \tag{40.14}$$

The dimensionless parameter Π_5 represents the ratio between the characteristic relaxation time of the food product and the characteristic time determined by its transport properties (i.e., mass, momentum or heat transport):

$$\Pi_5 = \frac{t_{\text{relax}}}{t_{\text{transport}}} = t_{\alpha,D,v}\omega_{\text{cr}} \tag{40.15}$$

Note that the term Π_3 is the viscous analog of Π_5:

$$\Pi_3 : \Pi_5 = \text{Re}^2 \tag{40.16}$$

Using $K = \alpha$, the following dimensionless number can be obtained:

$$\Pi_1 : \Pi_4 = \frac{C_p T t}{\alpha} \tag{40.17}$$

This relates the amount of heat generated in the system to the Fourier number of the food product.

High-Frequency Region ($\omega \gg \omega_{\text{cr}}$)

High-frequency electric-field oscillations result in rapidly alternating polarization processes in food products, increasing the influence of the dielectric properties of the food product on the performance of the process. A new set of variables for high-frequency electric fields is the following:

$$\Pi_\varepsilon = \text{K}^{e1}\text{L}^{e2}\rho^{e3}\varepsilon^{e4}\text{C}_p^{e5}(T, E, U, t)^{e6} \tag{40.18}$$

The set of dimensionless numbers obtained by solving the system of Equation 40.18 is as follows:

$$\Pi_{\varepsilon 1} = \frac{Kt}{L^2}; \quad \Pi_{\varepsilon 2} = \frac{K}{LU}; \quad \Pi_{\varepsilon 3} = \frac{L^2 E^2 \varepsilon}{K^2 \rho}; \quad \Pi_{\varepsilon 4} = \frac{C_p L^2 T}{K^2} \tag{40.19}$$

This is similar to the set in Equation 40.12 described above, and characterizes food-processing regimes where the dielectric permittivity of the food product is much greater than its conductivity.

Energy Absorption and Effective-Dose Calculation

The sets of dimensionless numbers derived above for the three frequency regions allow one to relate the processing conditions to the properties of the food product to be treated. However, these variables are difficult to use in the evaluation of the performance of a process, especially when two different processes need to be compared.

Clearly, the efficacy of a process should be a function of the total energy applied to the system, independent of the specific technology. Since the applied energy is dissipated because of the nonideal dielectric properties of the food product, the process performance function should include food-specific parameters. In the search for the best representative criterion for the efficacy of a process, it has been found that the electric-field/food-product interaction is most adequately described by a parameter called the "absorbed dose." This parameter, W_D, can be used to characterize both the ability of a foodstuff to absorb electric field energy, and the efficacy with which the energy is imparted (in combination with other dimensionless numbers). It is measured in $J \cdot kg^{-1}$ and has the dimensions $[L^2t^{-1}]$.

The absorbed dose is a function of the electric field intensity, the mass of the food product and its electrical properties. A combination of these variables results in the following closed system of five equations:

$$\Pi_{dose} = t^{e1} E^{e2} \rho^{e3} \sigma^{e4} W_D^{e5} \tag{40.20}$$

By solving this system using the rule of exponents, one can obtain a dimensionless number relating the absorbed dose to the electric field intensity and the material properties of the food product:

$$\Pi_{dose} = \frac{\rho W_D}{E^2 \sigma t} \tag{40.21}$$

By rearranging Equation 40.21 and using Π_1 given in Equation 40.12, an estimate of the absorbed dose as a function of the process and product parameters and the treatment time in the cases of active (low-frequency) and reactive (high-frequency) loads can be obtained:

$$W_D \sim \begin{cases} \dfrac{E^2 \sigma t}{\rho}, & \omega \ll \omega_{cr} \\[2ex] \dfrac{E^2 \varepsilon}{\rho}, & \omega \gg \omega_{cr} \end{cases} \tag{40.22}$$

Taking Equation 40.1 into account, one can determine the value of the relaxation time based on measurements of the electrical properties of the food (conductivity and dielectric permittivity) at low and high frequencies:

$$\tau \approx \left| \frac{\varepsilon_L - \varepsilon_H}{\sigma_L - \sigma_H} \right| \tag{40.23}$$

Theoretically, these frequency limits correspond to the conditions in Equation 40.2. However, in practice a narrower frequency range is used ($\omega_L < \omega_p$, $\omega_H > \omega_p$), where ω_p is the processing frequency. By substituting the value of the conductivity of the food defined by $\sigma' = 2\pi\omega\varepsilon''$ (Equation 40.1) into Equation 40.23, one can estimate the relaxation time and the characteristic (critical) frequency of the food product (Equation 40.13) as follows:

$$\tau = \frac{1}{\omega_{cr}} \approx \frac{1}{2\pi}\left|\frac{\varepsilon'_{\omega L} - \varepsilon'_{\omega H}}{\omega_L \varepsilon'_{\omega L} - \omega_H \varepsilon'_{\omega H}}\right| \tag{40.24}$$

The unified dimensionless representation of electric-field-based food processing developed above provides critical information for the scaling-up and validation of processes. It is important to find out what the major mechanism is for every combination of food product and treatment process. This knowledge makes it possible to select the most suitable model for the optimization and scaling-up of a process.

The principal mechanism of microbial inactivation at low frequencies (ohmic heating and medium electric field (MEF) treatment) is thermal, but for technologies using high-frequency electric fields (PEF, MW, RF and even MEF) it is not so obvious. Argument about the possible nonthermal effects (unrelated to inactivation caused by heat) of electric-field-based processing technologies has continued since these technologies were established (Geveke *et al.*, 2002; Banik *et al.*, 2003). Researchers have repeatedly reported nonthermal inactivation effects, although detailed descriptions or discussions of the mechanisms cannot easily be found in the literature. Several possible inactivation mechanisms have been proposed, including selective heating of microorganisms, electroporation, cell membrane rupture, and cell lysis caused by the coupling of electromagnetic energy.

Thermal microbial inactivation is mainly the result of denaturation of cell proteins. However, there is no generally accepted theory for the nonthermal mechanisms of inactivation of bacteria by electric fields. The mechanisms most often referred to are the dielectric breakdown or electroporation of cell membranes (Zimmermann, 1986; Castro *et al.*, 1993; Vega-Mercado *et al.*, 1996a). Dielectric breakdown of cell membranes is a result of a high electric field strength in the medium, and causes disruption of the cell membrane primarily at the sites of ion channels. Clearly, this mechanism is most important for systems where the dielectric interaction between the food product and the applied electric field is dominant. The second inactivation mechanism is electroporation of cell membranes, also known as ionic inactivation of cells. In a conductive medium (i.e., a food with significant ionic or salt contents), an applied high voltage leads to the buildup of a potential (concentration) gradient across the cell membrane (Schoenbach *et al.*, 1997). This gradient causes the expansion of ion channels (pore formation) and changes in the osmotic pressure. Cell membrane disruption occurs when the induced membrane potential exceeds a critical value of 1 V (Castro *et al.*, 1993). The existence of this mechanism is supported by data on cell membrane shrinkage and cytoplasm expulsion as a result of PEF treatment (Zhang *et al.*, 1997).

As follows from the proposed generalized representation of nonthermal food treatment processes, different food polarization regimes can easily be achieved by varying the process regime and/or the properties of the food product within one technology. The following criterion for determining the dominant mode in the interaction of a food product and an applied electric field (i.e., polarization) is proposed:

$$\omega_p \tau = \begin{cases} \ll 1 & \text{ionic} \\ \approx 1 & \text{mixed} \\ \gg 1 & \text{dielectric} \end{cases} \tag{40.25}$$

By substituting specific values of the relaxation time (a product factor) and the operating frequency (a process/design factor) into Equation 40.25, one can easily determine the type of food-product/electric-field interactions (polarization mode) that will dominate in a particular system. The process frequency ω_p can easily be determined for "single-frequency technologies," such as ohmic, microwave and RF heating. However, for multifrequency processes (which implies that the frequency of the applied field changes during the treatment) or technologies that use electrical pulses of complex shape, such as PEF and electric-arc discharge processes, the determination of ω_p requires a separate calculation.

In particular, electric-field pulses may be applied in the form of exponentially decaying, square-wave, oscillatory, bipolar or instant-reverse-charge pulses (Barbosa-Canovas *et al.*, 1999). In such cases, the specific process frequency is a function of the shape of the electrical impulse, its duration t_p, and the delay time t_d, i.e., the time interval between two pulses. Thus, the signal must be analyzed by a Fourier transform technique to determine the leading frequency in a PEF process. A rough estimate of the operating frequency range for any particular process is as follows:

$$\omega_p \in \left[\frac{1}{t_d}, \frac{1}{t_p}\right] \tag{40.26}$$

To elucidate the method of process analysis proposed here, examples of two nonthermal technologies are considered here.

Pulsed Electric Field Processing

Using the approach proposed above, values of the process frequency (Equation 40.3) and the corresponding absorbed dose (Equation 40.22) have been calculated using published data taken from 15 sources cited in CFSAN (2000). The resulting process efficacy (bacterial survival) is plotted in Figure 40.3 as a function of the total absorbed dose.

The typical values of the characteristic frequency for PEF treatment lie in the range of 1 kHz to 1 MHz (Zhang *et al.*, 1995b). On the other hand, the data available in the literature (Angersbach *et al.*, 1999; Ho *et al.*, 2003; Wang *et al.*, 2003) show that the

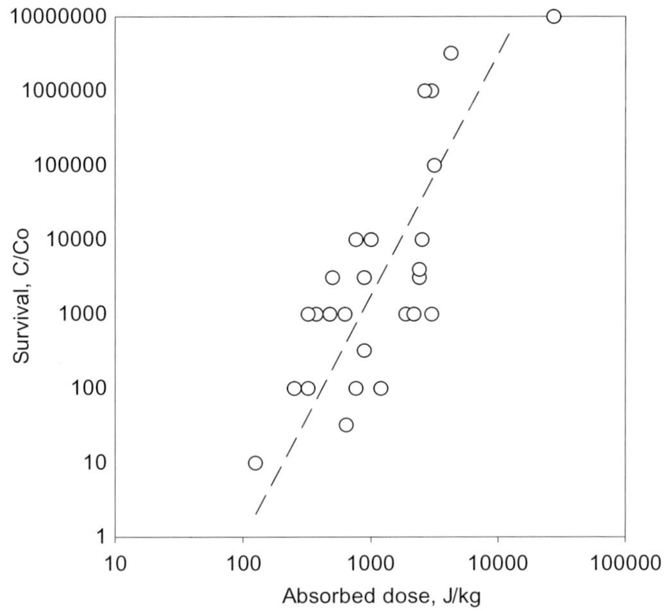

Figure 40.3 Bacterial survival as a function of absorbed dose for PEF-treated products and model solutions.

characteristic frequencies of many food products are also within the range defined by Equation 40.26, and therefore all process calculations should take both ionic and dielectric processes into account. Unfortunately, there is a significant gap in our knowledge about the electrical properties of foods, especially in the range of frequencies of the external electric field between 10^5 and 10^7 Hz. The reported data usually include values of the conductivity (Angersbach *et al.*, 1999) or the dielectric permittivity (Herve *et al.*, 1998) only, whereas data on the complex permittivity are vital for determining the dominant mode of interaction between a food product and the applied electric field in a particular PEF system.

Microwave Processing

Dielectric-permittivity and conductivity data for several food products (cheese sauce, liquid whey protein mixture, cooked macaroni noodles, macaroni and cheese, and whey protein gel), published by Wang *et al.* (2003), were used to calculate values of the critical process frequency and to determine the dominant polarization mode of the foodstuff during microwave processing. Data recalculated according to Equation 40.3 are depicted in Figure 40.4.

The critical process frequencies, calculated as functions of the product temperature, indicate that most food products can be considered as perfect dielectrics for 2.4 GHz microwave processing. However, for microwave processing at a lower working

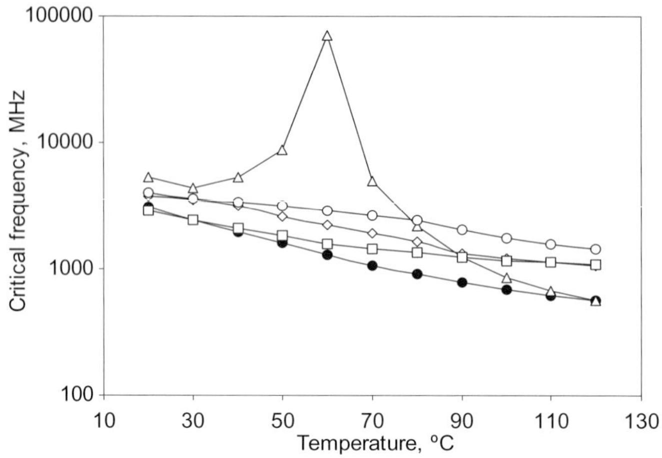

Figure 40.4 Critical process frequencies calculated for cheese sauce (•), liquid whey protein mixture (◇), cooked macaroni noodles (△), macaroni and cheese (○), and whey protein gel (□).

frequency (850 MHz), some products (cheese sauce and cooked macaroni noodles) should be analyzed as imperfect dielectrics, taking their conductivities into account.

Pulsed Electric Fields in Food Processing and Preservation

Pulsed electric field processing is a nonthermal method used to maintain food safety and increase the shelf-life of foods by inactivating spoilage-causing and pathogenic microorganisms. Many researchers have investigated this problem, including Sale and Hamilton (1967a,b), Mizuno and Hori (1988), Jayaram *et al.* (1992), Qin *et al.* (1994) and Pothakamury *et al.* (1996). PEF processing is advantageous over other methods because the changes in product color, flavor and nutritive value during treatment are minimized (Dunn and Pearlman, 1987; Jin *et al.*, 1998; Jia *et al.*, 1999; Jin and Zhang, 1999). High-intensity pulsed-electric-field processing involves the application of pulses of high voltage (typically 20–80 $kV \cdot cm^{-1}$) to a food material placed between two electrodes. PEF treatment is conducted at ambient, below ambient or slightly above ambient temperature for less than 1 s, and the energy loss due to heating of the food is minimized. With respect to food quality attributes, PEF technology is considered superior to traditional methods of heat treatment of foods because it avoids or greatly reduces detrimental changes in the sensory and physical properties (Quass, 1997). Although some studies have concluded that PEF treatment preserves the nutritional components of foods, the effects of PEF treatment on the chemical and nutritional aspects of foods must be better understood before PEF treatment can be used in food processing (Qin *et al.*, 1995a).

Some important aspects of PEF technology are the generation of a high electric field intensity, the design of chambers to impart a uniform treatment to the food with the minimum increase in temperature, and the design of electrodes that minimize the effect of electrolysis. The large field intensities are achieved by storing a large amount of energy in a capacitor bank (a series of capacitors) charged by a DC power supply, which is then discharged in the form of high-voltage pulses (Zhang *et al.*, 1995b). Studies of energy requirements have concluded that PEF treatment is an energy-efficient process compared with thermal pasteurization, particularly when a continuous system is used (Qin *et al.*, 1995b).

Treatment Chambers and Equipment

Currently, there are only two commercial systems available (one from PurePulse Technologies, Inc. and one from Thomson-CSF). Various laboratory- and pilot-scale treatment chambers have been designed and used for PEF treatment of foods. They can be classified as static/batch chambers (U-shaped polystyrene and glass-coil static chambers) or continuous chambers (chambers with ion-conductive membranes, chambers with baffles, enhanced-electric-field treatment chambers and coaxial chambers); see Figure 40.5.

A diagram of the PEF processing of foods is depicted in Figure 40.6. The apparatus consists of seven major components (Gongora-Nieto *et al.*, 2002b): a high-voltage power supply, an energy storage capacitor, a treatment chamber(s), a pump to move food though the treatment chamber(s), a cooling device, measuring devices (providing voltage, current and temperature measurements), and a computer to control operations.

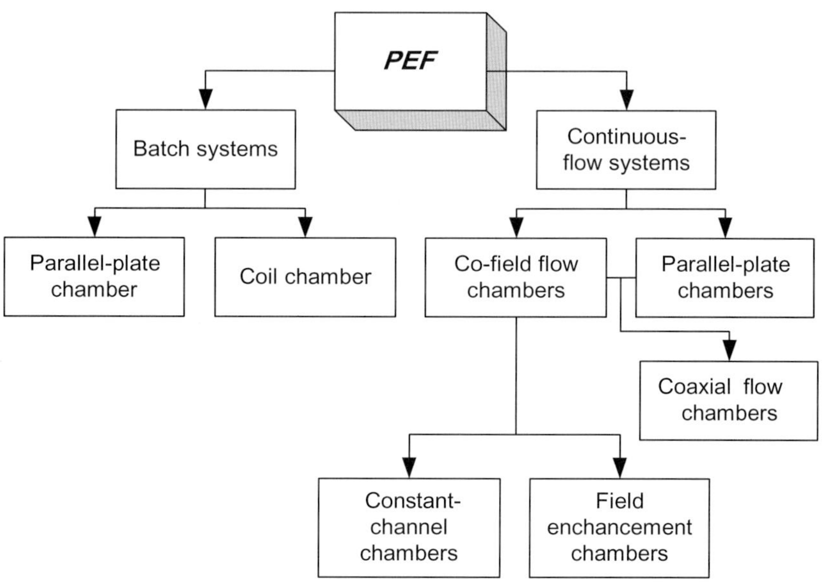

Figure 40.5 Classification of PEF treatment chambers.

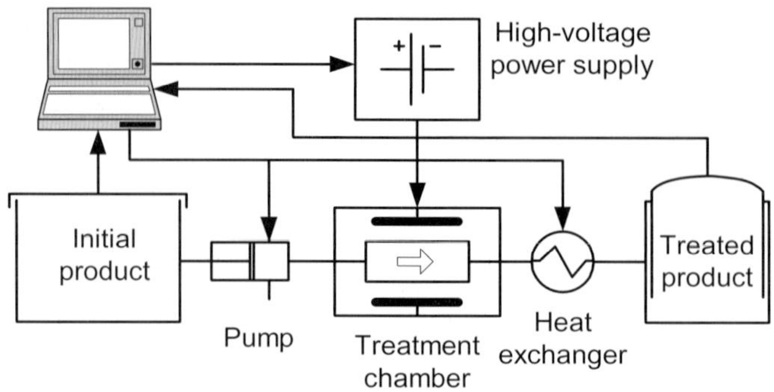

Figure 40.6 Flow chart of PEF food processing.

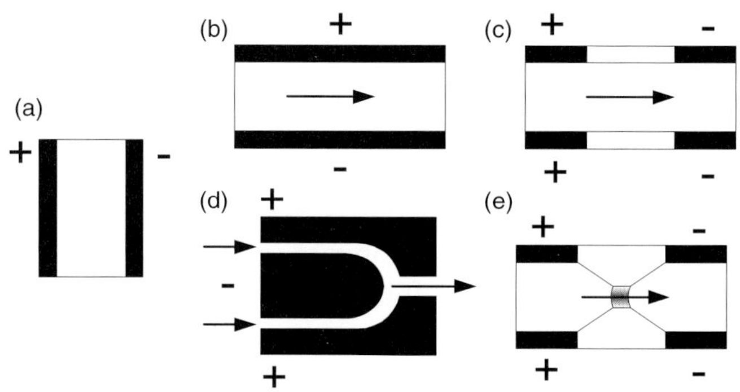

Figure 40.7 Different types of PEF treatment chamber: (a) parallel-plate chamber; (b) continuous-flow parallel-plate chamber; (c) co-field flow chamber; (d) coaxial continuous chamber; (e) enhanced-electric-field continuous-treatment chamber.

Batch-Type Processing ("Static") Chambers

Parallel-Plate Electrode Chambers

One example of this type consists of two carbon electrodes supported on brass blocks placed on a U-shaped polystyrene spacer (Figure 40.7a). Different spacers are used to regulate the electrode area and the amount of food to be treated. The brass blocks are provided with jackets for water recirculation and to control the temperature of the food during PEF treatment. This chamber can support a maximum electric field of $30\,kV{\cdot}cm^{-1}$. A second chamber model designed by Dunn and Pearlman (1987) consists of two stainless steel electrodes and a cylindrical nylon spacer. Another model (Barbosa-Canovas *et al.*, 1999) consists of two round-edged, disk-shaped stainless steel electrodes, with polysulfone used as the insulation material. The effective electrode area is $27\,cm^2$ and the gap between the electrodes can be selected as either 0.95 or 0.5 cm. This chamber can support up to $70\,kV{\cdot}cm^{-1}$. Water circulating at preselected temperatures through jackets built into the electrodes provides cooling of the chamber.

Glass-Coil Static Chambers

A model proposed by Lubicki and Jayaram (1997) uses a glass coil surrounding the anode. The volume of the chamber is $20\,cm^3$, which requires a filling liquid with a high conductivity and a permittivity similar to the sample to be used (the medium used was NaCl solution with $\sigma = 0.8$ to $1.3\,S{\cdot}m^{-1}$, and the filling liquid was water, with $\sigma \sim 10^{-3}\,S{\cdot}m^{-1}$), because there was no inactivation with a nonconductive medium (silicone oil).

Continuous-Flow PEF Treatment Chambers

Continuous-flow PEF treatment chambers (Qin *et al.*, 1995d) are suitable for large-scale operations and are more efficient than static chambers.

Parallel-Plate Chambers (Figure 40.7b)

The first experimental chambers were designed to treat a confined, static volume. Some of the first designs incorporated a parallel-plate geometry using flat electrodes separated by an insulating spacer. The major disadvantage of this type of design is the low productivity of these chambers. Because of limitations on the electric field strength, it is difficult to increase the product load and make this type of chamber more efficient.

Co-field Flow Chambers

Co-field chambers, described by Yin *et al.* (1997), have two hollow cylindrical electrodes separated by an insulator to form a tube through which the product flows (Figure 40.7c). The field distribution in a co-field chamber is not expected to be uniform, although some useful advantages may be gained by special shaping of the insulator. The primary advantage of co-field chambers is that they can be designed to operate in PEF systems at lower currents than for coaxial chambers.

Coaxial Continuous PEF Chambers

Coaxial chambers are generally composed of an inner cylinder surrounded by an outer annular cylindrical electrode that allows food to flow between the two electrodes (Figure 40.7d). A protruding outer electrode surface enhances the electric field within the treatment zones and reduces the field intensity in the remaining portion of the chamber. The electrode configuration shown here was obtained by optimizing the electrode design using a numerical electric-field computation. Using the optimized electrode shape, the prescribed field distribution along the fluid path without an electric-field enhancement point was determined. This treatment chamber has been used successfully for the inactivation of pathogenic and nonpathogenic bacteria, molds, yeasts, and enzymes present in liquid foods such as fruit juices, milk, and liquid whole eggs (Barbosa-Canovas *et al.*, 1999).

Enhanced-Electric-Field Continuous Treatment Chambers

Yin *et al.* (1997) applied the concept of an enhanced electric field in the treatment zones to develop a continuous-flow co-field PEF chamber with conical insulator shapes to eliminate deposits generated from gas in the treatment volume (Figure 40.7e). The conical regions were designed so that the voltage across the treatment zone could be almost equal to the voltage supplied.

Special-Design Flow-Through Chambers

Continuous Chamber with Ion-Conductive Membranes

One design by Dunn and Pearlman (1987) consists of parallel-plate electrodes and a dielectric spacer insulator (Figure 40.8a). The electrodes are separated from the food by conductive membranes made from sulfonated polystyrene and acrylic acid copolymers. An electrolyte is used to facilitate electrical conduction between the electrodes and the ion-permeable membranes.

Chamber with Electrode Reservoir Zones

Another continuous chamber described by Dunn and Pearlman (1987) contains electrode reservoir zones instead of electrode plates (Figure 40.8b). The dielectric spacer insulators have slot-like openings (orifices), between which the electric field is enhanced. The average residence time in each of these two reservoirs is less than 1 min.

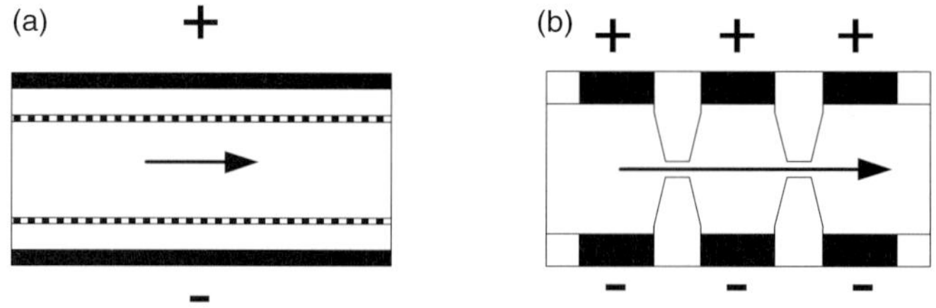

Figure 40.8 Special-design chambers: (a) continuous-treatment chamber with ion-conductive membranes separating the electrode and the food; (b) continuous-treatment chamber with electrode reservoir zones.

Mechanisms of Microbial Inactivation

The application of an electrical field to biological cells in a medium (e.g., water) causes a buildup of electric charges at the cell membrane (Schoenbach *et al.*, 1997). Membrane disruption occurs in many cellular systems when the induced membrane potential exceeds a critical value of 1 V, which, for example, corresponds to an external electric field of about $10\,kV{\cdot}cm^{-1}$ for *E. coli* (Castro *et al.*, 1993). Several theories have been proposed to explain microbial inactivation by PEF treatment (Schoenbach *et al.*, 1997; Velizarov *et al.*, 1998; Angersbach *et al.*, 1999; Bardos *et al.*, 2000; Schoenbach *et al.*, 2001a; Wouters Patrick *et al.*, 2001; Joshi *et al.*, 2002a; Joshi *et al.*, 2002b). Among them, the most studied (see Figure 40.9) are electrical breakdown, and electroporation or disruption of cell membranes (Zimmermann and Benz, 1980).

Electrical Breakdown

Zimmermann (1986) explained what electrical breakdown of a cell membrane entails. The membrane can be considered as a capacitor filled with a dielectric (Figure 40.9). The normal resting potential difference across the membrane V_m is 10 mV and leads to the buildup of a membrane potential difference V due to charge separation across the membrane. V is proportional to the field strength E and the radius of the cell. An increase in the membrane potential leads to a reduction in the thickness of the cell membrane. Breakdown of the membrane occurs if the critical breakdown voltage V_c (of the order of 1 V) is reached by a further increase in the external field strength. It is assumed that breakdown causes the formation of transmembrane pores (filled with conductive solution), which leads to an immediate discharge at the membrane and thus decomposition of the membrane. The breakdown is reversible if the pores produced are small compared with the total surface area of the membrane. With an

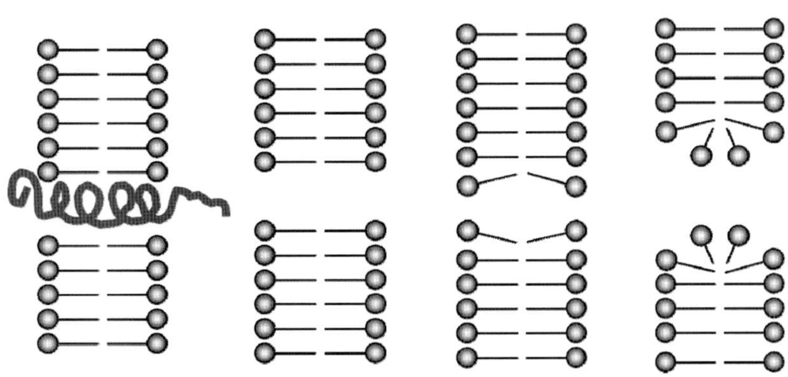

Figure 40.9 Schematic diagram of reversible and irreversible breakdown: pore development and cell membrane disruption.

electric field strength above the critical value and a long exposure time, larger areas of the membrane are subject to breakdown. If the size and number of pores become large in relation to the total surface area of the membrane, reversible breakdown turns into irreversible breakdown, which is associated with mechanical destruction of the cell membrane.

The corresponding electric field is given by $E_{critical} = V_{critical}/fa$, where a is the radius of the cell and f is a form factor that depends on the shape of the cell (Schoenbach *et al.*, 1997). For spherical cells, f is equal to 1.5; for cylindrical cells of length l with hemispheres of diameter d at each end, $f = l(l - d)/3$. Typical values of $V_{critical}$ required for the lysing of *E. coli* are of the order of 1 V. The critical field strength for the lysing of bacteria with dimensions of approximately 1 µm and a critical voltage of 1 V across the cell membrane is therefore on the order of $10\,kV{\cdot}cm^{-1}$ for pulses of 10 µs to several milliseconds in duration (Schoenbach *et al.*, 1997).

Electroporation

Electroporation is a phenomenon that occurs when high-voltage electric-field pulses temporarily destabilize the lipid bilayer and proteins of a cell membrane (Castro *et al.*, 1993; Zimmermann *et al.*, 2000; Joshi *et al.*, 2001; Schoenbach *et al.*, 2001b; Joshi *et al.*, 2002a; Joshi *et al.*, 2002b). The plasma membranes of cells become permeable to small molecules after being exposed to the electric field, and permeation then causes swelling and eventual rupture of the cell membrane (Kakorin *et al.*, 1998; Sukhorukov *et al.*, 1998; Teissie and Ramos, 1998; Schoenbach *et al.*, 2000). The main effect of an electric field on a microorganism cell is an increase in the permeability of the membrane due to compression and poration of the membrane (Winterhalter *et al.*, 1996; Wouters and Smelt, 1997; Ramos and Teissie, 2000; Wouters Patrick *et al.*, 2001; Huang *et al.*, 2003). Kinosita and Tsong (1977) demonstrated that an electric field of $2.2\,kV{\cdot}cm^{-1}$ induced pores in human erythrocytes approximately 1 nm in diameter. They suggested a two-step mechanism for pore formation in which the initial perforation is a response to an electrical suprathreshold potential, followed by a time-dependent expansion of the pore size (Figure 40.9). Larger pores are obtained by increasing the intensity of the electric field and the pulse duration or by reducing the ionic strength of the medium (Grahl and Maerkl, 1996; Joshi *et al.*, 2001).

Events in Electroporation and Microbial Lysis

Electric-Field-Induced Transmembrane Potential

It is known that when a cell (radius = R_{cell}) suspended in a medium is exposed to an external electric field (a DC field of strength E_{appl}), there is a rapid redistribution of cations in the vicinity of the plasma membrane, thus generating a transmembrane potential $\Delta\psi_{membr}$ with a rise time τ_{membr}:

$$\Delta\psi_{\text{membr}} = 1.5R_{\text{cell}}E_{\text{appl}}\cos\theta[1-\exp(-t/\tau_{\text{membr}})] \tag{40.27}$$

$$\tau_{\text{membr}} = R_{\text{cell}}C_{\text{membr}}(r_{\text{int}} + r_{\text{ext}}/2) \tag{40.28}$$

Here, θ is the angle between the field line and the normal from the center of the spherical cell to a point of interest on the membrane surface, and C_{membr}, r_{int} and r_{ext} are the membrane capacitance (per unit area), the resistivity of the cytoplasmic fluid and the resistivity of the external medium, respectively (Cole, 1972). For biological cells with diameters of micrometers, $\tau_{\text{membr}} < 1\,\mu s$ and the exponential term in Equation 40.27 approaches zero within 1 µs. Cells of larger diameter have $\tau_{\text{membr}} > 1\,\mu s$ (Kinosita *et al.*, 1988). The maximum transmembrane potential generated in a cell by a DC electric pulse a few times longer than τ_{membr} is

$$\Delta\psi_{\text{membr,max}} = 1.5R_{\text{cell}}E_{\text{appl}} \tag{40.29}$$

Kinetics of Electroporation in Cell Membranes

The plasma membrane of a cell is the first site of the interaction with the electric field. Beside lipids, there are proteins, carbohydrates and other types of molecules in the plasma membrane, most of which are either charged or polarizable. Channel proteins are especially sensitive to the value of $\Delta\psi_{\text{membr}}$, and each type of channel has a range of $\Delta\psi_{\text{membr}}$ in which it becomes conductive. The range of $\Delta\psi_{\text{membr}}$ for opening protein channels is approximately 50 mV, considerably smaller than the dielectric strength of the lipid bilayer, which is in the range of 150–400 mV. Like a lattice defect in the lipid bilayer, once a protein channel is forced to open, a strong current greatly exceeding the normal conduction of the channel will generate local heat sufficient to denature the protein. This denaturation may bc reversible or irreversible, depending on the extent of the temperature change and the properties of the channel. The time of opening and closing of the protein channel is in the submicrosecond range (Tsien *et al.*, 1987). Thermal denaturation of a protein takes milliseconds to seconds. Renaturation of a protein occurs in seconds (Kim and Baldwin, 1982). Electropores in lipid domains can reseal within seconds (Teissie and Tsong, 1981). Closing of PEF-perforated protein channels should occur in milliseconds. However, repair of a PEF-damaged cell membrane can take minutes to hours (Kinosita and Tsong, 1977).

Colloid Osmotic Lysis

A major difference between the electroporation of lipid vesicles and that of cells is the colloid osmotic lysis of cells (Kinosita and Tsong, 1977). A PEF-perforated cell membrane ceases to be a permeation barrier to ions and small molecules but not necessarily to proteins. The electroperforated membrane becomes semipermeable to cytoplasmic macromolecules. The osmotic pressure of these macromolecules causes the cell to swell. This process, known as colloidal swelling, eventually leads to rupture of the plasma membrane because of the excessive osmotic pressure imposed on the cell.

Colloidal osmotic pressure in PEF-treated red blood cells was identified as the main cause of electric-field-stimulated hemolysis. Colloidal swelling depends on the osmotic imbalance between the cytoplasm and the suspending medium. When the difference is large, PEF-treated cells will swell in times in the range of minutes. During this swelling phase, the electropores in the cell membranes also begin to reseal. If the resealing takes place faster than the swelling, the cells will shrink again and recover their original volume, thus averting membrane rupture. If, on the other hand, the resealing is slower than the swelling, the plasma membrane of the cells will be ruptured. Colloidal osmotic lysis may be prevented by balancing the osmotic pressures of the cytoplasm and the medium.

Electroosmosis in Electropores

An electric field parallel to a surface/liquid interface will cause a net hydrodynamic flow in the appropriate direction if there is an imbalance in the numbers of positive and negative charges in the layer of liquid adjacent to the charged surface. If electropores, which are expected to be induced closer to the "poles" of the cell that face the electrodes, are viewed as cylinders with an average net negative charge (from ionized headgroups of phospholipids and ionized amino acid side chains on integral proteins) on this surface and with their axes perpendicular to the plane of the membrane, then a hydrodynamic flow would be expected during an electric-field pulse (Figure 40.10). It has been predicted and experimentally demonstrated that the overall permeabilization difference between the two hemispheres is less than what was originally expected if an electroporation experiment is conducted taking electroosmosis into account (Sowers, 1988).

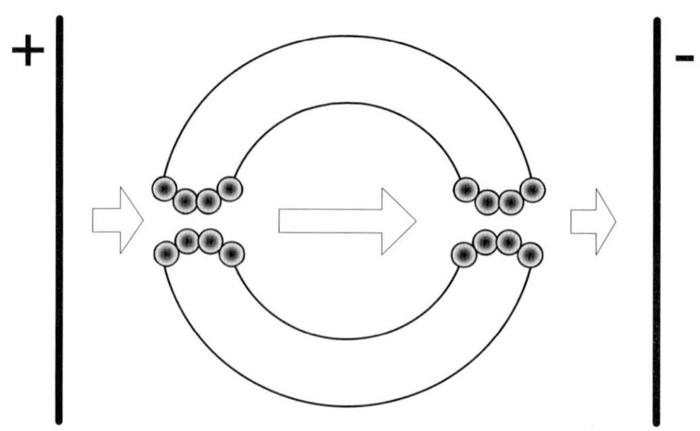

Figure 40.10 Electroosmosis-induced hydrodynamic flow toward the negative electrode, regardless of whether the electropores are in the positive-facing or negative-facing hemisphere.

Kinetics of Microbial Inactivation

Three types of factors that affect microbial inactivation by PEF treatment have been identified:

1. process factors (electric field intensity, pulse width, treatment time and temperature, and pulse waveshape);
2. microbial-entity factors (type, concentration and growth stage of microorganisms);
3. treatment-medium factors (pH, antimicrobial and ionic compounds, conductivity, and ionic strength).

Hülsheger and Niemann (1980) were the first to propose a mathematical model for inactivation of microorganisms by PEF treatment. Their model was based on representing the dependence of the survival ratio $S = N/N_0$ (the ratio of living-cell counts before and after PEF treatment) on the electric field intensity E by the following expression:

$$\ln S = -b_E(E - E_c) \tag{40.30}$$

where b_E is the regression coefficient, E is the applied electric field, and E_c is the critical electric field value obtained by extrapolating E to 100% survival.

The regression coefficient reflects the gradient of a straight survival curve and is a constant that depends on the microorganism and the medium. The critical electric field was found to be a function of the cell size and the pulse duration. Hülsheger *et al.* (1981) proposed an inactivation kinetic model that relates the microbial survival fraction S to the PEF treatment time t in the form

$$\ln S = -b_t \ln \frac{t}{t_c} \tag{40.31}$$

where b_t is the regression coefficient, t is the treatment time and t_c is the critical treatment time, or extrapolated value of t for 100% survival.

The model proposed by Peleg (1995) describes the sigmoid shape of the pathogen survival curves generated by PEF inactivation. This model represents the percentage of surviving organisms as a function of the electric field and the number of pulses applied. The model is defined by a critical electric field intensity that corresponds to 50% survival (E_d), and a kinetic constant K that is a function of the number of pulses, representing the steepness of the sigmoid curve. Generalized equations for the two models are shown in Table 40.2.

For both models, small values of the kinetic constants indicate a wide span in the inactivation rate curves and hence a lower sensitivity to PEF treatment, whereas large values imply a steep decline or a higher susceptibility. Lower E_c (or E_d) values indicate a lower resistance of the pathogens to the PEF treatment.

Table 40.2 Inactivation models.

Hülsheger's model (Hülsheger and Niemann, 1980)	**Peleg's model (Peleg, 1995)**
$S=\left(\frac{t}{t_c}\right)^{\frac{-(E-E_c)}{K}}$	$S=\frac{1}{1+e^{\frac{E-E_d}{K}}}$
E, electric field; t, treatment time; E_c, critical electric field; t_c, critical time; K, kinetic constant.	E_d, electric field when 50% of population is inactivated; K, kinetic constant.

Microbial Factors in the Efficacy of PEF Processing

Type of microorganism. Gram-positive bacteria ones are more resistant to PEF treatment than Gram-negative ones (Hülsheger *et al.*, 1983). In general, yeasts are more sensitive to electric fields than bacteria, owing to their larger size, although at low electric fields they seem to be more resistant than Gram-negative cells (Sale and Hamilton, 1967a; Qin *et al.*, 1995a). A comparison between the inactivation of two yeast species of different sizes showed that the electric field intensity needed to achieve the same inactivation level was inversely proportional to the cell size. These results are logical, but inconsistent with the results obtained by Hülsheger *et al.* (1983). Further studies are needed to better understand the effect of microorganism type on the effectiveness of microbial inactivation.

Growth stage of microorganism. Bacterial cells in the logarithmic phase are more sensitive to various stresses than are cells in the lag and stationary phases. Microbial growth in the logarithmic phase is characterized by a high proportion of cells undergoing division, during which the cell membrane is more susceptible to an applied electric field. Gaskova *et al.* (1996) reported that the killing effect of PEF treatment for *S. cerevisiae* in the logarithmic phase is 30% greater than in the stationary phase of growth.

Microbial Inactivation by PEF Treatment

Numerous publications on microbial inactivation have presented data on vegetative cells, the majority of them from a few genera. Extensive microbial-inactivation tests have been conducted to validate the concept of PEF treatment as a nonthermal food pasteurization process (Castro *et al.*, 1993; Qin *et al.*, 1995a,b; Zhang *et al.*, 1995b; Vega-Mercado *et al.*, 1996a; Vega-Mercado *et al.*, 1996b).

An applied intense pulsed electric field produces a series of degradative changes in blood, algae, bacteria and yeast cells (Castro *et al.*, 1993). These changes include electroporation and disruption of semipermeable membranes, which lead to cell swelling and/or shrinkage, and finally to lysis of the cells. The mechanisms for the inactivation of microorganisms include electric breakdown, an ionic punch-through effect and electroporation of cell membranes (Qin *et al.*, 1994). The inactivation of microorgan-

Table 40.3 Some bacteria effectively inactivated by PEF treatment.

Bacillus cereus (Rowan *et al.*, 2000; Pol *et al.*, 2001)	*Pseudomonas aeruginosa* (Lado and Yousef, 2002)
B. subtilis (Cserhalmi *et al.*, 2002)	*Pseudomonas fluorescens* (Vega-Mercado *et al.*, 1995; Bendicho *et al.*, 2002b)
Bacillus subtilis spores (Heinz *et al.*, 2001)	*Saccharomyces cerevisiae* (Zhang *et al.*, 1994)
Candida famata (Voronovsky *et al.*, 2002)	*Salmonella* (Jeantet *et al.*, 1999; Mattick *et al.*, 2001)
Escherichia coli (Unal *et al.*, 2001; Unal *et al.*, 2002)	*Staphylococcus aureus* (ATCC 25923) (Qin *et al.*, 1995d)
Listeria innocua (Calderon *et al.*, 1999a,b)	*Yersinia enterocolitica* (Lubicki and Jayaram, 1997)
Listeria monocytogenes (Scott A) (Dutreux *et al.*, 2000; Unal *et al.*, 2001; Unal *et al.*, 2002; Alvarez *et al.*, 2003)	
Lactobacillus leichmannii (Unal *et al.*, 2001; Wouters Patrick *et al.*, 2001; Rodrigo Ruiz *et al.*, 2003)	

isms is primarily caused by an increase in the permeability of their membranes due to compression and poration (Vega-Mercado *et al.*, 1996b).

Castro *et al.* (1993) reported a 5-log reduction in bacteria, yeast and mold counts suspended in milk, yoghurt, orange juice and liquid egg treated with a PEF. Zhang *et al.* (1995a) achieved a 9-log reduction in *E. coli* suspended in simulated milk ultrafiltrate (SMUF), treated with a PEF by applying a convergent electric field of strength $70\,kV{\cdot}cm^{-1}$, with a short treatment time of $160\,\mu s$. These processing conditions and results are adequate for commercial food pasteurization, which requires reduction by 6 to 7 log cycles. Table 40.3 presents the bacteria reported to be successfully inactivated by PEF treatment.

PEF Process Calculations and Variables

To treat foods with a PEF in a continuous system, a liquid food product is pumped through a series of treatment zones in a chamber with high-voltage electrodes on one side of each zone and a low-voltage electrode on the other side. The PEF process conditions are defined by the electric field strength applied and the treatment time.

Electric Field Intensity

This is one of the main factors influencing microbial inactivation (Hülsheger and Niemann, 1980; Dunne *et al.*, 1996). The microbial inactivation increases with an increase in the electric field intensity, above the critical transmembrane potential (Qin *et al.*, 1998). This is consistent with the electroporation theory, in which the induced potential difference across the cell membrane is proportional to the applied electric field. The critical electric field E_c (the electric field intensity below which inactivation does not occur) increases with the transmembrane potential of the cell. Transmembrane potentials, and consequently E_c, are larger for larger cells (Jeyamkondan *et al.*, 1999). The pulse width (or duration) also influences the critical electric field. For instance, for pulse widths greater than 50 μs, E_c was found to be 4.9 kV·cm^{-1}. but for pulse widths less than 2 μs, E_c was 40 kV·cm^{-1} (Schoenbach *et al.*, 1997).

Treatment Time

The treatment time is defined as the product of the number of pulses and the duration of a single pulse. An increase in either of these two variables improves microbial inactivation (Sale and Hamilton, 1967a). As noted above, the pulse width influences the microbial reduction by affecting E_c. Longer pulse widths decrease E_c, which results in higher inactivation; however, an increase in the pulse duration may also result in an undesirable temperature increase in the food. Optimum processing conditions therefore need to be established to obtain the highest inactivation rate with the lowest heating effect (Knorr *et al.*, 2001; Gongora-Nieto *et al.*, 2002a). The inactivation of microorganisms increases with the treatment time (Hülsheger *et al.*, 1983). In certain cases, however, the increase in the inactivation rate reaches saturation as the number of pulses increases. This is the case for the inactivation of *Saccharomyces cerevisiae* by PEF treatment, which reaches saturation with 10 pulses of an electric field of strength 25 kV·cm^{-1} (Barbosa-Canovas *et al.*, 1999).

The critical treatment time also depends on the electric field intensity applied (Zhang *et al.*, 1995b; Alvarez *et al.*, 2003; Rodrigo *et al.*, 2003). At electric field values above E_c, the critical treatment time decreases as the electric field increases. Barbosa-Cánovas *et al.* (1999) reported that for an electric field strength 1.5 times higher than E_c, the critical treatment time remained constant.

Pulse Waveshape

Electric-field pulses may be applied in the form of exponentially decaying, square-wave, oscillatory, bipolar or instant-reverse-charge pulses (Qin *et al.*, 1995d; Changjiang *et al.*, 2000). Oscillatory pulses are the least efficient for microbial inactivation, and square-wave pulses are more energy-efficient and lethal than exponentially decaying pulses (Zhang *et al.*, 1995b; Barbosa-Canovas and Zhang, 2001). Bipolar pulses are more lethal than monopolar pulses, because the PEF causes movement of charged molecules in the cell membranes, and a reversal in the orientation or polarity of the electric field causes a corresponding change in the direction of movement of the charged molecules (Qin *et al.*, 1994; Ho *et al.*, 1995). A difference between bipolar and monopolar pulses was reported in studies of the inactivation of *Bacillus* spp. (Ho and Mittal, 1997) and *E. coli* (Qin *et al.*, 1994). With bipolar pulses, the alternating changes in the movement of charged molecules cause a stress in the cell membrane and enhance its electric breakdown. Bipolar pulses also offer the advantages of minimum energy utilization, reduced deposition of solids on the electrode surface and decreased food electrolysis (Barbosa-Canovas *et al.*, 1999).

A study conducted by Zhang *et al.* (1997) showed the effects of square-wave, exponentially decaying and instant-charge-reversal pulses on the shelf-life of orange juice.

Square-wave pulses were more effective, yielding products with longer shelf-lives than those treated with exponentially decaying and charge-reversal pulses. In agreement with this study, Love (1998) quantitatively demonstrated the stronger inactivation effect of square-wave pulses over all other wave shapes.

Treatment Temperature

Experimental results have demonstrated that, in challenge tests, both treatment temperatures and process temperatures impact microbial survival and recovery (Zhang *et al.*, 1995b; Barsotti *et al.*, 1998).

PEF treatments at moderate temperatures (50 to 60 °C) have been shown to exhibit synergistic effects on the inactivation of microorganisms (Dunn and Pearlman, 1987; Jayaram *et al.*, 1992). At a constant electric field strength, pathogen inactivation increases with an increase in treatment temperature (Calderon *et al.*, 1999a). Since application of an electric field causes increase in the temperature of the food being treated, proper cooling of the treatment chamber is necessary to maintain food temperatures far below those that occur during a thermal pasteurization process (Bendicho *et al.*, 2002a; Lindgren *et al.*, 2002; Ohshima *et al.*, 2002).

Additional effects of a high treatment temperature include changes in the fluidity and permeability of the cell membrane, which increase the susceptibility of the cell to mechanical disruption (Hülsheger *et al.*, 1981). Also, a low transmembrane potential decreases E_c and therefore increases inactivation (Jeyamkondan *et al.*, 1999).

Electrochemistry of a Highly Polarized Electrode/Food Product Interface

Usually, PEF processing is considered as a "zero chemistry" treatment with no chemical reactions involved. However, the reported changes in the sensory and physical attributes of processed foods are not solely the result of Joule heating and the high electric current that passes through the food product. The treatment chambers in all existing PEF systems have an extremely high ratio of electrode surface area to treatment volume owing to limitations on the power supply and electric field strength. Therefore, the electrode materials are directly involved in the PEF treatment process. They interact with the treated food products by electrochemical reactions that occur at the surface of the highly polarized electrodes, and by electric-double-layer-assisted reactions of food particulates (the solid phase of the food product) with the electrode surface. These interactions include (Vetter, 1967; Bard and Faulkner, 1980; Bockris and Reddy, 1998):

- adsorption of organic and inorganic anions;
- changes in the chamber capacitance due to changes in the relaxation time of the electric double layer;
- electrophoretic deposition of food solids;
- electrocoagulation of solid phases at the electrode surface;

- electrodissolution of electrode material;
- hydrogen or oxygen evolution due to electrode reactions.

Interactions of the electrode surface with food matrix components have been previously underestimated by researchers and rarely investigated (Barbosa-Canovas and Zhang, 2001). Owing to the importance of the electrode surface properties to the behavior of food products in the vicinity of electrodes and therefore to the PEF treatment process, the electrochemical polarization characteristics of the electrode material play a critical role. This is also important because electrical impulses can accelerate electrochemically induced changes in the electrode polarization and the food properties.

Potential-scanning measurements of various food products in the range from ~0.5 to 3 V have been performed for two widely used electrode materials: aluminum (alloy 2024) and stainless steel. The electrochemical potential is the thermodynamic potential that characterizes the ability of an electrode to react in a solution. Increasing the thermodynamic potential by 1 V is equivalent to changing the reaction temperature by 10^3 K. An electrode polarization in the range between 2 and 3 V is considered as an extremely high polarization.

Despite their general similarity, these two electrodes demonstrate different behavior in an acid medium (orange juice) (Figure 40.11). In the low-polarization region, aluminum shows a more stable behavior and less corrosive activity. However, to increase the electric current through an aluminum electrode, one needs a very high voltage. A

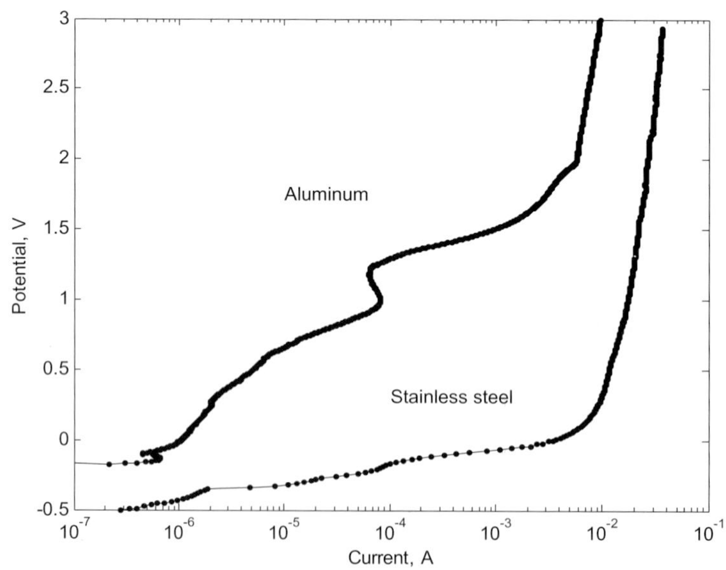

Figure 40.11 Polarization of stainless steel and aluminum electrodes in orange juice.

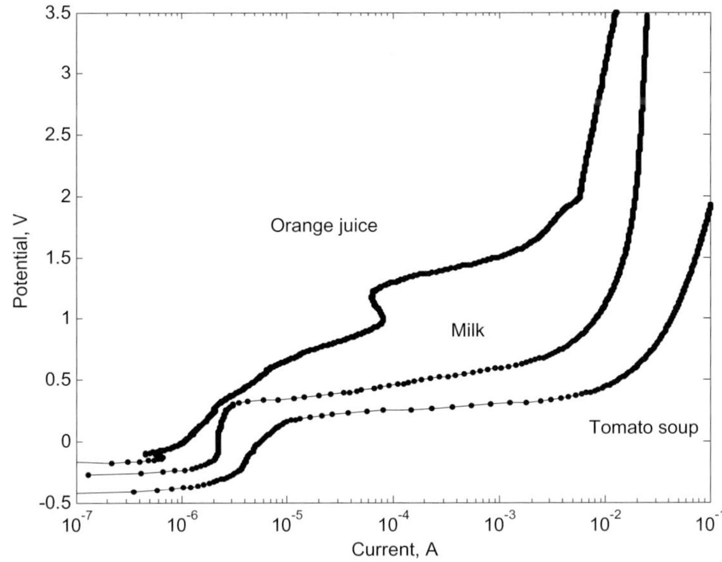

Figure 40.12 Potentiodynamic curves of a polarized stainless steel electrode for various food products.

stainless steel electrode can support a higher current owing to a lower adsorption of HO_3 ions on the electrode surface. In the high-polarization region, the two electrodes demonstrate similar volt–ampere characteristics. However, stainless steel has its potentiodynamic curve shifted in the direction of higher values of the electric current, and therefore is more suitable for PEF applications. For low-electric-field applications (ohmic heating), aluminum electrodes are preferred.

The composition of the food product is a significant factor influencing electrode processes. The potentiodynamic characteristics of stainless steel electrodes in orange juice, whole milk and tomato soup are depicted in Figure 40.12. Tomato soup has classical corrosion-type characteristics that include a Tafel region (adsorption) and electrodissolution (corrosion) of the electrode material (Bockris and Reddy, 1998). Orange juice has similar characteristics except for two regions of current–voltage instability, which can be explained by the aggregation of polarized pulp at the electrode surface. The most unusual potentiodynamic curve corresponds to the electrochemical treatment of milk. This type of curve usually corresponds to a passivated metal electrode. At low polarization potentials, the behavior of the stainless steel electrode in milk does not differ from that in the other food products. However, an increase in the potential leads to the deposition of milk constituents onto the electrode and blockage (passivation) of its surface. In the high-polarization regime, all three products behave similarly.

Mathematical Model of Continuous Operation (Esplugas *et al.*, 2001)

From an engineering point of view, it is of interest to differentiate between single-pass and recirculation modes of operation of a PEF treatment chamber. In both cases, the mathematical model consists of energy and mass balances, kinetic equations, and equilibrium conditions. It is possible to build a large and complicated mathematical model, but that would not be useful. In order to simplify the model, some assumptions may be made. Accordingly, plug-type flow in the PEF chamber, perfect mixing in the tank, and first-order kinetics for the inactivation of microorganisms were assumed in the work described here.

A simplified schematic illustration of a PEF installation operating in a single-pass mode is depicted in Figure 40.6. It can be assumed that the concentration of microorganisms in the feed tank c_T (microorganisms·L^{-1}) is the same as that at the PEF chamber inlet, and that the rate of microorganism destruction r (microorganisms·(L·s^{-1})$^{-1}$) follows first-order kinetics with respect to the microorganism concentration c (microorganisms/L):

$$r = -kc \tag{40.32}$$

where k (s^{-1}) is the kinetic constant of microorganism inactivation.

Assuming a stationary state and plug-type flow in the PEF chamber (Levenspiel, 1972), the microorganism balance gives the following equation:

$$q \ln(c/c_T) = -kV_r t \tag{40.33}$$

where q (L/s) is the fluid flow rate and V_r (L) is the volume of the PEF chamber.

According to Equation 40.33, the relation between the outlet microorganism concentration c (microorganisms/L) and the time t (s) is exponential:

$$c = c_T e^{-kV_r/q} \tag{40.34}$$

The energy balances are more complex. The energy E (J) dissipated during the discharge of the capacitor C (F) at a voltage V (V) is given by the following equation:

$$E = \frac{1}{2}CV^2 \tag{40.35}$$

Taking into account the frequency f (s^{-1}) of the pulses, the energy dissipation per second Q (J·s^{-1}) during the flow of the liquid through the chamber is

$$Q = \frac{1}{2}fCV^2 \tag{40.36}$$

However, only a fraction ϕ of this energy will heat the liquid food (flow rate q (L·s^{-1}), density ρ) that passes through the PEF chamber. This ratio ϕ must be less than 1, and depends strongly on the electrical conductivity of the food product. The energy balance for the PEF chamber after the stationary state is reached is represented by

$$q\rho C_p(T - T_T) = \phi Q \tag{40.37}$$

where q (L/s), ρ (kg·L^{-1}) and C_p (J·kg^{-1}·°C^{-1}) are the flow rate, density and specific heat capacity of the liquid food product, respectively; and T_{T} and T (°C) are the temperatures of the food in the feed tank and in the chamber, respectively. Consequently, the increase in temperature $T - T_{\mathrm{T}}$ of the food can be estimated as

$$(T - T_{\mathrm{T}}) = \frac{1}{2}\frac{\phi f C V^2}{q\rho C_p} \tag{40.38}$$

Process Calculations

The total possible temperature change per pair of electrodes in the treatment chamber (ΔT), the total energy input during the treatment per electrode pair (P), and the Reynolds number (N_{Re}) can be calculated using the following equations:

$$\begin{aligned} \Delta T &= \left(E^2 t\sigma / \rho C_p\right)/n \\ P &= E^2 t\sigma / n \\ N_{\mathrm{Re}} &= \rho D u / \mu \end{aligned} \tag{40.39}$$

The process variables used in these equations are described in Table 40.4.

Physical Properties of Food Products for PEF Processing

The physical properties of foods that are the most critical for the efficacy of PEF treatment are the electrical conductivity, density, specific heat capacity and viscosity of

Table 40.4 PEF process variables.

Process variable	Notation	Units
Electric field strength	E	V·m^{-1}
Total treatment time	t	s
Number of electrode pairs in treatment chamber	n	
Treatment zone diameter	D	m
Mean liquid velocity	μ	m·s^{-1}
Electrical conductivity of product	σ	S·m
Density of product	ρ	kg·m^{-3}
Specific heat capacity of product	C_p	kJ·kg^{-1}·°C^{-1}

the product. Some useful data can be found in Barbosa-Canovas and Zhang (2001). Liquid foods contain several ion species that carry an electrical charge and conduct electricity. At a given voltage, the electric current is directly proportional to the electrical conductivity of the food product (Zhang *et al.*, 1995b). An increase in the electrical conductivity causes an increase in the overall energy input and a change in the temperature of the product during processing.

The density and specific heat capacity of the food product affect the temperature change during PEF treatment. As the density of the product decreases, the total temperature change increases (Zhang *et al.*, 1995b). Similarly, a decrease in the specific heat capacity also increases the temperature change during PEF processing.

The viscosity of the product determines its flow characteristics, which can be calculated based on the Reynolds number. For a Reynolds number greater than 5000, the product flow is turbulent, which provides a uniform velocity profile in the treatment chamber, which, in turn, is likely to provide uniformity of the PEF process (Barsotti *et al.*, 1998; Lindgren *et al.*, 2002).

Conductivity, pH and Ionic Strength

The electrical conductivity of a medium (σ, $\Omega^{-1}\cdot m^{-1}$), which is defined as the ability to conduct electric current, is an important variable for PEF processing. The electrical conductivity is the reciprocal of the resistivity (r), which is measured in $\Omega\cdot m$. Foods with large electrical conductivities generate smaller peak electric fields across the treatment chamber, and therefore are not susceptible to PEF treatment (Barbosa-Canovas *et al.*, 1999). Studies of the inactivation of *Lactobacillus brevis* by PEF treatment showed that as the conductivity of the fluid increased, the resistance of the treatment chamber was reduced (Jayaram *et al.*, 1992), which in turn reduced the pulse width and decreased the rate of inactivation. Since an increase in the conductivity of the medium results from an increase in its ionic strength, the latter leads to a decrease in rate of inactivation of bacteria. Furthermore, an increased difference between the conductivities of the medium and the microbial cytoplasm weakens the membrane structure owing to an increased flow rate of ions across the membrane. Thus, the rate of inactivation of microorganisms increases with decreasing conductivity, even at equal input energy (Jayaram *et al.*, 1992). However, another study, performed by Dunne *et al.* (1996), showed that the resistivity had no influence on the effectiveness of inactivation of *E. coli* and *L. innocua* by PEF treatment. These contradictory results may be due to the microorganisms or medium used.

Vega-Mercado *et al.* (1996b) studied the effect of the pH and ionic strength of a medium (SMFU) during PEF treatment. The inactivation ratio increased from not detectable (zero) to 2.5 log cycles as the ionic strength of the solution was adjusted from 168 to 28 mM. At 55 $kV\cdot cm^{-1}$ (eight pulses), as the pH was reduced from 6.8 to 5.7, the inactivation ratio increased from 1.45 to 2.22 log cycles. The PEF treatment and the ionic strength of the solution were responsible for electroporation and compression of the cell membrane, whereas the pH of the medium affected the cytoplasm when the electroporation was complete. Dunne *et al.* (1996) reported that, depending on the microorganism, an acidic pH enhanced microbial inactivation, although no specific details were provided (what microorganisms were affected or what range of pH was used).

Particulate Foods

The inactivation of microorganisms in liquid–particulate systems was studied by Dunne *et al.* (1996). *E. coli*, *L. innocua*, *Staphyloccocus aureus* and *Lactobacillus acidophilus* were suspended in a 2 mm diameter alginate bead model system, and the effects of the PEF process variables on microbial inactivation were tested. The researchers concluded that the process was effective in killing microorganisms embedded in particulates. However, to achieve more than a 5-log-cycle reduction, high energy inputs were needed (70–100 $J \cdot ml^{-1}$, depending on the bacteria treated). Qin *et al.* (1995c) reported that dielectric breakdown occurred when air or vapor was present in the food because of the difference in dielectric constants between the liquid and the gas.

Application of PEF Treatment to Food Preservation

PEF treatment has been mainly applied to preserve the quality of foods, such as to improve the shelf-life of apple juice, cranberry juice, skim and chocolate milk, orange juice, liquid egg, and pea soup (Vega-Mercado *et al.*, 1996a).

Processing of Apple Juice and Cider

Simpson *et al.* (1995) reported that apple juice from concentrate treated with a PEF at $50\,kV \cdot cm^{-1}$ electric field strength, with 10 pulses of 2 μs pulse duration, and a maximum processing temperature of 45 °C, had a shelf-life of 28 days, compared with a shelf-life of 21 days for fresh-squeezed apple juice. There were no physical or chemical changes in the ascorbic acid or sugars in the PEF-treated apple juice, and a sensory panel found no significant differences between the untreated and electric-field-treated juices. PEF treatment extended the shelf-life of fresh apple juice; the treated apple juice had shelf-lives at 22 °C and 25 °C of more than 56 days and 32 days, respectively. There was no apparent change in the physicochemical and sensory properties of the juice. Evrendilek *et al.* (2001a) showed that PEF treatment of apple juice and PEF plus mild heat treatment of apple cider improved the shelf-life of the products compared with control samples at 4, 22 and 37 °C without degradation of vitamin C or any change in the color, measured by the L (white if $L = 100$, black if $L = 0$), a ($-a$ = green, a = red) and b ($-b$ = blue, $+b$ = yellow) values.

Processing of Orange Juice

A reduction in the native microbial flora of freshly squeezed orange juice by 3 log cycles was obtained with an applied electric field of 15 kV·cm^{-1} without significantly affecting the quality. The shelf-life of reconstituted orange juice treated in an integrated PEF pilot plant system consisting of a series of co-field chambers was evaluated by Qui *et al.* (1998) and Zhang *et al.* (1997). It was reported that the total aerobic counts were reduced by 3 to 4 log cycles for an electric field strength of under 32 kV·cm^{-1}. When stored at 4 °C, both heat- and PEF-treated juices had a shelf-life of more than 5 months. Vitamin C losses were lower and color was generally better preserved in PEF-treated juices than in heat-treated juices at up to 90 days (at a storage temperature of 4 or 22 °C) or 15 days (at a storage temperature of 37 °C) after processing. In the study by Yeom *et al.* (2000), orange juice was treated by PEF, and with an increase in electric field strength a longer shelf-life was obtained. Compared with heat treatment, more flavor components were retained in the PEF-treated orange juice.

Processing of Cranberry Juice

Cranberry juice was treated either by high-voltage PEF treatment at 20 kV·cm^{-1} and 40 kV·cm^{-1} for 150 μs or by thermal treatment at 90 °C for 90 s. Higher electric fields and longer treatment times reduced the number of viable microbial cells by a greater amount. The overall profile of volatile components was not affected by PEF treatment, but was affected by heat treatment. Compared with control samples, PEF treatment caused no color change in the samples. When the treatment conditions were 40 kV·cm^{-1} for 150 μs, there was no mold or yeast growth at either 22 and 4 °C and no bacterial growth at 4 °C (Jin and Zhang, 1999). When PEF treatment (32 kV·cm^{-1} electric field strength, 500 pulses per second (pps) frequency, 1.4 μs pulse duration and 47 μs total treatment time) and PEF plus heat treatment (60 °C for 32 s) were applied to cranberry juice, the shelf-life of juice stored at both 4 and 22 °C increased significantly (up to 197 days). The PEF and PEF plus heat treatments did not cause any significant differences in the color retention of the samples (Evrendilek *et al.*, 2001a).

Processing of Milk

A study of the inactivation of *Salmonella dublin* in homogenized milk and of the shelf-life of the product was performed using an electric field strength of 36.7 kV·cm^{-1} and a treatment time of over 25 min (Dunn and Pearlman, 1987). *S. dublin* was not detected after PEF treatment or after storage at 7–9 °C for 8 days. The naturally occurring bacterial population in the milk increased to 10^7 CFU·ml^{-1} in the untreated milk, whereas the treated milk showed approximately 4×10^2 CFU·ml^{-1}. Further studies by Dunn indicated less flavor degradation and no chemical or physical changes in the milk quality attributes for cheesemaking.

Studies were made of the shelf-life of raw skim milk (0.2% milk fat), treated with a PEF at 40 kV·cm^{-1}, with 30 pulses and 2 μs treatment time, using exponentially decaying pulses. The shelf-life of the milk was 2 weeks when it was stored at 4 °C; however, treatment of raw skim milk at 80 °C for 6 s followed by PEF treatment at 30 kV·cm^{-1}, with 30 pulses and 2 μs pulse width, increased the shelf-life to up to 22 days, with a total aerobic plate count of 3.6 log CFU·ml^{-1}. The inactivation of *Listeria monocytogenes* Scott A in pasteurized whole, 2% and skim milk by PEF treatment was also studied. *L. monocytogenes* was reduced by 1 to 3 log cycles at 25 °C and by 4 log cycles at 50 °C, with no significant differences being found among the three milks. The lethal effect of PEF treatment was a function of the field intensity and treatment time. In studies of the PEF inactivation of *Listeria innocua* suspended in skim milk and its subsequent sensitization to nisin the microbial population of *L. innocua* was reduced by 2.5 log cycles after PEF treatments at 30, 40 and 50 kV·cm^{-1}. The same PEF intensities and subsequent exposure to 10 IU·ml^{-1} of nisin achieved reductions of *L. innocua* by 2, 2.7 and 3.4 log cycles, respectively.

Similarly to the study of cranberry juice described above, chocolate milk was processed by PEF treatment (35 kV·cm^{-1} electric field strength, 600 pps frequency, 1.4 μs pulse duration and 45 μs total treatment time) and by PEF plus heat treatment (105 and 112 °C for 31.5 s, referred to in the following as "PEF + 105 °C" and "PEF + 112 °C," respectively) in a pilot-plant PEF processing system (Figure 40.4). Compared with control samples, the shelf-life of chocolate milk treated by PEF + 105 °C and PEF + 112 °C increased significantly at 4, 22 and 37 °C. The PEF plus heat treatments did not cause any significant differences in the color retention. Total plate counts during storage at 22 °C are shown in Figure 40.13.

Processing of Eggs

PEF studies on liquid egg, on heat-pasteurized liquid egg products, and on egg products with potassium sorbate and citric acid added as preservatives were conducted by Dunn

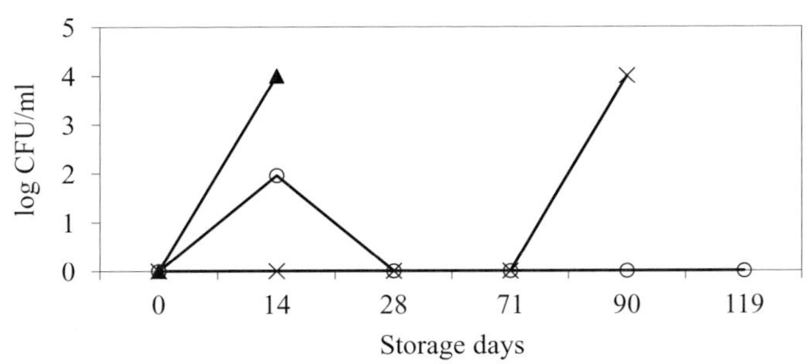

Figure 40.13 Total plate counts in chocolate milk during storage at 22 °C. ○ = PEF + 112 °C; X = PEF + 105 °C; ▲ = control.

and Pearlman (1987). Comparisons were made with regular heat-pasteurized egg products with and without the addition of preservatives when the products were stored at low (4 °C) and high (10 °C) refrigeration temperatures. The study focused on the importance of the hurdle approach in shelf-life extension. The effectiveness of this approach was even more evident from the results of storage at low temperatures, where the egg products had a final count of around 2.7 log CFU·ml^{-1} after storage at both 10 °C and 4 °C. The samples maintained a low count for 4 and 10 days, respectively, compared with a few hours for the heat-pasteurized samples.

In addition to color analysis of egg products, Ma *et al.* (1997) evaluated the density (an indicator of the foaming ability of egg protein) of fresh and PEF-treated liquid whole egg (LWE) and the strength of sponge cakes baked with PEF-treated eggs. The stepwise process used did not cause any difference in density or whiteness between the PEF-treated and the fresh LWE. The strength of the sponge cakes prepared with PEF-treated eggs was greater than that of cakes made with unprocessed eggs. This difference in strength was attributed to the lower volume obtained after baking with PEF-treated eggs. A statistical analysis of a sensory evaluation revealed no differences between cakes prepared from PEF-processed and fresh LWE.

A study reported by Hermawan (1999) indicated that there was a 90% reduction of *Salmonella enteritidis* inoculated into LWE with a circulation-mode fluid-handling system using a 200 pps pulse repetition rate, 2.12 μs pulse duration and 25 kV·cm^{-1} electric field strength.

Processing of Green Pea Soup

Vega-Mercado *et al.* (1996a) exposed pea soup to two steps of 16 pulses at 35 kV·cm^{-1} to prevent an increase in the temperature beyond 55 °C during PEF treatment. The shelf-life of the PEF-treated pea soup stored at refrigeration temperature (4 °C) exceeded 4 weeks, but 22 and 32 °C were found inappropriate for storing the product. There were no apparent changes in the physical and chemical properties or sensory attributes of the soup directly after PEF processing or during the 4 weeks of storage at refrigeration temperature.

Processing of Yoghurt-Based Products

PEF and mild heat (60 °C for 30 s) processing of yoghurt-based products similar to a dairy pudding dessert and a yoghurt-based drink were performed, and the results showed that the combination of PEF plus mild heat treatment significantly increased the microbial stability of the product at 4 and 22 °C without any difference in the sensory attributes. Sensory evaluation of the products indicated that there was no significant difference between the control and processed products. The color, pH and degrees Brix were not significantly affected by the processing conditions.

Processing of Rice Pudding

Owing to its higher viscosity, rice pudding was preheated to 55 °C for 30 s before PEF treatment. The processing conditions were 33 kV·cm^{-1} electric field strength, 100 L·h^{-1} flow rate, 1.47 μs pulse duration and 500 pps frequency. Monopolar negative pulses were applied. Shelf-life studies of the product showed that the total plate counts and the values of a color measurement of the PEF-treated and control samples were significantly different. The PEF-treated rice pudding had a shelf-life of 94 days, whereas the control samples were spoiled in 10 days.

PEF Treatment as a Hurdle Technology

In general, controlling a combination of factors (hurdles), such as pH, ionic strength and antimicrobial compounds in solution during PEF treatment can aid effectively in the inactivation of microorganisms. The term "hurdle technology" covers the intelligent use of multiple preservation procedures in combinations specifically relevant to particular types of foods. This concept is pertinent to the control of pathogenic and food spoilage microorganisms, and to almost all food commodities and products. Furthermore, hurdle technologies have been traditionally employed in all countries of the world, although with greatly differing emphases depending on the history and social characteristics of different cultures (Leistner and Gould, 2002).

Preservation technologies are based mainly on the inactivation of microorganisms or on the delay or prevention of microbial growth. Consequently, they must operate through those factors that most effectively influence the survival and growth of microorganisms (ICMSF 1980). Such factors are not numerous. They include a number of physical factors, some chemical ones, and some that are essentially microbial in that they depend on the nature of the microorganisms present in particular products. The most widely quoted classification of these factors derives from the original proposals of Mossel and Ingram (1955), updated by Mossel (1983). They include:

Intrinsic factors. Physical and chemical factors that exist within a food product, and with which contaminating microorganisms are inextricably in contact.

Processing factors. Procedures that are deliberately applied to foods in order to achieve improved preservation.

Extrinsic factors. Factors that influence microorganisms in foods, but which are applied from or exist outside the food; they also act during storage.

Implicit factors. Factors related to the nature of the microorganisms present in a food product, and to their interactions among themselves and with the environment during growth.

Net effects. These take into account the fact that many factors influence each other strongly, so that the overall effect of a combination of different factors may not be easily predictable, but may be usefully greater than the total of the expected effects of the individual factors.

Combination (hurdle) effects are the focus of many of the recent developments in the predictive modeling of microbial growth and survival in foods. The limits at which the various preservative factors inactivate or inhibit relevant microorganisms must be used to evaluate the effects of these factors on microorganisms that cause spoilage and food poisoning. However, it has to be remembered that these limits only apply if all other factors are optimal for the microorganisms in question. But this is hardly the case for any foodstuff. If more than one preservative factor (hurdle) is present, then an additive or even synergistic effect results, and this is the basis of the hurdle effect and the intentional use of hurdle technology.

The effective hurdle technologies typically employ multiple hurdles to preserve foods. In the use of such multiple hurdles, a consideration of the stress reactions and adaptations that microorganisms undergo underpins the logic of employing hurdles that affect different targets in the microbial cell. Ideally, the targets should be complementary, in order to gain synergism rather than simply additive effects (Leistner and Gould, 2002).

If mild heating can be applied to the food, then an injury that impairs the functionality of membranes may represent a further sensible target, which should amplify the effects of any previously applied hurdles that rely on properly functioning membranes. The potential value of the multitarget approach can therefore be easily appreciated, and perhaps built on more logically in the future. An example of a novel multitarget process is the use of PEF treatment, which damages cell membranes, in combination with the application of nisin, so that the membranes cannot be repaired, owing to the membrane-active action of nisin (Monticello, 1989; Calderon *et al.*, 1999a). Overall, therefore, homeostasis is interfered with by attacking two distinct targets.

It has been suspected that different hurdles might not just have an additive effect on microbial stability in food materials, but could act synergistically (Leistner, 1978). A synergistic effect could be achieved if the hurdles hit different targets within microbial cells at the same time, and thus disturb the overall homeostasis of the microorganisms present in several respects. If so, the repair of homeostasis and the activation of "stress shock proteins" in the microorganisms become more difficult. Therefore, employing different hurdles simultaneously for the preservation of a food should lead to optimal stability (Leistner, 1995). In practical terms, this could mean that it is more effective to employ several different preservative factors (hurdles) of small intensity than one preservative factor of larger intensity, because the different preservative factors might have a synergistic effect.

The multitarget preservation of foods is a promising research area, because if small hurdles with different targets are selected, a minimal treatment that provides the most effective preservation of foods could be obtained. It is anticipated that the targets of different preservative factors in microorganisms will be more fully elucidated, and that the hurdles could then be grouped into classes according to their targets. The mild but effective preservation of foods, i.e., a synergistic effect of hurdles, is likely to be achieved if the preservative measures are based on the intelligent selection and combination of hurdles taken from different target classes (Leistner, 1995). This approach is probably valid not only for traditional food preservation procedures, but also for the modern processes of irradiation, ultrahigh pressure, PEF and light beam treatment, in combination with conventional hurdles.

PEF Treatment Plus Hydraulic Pressing

Mechanical expression is widely used in the food industry for the extraction of fruit juices and vegetable oils, dewatering of fibrous materials, and other purposes (Schwartzberg, 1983). The efficacy of this process can be increased by plasmolysis, cellular damage or permeabilization of the raw material prior to its expression. Various methods are traditionally used to increase the degree of plasmolysis of the raw material, including heating, osmotic drying, freeze drying, alkaline breakage and enzymatic treatment (Barbosa-Canovas and Vega-Mercado, 1996). Earlier, the use of electric-field treatment (both DC and AC) was also proposed for plasmolysis of cellular material. Electroplasmolysis methods were shown to be good for the intensification of juice yield and for improving the product quality in juice production (McLellan *et al.*, 1991), in the processing of vegetables and plant-based raw materials (Grishko *et al.*, 1991), in the processing of various foodstuffs (Miyahara, 1985), in winemaking (Kalmykova, 1993), and in sugar production (Jemai, 1997). But these applications of electric fields were usually restricted by a high and uncontrolled increase in food temperature and by deterioration of the product quality because of electrolytic reactions with the electrode material and other factors.

Bazhal *et al.* (2001) investigated the influence of PEF treatment applied simultaneously with pressure treatment on the juice expression rate from fine-cut apple raw material. Three main phases of compression were observed in the case of mechanical expression. A unified approach was proposed for the analysis of data on the juice yield that allowed a reduction in the scatter of the data caused by differences in the quality of the samples. The application of PEF treatment at the moment when the specific electrical conductivity of the press cake reached its minimum value and the pressure reached a constant value was reported to provide the optimal results.

The combination of pressing and PEF treatment significantly enhanced the juice yield and improved the quality of the juice in comparison with samples not subjected to PEF treatment. The PEF treatment increased the yield from pressing whenever it was applied. The best results for the increase in juice yield at the lowest value of applied field were obtained when the PEF treatment was applied after a precompression period. Such pressure pretreatment before PEF application is necessary to increase the uniformity of the structure of the press cake, remove excess moisture and decrease the electrical conductivity of the material. In the study of Bazhal *et al.* (2001), a precompression period of 300–400 s and PEF treatment after that period were found to be optimal for the quality of the juice, which was confirmed by its coloration and transmittance. The simultaneous application of pressure and PEF treatment revealed the passive form of the PEF-induced cell plasmolysis, which developed very slowly under a low electric field without the application of pressure. The pressure promotes damage in defective cells, enhances the diffusion migration of moisture and depresses the resealing processes of the cells.

PEF Treatment Plus Bacteriocins

Microorganisms suffer cell membrane damage in the presence of a PEF, and nisin is a natural antimicrobial known to disrupt cell membrane integrity. Thus the combination of PEF and nisin represents a hurdle for the survival of *Listeria innocua* in liquid whole egg, which was investigated by Calderon *et al.* (1999a). *L. innocua* suspended in liquid egg was subjected to two different treatments: PEF treatment and PEF treatment followed by exposure to nisin. The frequency and pulse duration for the PEF treatment were 3.5 Hz and 2 ms, respectively. Electric field intensities of 30, 40 and 50 $kV{\cdot}cm^{-1}$ were used. The numbers of pulses applied to the liquid whole egg were 10.6, 21.3 and 32. The highest extent of microbial inactivation achieved with PEF treatment was 3.5 log cycles (U) for an electric field intensity of 50 $kV{\cdot}cm^{-1}$ and 32 pulses. The PEF treatment was conducted at relatively low temperatures, 36 °C being the highest. Exposure of *L. innocua* to nisin after the PEF treatment resulted in an additive effect on the inactivation of the microorganism. Moreover, a synergistic effect was observed as the electric field intensity, number of pulses and nisin concentration increased. *L. innocua* exposed to 10 $IU{\cdot}ml^{-1}$ of nisin after PEF treatment exhibited a decrease in population of 4.1 U for an electric field intensity of 50 $kV{\cdot}cm^{-1}$ and 32 pulses. Exposure of *L. innocua* to 100 $IU{\cdot}ml^{-1}$ of nisin following PEF treatment resulted in a decrease of 5.5 U for an electric field intensity of 50 $kV{\cdot}cm^{-1}$ and 32 pulses.

The model developed by Calderon *et al.* (1999a) for the inactivation of *L. innocua* by PEF treatment followed by exposure to nisin was shown to be successful in predicting the extent of microbial inactivation resulting from the combination of PEF treatment and nisin. The combination of these two preservation factors proved to be a hurdle against the survival of *L. innocua* in liquid whole egg. When energy conservation is a goal, inactivation of *L. innocua* in liquid egg products can be accomplished by selecting the proper combination of PEF intensity and nisin concentration.

Carvacrol was used in another study as a third preservative factor to further enhance the synergy between nisin and PEF treatment against vegetative cells of *Bacillus cereus* (Pol *et al.*, 2001). Applied simultaneously with nisin, carvacrol (0.5 mM) enhanced the synergy found between nisin and PEF treatment (16.7 $kV{\cdot}cm^{-1}$, 30 pulses) in potassium-*N*-2-hydroxy-ethylpiperazine-*N*-ethanesulfonic acid (HEPES) buffer. The influence of food ingredients on bactericidal activity was tested using skimmed milk that was diluted to 20% with sterile demineralized water. The efficacy of PEF treatment was not affected by the presence of proteins, and the results obtained in HEPES buffer correlated well with the results obtained in milk. Nisin showed less activity against *B. cereus* in milk, and carvacrol was not able to enhance the synergy between nisin and PEF treatment in milk, unless used in high concentrations (1.2 mM). This concentration did not in itself influence the viable count, but carvacrol did act synergistically with PEF treatment in milk, and not in HEPES buffer. This synergy was not influenced by milk proteins, since 5% milk still allowed synergy between carvacrol and PEF treatment to the same extent as in 20% milk.

References

Ade-Omowaye, B.I.O., Angersbach, A., *et al.* (2001) Use of pulsed electric field pre-treatment to improve dehydration characteristics of plant based foods. *Trends in Food Science & Technology* 12(8): 285–295.

Alvarez, I., Pagan, R., *et al.* (2003) The influence of process parameters for the inactivation of *Listeria monocytogenes* by pulsed electric fields. *International Journal of Food Microbiology* 87(1–2): 87–95.

Angersbach, A., Heinz, V., *et al.* (1999) Electrophysiological model of intact and processed plant tissues: cell disintegration criteria. *Biotechnology Progress* 15(4): 753–762.

Banik, S., Bandyopadhyay, S., *et al.* (2003). Bioeffects of microwave – a brief review. *Bioresource Technology* 87(2): 155–159.

Barbosa-Canovas, G. and Vega-Mercado, H. (1996) *Dehydration of Foods*. Chapman & Hall, New York.

Barbosa-Canovas, G. and Zhang, H. (2001) *PEF in Food Processing*. Technomic, Lancaster, UK.

Barbosa-Canovas, G., Gongora-Nieto, M., *et al.* (1999) *Preservation of Foods with Pulsed Electric Fields*. Academic Press, London.

Bard, A. and Faulkner, L. (1980) *Electrochemical Methods*. John Wiley & Sons, New York.

Bardos, D.C., Thompson, C.J., *et al.* (2000) Nonlinear cell response to strong electric fields. *Physics in Medicine and Biology* 45(7): 1965–1988.

Barsotti, L., Merle, P., *et al.* (1998) Food processing by pulsed electric fields. 1 – Physical aspects. *Sciences des Aliments* 18(6): 583–601.

Bazhal, M.I., Lebovka, N.I., *et al.* (2001). Pulsed electric field treatment of apple tissue during compression for juice extraction. *Journal of Food Engineering* 50(3): 129–139.

Bendicho, S., Barbosa-Canovas, G.V., *et al.* (2002a) Milk processing by high intensity pulsed electric fields. *Trends in Food Science & Technology* 13(6–7): 195–204.

Bendicho, S., Estela, C., *et al.* (2002b) Effects of high intensity pulsed electric field and thermal treatments on a lipase from *Pseudomonas fluorescens*. *Journal of Dairy Science* 85: 19–27.

Bockris, J. and Reddy, A. (1998) *Modern Electrochemistry*. Plenum, New York.

Calderon, L., Barbosa Canovas, V., *et al.* (1999a) Inactivation of *Listeria innocua* in liquid whole egg by pulsed electric fields and nisin. *International Journal of Food Microbiology* 51(1): 7–17.

Calderon, M., Barbosa-Canovas, G., *et al.* (1999b) Transmission electron microscopy of *Listeria innocua* treated by pulsed electric fields and nisin in skimmed milk. *International Journal of Food Microbiology* 51(1): 31–38.

Castro, A., Barbosa-Canovas, G., *et al.* (1993) Microbial inactivation of foods by pulsed electric fields. *Journal of Food Processing and Preservation* 17: 47.

CFSAN (2000) *Kinetics of Microbial Inactivation for Alternative Food Processing Technologies*. U. S. Food and Drug Administration, Washington, DC, p. 206.

Changjiang, W., Zhang, Q.H., *et al.* (2000) A 12 kV solid state high voltage pulse generator for a bench top PEF machine. In: *Proceedings of IPEMC* (00EX435), pp. 1347–1352.

Cole, K. (1972) *Membrane, Ions and Impulses*. University of California Press, Berkeley, CA.

Cserhalmi, Z., Vidacs, I., *et al.* (2002). Inactivation of *Saccharomyces cerevisiae* and *Bacillus cereus* by pulsed electric fields technology. *Innovative Food Science & Emerging Technologies* 3(1): 41–45.

Dunn, J. and Pearlman, J. (1987) Methods and apparatus of extending the shelf-life of fluid food products. US Patent 4695472.

Dunne, C., Dunn, J., *et al.* (1996) *Application of High Energy Electric Field Pulses to Preservation of Foods for Combat Rations*, Science and Technology for Force, Vol. XXI. Department of the Army, Norfolk, VA.

Dutreux, N., Notermans, S., *et al.* (2000) Pulsed electric fields inactivation of attached and free-living *Escherichia coli* and *Listeria innocua* under several conditions. *International Journal of Food Microbiology* 54(1–2): 91–98.

EPRI (1998) *Pulsed Electric Field Processing in the Food Industry: A Status Report on PEF*, CR-109742. Industrial and Agricultural Technologies and Services, Palo Alto, CA.

Esplugas, S., Pagan, R., *et al.* (2001) Engineering aspects of continuous treatment of fluid foods by PEF. In: *PEF in Food Processing* (eds G. Barbosa-Canovas and H. Zhang). Technomic, Lancaster, UK, p. 31.

Frubing, P. (2001) Dielectric spectroscopy. *University of Potsdam, Advanced Lab Experiments* M6: 1–22.

Gaskova, D., Sigler, K., *et al.* (1996) Effect of high-voltage electric pulses on yeast cells: factors influencing the killing efficiency. *Bioelectrochemical Bioenergetics* 39: 195.

Geveke, D.J., Kozempel, M., *et al.* (2002) Radio frequency energy effects on microorganisms in foods. *Innovative Food Science & Emerging Technologies* 3(2): 133–138.

Gongora-Nieto, M.M., Sepulveda, D.R., *et al.* (2002a) Food processing by pulsed electric fields: treatment delivery, inactivation level, and regulatory aspects. *Lebensmittel-Wissenschaft und-Technologie* 35(5): 375–388.

Gongora-Nieto, M.M., Younce, F., *et al.* (2002b) Metrology system for pulsed electric fields processing. *Innovative Food Science & Emerging Technologies* 3(4): 337–348.

Gongora-Nieto, M.M., Pedrow, P.D., *et al.* (2003) Impact of air bubbles in a dielectric liquid when subjected to high field strengths. *Innovative Food Science & Emerging Technologies* 4(1): 57–67.

Grahl, T. and Maerkl, H. (1996) Killing of microorganisms by pulsed electric fields. *Applied Microbiology and Biotechnology* 45(1–2): 148–157.

Grishko, A., Kozin, V., *et al.* (1991) Electroplasmolyzer for processing plant raw material. US Patent No. 5031521.

Heinz, V., Alvarez, I., *et al.* (2001) Preservation of liquid foods by high intensity pulsed electric fields – basic concepts for process design. *Trends in Food Science & Technology* 12(3–4): 103–111.

Herve, A.-G., Tang, J., *et al.* (1998) Dielectric properties of cottage cheese and surface treatment using microwaves. *Journal of Food Engineering* 37(4): 389–410.

Ho, S. and Mittal, G. (1997) *Analysis of 2 High Voltage Electric Pulse Systems for Batch and Continuous Pasteurization of Selected Food Products*. University of Guelph, Ontario, Canada.

Ho, S., Cross, J., *et al.* (1995) Inactivation of *Pseudomonas fluorescens* by high voltage electric pulses. *Journal of Food Science* 60(6): 1337.

Ho, S., Mittal, G., *et al.* (2003) Determination of electrical parameters of liquid foods for predicting waveforms of high-voltage low-energy pulses. *International Journal of Food Properties* 6(1): 115.

Huang, Y., Sekhon, N.S., *et al.* (2003) Instantaneous, quantitative single-cell viability assessment by electrical evaluation of cell membrane integrity with microfabricated devices. *Sensors and Actuators A: Physical* 105(1): 31–39.

Hülsheger, H. and Niemann, E. (1980) Lethal effect of high-voltage pulses on *E. coli* K12. *Radiation and Environmental Biophysics* 18(4): 281.

Hülsheger, H., Pottel, J., *et al.* (1981) Killing of bacteria with electric pulses of high field strength. *Radiation and Environmental Biophysics* 20: 53.

Hülsheger, H., Potel, J., *et al.* (1983) Electric field effects on bacteria and yeast cells. *Radiation and Environmental Biophysics* 22: 149–162.

ICMSF (1980) *Factors Affecting Life and Death of Microorganisms*. Academic Press, New York.

Jayaram, S., Castle, G., *et al.* (1992) Kinetics of sterilization of *Lactobacillus brevis* cells by the application of high voltage pulses. *Biotechnology and Bioengineering* 40: 1412–1420.

Jeantet, R., Baron, F., *et al.* (1999) High intensity pulsed electric fields applied to egg white: effect on *Salmonella enteridis* inactivation and protein denaturation. *Journal of Food Protection* 62: 1381–1386.

Jemai, A. (1997) *Contribution a l'etude de l'effect d'un traitement electrique sur les cossettes de betterave a sucre*. Universite de Technologic de Compiegne, Compiegne, France.

Jeyamkondan, S., D. S. Jayas, D.S., *et al.* (1999) Pulsed electric field processing of foods: a review. *Journal of Food Protection* 62(9): 1088–1096.

Jia, M., Zhang, Q., *et al.* (1999) Pulsed electric field processing effects on flavor compounds and microorganisms of orange juice. *Food Chemistry* 65(4): 445.

Jin, Z., Ruhlman, K., *et al.* (1998) Shelf life evaluation of pulsed electric fields treated aseptically packaged cranberry juice. In: *98 IFT Annual Meeting*, Atlanta, GA.

Jin, Z. and Zhang, Q. (1999) Pulsed electric field treatment inactivates microorganisms and preserves quality of cranberry juice. *Journal of Food Processing and Preservation* 23(6): 481–499.

Joshi, R.P., Hu, Q., *et al.* (2001) Self-consistent simulations of electroporation dynamics in biological cells subjected to ultrashort electrical pulses. *Physical Review E (Statistical, Nonlinear, and Soft Matter Physics)* 64(1): 011913/011911–011910.

Joshi, R.P., Hu, Q., *et al.* (2002a) Improved energy model for membrane electroporation in biological cells subjected to electrical pulses. *Physical Review E (Statistical, Nonlinear, and Soft Matter Physics)* 65(4): 041920/041921–041928.

Joshi, R.P., Hu, Q., *et al.* (2002b) Theoretical predictions of electromechanical deformation of cells subjected to high voltages for membrane electroporation. *Physical Review E (Statistical, Nonlinear, and Soft Matter Physics)* 65(2): 021913/021911–021910.

Kakorin, S., Redeker, E., *et al.* (1998) Electroporative deformation of salt filled lipid vesicles. *European Biophysics Journal* 27(1): 43–53.

Kalmykova, I. (1993) Application of electroplasmolysis for intensification of phenols extracting from the grapes in the technologies of red table wines and natural juice. [In Russian.] Odessa Technological Institute of Food Industry, Odessa, Ukraine.

Kim, P. and Baldwin, R. (1982) Specific intermediates in the folding reactions of small proteins and the mechanisms of protein folding. *Annual Review of Biochemistry* 51: 459.

Kinosita, K. and Tsong, T. (1977) Formation and resealing of pores of controlled sizes in human erythrocyte membrane. *Nature* 268: 438.

Kinosita, K., Ashikava, I., *et al.* (1988) Electroporation of cell membrane visualized under pulsed laser fluorescence microscope. *Biophysical Journal* 53: 1015.

Knorr, D., Angersbach, A., *et al.* (2001) Processing concepts based on high intensity electric field pulses. *Trends in Food Science & Technology* 12(3–4): 129–135.

Lado, B.H. and Yousef, A.E. (2002) Alternative food-preservation technologies: efficacy and mechanisms. *Microbes and Infection* 4(4): 433–440.

Lebovka, N.I., Praporscic, I., *et al.* (2003) Enhanced expression of juice from soft vegetable tissues by pulsed electric fields: consolidation stages analysis. *Journal of Food Engineering* 59(2–3): 309–317.

Leistner, L. (1978) Hurdle effect and energy saving. In: *Food Quality and Nutrition* (ed. W. Downey). Applied Science Publishers, London, pp. 553–557.

Leistner, L. (1995) Principles and applications of hurdle technology. In: *New Methods of Food Preservation* (ed. G. Gould). Blackie Academic & Professional, London, pp. 1–21.

Leistner, L. and Gould, G. (2002) *Hurdle Technologies*. Kluwer, New York.

Levenspiel, O. (1972) *Chemical Reaction Engineering*. John Wiley & Sons, New York.

Lindgren, M., Aronsson, K., *et al.* (2002) Simulation of the temperature increase in pulsed electric field (PEF) continuous flow treatment chambers. *Innovative Food Science & Emerging Technologies* 3(3): 233–245.

Love, P. (1998) Correlation of Fourier transforms of pulsed electric field waveform and microorganism inactivation. *IEEE Transactions on Dielectrics and Electrical Insulation* 5(1): 142.

Lubicki, P. and Jayaram, S. (1997) High voltage pulse application for the destruction of the Gram-negative bacterium *Yersinia enterocolitica*. *Bioelectrochemistry and Bioenergetics* 43(1): 135–141.

Ma, L., Chang, F.J. and Barbosa-Cánovas, G.V. (1997) Inactivation of E. coli in liquid whole eggs using pulsed electric fields technologies. New frontiers in food engineering. *Proceedings of the Fifth Conference of Food Engineering, American Institute of Chemical Engineers*, pp. 216–221.

Mattick, K.L., Legan, J.D., *et al.* (2001) Calculating *Salmonella* inactivation in nonisothermal heat treatments from isothermal nonlinear survival curves. *Journal of Food Protection* 64(5): 606–613.

McLellan, M., Kime, R., *et al.* (1991) Electroplasmolysis and other treatments to improve apple juice yield. *Journal of the Science of Food and Agriculture* 57: 303.

Miyahara, K. (1985) Methods and apparatus for producing electrically processed foodstuff. US Patent No. 4522834.

Mizuno, A. and Hori, Y. (1988) Destruction of living cells by pulsed high voltage application. *IEEE Transactions on Industry Applications* 24(3): 387–394.

Monticello, D. (1989) Control of microbial growth with nisin/lysozyme combinations. European Patent Application 89123445.2, 1989.

Mossel, D. (1983) Essentials and perspectives of the microbial ecology of foods. In: *Food Microbiology: Advances and Prospects* (eds T. Roberts and F. Skinner). Academic Press, London, pp. 1–45.

Mossel, D. and Ingram, M. (1955) The physiology of the microbial spoilage of foods. *Journal of Applied Bacteriology* 18: 232–268.

Ohshima, T., Okuyama, K. *et al.* (2002) Effect of culture temperature on high-voltage pulse sterilization of *Escherichia coli*. *Journal of Electrostatics* 55(3–4): 227–235.

Peleg, M. (1995) A model of microbial survival after exposure to pulse electric fields. *Journal of the Science of Food and Agriculture* 67(1): 93.

Pol, I., Mastwijk, H. *et al.* (2001) Influence of food matrix on inactivation of *Bacillus cereus* by combinations of nisin, pulsed electric field treatment, and carvacrol. *Journal of Food Protection* 64(7): 1012–1018.

Ponne, C.T. and Bartels, P.V. (1995) Interaction of electromagnetic energy with biological material – relation to food processing. *Radiation Physics and Chemistry* 45(4): 591–607.

Pothakamury, U., Vega-Mercado, H., *et al.* (1996) Effect of growth stage and processing temperature on the inactivation of *E. coli* by pulsed electric fields. *Journal of Food Protection* 59(11): 1167–1171.

Qin, B., Zhang, Q., *et al.* (1994) Inactivation of microorganisms by pulsed electric fields of different voltage waveforms. *IEEE Transactions on Dielectrics and Electrical Insulation* 1(6): 1047–1057.

Qin, B., Pothakamury, U., *et al.* (1995a) Food pasteurization using high intensity pulsed electric fields. *Journal of Food Technology* 49(12): 55.

Qin, B., Chang, F., *et al.* (1995b) Nonthermal inactivation of *S. cerevisiae* in apple juice using pulsed electric fields. *Lebensmittel-Wissenschaft und-Technologie* 28(6): 564.

Qin, B., Zhang, Q. *et al.* (1995c) Pulsed electric field treatment chamber design for liquid food pasteurization using a finite element method. *Transactions of the ASAE* 38(2): 557.

Qin, B.L., Barbosa Canovas, G.V., *et al.* (1995d) A continuous treatment system for inactivating microorganisms with pulsed electric fields. In: *Proceedings of IAS '95* (95CH35862), pp. 1345–1352.

Qin, B., Barbosa-Canovas, G., *et al.* (1998) Inactivating microorganism using a pulsed electric field continuous treatment system. *IEEE Transactions on Industrial Applications* 34(1): 43.

Quass, D. (1997) *Pulsed Electric Field Processing in the Food Industry*. Electric Power Research Institute, Palo Alto, CA.

Ramos, C. and Teissie, J. (2000) Electrofusion: a biophysical modification of cell membrane and a mechanism in exocytosis. *Biochimie* 82(5): 511–518.

Rodrigo, D., Ruiz, P., *et al.* (2003) Kinetic model for the inactivation of *Lactobacillus plantarum* by pulsed electric fields. *International Journal of Food Microbiology* 81(3): 223–229.

Ross, A.I.V., Griffiths, M.W., *et al.* (2003) Combining nonthermal technologies to control foodborne microorganisms. *International Journal of Food Microbiology* 89(2–3): 125–138.

Rowan, N.J., MacGregor, S.J., *et al.* (2000) Pulsed electric field inactivation of diarrhoeagenic *Bacillus cereus* through irreversible electroporation. *Letters in Applied Microbiology* 31(2): 110–114.

Sale, A. and Hamilton, W. (1967a) Effects of high electric fields on microorganisms I. Killing of bacteria and yeasts. *Biochimica et Biophysica Acta* 148(3): 781–788.

Sale, A. and Hamilton, W. (1967b) Effects of high electric fields on microorganisms II. Mechanism of action of the lethal effect. *Biochimica et Biophysica Acta* 148(3): 789–800.

Schoenbach, K.H., Peterkin, F.E., *et al.* (1997) The effect of pulsed electric fields on biological cells: experiments and applications. *IEEE Transactions on Plasma Science* 25(2): 284–292.

Schoenbach, K.H., Stark, R.H., *et al.* (2000) Biological/medical pulsed electric field treatments. *Conference Record of the 2000 Twenty-Fourth International Power Modulator Symposium. 50th Anniversary Edition* (00CH37049): 42–46.

Schoenbach, K.H., Beebe, S.J., *et al.* (2001a) Intracellular effect of ultrashort electrical pulses. *Bioelectromagnetics* 22(6): 440–448.

Schoenbach, K.H., Stark, R.H., *et al.* (2001b) Bioelectrics – new applications for pulsed power technology. *IEEE Conference Record Abstracts* PPPS(01CH37255): 56.

Schwartzberg, H. (1983) Expression-related properties. In: *Physical Properties of Food* (eds M. Peleg and E. Bagley). AVI, Westport, CT, p. 423.

Sowers, A. (1988) Fusion events and nonfusion contents mixing events induced in erythrocyte ghosts by an electric pulse. *Biophysical Journal* 54: 619.

Sukhorukov, V.L., Mussauer, H., *et al.* (1998) The effect of electrical deformation forces on the electropermeabilization of erythrocyte membranes in low- and high-conductivity media. *Journal of Membrane Biology* 163(3): 235–245.

Teissie, J. and Ramos, C. (1998) Correlation between electric field pulse induced long-lived permeabilization and fusogenicity in cell membranes. *Biophysical Journal* 74(4): 1889–1898.

Teissie, J. and Tsong, T. (1981) Electric field-induced transient pores in phospholipid bilayer vesicles. *Biochemistry* 20: 1548.

Thostenson, E.T. and Chou, T.-W. (1999) Microwave processing: fundamentals and applications. *Composites Part A: Applied Science and Manufacturing* 30(9): 1055–1071.

Tsien, R., Hess, P. *et al.* (1987) Calcium channels: mechanisms of selectivity, permeation, and block. *Annual Review of Biophysics and Biophysical Chemistry* 16: 265.

Unal, R., Kim Jin, G., *et al.* (2001) Inactivation of *Escherichia coli* O157:H7, *Listeria monocytogenes*, and *Lactobacillus leichmannii* by combinations of ozone and pulsed electric field. *Journal of Food Protection* 64(6): 777–782.

Unal, R., Yousef, A.E., *et al.* (2002) Spectrofluorimetric assessment of bacterial cell membrane damage by pulsed electric field. *Innovative Food Science & Emerging Technologies* 3(3): 247–254.

Vega-Mercado, H., Barbosa-Canovas, G., *et al.* (1995) Inactivation of a protease from *Pseudomonas fuorescens* M3/6 using high voltage pulsed electric fields. In: *Annual IFT Meeting*, Anaheim, CA.

Vega-Mercado, H., Martin-Belloso, O., *et al.* (1996a) Inactivation of *Escherichia coli* and *Bacillus subtilis* suspended in pea soup using pulsed electric fields. *Journal of Food Processing and Preservation* 20(6): 501.

Vega-Mercado, H., Pothakamury, U., *et al.* (1996b) Inactivation of *Escherichia coli* by combining pH, ionic strength and pulsed electric fields hurdles. *Food Research International* 29(2): 117.

Velizarov, S., Reitz, M., *et al.* (1998) Electropermeabilization and electrofusion of human cells modified by anaesthetic agents. *Bioelectrochemistry and Bioenergetics* 47(1): 89–96.

Venkatesh, M.S. and Raghavan, G.S.V. (2004) An overview of microwave processing and dielectric properties of agri-food materials. *Biosystems Engineering* 88(1): 1–18.

Vetter, K. (1967) *Electrochemical Kinetics*. Academic Press, New York.

Voronovsky, A.A., Abbas, C.A., *et al.* (2002) Development of a transformation system for the flavinogenic yeast *Candida famata*. *FEMS Yeast Research* 2(3): 381–388.

Wang, Y., Wig, T.D., *et al.* (2003) Dielectric properties of foods relevant to RF and microwave pasteurization and sterilization. *Journal of Food Engineering* 57(3): 257–268.

Winterhalter, M., Klotz, K.H., *et al.* (1996) On the dynamics of the electric field induced breakdown in lipid membranes. *IEEE Transactions on Industry Applications* 32(1): 125–130.

Wouters, P. and Smelt, J. (1997) Inactivation of microorganisms with pulsed electric fields: potential for food preservation. *Food Biotechnology* 11(3): 193–229.

Wouters Patrick, C., Bos Ad, P., *et al.* (2001) Membrane permeabilization in relation to inactivation kinetics of *Lactobacillus* species due to pulsed electric fields. *Applied and Environmental Microbiology* 67(7): 3092–3101.

Yin, Y., Zhang, Q., *et al.* (1997) High voltage pulsed electric field treatment chambers for the preservation of liquid food products. United States Patent 5690978.

Zhang, Q.H., Chang Fu, J., *et al.* (1994) Inactivation of microorganisms in a semisolid model food using high voltage pulsed electric fields. *Lebensmittel Wissenschaft and Technologie* 27(6): 538–543.

Zhang, Q., Qin, B., *et al.* (1995a) Inactivation of *E. coli* for food pasteurization by high-strength pulsed electric fields. *Journal of Food Processing and Preservation* 19(2): 103.

Zhang, Q., Barbosa-Canovas, G., *et al.* (1995b) Engineering aspects of pulsed electric field pasteurization. *Journal of Food Engineering* 25: 261–281.

Zhang, Q., Qiu, X., *et al.* (1997) *Recent Development in Pulsed Electric Field Processing*. National Food Processors Association, Washington, DC.

Zimmermann, U. (1986) Electrical breakdown, electropermeabilization and electrofusion. *Reviews of Physiology, Biochemistry and Pharmacology* 105: 176–256.

Zimmermann, U. and Benz, R. (1980) Dependence of the electrical breakdown voltage on the charging time in *Valonia utricularis*. *Journal of Membrane Biology* 53: 33.

Zimmermann, U., Friedrich, U., *et al.* (2000) Electromanipulation of mammalian cells: fundamentals and application. *IEEE Transactions on Plasma Science* 28(1): 72–82.

41

An Overview of Food Packaging: Material Selection and the Future of Packaging

Jasim Ahmed and Tanweer Alam

Introduction

Packaging is part of a coordinated system that starts from the preparation of goods for shipment and includes distribution, storage and merchandising at the optimum cost compatible with the requirements of the product. It can be considered as a combination of art, science and technology that is used in the transportation and selling of foods. Packaging materials have the four basic functions of providing protection, communication, convenience and containment (Paine, 1981; Robertson, 1993). The primary role of food packaging is to protect food products from the outside environment and from damage by abrasion, to contain the food, and to provide consumers with information about ingredients and nutrition (Dallyn and Shorten, 1998). The main requirement of food packaging is to maintain the safety, wholesomeness and quality of food. Packaging maintains the benefits of food processing after the process is complete, enabling foods to travel safely for long distances from their point of origin and still be in their natural state at the time of consumption. Traceability, convenience and tamper indication are secondary functions of increasing importance (Marsh and Bugusu, 2007). The objective of food packaging is to contain food in a cost-effective way that satisfies industry requirements, provides consumer satisfaction, maintains food safety and minimizes environmental impact. Packaging can be defined as a method to protect and contain foods with the aim of minimizing the environmental

Handbook of Food Process Design, First Edition. Edited by Jasim Ahmed and Mohammad Shafiur Rahman.

impact of our consumption. The ideal packaging would be comparable to that of natural products, such as banana peel, orange peel, coconut shells and eggshells.

The food processing and packaging industries spend an estimated 15% of their total variable costs on packaging materials (Esse, 2002). The opportunities for food packaging are astronomical. The developments in the packaging sector have been enormous, and processed food is now presented in various containers from metal cans to polyethylene terephthalate (PET) and microwavable polypropylene (PP) trays. Now, consumers can choose the same food in different packages. Microwaveable popcorn bags are a real innovation in the food packaging sector, with a self-opening stand-up feature. Food packaging has increased manifold in recent years for a number of reasons. Food is transported further than ever before and therefore requires more packaging as protection. There has been a rise in the number of ready meals sold; these tend to need larger amounts of packaging. Many foods are sold in smaller portions than in previous years, meaning even more packaging.

Consumer preferences and innovative technologies have always been the major forces that herd change in the food industry. In the past century particularly, the increases in population, urbanization, education, lifespan and communication drove the food industry to be large-scale and health-conscious, with the safety, attractiveness and convenience of foods being the primary motivation for technological transformations. Along with the concept of eating healthy foods to balance a declining medical system, there emerged a need for convenience foods that nourish, heal and fortify. Shifts in consumer preferences toward ready-to-cook meals, specialty foods and convenience foods have placed new demands on the food packaging industry. Food processors look for longer shelf-life, attractive packaging, and new materials for novel preparation methods such as microwaving, ohmic heating, aseptic processing (for particulate foods), high-pressure processing and high-pressure extrusion. At the same time, increasing emphasis is being placed on reducing the amount of packaging used, in response to the public's environmental concerns and to government regulations that have specified a target of a 50% staged reduction in packaging waste by the year 2015. These changes translate into a wide range of business opportunities for innovative approaches and solutions in the design and manufacture of packaging, processing and labeling equipment, and for new materials. Environmental factors, polymeric materials for manufacturing packaging materials, and attractive designs are some of the important aspects that have to be considered for a food packaging system.

Packaging offers the food industry an excellent marketing tool that is very useful for the growth and promotion of sales. Basically, a product and its packaging are an integrated unit. A manufacturer has to select appropriate packaging materials that can best represent the product, with appropriate customer appeal. A moisture/oxygen-sensitive product requires a package with adequate barrier properties. The development of both the product and the packaging should be carried out simultaneously and interactively. The most commonly used food packaging polymers are low-density polyethylene (LDPE), high-density polyethylene (HDPE), polypropylene (PP), polytetrafluoroethylene (PTFE) and nylon. Recently, there has been a focus on novel, advanced polymeric materials for enhanced food packaging. These materials are being developed based on existing polymer science methods, as well as newer technologies including biopolymers, nanotechnology and nanocomposites, and active and intelligent packaging. Proteins and polysaccharides are the biopolymers of prime interest, since they can be used effectively to produce edible and biodegradable films to replace petroleum-based packaging materials.

This chapter provides an overview of food packaging, including the basic need for packaging, the role of packaging in foods, the types of materials used in food packaging, and recent advances in food packaging. In addition, the suitability of various packaging materials for high-pressure processing is explored.

Why Do We Need Packaging?

Socioeconomic Factors in Food Packaging

The use of food packaging is a socioeconomic indicator of increased spending ability of the population, an increase in the gross domestic product or an increase in food availability (Brody *et al.*, 2008). Packaging technology must balance food protection with other social and environment issues, including energy and material costs, heightened social and environmental consciousness, and strict regulations on pollutants and disposal of municipal solid waste. Municipal solid waste (MSW) consists of items commonly thrown away, including packages, food scraps and trimmings. One of the U.S. Environmental Protection Agency (EPA)'s reports found that only approximately 31% of the MSW generated was from packaging-related materials, including glass, metal, plastic, paper and paperboard – a percentage that has remained relatively constant since the 1990s despite an increase in the total amount of MSW (Marsh and Bugusu, 2007). It has been reported that food packaging accounts for about two-thirds of total packaging waste by volume (Hunt *et al.*, 1990). Moreover, food packaging accounts for about 50% (by weight) of total packaging sales. Although the specific knowledge available has changed since the publication of the first Scientific Status Summary on the relationship between packaging and MSW (IFT, 1991), the issue remains poorly understood, complicating efforts to address the environmental impact of discarded packaging materials. Proper waste management is essential to protect human health and the environment and to preserve natural resources.

Packaging as a Preservation Tool

One of the main objectives of the packaging of food is to protect it against spoilage or deterioration due to physical damage, chemical changes or microbial growth. Physical protection protects food from mechanical damage during supply and distribution. Physical barriers resist impact, abrasion and crushing damage, so they can be used as shipping containers and as packaging for delicate foods (Marsh and Bugusu, 2007). On the other hand, chemical protection substantially decreases compositional changes influenced by environmental influences such as exposure to moisture, gases, or other environmental factors such as visible, infrared or ultraviolet radiation (Marsh and Bugusu, 2007). There are various packaging materials that can provide a chemical barrier for foods. Glass and metals provide a nearly absolute barrier to chemical and other environmental agents. Plastic packaging shows a large range of barrier properties, although it is more permeable than glass or metal. Biological protection provides a barrier to microorganisms (pathogens and spoiling agents), insects, rodents and other animals, thereby preventing disease and spoilage.

Foodstuffs may be divided into two main categories with regard to shelf-life: (i) fresh or perishable products, and (ii) shelf-stable or semiperishable products. Fresh products have a shelf-life ranging from a few hours to several days depending upon storage conditions. These products include refrigerated products (e.g., pasteurized milk). Shelf-stable products, on the other hand, keep well for up to several weeks or months at ambient temperatures. In either case, packaging has an important role to play in the product's shelf-life.

Besides the temperature of storage, the product's environment within the package is often crucial to its storage stability. This consideration is the basis of active packaging and modified-atmosphere packaging (MAP). Gas permeation properties are critical not only to MAP of foodstuffs such as fruit and vegetables, but also to long-life products obtained by in-package thermal processing or by aseptic packaging following sterilization treatment. Even intermediate-moisture foods rendered shelf-stable by manipulation of water activity rely heavily on packaging for their desired shelf-life.

Irrespective of the shelf-life-limiting changes, i.e., whether the product is subject to microbial spoilage or physicochemical deterioration, the characteristics of the package in terms of barrier properties, coupled with storage conditions, especially with regard to temperature, are the major determinants of shelf-life. Thus the shelf-life of a packaged food is a function of the temperature of storage and the properties of the packaging material. Kinetic models provide a useful tool for predicting shelf-life.

Function and Characteristics of Packaging

The basic function of packaging is to protect and preserve the contents during transit from the manufacturer to the ultimate consumer. Food packaging can retard product deterioration, retain the beneficial effects of processing, extend the shelf-life and increase the quality and safety of food. Packaging provides protection from three major classes of external influences: chemical, biological and physical. Chemical protection minimizes compositional changes triggered by environmental influences such as exposure to gases (typically oxygen), moisture (gain or loss) or light (visible, infrared or ultraviolet). Many different packaging materials can provide a chemical barrier. Glass and metals provide a nearly absolute barrier to chemical and other environmental agents, but few packages are purely glass or metal, since closure devices are added to facilitate both filling and emptying. Closure devices may contain materials that allow minimal levels of permeability. For example, plastic caps have some permeability to gases and vapors, as do the gasket materials used in caps to facilitate closure and in metal can lids to allow sealing after filling. Plastic packaging offers a large range of barrier properties but is generally more permeable than glass or metal. Biological protection provides a barrier to microorganisms (pathogens and spoiling agents), insects, rodents and other animals, thereby preventing disease and spoilage. In addition, biological barriers maintain conditions to control senescence (ripening and aging). Such barriers function via a multiplicity of mechanisms, including preventing access to the product, preventing odor transmission and maintaining the internal environment of the package. Physical protection shields food from mechanical damage and includes cushioning against the shock and vibration encountered during distribution. Typically developed from paperboard and corrugated materials, physical barriers resist impact, abrasion and crushing damage, so they are widely used as shipping containers and as packaging for delicate foods such as eggs and fresh fruit. Appropriate physical packaging also protects consumers from various hazards. For example, child-resistant closures hinder access to potentially dangerous products.

Packaging as a Marketing Tool

Packaging is an important tool for advertisement. The right packaging with the right message will rise above the competitive landscape. A package woos customers with an attractive label and design. It provides appropriate communication to the consumer about the product and how to use it, and other utility information. Packaging protects the interests of consumers. The information on the packaging includes quantity, price, inventory levels, lot number, size and weight; this information, together with the elapsed time since packaging and the color used, is very important for merchandising. Furthermore, innovative packaging can enhance sales in a competitive environment. The package also conveys important information about the product such as cooking instructions, brand identification and pricing. All of these enhancements may impact on waste disposal.

Convenience

Packaging should be convenient to the customer in ways such as ease of access, handling, and disposal; product visibility; resealability; and microwavability. Currently, the packaging of ready-to-eat foods is very convenient and customer-friendly. Oven-safe trays, boil-in bags and microwavable packaging make life easy and, in fact, they mean that no preparation is required for a meal. New closure designs provide ease of opening, resealability and special dispensing features. Advances in food packaging have facilitated the development of modern retail formats that offer consumers the convenience of one-stop shopping and the availability of food from around the world. These convenience features increase the value and competitive advantages of products but may also influence the amount and type of packaging waste requiring disposal.

Tamper Indication

Food tampering is the intentional contamination of a food product, with intent to cause harm to the consumer or to a private company (Canadian Food Inspection Agency, 2010). Food tampering may affect any part of the food product, such as the product itself, the packaging or the label. There are several measures to detect tampering, including banding, special membranes, breakaway closures, special printing on bottle liners or composite cans such as graphics or text that irreversibly changes upon opening, and special printing that cannot be easily duplicated (Marsh and Bugusu, 2007). Tamper-evident packaging usually requires additional packaging materials, which exacerbates disposal issues, but the benefits generally outweigh any drawbacks.

Mass Transfer and Food–Package Interactions

The quality of packaged food is directly related to the attributes of the food and packaging material (Cooksey, 2007; Lee *et al.*, 2008). Owing to the increasing awareness of health matters among consumers, the migration of substances from food packaging materials to foods has attracted the interest of the scientific and legislative communities (Fouad *et al.*, 1999; Lau and Wong, 2000). Mass transfer between food and packaging materials is an important consideration for food packaging. The quality of most packaged food deteriorates owing to mass transfer phenomena (e.g., moisture absorption, oxygen permeation, flavor loss, absorption of undesirable odors, and the migration of packaging components) (Kester and Fennema, 1986). In many cases, flavonoid components migrate to the packaging materials, leading to flavor alteration. A significant percentage of limonene and other volatiles are absorbed into the polymer when orange juice is stored inside an LDPE container. Migration may also result in mass transfer of an additive from the packaging material to the food. This transfer may increase the risk of chemical hazards and/or the formation of off-flavors. Any substance that migrates from a package material into a food is of concern if it could be harmful to the consumer. Even if the migrating substance is not potentially harmful, it could have an adverse effect on the flavor and acceptability of the food. The food–package interaction has been widely studied during the last few decades.

Several possibilities have been reported for the interaction between foods and packaging materials when they come into contact with each other (Gnanasekharan and Floros, 1997). These are the following. (i) Migration of volatile and nonvolatile compounds from packaging materials to the packaged food, including unreacted monomers or additives present in the polymerized packaging material. (ii) Sorption of components from the food or from the environment into the packaging material. The kind of molecules sorbed is dependent upon the type of interface between the food and the packaging. Some common examples are the sorption of fatty matter, pigments and vitamins into the packaging. (iii) Permeation of volatile compounds (flavors and water vapor) from the food through the packaging. The most studied aroma compound is D-limonene, which is present in orange juice. Interactions between D-limonene and packaging films have been studied extensively.

The migration of a compound from a packaging material into a food depends on the chemical and physical properties of the compound, the food and the polymer (Halek and Levinson, 1988). The migration processes are influenced by the concentration, molecular weight, solubility and diffusivity of the migrating compound; the partition coefficient between the food and the polymer; the temperature; the composition of the polymer and food; and structural properties (crystallinity, chain length and branching). Even though mathematical models are still in the process of development and are not very reliable for measuring the potential contamination of food with chemicals from packaging, they are still considered to be valuable tools (Arvanitoyannis and Bosnea 2004). The phenomenon of mass transfer from a packaging material into a foodstuff can be described by Fick's laws. For a reasonable prediction of migration using Fick's laws, two fundamental constants are needed: the partition coefficient of the migrating compound between the packaging material and the foodstuff or food simulant, and the diffusion coefficient of the compound in the packaging material. A simplified mass transfer operation is illustrated in Figure 41.1.

Basic Equations for Mass Transfer in Food Packaging

Diffusion is quantified by a kinetic parameter called the diffusion coefficient or diffusivity. For a macroscopically motionless medium (except for the diffusing compound) made up of one isotropic phase, such as a dense membrane or a packaging material, the driving force behind the transport process is the difference in chemical potential between the two phases on either side of the membrane. The transport process slowly tries to equalize the concentrations, partial pressures or chemical potentials of the penetrant in the phases separated by the membrane. A simplified illustration of steady-state mass transfer in monolayer packaging is shown in Figure 41.2.

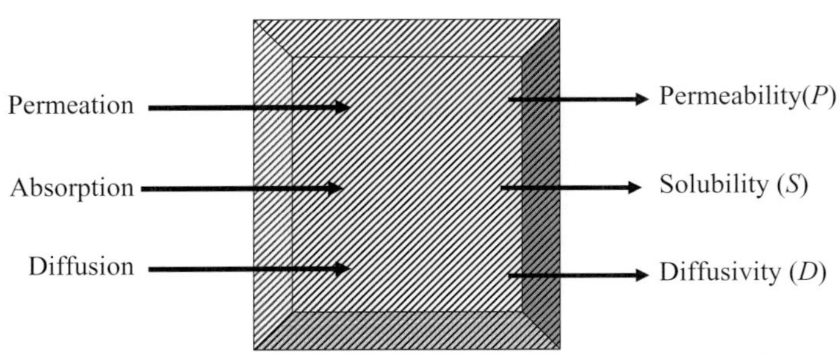

Figure 41.1 Mass transfer in food packaging.

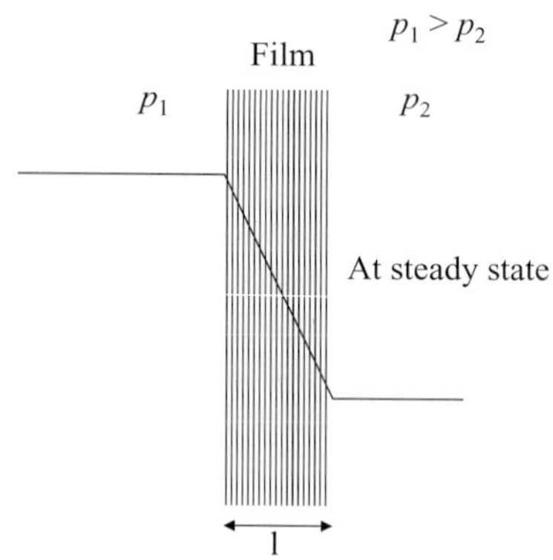

Figure 41.2 Diffusion through monolayer packaging.

The flux density J (kg·m^{-2}·s^{-1}) of a compound through a membrane of thickness l (m) and area A (m^2) in contact with the diffusing compound, with a concentration difference $C_1 - C_2$, where $C_1 > C_2$, can be written as

$$J = \frac{Q}{A \cdot t} \tag{41.1}$$

where Q (kg) is the quantity of the compound transferred through the membrane during a time t (s). The flux Q/t can be correlated with the average concentrations on the two sides of the membrane using the following mass transfer model:

$$\frac{Q}{t} = k \cdot (C_1 - C_2) \tag{41.2}$$

where k is the total mass transfer coefficient across the membrane (m^2·s^{-1}).

Fick's first law states that there is a linear relationship between the flux density of the compound and concentration difference between two sides of the membrane:

$$J = \frac{D}{l}(C_1 - C_2) \tag{41.3}$$

where D is the diffusivity, or diffusion coefficient (m^2·s^{-1}), for the compound that is diffusing in the membrane. The diffusion model is more appropriate when concentrations are measured with respect to position and time.

Furthermore, the concentration gradient between the membrane surfaces is related to the gradient of partial pressure in the vapor phase, which can be expressed by Henry's law as shown below:

$$(C_1 - C_2) = S(p_1 - p_2) \tag{41.4}$$

where S is the solubility coefficient ($kg{\cdot}m^{-3}{\cdot}Pa^{-1}$) of the compound in the membrane, and $(p_1 - p_2)$ is the pressure difference (Pa) between the two sides of the film. Combining Equations 41.3 and 41.4, we obtain

$$J = \frac{D \cdot S \cdot (p_2 - p_1)}{l} \tag{41.5}$$

At thermodynamic equilibrium, the gas permeability, or permeability coefficient P ($kg{\cdot}m^{-1}{\cdot}s^{-1}{\cdot}Pa^{-1}$), is the product of the diffusivity and the solubility coefficient, i.e., $P = DS$. The permeability coefficient defines the ease with which a compound can cross a membrane, whereas the solubility coefficient depends on the partial-pressure gradient of the compound. A wide range of values have been reported for diffusion coefficients in polymeric materials (10^{-19}–10^{-9} $m^2{\cdot}s^{-1}$). More details of the mass transfer between foods and packaging materials are described in a separate chapter.

Food Packaging Materials

The major categories of materials used for food packaging are glass, metals (aluminum, foils and laminates, tinplate, and tin-free steel), paper and paperboard, and plastics. A wide variety of plastics have been introduced in both rigid and flexible forms. Today's food packages often combine several materials to exploit each material's functional or aesthetic properties. There are many multilayered packaging materials containing either layers of different plastics or combinations of plastics with paper/board, metal or glass. The individual properties of the different materials are used to produce food packaging with the required characteristics. In many cases, a packaging material with two layers is chosen in such a way that one layer provides the basic strength and the second layer enables the packaging to be easily heat-sealed (Lord, 2008). Coatings are also often added to the basic plastic packaging material to provide additional barriers to the permeation of oxygen and water vapor. These coatings may be polymeric materials or vacuum-deposited aluminum. Detailed descriptions of some common packaging materials are given below.

Paper and Paperboard

The use of paper for food packaging dates back to the 17th century, with accelerated usage in the later part of the 19th century (Kirwan, 2003). Paper and paperboard are sheet materials produced from an interlaced network of cellulose fibers derived from wood by using sulfate and sulfite. The fibers are then pulped, bleached, and treated with chemicals and strengthening agents to produce the paper product. Paperboard is thicker than paper, with a higher weight per unit area, and is often made in multiple layers. It is commonly used to make containers for shipping, such as boxes, cartons, and trays, and is seldom used for direct food contact. There are several different types of paperboard, including white board, solid board, fiber board and chipboard (Soroka, 1999). Paper cartons are regularly used for the packaging of liquid foods. Tissue paper, paper plates and paper cups are other examples of paper and paperboard products. Plain paper is not suitable for food packaging, owing to its poor barrier properties, and it is not heat-sealable. Mostly, paper is coated with polyethylene, which permits much stronger heat-sealing and moisture protection. Paper is used in various forms in packaging, and these are briefly described below.

Kraft Paper

Kraft paper is made using a process that involves pulverizing the wood pulp and blending the material into large sheets of strong, brown wood filaments. It is an expensive option when it comes to paper products. The kraft process includes the use of sulfate in the conditioning of the wood pulp, which also helps to add to the overall strength of the finished paper. Kraft paper is available in several forms: natural brown, unbleached, heavy-duty and pure white. Natural brown kraft paper is the strongest of all types of paper and is commonly used for bags and wrapping. One of the most common uses of plain brown kraft paper is in the manufacture of paper bags for use in grocery stores.

Sulfite Paper

Lighter and weaker than kraft paper, sulfite paper is glazed to improve its appearance and to increase its wet strength and oil resistance. In the production process of sulfite paper, the wood pulp is treated with peroxide or hypochlorite and subjected to operations that yield a thick paper product. Repeated applications of the chemical treatment can be used to produce a lighter shade or a brilliant white appearance. From there, the sulfite paper can be processed further to any thickness that is desired, depending on how the end product is to be used. It can be coated for higher print quality, and is also used in laminates with plastic or foil. It is used to make small bags and wrappers for packaging biscuits and confectionery.

Greaseproof Paper

Greaseproof paper is made by a process known as beating, in which the cellulose fibers undergo a longer than normal hydration period that causes the fibers to break up and become gelatinous. These fine fibers thereafter pack densely to provide a surface that is resistant to oils but not to wet agents. Greaseproof paper is used to wrap snack foods, cookies, candy bars and other oily foods, a use that is being replaced by plastic films. Glassine is greaseproof paper with a highly smooth and glossy finish. It is used as a liner for biscuits, cooking fats, fast foods and baked goods.

Parchment Paper

Parchment paper is produced from acid-treated pulp (passed through a sulfuric acid bath). The acid modifies the cellulose to make it smoother and impervious to water and oil, which adds some wet strength. It does not provide a good barrier to air and moisture, is not heat-sealable, and is used mostly to package bakery products with a high fat content.

Glass

Glass has been used for food packaging for a long time. A great number of glass containers are currently manufactured for sealed packaging, since glass containers have several advantages for preserving foods. The production of glass containers involves heating a mixture of silica (the glass former), sodium carbonate (the melting agent), and limestone or calcium carbonate and alumina (stabilizers) to high temperatures until the materials melt into a thick liquid mass, which is then transferred to molds. Recycled broken glass (cullet) is also used in glass manufacture and may account for as much as 60% of the raw materials used. Glass possesses very good barrier properties, so it maintains product freshness for a long period of time without impairing the taste or flavor. The major limitations are the ratio of mass to volume, and the fact that it is brittle and nondegradable (although not harmful to the environment); it is mostly used for the production of bottles and jars. The ability to withstand high processing temperatures makes glass useful for heat sterilization of both low-acid and high-acid foods.

Plastics

Plastics are the most widely used packaging materials, and this is the category that has the largest numbers of variants (Lord, 2008). Plastics are synthesized by condensation, addition or crosslinking polymerization of monomer units. In condensation polymerization, the polymer chain grows by condensation reactions between molecules and is accompanied by the formation of water or alcohol.

A great advantage of plastics is the large variety of materials and compositions available, which makes it possible to adopt the most convenient packaging design for the very specific needs of each product. Some relevant characteristics of plastics are their low cost, lightness, thermosealability, good water resistance, ease of printing and microwavability (Kondo, 1990). They can also be formed into an unlimited variety of sizes and shapes, and converters can easily modify them. The optical properties (brightness and transparency) can also be adapted to the specific requirements of the product. This property allows the consumer to see the packaged product, providing it with visual appeal. The thermal and mechanical properties can also be partially modified in order, for instance, to manufacture retortable packages with plastics that have a high melting point, or thermosealable packages making use of plastics with a low melting point, and to develop very flexible structures (sachets and, wrappings), semi-rigid structures (trays and tubs) and rigid structures (bottles, closures and tanks). Some thermomechanical properties of selected polymers used in food packaging are listed in Table 41.1.

Polymers can be classified into two types according to their behavior on heating: thermoplastic and thermosetting polymers (Kondo, 1990). Thermoplastic polymers soften and melt on heating and solidify again on cooling. They are easily molded and

Table 41.1 Thermomechanical properties of polymers used in food packaging.

Polymer	Chemical structure	T_g (°C)	T_m (°C)
Polyethylene	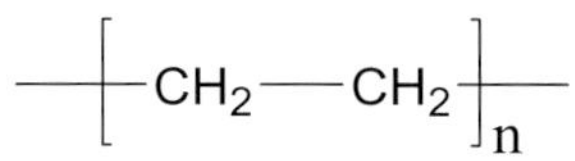		
LDPE		−103 to −133	125–136
HDPE		−110 to −120	108–134
Polypropylene	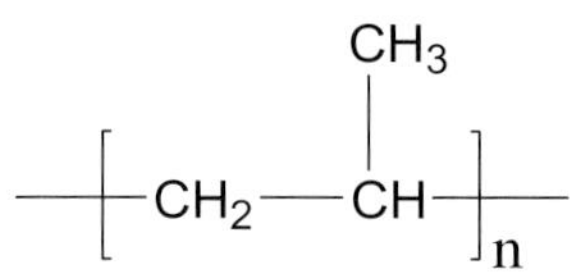	−20 to 5	160
Poly(vinyl dichloride)	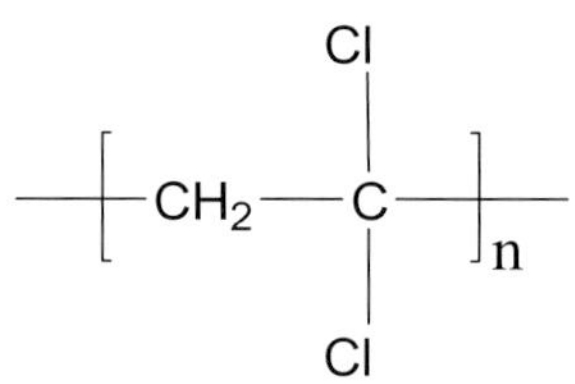	−17	
Poly(vinyl chloride)	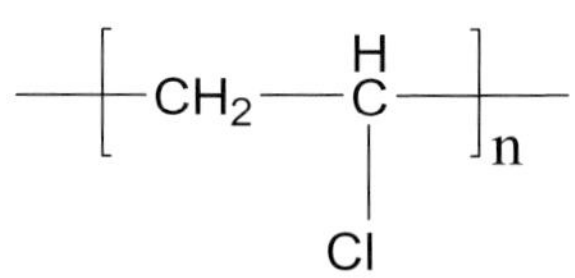	80	
Poly(vinyl acetate)	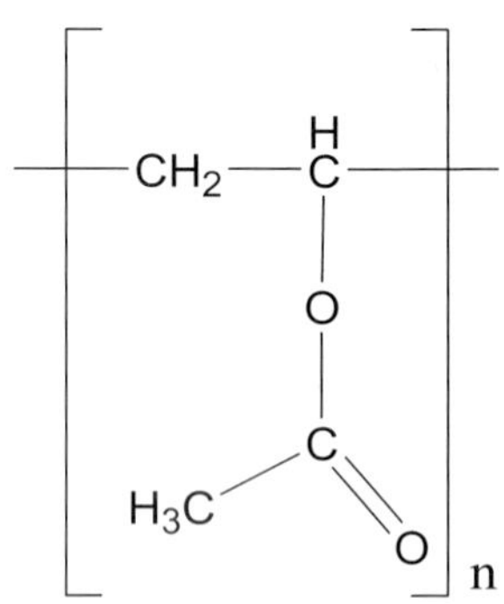	30	
Poly(vinyl alcohol)	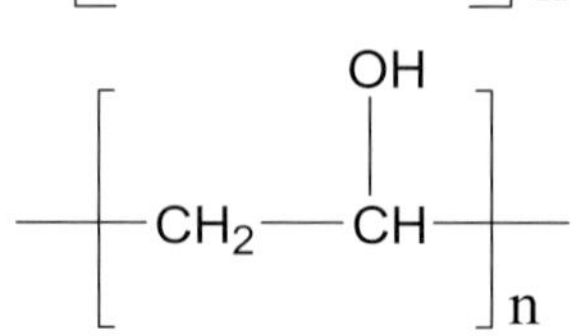	85	

Table 41.1 *(Continued)*

Polymer	Chemical structure	T_g (°C)	T_m (°C)
Polycarbonate	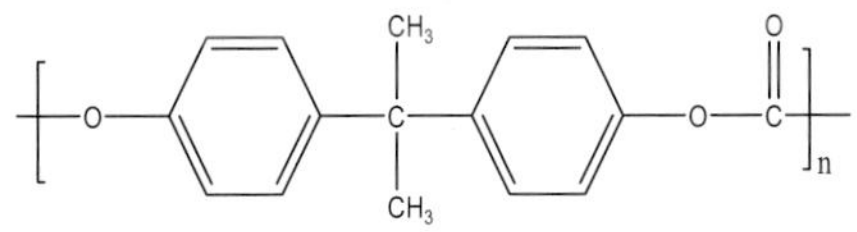	145	
Polystyrene	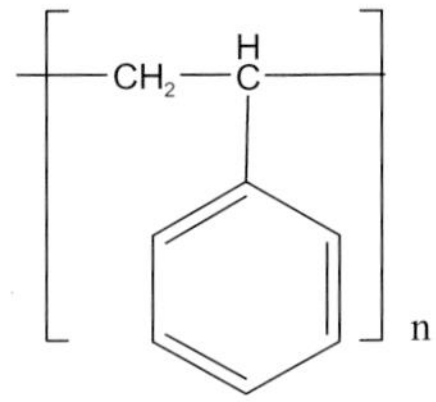	100	
Polyacrylonitrile	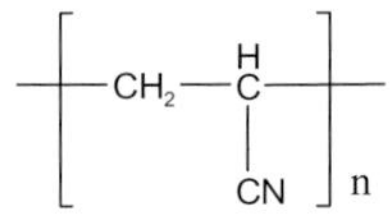	104	
Poly(ethylene terephthalate)	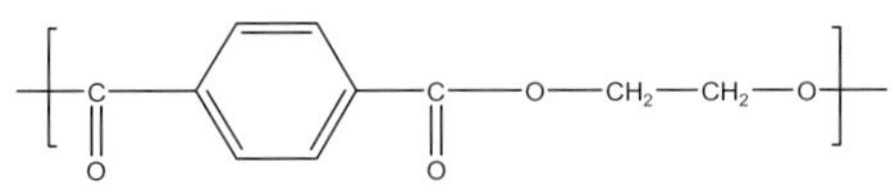	70	
Nylon-6	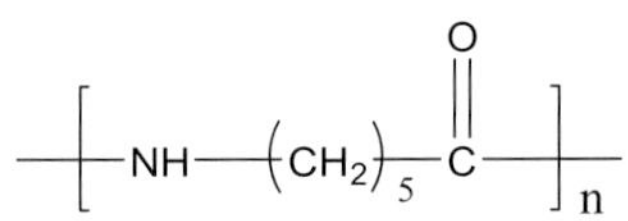	50	

extruded into films, fibers and packaging. Examples include polyethylene, polypropylene and polyvinyl chloride. Thermosetting polymers, in contrast, bccome hardened on cooling, and these plastics retain their shape and cannot return to their original form. They are hard and durable. Thermosets include polyurethanes, polyesters, epoxy resins and phenolic resins. Thermoplastics are less rigid than thermosets.

Various types of plastics, including polyolefins, polyesters, polyvinyl chloride, polyvinylidene chloride, polystyrene, polyamide and ethylene vinyl alcohol, are currently being used as packaging materials for food products. Although more than 30 types of plastics have been used as packaging materials (Lau and Wong, 2000), polyolefins and polyesters are the most common. A list of polymeric materials has been approved by Health Canada (see www.hc-sc.gc.ca) for food packaging and is listed in Table 41.2.

Moldable plastics, when molten, can be converted into sheets, various shapes, and structures, offering considerable design flexibility. Plastics are chemically resistant, inexpensive and lightweight, with a wide range of physical and optical properties. In fact, many plastics are heat-sealable and easy to print, and can be integrated into production processes where the package is formed, filled and sealed in the same production line. The major limitation of plastics is their variable permeability to light, gases, vapors and low-molecular-weight molecules.

Table 41.2 Permitted polymers for use in food packaging (Health Canada).

Name of polymer	Code
Polyethylene	PE
Polypropylene	PP
Poly(ethylene-vinyl acetate)	EVA
Polystyrene	PS
Polyvinyl chloride	PVC
Ionomer	I
Polyethylene terephthalate	PET
Polyvinyl acetate	PVAc
Polycarbonate	PC
Polyamide	PA
Polyvinyl alcohol	PVOH
Polyvinylidene chloride	PVDC

Polyolefins

The class of polyolefins represents the two most extensively used plastics in food packaging, namely polyethylene and polypropylene. Polyethylene and polypropylene both possess a unique combination of properties, including flexibility, lightness, strength, stability, moisture and chemical resistance, and easy processability. In addition, these polymers are recyclable and reusable.

The most well-known packaging material in the food industry that meets all desired criteria is polyethylene (PE), which has been in use by the food industry for over 50 years (Cutter, 2006). Polyethylene is the simplest, most versatile and most inexpensive plastic, and is synthesized by addition polymerization of ethylene. The properties of PE vary with the pressures and catalysts used in the polymerization process and with its density (Kondo,1990). According to the pressure applied during polymerization, PE is classified into low-pressure, medium-pressure and high-pressure PE. Based on its density, PE is classified into very low-density polyethylene (VLDPE), low-density polyethylene (LDPE), medium-density polyethylene (MDPE), and high-density polyethylene (HDPE) (Kondo, 1990). However, LDPE and HDPE are the forms most commonly used in food packaging.

LDPE softens at 100–105 °C and its crystallinity ranges from 60 to 70%. In LDPE, the polymer strands are entangled and loosely organized. It is flexible, easy to seal and resistant to moisture, but at the same time is a poor oxygen and carbon dioxide gas barrier. LDPE is relatively transparent, and therefore it is predominantly used in film applications and in applications where heat sealing is necessary. LDPE shows excellent cold resistance (up to −70 °C), and therefore is used in frozen-food packaging. HDPE is stiff, strong, tough, resistant to chemicals and moisture, permeable to gas, easy to process, and easy to form. HDPE is a harder plastic and has a higher melting point than LDPE, and it sinks in an alcohol–water mixture. Data on the oxygen and water vapor permeability of HDPE and LDPE are included in Table 41.3. Polyethylene

Table 41.3 Oxygen and water vapor permeability of selected polymers (100 μm thick) for food packaging (data from Utz, 1995).

Polymer	O_2 permeability at 23 °C and 0% RH [$cm^3(STP) \cdot m^{-2} \cdot d^{-1} \cdot bar^{-1}$]	Water vapor permeability at 40 °C and 90–0% RH ($g \cdot m^{-2} \cdot d^{-1}$)
EVOH	0.02–0.45	5.43–31.0
PVDC	0.31	0.23
PET	8.91	4.65
HDPE	581	1.48
BOPP	632	1.48
PS	1008	33.0
LDPE	2147	4.43

is used to make bottles for food products, juice and water; cereal box liners; margarine tubs; and grocery, trash and retail bags. Polyethylene bags are recycled. HDPE bottles are the most commonly recycled plastic packages.

Polypropylene (PP) is harder, denser and more transparent than polyethylene, has good resistance to chemicals and is effective at barring water vapor. The various forms of polypropylene have different melting points and hardnesses. Its high melting temperature (160 °C) makes it suitable for applications where thermal resistance is required, such as hot-filled and microwavable packaging. Popular uses include yoghurt containers and margarine tubs.

Polyesters

Polyethylene terephthalate (PET), polycarbonate (PC) and polyethylene naphthalate (PEN) are polyesters, which are obtained by condensation polymerization from ester monomers that result from reactions between a carboxylic acid and an alcohol. The most commonly used polyester in food packaging is PET.

Polyethylene Terephthalate

PET is a thermoplastic polyester that is synthesized by the condensation of terephthalic acid and ethylene glycol. There are several advantages in using PET as a packaging material. PET is almost unbreakable. Food products stored in PET have a long shelf-life, since PET is a good barrier to gases (oxygen and carbon dioxide) and moisture (Table 41.3). It shows good resistance to heat, mineral oils, solvents and acids, but not to bases. PET is the packaging material of first choice for beverages and mineral waters. The use of PET to make plastic bottles for carbonated beverages is gradually increasing (Van Willige *et al.*, 2002). The main reasons for its popularity are its glass-like transparency, gas barrier properties that allow retention of carbonation, light weight and shatter resistance. The major packaging applications of PET are in containers (bottles, jars and tubs), semirigid sheets for thermoforming (trays and blisters), and thin oriented films (bags and snack food wrappers). PET exists in both an amorphous (transparent) and a semicrystalline (opaque white) form. Amorphous PET possesses better ductility but less stiffness and hardness than its semicrystalline counterpart, which has good strength, ductility, stiffness and hardness.

Polycarbonate

PC is formed by polymerization of a sodium salt of bisphenol acid with carbonyl dichloride (phosgene). Its gas barrier properties are moderate, but it provides a very good barrier against flavors and aromas (Kondo, 1990). The novelty of PC over other types of plastic is its superior strength combined with light weight. PC is resistant to a wide range of temperatures (its melting point is 230 °C and its brittle temperature is −100 °C). Owing to its high-temperature resistance, PC containers can be used in process lines for pasteurization of puddings, rice cakes etc. In addition, it is transparent and durable. It is mainly used as a replacement for glass in products such as refillable water bottles and sterilizable baby bottles. Care must be taken during cleaning of PC, as harsh detergents catalyze the release of bisphenol A, a potential health hazard. Recently, many countries have put a ban on PC baby bottles.

Polyethylene Naphthalate

PEN is a condensation polymer of dimethyl naphthalene dicarboxylate and ethylene glycol. PEN is chemically similar to PET but more temperature-resistant (Guo and Zachmann, 1997). It exhibits excellent performance because of its high glass transition temperature. PEN shows excellent barrier properties for carbon dioxide, oxygen and water vapor, comparable to those of PET. It provides better performance at high temperatures, allowing hot refills, rewashing and reuse. It is more resistant to hydrolysis in alkaline and very hot aqueous conditions (Guo and Zachmann, 1997). PEN retains flavors and odors, and therefore it is well suited for manufacturing bottles for beverages such as beer. However, it is more expensive than PET.

Polyvinyl Chloride

Polyvinyl chloride (PVC) is obtained by radical polymerization or chain polymerization from vinyl chloride monomer (Kondo, 1990). PVC is heavy, stiff and ductile, and is a medium-strong, amorphous, transparent material. It has excellent resistance to chemicals (acids and bases), grease and oil; good flow characteristics; and stable electrical properties. Although PVC is primarily used in medical and other nonfood applications, its food uses include bottles and packaging films. Because it is easily thermoformed, PVC sheets are widely used for blister packs such as those for meat products and unit-dose pharmaceutical packaging. PVC can be transformed into materials with a wide range of flexibility by the addition of plasticizers such as phthalates, adipates, citrates and phosphates.

Polyvinylidene Chloride

Polyvinylidene chloride (PVdC) is an addition polymer of vinylidene chloride. It is heat-sealable and serves as an excellent barrier to water vapor, gases, and fatty and oily products. It is used in flexible packaging as a monolayer film, as a coating or as part of a co-extruded product. Major applications include the packaging of poultry, cured meats, cheese, snack foods, tea, coffee and confectionery. It is also used in hot filling, retorting, low-temperature storage and modified-atmosphere packaging. However, PVdC contains twice the amount of chlorine as PVC and therefore presents problems with incineration.

Polystyrene

Polystyrene (PS), an addition polymer of styrene, is clear, hard and brittle with a relatively low melting point. The glass transition temperature T_g of the polymer is about 100 °C. PS is a colorless polymer used extensively for low-cost applications. It is available commercially in both pellet and sheet form. It can be mono-extruded, co-extruded with other plastics, injection-molded or foamed to produce a range of products. Foaming produces an opaque, rigid, lightweight material with impact-protection and thermal-insulation properties. Typical applications of PS include protective packaging such as egg cartons, containers, disposable plastic silverware, lids, cups, plates, bottles and food trays. In expanded form, PS is used for nonfood packaging and cushioning, and it can be recycled or incinerated.

Polyamide (Nylon)

Nylon is a polyamide with an amide structure (–CO–NH–) in its main chain. Polyamide is synthesized by a condensation reaction between a diamine and a diacid, where the repeating units are held together by amide links. The various types of polyamides are characterized by a number that relates to the number of carbon atoms in the originating monomers. For example, nylon-6 has six carbon atoms in the monomers and is often used in packaging. It shows similar mechanical and thermal properties to PET and therefore it has similar usefulness, such as boil-in bag packaging. Nylon also offers good chemical resistance, toughness and low gas permeability.

Ethylene Vinyl Alcohol

Ethylene vinyl alcohol (EVOH) is a copolymer of ethylene and vinyl alcohol. It is an excellent barrier to oil, fat and oxygen. EVOH is a crystalline polymer. EVOH films show excellent gas barrier properties in dry conditions; however, EVOH is greatly affected by humidity and is therefore mostly used in multilayer co-extruded films in situations where it is not in direct contact with liquids.

Films Made by Lamination and Extrusion

Plastic materials can be manufactured into films containing a single material or a combination of more than one material. There are two ways of combining plastics: lamination and co-extrusion. Lamination involves bonding together two or more plastics or bonding a plastic to another material such as paper or aluminum. Bonding can be achieved by use of water-based, solvent-based or solids-based adhesives. After application of the adhesive, two or more films are passed between rollers to pressure-bond them together. Adhesives are currently being replaced by lasers for the lamination of thermoplastics (Kirwan and Strawbridge, 2003). Lamination enables reverse printing, in which the printing is buried between two layers and thus not subject to abrasion, and it can add or enhance heat-sealability. In co-extrusion, two or more layers of molten plastic are combined during film manufacture. The process is fast, but requires materials that have thermal characteristics that allow co-extrusion. Because co-extrusion and lamination combine multiple materials, recycling is complicated. However, combining materials results in the additive advantage of properties from each individual material and often reduces the total amount of packaging material required. Therefore, co-extrusion and lamination can be sources of packaging reduction.

Metals

Metals are the most versatile of all forms of packaging. They offer the combination of excellent physical protection and barrier properties, formability, decorative potential, recyclability, and consumer acceptance. Metal containers are vacuum-sealed and thermally sterilized under low oxygen pressure. The decomposition of nutrients is kept to a minimum in metal containers, since metals are a perfect barrier to oxygen, light and moisture. Double seaming of metal containers is very reliable. The lacquerability and printability of metals are superb. The major limitations of metal containers are cost, the weight of the containers and the fact that they are difficult to crush.

Aluminum and steel are the most predominantly used metals in food packaging. Even though many of the fundamental manufacturing processes, such as double seaming and body forming, were developed long ago, the evolution of can making still continues. A good combination of metallurgy and food engineering could produce more advanced metal containers with a lower price and lighter weight, with new applications.

Aluminum

Aluminum foil plays an important role in food packaging. Aluminum is a lightweight, silvery white metal derived from bauxite ore, where it exists in combination with oxygen as alumina. Magnesium and manganese are often incorporated into aluminum to improve its mechanical strength (Page *et al.*, 2003). Aluminum is highly resistant to most forms of corrosion; its natural coating of aluminum oxide provides a highly effective barrier to the effects of air, temperature, moisture and chemical attack. The mechanical, physical and chemical properties of aluminum foil such as its barrier effect, deadfold properties and suitability for food contact enable a wide range of applications in many different products and sectors (Lamberti and Escher, 2007). The material is light but strong, can be formed and converted into complex shapes, has a high thermal and electrical conductivity, and can be recycled without decrease in quality. Because aluminum foil is light in weight, it is energy-efficient for transportation. Moreover, aluminum foil can be considered as a modern packaging material since it is environmentally friendly. Aluminum foil is used for aseptic cartons, pouches, wrappings, bottle capsules, push-through blisters, laminated tubes, lids, trays and containers.

Laminates and Metallized Films

Lamination of packaging involves the binding of aluminum foil to paper or plastic film to improve the barrier properties. Thin gauges facilitate application. Although lamination to plastic enables heat-sealability, the seal does not completely bar moisture and air. Because laminated aluminum is relatively expensive, it is typically used to package high-value foods such as dried soups, herbs and spices. A less expensive alternative to laminated packaging is metallized film. Metallized films are plastics containing a thin layer of aluminum metal (Fellows and Axtell, 2002). These films have improved barrier properties to moisture, oils, air and odors, and the highly reflective surface of the aluminum is attractive to consumers. More flexible than laminated films, metallized films are mainly used to package snacks. Although the individual components of laminates and metallized films are technically recyclable, the difficulty of sorting and separating the material precludes economically feasible recycling.

Tinplate

Tinplate has been used for preserving food for well over a hundred years and today provides a robust form of packaging, allowing minimization of headspace oxygen and sterilization of the foodstuff within the hermetically sealed can, giving a long, safe ambient shelf-life with no or minimal use of preservatives (Blunden and Wallace, 2003). Produced from low-carbon steel (that is, black plate), tinplate is the result of coating both sides of black plate with thin layers of tin. The coating is achieved by dipping the sheets of steel in molten tin (hot-dipped tinplate) or by the electrodeposition of tin on the steel sheet (electrolytic tinplate). Although tin provides steel with some corrosion resistance, tinplate containers are often lacquered to provide an inert barrier between the metal and the food product. The commonly used lacquers are materials in the epoxy phenolic and oleoresinous groups, and vinyl resins. Tinplate has good ductility and formability, and therefore can be used for containers of many different shapes. Tinplate cans with a plain internal surface are used for low-acid foods, including tomato-based products, some fruits and some vegetables. The benefit provided by the bare tin surface inside the can is protection of the natural flavor and appearance of the food, through oxidation of the tin surface in preference to oxidative degradation of the food. This process retains the quality attributes that consumers expect from these products throughout the long shelf-life. Tinplate is used extensively in food and beverage cans. Europe produces and fills approximately 40000 million cans per annum for beverages (beers, carbonated soft drinks, water and wine), of which almost half are made of tinplate and all are internally lacquered (Blunden and Wallace, 2003). Its relatively low weight and high mechanical strength make it easy to ship and store. Finally, tinplate is easily recycled many times without loss of quality, and is significantly lower in cost than aluminum. A good review of the subject is available in the literature (Blunden and Wallace, 2003).

Tin-Free Steel

This is also known as electrolytic chromium-coated steel or chrome-oxide-coated steel. Tin-free steel requires a coating of an organic material to provide complete corrosion resistance. Although the chrome/chrome oxide makes tin-free steel unsuitable for welding, it also makes tin-free steel excellent for the adhesion of coatings such as paints, lacquers and inks. Like tinplate, tin-free steel has good formability and strength, but it is much cheaper than tinplate. Food cans, can ends, trays, bottle caps and closures can all be made from tin-free steel. Moreover, it can also be used to make large containers for bulk storage of ingredients or finished products (Fellows and Axtell, 2002).

Sterilization of Packaging Materials

Packaging makes food more convenient and provides the food with a higher level of safety from microbial and possible biochemical changes during transportation and storage. In order to meet the huge demand for processed foods with longer shelf-lives, packaging materials have to be sterile. There are many sterilization methods to choose from, such as the use of steam, sterile filtration, ethylene oxide gas, electron beam (E-beam) treatment and gamma radiation. Each technique has its own advantages and disadvantages.

According to the US Food and Drug Administration (FDA), sterilization is a process that is regulated only when it is used on low-acid foods (Code of Federal Regulations, 1986, Title 21, Part 113), and sterilants should not leave any residue on the food contact surface. However, when plastic packaging materials and chemical sterilants are used, the process is regulated as an indirect additive to food (Code of Federal Regulations, 1986, Title 21, Part 174). The process of obtaining FDA approval for the use of chemical sterilants in food packaging has eased significantly. Scientific evidence of the adequacy of the process must be filed by the processors before a system is put into commercial operation (Ansari and Datta, 2003).

Chemical processes using ethylene oxide, sodium hypochlorite, peracetic acid and hydrogen peroxide are used for sterilization of packaging. Ethylene oxide is slow in action and its desorption requires a very long time. Thus, it can be used for the pretreatment of packaging, but not for final sterilization of the packaging. Sodium hypochlorite and peracetic acid are effective sterilants; however, they have limitations in terms of the removal of residues. Alcohols such as glycols require high-temperature (≈100 °C) application for the desired sporicidal effect. Hydrogen peroxide (H_2O_2) has a high sporicidal effect at 80 °C, which makes it useful for packaging sterilization, although it shows poor effectiveness at ambient temperatures. The FDA regulations specify that a maximum concentration of 35% H_2O_2 may be used for sterilizing food contact surfaces, and the finished product must not exceed 0.5 ppm H_2O_2. It is applied to the material and then evaporated by heating using hot air or infrared radiation. Polyethylene was the first material to be approved for the use of hydrogen peroxide as a sterilant for packaging that was in direct contact with food (Federal Register, 1981, 46(6), 2341). Approval was later extended to other polymers (polyolefins, polystyrene, EVA, PET etc.) (Code of Federal Regulations, 1984, Title 21, Part 178, 1005; 1986, Title 21, Part 178, 1005).

Although steam or hot water is effective for the sterilization of tubes carrying food products, hot air (300 °C), with or without filtration, is commonly used for the sterilization of air injected into filling spaces. Air at 330–350 °C may also be used (for 30 min) for sterilization of tubes for food products. Sterilized air cooled to a temperature of 180–200 °C is used to evaporate H_2O_2, and when cooled to 50 °C can be employed for pressurizing filling chambers.

The use of UV radiation has several requirements, such as perpendicular incidence of the radiation, a dry atmosphere, a smooth surface, a low concentration of microorganisms, the absence of visible light to avoid reactivation of microorganisms, and shielding to protect the operator. If these requirements are satisfied, UV radiation can be used to provide a complementary treatment of already sterilized packaging.

The use of irradiation is becoming a common treatment for sterilizing packaging materials. This can be done in two ways: (i) sterilization of packaging materials for aseptic packaging, and (ii) radiation processing of prepackaged food. In both cases, the radiation stability of the packaging material is a key component for proper implementation of the technology. In addition, care must be taken that the irradiation does not significantly affect the physical and chemical properties of polymeric materials. Multilayer structures produced by co-extrusion are likely to satisfy the demands of radiation processing of prepackaged food, whereas single-layer plastics cannot meet the requirements of either of the types of treatment mentioned above (Chuaqui-Offermanns, 1989). Nowadays, packaging is mostly made of natural or synthetic plastics; therefore, the effect of irradiation on these materials is crucial for packaging engineering (Haji-Saeida *et al.*, 2007). As mentioned above, packaging materials can be irradiated either prior to or after filling; they are commonly irradiated prior to filling. In North America, the irradiation of empty packaging materials represents a significant activity in radiation facilities. In Europe, the packaging materials irradiated are mainly the following: bag-in-box materials; cardboard and plastic cups; plastic and paper bags; cardboard; tin cans; aluminum tubes; jerry cans; plastic tanks; rolls of plastic, aluminum and coated cardboard; and rubber and cork plugs (Neijssen, 1994). Plastic films laminated with aluminum foil are sterilized by radiation; they are used for hermetically sealed "bag-in-a-box" products, such as tomato paste, fruit juices and wines. Other aseptic packaging materials, including dairy product packaging, single-serving containers (e.g., for cream) and wine bottle corks, are also sterilized by irradiation prior to filling and sealing to prevent product contamination. With certain irradiation facilities, it is possible to determine and specify the most suitable shapes and sizes of the unit packs (Kubera, 1994).

In commercial aseptic processing systems, sterile food products are filled into sterilized packages or containers in a sterile environment, and the packages are then hermetically sealed to prevent recontamination of the product. This can be achieved in two ways: (i) using presterilized preformed containers such as bottle and cans, or (ii) sterilizing the packaging material, forming it into suitable containers, filling them with the sterile product and sealing the package in "form-fill-and-seal" (FFS) machines. In aseptic systems, the packaging material is usually sterilized either inside the packaging machine, or externally and then introduced aseptically into the aseptic zone of the packaging machine. The sterilization of packaging materials in aseptic systems is well described in a review paper (Ansari and Datta, 2003).

Packaging Design

Consumer packaging serves to contain and communicate. The main drivers behind packaging design used to be the cost to the manufacturer and the potential impact on consumers. However, there are several factors to be considered in packaging design today. The factors that concern food packaging design are:

- the physical/mechanical properties of the packaging materials used;
- the barrier properties of the packaging materials;
- hazards in distribution;
- printing characteristics;
- productivity, machinery, automation and packaging speed;
- sealability of the packaging materials;
- hygiene, sanitation and ecological factors;
- the strength of the external packaging;
- environmental conditions;
- effectiveness for display.

A package should protect the product, be suitable for production-line speeds, sell the item, provide reusable value to the user, satisfy legal requirements and keep packaging-related expenses low. More recently, however, the additional factor of openability has become more important (Yoxall *et al.*, 2006). Package design can be considered from two different viewpoints: first, design based on functional requirements, and second, design based on sales requirements.

Design Based on Functional Requirements

Packaging design must satisfy various functional criteria: in-home, in-store, production, distribution, safety and legal. For consumers (in-home), packaging should be convenient to use and store, reinforce the consumers' expectations of the product, and guide them in using the product safely and effectively. In addition, packaging should be recyclable and environmentally friendly.

For the purposes of sale (in-store), packaging should be attractive and visible on the shelf, identify the product easily and differentiate the product from others, communicate benefits and uses, and attract customers to actually purchase the item. The product must also be easy for retailers to store and to stock on shelves or the floor, and simple to process at a checkout counter or other final point of distribution.

From the point of view of production, packaging is assessed on cost. A design is only successful when the cost of the packaging materials is economically feasible for the company. A container that is too large or too small is not suitable for a process line, since it could significantly slow the speed of the production machines.

Packaging considerations related to distribution and safety are very important. A package should retain its shape during storage, shipment and distribution of the product. It should not hurt consumers. Examples of packages that might result in harm to consumers include those with sharp edges, such as some pull-top canisters; glass containers; and heavy boxes that might break when the consumer is carrying them or cause strain or injury to the consumer when picked up or set down. Packaging developers have to be very careful during the design process about any decisions that have an impact on distribution and safety. For example, packages must be able to withstand the pressure of crates stored on top of them. Packages should be able to resist moisture, withstand temperature changes and withstand rough handling. From a cost standpoint, packages must also be designed to suit standardized transportation requirements related to weight, size and durability. Finally, they should be designed so that the bar code on the package is easily scanned.

The packaging should follow rules and legislation as directed by government and regulatory agencies. Numerous laws have been passed to protect consumers from adulterants and unsafe products. For instance, some laws require that potentially dangerous goods, such as gasoline and drugs, be stored in specially constructed containers. Other laws forbid misrepresenting the quality or quantity of a product through misleading packaging. The most influential class of laws that affect packaging is that related to labeling. Under various laws, labels must provide the sodium content if other nutritional information is shown. They must also show the ingredients, in descending order from the one of highest quantity to the one of lowest quantity.

Certain food items, such as beef, may also be required to display qualitative "grade labels" or inspection labels. Some laws are intended primarily to discourage misleading labeling related to health benefits of food items. Specifically, many package labels have claimed, subjectively, that their contents were "low-sugar" or "high-fiber," or possessed some other health benefits, when the facts indicated otherwise. Basically, the new laws require most food labels to specify values such as calorie and cholesterol contents, fat and saturated-fat percentages, and sodium levels.

Design Based on Marketing Demand

In addition to the functional requirements, product packaging must be designed in such a way that it will attract customers. The major areas of marketing requirements are apparent size, attention-drawing power, an impression of quality, and clear readability of the brand name.

Design for apparent size entails designing the packaging to look as large as possible without misrepresenting the actual contents. This objective can be achieved by ensuring that the panels or dimensions of the package most likely to be viewed by the consumer are the largest; the product or brand name should be clearly visible.

An aesthetic and obtrusive packaging design definitely attracts the attention of consumers. Mostly, the product is displayed on the front of the package in the form of a picture, artwork or a see-through window and thereby woos consumers. In addition, bright colors, glossy stock, obtrusive carton displays, and other elements can garner positive attention if used prudently.

An impression of quality is an important sales requirement for packaging because items that are perceived to be of low quality are usually assumed to be poor value, regardless of price. Readability is the fourth basic sales requirement for successful package design. This element is of paramount importance for products such as breakfast cereals that are shelved next to several competing brands and products. Among other guidelines, letters and logos should be large and printed in the same type style as that used in complementary print and television advertising. The requirement of readability contributes to the difficulties of packaging completely new products.

Packaging for Nonthermal Processes

Nonthermal processing technologies are currently creating interest among food scientists, manufacturers and consumers because they have a minimal impact on the nutritional and sensory properties of foods, and extend shelf-life by inhibiting or killing microorganisms. These preservation processes are considered to preserve quality attributes better than conventional processes do. There are various nonthermal processing technologies are currently being used or in a developmental stage, including processes using high pressure, ionizing radiation, ultrasound, pulsed electric fields, pulsed light, high-voltage arc discharges and magnetic fields. In most cases, the food materials are packed in suitable packaging materials before being processed. Appropriate food packaging is a critical element in ensuring food quality and safety, especially for new, innovative processes. However, a crucial question concerns the possible effects of nonthermal processing on the packaging materials, and thus on the quality and safety of the packaged foods.

Among the nonthermal processing technologies, high-pressure processing is the front runner in terms of commercial success, and various pressure-treated foods are already commercially available in Europe, North America and Japan. In this section, the effects of high-pressure processing on packaging materials and possible interactions between foods and packaging materials are discussed.

Packaging for High-Pressure Processing (HPP)

The preservation of foods by high hydrostatic pressure (up to 1000 MPa) is a promising technique for food processing. The food materials are generally packed in a flexible packaging material and put in a pressure chamber for pressure treatment. The pressure is usually increased over a period of minutes, held at the final pressure for varying amounts of time and then quickly released, usually within a few seconds. The headspace should be minimized to reduce the package volume and reduce the time required to reach the target pressure (Caner *et al.*, 2004). Heat is generated owing to adiabatic compression during pressurization. The packages must be able to withstand the operating conditions of high pressure and temperature, and they must have sufficient mechanical, sealing and gas barrier properties (Galotto *et al.*, 2009). Furthermore, the packages should ensure the maintenance of the quality of the food throughout the entire marketing channel.

Plastic bottles and semirigid plastics can be used if they are properly sealed; for example, such packs are used for fruit purees, pastes and juices (Le-Bail *et al.*, 2006). High-pressure processing (HPP) can be used in a continuous process for liquids and can also process foods in their packaging (Morris *et al.*, 2007). The increasing use of plastic structures in food packaging has led to an increased interest in the area of mass transport. Polymeric materials are not perfectly inert in direct contact with food materials. Food–packaging interactions include transport of gases, vapors, water and other low-molecular-weight compounds, and also include chemical changes in the food, the package or both (Calvert and Billingham, 1979; Bieber *et al.*, 1984). Moreover, the behavior of packaging materials during high-pressure treatment, especially during the pressurization and depressurization steps, has not yet been fully studied.

In the last few years, some research work has been carried out on various aspects of packaging materials under high pressure (Caner *et al.*, 2000; Dobiáš *et al.*, 2004; Le-Bail *et al.*, 2006; Galotto *et al.*, 2008, 2009; Fairclough and Conti, 2009; Bull *et al.*, 2010; Mauricio-Iglesias *et al.*, 2010). However, the information that has been obtained is limited and is yet to be validated. It is worth mentioning that the packaging issue is very crucial to establishing HPP as an effective nonthermal processing technology. A brief overview of the research findings is presented in Table 41.4.

Table 41.4 Effects of high pressure on packaging materials

Materials used[a]	Process parameters	Properties studied	Results	Reference
PA/PE	500 MPa	Structure using SEM	Significant dependence of permeation rate	Goetz and Weisser (2002)
PP/EVOH/PP, OPP/PVOH/PE, PET/Al/CPP	400–600 MPa, 10 min, 20–40 °C	Water transmission rate (WTR, oxygen transfer rate (OTR), tensile stress (TS)	Not significantly affected	Masuda *et al.* (1992)
PE, PP, BOPP; PA/PE; PE/PA/EVOH/PE; PET/PE/EVOH/PE; LDPE/PA/LDPE; LDPE/EVOH/LDPE/APET/Exp.PET/APET	600 MPa, 60 min at room temperature	Mechanical properties, transparency, WTR, OTR, migration characteristics	HP affected functional properties of tested films, loss of heat sealability, migration levels affected significantly	Dobiáš *et al.* (2004)
PE/nylon/Al/PP Nylon/EVOH/PE	200, 400, 690 and 827 MPa, 30–95 °C	Migration, delamination	Delamination observed between PP and Al layers	Schauwecker *et al.* (2002)
Four packaging materials, including PE/EVOH/PE and met-PET/PE.	400 MPa, 60 °C	Mechanical properties	Delamination and wrinkling observed in packaging materials; no significant changes in mechanical properties	Galotto *et al.* (2008)
Ethylene–vinyl alcohol copolymers (EVOH): EVOH26 and EVOH48	400 and 800 MPa, 5 and 10 min, 40 and 75 °C	Structure	No effect on structure of copolymers with high ethylene content (EVOH48); improvement in crystalline morphology of EVOH26; improves barrier properties	López-Rubio *et al.* (2005)
LDPE/HDPE/LDPE; PET/LDPE/HDPE	500 MPa 30 min, 25 °C	Sorption: *p*-cymene, acetophenone	Structures remained unaffected	Kübel *et al.* (1996)
LLDPE/Irganox 1076; LLDPE/Uvitex OB; PLA/Uvitex OB	HP/HT: 800 MPa, 5 min, 115 °C HP/LT: 800 MPa, 5 min, 40 °C	Migration of polymer	No effect of HP/T treatments on Uvitex OB and Irganox 1076 migration from LLDPE and PLA. Diffusivity of Uvitex OB in PLA is much lower than in LLDPE and shows good barrier properties. HP/HT clearly affects the structure of PLA	Mauricio-Iglesias *et al.* (2010)

PLASiOx/PLA and PET-AlOx	500 MPa, 15 min, 50 °C	Mechanical, thermal, and gas barrier properties	HP treatment significantly affected the mechanical, thermal and gas barrier properties of both films. In the PETAlOx film, many pinholes and cracks were formed. In the PLASiOx-PLA film, an abrupt change in properties occurred when the film was in contact with water	Galotto *et al.* (2009)
High-barrier laminated films, including PET/AlO_x/LDPE	800 MPa, 50 °C;	Permeability	Most films showed significant, albeit small, increases in permeance after pressure treatment, but metallized PET/EVA/linear LDPE suffered significant increases in permeance to oxygen, carbon dioxide and water by up to 150%	Caner *et al.* (2000)
PET/ON/R-CPP; PET/PVDC/CPP; PET/ON/Al/R-CPP; PET/Al/CPP; PET/Al/PP; PET–SiO_x/OPA; PET–AlO_x/ON	600 MPa, 5–10 min, 115 °C			Bull *et al.* (2010)
PLA isomers (L, D and DL)	350, 450 and 650 MPa, 15 min, 22–26 °C		T_g decreased; T_m remained unaffected. Partial shift in crystallinity of L-isomer at 650 MPa, and enthalpy dropped	Ahmed *et al.* (2009)
PP/Irganox 1076	800 MPa, 60 °C.	Migration behavior of Irganox in PP	No significant migration of Irganox 1076 from PP	Caner and Harte (2005)
PP	695 MPa, 10 min, 70 °C	Effect of varying the air headspace on the structure of films	Increasing the amount of air headspace increased the prevalence of opaque areas.	Fairclough and Conti (2009)

[a] See the original references for explanation of the acronyms in this column.

The barrier properties of packaging materials after high-pressure treatment have been studied by various researchers. Masuda *et al.* (1992) found that the barrier behavior of flexible plastic structures was not significantly affected by HPP. These authors studied the permeation properties of the following flexible structures: PP/EVOH/PP, OPP/EVOH/PE, PVDC-coated OPP/CPP and PET/Al/CPP. After pressure treatment (400 MPa and 600 MPa for 10 min), the permeability to water vapor at 40 °C and 90% RH and the oxygen permeability at 23 °C and 90% RH did not change relative to the material's initial water and oxygen barrier properties. Ochiai and Nakagawa (1992) expressed a similar view after measurement of the barrier properties of laminated plastic package structures filled with water and pressure-treated at 400 MPa for 10 min: the oxygen and water permeabilities did not change after HPP.

The effects of HPP on the mechanical and physical characteristics of eight high-barrier multilayer films (PET/SiO_x/LDPE, PET/Al_2O_3/LDPE, PET/PVDC/nylon/HDPE/PP, PE/nylon/EVOH/PE, PE/nylon/PE, metallized PET/EVA/LLDPE, PP/nylon/PP and PET/PVDC/EVA) and a single-layer polypropylene film were investigated by Caner *et al.* (2003). Pouches made from these films were filled with distilled water, sealed and then pressure-treated at 600 and 800 MPa for 5, 10 and 20 min at a process temperature of 45 °C. The results indicated that there were no significant changes in the tensile strength, elongation or modulus of elasticity of any of the films after HPP. However, significant physical damage to metallized PET (MET-PET) was identified by SEM and and C-mode scanning acoustic microscopy. Thus it could be concluded that MET-PET is not suitable for batch-type HPP. Similar experiments were carried out by Le-Bail *et al.* (2006). These authors investigated the effects of HPP (at 200, 400 and 600 Pa) on the mechanical properties and the water vapor permeability of seven selected packaging materials (NOD 259 (PA-PE), BB4L, PET/BOA/PE, PET/PVDC/PE, PA/SY, LDPE and EVA/PE). The results obtained indicated that HPP affected the mechanical strength of the packaging material minimally. The depressurization rate did not have any significant influence. The barrier properties to water vapor were not significantly affected and were even slightly enhanced for LDPE, which is a packaging material commonly used for HPP applications.

Goetz and Weisser (2002) measured the permeation rate of LDPE/HDPE/LDPE (12 μm/12 μm/12 μm) polymer films for an aroma compound, *p*-cymene (0.25 volume %), at a pressure level of 50 MPa at 23 °C. The permeation rate decreased with increasing pressure. Caner and Harte (2005) studied the migration of Irganox 1076 from polypropylene in high-pressure-treated and untreated polypropylene pouches containing either 95% or 10% ethanol as aqueous food-simulating liquids (FSLs) for 20 days at 40 °C and 60 °C. After contact, the concentrations of Irganox in the PP and the FSL were measured to study the migration behavior. The results indicated that migration of additives from plastics into food simulants is possible only when the contact period is too long. No significant differences in the migration level of Irganox 1076 compared with controls into either of the FSLs were observed after HPP treatment. Increasing the HPP temperature increased the migration of Irganox from the PP into the FSL.

Overall, the data indicated that the migration from a single-layer PP material did not significantly increase as a result of HPP.

The effect of HPP (695 MPa for 10 min at 70 °C) on the structure of polypropylene pouches and the effect of an air headspace on a PP heat-sealing inner layer was reported by Fairclough and Conti (2009). With increasing pressure, both nitrogen and oxygen become increasingly soluble in the polymer layer. If the pressure was rapidly released, this caused voids and pits to form in the inner layer. The films showed opaque areas that scattered light well and appeared white. These areas came in two forms: one associated with folds in the material, and the other distributed throughout the film. Microscopic study revealed these opaque areas to be pits and craters in the film. An estimate of the increase in solubility (70-fold) was calculated from regular-solution theory. Fairclough and Conti recommended slow decompression to avoid the formation of these features. The most practical solution would be to avoid the use of polypropylene in ultrahigh-pressure processes.

Ahmed *et al.* (2009) studied the effect of high-pressure treatment on the thermal properties of amorphous and semicrystalline polylactide (PLA). Three different PLA isomers (D, L and DL) with molar masses in the range of 3800–6200 were pressurized (at 350, 450 and 650 MPa) for a holding time of 15 min, and the process temperature was maintained in the range of 22–26 °C. The thermal properties (glass transition temperature, melting behavior and crystallinity) of postprocess samples were analyzed by differential scanning calorimetry and compared with those of the untreated samples. The glass transition temperature T_g was found to decrease as pressure was applied to the lactides. During pressure treatment, it was observed that both the melting and the crystallization peak of the L-isomer were significantly reduced at 650 MPa, and this observation was quantified by measuring a decrease in the enthalpies of fusion (ΔH_m) and crystallization (ΔH_c). Fourier transform infrared spectroscopy could not detect any change in the crystalline band (1300–1150 cm^{-1}) of the pressure-treated L-isomer.

High-pressure thermal (HPT) processing has the potential to deliver quality benefits for a range of processed foods. By exploiting the rapid temperature increase and decrease that accompany pressurization and depressurization, commercial sterilization of foods can potentially be achieved by HPT processing with an overall reduced thermal exposure compared with conventional thermal processing technologies. Bull *et al.* (2010) studied the suitability of packaging materials for high-pressure-assisted pasteurization and sterilization in combination with HPT processing. These authors studied the barrier properties after HPT processing (at 600 MPa and 110–115 °C for 5–10 min) of 11 commercially available packaging materials developed for conventional thermal sterilization processes. The structures encompassed barriers based on aluminum (Al) foil, vapor-deposited silicon and aluminum oxides (SiO_x and AlO_x), oriented nylon/polyamide (ON/OPA) and PVDC–methyl acrylate (PVDC–MA). It was observed that the barrier properties of films containing vapor-deposited oxide and nylon were compromised by the combination of high pressure (600 MPa) and high temperature (~110 °C) that would reasonably be expected to be required to render food commercially sterile by HPT processing. However, the barrier properties of films containing aluminum foil and PVDC–MA were not significantly affected by HPT processing. All of the materials suffered cosmetic deformation of the outer surface to some degree, and mechanisms for these changes were proposed.

In similar studies of HPT processing intended for a pasteurization treatment (800 Pa for 5 min, from 20 to 40 °C) and a sterilization treatment (800 MPa for 5 min, from 90 to 115 °C), Mauricio-Iglesias *et al.* (2010) evaluated the migration of additives (Irganox 1076 and Uvitex OB) from one synthetic common packaging material (LLDPE) and one biosourced material (PLA) in contact with four food-simulating liquids during and after HPT treatments, and compared the results with those obtained with conventional pasteurization and sterilization. LLDPE withstood the high-pressure sterilization, whereas it melted during the conventional sterilization. No difference was observed in the migration from LLDPE for either of the treatments. In the case of PLA, the migration of Uvitex OB was very low or not detectable in all the cases studied. However, the effect of HPT processing on PLA could not be properly assessed, since the migration was too low to be detected within the experimental error. The diffusivity of Uvitex OB in PLA was much lower than in LLDPE, and PLA therefore showed good barrier properties. However, both of the sterilization processes (HPT and conventional) clearly affected the structure of the PLA, which suggests that PLA should be restricted to low-temperature treatments.

Biodegradable Packaging

The present global concern about petrochemical-based plastic materials has generated much interest in biodegradable, or "green," packaging materials. Biodegradable plastics have matured from their infancy, and polymers with true susceptibility to microbial degradation are a reality today (Ahmed and Varshney, 2011). The attribute of compostability is very significant for biopolymer materials because whereas recycling is energy-expensive, composting allows disposal of packages in the soil, where they are transformed into water, carbon dioxide and inorganic compounds (Urayama *et al.*, 2003). Biodegradation standards assess the propensity of a material to degrade biologically. A few years ago, the U.S. Composting Council and the Biodegradable Products Institute initiated certification for materials suitable for composting, following ASTM guidelines (ASTM, 2003). According to these guidelines, a "biodegradable plastic" is defined as a degradable plastic in which the degradation results from the action of naturally occurring microorganisms such as bacteria, fungi and algae.

Biodegradable or green packaging must satisfy some basic requirements to be an ideal candidate for food packaging. These requirements include barrier properties (to water vapor, gases, light and aromas), optical properties (transparency), strength, welding and molding properties, printing properties, migration resistance, chemical and temperature resistance, the ability to satisfy disposal requirements, antistatic

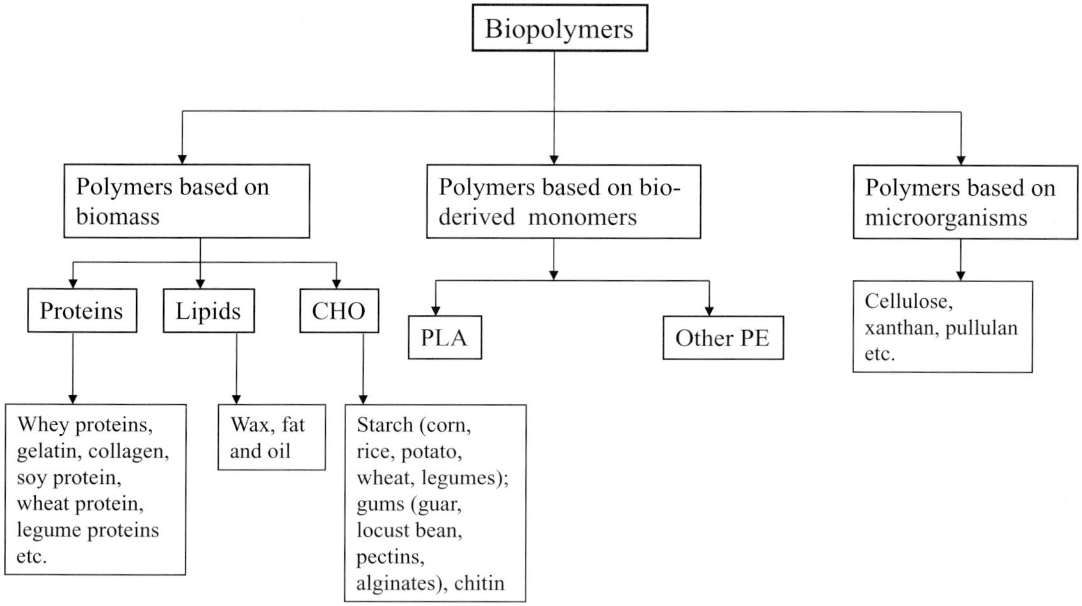

Figure 41.3 The various categories of bio-based materials. (CHO, carbohydrates; PLA, polylactide; PE, polyesters.)

properties, and the ability to retain sensory properties. And, above all, it must strictly follow food safety regulations.

Bio-based polymers, or biopolymers, are obtained from renewable resources (Weber *et al.*, 2002), as illustrated in Figure 41.3. These renewable resources consist of proteins (whey protein, soy protein, collagen, gelatin, wheat protein etc.), polysaccharides (starch, alginates, pectin, carrageenans and chitosan/chitin) and lipids (fats, waxes and oils) (Comstock *et al.*, 2004). Polymers such as polylactide have gained growing attention in the last decade as food packaging materials because they can easily be obtained from renewable resources, their production consumes quantities of carbon dioxide, they can be recycled and composted, and their physical and mechanical properties can be tailored through the polymer architecture (Sinclair, 1996; Siracusa *et al.*, 2008). Although the production costs of PLA are relatively high compared with conventional plastics, there are predictions that this will change with time as production volume and demand both increase (Auras *et al.*, 2004). It is important to note that while some bio-based packaging materials may be biodegradable, not all biodegradable materials are bio-based (Weber *et al.*, 2002).

Recent technological advances have also allowed biopolymers to be processed similarly to petroleum-based plastics, whether in sheet form or by extrusion, spinning, injection molding or thermoforming (Comstock *et al.*, 2004). Some of the beneficial characteristics of PLA fiber products include their natural soft feel, ease of processing, and unique stain and soil resistance. PLA has already been approved for its intended use in fabricating articles for contact with food by the US Food and Drug Administration, and it is already being exploited for short-shelf-life food packaging such as containers, drinking cups, sundae and salad cups, overwrap and lamination films, and blister packages (Plackett *et al.*, 2006). PLA-based packaging materials are currently being used for supermarket products in Europe and North America for packaging bottled water, juices and yoghurts. The containers satisfy German and European Union (EU) food standards. PLA excels at resistance to staining in standard tests with coffee, tea, cola, catsup and other food products. Therefore, PLA is gradually moving toward being a "green" food packaging material.

The Future of Packaging

With consumer demands for quality and for fresh-like food, traditional packaging is no longer acceptable in today's food industry. Innovative packaging with enhanced functions is constantly being sought in response to the demands of consumers for minimally processed foods with less preservatives, and in response to strict regulatory requirements, market globalization and the recent threat of food bioterrorism (Yam *et al.*, 2005). To meet these challenges, active packaging and intelligent packaging can play important roles in future food packaging.

Active Packaging

Active packaging is an innovative concept that can be defined as a mode of packaging in which the package, the product and the environment interact to prolong shelf-life or enhance safety or sensory properties, while maintaining the quality of the product (Suppakul *et al.*, 2003). It allows the active preservation of foods, according to their needs, by modification of the environment inside the package by removing undesired gases or by regulating the composition of the gas in the package headspace.

Active systems can be classified according to their functionality as scavengers, regulators and emitters, and their action can be specific for several substances (O_2, CO_2, ethylene etc.). The internal atmosphere may be regulated by substances that absorb (scavenge) or release (emit) gases or vapors. Some examples of active packaging systems are oxygen scavengers, carbon dioxide scavengers and emitters, humidity adsorbers and controllers, ethylene scavengers, aroma emitters and absorbers, enzymatically active systems, and antimicrobial systems (Lopez-Rubio *et al.*, 2004). More details of active packaging are available in the literature (Lopez-Rubio *et al.*, 2004; Kerry *et al.*, 2006; Brody *et al.*, 2008).

The next generation of food packaging may include materials with antimicrobial properties to prevent food-borne microbial outbreaks while maintaining desired quality attributes and safety. Such packaging could play a role in extending the shelf-life of foods and reduce the risk from pathogens (Appendini and Hotchkiss, 2002). Antimicrobial polymers may find use in other food contact applications as well.

Table 41.5 Examples of applications of active packaging for use in the food industry (adapted from Kerry *et al.*, 2006).

Operations	Details
Absorbing/scavenging properties	O_2, CO_2, moisture, ethylene, flavors, taints, UV light
Releasing/emitting properties	Ethanol, CO_2, preservatives, antioxidants, SO_2, flavors, pesticides
Removal properties	Catalyzing food component removal: lactose, cholesterol
Temperature control	Insulating materials, self-heating and self-cooling packaging, microwave susceptors and modifiers, temperature-sensitive packaging
Microbial and quality control	UV- and surface-treated packaging materials

Antimicrobial packaging is a form of active packaging. It acts as an additional safety measure by reducing, inhibiting or retarding the growth of microorganisms that may be present in the packaged food or packaging material itself.

There are various types of antimicrobial packaging currently available on the market (Appendini and Hotchkiss, 2002). These include (i) the addition of sachets or pads containing volatile antimicrobial agents into packages, (ii) incorporation of volatile and nonvolatile antimicrobial agents directly into polymers, (iii) coating or adsorbing antimicrobials onto polymer surfaces, (iv) immobilization of antimicrobials in polymers by ionic or covalent linkages, and (v) the use of polymers that are inherently antimicrobial. A few examples of applications of active packaging for use in the food industry are listed in Table 41.5.

Although active-packaging technologies are used commercially in Japan, the US, Europe and Australia, no specific regulations exist to date for active packaging. Most of the active agents are considered as constituents of food-contact materials, and thus these systems should comply with the present regulations regarding migration. Moreover, a more exhaustive study of the chemical, microbiological and physiological effects of the technologies applied must be carried out before full implementation of the technology.

Edible Packaging

Edible packaging is defined as a thin layer of edible material formed on a food as a coating or placed (preformed) on or between food components (Pagella *et al.*, 2002). Edible films and coatings have received increasing attention recently as an interesting alternative to food packaging. Considerable research has been conducted to develop and apply bio-based polymers made from a variety of agricultural commodities and food industry waste. In view of their beneficial impact on the environment, natural polymers have been studied extensively for the development of edible packaging. A variety of polysaccharides (starch and hydrocolloids), proteins (whey proteins, soybean proteins and fish proteins) and lipids have been used, either individually or in mixtures, to produce edible films. These materials present the possibility of obtaining thin films and coatings to cover fresh or further-processed foods to extend their shelf-life (Baldwin, 1994).

Edible films and coatings have some advantages such as edibility, biocompatibility, barrier properties, absence of toxicity, the fact that they are nonpolluting, and low cost (Han, 2000). Moreover, biofilms and coatings, by themselves or acting as carriers of food additives (i.e., antioxidants and antimicrobials), have been considered particularly for food preservation because of their ability to extend the shelf-life (Franssen and Krochta, 2003).

Proteins have been demonstrated to be excellent candidates for applications as edible films and coatings. Edible films based on milk proteins have good gas barrier properties and mechanical strength (Chen, 1995). Their water vapor resistance can be improved by forming emulsions or multiple-layer, multiple-component films. Edible films based on milk proteins are bland in taste, with an excellent capacity to integrate sensory enhancers. An edible film based on squid mantle muscle protein with 0.5% sodium citrate gave a transparent film with the highest tensile strength and least degradation of myosin heavy chain (Leerahawong *et al.*, 2011). The film showed excellent UV barrier properties.

The potential of polysaccharides as edible films has long been recognized (Guilbert, 1986; Cuq *et al.*, 1995). Their application in agricultural products for extending the shelf-life of fresh fruits and vegetables (Baldwin, 1994; Park, 1999) and their use in encapsulation processes for retarding the loss of flavor have become widespread. Their use in breading and batters for reducing oil uptake during frying is being investigated (Park *et al.*, 1993; Albert and Mittal, 2002).

Starches can be used to form edible and biodegradable films and, to date, high-amylose starch and hydroxylpropylated high-amylose starch have been used to form self-supporting films by casting from aqueous solution (Mark *et al.*, 1966).

Intelligent or Smart Packaging

Intelligent, or smart, packaging is basically designed to monitor and communicate information about food quality (Kerry *et al.*, 2006). A package can be made smart through functional attributes that add benefits to the food and hence to consumers. It is essentially an integrating method that deals with mechanical, chemical, electrical and/or electronically driven functions that enhance the usability or effectiveness of the food product in a proven way (Mahalik and Nambiar, 2010). Some common examples of intelligent packaging are time–temperature indicators (TTIs), ripeness indicators, biosensors and radio frequency identification (RFID). In addition, self-heating and self-cooling containers with electronic displays indicating use-by dates and information regarding the nutritional qualities and origin of the product in numerous languages are available in smart packaging (Mahalik and Nambiar, 2010). These smart devices may be incorporated into packaging materials or attached to the inside or outside of a package. The FDA recognizes TTIs for fish products, so their importance may increase in the seafood industry. Some superstores have already used RF sensors for tracking and tracing produce and other perishable commodities. The architecture and working principles of RFID devices have been described by Brody *et al.* (2008).

RFID uses radio waves to track items wirelessly. It makes use of tags or transponders (data carriers), readers (receivers), and computer systems (software, hardware, networking and a database). The tags consist of an integrated circuit, a tag antenna, and a battery if the tag is passive (most active tags do not require battery power). The integrated circuit includes a nonvolatile memory microchip for data storage, an AC/DC converter, encode/decode modulators, logic control and antenna connectors. The wireless data transfer between a transponder/tag and a reader makes RFID technology far more flexible than other methods of contact identification (Finkenzeller, 2003; RFID Journal, Inc., 2005), and thus makes RFID ideal for food packaging. The working principles of an RFID system are as follows (Brody *et al.*, 2008):

1. Data stored in a tag are activated by a reader when an object with an embedded tags enters the electromagnetic zone of a reader.
2. The data are transmitted to a reader for decoding.
3. The decoded data are transferred to a computer system for further processing.

A few examples of applications of intelligent packaging for use in the food industry are listed in Table 41.6.

Nano-Packaging

Nanotechnology has been significantly increasing its impact on the food and beverage packaging industry and has the potential to transform food packaging materials in the future. Research on the use of nanocomposites for food packaging began in the 1990s. Since then, sales of nano-related packaging products have been increasing every year worldwide. Nanoscale innovation could potentially introduce many amazing improvements to food packaging in the form of barrier and mechanical properties, detection of pathogens, and smart and active packaging with food safety and quality benefits (Brody *et al.*, 2008). Nanotechnology enables designers to alter the structure of packaging materials on the molecular scale, in order to give the material the desired properties. With different nanostructures, plastics can be given various gas and water vapor

Table 41.6 Examples of applications of intelligent packaging for use in the food industry (adapted from Kerry *et al.*, 2006).

Operations	Details
Tamper evidence and pack integrity	Detection of breach of pack containment
Indicators of product safety/quality	Time–temperature indicators (TTIs), gas-sensing devices, microbial growth and pathogen detection
Traceability/antitheft devices	Radio frequency identification (RFID) labels, tags, chips
Product authenticity	Holographic images, logos, hidden-design print elements, RFI

permeabilities to fit the requirements of various foods. By adding nanoparticles, one can achieve packages with more resistance to light and fire, better mechanical and thermal performance, and less gas absorption. These properties can significantly increase the shelf-life and sensory characteristics of food products, and facilitate transportation and usage. The addition of nanosensors to food packages is also anticipated in the future. Nanosensors could be used to detect chemicals, pathogens and toxins in foods (Brody *et al.*, 2008).

Biopolymer–clay nanocomposites have created new interest because of the significant property enhancements that can be obtained by incorporating only a few weight percent of rigid particles (nanofillers) with sizes in the nanometer range (approximately 100 nm). These nanofillers have been found to improve the mechanical, barrier, electrical, gas permeability and thermal properties of the base polymer significantly at very low filler concentrations with respect to their conventional microcomposite counterparts (Sinha *et al.*, 2002; Angellier *et al.*, 2005; Joshi *et al.*, 2006; Ginzburg *et al.*, 2009; Ahmed *et al.*, 2010a,b). These remarkable property enhancements make nanocomposites superior candidates for application as materials for food packaging. The properties of nanocomposites depend upon the state of dispersion of the nanoparticles in the polymer matrix (Mackay *et al.*, 2006). However, the relationship between the mechanical behavior and the state of dispersion of the nanoparticles remains unresolved (Akcora *et al.*, 2010). Occasionally, clay platelets fail to disperse, and remain aggregated into large "stacks"; in this situation, the modulus and strength of the composite may be comparable to those of the matrix polymer but the toughness and ultimate elongation may even be worse (Ginzburg *et al.*, 2009).

In general, polymer nanocomposites are made by dispersing inorganic or organic nanoparticles into either a thermoplastic or a thermoset polymer. The nanoparticles can be three-dimensional spherical or polyhedral particles (e.g., colloidal silica), two-dimensional nanofibers (e.g., nanotubes or whiskers), or one-dimensional disk-like particles (e.g., clay platelets). Such nanoparticles offer enormous advantages over traditional macroparticles and microparticles (e.g., talc, glass and carbon fibers) owing to their higher surface area and aspect ratio, the improved adhesion between the nanoparticles and the polymer, and the lower amount of loading required to achieve equivalent properties (Zeng *et al.*, 2005). Details of clay-based nanocomposites are available in a review article by Zeng *et al.* (2005).

There are several different techniques for making clay-based polymer nanocomposites, including in-situ polymerization, solution exfoliation and melt intercalation. The most successful technique for preparing polymer/layered-silicate nanocomposites is to intercalate polymers into the galleries of the silicate structure (Sinha *et al.*, 2002). Usually, the intercalation of polymer chains into the silicate galleries is performed either by insertion of suitable monomers into the galleries and subsequent polymerization (Figure 41.4) or by direct insertion of polymer chains into the galleries from either a solution or a melt. Basically, the process involves the annealing and mixing of the polymer and the layered silicate nanomaterial above the softening temperature

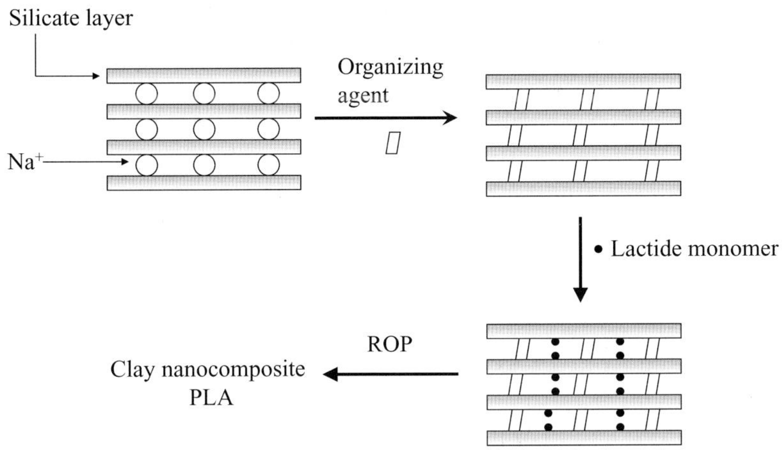

Figure 41.4 Schematic diagram of synthesis of nanoclay PLA (adapted from Ahmed and Varshney, 2011). (ROP, ring-opening polymerization.)

of the polymer. During annealing, the polymer chains diffuse from the melted bulk into the galleries.

Various biopolymers have been experimented on to produce nanocomposites. Polylactide is an emerging biopolymers with potential applications in food packaging. To enhance the barrier and mechanical properties, other synthetic polymers have been incorporated into PLA or it has been mixed with fillers to reduce the cost and/or improve performance (e.g., proteins, starch, inorganic fillers, and natural flax fiber have been used). It has also been chemically modified with the aim of extending its applications in more specialized or severe circumstances. Polymer nanocomposites represent a better alternative to traditional polymer composites (Petersen *et al.*, 1999; Alexandre and Dubois, 2000). Currently, montmorillonite (MMT) has been used in the formation of nanocomposites because of its high surface area and aspect ratio. MMT has a very high elastic modulus (178 GPa) as compared with most biopolymers. The high value of the elastic modulus enables MMT to improve the mechanical properties of biopolymers by carrying a significant portion of the applied stress (Fornes and Paul, 2003).

Starch–clay combinations are another type of biodegradable nanocomposite that has been tested for various applications, including food packaging (Huang *et al.*, 2005; Yoon and Deng, 2006; Cyras *et al.*, 2008). Significant improvements in mechanical properties were noticed; both the Young's modulus and the tensile strength were increased with the addition MMT. Barrier properties are of prime importance in bottling and food packaging. Cyras *et al.* (2008) reported that the effective diffusion coefficients for nanocomposites were lower than for starch alone. This suggests that the addition of MMT reduced the water uptake of the starch films, possibly owing to the tortuous structure formed by the exfoliated clay.

Protein isolates have been used to synthesize nanomaterials that may lead to food packaging applications. Whey protein isolate (WPI) acts as an edible film and coating material. TiO_2 was incorporated into WPI films to form a nanocomposite with improved antimicrobial properties (Sothornvit and Krochta, 2005). Zhou *et al.* (2009) recommended whey–TiO_2 nanocomposites to be used as food-grade, biodegradable packaging materials. The addition of small amounts (<1 wt%) of TiO_2 nanoparticles significantly increased the tensile strength of the WPI film (from 1.69 to 2.38 MPa).

Soy protein isolate (SPI) is a good candidate for a biodegradable plastic owing to its thermoplastic properties. However, it has a poor response to moisture and a high rigidity. Plasticization of SPI further decreases the barrier properties and tensile strength of films made from this material. Soy protein nanocomposite films showed reduced water vapor permeability, and improved elastic modulus and tensile strength compared with counterparts without fillers (Dean and Yu, 2005; Yu *et al.*, 2007).

Cellulose nanofibers (CNF) are another potential source of nanoparticles to improve the mechanical properties of packaging materials. It was found that incorporation of CNF improved the tensile properties, water vapor permeability and glass transition temperature of fruit-puree-based films (Azeredo *et al.*, 2009). The tensile strength increased (from 4.09 to 8.76 MPa) with an increase in the CNF concentration from 0% to 36%. It was proposed that there was an effective increase in the tensile strength and Young's modulus owing to the formation of a fibril network within the matrix, especially at higher concentrations of CNF.

The technology of nanocomposites is still in its infancy, and there are several issues that have raised concerns about food safety. Speculation is rife that nanotechnology may be creating strange new substances that should not be commercialized until further research has been completed. Any manufacturer can rely on past decisions approved by the FDA for any chemicals to go to market without involving the FDA, if the chemical it proposes to market is the same as the listed chemical, if it is sold for the same intended purpose, and if the product complies with any specifications and limitations placed on the intended use when the decision was made to permit marketing.

Packaging Safety, Legislation and Regulations

There are some health concerns regarding residual monomers and other components in petroleum-based packaging materials, including stabilizers, plasticizers, and condensation components such as bisphenol A. To ensure public safety in the United States, the FDA reviews packaging materials carefully and regulates those materials. Any substance that can sensibly be expected to migrate into food is classified as an indirect food additive and is subject to FDA regulations. There is a threshold of regulation, which is defined as a specific level of dietary exposure that typically induces toxic effects. If the level of a substance is below the threshold limit and therefore poses negligible safety concerns, the substance may be exempted from regulation as a food additive. The FDA revisits the threshold level if new scientific information raises concerns. Furthermore, the FDA advises consumers to use plastics for their intended purposes in accordance with the manufacturer's directions to avoid unintentional safety concerns. Despite these safety concerns, the use of plastics in food packaging has continued to increase owing to the low cost of the materials and their functional advantages (such as thermosealability, microwavability, their optical properties, and their ability to be formed into unlimited sizes and shapes) over traditional materials such as glass and tinplate (López-Rubio *et al.*, 2004).

In order to protect consumers against potential hazards from oral exposure to packaging materials, different countries have adopted several different frameworks. The history of formal food packaging regulation in the United States began with the passage of the Food Additives Amendment of 1958, and although some statutes in other countries dealt with packaging in general terms prior to that time, it is believed that this 1958 law set in motion all of the modern thinking on the subject (Heckman, 2005). The European Commission adopted a Framework Directive in November 1976, which established the general principles for all food-contacting materials and objects, as well as criteria and procedures that should be followed for the elaboration of specific Directives (Robertson, 1993).

European Union legislation has five main instruments: Regulations, Directives, Decisions, Recommendations and Opinions. The safety of food packaging materials is therefore generally based on the lack of potential toxic substances (from toxicological data) and the absence of migration from such substances (migration testing). Directive 76/893/EEC (European Union, 1976) established the principle that all materials and objects intended to come into contact directly or indirectly with foodstuffs must not transfer any of their constituents to the food product in quantities that could endanger human health, produce an unacceptable change in the composition of the food or alter its sensory characteristics (Heckman, 2005; Robertson, 1993).

EU Directive 2002/72/EC lays down limits with respect to the concentrations of certain substances in packaging and of migrating substances in foodstuffs or corresponding food simulants. These regulations are based on toxicological data on substances. EU Directive 2002/72/EC stipulates that a maximum of $10\,mg{\cdot}dm^{-2}$ or $60\,mg{\cdot}kg^{-1}$ of physiologically nonhazardous substances may transfer from the packaging to the foodstuff (global migration). Directive 89/109/EEC (European Union, 1989) was substituted in October 2004 by Regulation (EC) 1935/2004 (European Union, 2004) of the European Parliament and Council, and therefore converted into European Law. It contains general provisions on the safety of active and intelligent (A&I) packaging and sets out the framework for the safety evaluation process. A new Regulation specific to A&I packaging, published in 2009 (Regulation (EC) 450/2009), sets out additional requirements in order to ensure the safe use of A&I packaging and introduces an authorization scheme for substances used for active and intelligent functions in food contact materials.

In the US, the Food and Drug Administration proposed procedural regulations to govern the administration of the food additives law; these were published on December 9, 1958, and the rules were finalized on March 28, 1959. At the time, it was generally understood that the FDA would consider data derived from extraction studies and would concur in "nonadditive" status where such data indicated that there would be no detectable extraction with analytical methodology generally sensitive to a part or two per million. More than 250 pages of the Code of Federal Regulations set forth the Indirect Food Additive Regulations put into effect before 1997. They are effective at this time and will remain so. The Pre-Market Notification system now mandated by Congress in the Food and Drug Modernization and Accountability Act of 1997 effectively replaces the Petition-Regulation system put into play in 1958. Nevertheless, it is important to know what needed to be included in the Food Additive Petitions to bring about favorable FDA action, since the regulations adopted under the new law require that essentially everything required to be submitted in petitions needs to be provided in an acceptable Pre-Market Notification. The FDA Form 3480 is used to prepare and submit Food Contact Notifications, and FDA's Guidance Documents 44 make the data requirements quite clear. In general, the new law requires the same data as was required for a successful Food Additive Petition but it does so in a more systematic way. A detailed discussion of US and European regulations is available elsewhere (Heckman, 2005).

Conclusions

Foods are packaged to preserve their qualities. Processed foods pass through a distribution process and there are many hazards in the distribution system. Packaging is the best way to prevent hazards in the distribution systems. The food industry has seen great advances in the packaging sector since its inception. These advances have led to improved food quality and safety. Packaging materials have changed with time and demands. The interaction between foods and package materials is a real challenge in food package design. New advances have focused on controlling moisture migration, microbial growth, and volatile flavors and aromas. Currently, biopolymers have an edge over petroleum-based polymers from an environmental point of view. However, biodegradable packaging materials must satisfy the critical requirements on mechanical, thermal and barrier properties for the intended application. With the introduction of nonthermal processing, there is a great demand for packaging materials suitable for such processes. In general, it has been observed that most conventional packaging materials remain unaffected by high-pressure processing. Nanotechnology has the potential to influence the packaging sector significantly. Nanoscale innovations in the form of pathogen detection, active packaging and barrier formation are poised to take food packaging in a new direction.

References

Ahmed, J. and Varshney, S.K. (2011) Polylactides – chemistry, properties and green packaging – a review. *International Journal of Food Properties* 14: 37–58.

Ahmed, J., Varshney, S.K. and Ramaswamy, H.S. (2009) Effect of high pressure treatment on thermal and rheological properties of lentil flour slurry. *Lebensmittel-Wissenschaft und-Technologie* 42: 1538–1544.

Ahmed, J., Varshney, S.K. and Auras, R. (2010a) Rheological and thermal properties of polylactide/silicate nanocomposites films. *Journal of Food Science* 75: N17–N24.

Ahmed, J., Varshney, S.K., Auras, R. and Hwang, S.W. (2010b) Thermal and rheological properties of L-polylactide/polyethylene glycol/silicate nanocomposites films. *Journal of Food Science* 75: N97–N108.

Albert, S. and Mittal, G.S. (2002) Comparative evaluation of edible coatings to reduce fat uptake in a deep-fried cereal product. *Food Research International* 35(5): 445–458.

Akcora, P., Kumar, S.K., Moll, J., Lewis, S., Schadler, L.S., Li, Y., Benicewicz, B.C., Sandy, A., Narayanan, S., Ilavsky, J., Thiyagrajan, P., Colby, R.H. and Douglas, J. (2010) Gel-like mechanical enforcement in polymer nanocomposite melts. *Macromolecules* 43: 1003–1010.

Alexandre, M. and Dubois, P. (2000) Polymer-layered silicate nanocomposites: preparation, properties and uses of a new class of materials. *Materials Science and Engineering* 28: 1–63.

Angellier, H., Molina-Boisseau, S., Lebrun, L. and Dufresne, A. (2005) Processing and structural properties of waxy maize starch nanocrystals reinforced natural rubber. *Macromolecules* 38: 3783–3792.

Ansari, M.I.A. and Datta, A.K. (2003) An overview of sterilization methods for packaging materials used in aseptic packaging systems. *Transactions of the Institute of Chemistry* E 81, Part C: 57–65.

Appendini, P. and Hotchkiss, J.H. (2002) Review of antimicrobial food packaging. *Innovative Food Science & Emerging Technologies* 3: 113–126.

Arvanitoyannis, I.S. and Bosnea, L. (2004) Migration of substances from food packaging materials to foods. *Critical Reviews in Food Science & Nutrition* 44: 63–76.

ASTM (2003) *Standard Specification for Compostable Plastics*, D6400-04. ASTM International, West Conshohocken, PA. DOI: 10.1520/D6400-04, www.astm.org.

Auras, R., Harte, B. and Selke, S. (2004) Effect of water on the oxygen barrier properties of polyethylene terephthalate and polylactide films. *Journal of Applied Polymer Science* 92: 1790–1803.

Azeredo, H.M.C., Mattoso, L.H.C., Wood, D., Williams, T.G., Avena-Bustillos, R.J. and McHugh, T.H. (2009) Nanocomposite edible films from mango puree reinforced with cellulose nanofibers. *Journal of Food Science* 74(5): N31–N35.

Baldwin, E.A. (1994) Edible coatings for fresh fruits and vegetables: past, present, and future. In: *Edible Coatings and Films to Improve Food Quality* (eds J.M. Krochta, E.A. Baldwin and M. NisperosCarriedo). Technomic, Lancaster, PA, p. 25.

Bieber, W.D., Fritag, W., Figge, K., VomBruck, C.G. and Rossi, L. (1984) Transfer of additives from plastic materials into foodstuffs and into food stimulants –compression. *Food Chemistry and Toxicology* 22: 737–742.

Blunden, S. and Wallace, T. (2003) Tin in canned food: a review and understanding of occurrence and effect. *Food Chemistry and Toxicology* 41: 1651–1662.

Brody, A.L., Bugusu, B., Han, J.H., Sand, C.K. and Mchugh, T.H. (2008) Innovative food packaging solutions. *Journal of Food Science* 73, R107–R116.

Bull, M.K., Steele, R.J., Kelly, M.S.A. and Olivier, B. (2010) Chapman Packaging under pressure: effects of high pressure, high temperature processing on the barrier properties of commonly available packaging materials. *Innovative Food Science and Emerging Technologies* 11: 133–137.

Calvert, P.D. and Billingham, N.C. (1979) Loss of additives from polymer: a theoretical model. *Journal of Applied Polymer Science* 24: 357–370.

Canadian Food Inspection Agency (2010). http://www.inspection.gc.ca/english/fssa/concen/tipcon/tamalte.shtml.

Caner, C. and Harte, B. (2005) Effect of high-pressure processing on the migration of antioxidant Irganox 1076 from polypropylene film into a food stimulant. *Journal of the Science of Food and Agriculture* 85: 39–46.

Caner, C., Hernandez, R.J. and Pascall, M.A. (2000) Effect of high pressure processing on selected high banier laminated films used for food packaging. *Packaging Technology and Science* 13: 183–195.

Caner, C., Hernandez, R.J., Pascall, M.A. and Reimer, J. (2003) The use of mechanical analyses, scanning electron microscopy and ultrasonic imaging to study the effects of high-pressure processing on multilayer films. *Journal of the Science of Food and Agriculture* 83: 1095–1103.

Caner, C., Hernandez, R.J. and Harte, B.R. (2004) High-pressure processing effects on the mechanical, barrier and mass transfer properties of food packaging flexible structures: a critical review. *Packaging Technology and Science* 17: 23–29.

Chen, H. (1995) Functional properties and applications of edible films made of milk proteins. *Journal of Dairy Science* 78: 2563–2583.

Chuaqui-Offermanns, N. (1989) Food packaging materials and radiation processing of food: a brief review. *International Journal of Radiation Applications Institute, Part C. Radiation Physics and Chemistry* 34: 1005–1007.

Comstock, K., Farrell, D., Godwin, C. and Xi, Y. (2004) From hydrocarbons to carbohydrates: food packaging of the future. http://depts.washington.edu/poeweb/gradprograms/envmgt/2004symposium/GreenpackagingReport. pdf (accessed March 28, 2006).

Cooksey, K. (2007) Interaction of food and packaging contents. In: *Intelligent and Active Packaging for Fruits and Vegetables* (ed. C. Wilson). CRC, Boca Raton, FL, pp. 201–237.

Cuq, B., Aymard, C., Cuq, J.L. and Guilbert, S. (1995) Edible packaging films based on fish myofibrillar proteins: formulation and functional properties. *Journal of Food Science* 60: 1369–1374.

Cutter, C.N. (2006) Opportunities for bio-based packaging technologies to improve the quality and safety of fresh and further processed muscle foods. *Meat Science* 74: 131–142.

Cyras, V.P., Manfredi, L.B., Ton-That, M.T. and Vazquez, A. (2008) Physical and mechanical properties of thermoplastic starch/montmorillonite nanocomposite films. *Carbohydrate Polymers* 73(1): 55–63.

Dallyn, H. and Shorten, D. (1998) Hygiene aspects of packaging in the food industry. *International Biodeterioration* 24: 392–398.

Dean, K. and Yu, L. (2005) *Biodegradable Polymers for Industrial Application*. CRC, Boca Raton, FL, pp. 289–309.

Dobiáš, J., Voldřich, M., Marek, M. and Chudáčková, K. (2004) Changes of properties of polymer packaging films during high pressure treatment. *Journal of Food Engineering* 61: 545–549.

Esse, R. (2002) *Flexible Packaging End-Use Market Analysis*. Flexible, Linthicum, MD.

European Union (1976) http://eur-lex.europa.eu/JOHtml.do?uri=OJ:L:1976:340:SOM:EN:HTML.

European Union (1989) http://www.food.gov.uk/multimedia/pdfs/fcmcons2004ecdoc.pdf.

European Union (2004) http://ec.europa.eu/food/food/chemicalsafety/foodcontact/framework_en.htm.

Fairclough, J.P.A. and Conti, M. (2009) Influence of ultra-high pressure sterilization on the structure of polymer films. *Packaging Technology Science* 22: 303–310.

Fellows, P. and Axtell, B. (2002) Packaging materials. In: *Appropriate Food Packaging: Materials and Methods for Small Businesses* (eds P. Fellows P and B. Axtell). ITDG, Rugby, UK, pp. 25–77.

Finkenzeller, K. (2003) *RFID Handbook: Fundamentals and Applications*, 2nd edn. John Wiley & Sons, Chichester, UK.

Fornes, T.D. and Paul, D.R. (2003) Modeling properties of nylon 6/clay nanocomposites using composite theories. *Polymer* 44: 4993–5013.

Fouad, M.M.K., Sayed, A.M. El and Maledy, A.N. (1999) Migration of DINP and DOP plasticisers from PVC sheets into food. *Environmental Management and Health* 10: 297–302.

Franssen, L.R. and Krochta, J.M. (2003) Edible coatings containing natural antimicrobials for processed foods. In: *Natural Antimicrobials for the Minimal Processing of Foods* (ed. S. Roller). CRC, Boca Raton, FL.

Galotto, M.J., Ulloa, P.A., Hernández, D., Fernández-Martín, F., Gavara, R. and Guarda, A. (2008) Mechanical and thermal behaviour of flexible food packaging polymeric films materials under high pressure/temperature treatments. *Packaging Technology and Science* 21: 297–308.

Galotto, M.J., Ulloa, P.A., Guarda, A,, Gavara, R. and Miltz, J. (2009) Effect of high-pressure food processing on the physical properties of synthetic and biopolymer films E304. *Journal of Food Science* 74: 304–311.

Ginzburg, V.V., Weinhold, J.D., Jog, P.K. and Srivastava, R. (2009) Thermodynamics of polymer–clay nanocomposites revisited: compressible self-consistent field theory modeling of melt-intercalated organoclays. *Macromolecules* 42: 9089–9095.

Gnanasekharan, V. and Floros, J.D. (1997) Migration and sorption phenomena in packaged foods. *Critical Reviews in Food Science and Nutrition* 37: 519–559.

Goetz, J. and Weisser, H. (2002) Permeation of aroma compounds through plastic films under high pressure: in-situ measuring method. *Innovative Food Science & Emerging Technologies* 3: 25–31.

Guilbert, S. (1986) Technology and application of edible protective film. In: *Food Packaging and Preservation* (ed. M. Matathoulthi). Elsevier Applied Science, New York, pp. 371–394.

Guo, M. and Zachmann, H.G. (1997) Structure and properties of naphthalene-containing polyesters. 2. Miscibility studies of poly(ethylene naphthalene-2,6-dicarboxylate) with poly(butylene terephthalate) by 13C CP/MAS NMR and DSC. *Macromolecules* 30: 2746–2750.

Haji-Saeida, M,, Sampaa, M.H.O. and Chmielewski, A.G. (2007) Radiation treatment for sterilization of packaging materials. *Radiation Physics and Chemistry* 76: 1535–1541.

Halek, G.W. and Levinson, J.J. (1988) Partitioning behavior and off-flavor thresholds in cookies from plastic packaging film printing ink compounds. *Journal of Food Science* 53: 1806–1822.

Han, J.H. (2000) Antimicrobial food packaging. *Food Technology* 54: 56–65.

Heckman, J.H. (2005) Food packaging regulation in the United States and the European Union. *Regulatory Toxicology and Pharmacology* 42: 96–122.

Huang, J.C., He, C.B., Liu, X.M., Xu, J.W., Tay, C.S.S. and Chow, S.Y. (2005) Organic–inorganic nanocomposites from cubic silsesquioxane epoxides: direct characterization of interphase, and thermomechanical properties. *Polymer* 46: 7018–27.

Hunt, R.G., Sellers, V.R., Frankalin, W.E., Nelson, J.M., Rathje, W.L., Hughes, W.W. and Wilson, D.C. (1990) *Estimates of the Volume of MSW and Selected Components in Trash Cans and Land Fills*. Report prepared by The Garbage Project and Franklins Assn. Ltd. for the Council for Solid Waste Solutions, Tucson, AZ.

IFT (1991) Effective management of food packaging: from production to disposal. In: IFT scientific status summary (by K.S. Marsh). *Food Technology* 45: 225–234.

Joshi, M., Butola, B.S., Simon, G. and Kukaleva, N. (2006) Rheological and viscoelastic behavior of HDPE/octamethyl–POSS nanocomposite. *Macromolecules* 39: 1839–1849.

Kerry, J.P., O'Grady, M.N. and Hogan, S.A. (2006) Past, current and potential utilization of active and intelligent packaging systems for meat and muscle-based products: a review. *Meat Science* 74: 113–130.

Kester, J.J. and Fennema, O.R. (1986) Edible films and coatings: a review. *Food Technology* 40: 47–59.

Kirwan, M.J. (2003) Paper and paperboard packaging. In: *Food Packaging Technology* (eds R. Coles, D. McDowell and M.J. Kirwan). Blackwell Publishing Ltd., Oxford, London, pp. 241–281.

Kirwan, M.J. and Strawbridge, J.W. (2003) Plastics in food packaging. In: *Food Packaging Technology* (eds R. Coles, D. McDowell and M.J. Kirwan). Blackwell Publishing Ltd., Oxford, London, pp. 174–240.

Kondo, K. (1990) Plastic containers. In: *Food Packaging* (ed. T. Kadoya). Academic Press, San Diego, CA, pp. 117–142.

Kübel, J., Ludwig, H., Marx, H. and Tauscher, B. (1996) Diffusion of aroma compounds into packaging films under high pressure. *Packaging Technology and Science* 9: 143–152.

Kubera H. (1994) Effect of ionizing radiation on packaging materials and packagings for irradiated food. *Roczniki Naukowe Akademii Rolniczej w Poznaniu*, Seria II.

Lamberti, M. and Escher, F. (2007) Aluminium foil as a food packaging material in comparison with other materials. *Food Reviews International* 23: 407–433.

Lau, O.-W. and Wong, S.K. (2000) Contamination in food from packaging material. *Journal of Chromatography A* 882: 225–270.

Le-Bail, A., Hamadami, N. and Bahuaud, S. (2006) Effect of high-pressure processing on the mechanical and barrier properties of selected packagings. *Packaging Technology and Science* 19: 237–243.

Lee, S., Yim, M.J., Master, R.N., Wong, C.P. and Baldwin, D.F. (2008) Void formation study of flip chip in package using no-flow underfill. *IEEE Transactions on Electronics Packaging Manufacturing* 31: 297–305.

Leerahawong, A., Arii, R., Tanaka, M. and Osako, K. (2011) Edible film from squid (*Todarodes pacificus*) mantle muscle. *Food Chemistry* 124: 177–182.

López-Rubio, A., Almenar, E., Hernandez-Munoz, P., Lagaron, J.M., Catala, R. and Gavara, R. (2004) Overview of active polymer-based packaging technologies for food applications. *Food Review International* 2: 357–387.

López-Rubio, A., Lagaron, J.M., Hernandez-Munoz, P., Almenar, E., Catala, R., Gavara, R. and Pascall, M.A. (2005) Effect of high-pressure treatments on the properties of EVOH-based food packaging materials. *Innovative Food Science & Emerging Technologies* 6: 51–58.

Lord, J.B. (2008) The food industry in the United States. In: *Developing New Food Products for a Changing Marketplace*, 2nd edn (eds A.L. Brody and J. Lord). CRC, Boca Raton, FL, pp. 1–23.

Mackay, M.E., Tuteja, A., Duxbury, P.M., Hawker, C.J., Van Horn, B., Guan, Z.B., Chen, G.H. and Krishnan, R.S. (2006) General strategies for nanoparticle dispersion. *Science* 311: 1740–1743.

Mahalik, N.P. and Nambiar, A.N. (2010) Trends in food packaging and manufacturing systems and technology. *Trends in Food Science & Technology* 21(3): 117–128.

Mark, A.M., Roth, W.B., Mehltretter, C.L. and Rist, C.E. (1966) Oxygen permeability of amylomaize starch films. *Food Technology* 20: 75–80.

Marsh, K.S. and Bugusu, B. (2007) Food packaging – roles, materials and environmental issues. *Journal of Food Science* 72: R39–R55.

Masuda, M., Saito, Y., Iwanami, T. and Hirari, Y. (1992) Effect of hydrostatic pressure on packaging materials for food. In: *High Pressure and Biotechnology* (eds C. Balny, R. Hayashi, K. Heremans and P. Masson). Colloque INSERM/John Libbey Eurotext, London, pp. 545–547.

Mauricio-Iglesias, M., Jansana, S., Peyron, S., Gontard, N. and Guillard, V. (2010) Effect of high-pressure/temperature (HP/T) treatments of in-package food on additive migration from conventional and bio-sourced materials. *Food Additives and Contaminants* 27: 118–127.

Morris, C., Brody, A.L. and Wicker, L. (2007) Non-thermal food processing/preservation technologies: a review with packaging implications. *Packaging Technology and Science* 20: 275–286.

Neijssen, P. (1994) Gamma sterilization of packaging. *Scandinavian Dairy Information* 8: 40–42.

Ochiai, S. and Nakagawa, Y. (1992) Packaging for high pressure food processing. *Conferences – National Institute of the Health and Medical Research Symposia and Seminars* 224: 515–521.

Page, B., Edwards, M. and May, N. (2003) Metal cans. In: *Food Packaging Technology* (eds R. Coles, D. McDowell and M.J. Kirwan). Blackwell Publishing Ltd., Oxford, London, pp. 121–151.

Pagella, C., Spigno, G. and De Faveri, D.M. (2002) Characterization of starch based edible coatings. *Transactions of the IChemE, Part C* 80: 193–197.

Paine, F.A. (1981) *Fundamentals of Packaging,* 1st revised edn. Brookside, Leicester, UK.

Park, H.J. (1999) Development of advanced edible coatings for fruits. *Trends in Food Science and Technology* 10: 254–260.

Park, H.J., Weller, C.L., Vergano, P.J. and Testin, R.F. (1993) Permeability and mechanical properties of cellulose based edibles films. *Journal of Food Science* 58: 1361–1364.

Petersen, K., Nielsen, P.V., Bertelsen, G., Lawther, M., Olsen, M.B. and Nilssonk, N.H. (1999) Potential of biobased materials for food packaging. *Trends in Food Science and Technology* 10: 52–68.

Plackett, D.V., Holm V.K., Johansen, P., Ndoni, S., Nielsen, P.V., Sipilainen-Malm, T., Södergård, A. and Verstichel, S. (2006) Characterization of L-polylactide and L-polylactide–polycaprolactone co-polymer films for use in cheese-packaging applications. *Packaging Technology and Science* 19: 1–24.

RFID Journal, Inc. (2005) What is RFID? *RFID Journal* [Internet magazine]. Available from: http://www.rfidjournal.com/article/articleview/1339/1/129/. Accessed March 13, 2008.

Robertson, G.L. (1993) *Food Packaging: Principles and Practice.* Marcel Dekker, New York.

Schauwecker, A., Balasubramaniam, V.M., Sadler, G., Pascall M.A. and Adhikari, C. (2002) Influence of high-pressure processing on selected polymeric materials and on the migration of a pressure-transmitting fluid. *Packaging Technology and Science,* 15: 255–262.

Sinclair, R.G. (1996) The case for polylactic acid as a commodity packaging plastic. *Journal of Macromolecular Science Part A: Pure and Applied Chemistry* 33: 585–597.

Sinha, R.S., Maiti, P., Okamoto, M., Yamada, K. and Ueda, K. (2002) New polylactide/layered silicate nanocomposites. 1. Preparation, characterization and properties. *Macromolecules* 35: 3104–3110.

Siracusa, V., Rocculi, P., Romani S. and Rosa, M.D. (2008) Biodegradable polymers for food packaging: a review. *Trends in Food Science and Technology* 19: 634–643.

Soroka, W. (1999) Paper and paperboard. In: *Fundamentals of Packaging Technology*, 2nd edn (eds A. Emblem and H. Emblem). Institute of Packaging Professionals, Herndon, VA, pp. 95–112.

Sothornvit, R. and Krochta, J.M. (2005) Plasticizers en edible films and coatings. In: *Innovation in Food Packaging* (ed. J.H. Han). Elsevier, New York, pp. 40–433.

Suppakul, P., Miltz, J., Sonneveld, K. and Bigger, S.W. (2003) Active packaging technologies with an emphasis on antimicrobial packaging and its applications. *Journal of Food Science* 68: 408–420.

Urayama, H., Moon, S.I. and Kimura, Y. (2003) *Macromolecules Materials Engineering* 288: 137–143.

Utz, H. (1995) *Barriereeigenschaften aluminiumbedampfter Kunststoffolien.* Dissertation, Technischen Universität München., Munich.

Van Willige, R.W.G., Linssen, J.P.H., Meinders, M.B.J., Van Der Steger, H.J. and Voragen, A.G.J. (2002) Influence of flavor absorption on oxygen permeation through LDPE, PP, PC and PET plastics food packaging. *Food Additives and Contaminants* 19: 303–313.

Weber, C.J., Haugaard, V., Festersen, R. and Bertelsen, G. (2002) Production and applications of biobased packaging materials for the food industry. *Food Additives and Contaminants* 19: 172–177.

Yam, K.L., Takhistov, P.T. and Miltz, J. (2005) Intelligent packaging: concepts and applications. *Journal of Food Science* 70: R1–R10.

Yoon, S. and Deng, Y. (2006) Clay–starch composites and their application in papermaking. *Journal of Applied Polymer Science* 100: 1032–1038.

Yoxall, A., Janson, R., Bradbury, S.R., Langley, J., Wearn, J. and Hayes, S. (2006) Openability: producing design limits for consumer packaging. *Packaging Technology and Science* 19(4): 219–225.

Yu, J., Cui, G., Wei, M. and Huang, J. (2007) Facile exfoliation of rectorite nanoplatelets in soy protein matrix and reinforced bionanocomposites thereof. *Journal of Applied Polymer Science* 104(5): 3367–3377.

Zeng, Q.H., Yu, A.B., Max Lu, G.Q. and Paul, D.R. (2005) Clay-based polymer nanocomposites: research and commercial development. *Journal of Nanoscience and Nanotechnology* 5: 1574–1592.

Zhou, J.J., Wang, S.Y. and Gunasekaran, S. (2009) Preparation and characterization of whey protein film incorporated with TiO_2 nanoparticles. *Journal of Food Science* 74: N50–N56.

42

Mass Transport Phenomena in Food Packaging Design

Marcella Mastromatteo, Amalia Conte and Matteo Alessandro Del Nobile

Introduction

The Purpose of Packaging

Food is packaged to preserve its quality and freshness, to add appeal to consumers and to facilitate storage and distribution (as stated by the Codex Alimentarius Commission in 1985). To accomplish this goal, the package must serve the important functions of containing and protecting the food, providing convenience, and conveying product information. The package protects the food against physical, chemical and biological damage. It also acts as a physical barrier to oxygen, moisture, volatile chemical compounds and microorganisms that are detrimental to the food (Brown, 1992). The package has to be considered as an integral part of the preservation system because it provides a barrier between the food and the external environment. It is usually a composite item meeting several different needs (da Cruz *et al.*, 2007). What we call the preservation role is a fundamental requirement of food packaging, since it is directly related to the safety of the consumer.

Besides the preservation function, packaging has several other important roles to play in delivering safe, wholesome and attractive foods to the market and in doing this economically and with minimal environmental impact. The package affects, to a great extent, the purchase decisions of the consumer (Grorski-Berry, 1999). In fact,

Handbook of Food Process Design, First Edition. Edited by Jasim Ahmed and Mohammad Shafiur Rahman.

if the product draws the attention of the consumer, the product will be purchased. If the product is good and lives up to expectations, the consumer will be satisfied and will buy it again. The packaging not only has to look good, it must also appeal to at least one sense by smelling good, feeling exceptional or giving off an irresistible sound when it is opened. The message conceived by the producer of the packaging is not received in the same way by every consumer, but it is revised by each in light of their personal needs and specific sensibilities. Apart from obvious reasons such as necessity or utility, the consumer desires and buys an object for the image that it provides or for the identity that it is perceived to have in society. The packaging also has a meaning of a promotional kind, which is perceived both in the place where the product is sold and in the place where it is used, and it is able to pass on information and feelings about brand and corporate image using, essentially, well-structured iconic and linguistic codes (Brown, 1992). The information on labels, including indications of preservation methods and expiry date, should be always considered sufficiently clear and complete by the consumer.

Extra and enhanced functions have been sought by the food packaging sector to meet consumer demands: packaging that improves the service provided by a product, packaging that communicates and simplifies the new uses of a product, and packaging that simplifies an offer of complex solutions (Lee *et al.*, 2008). Efforts to improve the performance of packaging solutions and to control fresh foods can be directed towards many areas. Active packaging is the most relevant innovative idea that has been applied to increase consumer satisfaction (Appendini and Hotchkiss, 2002; Lopez-Rubio *et al.*, 2004). Active packaging has been defined as a system in which the product, the package and the environment interact in a positive way to extend the shelf-life of the product or to control some characteristics that cannot be obtained otherwise (Cutter, 2002). This technology could be described as food packaging with an enhanced protection function. In recent years, an additional step forward has been taken by developing active packaging with antimicrobial properties that contain natural instead of synthetic additives (Suppakul *et al.*, 2003).

Low-environmental-impact polymers represent another interesting category of new food packaging materials, since they are in accordance with some current themes, such as the prevention of environmental pollution, and also convenience and food quality (Petersen *et al.*, 1999; Siracusa *et al.*, 2008). To develop environmentally sustainable solutions, innovations in the characteristics of polymeric materials are needed (Karbowiak *et al.*, 2009). In the near future, the aim is to develop eco-friendly packaging solutions. The solution will lie in the implementation of preventive measures and, above all, in the adoption of policies shared by all actors in the production chain.

Another interesting function that is extremely appreciated by the consumer is intelligent packaging. This represents a technology that uses the communication function of the package to facilitate decision making, in order to achieve the benefits of enhanced food quality and safety (Yam *et al.*, 2005). Researchers have achieved good results for the final setup of these devices, but these devices have not yet come into widespread use in fresh products at the level of primary packaging. Conducting such research frequently requires a systems approach involving interactions between researchers in food packaging, food engineering, nanotechnology and other disciplines. Research opportunities also exist in integrating data carriers and package indicators into the same device. Combining intelligent packaging and active packaging offers many intriguing possibilities, thus allowing the development of more sophisticated packaging systems (Kerry *et al.*, 2006).

Correlation between Mass Transport Properties and Shelf-Life

The shelf-life of a product is defined as the period of time during which the quality of the packaged food remains acceptable. This period may range from a few days to more than one year. Thus, the packaging must have properties that ensure that the shelf-life is not compromised (Kilkast and Subramaniam, 2000; da Cruz *et al.*, 2007). The properties of interest vary between applications: what is sufficient for a short-shelf-life product may be completely unsatisfactory for a food with a longer shelf-life (Paine and Paine, 1983). There are numerous variables that can play a significant role in establishing package performance, such as the initial food quality, the processing operations, the size, shape and appearance of the package, the distribution method, and the method of disposal of the package. Generally speaking, the properties that determine the adequacy of packages to meet performance requirements can be grouped into the following categories: mechanical, thermal, optical and mass transport properties (Brown, 1992). In spite of this, the extent to which packaging fulfills its preservation role is greatly dependent on the ability of the material to provide a barrier to the environmental factors that cause spoilage. The factors with which we are concerned are water vapor, gases and light. When considering the barrier properties of the common packaging materials to low-molecular-weight compounds, we can divide them into total barriers (glass, tin-plated and other coated steel, and aluminum above 17 μm in thickness) and partial barriers (aluminum foil below 17 μm, and paper-based and plastic materials).

Mass transport phenomena are of great importance in food packaging using plastics, since polymeric matrices are permeable to moisture, oxygen, carbon dioxide, nitrogen and other low-molecular-weight compounds. Glass and metal packaging materials are not permeable to low-molecular-weight compounds, whereas paper-based materials are too permeable. Hence, these last types of materials do not provide an opportunity for the designer to optimize the barrier properties for various applications. Polymers can provide a wide range (covering three or four orders of magnitude) of permeabilities for different applications, thus justifying studies aimed at ensuring adequate barrier protection.

For foods that are sensitive to oxygen or moisture, barrier protection against low-molecular-weight compounds is the major function of the package in terms of providing adequate protection. For example, moisture can move from the external environment through the package into the headspace, thus increasing the relative humidity and causing a moisture-sensitive food to have less crispness and consequently a shorter shelf-life (Sanches Silva *et al.*, 2004). Moisture loss can also be critical: water vapor escaping from the package can provoke undesirable textural changes in food (Roca *et al.*, 2006). As regards the function as an oxygen barrier, whenever a product contains fat we can assume that an oxygen barrier is required. For most foods, packaging technologists are concerned with keeping oxygen out of the pack (Del Nobile *et al.*, 2003c). However, there are some exceptions, such as fresh meat, that need a certain amount of oxygen to maintain freshness (Quintavalla and Vicini, 2002). There are also occasions when the transport of gases is desirable, as happens with fresh-cut produce, where an exchange of gas through the package is necessary to accommodate respiration and transpiration; and to maintain an optimum gas composition in the headspace (Del Nobile *et al.*, 2006; Rico *et al.*, 2007; Conte *et al.*, 2009a).

On the basis of the considerations above, we can sum up by saying that the mass barrier properties of packaging can play a key role in determining product shelf-life. In several cases, in fact, shelf-life calculations can be based solely on the barrier properties of the packaging material. Conversely, the desired shelf-life can dictate the barrier specification of the polymeric matrix of the packaging.

Design of Packaging Systems Intended for Food Applications

As described above, in situations where food deterioration is driven by either gas or moisture permeation from the ambient environment, an accurate choice of the mass transport properties of the packaging may bring about an increase in the product shelf-life. Each category of food has its specific quality attributes, storage conditions and expected shelf-life, and specific packaging tools are applied to it. For a specific shelf-life, defined on the basis of marketing requirements, consumer expectations and distribution organization, the required characteristics of the package can be found from the interactions between the food, the packaging and the environment. Packaging design is clearly a fundamental part of any newly launched product. Considering the importance of packaging in determining product shelf-life, the correct approach is one in which the development of the product and of its packaging system are considered at the same level of importance (Cleland, 1996). An optimization process by means of experimental evaluation (the trial and error approach) would take a long time, and it is usually very expensive. It would be necessary to invest long, laborious effort in the experimental determination of the acceptability of a given plastic for containing a specific food. Moreover, long-shelf-life studies do not fit with the speed requirements of the food industry; therefore, many efforts have been made to develop shelf-life simulation tools. Even though package design by the trial and error method is possible, the most suitable approach is to deploy design-engineering methods based on quantitative models that relate the relevant factors influencing quality loss and gas transfer systems. Packaging design has gained great importance over the last few decades because of the numerous options available for packaging materials that offer alternative ways to reduce cost and time (Rodriguez-Aguilera and Oliveira 2009).

The idea behind designing barriers to protect food is to select the packaging in terms of materials and dimensions to provide adequate protection to the packaged food. To accomplish this goal, the following steps must be carried out. First of all, for a given food product, the main phenomena that are detrimental to quality, as well as their kinetics, have to be assessed. Many foods are sensitive to more than one spoilage factor. For example, whereas for raw and cooked meats taint and physical damage are the main deterioration indices, oxidation and moisture changes are specific problems for breakfast cereals. Once the main factor responsible for product unacceptability has been identified, it is of prime importance to create conditions in the package that will slow down the relevant detrimental process. Mathematical models can be developed to describe the chosen spoilage phenomenon and predict product shelf-life under real working conditions. From all of the assembled data, the equations defining the decay of quality during storage of the packaged food can be solved. Once potentially suitable packaging solutions have been determined, it is essential to carry out only limited storage trials to validate the mathematical approach. This involves shelf-life tests carried out using a pilot plant to pack and store the food.

Packaging design today involves strategies and solutions for products and services. Planning in a strategic way means understanding in what way innovation can be part of a company's targets.

Barrier Properties: Steady State

Basic Theoretical Principles, Problems and Limitations of Using the Permeability Coefficient

In food science applications, a knowledge of the permeability coefficients of packaging materials is of great importance, since they are used for predicting the shelf-life of packaged foods. Permeation is defined as the transfer of molecules through a barrier formed by a film, without consideration of what mechanism is involved. There are usually two possible models of mass transfer: a capillary flow type of model, and an activated-diffusion type. Capillary flow is the dominant mechanism in flexible films that either consist of highly porous media or have defects such as cracks, folds and micropores (Heiss, 1980; Chao and Rizvi, 1988). Of much more importance to the field of food packaging is the activated-diffusion type, which describes the transport of molecules through homogeneous bulk polymeric materials, which may be either amorphous or semicrystalline (Rogers *et al.*, 1972; Crank, 1975). For the activated-diffusion type of mass transfer, the permeation process is described by means of the following equation:

$$J = P \cdot \frac{(p_1 - p_2)}{\ell} \tag{42.1}$$

where J is the steady-state mass flux, P is the permeability coefficient, p_1 and p_2 are the partial pressures of the permeant on the upstream and downstream sides of the film, and ℓ is the film thickness. In fact, the permeability coefficient is the proportionality constant between the partial-pressure gradient of the permeant through the film and the mass flux in the steady state. Alternatively, it can be taken to mean the proportionality constant between a cause (i.e., the partial-pressure gradient of the permeant through the film) and an effect (the mass flux of the permeant). Permeation through bulk polymers is generally described as a three-step process during which molecules move along a partial-pressure gradient. In the first step, the permeant is dissolved into the amorphous phase of the film on the upstream side. In the second step, once inside the film, each molecule moves randomly through the polymeric matrix. In the third step, the molecules desorb from the film on the downstream side. Absorption and desorption depend on the solubility of the penetrant in the polymer. On the other hand, the mobility of the penetrant in the polymeric matrix is related to its diffusivity (Rogers *et al.*, 1972; Crank, 1975; Comyn, 1985; DeLassus and Hilker, 1987). Therefore, the barrier properties of the film, which can be expressed and described by the permeability coefficient, are related to the diffusivity and solubility of the film–penetrant system. As reported by Miller and Krotcha (1997), the diffusion coefficient describes the movement of permeant molecules through the polymer, and thus represents a kinetic property of the polymer–permeant system. On the other hand, the solubility coefficient describes the dissolution of the permeant in the polymer, and thus represents a thermodynamic property of the polymer–permeant system; the permeability coefficient, being a function of both the solubility of the absorbed low-molecular-weight compound and its diffusion rate through the film, incorporates both kinetic and thermodynamic properties.

The solubility coefficient is defined as the ratio between the activity of the permeant and its concentration at equilibrium inside the polymer:

$$S = \left(\frac{C}{a}\right)_{eq} = \left(\frac{C}{a_{ext}}\right)_{eq} \cong \left(\frac{C}{C_{ext}}\right)_{eq} \tag{42.2}$$

where C and C_{ext} are the permeant concentrations in the membrane and in the external phase, respectively, and a and a_{ext} are the permeant activities in the membrane and in the external phase, respectively. In the case of a gaseous permeant, the above equation can be written as follows:

$$S = \left(\frac{C}{p}\right)_{eq} \tag{42.3}$$

where p is the permeant partial pressure in the external phase. Generally, the amount of permeant absorbed by a polymer at equilibrium and the shape of the sorption isotherm depend on the magnitude and type of the intermolecular forces between the polymer and the permeant. The sorption isotherm describes the relationship between the content of a low-molecular-weight compound sorbed in a material and its activity at equilibrium. In an ideal system, C is proportional to a, which means a linear sorption isotherm (Henry's law). This is the behavior exhibited by a polyamide membrane for the gases Ar and N_2 (Motamedian *et al.*, 1998). In the case of nonideal polymer–permeant systems, such as cellulose acetate and H_2 gas (Motamedian *et al.*, 1998), the sorption isotherm is nonlinear.

The solubilization process of low-molecular-weight compounds in semicrystalline polymers is strictly related to that in the corresponding totally amorphous polymer (Michaels and Parker, Jr., 1959; Michaels and Bixler, 1961). It is generally assumed that the crystal phase acts as an impervious region to low-molecular-weight compounds. Under this hypothesis, the solubility coefficient of a semicrystalline polymer can be written as a function of the solubility coefficient as in the following equation:

$$S = \beta \cdot S^{*} \tag{42.4}$$

where S is the solubility of the semicrystalline polymer, β is the amorphous volume fraction of the polymer and S^* is the solubility coefficient of the totally amorphous polymer. The diffusion of low-molecular-weight compounds in polymers is generally governed by two simultaneously occurring phenomena: a substantially stochastic phenomenon (related to Brownian motion), where the penetrant flow is driven exclusively by a concentration gradient and, only in the case of permeant-induced swelling of the polymeric matrix, a relaxation phenomenon that leads the polymer to swell to equilibrium (Del Nobile *et al.*, 1994). When the mass transfer takes place in a substantially unperturbed polymer matrix, as in the case of gas diffusion in rubbery polymers (or whenever the solvent-induced polymer swelling is negligible), the diffusion process is substantially controlled by a stochastic phenomenon. The diffusion related to Brownian motion is generally described by means of Fick's first law. In the specific case of diffusion through a plane sheet, the Fick model reduces to the following equation:

$$J = -D_{\mathrm{F}} \cdot \frac{\partial C}{\partial x} \tag{42.5}$$

where D_{F} is the diffusion coefficient and x is the spatial coordinate.

It is possible to distinguish between two different behaviors, ideal and nonideal Fickian diffusion. In the case of ideal behavior, the diffusion coefficient D_{F} is constant, whereas in the case of nonideal behavior, the diffusion coefficient is a function of the local permeant concentration ($D_{\mathrm{F}} = f(C)$). If one integrates both sides of Equation 42.5 over the film thickness, one obtains

$$\int_0^{\ell} J = -\int_0^{\ell} D_{\mathrm{F}} \frac{\partial C}{\partial x} dx = -\int_{C_1}^{C_2} D_{\mathrm{F}} dC$$

and also

$$\int_{0}^{\ell} J dx = \int_{C_2}^{C_1} D_F dC \tag{42.6}$$

where C_1 and C_2 are the permeant concentrations on the upstream and downstream sides of the film.

In the steady state, Equation 42.6 can be rewritten as follows:

$$J \cdot \ell = \int_{C_2}^{C_1} D_F dC \tag{42.7}$$

By rearranging the steady-state gas flow equation (Equation 42.1), one obtains

$$J \cdot \ell = P \cdot (p_1 - p_2) \tag{42.8}$$

Substituting Equation 42.7 into Equation 42.8, the following equation is obtained:

$$P \cdot (p_1 - p_2) = \int_{C_2}^{C_1} D_F dC$$

and also

$$P = \frac{1}{(p_1 - p_2)} \int_{C_2}^{C_1} D_F dC \tag{42.9}$$

From Equation 42.9, it emerges that the permeability coefficient can be a function of the boundary conditions (the permeant partial pressures on the upstream and downstream sides of the film). Generally, the barrier properties of polymeric films are determined simply by evaluating the permeability coefficient of a given polymeric diffusing system for a single set of boundary conditions (Del Nobile *et al.*, 1996a,b). This approach can be used successfully whenever the permeability coefficient does not depend on the boundary conditions. In fact, in these cases (ideal cases), the permeation coefficient is constant because the permeating molecules do not change the free volume of the polymeric matrix much (Myers *et al.*, 1961). In the ideal case, D_F does not depend on the local permeant concentration, and Equation 42.9 becomes

$$P = \frac{D_F}{p_1 - p_2} \cdot (C_1 - C_2)$$

and, also, S does not depend on the permeant partial pressure:

$$P = \frac{D_F}{p_1 - p_2} \cdot S(p_1 - p_2)$$

From the above equation, the following equation is obtained:

$$P = D_F \cdot S \tag{42.10}$$

Therefore, in the particular case where both the diffusion coefficient and the solubility coefficient are constant, P is simply the product of D_F and S. When the permeability coefficient depends on the partial pressures of the diffusing molecules on the upstream and downstream sides of the film, the permeability coefficient determined for a single set of boundary conditions is useless for predicting the barrier properties of packaging under real working conditions. According to Equation 42.9, in these cases the relationship between the barrier properties of the film and the boundary conditions can be addressed by determining $C = f_1(p)$ and $D_F = f_2(C)$. In general, the sorption isotherm of the permeant is used to evaluate the dependence of the solubility coefficient on the partial pressure of the permeant, whereas the relationship between the diffusion coefficient and the local permeant concentration can be determined by evaluation of the sorption kinetics of the permeant. These techniques to assess the dependence of the permeability coefficient on the boundary conditions will be discussed in more detail later on. It is worth noting that water is the main low-molecular-weight compound that plasticizes the polymeric matrix, and it could make the barrier properties of a package depend on the relative humidity on both side of the film. For this reason, when we discuss the issue of determining the dependence of the permeability coefficient on the boundary conditions, we shall focus on this molecule. In fact, water molecules, acting as a plasticizer, increase the free volume of the polymeric matrix and consequently its macromolecular mobility. Therefore, the solubility coefficient depends on the permeant activity, whereas the diffusivity coefficient depends on the local permeant concentration. As a result, the permeability coefficient depends on the upstream and downstream partial pressures of water vapor (Hernandez 1994). This topic will be treated in detail in the "Case Studies" section below. In particular, the work of Buonocore *et al.* (2005) will be presented and discussed. These authors performed water permeation tests on water-sensitive nylon films, and the experimental data obtained were used to predict the water permeability coefficient by means of an appropriate model. The results of model validation will be also reported.

The permeability coefficient also depends on the temperature according to the Arrhenius equation (Williams *et al.*, 1955):

$$P = P_0 \exp\left(\frac{-E_0}{RT}\right) \tag{42.11}$$

where P_0 is the Arrhenius constant, E_0 is the activation energy for permeation, R is the universal gas constant and T is the absolute temperature.

Permeability Coefficients of Multilayer Packaging Systems

The conventional approach to producing high-performance films for foodstuff packaging is to combine films made of different materials into multilayer structures to achieve the required properties. The strategy of combining different polymers with various gas or water barrier properties is a useful tool for reducing the sensitivity of a polymeric system to low-molecular-weight compounds and to improve the performance of a packaging material. Multilayer structures are also used to optimize the properties of packages from other points of view. This technique permits one to reach a good compromise between other fundamental characteristics of a packaging material such as mechanical properties, weldability, printability and cost.

On the basis of the above, it is clear that the evaluation of the permeability coefficient of every multilayer film would require numerous measurements, owing to the enormous number of film structures that could be made by combining the commercially available films. Therefore, it is necessary to find a relationship that relates the permeability of each individual layer and its thickness to that of the multilayer film. If a three-layer structure is considered, the mass flux through the multilayer film is

$$J = P_{TOT} \cdot \frac{p_1 - p_3}{\ell_{TOT}} \tag{42.12}$$

where p_1 and p_3 are the permeant partial pressures on the upstream and downstream sides of the film, respectively, ℓ_{TOT} is the thickness of the multilayer film, and P_{TOT} is the permeability coefficient of the multilayer structure. As an example, Figure 42.1 shows a schematic illustration of a three-layer structure. The mass fluxes through each film are:

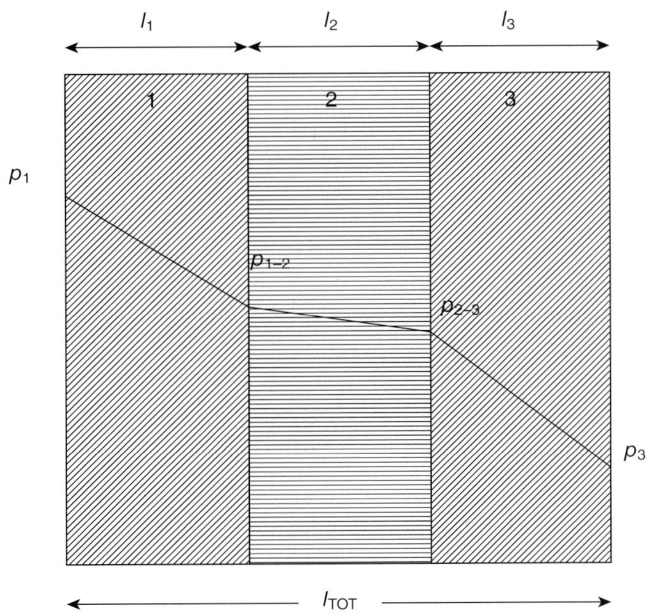

Figure 42.1 Representation of a three-layer structure.

$$J_1 = P_1 \cdot \frac{p_1 - p_{1\text{-}2}}{\ell_1} = P_1 \cdot \frac{\Delta p_1}{\ell_1} \tag{42.13}$$

$$J_2 = P_2 \cdot \frac{p_{1\text{-}2} - p_{2\text{-}3}}{\ell_2} = P_2 \cdot \frac{\Delta p_2}{\ell_2} \tag{42.14}$$

$$J_3 = P_3 \cdot \frac{p_{2\text{-}3} - p_3}{\ell_3} = P_3 \cdot \frac{\Delta p_3}{\ell_3} \tag{42.15}$$

where $p_{1\text{-}2}$ and $p_{2\text{-}3}$ are the permeant partial pressures in equilibrium with the permeant concentrations at the interfaces between layers 1 and 2 and between layers 2 and 3, respectively. Rearranging Equations 42.13 and 42.15, one obtains

$$\Delta p_1 = \frac{J_1 \cdot \ell_1}{P_1} \tag{42.16}$$

$$\Delta p_2 = \frac{J_2 \cdot \ell_2}{P_2} \tag{42.17}$$

$$\Delta p_3 = \frac{J_3 \cdot \ell_3}{P_3} \tag{42.18}$$

By adding Equation 42.16 to Equation 42.18 and keeping in mind that, according to the steady-state hypothesis, the mass flux through all of the films must be the same (i.e., $J_1 = J_2 = J_3 = J$), and also recalling that, by definition, $\Delta p_{\text{TOT}} = \Delta p_1 + \Delta p_2 + \Delta p_3$, the following equation is obtained:

$$\Delta p_{\text{TOT}} = J \cdot \left[\frac{\ell_1}{P_1} + \frac{\ell_2}{P_2} + \frac{\ell_3}{P_3} \right] \tag{42.19}$$

Rearranging Equation 42.19, one obtains

$$J = \frac{1}{\left[\frac{\ell_1}{P_1} + \frac{\ell_2}{P_2} \cdot \frac{\ell_3}{P_3} \right]} \cdot \Delta p_{\text{TOT}} \tag{42.20}$$

Comparing Equations 42.20 and 42.12, one obtains

$$P_{\text{TOT}} = \frac{1}{\left(\frac{\ell_1}{\ell_{\text{TOT}}} \cdot \frac{1}{P_1} + \frac{\ell_2}{\ell_{\text{TOT}}} \cdot \frac{1}{P_2} + \frac{\ell_3}{\ell_{\text{TOT}}} \cdot \frac{1}{P_3} \right)} \tag{42.21}$$

The above equation can be written in a generalized way as follows:

$$P_{\text{TOT}} = \sum_{i=1}^{i=n} \frac{1}{\dfrac{\ell_i}{\ell_{\text{TOT}}} \cdot \dfrac{1}{P^i}} \tag{42.22}$$

where ℓ_i is the thickness of the *i*th layer in the multilayer structure, and P^i is the permeability coefficient of the *i*th layer.

Macroperforations and Perforation-Mediated Packaging Systems

Products such as fruit and vegetables with a high respiratory activity require the use of packaging with a high permeability to avoid extremely low levels of O_2 and/or high levels of CO_2 in the package headspace arising in the steady-state condition from the balance between respiration and permeation. The permeability of commercially available plastic films is generally far too low for the optimal requirements of many commodities. For this reason, perforations or channels in the package are used to widen the range of mass transport properties of packages made from commercially available films. The number of perforations, cross-sectional area of the perforations and length of the diffusion channel affect the barrier properties of the package and thus can be used as control variables to obtain the desired mass transport properties. Another type of structure is called the perforation-mediated packaging system, which is a system where tubes, which may be packed with an inert filling, are inserted in an otherwise airtight package (Fonseca *et al.*, 2000).

According to what was said earlier, in the case of fruit and vegetables two main processes contribute to the changes in the amount of a low-molecular-weight compound, q, inside a package containing fresh produce: physiological activity (respiration, resulting in oxygen consumption and carbon dioxide production, and transpiration, resulting in emission of water vapor), with a total rate f, and permeation of gases through the packaging film, with a rate F:

$$\frac{dq}{dt} = f + F \tag{42.23}$$

The amount of a low-molecular-weight compound is usually represented by the product of its partial concentration (denoted by C_{O_2} for oxygen and C_W for water vapor) and the free volume within the package, V_{st}:

$$q = V_{st} \cdot C_{O_2} \tag{42.24}$$

The total amount of low-molecular-weight compound that permeates through a macroscopically homogeneous film is given by the mass flux multiplied by the film area $A: F = J \cdot A$. If the film is perforated, the perforations represent an alternative route for gas transport that is in parallel to the barrier formed by the plastic material. In this case, the total flow is

$$F = J \cdot A + J_h \cdot A_h \tag{42.25}$$

where A_h is the total area of the holes and J_h is the flux of gas per unit area of the holes. A similar equation can be written for the water vapor flow F_W. The steady-state diffusion flux of a gas through a nonperforated film obeys Equation 42.1 (Cussler, 1984). This can be rewritten as follows:

$$J = P \cdot \frac{(C_A - C_{in})}{\ell} \tag{42.26}$$

where C_A and C_{in} are the concentrations of the gas in the ambient atmosphere and inside the package, respectively. Macroscopic perforations in polymeric films have diameters of the order of 10^{-4} m or greater, whereas the mean free path of gas molecules at atmospheric pressure is much less, being about 1 to 2×10^{-7} m (Kesting and Fritzsche, 1993). Therefore, the transport through a perforation may be treated as macroscopic diffusion in a cylindrical pathway filled with air. The diffusive flux in this case obeys Fick's law:

$$J_h = -D_F^{air} \frac{(C_{in} - C_A)}{L_h} \tag{42.27}$$

with the diffusion coefficient of the gas in air, D_F^{air}, replacing the permeability coefficient, P, of Equation 42.26. The diffusive path length is denoted here by L_h. Its evaluation is a complicated diffusion problem. If the distance between the perforations is much greater than their radius, estimation of L_h may be done approximately by using the model employed by Meidner and Mansfield (1968) and Nobel (1974) for computation of stomatal resistance. According to this approximation, L_h equals the length of the cylindrical pore (L), increased by the radius of the hole (z_h):

$$L_h = L + z_h \tag{42.28}$$

Therefore, Equation 42.28 has been used in subsequent computations. In particular, taking Equations 42.25–42.28 into account, the total flux through a plastic film of area A with N perforations of radius z_h (total area $A_h = \pi z_h^2 N$) may be represented as

$$F = (C_A - C_{in}) \cdot \left[\frac{A \cdot P}{\ell} + \frac{A_h \cdot D_F}{(L + z_h)} \right] \tag{42.29}$$

The water vapor flux through the perforated film is represented in a similar way, with $\theta \cdot H$ replacing the concentration:

$$F_W = \theta \cdot (H_A - H) \cdot \left[\frac{A \cdot P_W}{\ell} + \frac{A_h \cdot D_W}{(L + z_h)} \right] \tag{42.30}$$

where H_A is the relative humidity in the ambient atmosphere, H is the relative humidity in the packaging, D_W is the diffusion coefficient of water vapor in air and θ is the water vapor concentration at the saturation vapor pressure. Fishman *et al.* (1996) validated the above mathematical model by comparison with experiments on modified-atmosphere packages containing fresh fruit, as will be discussed later on.

Case Studies

Evaluation of the Water Permeability Coefficient as a Function of Boundary Conditions

Buonocore *et al.* (2005) studied the water transport properties of a commercially available moderately hydrophilic nylon film. As was said above, a more complex analysis is necessary to determine the relationship between the permeability coefficient and the water vapor partial pressures on the upstream and downstream sides of the film. In this case Equation 42.9 can be rewritten as follows:

$$P(a_W^1, a_W^2) = \frac{\int_{C_W^2(a_W^2)}^{C_W^1(a_W^1)} D_F(C_W) \cdot dC_W}{p_O \cdot (a_W^1 - a_W^2)} \tag{42.31}$$

As reported by Del Nobile *et al.* (1997a), a simple equation to relate D_F to the local penetrant concentration can be obtained by rearranging the relationship proposed by Fujita (1961) to describe the dependence of the thermodynamic diffusion coefficient on the local penetrant volume fraction:

$$D_F(C_W) = B_1 \cdot \exp\left(\frac{-1}{B_2 + B_3 \cdot C_W} \right) \tag{42.32}$$

where the B_i 's are constants to be determined by fitting the model to experimental data.

By substituting the expression proposed for $D_F(C_W)$ into Equation 42.31, one obtains

$$P(a_W^1, a_W^2) = \frac{\int_{C_W^2(a_W^2)}^{C_W^1(a_W^1)} B_1 \cdot \exp\left(\frac{-1}{B_2 + B_3 \cdot C_W} \right) \cdot dC_W}{p_O \cdot (a_W^1 - a_W^2)} \tag{42.33}$$

According to Equation 42.33, the relationship between the water absorbed at equilibrium into the polymer and the water activity, as well as the dependence of the diffusion coefficient on the local water concentration (i.e., the constants B_i) needs to be determined to address the dependence of the permeability coefficient on the water activity on the upstream and downstream sides of the film.

Concerning the dependence of the solubility coefficient of water on the water activity, an equation proposed by Flory (1953) to describe the process of mixing of a linear polymer with a low-molecular-weight compound was used. In particular, owing to the presence of specific interactions between water molecules and hydrophilic sites on the polymer backbone, it was hypothesized that the absorbed molecules are in part randomly dispersed in the polymer matrix (dissolved water or free water) and in part physically bonded to the hydrophilic sites (adsorbed water or bound water) (Netti *et al.*, 1996). Therefore, the total amount of absorbed water was expressed as follows:

$$C_W = C_W^D + C_W^H \tag{42.34}$$

where C_W is the total concentration of total absorbed at equilibrium into the polymer, C_W^D is the concentration of dissolved water and C_W^H is the concentration of adsorbed water. The Flory equation that relates C_W^D to the water activity a_W is the following:

$$\ln(a_W) = \ln\left(\frac{C_W^D}{\frac{\delta_W}{\delta_p} + C_W^D}\right) + \left(1 - \frac{C_W^D}{\frac{\delta_W}{\delta_p} + C_W^D}\right) + \varphi \cdot \left(1 - \frac{C_W^D}{\frac{\delta_W}{\delta_p} + C_W^D}\right)^2 \tag{42.35}$$

where δ_W is the density of water, δ_p is the density of the dry polymer and φ is the Flory interaction parameter. The dependence of C_W^H on the water activity was described by means of the Langmuir equation (Netti *et al.*, 1996),

$$C_W^H = \frac{C_H' \cdot b \cdot a_W}{1 + b \cdot a_W} \tag{42.36}$$

where C_H' is the adsorption capacity and b is a constant that takes into account the affinity between the adsorption sites and water molecules. By combining Equations 42.35 and 42.36, it is possible to relate C_W to the water activity. However, owing to the form of Equation 42.35, it is not possible to combine the above equations to obtain a relationship where C_W is an explicit function of a_W. In fact, for a given a_W , the corresponding value of C_W was evaluated by first numerically solving Equation 42.35 to determine C_W^D, then evaluating C_W^H using Equation 42.36 and finally summing these two terms.

The constants B_i appearing in Equation 42.33 are generally determined by a study of the water sorption kinetics. When mass transfer takes place in a substantially unperturbed polymer matrix, the diffusion process is substantially controlled by a stochastic phenomenon. The other limiting behavior is encountered when a very thin slab of polymer is put in contact with a swelling penetrant. In this case the characteristic diffusion time is much lower than the relaxation time of the polymer; hence, polymer relaxation becomes the limiting phenomenon controlling the solvent uptake kinetics (Del Nobile *et al.*, 1994). In the case of water diffusion in water-sensitive polymers, the experimental observations range between these two limiting phenomena. Several approaches have been reported in the literature to describe solvent-induced polymer relaxation (Del Nobile *et al.*, 1994). Among them, one of the simplest is that proposed by Long and Richman (1960). These authors assumed that when the external water activity is suddenly changed, the solvent concentration at the polymer surface does not reach the equilibrium value instantaneously (as it would do if only stochastic diffusion were considered), but it first rapidly increases to a value lower than the equilibrium value. It then increases gradually until it reaches equilibrium. The instantaneous response of the system to the increase in the external water activity represents the elastic response of the polymer matrix to the external perturbation. The value of $C_W^B(0)$, the initial water concentration at the boundaries of the film, depends on both the initial macromolecular mobility, which in turn depends on the initial water concentration (i.e., $C_W(0)$), and the extent of perturbation (i.e., the difference between the final and initial water concentrations at the film boundaries, $C_W(\infty) - C_W(0)$). Owing to the complexity of the phenomena involved, the following empirical relationship was proposed to relate $C_W^B(0)$ to $C_W(0)$ and $C_W(\infty) - C_W(0)$:

$$K = \frac{1-[C_W(0)]^{K_1} \cdot \exp[-K_2 \cdot C_W(0)]}{\exp\{K_3 \cdot [C_W(0) - C_W(\infty)]^2\}} \tag{42.37}$$

where K is the normalized initial water concentration at the boundaries of the film, which is given by

$$K = \frac{C_W^B(0) - C_W(0)}{C_W(\infty) - C_W(0)}$$

and spans the range from zero to one, and the K_i's are constants to be evaluated by fitting the model to the experimental data. The rate at which the water concentration at the boundaries gradually increases to reach equilibrium is directly related to the relaxation kinetics of the polymeric matrix. To describe the boundary condition relaxation rate, the following empirical equation was proposed:

$$\frac{d\alpha(t)}{dt} = \{\alpha_1 \cdot \exp[\alpha_2 \cdot C_W^B(t)] \cdot \sqrt{\alpha(t)}\} \cdot \{1 - \exp[-(1-\alpha(t))]\} \tag{42.38}$$

where $\alpha(t)$ is defined by the following equation:

$$\alpha(t) = \frac{C_W^B(t) - C_W(0)}{C_W(\infty) - C_W(0)}$$

This is the normalized water concentration at the boundaries of the film at time t; it spans the range from zero to one, and represents the driving force for the polymer relaxation process. The α_i's are constants to be evaluated by fitting the model to the experimental data. As can be seen in Equation 42.38, $d\alpha(t)/dt$ is expressed as the product of two terms. The first term, which prevails at the early stages of the sorption process, is the kinetic constant of the polymer relaxation phenomenon; therefore, it increases as the macromolecular mobility of the polymer increases. The second term, which prevails at later stages of the sorption process, is a measure of the extent of the driving force for the polymer relaxation (i.e., the distance of the system from equilibrium). Therefore, it is a decreasing function of $\alpha(t)$.

In order to determine the dependence of the diffusion coefficient on the local water concentration (i.e., for the assessment of the constants B_i), water permeation tests were performed by Buonocore *et al.* (2005). and the experimental data obtained were used to verify the ability of the above equations to fit them. In particular, Equations 42.5 and 42.32 were used to describe the stochastic diffusion, and Equations 42.37 and 42.38 were used to describe the relaxation kinetics of the polymer matrix. The water uptake kinetics were obtained by numerically solving Fick's second law for a plane sheet with the appropriate initial and boundary conditions:

$$\frac{\partial C}{\partial t} = \frac{\partial}{\partial x}\left(D_{\mathrm{F}} \cdot \frac{\partial C}{\partial x}\right) \tag{42.39}$$

Two different boundary conditions were chosen, and thus two different models were proposed. The first model, which will be referred to as the "anomalous diffusion model," was derived taking the polymer matrix relaxation into account. Therefore, the following initial and boundary conditions were used:

$$\begin{cases} C_{\mathrm{W}} = C_{\mathrm{W}}(0) & \Rightarrow \quad t = 0; \quad 0 < x < \ell \\ C_{\mathrm{W}} = C_{\mathrm{W}}^{\mathrm{B}}(t) & \Rightarrow \quad \forall t; \quad x = 0, x = \ell \end{cases}$$

The second model, which will be referred to as the "nonideal Fickian model," was derived neglecting the polymer matrix relaxation. Therefore, the following initial and boundary conditions were used:

$$\begin{cases} C_{\mathrm{W}} = C_{\mathrm{W}}(0) & \Rightarrow \quad t = 0; \quad 0 < x < \ell \\ C_{\mathrm{W}} = C_{\mathrm{W}}(\infty) & \Rightarrow \quad \forall t; \quad x = 0, x = \ell \end{cases}$$

The nylon film used to validate the models was a commercially available film and had an average thickness of 70 μm. In order to determine the relationship between the water concentration and the water activity, water sorption isotherm tests were run by Buonocore *et al.* (2005), and the experimental data obtained were used to verify the ability of Equation 42.34 to fit them. Buonocore *et al.* (2005) found good agreement between predicted (Equation 42.34) and experimental data, suggesting that the process

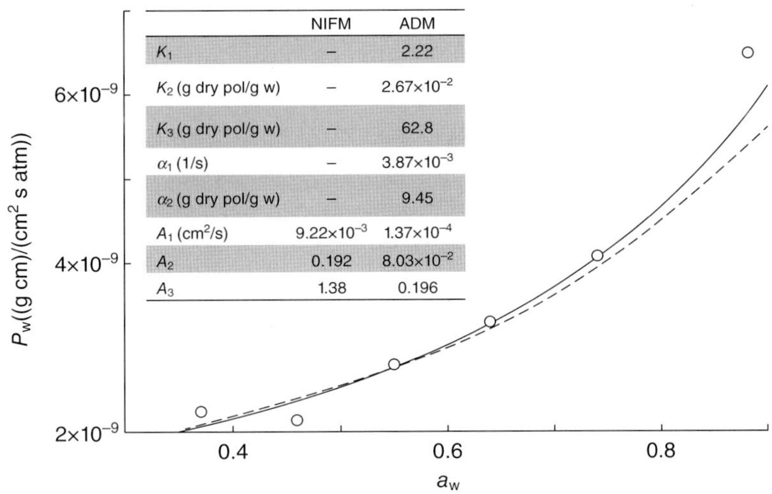

	NIFM	ADM
K_1	–	2.22
K_2 (g dry pol/g w)	–	2.67×10^{-2}
K_3 (g dry pol/g w)	–	62.8
α_1 (1/s)	–	3.87×10^{-3}
α_2 (g dry pol/g w)	–	9.45
A_1 (cm²/s)	9.22×10^{-3}	1.37×10^{-4}
A_2	0.192	8.03×10^{-2}
A_3	1.38	0.196

Figure 42.2 Water permeability of nylon film plotted as a function of water vapor activity on the upstream side of the film (the water activity on the downstream side of the film is equal to zero), along with fitting parameters. (○) Experimental data; (——) "anomalous diffusion model" (ADM); (— —) "nonideal Fickian model" (NIFM).

of sorption of water into the film investigated could be successfully described by the proposed equation.

The water permeability of the film investigated, as determined by means of permeation tests, is plotted as a function of the water activity on the upstream side of the film in Figure 42.2. The curves shown in this figure were obtained by predicting the water permeability coefficient by means of the proposed models using the fitting parameters also listed in the figure. As can be inferred from the data shown in Figure 42.2, the ability of the "anomalous diffusion model" to predict the experimental data is quite satisfactory. The predictive ability of the "nonideal Fickian model" decreases as the water activity increases (i.e., as the influence of the polymer matrix relaxation process on diffusion becomes increasingly significant), being quite satisfactory at low water activity, and only just acceptable at high water activity. The above results suggest that at low water activity, both models can successfully used to predict the water barrier properties of nylon films. In contrast, at high water activity, the superposition of solvent-induced polymer relaxation on stochastic diffusion cannot be neglected, and only the "anomalous diffusion model" is able to satisfactorily predict the water permeability coefficient.

Evaluation of the Permeability Coefficient of a Multilayer System

Del Nobile *et al.* (2004) used a multilayer film made up of five polyolefin layers and a gas barrier layer (made of EVOH), and water as a permeant to validate the model

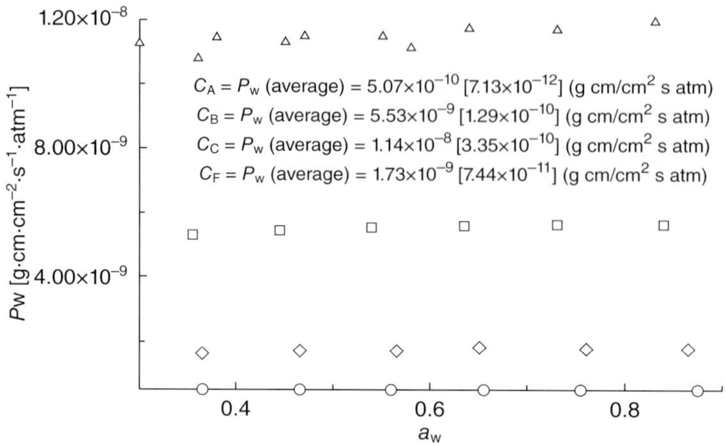

Figure 42.3 Water permeability coefficients of the four polyolefin films investigated, plotted as a function of the value of a_w on the upstream side of the film. The value of a_w on the downstream side of the film was set to zero. (○) Film A, (□) film B, (△) film C, (◇) film F. The mean water permeability coefficients of each polyolefin layer are also shown.

above that relates the permeability of a multilayer film to those of its constituent layers. As far as the polyolefin layers are concerned, the water transport properties are not expected to depend on the boundary conditions. Figure 42.3 shows the water permeability coefficients of the polyolefin layers investigated. As one would expect, the water permeability coefficient does not depend on the external humidity. Therefore, in these cases the water barrier properties of the film can be determined by a single water permeation test. The mean water permeability coefficients of each polyolefin layer, calculated by averaging the values shown in Figure 42.3, are also listed in the figure. Both the three polyolefin films on one side of the gas barrier layer and the two polyolefin films on the other side were considered as a single film, since their permeability does not depend on the water activity.

EVOH is a moderately hydrophilic polymer, whose water transport properties are expected to depend on the external humidity. Figure 42.4 shows the water permeability of the EVOH film investigated , as determined by permeation tests, plotted as a function of the water activity on the upstream side of the film (the water activity on the downstream side of the film was set equal to zero). The curve shown in the figure was obtained by predicting the water permeability coefficient by means of Equation 42.33 using the parameter values also shown in the figure. Del Nobile *et al.* (2004) evaluated the goodness of fit by means of the relative percentage difference (denoted by $E\%$) (Boquet *et al.*, 1978) calculated using the experimental and predicted values. They found that the ability of Equation 42.32 to predict the experimental data was quite satisfactory, with an $E\%$ value equal to 5.74.

As said above, the multilayer films intended for food-packaging applications commercially available have one moderately hydrophilic film (i.e., EVOH), which serves

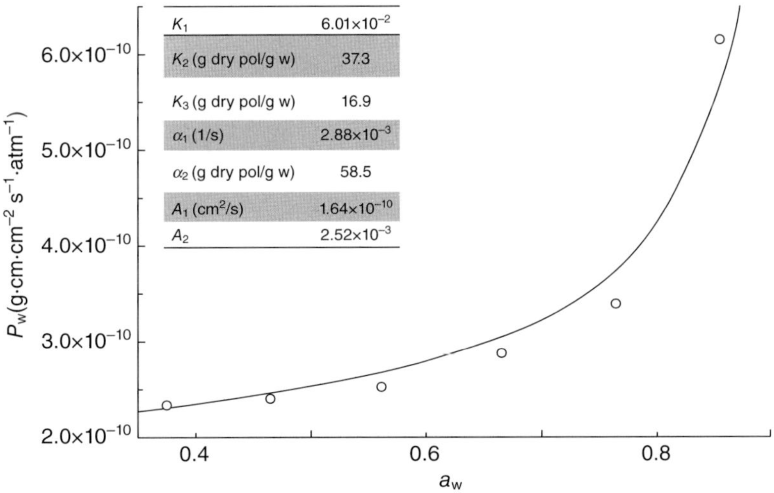

Figure 42.4 Water permeability coefficient of the EVOH film investigated, plotted as a function of the value of a_w on the upstream side of the film. The value of a_w on the downstream side of the film was set to zero. (○) Experimental data, (——) data trend as predicted by means of Equation 42.33. The values of the fitting parameters are also shown.

as a barrier film to gases, placed between two hydrophobic films (such as polyolefin films). In these cases Equation 42.22 can be written as follows:

$$P_W^{TOT}\left(a_W^u, a_W^d\right) = \frac{1}{\frac{\ell_1}{\ell}\cdot\frac{1}{P_W^1} + \frac{\ell_2}{\ell}\cdot\frac{1}{P_W^2\left(a_W^{1-2}, a_W^{2-3}\right)} + \frac{\ell_3}{\ell}\cdot\frac{1}{P_W^3}} \tag{42.40}$$

where a_W^{1-2} is the water activity in equilibrium with the permeant concentration at the interface between the first and second layers in the multilayer structure, and a_W^{2-3} is the water activity in equilibrium with the permeant concentration at the interface between the second and third layers. As can be inferred from Equation 42.40, to calculate $P_W^{TOT}(a_W^u, a_W^d)$, the values of a_W^{1-2} and a_W^{2-3} need to be calculated first. These two values can be determined by numerically solving the following set of nonlinear equations:

$$\begin{cases} J = P_W^1 \cdot \dfrac{p_W^0\left(a_W^u - a_W^{1-2}\right)}{\ell_1} \\ J = P_W^2\left(a_W^{1-2}, a_W^{2-3}\right)\cdot\dfrac{p_W^0\left(a_W^{1-2} - a_W^{2-3}\right)}{\ell_2} \\ J = P_W^3 \cdot \dfrac{p_W^0\left(a_W^{2-3} - a_W^d\right)}{\ell_3} \end{cases} \tag{42.41}$$

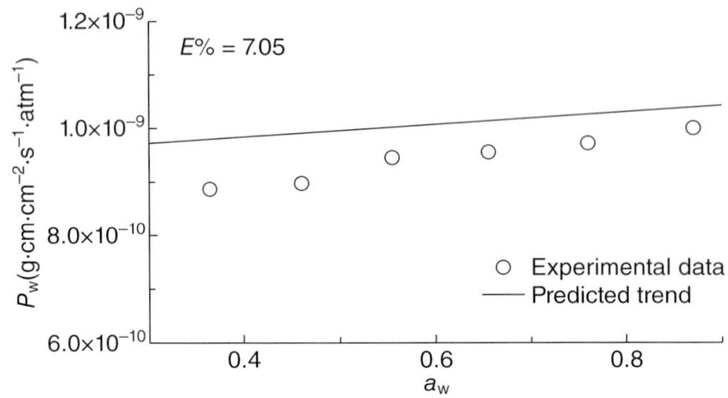

Figure 42.5 Water permeability coefficient of the multilayer film plotted as a function of the value of a_w on the upstream side of the film. The value of a_w on the downstream side of the film was set to zero. The relative percentage difference (E%) is also shown.

The set of nonlinear equations above was solved in the following way. The flux J was derived from the first equation by choosing an arbitrary value for a_W^{1-2}. By replacing J in the third equation, a_W^{2-3} was obtained. Therefore, it was possible to obtain the permeability coefficient $P_W^2(a_W^{1-2}, a_W^{2-3})$ by means of Equation 42.33. Then, a_W^{1-2}, a_W^{2-3} and $P_W^2(a_W^{1-2}, a_W^{2-3})$ were used to calculate the flux from the second equation. This new value of J was substituted in the first equation in order to evaluate a new value of a_W^{1-2}, which was then compared with the value previously calculated. The iteration ended when these two values converged.

Figure 42.5 shows the water permeability coefficient of the multilayer film plotted as a function of the water activity on the upstream side of the film (the water activity on the downstream side of the film was set equal to zero). The water permeability coefficient, as predicted by means of Equation 42.40, is also shown. The values of P_W^1 and P_W^3 were calculated by applying Equation 42.22 to the first three layers of the multilayer structure (total thickness $l_1 = l_A + l_B + l_C$) and to the last two layers (total thickness $l_2 = l_E + l_F$).

The value of E%, equal to 7.05, suggests that the approach proposed by Del Nobile *et al.* (2004) was successful in predicting the water permeability coefficient of the multilayer film from a knowledge of the water transport properties of each constituent layer. This approach could be used advantageously in the field of food packaging to design multilayer structures. In fact, the model could be used in combination with a database of the transport properties of the main polymeric films used for food packaging applications to specifically design a multilayer film, in terms of the materials and thickness of each layer, for a given application.

Packaging Design for Potato Chips

As stated above, the control of the initial composition of the headspace could represent a challenge for scientific research aimed at extending the shelf-life of a packaged product. An estimate of the best packaging conditions can be obtained by using a mathematical approach. In order to show the effectiveness of such an approach, a determination of the optimal initial headspace composition, in terms of nitrogen and water vapor, was carried out by Del Nobile (2001) for potato chips. As is known, potato chips are packaged in bags containing nitrogen. The packaging films used for this application have a low permeability to oxygen and water vapor because the former is responsible for lipid oxidation (a quality subindex) and the latter is responsible for crispness reduction (another quality subindex) (Sanches Silva *et al.*, 2004). Two different multilayer systems were considered by Del Nobile (2001): (a) a polypropylene (PP) film laminated with a metallized polypropylene (PPM) film (PP/PPM), and (b) two laminated films of polyethylene (PE) coated with polyvinylidene chloride (PVdC) (PVdC/PE/PE/PVdC). The mathematical model was based on the consideration that lipid oxidation depends on the oxygen partial pressure, the water vapor partial pressure and the extent of the lipid oxidation reaction, according to an empirical equation proposed by Quast and Karel (1972):

$$\frac{d\,Ext(t)}{dt} = mp \cdot \left(Ext + \frac{M_1 + M_2 \cdot Ext(t)}{\sqrt{\frac{p_{H_2O}^{in}(t)}{p_{H_2O}^{*}} \cdot 100}} \right) \cdot \left(\frac{p_{O_2}^{in}(t)}{M_3 + M_4 \cdot p_{O_2}^{in}(t)} \right) \tag{42.42}$$

where $Ext(t)$ is the extent of the lipid oxidation reaction at time t, mp is the mass of the packaged product, $p_{H_2O}^{in}(t)$ is the water vapor partial pressure in the package headspace at time t, $p_{H_2O}^{*}$ is the saturated water vapor pressure at the test temperature, $p_{O_2}^{in}(t)$ is the oxygen partial pressure in the package headspace at time t, and the M_i's are the model parameters. The M_i's typify the packaged food, and are generally determined by fitting Equation 42.42 to experimental data. The water sorption isotherm of potato chips can be successfully described by the Khun isotherm (Quast and Karel, 1972):

$$c = \frac{M_5}{100 \cdot \ln\left(\frac{p_{H_2O}^{in}}{p_{H_2O}^{*}} \right)} + M_6 \tag{42.43}$$

where c represents the absorbed water concentration, and M_5 and M_6 are the parameters of the model, evaluated by fitting the experimental data.

By considering the mass balance of the oxygen and the water vapor contained in the package headspace, the following equations can be obtained:

$$\frac{dp_{H_2O}^{in}(t)}{dt} = \frac{A \cdot P_{H_2O} \cdot \frac{p_{H_2O}^{o} - p_{H_2O}^{in}(t)}{\ell}}{\frac{V_{st}}{R \cdot T} \quad \frac{M_5 \cdot mp}{100 \cdot mp \cdot p_{H_2O}^{in}(t) \cdot \left[\ln\left(\frac{p_{H_2O}^{in}(t)}{p_{H_2O}^{*}} \right) \right]^2}} \tag{42.44}$$

$$\frac{dp_{O_2}^{in}(t)}{dt} = \frac{R \cdot T}{V_{st}} \cdot \left(A \cdot P_{O_2} \cdot \frac{p_{O_2}^{o} - p_{O_2}^{in}(t)}{\ell} \right) + \\ + \frac{R \cdot T}{V_{st}} \cdot \left[\frac{MW \cdot 44.615 \cdot 10^{-9}}{3600} \cdot \left(Ext + \frac{M_1 + M_2 \cdot Ext(t)}{\sqrt{\frac{p_{H_2O}^{in}(t)}{p^{*}_{H_2O}} \cdot 100}} \right) \cdot \left(\frac{p_{O_2}^{in}(t)}{M_3 + M_4 \cdot p_{O_2}^{in}(t)} \right) \right] \tag{42.45}$$

where MW is the molecular weight of water, P_{O_2} is the oxygen permeability of the packaging film, P_{H_2O} is the water vapor permeability of the packaging film, $p_{H_2O}^{o}$ is the water vapor partial pressure outside the package and $p_{O_2}^{o}$ is the oxygen partial pressure outside the package. Equations 42.42, 42.44 and 42.45 represent a set of three ordinary differential equations with three unknown functions (i.e., $Ext(t)$, $p_{H_2O}^{in}(t)$, $p_{O_2}^{in}(t)$).

Once the initial conditions are known, these differential equations can be solved numerically (Press *et al.*, 1989). In this way, it is possible to predict the kinetics of the decay of the two quality subindices during storage, and consequently the shelf-life of the product. In fact, $Ext(t)$ is directly related to lipid oxidation phenomena, whereas $p_{H_2O}^{in}(t)$ is related to the amount of water sorbed into the packaged food, and consequently to its crispness. The storage and packaging conditions considered for the simulations were: storage temperature 37 °C, oxygen partial pressure outside the package 0.21 atm, water vapor partial pressure outside the package 24.8×10^{-3} atm (RH = 40%), the gas flushed into the package was composed of nitrogen and water vapor, the initial humidity value was considered as a design variable, and the initial value of Ext was equal to zero. As stated earlier, the total quality of the product can be described by two normalized quality subindices, one related to the amount of absorbed water and the other to the extent of lipid oxidation (Quast and Karel, 1972):

$$Q_w(t) = 1 - \frac{p_{H_2O}^{in}(t)}{p_{H_2O}^{max}} \tag{42.46}$$

$$Q_0(t) = 1 - \frac{Ext(t)}{Ext^{max}} \tag{42.47}$$

where $Q_W(t)$ is the quality subindex related to absorbed water at time t; $p_{H_2O}^{max}$ is the threshold value of $p_{H_2O}^{in}(t)$, equal to 20×10^{-3} atm (Quast and Karel, 1972); $Q_0(t)$ is the quality subindex related to the extent of the lipid oxidation reaction at time t; and Ext^{max} is the threshold value of $Ext(t)$, equal to 1200 μL O_2(STP)·g^{-1} (Quast and Karel, 1972). The product becomes unacceptable whenever one of the two subindices exceeds its threshold value. It is worth noting that in the hypothetical case where the product quality was related only to $Q_W(t)$, the shelf-life of the product would decrease as the initial value of the water vapor partial pressure in the package headspace, $p_{H_2O}^{in}$, increases (Equation 42.46), whereas, if the product quality was related solely to lipid oxidation, the shelf-life would increase with an increase in $p^{in}_{H_2O}$ (Equation 42.47). The "real" shelf-life vs. $p^{in}_{H_2O}$ is obtained, by definition, by superimposing the two above conditions to determine a curve, which is expected to have a maximum. Therefore, substituting nitrogen with a mixture of water vapor and nitrogen in such a way as to reach the optimal value of the water concentration leads to an increase in the product shelf-life.

The optimal value of $p^{in}_{H_2O}$, at which the shelf-life reaches its highest value, depends on the initial and storage conditions, and on the water vapor and oxygen permeabilities of the packaging film. The optimal value of $p^{in}_{H_2O}$ (i.e., $(p^{in}_{H_2O})_{opt}$) was evaluated for ℓ_{PVdC} ranging from 1 μm to 6 μm. When the value of ℓ_{PVdC} is increased, the permeability of the multilayer film decreases, but its mechanical properties and selectivity toward oxygen and water vapor are substantially unmodified. To determine the water vapor and oxygen permeability of the PVdC/PE/PE/PVdC multilayer, Equation 42.22 was used:

$$P = \frac{1}{\frac{\ell_{PVdC}}{\ell} \cdot \frac{1}{P_{PVdC}} + \frac{\ell_{PE}}{\ell} \cdot \frac{1}{P_{PE}}} \tag{42.48}$$

where ℓ_{PVdC} is the total thickness of PVdC, ℓ_{PE} is the total thickness of PE, P_{PVdC} is the permeability of PVdC and P_{PE} is the permeability of PE. To determine the water vapor permeability of the PP/PPM multilayer, Del Nobile (2001) assumed that $P^{H_2O}_{PP} / P^{H_2O}_{PE} = P^{N_2}_{PP} / P^{N_2}_{PE}$ and $P^{H_2O}_{PP} / P^{H_2O}_{PP+PPM} = P^{O_2}_{PP} / P^{O_2}_{PP+PPM}$, where $P^{H_2O}_{PP}$ is the permeability of polypropylene to water, $P^{O_2}_{PP}$ is the permeability of polypropylene to oxygen, $P^{N_2}_{PP}$ is the permeability of polypropylene to nitrogen, $P^{H_2O}_{PE}$ is the permeability of polyethylene to water, $P^{N_2}_{PE}$ is the permeability of polyethylene to nitrogen, $P^{O_2}_{PP+PPM}$ is the permeability of PP/PPM to oxygen, and $P^{H_2O}_{PP+PPM}$ is the permeability of PP/PPM to water. The values of $P^{N_2}_{PP}$, $P^{H_2O}_{PE}$, $P^{O_2}_{PP+PPM}$, $P^{N_2}_{PE}$ and $P^{O_2}_{PP}$ were taken from data reported in the literature (Del Nobile, 2001).

Figure 42.6 shows two curves: the first represents the shelf-life of the product (SL) plotted as a function of ℓ_{PVdC} when $p^{in}_{H_2O}$ is equal to 62×10^{-6} atm (RH = 0.1%, i.e., technical nitrogen), and the second represents the shelf-life of the product (OSL) plotted as a function of ℓ_{PVdC} when $p^{in}_{H_2O}$ is equal to $(p^{in}_{H_2O})_{opt}$. A percentage increment in shelf-life (ΔSL%), defined by ΔSL% = ((OSL − SL)/SL) × 100, as high as 48% was obtained for a value of ℓ_{PVdC} equal to 4.4×10^{-4} cm.

The same approach was used for the PP+PPM multilayer. In Figure 42.7, ΔSL% is shown as a function of the metallization level χ. The metallization level is a measure of the thickness of the aluminum coating; it is defined by the relationship $P_{met} = P^{\star}_{met}/\chi$, where $P^{\star}_{met}$ represents the permeability of the metallized film, and P_{met} represents the permeability of a hypothetical metallized film with a different thickness of the aluminum coating. Similarly to what was found for the PVdC/PE/PE/PVdC multilayer, OSL was always higher than SL, with a value of ΔSL% as high as 80%.

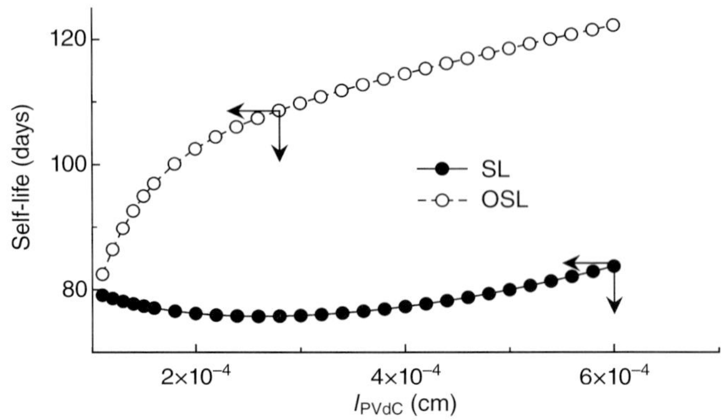

Figure 42.6 SL and OSL as a function of ℓ_{PVdC} for the PVdC/PE/PE/PVdC multilayer film.

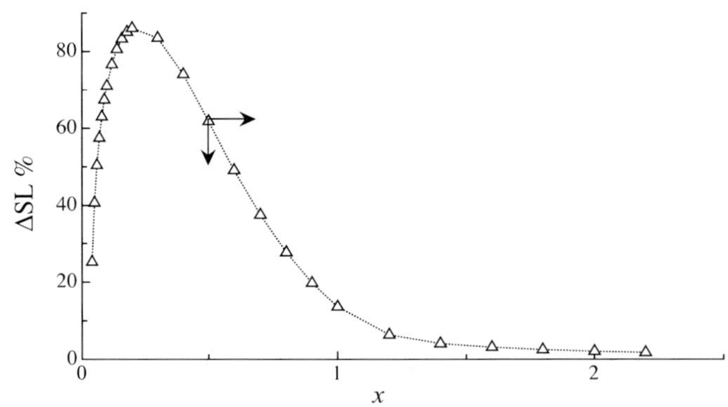

Figure 42.7 ΔSL% as a function of the metallization level χ for the PP/PPM multilayer film.

The behavior of SL in both of the above cases shows a minimum. This means that when the value of either ℓ_{PVdC} or χ is increased, the water vapor and oxygen permeabilities are reduced, and consequently the rates at which the respective partial pressures increase inside the package during storage are reduced. Hence, as is evident from Equation 42.44, either an increase or a decrease in the lipid oxidation rate can occur, thus suggesting that in the range investigated for both ℓ_{PVdC} and χ, $Q_0(t)$ determines the unacceptability of the product. To sum up, it is interesting to note that the mathematical approach allowed the effects of both the thickness and the metallization level on the optimal initial humidity for potato chips to be determined, since the results are not intuitive. In particular, the author of that study found that the effect related to water vapor was dominant for values of ℓ_{PVdC} lower than 2.5 μm and values of χ lower than 0.5.

Prediction of Shelf-Life of Cereal-Based Dry Products

The shelf-life of cereal-based dry products is strictly related to the gain of moisture from the external environment, which causes loss of crispness of the packaged food. The water concentration of the packaged food depends on the affinity between the food and water, and on the water activity in the package headspace, which in turn depends on the barrier properties of the packaging material. Generally, the shelf-life of such a type of product is several months and, owing to this long time period, an experimental evaluation of the optimal packaging conditions is generally unfeasible. Rough shelf-life estimation can be done by empirical methods; most of them have been proposed in the last 30 years. In the following, the most relevant are reported.

Azanha and Faria (2005), for example, evaluated the ability of three different empirical equations (a linear equation, a midpoint equation and a logarithm interval equation) to predict the stability of commercial cornflakes in flexible packaging. To this end, the initial moisture contents, moisture sorption isotherms, moisture content changes, water vapor transmission rates and water vapor permeability coefficients of HDPE-based films of different thicknesses (20, 40 and 50 μm) were determined. The mathematical simulation demonstrated that the midpoint and logarithm interval models were suitable for describe the percentage moisture content in the two thickest bags, whereas the linear model was good only for the 20 μm bag, thus suggesting that the higher the precision required, the higher the complexity of the model. To better simulate the stability of cornflakes, exact experimental points on the sorption isotherms and the correct model to fit these data are necessary. Moreover, a critical sensory evaluation is of crucial importance for confirming the model, thus demonstrating that empirical mathematical models cannot be applied to a complex case study.

A similar approach, based on the relationship between the barrier properties of the packaging material and the moisture absorbed by the dry product, was used by Siripatrawan (2009) for two types of rice crackers. The Guggenheim–Anderson–de Boer (GAB) model was fitted to experimental data for the moisture sorption isotherm, as this is the most versatile sorption model available in the literature. The critical water activity for product acceptability was judged by a panel test. The shelf-life simulation model was developed on the basis of a simple mass balance:

$$\int_{t=0}^{t=t} dt = \frac{\ell \cdot wp}{P_{H_2O} \cdot A} \int_{m_1}^{m_c} \frac{dm}{\left[p_{H_2O}^{o} - p_{H_2O}^{in}\right]} \tag{42.49}$$

where t is the storage time (days), wp is the dry weight of the packaged food (g), m is the moisture content of the product on a percentage dry basis, m_1 is the initial moisture content of the sample and m_c is the critical moisture content, determined at the critical water activity from the sorption isotherm, as determined by the GAB model. The performance of the shelf-life simulation was measured using the coefficient of determination and the root mean square error. The author of the study stated that in this case also, the simulation model tested had limited applicability, being suitable for predicting the shelf-life of only one type of cracker, those containing a smaller fat content. The difference between the experimental and predicted shelf-life suggested that the assumption of a constant water vapor permeability is not very accurate, thus leading to an overestimation of the shelf-life.

In the literature, there are numerous other mathematical models (Iglesias *et al.*, 1979; Labuza and Contreras-Medellin, 1981; Tubert and Iglesias, 1985) that attempt to predict the stability of dry products to a good approximation, but in all cases an overestimation of the shelf-life is made if the dependence of the water permeability coefficient on the water activity inside and outside the package is neglected (Del Nobile *et al.*, 2003a). In fact, when a multilayer system includes a moderately hydrophilic film, the water barrier properties of the package depend on the water activity on the upstream and downstream sides of the film (Hernandez, 1994). Del Nobile *et al.* (2003a) developed a mathematical model that takes this specific aspect into account, and also evaluates the error made in the shelf-life prediction if the dependence of the barrier properties of the film on the boundary condition is not taken into account. The proposed model consists of two parts; the first aims to predict the dependence of the water permeability coefficient on the water activity on the upstream and downstream sides of the film, and the second part aims to predict the time evolution of the water activity inside the package during storage by considering the water mass balance inside the bag. The model is based on the following assumptions: the water vapor in the package headspace is always in equilibrium with the water absorbed in the packaged product, and the water vapor in the package headspace can be considered as an ideal gas (Quast and Karel, 1972; Bell and Labuza, 2000). The mass balance of water in the package can be written as follows:

$$\frac{dn_{\mathrm{TOT}}(t)}{dt} = P_{\mathrm{W}}^{\mathrm{M}}\left(a_{\mathrm{W}}^{\mathrm{est}}, a_{\mathrm{W}}^{\mathrm{in}}(t)\right)\cdot A\cdot p_{\mathrm{H_2O}}^{\mathrm{o}}\cdot\frac{a_{\mathrm{W}}^{\mathrm{est}} - a_{\mathrm{W}}^{\mathrm{in}}(t)}{\ell} \tag{42.50}$$

The water permeability coefficient of a bilayer film obtained by laminating a moderately hydrophilic film, such as a polyamide, with a polyolefinic film ($P_{\mathrm{W}}^{\mathrm{M}}(a_{\mathrm{W}}^{\mathrm{est}}, a_{\mathrm{W}}^{\mathrm{in}}(t))$) is given by the following equation, based on Equation 42.22:

$$P_{\mathrm{W}}^{\mathrm{M}}\left(a_{\mathrm{W}}^{\mathrm{ext}}, a_{\mathrm{W}}^{\mathrm{in}}\right) = \frac{1}{\frac{\ell_1}{\ell}\cdot\frac{1}{P_{\mathrm{w}}\left(a_{\mathrm{W}}^{\mathrm{in}}, a_{\mathrm{W}}^{1-2}\right)} + \frac{\ell_2}{\ell}\cdot\frac{1}{P_{\mathrm{W}}^{2}}} \tag{42.51}$$

a_{W}^{1-2} was evaluated from Equation 42.41. The value of $n_{\mathrm{TOT}}(t)$ given by Equation 42.50 is the sum of two contributions: the number of moles of water present in the package headspace and the number of moles of water absorbed into the packed product:

$$n_{\mathrm{TOT}}(t) = n_{\mathrm{st}}(t) + n_{\mathrm{ass}}(t) \tag{42.52}$$

$n_{st}(t)$ depends on the water activity inside the package according to the following equation:

$$n_{st}(t) = \frac{V_{st} \cdot p^{o}_{H_2O} \cdot a^{in}_{w}(t)}{R \cdot T} \tag{42.53}$$

According to the water sorption isotherm as described by the GAB model, the number of moles of water absorbed into the packaged food depends on the internal water activity according to the following equation:

$$n_{ass}(t) = \frac{mp}{18} \cdot \frac{M_0 \cdot G \cdot W \cdot a^{in}_{w}(t)}{(1 - W \cdot a^{in}_{w}(t)) \cdot (1 - W \cdot a^{in}_{w}(t) + G \cdot W \cdot a^{in}_{w}(t))} \tag{42.54}$$

By substituting Equations 42.53 and 42.54 into Equation 42.50, the following equation is obtained:

$$\begin{aligned} &\frac{d}{dt}\left[\frac{mp}{18} \cdot \frac{M_0 \cdot G \cdot W \cdot a^{in}_{w}(t)}{(1 - W \cdot a^{in}_{w}(t)) \cdot (1 - W \cdot a^{in}_{w}(t) + G \cdot W \cdot a^{in}_{w}(t))} + \frac{V_{st} \cdot p^{o}_{H_2O} \cdot a^{in}_{w}(t)}{R \cdot T}\right] \\ &= P^{M}_{W}\left(a^{est}_{w}, a^{in}_{w}(t)\right) \cdot A \cdot p^{o}_{H_2O} \cdot \frac{a^{est}_{w} - a^{in}_{w}(t)}{\ell} \end{aligned} \tag{42.55}$$

which can be rearranged into the following form:

$$\frac{da^{in}_{w}(t)}{dt} = \frac{P^{M}_{W}\left(a^{est}_{w}, a^{in}_{w}(t)\right) \cdot A \cdot p^{o}_{H_2O} \cdot \dfrac{a^{est}_{w} - a^{in}_{w}(t)}{\ell}}{\dfrac{V_{st} \cdot p^{o}_{H_2O}}{R \cdot T} + \dfrac{mp \cdot M_0 \cdot G \cdot W \cdot \left[1 - \left(W \cdot a^{in}_{w}(t)\right)^2 \cdot (G-1)\right]}{18 \cdot \left[(1 - W \cdot a^{in}_{w}(t)) \cdot (1 - W \cdot a^{in}_{w}(t) + G \cdot W \cdot a^{in}_{w}(t))\right]^2}} \tag{42.56}$$

Equation 42.56 is an ordinary differential equation, whose unknown function is $a_W(t)$. It was integrated numerically using a fourth-order Runge–Kutta formula (Press *et al.*, 1989). Once the function $a_W(t)$ is known, the amount of water absorbed into the packed product during storage can be easily determined by means of the GAB model. As pointed out by Del Nobile *et al.* (2003a), the major limitation on an extension of the proposed model to the general case is that it makes the hypothesis that there is an instantaneous equilibrium between the water vapor present in the package headspace and the water absorbed in the packaged product. Wherever this assumption is satisfied, the proposed model can be used advantageously to predict the shelf-life of cereal-based dry foods or to design a package properly.

This model was used to simulate the storage behavior of three dry foods, named A, B and C, packaged with three hypothetical bilayer films based on polyamides, named 1, 2 and 3. For the sake of simplicity, food A packaged using the bilayer film 1 will be referred to as sample A1, and a similar convention will be used to refer to the other samples. The shelf-life was calculated assuming the critical moisture content to be 0.04 $g_{water}/g_{dry\ matter}$, which corresponds to a water activity of around 0.4 (Labuza and Contreras-Medellin, 1981). In order to highlight the error made when the permeability coefficient is considered to be constant, the shelf-life was also calculated according to the following equation:

$$\frac{da_w^{in}(t)}{dt} = \frac{P_{W,Av.}^{M} \cdot A \cdot p_{H_2O}^{o} \cdot \dfrac{a_w^{est} - a_w^{in}(t)}{\ell}}{\dfrac{V_{st} \cdot p_{H_2O}^{o}}{R \cdot T} + \dfrac{mp \cdot M_0 \cdot G \cdot W \cdot \left[1 - \left(W \cdot a_w^{in}(t)\right)^2 \cdot (G-1)\right]}{18 \cdot \left[\left(1 - W \cdot a_w^{in}(t)\right) \cdot \left(1 - W \cdot a_w^{in}(t) + G \cdot W \cdot a_w^{in}(t)\right)\right]^2}} \tag{42.57}$$

The mean value of the water permeability coefficient, $P_{W,Av}^{M}$, was calculated from the following equation:

$$P_{W,Av.}^{M} = \frac{\displaystyle\int_{a_W=0.1}^{a_W=0.7} P_W^M\left(a_w^{est}, a_w^{in}(t)\right) da_w}{0.6} \tag{42.58}$$

Equation 42.58 was solved numerically using the extended Simpson's rule (Press *et al.*, 1989). The data for the shelf-life calculated for six of the nine sample combinations (three foods with three packaging materials), as predicted by means of both Equation 42.56 and Equation 42.57, is shown in Figure 42.8. This figure clearly evidences that

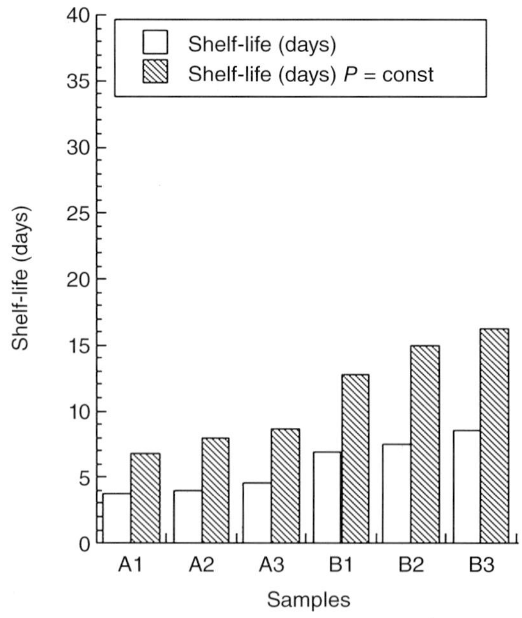

Figure 42.8 Shelf-life of the samples investigated, as predicted by means of the proposed model.

the predicted shelf-life depends strongly on the type of food, whereas the type of polyamide used for the package has only a slight influence. As can be seen, the shelf-life predicted by Equation 42.57 is about 90% greater than the shelf-life predicted by means of Equation 42.56; the same trend was also observed for food C. The reason for the observed large difference has to be ascribed to the fact that a substantial difference exists between the average water permeability coefficient of the bilayer and the water permeability coefficient under real working conditions. To sum up, it is not possible to determine the water barrier properties of a given moderately hydrophilic material by simply evaluating the permeability coefficient for a single set of boundary conditions. In fact, water molecules, acting as plasticizers, increase the macromolecular mobility of the polymer; thus both the solubility and the diffusivity, and consequently the permeability coefficient, depend on the local water concentration. As a consequence, in these cases, the permeability coefficient cannot be determined by a single measurement; instead, a more complex analysis is necessary to understand all the phenomena involved in the permeation process.

Design of Plastic Packages for Minimally Processed Foods

Minimally processed crops consist of washed, cut and packaged fruit and vegetables. The shelf-life of these products is very short, owing to the cutting operations, which provoke a physiological disorder (Soliva-Fortuny and Martìn-Belloso, 2003; Martin-Diana *et al.*, 2007; Olivas *et al.*, 2007). Among the various alterations in transpiration, enzymatic activity and microbial proliferation that occur, for most products the accelerated respiration rate represents the main factor that provokes food unacceptability (Ragaert *et al.*, 2007; Rico *et al.*, 2007). Thus, strategies aimed at slowing down respiration activity are generally successful for such foods. One of the most widespread preservation techniques is the substitution of air by a gas with a different composition at the moment of packaging (active modified-atmosphere packaging, or active MAP). The selection of a gas combination that avoids the usual transient state before the equilibrium state of the gas in the bag is reached can reduce the respiration rate during the transient state, thus promoting product preservation and consequently, shelf-life prolongation (Lee *et al.*, 1996; Rodriguez-Aguilera and Oliveira, 2009). The appropriate atmosphere within a package depends on various factors, such as the characteristics and weight of the product, the respiring surface area, the storage conditions and the barrier properties of the packaging (Mahajan *et al.*, 2007). The natural variability of raw material and its dynamic response to processing and storage conditions may render it impossible to identify a truly optimal atmosphere by using general empirical methods, thus suggesting that significant advances in the packaging of minimally processed foods may require the development of complex mathematical models that incorporate the dynamic response of the product to the environment (Jacxsens *et al.*, 1999; Saltveit, 2003).

As stated earlier, as the quality loss of many types of fresh-cut produce is strictly related to the respiration rate, packaging design could be assessed by taking only respiration activity into account (Fonseca *et al.*, 2002). Predictive models of the respiration rate of minimally processed foods have been used to optimize the packaging characteristics and/or the gas composition (Fonseca *et al.*, 2002; Rocculi *et al.*, 2006; Del Nobile *et al.*, 2008, 2009a,b; Conte *et al.*, 2009a,b). Del Nobile *et al.* (2007) used a simple mathematical model to design the packaging of three fruits: bananas, prickly pears and kiwi fruit. Del Nobile *et al.* (2007) validated the model by monitoring the gas concentrations in the package headspace during the entire observation period. Two multilayer films were used, a laminated high-barrier film (PE/aluminum/PET) and a traditional co-extruded low-barrier polyolefin film (Coex). The model used to describe the respiration rate of the packaged food was that proposed by Del Nobile *et al.* (2006); it is based on the Michaelis–Menten type of enzyme kinetics, as follows:

$$\begin{cases} rO_2 = A_1 \cdot \exp\{-A_2 \cdot [CO_2]\} \cdot [O_2] \\ rCO_2 = k_1 \cdot \{A_1 \cdot \exp\{-A_2 \cdot [CO_2]\} \cdot [O_2]\} \end{cases} \tag{42.59}$$

where rO_2 is the oxygen consumption rate expressed in mL·kg^{-1}·h^{-1}; $[O_2]$ is the percentage headspace oxygen concentration; $[CO_2]$ is the percentage headspace carbon dioxide concentration; A_1 is the preexponential term expressed in mL·kg^{-1}·h^{-1}, which is the maximum oxygen consumption rate; A_2 is the exponential factor and accounts for the carbon dioxide-induced respiration inhibition, which is dimensionless; rCO_2 is the carbon dioxide production rate expressed in mL·kg^{-1}·h^{-1}; and k_1 is the ratio between the numbers of moles of carbon dioxide produced and moles of oxygen consumed, which is dimensionless. To describe the time evolution of the oxygen and carbon dioxide concentrations inside the package during storage, the mass balance on these two substances in the package headspace was written as

$$\frac{d(n_{O_2}(t))}{dt} = A \cdot P_{O_2} \cdot \frac{\left[p^{0}_{O_2} - \dfrac{n_{O_2}(t) \cdot R \cdot T}{V_{st}}\right]}{\ell} + \\ - mp \cdot 4.615 \cdot 10^{-6} \cdot \{A_1 \cdot [O_2] \cdot \exp\{-A_2 \cdot [CO_2]\}\} \tag{42.60}$$

$$\frac{d(n_{CO_2}(t))}{dt} = A \cdot P_{O_2} \cdot \frac{\left[p^{0}_{CO_2} - \dfrac{n_{CO_2}(t) \cdot R \cdot T}{V_{st}}\right]}{\ell} + \\ - mp \cdot 4.615 \cdot 10^{-6} \cdot \{k_1 \cdot \{A_1 \cdot [O_2] \cdot \exp\{-A_2 \cdot [CO_2]\}\}\} \tag{42.61}$$

where $n_{O_2}(t)$ is the number of moles of oxygen in the package headspace at time t, $n_{CO_2}(t)$ is the number of moles of carbon dioxide in the package headspace at time t, P_{CO_2} is the carbon dioxide permeability of the package, and $p^{0}_{CO_2}$ is the external carbon dioxide partial pressure. Equations 42.60 and 42.61 represent a set of two ordinary differential equations, which was integrated numerically using a fourth-order Runge–Kutta formula (Press *et al.*, 1989).

Del Nobile *et al.* (2007) found a decrease in the headspace oxygen concentration along with an increase in the headspace carbon dioxide concentration for all of the packaged products during the observation period. The set of ordinary differential equations in Equations 42.60 and 42.61 was fitted simultaneously to the data for the oxygen and carbon dioxide headspace concentrations and a good fit was obtained, thus suggesting that the proposed respiration model accurately described the behavior of the fresh-cut produce studied. On the basis of the calculated fitting parameters (A_1, A_2 and k_1), a single package variable was optimized: the film thickness. In particular, the film used for packaging design was the co-extruded film, and its thickness was increased until the simulation indicated that the steady-state oxygen concentration in the package headspace had reached a value ranging between 7% and 8%, which guaranteed the lowest produce respiration rate, while avoiding the onset of anaerobic respiration. The study demonstrated that an increase in the film thickness corresponded to a decrease in the equilibrium oxygen concentration in the package headspace and an increase in the equilibrium carbon dioxide concentration. Del Nobile *et al.* (2007) introduced the total amount of oxygen consumed as a produce quality subindex, as this is directly related to the extent of metabolic activity (the senescence level) associated with product respiration (Böttcher *et al.*, 2003). Figure 42.9 shows the total amount of oxygen consumed by the three fruits packaged in two different systems, a traditional bag and a new, thicker bag, as predicted by the model. The values of the total amount of oxygen consumed demonstrated that the commercially available film did not represent the optimal packaging solution, whereas increasing the film thickness led to a lower senescence level and, as a consequence, to shelf-life prolongation in all cases investigated.

A different mathematical approach (which neglects the transient state) was used by Jacxsens *et al.* (1999) to design equilibrium modified-atmosphere (EMA) packages (1–5 mL/100 mL oxygen, 3–10 mL/100 mL carbon dioxide, balance nitrogen) for several types of fresh-cut produce. To this end, the respiration rate was calculated with two mathematical models (Gong and Corey, 1994; Haggar *et al.*, 1992). When the respiration rate, fill weight and package dimensions were known, the required transmission rates of the film for oxygen and carbon dioxide could be calculated based on the following steady-state equations:

$$\begin{cases} rO_2 \cdot mp = \mathrm{OTR} \cdot A \cdot \{[O_2]_{out} - [O_2]_{in}\} \\ rCO_2 \cdot mp = \mathrm{CO_2TR} \cdot A \cdot \{[CO_2]_{in} - [O_2]_{out}\} \end{cases} \tag{42.62}$$

where A is the area of the film (m^2) = length × width × 2, mp is the fill weight (kg), OTR is the oxygen transmission rate ($mL\,O_2/(m^2\,24\,h\ atm)$) and CO_2TR is the carbon dioxide transmission rate ($mL\,CO_2/(m^2\,24\,h\,atm)$). To validate this method, seven minimally processed vegetables were tested, thus demonstrating that this systematic

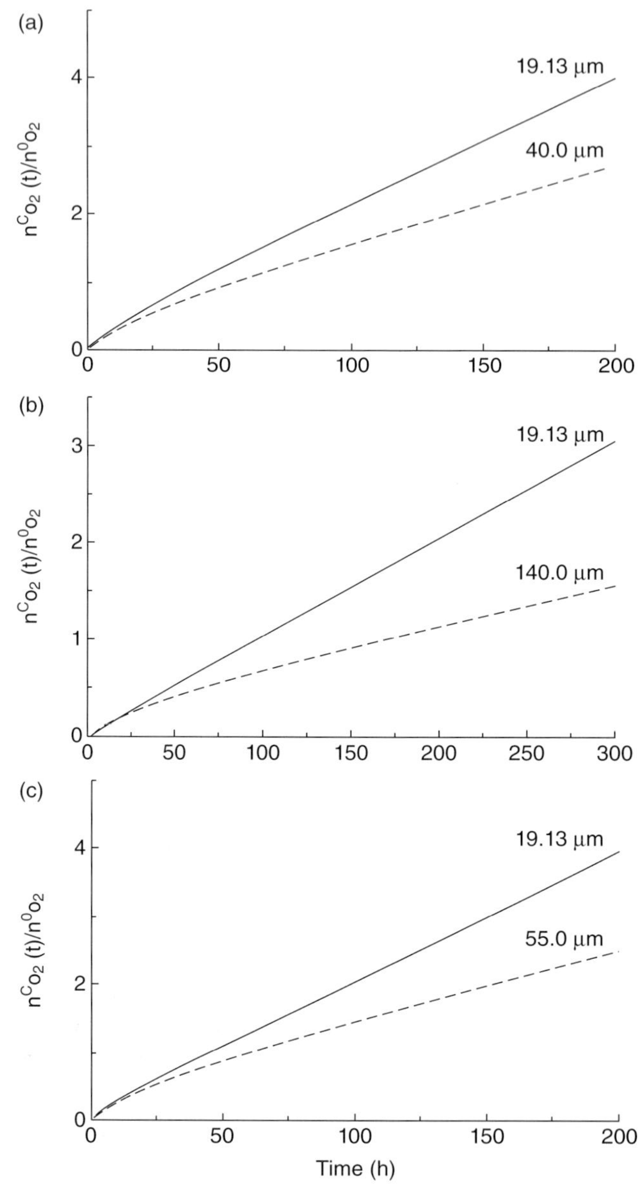

Figure 42.9 Variation during storage of the ratio between the numbers of moles of oxygen consumed and moles of oxygen initially present in the package headspace for prickly pear (a), kiwi (b) and banana (c), packaged in two different types of bag.

approach can work satisfactorily. However, the microbiological quality, the temperature, and the oxygen and carbon dioxide dependence of the respiration rate were not taken into account, even though these factors usually represent strategic variables in the packaging design.

For this reason, the same authors developed an integrated mathematical system to design EMA packages for fresh-cut produce to be stored at temperatures between 2 and 15 °C (Jacxsens *et al.*, 2000). In another publication, these authors used the same predictive model to evaluate the impact of temperature fluctuations in a simulated cold distribution chain on EMA packages of fresh-cut vegetables (Jacxsens *et al.*, 2002). In this integrated system, the effect of temperature on produce respiration and film permeability was described by an Arrhenius-type equation, while the effect of oxygen and carbon dioxide on the respiration rate was described by Michaelis–Menten kinetics. Prediction using this integrated approach and validation of it showed that the optimal EMA conditions could be generated between 2 and 10 °C. Above 10 °C, an overestimation of the internal oxygen concentration occurred, probably because oxygen consumption provoked by the microbial activity that was generally promoted at higher temperatures was not considered.

Effect of Perforation on Oxygen and Water Vapor Dynamics under MAP

As said above, in the case of products with a high respiratory activity, perforations or channels in the package can be used to obtain the desired transport properties of the package. Fishman *et al.* (1996) developed a model to calculate the oxygen concentration and relative humidity in a perforated-film modified-atmosphere package containing mango fruit. The package parameters, such as the number and area of the perforations, and the surface area of the film, were used as design variables. The model included equations describing fruit respiration and transpiration, and the permeation of oxygen and water vapor through the perforation holes (Equations 42.29 and 42.30). The respiration rate, rO_2, was used to describe changes in the headspace gas concentration due to fruit respiration. Multiplication of rO_2 by the weight of the fruit, mp, gives the contribution from respiration to changes in the amount of gas in the headspace, V_{st}. The term f in Equation 42.23 is given by

$$f = -rO_2 mp \tag{42.63}$$

A linear model with one parameter was used to describe the respiration rate as a function of the concentration in the modified-atmosphere package:

$$rO_2 = \psi \cdot O_2 \tag{42.64}$$

Combining Equations 42.23, 42.24, 42.63 and 42.64, subject to $F = 0$, results in the following linear differential equation for the closed system:

$$\frac{dO_2}{dt} = -\psi \left(\frac{mp}{V_{st}} \right) O_2 \tag{42.65}$$

which has a solution

$$O_2(t) = O_{2(0)} \exp\left(-\psi \frac{mp \cdot t}{V_{st}}\right) \tag{42.66}$$

The transpiration flow f_W replaces the flow f in Equation 42.23 in the case of water vapor. This is the amount of water vapor that passes away from the total fruit surface per unit of time. Denoting the water vapor efflux from a unit surface area of the fruit per unit of time by ϕ_W, and the area of the fruit by A_f, the total flow f_W is calculated as

$$f_W = \phi_W \cdot A_f \tag{42.67}$$

Fishman *et al.* (1996) hypothesized that ϕ_W is proportional to the difference between the relative humidity of the internal atmosphere of the fruit and that of the modified atmosphere in the package:

$$\phi_W = \rho \cdot \theta \cdot (H_f - H) \tag{42.68}$$

where H_f is the relative humidity of the internal atmosphere of the fruit and ρ is the permeance of the fruit surface to water vapor. The relative humidity of the internal atmosphere of the fruit was considered, as a first approximation, to be 1.0 (or 100%). This parameter depends on the solute content of the fruit and is in fact slightly less than 1.0 (Patterson *et al.*, 1993). Inserting Equations 42.24, 42.29, 42.63 and 42.64 into Equation 42.23 gives the following equation describing the dynamics of oxygen in the modified atmosphere:

$$\frac{dO_2}{dt} = -\psi \frac{O_2 \cdot mp}{V_{st}} + \frac{(O_{2(A)} - O_2)\left[\dfrac{A \cdot P}{\ell} + \dfrac{A_h \cdot D_F}{(L + z_h)}\right]}{V_{st}} \tag{42.69}$$

where $O_{2(A)}$ is the oxygen concentration in the ambient atmosphere. Combining Equations 42.24, 42.30, 42.67 and 42.68 with Equation 42.23 leads to an equation describing the dynamics of water vapor in the modified atmosphere:

$$\frac{dH}{dt} = A_f \rho \frac{(H_f - H)}{V_{st}} - \frac{(H - H_A)\left[\dfrac{A \cdot P_W}{\ell} + \dfrac{A_h \cdot D_W}{(L + z_h)}\right]}{V_{st}} \tag{42.70}$$

The steady-state solutions of Equations 42.69 and 42.70 (denoted by an asterisk) are obtained by equating the terms dO_2/dt and dH/dt to zero:

$$O_2^{*} = O_{2(A)} \frac{\left[\frac{A \cdot P}{\ell} + \frac{A_h \cdot D_F}{(L + z_h)}\right]}{\left[\psi m p + \frac{A \cdot P}{\ell} + \frac{A_h \cdot D_F}{(L + z_h)}\right]} \tag{42.71}$$

$$H^{*} = \frac{\left\{A_f \cdot \rho \cdot H_f + \left[\frac{A \cdot P_W}{\ell} + \frac{A_h \cdot D_W}{(L + z_h)}\right] H_A\right\}}{\left[A_f \cdot \rho + \frac{A \cdot P_W}{\ell} + \frac{A_h \cdot D_W}{(L + z_h)}\right]} \tag{42.72}$$

The general solution of the linear differential equations in Equations 42.69 and 42.70, subject to the initial conditions of the ambient atmosphere, i.e., $O_2 = O_{2(A)}$ and $H = H_A$ at $t = 0$, is

$$O_2 = O_2^{*} - (O_2^{*} - O_{2(A)}) \exp(-t/t^{*}) \tag{42.73}$$

$$H = H^{*} - (H^{*} - H_A) \exp(-t/t_W^{*}) \tag{42.74}$$

where $t^{\star}$ and t_W^{*} are characteristic times for the dynamics of the oxygen concentration and relative humidity, respectively. The characteristic times in Equations 42.73 and 42.74 represent the following combinations of parameters:

$$t^{*} = \frac{V_{st}}{\left[\psi m p + \frac{A \cdot P}{\ell} + \frac{A_h \cdot D_F}{(L + z_h)}\right]} \tag{42.75}$$

$$t_W^{*} = \frac{V_{st}}{\left[A_f \rho + \frac{A \cdot P_W}{\ell} + \frac{A_h \cdot D_W}{(L + z_h)}\right]} \tag{42.76}$$

Equations 42.73 and 42.74 represent the solution of Equations 42.69 and 42.70 on the assumption that the coefficients of the equations are constant in time. For climacteric fruits, such as mango, the parameters of respiration and transpiration are known to change with time. Fishman *et al.* (1996) stated that Equations 42.73 and 42.74 provide good approximations when the climacteric changes are slower than the characteristic times of the system, so that the coefficients may be considered as constant within the time required to reach a steady state.

To validate the model, Fishman *et al.* (1996) compared the predictions with experiments on mango fruit in simulated passive modified-atmosphere packages with perforated and nonperforated films. The observed (circles) and calculated (curves) time evolutions of the oxygen concentration in the modified atmosphere were compared

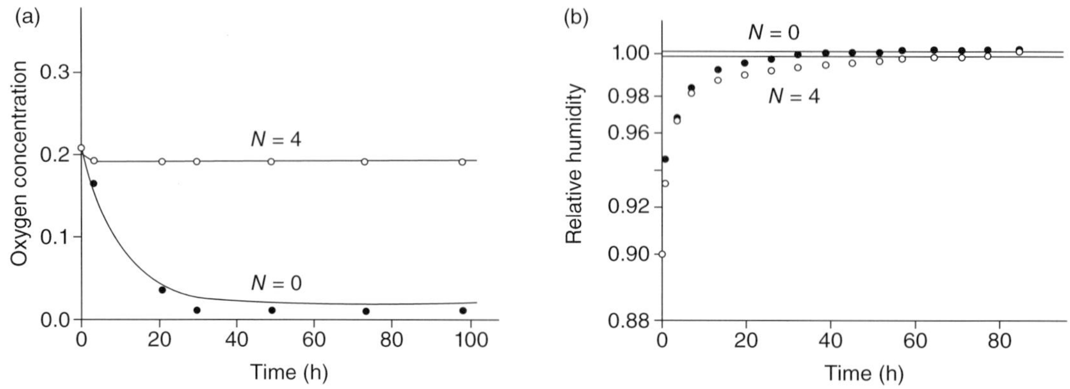

Figure 42.10 Model validation. (a) Effect of perforations on the change of O_2 concentration in modified-atmosphere packages with time. Calculated curves (Equation 42.73) for films with ($N = 4$) and without ($N = 0$) perforations, compared with experimental data at 20 °C (circles). (b) Effect of perforations on the change of in-package relative humidity with time. Model prediction (Equation 42.74) for films with ($N = 4$) and without ($N = 0$) perforations (curves) compared with experimental data at 20 °C (circles).

for both $N = 0$ and $N = 4$ perforations (Figure 42.10a). The predicted and observed results showed similar patterns. Fishman *et al.* (1996) also found a good correspondence between the predicted and observed steady-state relative humidity levels during the same period (Figure 42.10b). The results showed that perforations affected the dynamics and steady-state concentration of oxygen in the modified-atmosphere package much more strongly than those of water vapor. Moreover, in the experimental system of Fishman *et al.* (1996), the oxygen concentrations in the perforated and non-perforated packages differed by an order of magnitude, whereas the in-package relative humidity levels were practically equal. The model developed by Fishman *et al.* (1996) could be applied to design packaging for other commodities, once the transpiration coefficients are known.

Barrier Properties: Transient State

Basic Theoretical Principles, Problems and Limitations

As mentioned earlier in this chapter, the diffusion and solubility coefficients are the fundamental parameters that are needed to evaluate the flow of gases through plastic films. Generally, the determination of these coefficients and therefore of their product, the permeability coefficient, is used to predict the mass flux of low-molecular-weight compounds through plastic films in the steady state. In the case where the mass exchanged between the interior of the package and the outer environment during the transient state cannot be neglected, as in the case of rigid containers such as bottles, a different approach must be used do determine the permeant mass flux. Much scientific evidence (Barrer *et al.*, 1958; Michaels *et al.*, 1963; Vieth *et al.*, 1966; Paul, 1979) has shown that the use of a simple permeability coefficient is not adequate for judging the barrier provided by plastic materials in the transient state. This procedure, in fact, implicitly assumes that a steady-state treatment will suffice to predict the gas loss from a bottle, whereas it has been proven that the dynamics of the transport process must not be ignored. On the other hand, it has been demonstrated that the transport model used to describe the barrier properties of bottle materials has a great influence on the prediction of the shelf-life of packaged foods.

The transient state can be explained as follows. A film, when challenged by a pressure difference between the upstream and downstream sides, will go through a transient period during which its concentration profile changes with time. During this time, the concentration profile gradually adjusts itself until the steady state is achieved; this transient period is sometimes referred to as the conditioning of the film. In order to estimate the mass exchanged during the transient state, it is necessary to evaluate the concentration gradient at the interface between the packaged food and the polymer phase. In particular, this can be done by writing the mass balance differential equation for a permeant dissolved inside a bottle wall, assuming that the flux is only diffusive and is described by Fick's first law. The differential equation obtained, i.e., Fick's second law (Equation 42.39), can be solved to provide the evolution of the permeant concentration profile in the bottle wall during storage. The permeant concentration profile is then used to estimate the permeant mass flux at the interface between the packaged food and the bottle wall using Fick's first law. This approach can be used successfully to describe the unsteady state of a rigid package, and can then be an appropriate guide to the design of such a package.

Case Studies

Prediction of Shelf-Life of Soft Drinks as Affected by Thermal History

The carbon dioxide content is the quality index of many soft drinks. In fact, in most cases, a decrease in the carbon dioxide concentration by as little as 10% causes their taste to become flat and hence unacceptable to the consumer (Fenelon, 1973). The carbon dioxide content of these beverages decreases with time owing to gas permeation through the plastic bottle. The simple approach usually used in practical applications assumes that the partitioning of the gas between the gas phase and the carbonated beverage is described by Henry's law, and that the mass transfer through the plastic bottle wall is governed by the "dual-sorption–dual-mobility" model.

Del Nobile *et al.* (1997b) presented a study of the influence of the thermal history of the bottle during the period of time that elapses between capping and consumption. They considered an aspect that is often neglected in predicting the shelf-life of carbonated beverages bottled in glassy polymer containers, and showed the error that one can make by underestimating the importance of the thermal history of the bottle.

Test parameters:	
Reference temperature	5 °C
Initial headspace pressure	4 atm
Gas phase	CO_2
Liquid phase	Carbonated water
Container:	
Thickness	0.04 cm
Material	PET
Gas phase:	
surface area	55 cm^2
volume	119 cm^3
Liquid phase:	
surface area	931.7 cm^2
volume	2000 cm^3

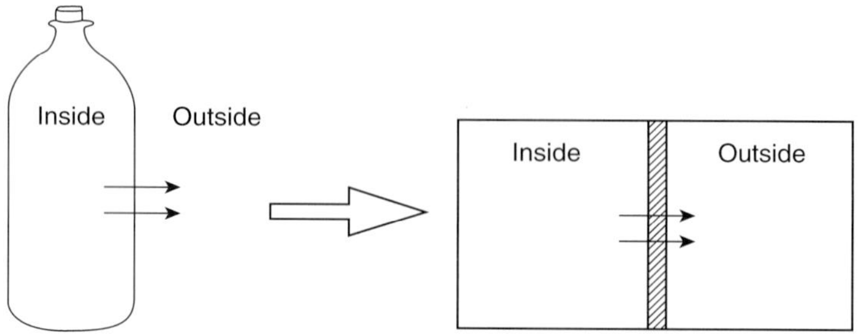

Figure 42.11 Representation of a plastic beverage container and equivalent cylinder. The parameters apply to a typical 2L soft drink bottle.

Figure 42.11 shows a schematic illustration of a typical 2L plastic soft-drink bottle, which is assumed to be made of polyethylene terephthalate (PET). To facilitate the mathematical analysis, the bottle containing the carbonated beverage is represented by a cylinder composed of an outer shell (made of plastic or glass) of constant thickness ℓ and an internal beverage core. Moreover, it is assumed that mass diffusion of the gas occurs only in the radial direction and through the lateral surface. Such a container represents a closed system where a liquid phase of volume V_1 is at all times in equilibrium with a gas phase of volume V_2. The equilibrium existing between the phases is described by Henry's law:

$$C_{CO_2} = Y \cdot p_{CO_2} \qquad (42.77)$$

where C_{CO_2} is the molar concentration of carbon dioxide in the liquid phase, p_{CO_2} is its partial pressure in the gas phase and Y is the partition coefficient. The proposed mechanism for carbon dioxide loss is also shown in Figure 42.11. The driving force for carbon dioxide permeation is assumed to be uniform over the container's internal surface and, at all times, equal to the partial pressure of carbon dioxide in the container headspace (the partial pressure of carbon dioxide in the external atmosphere is set equal to zero); it is a decreasing function of time. In addition, equilibrium between the phases is assumed to be established continuously over the entire period of storage, in accordance with Equation 42.77. The carbon dioxide losses are evaluated via solution of the differential equation that describes the carbon dioxide mass balance in the bottle wall, assuming a one-dimensional geometry together with the proper initial and boundary conditions. Accordingly, the carbon dioxide concentration profile $C_{CO_2}(t, x)$ is evaluated at each time, and from it the total flux of gas leaving the bottle is estimated. To compute the amount of gas that has traversed the surface at $x = 0$ (i.e., the interface between the beverage and the bottle wall) and time t requires the use of a mathematical model to represent the equilibrium between the gas and the polymer phase and the subsequent transport away from this surface. Many investigators have demonstrated that a more appropriate model to describe gas transport through glassy polymers is the dual-sorption–dual-mobility model (Vieth *et al.*, 1966; Koros, 1980; Chern *et al.*, 1983; Koros and Hellums, 1990). According to this model, the total gas concentration that is absorbed into the polymer is the result of two independent mechanisms. One mechanism follows Henry's law and corresponds to the expectation for liquids and rubbery polymers. The additional model has a Langmuir form and is attributed to sorption into fixed sites or frozen microvoids; it is analogous to the sorption isotherms for gases and vapors in zeolites.

According to the "dual-sorption–dual-mobility" model, the equilibrium relationship that describes the partitioning of the gas at the gas–polymer interface is

$$C_{\mathrm{TOT}} = C_{\mathrm{D}} + C_{\mathrm{H}} = k_{\mathrm{D}} \cdot p + \frac{C'_{\mathrm{H}} \cdot b \cdot p}{1 + b \cdot p} \tag{42.78}$$

where C_{TOT} is the total molar concentration of gas molecules inside the polymer; the subscripts H and D identify the two different populations of molecules, the former being adsorbed in holes, and the latter being dissolved in the polymer structure; k_{D} is the partition coefficient relative to the gas population dissolved according to Henry's law; b is a parameter that describes the affinity between the adsorbed molecules and the polymer; C'_{H} is the adsorption capacity of the polymer; and p is the partial pressure of the gas in contact with the polymer. The two populations of penetrant molecules have different mobilities and are always in local equilibrium. The modified form of Fick's first law that applies to this model for a plane sheet is

$$J_x = -D_{\mathrm{D}} \cdot \frac{\partial C_{\mathrm{D}}}{\partial x} - D_{\mathrm{H}} \cdot \frac{\partial C_{\mathrm{H}}}{\partial x} \tag{42.79}$$

where J_x is the total mass flux of the gas through the polymer in the x direction, and D_{D} and D_{H} are the diffusivities of the molecules that are dissolved and adsorbed, respectively. The assumption of local equilibrium between the two absorbed populations permits the writing of Equation 42.79 in a more useful way:

$$J_x = -D_{\mathrm{eff}} \cdot \frac{\partial C_{\mathrm{TOT}}}{\partial x} \tag{42.80}$$

where the "effective" Fickian diffusivity, D_{eff}, is a concentration-dependent coefficient given by

$$D_{\mathrm{eff}} = D_{\mathrm{D}} \cdot \left[\frac{1 + \dfrac{\left(\dfrac{D_{\mathrm{H}}}{D_{\mathrm{D}}}\right) \cdot \left(\dfrac{C'_{\mathrm{H}} \cdot b}{k_{\mathrm{D}}}\right)}{(1 + b \cdot p)^2}}{1 + \dfrac{\dfrac{C'_{\mathrm{H}} \cdot b}{k_{\mathrm{D}}}}{(1 + b \cdot p)^2}} \right] \tag{42.81}$$

The differential equation derived from the carbon dioxide mass balance in the bottle wall has the form

$$\frac{\partial C_{TOT}}{\partial t} = \frac{\partial}{\partial x} \cdot \left(D_{eff} \cdot \frac{\partial C_{TOT}}{\partial x} \right) \tag{42.82}$$

According to the proposed mechanism, the appropriate boundary conditions that apply to the case of a carbonated-beverage bottle are

$$t = 0, \quad 0 < x < \ell \rightarrow p_{CO_2} = 0$$

$$t \geq 0, \quad x = \ell \quad \rightarrow p_{CO_2} = 0$$

$$t \geq 0, x = 0 \quad \rightarrow p_{CO_2} = \frac{n_{CO_2}(t)}{\dfrac{V_2}{R \cdot T} + Y \cdot V_1} \tag{42.83}$$

where n_{CO_2} is the total number of moles of carbon dioxide inside the container. Generally speaking, the modeling procedure set up to predict the shelf-life of a carbonated beverage does not change substantially when Equations 42.78, 42.80 and 42.81 are used to describe gas sorption and transport inside the polymer wall when Henry's law and Fick's first law are used. However, there is a fundamental difference between the two approaches that has a practical implication. The dual-sorption–dual-mobility model implies that the gas permeability through the polymer wall is not constant, but is dependent on the upstream and downstream pressures. This is of great importance when one is dealing with shelf-life prediction for carbonated beverages bottled in plastic containers in which the internal pressure is much higher than atmospheric pressure and changes with time during storage.

Del Nobile *et al.* (1997b) solved Equation 42.82 using the parameters listed in Figure 42.11 and the boundary conditions set out above by means of a computer program based on an implicit finite difference discretization scheme. The total number of moles (moles of carbon dioxide in the headspace + moles of carbon dioxide dissolved in the beverage) inside the container, $n_{CO_2}(t)$, was found at any given time by subtracting from the initial number of moles the amount of carbon dioxide that had left the interior of the container. In simulations in which the temperature varied continuously during storage, the parameters needed at each step to describe gas sorption and transport into the polymer wall were updated by considering their dependence on the temperature. Figure 42.12 shows the relationships between the sorption and transport parameters and temperature.

In order to reproduce the thermal history experienced by a bottle during transportation and outdoor storage without any protection, Del Nobile *et al.* (1997b) assumed that the daily average temperature of the soft drink changes during the year according to a sinusoidal function having an average value equal to 17.5 °C, a fluctuation ampli-

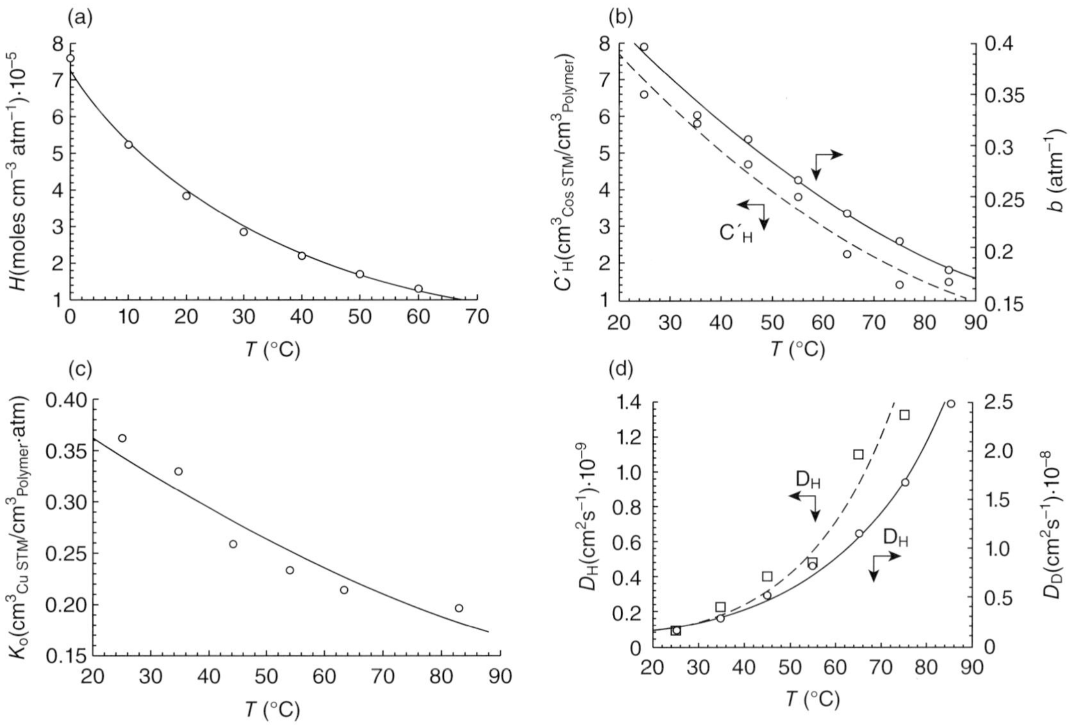

Figure 42.12 (a) Influence of temperature on the Henry's law constant for the CO_2–water system; (b) and (c) influence of temperature on the sorption parameters b, C'_H and k_D for the CO_2–PET system (after Koros and Paul, 1978); (d) influence of temperature on the transport parameters D_D and D_H for the CO_2–PET system (after Koros and Paul, 1978).

tude of ±12.5 °C and a period of 365 days. It was also assumed that the temperature varied during the day around the average temperature according to an asymmetric sine function with a period of 24 h and an amplitude of 25 °C. The asymmetry was introduced to account for thermal spikes due to exposure to sunlight. Figure 42.13 shows the thermal history of a soft-drink bottle over a limited time window of 10 days as described by the following equation:

$$T = 17.5 + 12.5 \cdot \sin\left(\frac{2\pi t}{31536000}\right) + 12.5 \cdot \sin\left(\frac{2\pi t}{86400}\right) + \left|7.5 \cdot \sin\left(\frac{2\pi t}{86400}\right)\right| \tag{42.84}$$

Two other cases were considered in order to highlight the error that would be made in the model by neglecting the temperature fluctuations. In the first case, the temperature of the carbonated beverage was set equal to 20 °C, which was the temperature of a climate chamber; in the other case, it was set equal to 28.9 °C, which is the temperature averaged over a period of 70 days according to Equation 42.84.

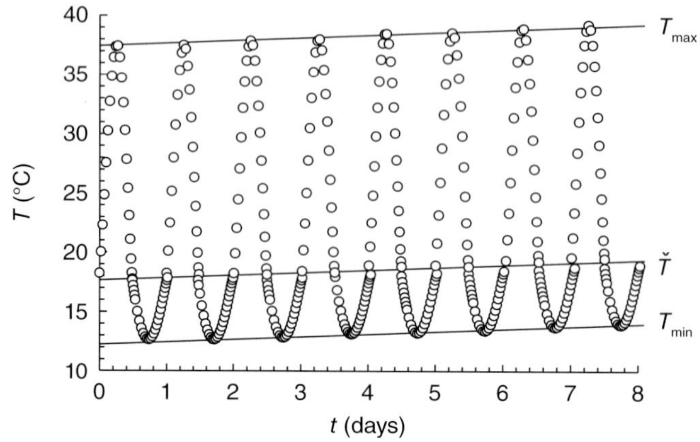

Figure 42.13 Thermal history of a bottle of soft drink as predicted by Equation 42.84. $\check{T}$ is the indoor daily average temperature.

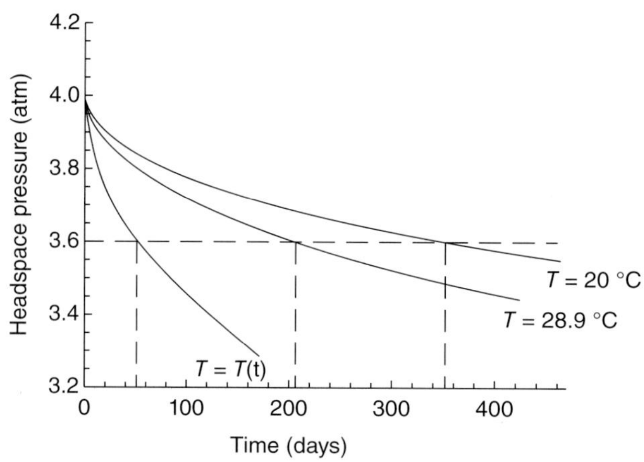

Figure 42.14 Effect of thermal history on the decay of the headspace pressure inside a soft-drink bottle.

Figure 42.14 shows the decay of the pressure inside a plastic bottle as predicted in the three cases considered. One can observe that there are remarkable difference in the shelf-life among the three situations, and this confirms the concern about the adequacy of some assumptions about the storage temperature that are generally made in designing carbonated-beverage containers. If the temperature of the soft drink is assumed to remain constant and equal to room temperature for the entire period of time between bottling and consumption, its shelf-life is longer than 1 year. Therefore, Del Nobile *et al.* (1997b) concluded that the PET bottles commonly used for this application provide adequate protection for the carbonated beverage over a period of time that is more than enough for normal distribution cycles. A similar conclusion was reached if, instead, it was assumed that the temperature is constant and equal to the average temperature of the distribution and storage period. The temperature used to perform the calculations in the second case was 9 °C higher than the temperature used in the first case, and the shelf-life of the carbonated beverage was reduced from 365 to 206 days.

Although the shelf-life reduction is quite significant, it may not be considered relevant from a practical point of view. In fact, a shelf-life of 7 months is still more than adequate for commercial purposes. However, both of these results are based on approximations that are false, and the predicted shelf-life is overestimated. In the first case, where the storage temperature was constant and equal to room temperature for the entire storage period, proper account was not taken of the fact that during storage and distribution, the temperature of soft drinks can often fluctuate in a significant manner. Secondly, the influence of temperature on the sorption and transport parameters is underestimated. In fact, by averaging the temperature and using the corresponding parameters in the calculations, it is implicitly assumed that the transport and sorption parameters change linearly with temperature, and this is far from true. When the thermal history of the bottle is considered and the proper relationships that describe the dependence of the physical parameters on the temperature are used, one finds that under conditions comparable to those occurring during distribution and outdoor storage, the shelf-life of the carbonated beverage is less than 2 months, which is remarkably less than that predicted in the other two cases. Moreover, the shelf-life that has been predicted in this case is comparable to the time that is usually required to distribute and sell soft drinks. Therefore, the results found by Del Nobile *et al.* (1997b) clearly demonstrate that there is a real risk that many bottles of soft drinks may reach the consumer at the very end of their shelf-life.

The approach proposed by Del Nobile *et al.* (1997b) can be used successfully when one is dealing with prediction of the shelf-life of carbonated beverages bottled in plastic containers in which the internal pressure is much higher than atmospheric pressure and changes with time during storage.

Prediction of Shelf-Life of Bottled Virgin Olive Oil

The shelf-life of a bottled vegetable oil is limited by the autoxidation of unsaturated fatty acids, with the formation of hydroperoxides. The decomposition of hydroperoxides gives rise to various compounds, some of which are volatile and responsible for the sensory degradation of the oil (Frankel, 1998). The work presented by Del Nobile *et al.* (2003b) gives evidence of the limits, in terms of performance, of the use of plastic containers, and aimed to evaluate the efficiency of innovative containers for prolonging the shelf-life of virgin olive oil by simulating the behavior of bottled oil using a mathematical model. In particular, the influence of both the oxygen diffusivity and the wall thickness of the plastic container on the kinetics of the quality decay of bottled virgin olive oil was addressed.

In the work presented by Del Nobile *et al.* (2003b), the model was derived by assuming the average hydroperoxide concentration to be a measure of the quality of virgin olive oil. The mathematical model was obtained by combining the mass balance equations for oxygen and hydroperoxides with an equation describing the rate of hydroperoxide formation and decomposition. To validate the model, Del Nobile *et al.* (2003b) monitored the average hydroperoxide concentration of virgin olive oil bottled in glass, PET and starch–PCL blend containers stored at 40 °C.

To this aim, it was assumed that (i) the bottle containing the oil could be represented by a cylinder composed of an outer shell (made of plastic or glass) and an internal core of oil, (ii) oxygen mass diffusion occurred only in the radial direction and through the lateral surface, and (iii) the diffusive mass fluxes of hydroperoxides through both the oil and the container wall were negligible. Under the above restrictions, the mass balance equation for the hydroperoxides dissolved in the bottled oil can be expressed as follows:

$$\frac{\partial C_{\mathrm{ROOH}}}{\partial t} = R_{\mathrm{F}} - R_{\mathrm{D}} \tag{42.85}$$

where C_{ROOH} is the local hydroperoxide concentration in the bottled oil, R_{F} is the rate at which hydroperoxides are formed and R_{D} is the rate at which hydroperoxides are decomposed.

Several models have been reported in the literature to describe the rate at which hydroperoxides are formed by the oxidation of unsaturated fatty acids (Quast *et al.*, 1972; Quast and Karel, 1972). The model proposed by Quast and Karel (1972) to describe lipid oxidation in potato chips was used by Del Nobile *et al.* (2003b). Assuming that the relative humidity is constant during storage, the model proposed by Quast and Karel (1972) is further simplified to the following relationship:

$$R_{\mathrm{F}} = (E_1 + E_2 \cdot C_{\mathrm{ROOH}}) \cdot \left(\frac{p_{\mathrm{O_2}}^{\mathrm{Oil}}}{E_3 + E_4 \cdot p_{\mathrm{O_2}}^{\mathrm{Oil}}} \right) \tag{42.86}$$

where the E_i's are constants and have to be regarded as fitting parameters, and $p_{\mathrm{O_2}}^{\mathrm{Oil}}$ is the oxygen partial pressure. Assuming that the solubilization process of oxygen into the oil is governed by Henry's law, $p_{\mathrm{O_2}}^{\mathrm{Oil}}$ is related to the oxygen concentration in the oil by the following relationship:

$$p_{\mathrm{O_2}}^{\mathrm{Oil}} = \frac{C_{\mathrm{O_2}}^{\mathrm{Oil}}}{S_{\mathrm{O_2}}^{\mathrm{Oil}}}$$

where $C_{\mathrm{O_2}}^{\mathrm{Oil}}$ is the oxygen concentration and $S_{\mathrm{O_2}}^{\mathrm{Oil}}$ is the solubility of oxygen in virgin olive oil. For the sake of simplicity, Del Nobile *et al.* (2003b) assumed that at a given temperature, the overall rate at which hydroperoxides decompose depends only on their concentration according to the following equation:

$$R_D = E_5 \cdot C_{ROOH} \tag{42.87}$$

By substituting Equations 42.86 and 42.87 into Equation 42.85, the following equation is obtained:

$$\frac{\partial C_{ROOH}}{\partial t} = (E_1 + E_2 \cdot C_{ROOH}) \cdot \left(\frac{p_{O_2}^{Oil}}{E_3 + E_4 \cdot p_{O_2}^{Oil}} \right) - E_5 \cdot C_{ROOH} \tag{42.88}$$

The mass balance for the oxygen dissolved in the bottled oil can be expressed as follows:

$$\frac{\partial C_{O_2}^{Oil}}{\partial t} = \zeta \cdot \left[\frac{D_{O_2}^{Oil}}{r} \cdot \frac{\partial}{\partial r} \cdot \left(r \cdot \frac{\partial C_{O_2}^{Oil}}{\partial r} \right) \right] - (E_1 + E_2 \cdot C_{ROOH}) \cdot \left(\frac{p_{O_2}^{Oil}}{E_3 + E_4 \cdot p_{O_2}^{Oil}} \right) \tag{42.89}$$

where r is the spatial coordinate, $D_{O_2}^{Oil}$ is the oxygen diffusivity in the oil, and ζ is a constant equal to 1 for plastic containers and to 0 for glass containers. The term on the right-hand side of Equation 42.89 enclosed in the square brackets is related to the diffusive mass flux of oxygen, and was obtained by assuming that the diffusion and solubilization processes of oxygen are governed by Fick's first law and Henry's law, respectively. In the case of gas-permeable containers, such as plastic containers, to evaluate the amount of oxygen permeating through the container wall, it is necessary to write the mass balance equation for the oxygen dissolved in the container wall, which, in the case under investigation, can be expressed as follows:

$$\frac{\partial C_{O_2}^{Polym}}{\partial t} = \frac{1}{r} \cdot \frac{\partial}{\partial r} \cdot \left(r \cdot D_{O_2}^{Polym} \cdot \frac{\partial C_{O_2}^{Polym}}{\partial r} \right) \tag{42.90}$$

where $C_{O_2}^{Polym}$ is the concentration of oxygen dissolved in the container wall and $D_{O_2}^{Polym}$ is the diffusivity of oxygen through the container wall. According to Equation 42.81, $D_{O_2}^{Polym}$ can be rewritten as follows (Paul and Koros, 1976):

$$D_{O_2}^{Polym} = D_D \cdot \left[\frac{1 + \dfrac{F \cdot X}{\left(1 + b \cdot p_{O_2}^{Polym}\right)^2}}{1 + \dfrac{X}{\left(1 + b \cdot p_{O_2}^{Polym}\right)^2}} \right] \tag{42.91}$$

where $p_{O_2}^{Polym}$ is the oxygen partial pressure in the container wall in equilibrium with the dissolved oxygen. F and X are defined as follows:

$$F = \frac{D_D}{D_H}$$

and

$$X = \frac{C'_{\mathrm{H}} \cdot b}{k_{\mathrm{D}}}$$

At low oxygen partial pressures, as in the case under investigation, Equation 42.91 becomes

$$D_{\mathrm{O_2}}^{\mathrm{Polym}} = D_{\mathrm{D}} \cdot \left[\frac{1 + F \cdot X}{1 + X} \right] \tag{42.92}$$

In this case $D_{\mathrm{O_2}}^{\mathrm{Polym}}$ is constant, and hence Equation 42.90 can be rewritten as

$$\frac{\partial C_{\mathrm{O_2}}^{\mathrm{Polym}}}{\partial t} = \frac{D_{\mathrm{O_2}}^{\mathrm{Polym}}}{r} \cdot \frac{\partial}{\partial r} \cdot \left(r \cdot \frac{\partial C_{\mathrm{O_2}}^{\mathrm{Polym}}}{\partial r} \right) \tag{42.93}$$

Interfacial conditions were imposed to ensure chemical and physical equilibrium at the interface between the oil and the plastic. Hence, both the mass flow and the oxygen partial pressure of the juxtaposed substances were required to be equal at the interface, that is,

$$\left(J_{\mathrm{O_2}}^{\mathrm{Polym}} \right)_{\mathrm{int}} = \left(J_{\mathrm{O_2}}^{\mathrm{Oil}} \right)_{\mathrm{int}} \tag{42.94}$$

$$\left(p_{\mathrm{O_2}}^{\mathrm{Polym}} \right)_{\mathrm{int}} = \left(p_{\mathrm{O_2}}^{\mathrm{Oil}} \right)_{\mathrm{int}} \tag{42.95}$$

where $\left(J_{\mathrm{O_2}}^{\mathrm{Polym}} \right)_{\mathrm{int}}$ is the oxygen mass flux at the interface in the container wall, $\left(J_{\mathrm{O_2}}^{\mathrm{Oil}} \right)_{\mathrm{int}}$ is the oxygen mass flux at the interface in the oil, $\left(p_{\mathrm{O_2}}^{\mathrm{Polym}} \right)_{\mathrm{int}}$ is the oxygen partial pressure at the interface in the plastic and $\left(p_{\mathrm{O_2}}^{\mathrm{Oil}} \right)_{\mathrm{int}}$ is the oxygen partial pressure at the interface in the oil. Equations 42.88 and 42.89 and, only for gas-permeable containers, 42.93, 42.94 and 42.95 form a set of differential equations, which, using the proper initial and boundary conditions, were solved simultaneously by numerical methods to predict the evolution of oxygen and hydroperoxides inside the bottled oil during storage. The average hydroperoxide concentration (denoted by $C_{\mathrm{ROOH}}^{\mathrm{AV}}$) was obtained by averaging C_{ROOH} over the volume of the bottled olive oil.

Del Nobile *et al.* (2003b) used three different types of cylindrically shaped bottles to run the tests: commercially available glass bottles, commercially available PET bottles and prototype bottles obtained by using an experimental material based on a starch–PCL polymeric blend. Figure 42.15 shows $C_{\mathrm{ROOH}}^{\mathrm{AV}}$ plotted as a function of storage time for the three samples. The differences among the behavior of the samples examined can be ascribed to the different amounts of oxygen available for the autoxidation reactions of the unsaturated fatty acids. PET is characterized by a low oxygen diffusivity ($8.8 \times 10^{-9}\,\mathrm{cm^2\,s^{-1}}$). Therefore, the amount of oxygen permeating through the bottle

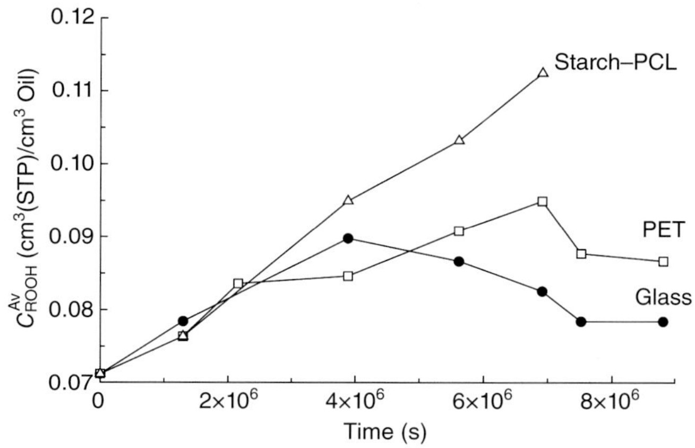

Figure 42.15 Average hydroperoxide concentration during the storage period.

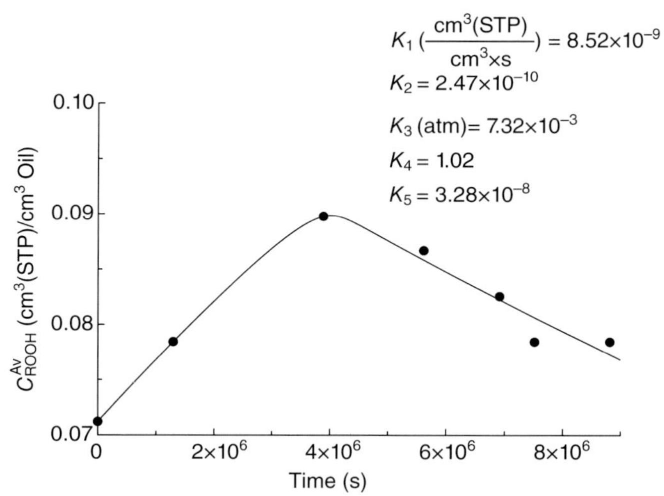

Figure 42.16 Average hydroperoxide concentration of oil packaged in a glass bottle. The curve is the best fit to the experimental data. The results of fitting parameters are also shown.

wall is relatively small, and the behavior of the oil in PET is close to that of the sample in glass. By contrast, the starch–PCL blend is characterized by a relatively high oxygen diffusivity ($1.26 \times 10^{-7}\,cm^2\,s^{-1}$). As a consequence, this sample shows a markedly different behavior from the other two samples investigated.

The constants E_1, E_2, E_3, E_4 and E_5 were obtained by fitting Equations 42.81 and 42.82 to the data for the sample packaged in the glass bottle. The results of this fitting, along with the values obtained for the fitting parameters, are shown in Figure 42.16.

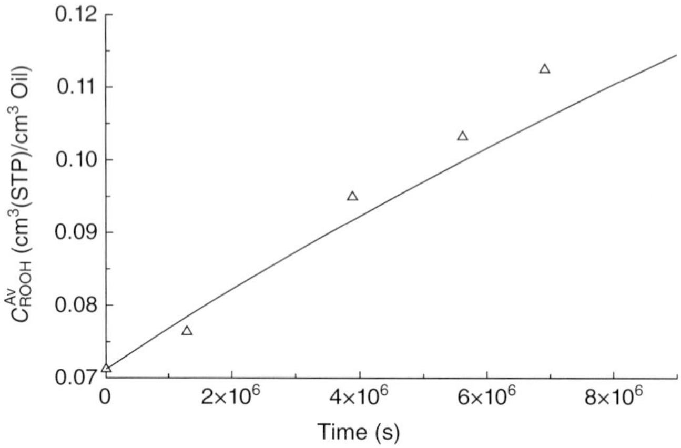

Figure 42.17 Average hydroperoxide concentration of oil stored in a starch–PCL polymeric blend bottle; (—) model prediction.

To assess the predictive ability of the model developed, the fitting parameters were also used to predict the evolution of the average hydroperoxide concentrations in samples packaged in PET and in the starch–PCL polymeric blend. As an example, Figure 42.17 shows the average hydroperoxide concentration plotted as a function of storage time as predicted by the model, along with the experimental data for oil in the starch–PCL blend; similar results were also recorded with the other sample. Despite the restrictions imposed to derive the model, its predictive ability appears to be quite satisfactory. The major limitations on extending the proposed model to the general case are (i) the use of the empirical Equations 42.86 and 42.87 to describe the hydroperoxide formation and breakdown reactions, and (ii) that the oxygen mass flux in the axial direction of the bottle is considered negligible. Whenever these assumptions are satisfied, the model can be used satisfactorily to fit and/or predict the quality decay kinetics of bottled olive oil. The model developed by Del Nobile *et al.* (2003b) could be used in many practical applications to manufacture well-designed plastic bottles in order to control the oxidation kinetics during the storage of bottled oil.

References

Appendini, P. and Hotchkiss, J.H. (2002) Review of antimicrobial food packaging. *Innovative Food Science and Emerging Technologies* 3: 113–126.

Azanha, A.B. and Faria, J.A.F. (2005) Use of mathematical models for estimating the shelf life of cornflakes in flexible packaging. *Packaging Technology and Science* 18: 171–178.

Barrer, R.M., Barrie, J.A. and Slater, J. (1958) Sorption and diffusion of ethyl cellulose. Part III. Comparison between ethyl cellulose and rubber. *Journal of Polymer Science* 27: 177–197.

Bell, L.N. and Labuza, T.P. (2000) *Moisture Sorption: Practical Aspects of Sorption Isotherm Measurement and Use*, 2nd edn. American Association of Cereal Chemists, St. Paul, MN.

Boquet, R., Chirifie, J. and Iglesias, H.A. (1978). Equations for fitting water sorption isotherms of foods. II. Evaluation of various two-parameters models. *Journal of Food Technology* 13: 319–327.

Böttcher, H., Günther, I. and Kabelitz, L. (2003) Physiological postharvest responses of Common Saint-John's wort herbs (*Hypericum perforatum* L.). *Postharvest Biology and Technology* 29: 342–350.

Brown, W.E. (1992) *Plastics in Food Packaging: Properties, Design, and Fabrication.* Marcel Dekker, New York.

Buonocore, G.G., Conte, A., Nicolais, L. and Del Nobile, M.A. (2005) Modelling the water transport properties of traditional and innovative polymeric films for food packaging applications. *Research Advances in Food Science* 5:35–52.

Chao, R.R. and Rizvi, S.S.H. (1988) Oxygen and water vapor transport through polymeric film – a review of modeling approaches. In: *Food and Packaging Interactions* (ed. J.H. Hotchkiss), ACS Symposium Series 365. American Chemical Society, Washington, DC, p. 217.

Chern, R.T., Koros, W.J., Sanders, E.S. and Yui, R. (1983) Second component. Effects in sorption and permeation of gases in glassy polymers. *Journal of Membrane Science* 15: 157–169.

Cleland, A.C. (1996) Package design for refrigerated food: the need for multidisciplinary project teams. *Trends in Food Science and Technology* 7: 269–271.

Comyn, J. (1985) Introduction to polymer permeability and the mathematics of diffusion. In: *Polymer Permeability* (ed. J. Comyn). Elsevier, London, pp. 1–10.

Conte, A., Scrocco, C., Brescia, I. and Del Nobile, M.A. (2009a) Different packaging strategies for fresh-cut "barattiere" melon cultivar (*Cucumis melo* L.). *International Journal of Food Science and Technology* 44: 1422–1428.

Conte, A., Scrocco, C., Brescia, I. and Del Nobile, M.A. (2009b) Packaging strategies to prolong the shelf life of minimally processed lampascioni (*Muscari comosum*). *Journal of Food Engineering* 90: 199–206.

Crank, J. (1975) *The Mathematics of Diffusion*, 2nd edn. Clarendon Press, Oxford.

Cussler, E.L. (1984) *Diffusion: Mass Transfer in Fluid Systems*. Cambridge University Press.

Cutter, C.N. (2002) Microbial control by packaging: a review. *Critical Reviews in Food Science and Nutrition* 42: 151–161.

da Cruz, A.G., Faria, J.A.F. and Van Dender, A.G.F. (2007) Packaging systems and probiotic dairy foods. *Food Research International* 40: 951–956.

Del Nobile, M.A. (2001) Packaging design for potato chips. *Journal of Food Engineering* 47: 211–215.

Del Nobile, M.A., Mensitieri, G., Netti, P.A. and Nicolais, L. (1994) Anomalous diffusion in poly-ether-ether-ketone (PEEK). *Chemical Engineering Science* 49(5): 633–644.

Del Nobile, M.A., Mensitieri, G., Manfredi, C., Arpaia, A. and Nicolais, L. (1996a) Low molecular weight molecules diffusion in advanced polymers for food packaging applications. *Polymers for Advanced Technologies* 7: 409–417.

Del Nobile, M.A., Mensitieri, G. and Nicolais, L. (1996b) Effect of chemical composition on gas transport properties of ethylene based ionomers. *Polymer International* 41: 73–78.

Del Nobile, M.A., Mensitieri, G., Ho, L.H., Huang, S.J. and Nicolais, L. (1997a) Moisture transport properties of a degradable nylon for food packaging. *Packaging Technology and Science* 10: 311–330.

Del Nobile, M.A., Mensitieri, G., Nicolais, L. and Masi, P. (1997b) The influence of the thermal history on the shelf life of carbonated beverages bottled in plastic containers. *Journal of Food Engineering* 34: 1–13.

Del Nobile, M.A., Buonocore, G.G., Limbo, S. and Fava, P. (2003a) Shelf life prediction of cereal-based dry foods packed in moisture-sensitive films. *Journal of Food Engineering* 68: 1292–1300.

Del Nobile, M.A., Ambrosino, M.L., Sacchi, R. and Masi, P. (2003b) Design of plastic bottles for packaging of virgin olive oil. *Journal of Food Science* 68(1): 170–175.

Del Nobile, M.A., Bove, S., La Notte, E. and Sacchi, R. (2003c) Influence of packaging geometry and material properties on the oxidation kinetic of bottles virgin olive oil. *Journal of Food Engineering* 57: 189–197.

Del Nobile, M.A., Buonocore, G.G., Dainelli, D. and Nicolais, L. (2004) A new approach to predict the water transport properties of multi-layer films intended for food packaging applications. *Journal of Food Science* 69(3): 85–90.

Del Nobile, M.A., Baiano, A., Benedetto, A. and Massignan, L. (2006) Respiration rate of minimally processed lettuce as affected by packaging. *Journal of Food Engineering* 74: 60–69.

Del Nobile, M.A., Licciardello, F., Scrocco, C., Muratore, G. and Zappa, M. (2007) Design of plastic packages for minimally processed fruits. *Journal of Food Engineering* 79: 217–224.

Del Nobile, M.A., Sinigaglia, M., Conte, A., Speranza, B., Scrocco, C., Brescia, I., Bevilacqua, A., Laverse, J., La Notte, E. and Antonacci, D. (2008) Influence of postharvest treatments and film permeability on quality decay kinetics of minimally processed grapes. *Postharvest Biology and Technology* 47: 389–396.

Del Nobile, M.A., Conte, A., Scrocco, C. and Brescia, I. (2009a) New strategies for minimally processed cactus pear packaging. *Innovative Food Science and Emerging Technologies* 10: 356–362.

Del Nobile, M.A., Conte, A., Scrocco, C., Laverse, J., Brescia, I., Conversa, G. and Elia, A. (2009b) New packaging strategies to preserve fresh-cut artichoke during refrigerated storage. *Innovative Food Science and Emerging Technologies* 10: 128–133.

DeLassus, P.T. and Hilker, B.L. (1987) Interaction of high barrier plastic with food: permeation and sorption. In: *Food Product–Package Compatibility* (eds J.I. Gray, B.R. Harte and J. Miltz). Technomic, Lancaster, pp. 229–244.

Fenelon, P.J. (1973) Prediction of pressure loss in pressurized plastic containers. *Polymer Engineering and Science* 13: 440–451.

Fishman, S., Rodov, V. and Ben-Yehoshua, S. (1996). Mathematical model for perforation effect on oxygen and water vapor dynamics in modified-atmosphere packages. *Journal of Food Science* 61(5): 957–961.

Flory, P.J. (1953) *Principles of Polymer Chemistry*. Cornell University Press, Ithaca, NY.

Fonseca, S.C., Oliveira, F.A.R., Lino, I.B.M., Brecht, J.K. and Chau, K.V. (2000) Modeling O_2 and CO_2 exchange for development of perforation-mediated modified atmosphere packaging. *Journal of Food Engineering* 43(1): 9–15.

Fonseca, S.C., Oliveira, F.A.R. and Brecht, J.K. (2002) Modelling respiration rate of fresh fruits and vegetables for modified atmosphere packages: a review. *Journal of Food Engineering* 52: 99–119.

Frankel, E.N. (1998) *Lipid Oxidation*. The Oily Press, Dundee.

Fujita, H. (1961) Diffusion in polymer–diluent systems. *Fortschritte Hochpolymer-Forschung* 3: 1–47.

Gong, S. and Corey, K.A. (1994) Prediction of steady-state oxygen concentration in modified-atmosphere packages of tomatoes. *Journal of American Society for Horticultural Science* 119(3): 546–550.

Grorski-Berry, D.M. (1999) Wrapping it all up the value of packaging. *Journal of Dairy Science* 82: 2257–2258.

Haggar, P.E., Lee, D.S. and Yam, K.L. (1992) Application of an enzyme kinetics based respiration model to closed system experiments for fresh produce. *Journal of Food Process Engineering* 15: 143–157.

Heiss, R. (1980) *Verpackung von Lebensmitteln*. Springer, Berlin.

Hernandez, R.J. (1994) Effect of water vapor on the transport properties of oxygen through polyamides packaging materials. *Journal of Food Engineering* 22: 495–507.

Iglesias, H.A., Viollaz, P. and Chirife, J. (1979) Technical note: a technique for predicting moisture transfer in mixtures of packaged dehydrated foods. *Journal of Food Science* 14: 89.

Jacxsens, L., Devlieghere, F. and Debevere, J. (1999) Validation of a systematic approach to design equilibrium modified atmosphere packages for fresh-cut produce. *Lebensmittel-Wissenschaft & Technologie* 32: 425–432.

Jacxsens, L., Devlieghere, F., De Rudder, T. and Debevere, J. (2000) Designing equilibrium modified atmosphere packages for fresh-cut vegetables subjected to changes in temperature. *Journal of Food Science* 33: 178–187.

Jacxsens, L., Devlieghere, F. and Debevere, J. (2002) Predicting modeling for packaging design: equilibrium modified atmosphere package of fresh-cut vegetables subjected to a simulated distribution chain. *International Journal of Food Microbiology* 73: 331–341.

Karbowiak, T., Debeaufort, F., Voilley, A. and Trystram, G. (2009) From macroscopic to molecular scale investigations of mass transfer of small molecules through edible packaging applied at interfaces of multiphase food products. *Innovative Food Science and Emerging Technologies* 10: 116–127.

Kerry, J.P., O'Grady, M.N. and Hogan, S.A. (2006) Past, current and potential utilization of active and intelligent packaging systems for meat and muscle-based products: a review. *Meat Science* 74: 113–130.

Kesting, R.E. and Fritzsche, A.K. (1993) *Polymeric Gas Separation Membranes*. John Wiley & Sons, New York.

Kilkast, D. and Subramaniam, P. (2000) The stability and shelf life of foods. In: *Biobased Packaging Materials for the Food Industry* (ed. C. Weber). CRC, Boca Raton, FL, pp. 1–18.

Koros, W.J. (1980) Model for sorption of mixed gases in glassy polymers. *Journal of Polymer Science: Polymer Physics Edition* 18: 981–992.

Koros, W.J. and Hellums, M.W. (1990) *Encyclopedia of Polymer Science and Engineering, Supplement Volume*. John Wiley & Sons, New York, pp. 724–802.

Koros, W.J. and Paul, D.R. (1978) Sorption in poly(ethylene terephthalate) above and below the glass transition. *Journal of Polymer Science* 16: 1947–1963.

Labuza, T.P. and Contreras-Medellin, R. (1981) Prediction of moisture protection requirements for foods. *Cereal Foods World* 26: 335–343.

Lee, L., Arult, J., Lencki, R. and Castaigne, F. (1996) A review on modified atmosphere packaging and preservation of fresh fruit and vegetables: physiological basis and practical aspects – Part II. *Packaging Technology and Science* 9: 1–17.

Lee, D.S., Yam, K.L. and Piergiovanni, L. (2008) *Food Packaging Science and Technology*. CRC, Boca Raton, FL.

Long, F.A. and Richman, D. (1960) Concentration gradients for diffusion of vapors in glassy polymers and their relation to time dependent diffusion phenomena. *Journal of the American Chemical Society* 82: 513–519.

Lopez-Rubio, A., Almenar, E., Hernandez-Munoz, P., Lagaron, J.M., Catala, R. and Gavara, R. (2004) Overview of active polymer-based packaging technologies for food applications. *Food Reviews International* 20: 357–387.

Mahajan, P.V., Oliveira, F.A.R., Montanez, J.C. and Frias, J. (2007) Development of user-friendly software for design of modified atmosphere packaging for fresh and fresh-cut produce. *Innovative Food Science and Emerging Technologies* 8: 84–92.

Martin-Diana, A.B., Rico, D., Frias, J.M., Barat, J.M., Henehan, G.T.M. and Barry-Ryan, C. (2007) Calcium for extending the shelf life of fresh whole and minimally processed fruits and vegetables: a review. *Trends in Food Science & Technology* 18: 210–218.

Meidner, H. and Mansfield, T.A. (1968) *Physiology of Stomata*. McGraw-Hill, New York.

Michaels, A.S. and Bixler, H.J. (1961) Solubility of gases in polyethylene. *Journal of Polymer Science* 154(50): 393–412.

Michaels, A.S. and Parker, R.B., Jr. (1959) Sorption and flow of gases in polyethylene. *Journal of Polymer Science* 138(41): 53–71.

Michaels, A.S., Vieth, W.R. and Barrie, J.A. (1963) Solution of gases in polyethylene terephtalate. *Journal of Polymer Science* 34: 1–12.

Miller, K.S. and Krotcha, J.M. (1997) Oxygen and aroma barrier properties of edible films: a review. *Trends in Food Science & Technology* 8: 228–237.

Motamedian, S., Pusch, W., Tanioka, A. and Becker, F. (1998) Sorption isotherms of gases by polymer membranes in the glassy state: an explanation based on the non-equilibrium thermodynamics. *Journal of Colloid and Interface Science* 204: 135–142.

Myers, A.W., Meyer, J.A., Rogers, C.E., Stannett, V. and Szwarc, M. (1961) *Tappi* 44(1): 58–64.

Netti, P.A., Del Nobile, M.A., Mensitieri, G., Ambrosio, L. and Nicolais, L. (1996) Water sorption kinetics in hyaluronic acid esters. *Bioactive and Compatible Polymers* 11(4): 312–327.

Nobel, P.S. (1974) *Introduction to Biophysical Plant Physiology*. W.H. Freeman, San Francisco.

Olivas, G.I., Mattinson, D.S. and Barbosa-Canovas, G.V. (2007) Alginate coatings for preservation of minimally processed "Gala" apples. *Postharvest Biology & Technology* 45: 89–96.

Paine, F. and Paine, Y. (1983) *A Handbook of Food Packaging*. Blackie, London.

Patterson, B.D., Jobling, J.J. and Moradi, S. (1993) Water relations after harvest – new technology helps translate theory into practice. In: *Proceedings of the Australasian Postharvest Conference*, 1993, University of Queensland, Getton, Queensland, pp. 99–102.

Paul, D.R. (1979) Gas sorption and transport in glassy polymers. *Berichte der Bunsengesellschaft: Physical Chemistry* 83: 294–302.

Paul, D.R. and Koros, W.J. (1976) Effect of partially immobilizing sorption on permeability and diffusion time lag. *Journal of Polymer Science* 14: 675–685.

Petersen, K., Nielsen, P.V., Bertelsen, G., Lawther, M., Olsen, M.B., Nilsson, N.H. and Mortensen, G. (1999) Potential of bio-based materials for food packaging. *Trends in Food Science & Technology* 10: 52–68.

Press, W.H., Flannery, B.P., Teukolsky, S.A. and Vetterling, W.T. (1989) *Numerical Recipes in Pascal*. Cambridge University Press, Cambridge.

Quast, D.G. and Karel, M. (1972) Computer simulation of storage life of foods undergoing spoilage by two interacting mechanisms. *Journal of Food Science* 37: 679–683.

Quast, D.G., Karel, M. and Rand, M. (1972) Development of a mathematical model for oxidation of potato chips as a function of oxygen pressure, extent of oxidation and equilibrium relative humidity. *Journal of Food Science* 37: 673–678.

Quintavalla, S. and Vicini, L. (2002) Antimicrobial food packaging in meat industry. *Meat Science* 62: 373–380.

Ragaert, P., Devlieghere, F. and Debevere, J. (2007) Role of microbiological and physiological spoilage mechanisms during storage of minimally processed vegetables. *Postharvest Biology & Technology* 44: 185–194.

Rico, D., Martín-Diana, A.B., Barat, J.M. and Barry-Ryan, C. (2007) Extending and measuring the quality of fresh-cut fruit and vegetables: a review. *Trends in Food Science & Technology* 18: 373–386.

Roca, E., Guillard, V., Guilbert, S. and Gontard, N. (2006) Moisture migration in a cereal composite food at high water activity: effects of initial porosity and fat content. *Journal of Cereal Science* 43: 144–151.

Rocculi, P., Del Nobile, M.A., Romani, S., Baiano, A. and Dalla Rosa, M. (2006) Use of a simple mathematical model to evaluate dipping and MAP effects on aerobic respiration of minimally processed apples. *Journal of Food Engineering* 76: 334–340.

Rodriguez-Aguilera, R. and Oliveira, J.C. (2009) Review of design engineering methods and applications of active and modified atmosphere systems. *Food Engineering Review* 1: 66–83.

Rogers, C.E., Fels, M. and Li, N.N. (1972) Separation by permeation through polymeric membranes. In: *Recent Developments in Separation Science* (ed. N.N. Li), Vol. 2. CRC, Cleveland, OH, pp. 107–155.

Saltveit, M.E. (2003) Is it possible to find an optimal controlled atmosphere? *Postharvest Biology and Technology* 27: 3–13.

Sanches Silva, A., López Hernández, J. and Paseiro Losada, P. (2004) Modified atmosphere packaging and temperature effect on potato crisps oxidation during storage. *Analytica Chimica Acta* 524: 185–189.

Siracusa, V., Rocculi, P., Romani, S. and Dalla Rosa, M. (2008) Biodegradable polymers for food packaging: a review. *Trends in Food Science & Technology* 19: 634–643.

Siripatrawan, U. (2009) Shelf life simulation of packaged rice crackers. *Journal of Food Quality* 32: 224–239.

Soliva-Fortuny, R.C. and Martìn-Belloso, O. (2003) New advances in extending the shelf-life of fresh-cut fruit: a review. *Trends in Food Science and Technology* 14: 341–353.

Suppakul, P., Miltz, J., Sonneveld, K. and Bigger, S.W. (2003) Active packaging technologies with an emphasis on antimicrobial packaging and its applications. *Journal of Food Science* 68: 408–420.

Tubert, A.H. and Iglesias, H.A. (1985) Water sorption isotherms and prediction of moisture gain during storage of packaged cereal crackers. *Lebensmittel-Wissenschaft & Technologie* 19: 365–368.

Vieth, W.R., Tam, P.M. and Michaels, A.S. (1966) Dual sorption mechanism in glassy polystyrene. *Journal of Colloid and Interface Science* 22: 360–370.

Williams, M.L., Landel, R.F. and Ferry, D.J. (1955) The temperature dependence of relaxation mechanisms in amorphous polymers and other glass-forming liquids. *Journal of the American Chemical Society* 77: 3701–3707.
Yam, K.L., Takhistov, P.T. and Miltz, J. (2005) Intelligent packaging: concepts and applications. Review. *Journal of Food Science* 70: 1–10.

43

Design of Modified and Controlled Atmospheres

Gurbuz Gunes and Celale Kirkin

Introduction

Packaging is an unavoidable unit operation in food manufacturing. The function of packaging includes containment, protection, communication and convenience (Robertson, 1993). Food products deteriorate by chemical, physical and microbiological means. These reactions in food are controlled by various processing/preservation methods. Products are usually packaged and stored under particular conditions after processing.

Alteration of the gaseous environment around food products can decrease the rate of quality degradation, which is affected by the gaseous environment. Special gas mixtures can be applied to foods by using modified-atmosphere packaging (MAP) or controlled-atmosphere (CA) storage for preservation of quality. MAP is the enclosure of a food product in a package with a composition of the headspace gas different from that of air. Although vacuum packaging also fits into this definition, it is usually designated separately as "vacuum packaging," whereas any packaging with a special gas mixture is usually designated as MAP. The most common gases used in MAP are CO_2, O_2 and N_2. There are also gases such as argon (Ar), nitrous oxide (N_2O), helium (He), xenon (Xe) and carbon monoxide (CO) that have been the subject of interest for MAP.

Handbook of Food Process Design, First Edition. Edited by Jasim Ahmed and Mohammad Shafiur Rahman.

Controlled-atmosphere storage involves storage of food products in bulk form in a storage room whose gaseous environment is constantly controlled and set at specific levels during the storage period. These storage facilities have a special construction to prevent gas leakage in and out through the walls and doors. CA storage is widely used in the storage of fresh horticultural commodities.

Gases Used in Modified Atmospheres

There are three main gases, namely CO_2, O_2 and N_2, that are used in MAP. In addition, other gases such as Ar, N_2O, He, Xe and CO have been tested in MAP applications. The use and function of these gases will be briefly described in the following sections.

Carbon Dioxide

Carbon dioxide is used in modified atmospheres to inhibit bacterial and fungal growth in packaged food products (Farber, 1991). It also inhibits respiration and other metabolic activities in fresh produce (Kader *et al.*, 1998). The concentration used depends on the food product, as different products tolerate different levels of CO_2. Carbon dioxide is soluble in both aqueous and lipid phases, and its solubility is inversely proportional to temperature (Church and Parsons, 1995). For this reason, its antimicrobial and antimetabolic activities are higher at low temperatures. Dissolved CO_2 forms carbonic acid in tissues, resulting in a pH reduction, which is thought to be responsible for its effects (Church and Parsons, 1995). Carbon dioxide extends both the lag phase and the logarithmic-growth phase of microorganisms (Farber, 1991). The amount of CO_2 used in packages is sometimes limited by its high solubility in the food, as this can cause package collapse due to volume reduction in the headspace.

Oxygen

Oxygen has both desirable and undesirable effects on the quality of packaged food, depending on its concentration and the product. Oxygen induces oxidative reactions, resulting in degradation of flavor and color. It also stimulates the growth of aerobic microorganisms, causing spoilage and safety problems. On the other hand, O_2 is required at low levels to maintain aerobic respiration of fresh produce. It is also required at low levels to inhibit the growth of anaerobic pathogens such as *C. botulinum* in the product. MAP generally involves an O_2- free atmosphere or a low O_2 level for most food products. The use of high-O_2 MAP has been shown to have beneficial effects on the quality of some food products (Day, 2003). Super atmospheric O_2 levels can inhibit microbial growth and some oxidative reactions, such as enzymatic browning through substrate inhibition (Artés *et al.*, 2009; Day, 2003).

Nitrogen

Nitrogen is chemically inert and is generally used as a filler gas in MAP (Church and Parsons, 1995). 100% N_2 is used as an alternative to vacuum packaging for some products. N_2 is also used to prevent package collapse in high-CO_2 MAP owing to its low solubility in food products.

Other Gases

Novel MAP gases such as Ar, N_2O, He, Xe and CO have also been tested in MAP applications. These gases have been shown to have inhibitory effects on microbial growth, oxidative reactions and color changes in some products. For instance, the use of Ar in MAP was effective for quality retention and shelf-life extension of trout fillets (Choubert *et al.*, 2008). Argon also inhibited enzyme activity and respiration in fresh produce, and stimulated accumulation of phenolics (Day, 2003; Jamie and Saltveit, 2002). The use of nitrous oxide in MAP decreased browning, ripening and softening of spinach and fresh-cut kiwi fruit (Rodríguez-Hidalgo *et al.*, 2010; Rocculi *et al.*, 2005). The use of Xe in MAP improved retention of the quality of asparagus spears (Zhang *et al.*, 2008). The use of helium in MAP reduced chlorophyll loss in broccoli florets and decreased chilling injury in sensitive plants (Jamie and Saltveit, 2002). Carbon monoxide is used in MAP to stabilize the bright red color of meat and to inhibit oxidation caused by high-O_2 MAP or irradiation treatment (Seyfert *et al.*, 2007; Stetzer *et al.*, 2007; John *et al.*, 2005; Kusmider *et al.*, 2002).

Packaging Materials

The packaging materials used in MAP can theoretically be glass, metal or plastics. However, plastics are mostly used in MAP because they are flexible and have specific permeabilities to gases. Plastics are made of high-molecular-weight polymers and can easily be produced in different forms, such as films, trays and bottles. The most common plastics used in food packaging are polyethylene (PE), polypropylene (PP), polyethylene terephthalate (PET), polyvinyl chloride (PVC), ethylene vinylacetate (EVA), ethylene vinyl alcohol (EVOH), polyamides (nylon), ionomers, polycarbonates, polystyrene and cellulose acetate (Mangaraj *et al.*, 2009). These are generally used in laminated or co-extruded forms to obtain the desired properties of the final materials used in the packaging. Modified-atmosphere packaging requires materials with different gas permeabilities, depending on the product being packaged. Multilayered materials with an aluminum or EVOH layer are used for products requiring packages with high gas barrier properties. Some products, such as fresh produce, require specific O_2 and CO_2 permeabilities. Low-density polyethylene, polypropylene or perforated materials are used for such products. The permeability coefficients of some common packaging materials used in MAP are given in Table 43.1.

Table 43.1 Permeability coefficients of packaging films commonly used in MAP. (Adapted from Exama *et al.* (1993) and Mangaraj *et al.* (2009)).

Film	Permeability coefficient ($mL{\cdot}mil{\cdot}cm^{-2}{\cdot}h^{-1}{\cdot}atm^{-1}$)	
	O_2	CO_2
Cellulose acetate[a]	3.90×10^{-3}	3.14×10^{-2}
Ethyl cellulose[a]	1.55×10^{-1}	3.78×10^{-1}
Ethylene–vinyl alcohol (EVOH) (32 mol% ethylene)[b]	1.28×10^{-6}	3.97×10^{-5}
Ethylene–vinyl alcohol (EVOH) (44 mol% ethylene)[b]	4.92×10^{-6}	1.48×10^{-4}
High-density polyethylene (HDPE)[b]	$(6.56–11.97) \times 10^{-3}$	$(3.93–7.17) \times 10^{-2}$
Low-density polyethylene (LDPE)[b]	$(2.62–3.44) \times 10^{-2}$	$(1.64–2.15) \times 10^{-1}$
Natural rubber[a]	9.50×10^{-2}	6.72×10^{-1}
Polybutadiene[a]	8.39×10^{-2}	7.73×10^{-1}
Polyethylene terephthalate (PET) (unoriented)[b]	$(1.97–3.94) \times 10^{-4}$	$(1.00–2.01) \times 10^{-3}$
Polyethylene terephthalate (PET) (oriented)[b]	1.77×10^{-4}	8.70×10^{-4}
Polypropylene (PP)[b]	$(8.20–15.42) \times 10^{-3}$	$(4.61–8.66) \times 10^{-2}$
Polyvinyl chloride (PVC)[b]	$(6.06–393.70) \times 10^{-4}$	$(3.70–240.16) \times 10^{-3}$

[a]At 4 °C. [b]At 25 °C.

Design of Modified-Atmosphere Packaging for Foods

Packaging requirements differ between foods. Therefore, the optimum packaging condition (gas requirement) for each particular product needs to be determined first. Recommended modified-atmosphere conditions for selected respiring and nonrespiring foods are given in Tables 43.2 and 43.3, respectively. Depending on the size of the product that needs to be packaged, the characteristics of the packaging materials need to be determined and compared with the materials commercially available. If the required material is not available, then the size and shape of the package can be changed so that some commercially available material might meet the requirements. The design of MAP is discussed separately for each category of food in the following sections.

Fresh Fruit and Vegetable Products

Fresh whole or minimally processed (fresh-cut) fruits and vegetables can be packaged in a modified atmosphere for shelf-life extension. These products respire, i.e., they consume O_2 and produce CO_2. The shelf-life of fresh fruits and vegetable products can be extended by reducing their respiration rate. Reduced O_2 and elevated CO_2 generally decrease the respiration rate of fresh produce. However, the levels of O_2 and CO_2 in the package must be in accordance with the specific low O_2 and elevated CO_2 limits for each produce item (Kader *et al.*, 1998). Oxygen and CO_2 levels beyond these limits result in anaerobic respiration and tissue damage, which cause rapid quality degradation. Each type of fresh produce has a different optimum gaseous environment for quality retention and shelf-life extension (Table 43.2). These gaseous atmospheres are

Table 43.2 Recommended modified-atmosphere (MA) conditions for selected fruits and vegetables. (Adapted from Kader *et al.* (1998) and Gorris and Peppelenbos (2007).)

Product	Temperature (°C)	Recommended MA		Potential
		O_2 (%)	CO_2 (%)	
Fruits				
Apple	0–5	1–3	1–5	Excellent
Apricot	0–5	2–3	2–3	Fair
Avocado	5–13	2–5	3–10	Good
Banana	12–15	2–5	2–5	Excellent
Cherry, sweet	0–5	3–10	10–15	Good
Fig	0–5	5–10	15–20	Good
Grape	0–5	2–5	1–3	Fair
Grapefruit	10–15	3–10	5–10	Fair
Kiwi fruit	0–5	1–2	3–5	Excellent
Lemon and lime	10–15	5–10	0–10	Good
Mango	10–15	3–5	5–10	Fair
Olive	5–10	2–3	0–1	Fair
Orange	5–10	5–10	0–5	Fair
Peach and nectarine	0–5	1–2	3–5	Good
Pear	0–5	1–3	0–3	Excellent
Persimmon	0–5	3–5	5–8	Good
Plum	0–5	1–2	0–5	Good
Raspberry and other cane berries	0–5	5–10	15–20	Excellent
Red currant	0–2	5–10	15–20	Excellent
Strawberry	0–5	5–10	15–20	Excellent
Vegetables				
Asparagus	0–5	air	5–10	Excellent
Bean	5–10	2–3	4–7	Fair
Broccoli	0–5	1–2	5–10	Excellent
Brussels sprout	0–1	2–4	4–6	Excellent
Cabbage	0–5	2–3	3–6	Excellent
Cantaloupe	3–7	3–5	10–15	Good
Celery	3–5	1–4	0–5	Good
Chicory	5	2–3	5–10	Excellent
Corn (sweet)	0–5	2–4	5–10	Good
Cucumber	8–12	3–5	0	Fair
Leek	0–5	1–2	3–5	Good
Lettuce	0–5	1–3	0	Good
Onion	0–5	1–2	2–5%	Good
Pepper	8–12	3–5	0	Fair
Spinach	0–5	air	0–20	Good
Tomato	12–20	3–5	0–3	Good

low in O_2 and high in CO_2 compared with air. However, recent studies have indicated that superatmospheric O_2 levels can also be useful for the packaging of some whole and fresh-cut commodities (Artés *et al.*, 2009; Day, 2003). The respiration rates of fresh-cut products are much higher, and they can tolerate higher CO_2 and lower O_2 levels than their whole forms (Brecht, 1995). MAP is unavoidable for fresh-cut produce, because it is more vulnerable to biochemical and microbial quality degradation.

Table 43.3 Recommended modified-atmosphere (MA) conditions for selected nonrespiring foods.

Product	Recommended MA[a]		Reference
	O_2 (%)	CO_2 (%)	
Muscle foods			
Chilled chicken breast	0	70	Al-Haddad *et al.* (2005)
Fish (fatty)	0	60	Parry (1993)
Fresh meat	0	100	Skandamis and Nychas (2002)
Meatballs	1–3	33	Ozturk *et al.* (2010)
Poultry	0–80	20–100	Philips (1996)
Rainbow trout	0–2.5, 80	20–90	Yılmaz *et al.* (2009)
Salmon	20	60	Parry (1993)
Scallops	10	60	Simpson *et al.* (2007)
Sea bass fillets	15	63	Mercogliano *et al.* (2009)
White fish (nonfatty)	30	40	Parry (1993)
Dairy foods			
Hard cheese	0	100	Parry (1993)
Soft cheese	0	30	Parry (1993)
Whey cheese	0	60–70	Dermiki *et al.* (2008), Temiz *et al.* (2009)
White cheese	0–10	0–75	Kırkın (2009)
Bakery foods			
Fresh pasta	0	70–100	Del Nobile *et al.* (2009), Parry (1993)
Wheat bread	0	100	Rasmussen and Hansen (2001)
Cake	0	70–100	Guynot *et al.* (2004)
Other foods			
Eggs	0	20	Yalamanchili (2009)
Hazelnut	<10	0	Martin *et al.* (2001)
Peanut	0	65	Ellis *et al.* (1994)
Steamed rice	>10		Kasai *et al.* (2005)

[a] N_2 was used to balance.

Reduced O_2 levels with elevated CO_2 decreased enzymatic browning at cut surfaces, tissue softening, respiration, ethylene production and microbial growth in fresh-cut produce (Gunes *et al.*, 2001).

Typical respiration rates of fresh produce are given in Table 43.4. The respiration rate of a fresh product is measured in different O_2- and CO_2-containing environments, and a suitable mathematical model is fitted to the data. The respiration models are used in the design equations for MAP to predict the changes in the gas concentrations in the package headspace with time. Several mathematical models have been used to fit the respiration rates of fresh commodities as a function of the O_2 and CO_2 concentrations (Cameron *et al.*, 1989; Iqbal *et al.*, 2009; Lee *et al.*, 1991; Ravindra and Goswami, 2008; Renault *et al.*, 1994). Enzyme-kinetics-based models (based on the Michaelis–Menten model with several different types of inhibition) have been used extensively in modeling the respiration rate of fresh produce as a function of the O_2 and CO_2 concentrations (Table 43.5). The model parameters have been determined for a variety of fresh produce and reported in the literature (Table 43.6).

Table 43.4 Respiration rates of selected fruits and vegetables at 5 °C. (Adapted from Kader and Saltveit (2003) and Mahajan *et al.* (2006)).

Product	Respiration rate	
	R_{O2} ($mL\,O_2 \cdot g^{-1} \cdot h^{-1}$)	R_{CO2} ($mL\,CO_2 \cdot kg^{-1} \cdot h^{-1}$)
Apple (whole)	8.42–29.07	10.31–37.79
Asparagus	41.76	36.33
Banana	14.14	15.01
Blueberry	3.14–365.62	3.21–290.16
Cabbage	112.56	90.05
Carrot	26.05	14.58
Cauliflower	9.80	13.88
Cherry	13.77–48.32	12.66–38.66
Cucumber	24.39	16.52
Garlic	31.53	23.46
Kale		22[a]
Lettuce		30[a]
Litchi	13.90	11.60
Mango	19.28–22.10	17.53–17.74
Mushroom	118.36	106.34
Onion (green)	50.92–51.38	40.97–47.80
Pepper	19.38	12.00
Raspberry	43.46	48.63
Red pepper	15.15	14.09
Soybean	50.52	14.22
Strawberry	21.62	18.11
Tomato	13.80–36.01	12.84–18.30

[a] $mg\,CO_2 \cdot kg^{-1} \cdot h^{-1}$.

Table 43.5 Respiration rate equations based on enzyme inhibition kinetics models. (Adapted from Aguilera and Oliveira (2009) and Torrieri *et al.* (2009)).

Equation	Inhibition type
$R_{O2} = \dfrac{V_m . y_{O2}}{K_m + y_{O2}}$	No inhibition
$R_{O2} = \dfrac{V_m . y_{O2}}{K_m . \left(1 + \dfrac{y_{CO2}}{K_i}\right) + y_{O2}}$	Competitive
$R_{O2} = \dfrac{V_m . y_{O2}}{K_m + y_{O2} . \left(1 + \dfrac{y_{CO2}}{K_i}\right)}$	Uncompetitive
$R_{O2} = \dfrac{V_m . y_{O2}}{(K_m + y_{O2}) . \left(1 + \dfrac{y_{CO2}}{K_i}\right)}$	Noncompetitive

Table 43.6 Enzyme-kinetics-based model parameters for respiration rate (R_{O2} and R_{CO2}) of selected fruits and vegetables

Product	Temperature (°C)	Inhibition type		V_m	K_m	K_i	Reference
Apple	7	Uncompetitive	R_{O2}	15.55[a]	5.59[b]	10.87[c]	Mahajan and Goswami (2001)
			R_{CO2}	21.62[a]	13.21[b]	13.46[c]	
Banana (green and fresh)	10	Uncompetitive	R_{O2}	16.36[a]	6.4[b]	10.11[c]	Bhande *et al.* (2008)
			R_{CO2}	26.13[a]	10.42[b]	12.75[c]	
	30		R_{O2}	43.76[a]	8.92[b]	6.13[c]	
			R_{CO2}	56.43[a]	14.12[b]	7.79[c]	
Banana	21	Uncompetitive	R_{O2}	53.79[a]	3.427[b]	102.672[c]	Heydari *et al.* (2010)
			R_{CO2}	81.95[a]	11.124[b]	70.491[c]	
Blueberry (Duke)	15	Uncompetitive	R_{O2}	22.71[a]	7.63[b]	14.42[c]	Song *et al.* (2002)
			R_{CO2}	17.64[a]	5.08[b]	11.99[c]	
	25		R_{O2}	28.20[a]	0.12[b]	16.65[c]	
			R_{CO2}	21.09[a]	0.09[b]	52.4[c]	
Blueberry (Elliot)	15	Uncompetitive	R_{O2}	27.44[a]	6.96[b]	12.79[c]	Song *et al.* (2001)
			R_{CO2}	29.77[a]	9.35[b]	8.47[c]	
	25		R_{O2}	44.95[a]	5.94[b]	14.81[c]	
			R_{CO2}	48.06[a]	2.49[b]	9.96[c]	
Broccoli	3	Competitive	R_{O2}	2.4×10^{-7}[d]	0.8[b]	3[c]	Torrieri *et al.* (2010)
	20			1.7×10^{-6}[d]	0.5[b]	1[c]	
Carrot (cut)	10	Uncompetitive	R_{O2}	27.7[a]	1.15[b]	3.30[c]	Lee *et al.* (1996)
			R_{CO2}	15.2[a]	0.82[b]	8.32[c]	
Cherry	2	No inhibition	R_{O2}	4–8[a]	16–46[b]		Jamie *et al.* (2001)
	20			49–103[a]	29–32[b]		

(*Continued*)

Table 43.6 (*Continued*)

Product	Temperature (°C)	Inhibition type		V_m	K_m	K_i	Reference
Coleslaw	5	Uncompetitive	R_{O2}	22.72[a]	1.083[b]	23.16[c]	McLaughlin and O'Beirne (1999)
Cucumber (fresh-cut)	10	Uncompetitive	R_{O2}	30.0[a]	4.38[b]	1.27[c]	Lee *et al.* (1996)
			R_{CO2}	19.9[a]	3.45[b]	0.71[c]	
Curled lettuce	3	Uncompetitive	R_{O2}	0.354[a]	0.84[b]	3.41[c]	An *et al.* (2009)
			R_{CO2}	0.190[a]	2.85[b]	44.8[c]	
Garlic (cut)	10	Uncompetitive	R_{O2}	35.1[a]	2.30[b]	6.00[c]	Lee *et al.* (1996)
			R_{CO2}	24.7[a]	1.06[b]	9.91[c]	
Green onion (fresh-cut)	0	No inhbition	R_{O2}	11.75[a]	1.77[b]		Hong and Kim (2001)
			R_{CO2}	9.19[a]	1.34[b]		
	20		R_{O2}	91.74[a]	2.39[b]		
			R_{CO2}	78.74[a]	1.29[b]		
Green pepper (whole)	10	Uncompetitive	R_{O2}	24.4[a]	5.34[b]	3.70[c]	Lee *et al.* (1996)
			R_{CO2}	12.7[a]	1.00[b]	2.74[c]	
Mango (green mature)	5	Uncompetitive	R_{O2}	29.48[a]	12.85[b]	8.50[c]	Ravindra and Goswami (2008)
			R_{CO2}	31.25[a]	12.38[b]	6.50[c]	
	30		R_{O2}	68.10[a]	17.15[b]	4.85[c]	
			R_{CO2}	72.80[a]	18.26[b]	3.85[c]	
Onion (fresh-cut)	10	Uncompetitive	R_{O2}	4.20×10^{-4}[e]	1.54×10^{-3}[f]	0.449[f]	Lee and Renault (1998)
			R_{CO2}	3.82×10^{-4}[e]	7.93×10^{-5}[f]	0.896[f]	
Pomegranate arils	4	Competitive	R_{O2}	3.1[a]	3.8[b]	115.2[c]	Ersan *et al.* (2010)
Strawberry	3	Uncompetitive	R_{O2}	0.44[g]	0.99[h]	8.80[h]	An *et al.* (2009)
			R_{CO2}	0.278[g]	1.79[h]	8.07[h]	

[a] $mL \cdot kg^{-1} \cdot h^{-1}$. [b] $\%O_2$. [c] $\%CO_2$. [d] $mol \cdot kg^{-1} \cdot s^{-1}$. [e] $mol \cdot kg^{-1} \cdot h^{-1}$. [f] mole fraction. [g] $mmol \cdot kg^{-1} \cdot h^{-1}$. [h] kPa.

The respiration of fresh produce makes MAP a dynamic system in which the headspace gaseous environment is affected by the rate of respiration and the rate of gas transmission through the package. As the O_2 in the headspace is consumed by the produce through respiration, O_2 from the outside atmosphere diffuses into the headspace owing to the concentration gradient. Similarly, as CO_2 is emitted by the produce through respiration into the headspace, it diffuses out of the package owing to the concentration gradient (Figure 43.1). An equilibrium condition with constant headspace O_2 and CO_2 levels is reached when the respiration rate of the packaged produce becomes equal to the transmission rate of the gases through the package. Packaging that provides this condition is referred to as equilibrium modified-atmosphere packaging (EMAP).

The modified atmosphere in the headspace can be created passively or actively. In passive modification, the fresh produce is placed in the package and sealed without any gas flushing. The headspace atmosphere is modified by respiration and permeation, reaching equilibrium. The time required to reach the equilibrium atmosphere is relatively long and depends on the respiration rate of the produce. Passive modification is used for fresh produce for which there is no significant quality degradation during the time taken to reach equilibrium. Active modification involves an initial flushing of the package with gas before sealing. When the produce to be packaged shows rapid quality degradation in air, active modification is used. A gas mixture with the optimum gas composition for the product is usually flushed into the headspace before sealing. Active modification is especially required for fresh-cut produce, as the quality of products of this type degrades rapidly in air.

The design of MAP for respiring products is more challenging than the design of MAP for nonrespiring foods because of this dynamic gas system. The first step in the design of MAP for fresh produce is to determine the optimum gaseous environment. Next, the respiration rate of the fresh produce as affected by the gaseous environment (i.e., the CO_2 and O_2 concentrations) is determined. The respiration rate in the optimum gaseous environment is required to design the MAP.

Mathematical models describing the gas exchange and the headspace O_2 and CO_2 concentrations as a function of the respiration rate and the package properties are used in MAP design. To develop a model, the material balance for both O_2 and CO_2 in the package headspace is considered, as follows:

$$\begin{bmatrix}\text{Rate of gas accumulation}\\ \text{in the headspace}\end{bmatrix} = \begin{bmatrix}\text{Rate of gas permeated}\\ \text{into the headspace}\end{bmatrix} + \begin{bmatrix}\text{Rate of gas generated}\\ \text{through respiration}\end{bmatrix}$$

The design equations for O_2 and CO_2 are as follows (Aguilera and Oliveira, 2009):

$$V_f \frac{dy_{O2}}{dt} = \frac{P_{O2}}{x} P_{atm} A_f \left[y_{O2,out} - y_{O2,in} \right] - R_{O2} W \tag{43.1}$$

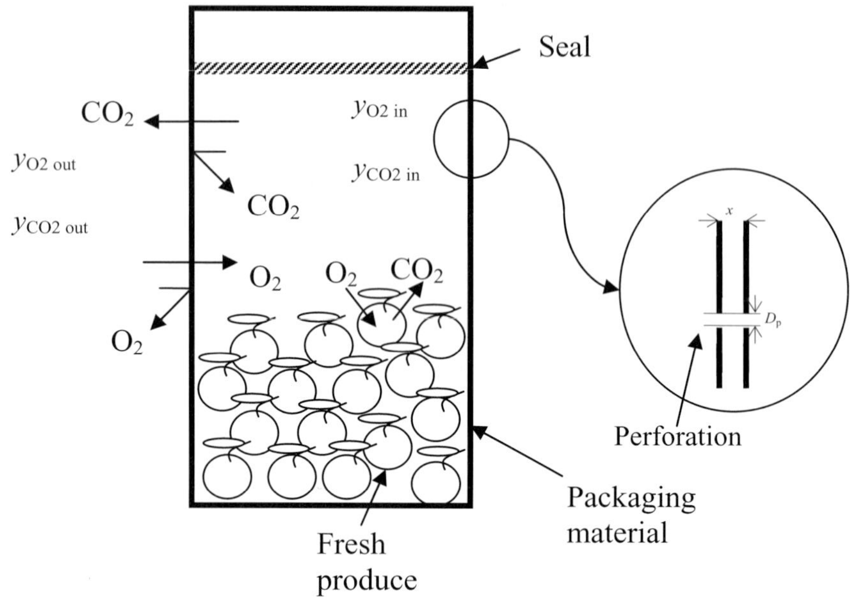

Figure 43.1 Gas exchange in a modified-atmosphere package containing fresh produce; x is the film thickness, D_p is the diameter of the perforations if they exist, y_{O2} is the partial pressure of O_2 and y_{CO2} is the partial pressure of CO_2.

$$V_f \frac{dy_{CO2}}{dt} = \frac{P_{CO2}}{x} P_{atm} A_f \left[y_{CO2,out} - y_{CO2,in} \right] + R_{CO2} w \tag{43.2}$$

where V_f is the free volume (the headspace); y is the partial pressure of each gas (y_{O2} = O_2 partial pressure and y_{CO2} = CO_2 partial pressure, inside or outside of the package); x is the thickness of the film; P_{O2} and P_{CO2} are the O_2 and CO_2 permeability coefficients, respectively, of the film; A_f is the surface area of the film; R is the respiration rate (R_{O2} = O_2 consumption rate and R_{CO2} = CO_2 production rate); P_{atm} is the atmospheric pressure; and w is the weight of the product. Unsteady-state gas concentrations in the package headspace can be determined by inserting equations for the respiration rates R as a function of the O_2 and CO_2 partial pressures into Equations 43.1 and 43.2. In the steady state, an equilibrium modified atmosphere is attained, in which the gas concentrations in the headspace do not change with time. The steady-state equations are:

$$\frac{dy_{O2}}{dt} = 0 \Rightarrow \frac{P_{O2} A_f}{x} P_{atm} (y_{O2,out} - y_{O2,eq}) = R_{O2,eq} w \tag{43.3}$$

$$\frac{dy_{CO2}}{dt} = 0 \Rightarrow \frac{P_{CO2} A_f}{x} P_{atm} (y_{CO2,eq} - y_{CO2,out}) = R_{CO2,eq} w \tag{43.4}$$

where $y_{O2,eq}$ and $y_{CO2,eq}$ are the partial pressures of O_2 and CO_2, respectively, in the headspace at equilibrium, and $R_{O2,eq}$ and $R_{CO2,eq}$ are the O_2 consumption rate and CO_2 production rate, respectively, in the equilibrium gaseous atmosphere. These equations can be solved for the required area A_f, given the quantity w of the commodity and the permeability coefficients of the packaging materials. They can also be solved for the permeability coefficients P_{O2} and P_{CO2}, given the quantity w of the commodity and the area A_f. The calculated properties of the packaging material are compared with the commercially available materials. If there is a commercially available packaging material with properties similar to the calculated ones, successful MAP is possible for the product.

If Equations 43.3 and 43.4 are combined (by dividing one by the other), we get

$$\frac{(y_{O2,out} - y_{O2,eq})}{(y_{CO2,eq} - y_{CO2,out})} = \frac{\beta}{RQ} \tag{43.5}$$

where β is the selectivity of the film (the ratio of P_{CO2} to P_{O2}), and RQ is the respiratory quotient (the ratio of R_{CO2} to R_{O2}). RQ is usually assumed to be equal to 1, although it can be between 0.7 and 1.2. The selectivity of common packaging materials can range from 2.2 to 10.2 (Exama *et al.*, 1993).

Example 1: Calculation of permeability requirements for pomegranate arils

Given:
Desired weight of arils in package: 500 g.
Desired equilibrium gas concentrations: 2% O_2 and 10% CO_2.
Respiration rate measured in the equilibrium condition: $1.01\,cm^3\,O_2 \cdot kg^{-1} \cdot h^{-1}$.
Desired package form: tray with a permeable film on the top side.
Density of arils: $0.55\,g \cdot cm^{-3}$.
Desired fraction of free volume in package: 30%.

Calculations:
Volume of package = volume of arils + 30% free volume:

$$V = 500\ g \times (1\ cm^3/0.55\ g) + 0.30 \times V$$

$$V = 1299\ cm^3$$

The dimensions of the tray are selected to obtain the total volume calculated above:

$$\text{Length} = 10\ cm;\ \text{width} = 10\ cm;\ \text{height} = 13\ cm.$$

Thus, permeation area $A = 10 \times 10 = 100\,cm^2$.

$$\frac{P_{O2}A_f}{X}P_{atm}(y_{O2,out} - y_{O2,eq}) = R_{O2,eq}w$$

$$\frac{P_{O2} \times 0.01}{X} \times 1 \times (0.208 - 0.02) = 1.01 \times 0.5$$

$P_{O2}/X = 268.6\,\text{mL}\,O_2 \cdot (m^2 \cdot h \cdot atm)^{-1}$. This is the required O_2 permeability for the package. For CO_2 permeability, assume $R_{CO2} = R_{O2} = 1.01\,cm^3\,CO_2 \cdot kg^{-1} \cdot h^{-1}$:

$$\frac{P_{CO2}A_f}{X}P_{atm}(y_{CO2,eq} - y_{CO2,out}) = R_{CO2,eq}w$$

$$\frac{P_{CO2} \times 0.01}{X} \times 1 \times (0.10 - 0) = 1.01 \times 0.5$$

$P_{CO2}/X = 505.0\,\text{mL}\,CO_2 \cdot (m^2 \cdot h \cdot atm)^{-1}$. This is the required CO_2 permeability for the package.

For many fresh commodities, these calculations result in impractical solutions for real-life applications. For example, the calculated area may be too big or too small, so that it may not be possible to have a reasonable commercial package. In other cases, there may be no commercial packaging materials with the calculated permeabilities. It is very difficult to obtain the optimum O_2 and CO_2 conditions simultaneously in the headspace at equilibrium for many products, because the CO_2 permeabilities are much higher than the O_2 permeabilities for plastic packaging films (Exama *et al.*, 1993). Perforations in existing packaging materials can solve these problems to some extent.

Perforation is simply making holes of various sizes and numbers in an existing commercial plastic film. Design equations for MAP using perforated films are derived in a similar way to that above, using material balances for each gas. Transmission of gases occurs through both the perforations and the film. Design equations for macroperforated and microperforated packages have been derived in the literature (Techavises and Hikida, 2008; González *et al.*, 2008). These equations are reproduced, using a modified notation, as follows:

$$\frac{dV_{O2}}{dt} = \left(n_p \frac{D_{O2}A_h}{L_h} + A_f \frac{P_{O2}}{X}\right)\left(P_{O2,out} - P_T \frac{V_{O2}(t)}{V_{O2}(t) + V_{CO2}(t) + V_{N2}(t) + V_{H2O}(t)}\right) - wR_{O2} \tag{43.6}$$

$$\frac{dV_{CO2}}{dt} = \left(n_p \frac{D_{CO2}A_h}{L_h} + A_f \frac{P_{CO2}}{X}\right)\left(P_{CO2,out} - P_T \frac{V_{CO2}(t)}{V_{O2}(t) + V_{CO2}(t) + V_{N2}(t) + V_{H2O}(t)}\right) + wR_{CO_2} \tag{43.7}$$

where V is the volume fraction of the gas, A_f is the area of the film, n_p is the number of perforations, D is the diffusion coefficient of O_2 or CO_2 in air ($m^2 \cdot s^{-1}$), A_h is the area of a perforation (m^2), and L_h is the diffusion path length (m), which is equivalent to the film thickness x. The term $(DA_h)/L_h$ is the effective permeability of one perforation to O_2 or CO_2 in $m^3 \cdot s^{-1}$. Under steady-state conditions, the equations for MAP with a perforated film become

$$wR_{O2} = \left(n_p \frac{D_{O2} A_h}{L_h} + A_f \frac{P_{O2}}{x}\right)\left(P_{O2,out} - P_T \frac{V_{O2}}{V_{O2} + V_{CO2} + V_{N2} + V_{H2O}}\right) \quad (43.8)$$

$$\Rightarrow wR_{O2} = \left(n_p \frac{D_{O2} A_h}{L_h} + A_f \frac{P_{O2}}{x}\right)(P_{O2,out} - P_{O2,eq}) \quad (43.9)$$

$$-wR_{CO2} = \left(n_p \frac{D_{CO2} A_h}{L_h} + A_f \frac{P_{CO2}}{x}\right)\left(P_{CO2,out} - P_T \frac{V_{CO2}}{V_{O2} + V_{CO2} + V_{N2} + V_{H2O}}\right) \quad (43.10)$$

$$\Rightarrow wR_{O2} = \left(n_p \frac{D_{CO2} A_h}{L_h} + A_f \frac{P_{CO2}}{x}\right)(P_{CO2,out} - P_{CO2,eq}) \quad (43.11)$$

These equations can be solved for any given situation to design MAP using a perforated packaging material. For example, if there are no commercially available packaging materials that satisfy the requirements for a particular type of fresh produce, the perforation area A_h and the number of perforations in the material can be calculated using Equation 43.11 to obtain the desired equilibrium gas composition in the package headspace.

Example 2: Calculation of the number of perforations required in a package of the pomegranate arils considered in Example 1

$$wR_{O2} = \left(n_p \frac{D_{O2} A_h}{L_h} + A_f \frac{P_{O2}}{x}\right)(P_{O2,out} - P_{O2,eq})$$

$$\Rightarrow 0.5 \times 1.01 = \left(n_p \frac{D_{O2} A_h}{L_h} + A_f \frac{P_{O2}}{x}\right)(0.208 - 0.02)$$

$$\Rightarrow \left(n_p \frac{D_{O2} A_h}{L_h} + A_f \frac{P_{O2}}{x}\right) = 2.686$$

Given the characteristics of the packaging film (P_{O2} and x) and the surface area of the package for gas transmission, A_f, the number of perforations n_p can be calculated for any selected size of perforations.

If we use a 71 µm thick polypropylene film with a permeability of 0.006 $mL{\cdot}cm^{-2}{\cdot}h^{-1}$ at 4 °C and 1 atm, and if a perforation diameter of 10 µm is chosen, then the number of perforations in a film of area 100 cm^2 is calculated as follows.

The diffusion coefficient D_{O2} of O_2 in air is assumed to be 0.143 $cm^2{\cdot}s^{-1}$ = 514.8 $cm^2{\cdot}h^{-1}$. This gives

$$\left(n_p \frac{514.8 \times \left(\pi \times (10 \times 10^{-4})^2 / 4 \right)}{71 \times 10^{-4}} + 100 \times 0.006 \right) = 2.686$$

$$\Rightarrow n_p = 36.6 \cong 37$$

That is, the number of perforations required is 37.

The design of MAP involves the use of permeability and respiration data, which are affected by temperature. The permeability of the packaging film and the respiration rate are taken at a specific temperature in the solution of the design equations. Both the respiration rate and the permeability increase with an increase in temperature, but at different rates. The temperature dependences of the respiration rate and permeability are described by Arrhenius-type equations. The activation energies E_a for most fruits and vegetables in air have been reported to be between 29.0 and 92.9 $kJ{\cdot}mol^{-1}$ (Aguilera and Oliveira, 2009). Activation energies for common packaging materials in the range of 8.4 to 66.5 $kJ{\cdot}mol^{-1}$ have been reported (Mahajan *et al.*, 2006). The respiration rate increases faster than the permeability with temperature. Therefore, the exposure of packaged fruits and vegetables to elevated temperatures (temperature abuse) would result in formation of an anaerobic atmosphere with an elevated CO_2 level, which would cause rapid quality degradation. Thus, temperature abuse is one of the main risk factors in MAP design, and strict temperature control during storage, distribution and marketing is required for fresh produce in modified-atmosphere packages.

Muscle Foods

Muscle foods can be preserved by using a modified atmosphere. Recommended modified-atmosphere conditions for some selected products are given in Table 43.3. The use of MAP will be discussed separately for different types of meat and meat products, as the required atmospheres vary significantly.

Fresh Red Meat

Red meat can deteriorate through microbial growth and oxidation (flavor, odor and color degradation). A bright red color is a very important quality parameter for fresh red meat, as it is associated with freshness and good eating quality by consumers. The red color of meat is related to myoglobin, the principal pigment in red meat. The presence of various chemical derivatives of myoglobin affects the red color. Reduced myoglobin (Mb) is dominant in the absence of O_2 and is purple in color. Oxymyoglobin (O_2Mb) forms when Mb is exposed to O_2, and is bright red. Metmyoglobin (MetMb) is brown in color, and it is formed when Mb is exposed to low O_2 concentrations (0.5–1%) or to air for long periods of time (McMillin, 2008). MetMb cannot bind to O_2 even in a high-O_2 environment. Myoglobin can also be converted to choleglobin and sulfmyoglobin, which are green in color, by bacterial by-products (Robertson, 1993). Therefore, high O_2 is required for the stability of the bright red color of red meat. On the other hand, O_2 can promote the oxidation of fats and the growth of many spoilage microorganisms in meat.

High-O_2 MAP is the most common type of packaging for fresh chilled red meat in display packages. The most common gas mixture used is 80% O_2 : 20% CO_2, although 25–90% O_2 and 15–80% CO_2 can be used (McMillin, 2008). High-O_2 MAP inhibits the growth of surface microorganisms and maintains the bright red color of the meat for an extended period of time during storage. High O_2 also inhibits the growth of anaerobic organisms, including lactic acid bacteria, in meat. The formation of off-odors and rancidity in meat can limit the shelf-life of meat in high-O_2 MAP, depending on its fat content. High-barrier films are required in high-O_2 packages so that the gas mixture used for flushing is retained in the headspace for as long as possible during storage. The amount of O_2 in the headspace usually decreases because of permeation from the package and respiration of tissues.

Low-O_2 packaging, in the form of either vacuum packaging or MAP, is also used for chilled red meat. Vacuum packaging is a practical and cost-effective packaging method, but it has the disadvantage that the package and the meat are subjected to mechanical strain. This can deform the package and impair its integrity, as well as cause increased drip loss from the meat, and even cause the package to rupture if bone is present (Robertson, 1993). Some of these problems can be eliminated by vacuum skin packaging (VSP), in which the product acts as a forming mold. Here, a soft film molds itself to the shape of the product and seals to a base film, creating an anaerobic medium for the meat (Robertson, 1993).

MAP with low O_2 or without O_2 can be used as an alternative to vacuum packaging. In this case, elevated CO_2 (up to 100%) balanced with N_2 is used. This type of MAP has the advantages that the antimicrobial activity of CO_2 is used, and that the issue of mechanical damage in vacuum packaging is eliminated. Anaerobic aerotolerant lactobacilli are selected by anaerobic atmospheres. Aerobic spoilage organisms are effectively inhibited by CO_2 at levels of 20–60% (McMillin, 2008). Carbon dioxide is soluble in both the aqueous and the lipid phases in meat, and it shows its maximum inhibitory effect on the growth of microorganisms above the saturation level. Owing to the solubility of CO_2 in meat, package collapse can be an issue for packages with high CO_2 levels. In such cases, the meat can be exposed to CO_2 to achieve saturation prior to packaging so that little or no further dissolution of CO_2 takes place in the meat after packaging (Jeremiah, 2001). Both low-O_2 MAP and vacuum packaging have the disadvantage that they induce a dark color due to the unoxygenated state of the myoglobin. However, low-O_2 packages can include both an outer barrier film and a permeable overwrapped inner film so that when the outer film is peeled away during retail display, the meat can be oxygenated, causing it to "bloom."

Carbon monoxide can be used in low-O_2 packaging for red meat to provide color stability. Either the meat can be exposed to CO prior to packaging, or CO can be flushed into the a low-O_2 package prior to sealing. CO, along with either an anaerobic atmosphere or a high-O_2 atmosphere, has been reported to inhibit oxidation and the growth of spoilage organisms in meat (John *et al.*, 2005; Kusmider *et al.*, 2002; McMillin, 2008). However, although a small amount of CO is effective for color stability, its commercial use is limited owing to concern about its safety for employees.

Fresh Poultry and Fish

Vacuum packaging can extend the shelf-life of poultry meat at refrigeration temperatures. However, the disadvantages of vacuum packaging described above for red meat also apply to poultry meat. Oxygen-free or low-O_2 MAP with elevated CO_2 results in better preservation of poultry meat without the disadvantages of vacuum packaging. Carbon dioxide at levels above 20% inhibits the growth of various microorganisms and thus extends the shelf-life of poultry meat. MAP using 25–70% CO_2 balanced with N_2 is recommended for chicken (Saucier *et al.*, 2000). Higher CO_2 levels result in longer shelf-life (Sawaya *et al.*, 1995; Vongsawadi *et al.*, 2008). Treatment of chicken with lactic or acetic acid can further extend the shelf-life under high-CO_2 MAP (Rao and Sachindra, 2002). However, temperature abuse of chicken packaged in an anaerobic atmosphere can result in toxin production by *C. botulinum*. Extension of the shelf-life by use of anaerobic MAP may provide sufficient time for the spores of the pathogen to germinate and produce toxin (Rao and Sachindra, 2002). The use of additional hurdles and the inclusion of low levels of O_2 can reduce this risk in packaged chicken.

The shelf-lives of various fresh fishery products, taken from the recent literature, have been tabulated by Sivertsvik *et al.* (2002). Elevated CO_2 (40–100%) in anaerobic MAP increased the shelf-life of many fish and seafood varieties significantly in comparison with air or vacuum packaging (Yesudhason *et al.*, 2009; Sivertsvik *et al.*, 2002). Increased CO_2 levels generally increased the shelf-life by inhibiting microbial spoilage. However, a potential safety issue due to *C. botulinum* exists when anaerobic MAP is used for fishery products, especially if the product is exposed to elevated storage temperatures of 8–16 °C (Sivertsvik *et al.*, 2002). The safest packaging with regard to toxin production by *C. botulinum* was suggested to be MAP with the inclusion of equal amounts of CO_2 and O_2 (Sivertsvik *et al.*, 2002). This can still give sufficient shelf-life for nonfatty fishery products, but the shelf-life will be limited by oxidative degradation in the case of fatty products. Strict temperature control below 4 °C and the use of additional hurdles can also significantly decrease the safety issue. For fatty fish, anaerobic MAP with 60% CO_2 has been suggested, whereas for nonfatty fish, an aerobic modified atmosphere (30% O_2:40% CO_2 balanced with N_2) is recommended (Parry, 1993).

Meat Products

Fresh red meat can be processed into many different foods, such as fermented products (sausages, hotdogs, salami, pastrami, etc.), seasoned ready-to-cook meatballs and marinated ready-to-cook products. Poultry and fishery meats can also be processed into similar products. These products, either uncooked or precooked, can be preserved by anaerobic MAP with elevated CO_2 (20–35%) (Parry, 1993). Maintenance of a bright red color in beef products is not critical for such products, because the colors of seasoned meatballs and of marinated products in various sauces are modified by the ingredients used. The characteristic red colors of fermented meat products are stabilized by additives, so oxygenation of myoglobin is not critical in such products. However, the growth of *C. botulinum* and toxin production by it can be a safety issue in such products, especially if they are subjected to temperature abuse. There have been studies in which the inclusion of low O_2 levels (up to 5%) in MAP for meatballs (seasoned ground beef) and marinated chicken did not cause significant oxidative quality degradation compared with anaerobic MAP (Doğu, 2009; Ozturk *et al.*, 2010). *C. botulinum* growth and toxin production can be inhibited by O_2 levels of 2% or higher (Rao and Sachindra, 2002). Therefore, the inclusion of a small amount of O_2 (up to 5%) can decrease the *C. botulinum* safety issue without significant quality degradation.

Cereal-Based Food Products

Cereals, such as bread, cakes, biscuits, pasta and dough, can also be preserved by MAP. The shelf-life of these products is affected by the processing conditions, the characteristics of the product, the storage conditions and the packaging (Galic *et al.*, 2009). Moisture content, water activity, pH, and antimicrobial and antioxidant additives affect the shelf-life of cereal-based products. Degradation of these products usually occurs through biological processes (by microbial and, especially, fungal activity, and because of insects), through oxidative reactions and through physical spoilage (staling, moisture loss and texture). MAP can protect products by controlling these degradation mechanisms. Recommended modified-atmosphere conditions for some selected products are given in Table 43.3. Fresh pasta, pies, cakes, bread and other bakery products should be packaged in high-barrier materials using an anaerobic atmosphere with elevated CO_2 to inhibit microbial, oxidative and textural quality loss processes (Del Nobile *et al.*, 2009; Galic *et al.*, 2009; Guynot *et al.*, 2004; Rasmussen and Hansen, 2001). Bakery products are usually packaged in an anaerobic atmosphere with 60% CO_2 (balanced with N_2), although higher CO_2 levels can also be used (Galic *et al.*, 2009). Exclusion of O_2 and elevated CO_2 can also inhibit insects in packaged cereals (Jayas and Jeyamkondan, 2002). Carbon dioxide in MAP may also delay staling of bread, although conflicting results have been reported in the literature (Galic *et al.*, 2009). The exclusion of O_2 from the package atmosphere is critical for shelf-life extension. Active packaging, including use of O_2 absorbers, ethanol emitters and antimicrobial agents such as calcium propionate, can further increase the benefits of MAP for some cereal-based products (Galic *et al.*, 2009).

Snack Foods, Tree Nuts and Dried Fruits

Snack foods include chips (potato, corn and tortilla chips), pretzels, popcorn, extruded foods, tree nuts and dried fruits. Insects, molds and oxidation are the most important problems that limit the shelf-life of these foods. Sun-dried raisins, apricots, figs and dates are susceptible to attack by various insects and fungi. These products can be preserved by MAP. Recommended modified-atmosphere conditions for some selected products are given in Table 42.3. It has been shown that anaerobic or low-O_2 MAP with elevated CO_2 successfully inhibited insect infestations in dried fruits and tree nuts (Kader *et al.*, 1998; Jayas and Jeyamkondan, 2002). MAP with elevated CO_2 can also reduce darkening of dried fruits during storage. Tree nuts contain large amounts of polyunsaturated lipids, making them susceptible to oxidative rancidity. The deshelling and roasting applied to tree nuts increase further their susceptibility to oxidation. Vacuum packaging and N_2 packaging are commonly used for tree nuts to delay oxidative rancidity. MAP is usually preferable to vacuum packaging, which can mechanically damage these sensitive products.

Chips, popcorn and extruded snacks also contain large amounts of fat, making them susceptible to oxidation. All of these foods are also susceptible to mechanical damage, and thus special cushioned packages are needed. Flushing the package with gas with a positive headspace pressure also serves to provide cushioning (by the ballooning effect) for fragile contents. Packaging materials with high barrier properties for gases and water vapor should be used in order to maintain the target modified atmosphere and prevent migration of O_2 into the headspace. Pillow pouches and bags are the most common types of packages used for snack foods. The package should not be transparent, to prevent exposure of the contents to light, which can catalyze fat oxidation. Active packaging systems containing O_2 absorbers can further increase the benefits of MAP for tree nuts, as even residual O_2 in the headspace can cause oxidation in the product.

Dairy Foods

MAP can delay microbial, chemical and physical spoilage and extend the shelf-life of dairy products (Ooraikul, 2003). The modified atmospheres used for cheese products are generally a combination of CO_2 and N_2, but low amounts of O_2 may also be used. Recommended modified-atmosphere conditions for some selected products are given in Table 43.3. High levels of CO_2 can effectively inhibit the growth of yeasts and molds. Shelf-life extension of various cheeses by MAP has been reported in the literature (Dermiki *et al.*, 2008; Favati *et al.*, 2007; Gonzalez-Fandos *et al.*, 2000; Papaioannou *et al.*, 2007). Packaging with 30% CO_2:70% N_2 extended the shelf-life of Provolone cheese by 50% compared with vacuum packaging (Favati *et al.*, 2007). MAP with 60% CO_2:40% N_2 extended the shelf-life of whey cheese (Erkan *et al.*, 2007). Anaerobic MAP with 70% or 30% CO_2 resulted in a longer shelf-life of cheese compared with vacuum packaging (Papaioannou *et al.*, 2007).

The extension of the shelf-life of cheese by MAP is achieved mainly through inhibition of microbial growth. The growth of staphylococci, molds, yeasts, lactic acid bacteria and overall mesophilic bacteria on cheese was suppressed by MAP (Eliot *et al.*, 1998; Erkan *et al.*, 2007; Oyugi and Buys, 2007; Papaioannou *et al.*, 2007). Microbial growth decreased with increased CO_2 concentration (Westall and Filtenborg, 1998). Packaging in 20–80% CO_2 with 0.5% or less O_2 suppressed the growth of fungi in sliced Cheddar cheese (Taniwaki *et al.*, 2001). Packaging with 50% CO_2:50% N_2 and 95% CO_2:5% N_2 preserved the microbial quality of Stracciatella cheese (Gammariello *et al.*, 2009).

MAP may also reduce proteolysis and lipid oxidation in whey cheese (Dermiki *et al.*, 2008). It was observed that proteolysis and lipolysis reactions were higher in packages containing air than in anaerobic packages with 20% or higher CO_2 (Gonzalez-Fandos *et al.*, 2000). Packaging with 75% CO_2 and up to 10% O_2 delayed microbial spoilage of precut white cheese, reduced proteolysis, lipolysis and oxidation reactions, and preserved the color, texture and sensory properties (Kırkın, 2009). Exposure to light may also influence the quality of packaged products. Changes in sensory properties, color and riboflavin content were observed in Havarti cheese packaged in a modified atmosphere and stored under light (Kristensen *et al.*, 2000).

Elevated CO_2 in cheese packaging can also decrease microbial safety issues. Pure CO_2 (100%) inhibited the growth of staphylococci, lactobacilli and *Bacillus* for two days at 4 °C (Pintado and Malcata, 2000). Inclusion of high CO_2 levels in MAP reduced the risk of botulism and listeriosis in cottage cheese (Chen and Hotchkiss, 1993). MAP may also be applied to other dairy foods, such as yoghurt, fermented milk beverages and milk (Hotchkiss *et al.*, 2006). The body and texture of sour cream were preserved by packaging under N_2, and the flavor was preserved by packaging under CO_2 (Kosikowski and Brown, 1973). Hotchkiss *et al.* (1999) suggested that CO_2 could increase the shelf-life of pasteurized refrigerated milk by up to twofold. Nitrogen flushing of whole milk powder resulted in lower peroxide values, lipid oxidation and off-odor compared with air storage (Lloyd *et al.*, 2009).

Equipment for MAP

MAP can be done by packaging machines using gas flushing or vacuum compensation. The replacement of air by the desired gas mixture can be achieved by flushing the headspace with a continuous gas stream to dilute the air around the product, followed by hermetic sealing of the package. This system is fast but allows residual O_2 remain

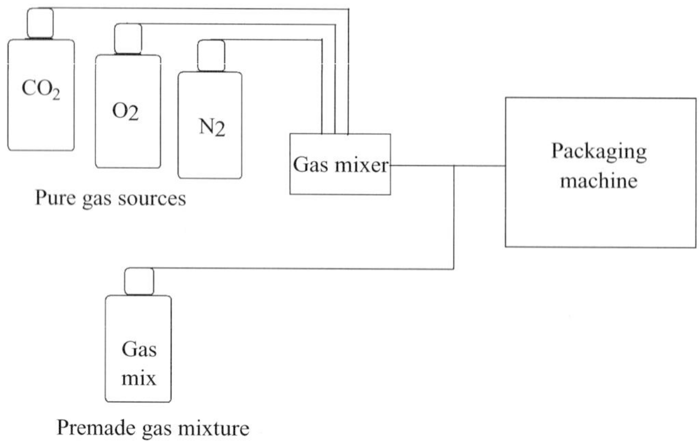

Figure 43.2 Typical flow diagram of a MAP process.

in the package, and thus is not suitable for strictly anaerobic MAP. The vacuum compensation method removes air from the headspace by creating a vacuum and then breaking the vacuum with the designated gas mixture. This two-step process is very effective in removing residual oxygen and thus is suitable for packaging of O_2-sensitive products, but it is relatively slow.

MAP machines are available for various types of packages, including trays, pillow pouches and bags. These packages can be fed preformed to the packaging machine in the case of machines such as tray sealers, vacuum chamber machines and bag sealer machines. The packages can also be formed inside the packaging machine in the case of machines such as form–fill–seal and deep-drawing machines. The gas mixtures used in MAP can be obtained premade or be obtained on site by the use of gas mixers. Gas concentrations are monitored using gas sensors integrated into the packaging machine, and in laboratories for quality control purposes. A typical schematic diagram of a MAP process is shown in Figure 43.2.

Controlled-Atmosphere Storage

Controlled-atmosphere storage involves an initial alteration of the atmospheric composition to specific levels in a storage room or large storage bins, followed by strict control of the gas composition to maintain it at the initial levels during the storage period. It is used commercially for the bulk storage of fresh produce and grain. The gas compositions used in controlled-atmosphere storage usually have reduced O_2 and elevated CO_2 levels, which are specific to each commodity. The recommended conditions for controlled-atmosphere storage are similar to those described for MAP earlier in this chapter. The mode of action of controlled-atmosphere storage is similar to that of MAP: the respiration and ethylene production rates of the stored commodity are decreased, and insect infestation and fungal growth are inhibited. Thus, the ripening of climacteric fruits during storage can be controlled, resulting in an extension of storage life. Moreover, the activities of postharvest pathogens, insects and their eggs and larvae are inhibited by the controlled atmosphere, resulting in prevention of decay and infestation during storage.

Controlled-atmosphere storage of fresh fruits and vegetables and the potential storage lives that can be achieved have been evaluated extensively by Thompson (1998) and Kader *et al.* (1998). Controlled-atmosphere storage is commonly practiced for the storage of apples and pears worldwide, proving a storage life of 6 to 12 months. Tree nuts and dried fruits and vegetables can be stored in bulk in a controlled atmosphere with a storage life of more than 12 months. Cabbages, kiwi fruits, persimmons, pomegranates, avocados, bananas, cherries, grapes, mangoes, olives, onions, nectarines, peaches, plums and tomatoes can be stored in a controlled atmosphere for 1 to 6 months. Controlled atmospheres can also be applied during long-distance transport of fresh produce with a limited storage life. Asparagus, broccoli, pineapples, papayas, figs, lettuces, strawberries, sweetcorn and muskmelons are commodities that have a storage life under a controlled atmosphere of less than 1 month. Controlled atmospheres with no O_2 and elevated CO_2 also have great potential for the storage of grain, mainly to prevent insect and mold damage.

Controlled-atmosphere storage requires gas-tight rooms, sealed with metal cladding and with carefully sealed doorways (Fellows, 2000). The gas levels are continuously monitored using special gas sensors and are maintained at the designated levels using special systems including CO_2 scrubbers, solid or liquid CO_2, N_2 purging, O_2 purging, and humidity control (Fellows, 2000).

Nomenclature

A_f: film area, m^2
A_h: perforation area, m^2
D: diffusion coefficient, $m^2 \cdot s^{-1}$
D_p: perforation diameter, m
K_i: Inhibition constant, $\%CO_2$
L_h : diffusion path length, m
n_p: number of perforations
P_{atm}: atmospheric pressure, atm
P_{CO2}: CO_2 permeability coefficient, $mL \cdot mil \cdot cm^{-2} \cdot h^{-1} \cdot atm^{-1}$
P_{O2}: O_2 permeability coefficient, $mL \cdot mil \cdot cm^{-2} \cdot h^{-1} \cdot atm^{-1}$
R_{CO2}: CO_2 production rate, $mL\ CO_2 \cdot kg^{-1} \cdot h^{-1}$
$R_{CO2,eq}$: CO_2 production rate in equilibrium gaseous atmosphere, $mL\ O_2 \cdot kg^{-1} \cdot h^{-1}$
R_{O2}: O_2 consumption rate, $mL\ O_2 \cdot kg^{-1} \cdot h^{-1}$
$R_{O2,eq}$: O_2 consumption rate in equilibrium gaseous atmosphere, $mL\ O_2 \cdot kg^{-1} \cdot h^{-1}$
RQ: respiratory quotient, dimensionless

V:	volume fraction of a gas, m^3
V_f:	free volume, m^3
V_m:	maximum respiration rate ($mL \cdot kg^{-1} \cdot h^{-1}$)
w:	product weight, kg
x:	film thickness, m
y_{CO2}:	CO_2 partial pressure, atm
$y_{CO2,eq}$:	partial pressure of CO_2 in the headspace at equilibrium, atm
y_{O2}:	O_2 partial pressure, atm
$y_{O2,eq}$:	partial pressure of O_2 in the headspace at equilibrium, atm
β:	selectivity of the film, dimensionless

References

Al-Haddad, K.S.H., Al-Qassemi, R.A.S. and Robinson, R.K. (2005) The use of gaseous ozone and gas packing to control populations of *Salmonella infantis* and *Pseudomonas aeruginosa* on the skin of chicken portions. *Food Control* 16: 405–410.

An, D.S., Park, E. and Lee, D.S. (2009) Effect of hypobaric packaging on respiration and quality of strawberry and curled lettuce. *Postharvest Biology and Technology* 52: 78–83.

Artés, F., Gómez, P., Aguayo, E., Escalona, V. and Artés-Hernández, F. (2009) Sustainable sanitation techniques for keeping quality and safety of fresh-cut plant commodities. *Postharvest Biology and Technology* 51: 287–296.

Aguilera, R.R. and Oliveira J.C. (2009) Review of design engineering methods and applications of active and modified atmosphere packaging systems. *Food Engineering Review* 1: 66–83.

Bhande, S.D., Ravindra, M.R. and Goswami, T.K. (2008) Respiration rate of banana fruit under aerobic conditions at different storage temperatures. *Journal of Food Engineering* 87: 116–123.

Brecht, J.K. (1995) Physiology of lightly processed fruits and vegetables. *HortScience* 30: 18–22.

Cameron, A.C., Boylan-Pett, W. and Lee, J. (1989) Design of modified atmosphere packaging systems: modeling oxygen concentrations within sealed packages of tomato fruits. *Journal of Food Science* 54(6): 1413–1421.

Chen, J.H. and Hotchkiss, J.H. (1993) Growth of *Listeria monocytogenes* and *Clostridium sporogenes* in cottage cheese in modified atmosphere packaging. *Journal of Dairy Science* 76: 972–977.

Choubert, G., Brisbarre, F., Parfouru, D. and Baccaunaud, M. (2008) Argon modified atmosphere packaging for fillets of rainbow trout (*Oncorhynchus mykiss*) fed astaxanthin or canthaxanthin. *Journal of Aquatic Food Product Technology* 17(2): 117–136.

Church, I.J. and Parsons, A.L. (1995) Modified atmosphere packaging technology: a review. *Journal of the Science of Food and Agriculture* 67: 143–152.

Day, B.P.F. (2003) Novel MAP applications for freshprepared produce. In: *Novel Food Packaging Techniques* (ed. R. Ahvenainen). CRC, Boca Raton, FL, pp. 189–207.

Del Nobile, M.A., Di Benedetto, N., Suriano, N., Conte, A., Corbo, M.R. and Sinigaglia, M. (2009) Combined effects of chitosan and MAP to improve the microbial quality of amaranth homemade fresh pasta. *Food Microbiology* 26: 587–591.

Dermiki, M., Ntzimani, A., Badeka, A., Savvaidis, I.N. and Kontominas, M.G. (2008) Shelf-life extension and quality attributes of the whey cheese "Myzithra Kalathaki" using modified atmosphere packaging. *LWT Food Science and Technology* 41: 284–294.

Doğu, E. (2009) *Marine edilmiş pişirmeye hazır tavuk etlerinin modifiye atmosfer paketleme ile muhafazası*. MSc thesis, Istanbul Technical University, Turkey.

Eliot, S.C., Vuillemard, J.C. and Emond, J.P. (1998) Stability of shredded Mozzarella cheese under modified atmospheres. *Journal of Food Science* 63(6): 1075–1080.

Ellis, W.O., Smith, J.P., Simpson, B.K., Ramaswamy, H. and Doyon, G. (1994) Effect of gas barrier characteristics of films on aflatoxin production by *Aspergillus flavus* in peanuts packaged under modified atmosphere packaging. *Food Research International* 27(6): 505–512.

Erkan, M.E., Vural, A., Çiftçioğlu, G., Aydın, A and Aksu, H. (2007) Comparison for the effect of MAP on Lor whey cheeses with two different initial microflora. *Archiv für Lebensmittelhygiene* 58: 51–56.

Ersan, S., Gunes, G. and Zor, A.O. (2010) Respiration rate of pomegranate arils as affected by O_2 and CO_2, and design of modified atmosphere packaging. *Acta Horticulturae* 876: 189–196.

Exama, A., Arul, J., Lencki, R.W., Lee, L.Z. and Toupin, C. (1993) Suitability of plastic films for modified atmosphere packaging of fruits and vegetables. *Journal of Food Science* 58(6): 1365–1370.

Farber, J.M. (1991) Microbiological aspects of modified atmosphere packaging – a review. *Journal of Food Protection* 54: 58–70.

Favati, F., Galgano, F. and Pace, A.M. (2007) Shelf-life evaluation of portioned Provolone cheese packaged in protective atmosphere. *LWT Food Science and Technology* 40: 480–488.

Fellows, P. (2000) *Food Processing Technology: Principles and Practice*, 2nd ed. CRC, Boca Raton, FL.

Galic, K., Curic, D. and Gabric, D. (2009) Shelf life of packaged bakery goods – a review. *Critical Reviews in Food Science and Nutrition* 49: 405–426.

Gammariello, D., Conte, A., Di Giulio, S., Attanasio, M. and Del Nobile, M.A. (2009) Shelf life of Stracciatella cheese under modified-atmosphere packaging. *Journal of Dairy Science* 92: 483–490.

González, J., Ferrer, A., Oria, R. and Salvador, M.L. (2008) Determination of O_2 and CO_2 transmission rates through microperforated films for modified atmosphere packaging of fresh fruits and vegetables. *Journal of Food Engineering* 86: 194–201.

Gonzalez-Fandos, E., Sanz, S. and Olarte, C. (2000) Microbiological, physicochemical and sensory characteristics of Cameros cheese packaged under modified atmospheres. *Food Microbiology* 17: 407–414.

Gorris, L.G.M. and Peppelenbos, H.W. (2007) Modified-atmosphere packaging of produce. In: *Handbook of Food Preservation* (ed. M.S. Rahman). Marcel Dekker, New York, pp. 315–333.

Gunes, G., Watkins, C.B. and Hotchkiss, J.H. (2001) Physiological responses of fresh-cut apple slices under high CO_2 and low O_2 partial pressures. *Postharvest Biology and Technology* 22: 197–204.

Guynot, M.E., Marín, S., Sanchis, V. and Ramos, A.J. (2004) An attempt to minimize potassium sorbate concentration in sponge cakes by modified atmosphere packaging combination to prevent fungal spoilage. *Food Microbiology* 21: 449–457.

Heydari, A., Shayesteh, K., Eghbalifam, N., Bordbar, H. and Falahatpisheh, S. (2010) Studies on the respiration rate of banana fruit based on enzyme kinetics. *International Journal of Agriculture and Biology* 12: 145–149.

Hong, S.I. and Kim, D.M. (2001) Influence of oxygen concentration and temperature on respiratory characteristics of fresh-cut green onion. *International Journal of Food Science and Technology* 36: 283–289.

Hotchkiss, J.H., Chen, J.H. and Lawless, H.T. (1999) Combined effects of carbon dioxide addition and barrier films on microbial and sensory changes in pasteurized milk. *Journal of Dairy Science* 82: 690–695.

Hotchkiss, J.H., Werner, B.G. and Lee, E.Y.C. (2006) Addition of carbon dioxide to dairy products to improve quality: a comprehensive review. *Comprehensive Reviews in Food Science and Food Technology* 5: 158–168.

Iqbal, T., Rodrigues, F.A.S., Mahajan, P.V. and Kerry, J.P. (2009) Mathematical modeling of the influence of temperature and gas composition on the respiration rate of shredded carrots. *Journal of Food Engineering* 91: 325–332.

Jamie, P. and Saltveit, M.E. (2002) Postharvest changes in broccoli and lettuce during storage in argon, helium, and nitrogen atmospheres containing 2% oxygen. *Postharvest Biology and Technology* 26: 113–116.

Jamie, P., Salvador, M.L. and Oria, R. (2001) Respiration rate of sweet cherries: "Burlat", "Sunburst" and "Sweetheart" cultivars. *Journal of Food Science* 66(1): 43–47.

Jayas, D.S. and Jeyamkondan, S. (2002) Modified atmosphere storage of grains meats fruits and vegetables. *Biosystems Engineering* 82(3): 235–251.

Jeremiah, L.E. (2001) Packaging alternatives to deliver fresh meats using short- or long-term distribution. *Food Research International* 34: 749–772.

John, L., Cornforth, D., Carpenter, C.E., Sorheim, O., Pettee, B.C. and Whittier, D.R. (2005) Color and thiobarbituric acid values of cooked top sirloin steaks packaged in modified atmospheres of 80% oxygen, or 0.4% carbon monoxide, or vacuum. *Meat Science* 69: 441–449.

Kader, A.A. and Saltveit, M.E. (2003) Respiration and gas exchange. In: *Postharvest Physiology and Pathology of Vegetables*, 2nd edn (eds J.A. Bartz and J.K. Brecht). Marcel Dekker, New York, pp. 7–29.

Kader, A.A., Singh, P.R. and Mannapperuma, J.D. (1998) Technologies to extend the refrigerated shelf life of fresh fruits. In: *Food Storage Stability* (eds I.A. Taub and R.P. Singh). CRC, Boca Raton, FL, pp. 419–435.

Kasai, Y., Kimura, B., Kawasaki, S., Fukaya, T., Sakuma, K. and Fuljii, T. (2005) Growth and toxin production by *Clostridium botulinum* in steamed rice aseptically packed under modified atmosphere. *Journal of Food Protection* 68(5): 1005–1011.

Kırkın, C. (2009) *Tüketime hazır, doğranmış beyaz peynirlerin modifiye atmosferde paketleme ile muhafazası.* MSc thesis, Istanbul Technical University, Turkey.

Kosikowski, F.V. and Brown, D.P. (1973) Influence of carbon dioxide and nitrogen on microbial populations and shelf life of cottage cheese and sour cream. *Journal of Dairy Science* 56: 33–38.

Kristensen, D., Orlien, V., Mortensen, G., Brockhoff, P. and Skibsted, L.H. (2000) Light-induced oxidation in sliced Havarti cheese packaged in modified atmosphere. *International Dairy Journal* 10: 95–103.

Kusmider, E.A., Sebranek, J.G., Lonergan, S.M. and Honeyman, M.S. (2002) Effects of carbon monoxide packaging on color and lipid stability of irradiated ground beef. *Journal of Food Science* 67(9): 3464–3468.

Lee, D.S. and Renault, P. (1998) Using pinholes as tools to attain optimum modified atmospheres in packages of fresh produce. *Packaging Technology and Science* 11: 119–130.

Lee, D.S., Haggar, P.E., Lee, J. and Yam, K.L. (1991) Model for fresh produce respiration in modified atmospheres based on principles of enzyme kinetics. *Journal of Food Science* 56(6): 1580–1585.

Lee, K.S., Park, I.S. and Lee, D.S. (1996) Modified atmosphere packaging of a mixed prepared vegetable salad dish. *International Journal of Food Science and Technology* 31: 7–13.

Lloyd, M.A., Hess, S.J. and Drake, M.A. (2009) Effect of nitrogen flushing and storage temperature on flavor and shelf-life of whole milk powder. *Journal of Dairy Science* 92: 2409–2422.

Mahajan, P.V. and Goswami, T.K. (2001) Enzyme kinetics based modeling of respiration rate for apple. *Journal of Agricultural Engineering Research* 79(4): 399–406.

Mahajan, P.V., Oliveira, F.A.R., Sousa, M.J., Fonseca, S.C. and Cunha, L.M. (2006) An interactive design of MA-packaging for fresh produce. In: *Handbook of Food Science, Technology, and Engineering*, Vol. 3 (ed. Y.H. Hui). CRC, Boca Raton, FL, pp. 119-1–119-16.

Mangaraj, S., Goswami, T.K. and Mahajan, P.V. (2009) Applications of plastic films for modified atmosphere packaging of fruits and vegetables: a review. *Food Engineering Reviews* 1: 133–158.

Martin, M.B.S., Fernández-García, T., Romero, A. and López, A. (2001) Effect of modified atmosphere storage on hazelnut quality. *Journal of Food Processing Preservation* 25: 309–321.

McLaughlin, C.P. and O'Beirne, D. (1999) Respiration rate of dry coleslaw mix as affected by storage temperature and respiratory gas concentrations. *Journal of Food Science* 64(1): 116–119.

McMillin, K.W. (2008) Where is MAP going? A review and future potential of modified atmosphere packaging for meat. *Meat Science* 80: 43–65.

Mercogliano, R., De Felice, A., Panzardi, M. and Anastasio, A. (2009) Evaluation of the shelf life of vacuum- and modified atmosphere packaged *Dicentrarchus labrax* fillets. *Veterinary Research Communications* 33 (Supplement 1): S245–S247.

Ooraikul, B. (2003) Modified atmosphere packaging. In: *Food Preservation Techniques* (eds P. Zeuthen and L. Bøgh-Sorensen). Woodhead, Boca Raton, FL, pp. 338–355.

Oyugi, E. and Buys, E.M. (2007) Microbiological quality of shredded Cheddar cheese packaged in modified atmospheres. *International Journal of Dairy Technology* 60(2): 89–95.

Ozturk, A., Yilmaz, N. and Gunes, G. (2010) Effect of different modified atmosphere packaging on microbial quality, oxidation and colour of a seasoned ground beef product (meatball). *Packaging Technology and Science* 23: 19–25.

Papaioannou, G., Chouliara, I., Karatapanis, A.E., Kontominas, M.G. and Savvaidis, I.N. (2007) Shelf-life of a Greek whey cheese under modified atmosphere packaging. *International Dairy Journal* 17: 358–364.

Parry, R.T. (1993) Introduction. In: *Principles and Applications of Modified Atmosphere Packaging of Food* (ed. R.T. Parry). Blackie Academic & Professional, Glasgow, pp. 1–18.

Philips, C.A. (1996) Review: modified atmosphere packaging and its effects on the microbiological quality and safety of produce. *International Journal of Food Science and Technology* 31: 463–479.

Pintado. M.E. and Malcata, F.X. (2000) The effect of modified atmosphere packaging on the microbial ecology in *Requeijão*, a Portuguese whey cheese. *Journal of Food Processing Preservation* 24: 107–124.

Rao, D.N. and Sachindra, N.M. (2002) Modified atmosphere and vacuum packaging of meat and poultry products. *Food Reviews International* 18(4): 263–293.

Rasmussen, P.H. and Hansen, A. (2001) Staling of wheat bread stored in modified atmosphere. *LWT Food Science and Technology* 34: 487–491.

Ravindra, M.R. and Goswami, T.K. (2008) Modelling the respiration rate of green mature mango under aerobic conditions. *Biosystems Engineering* 99: 239–248.

Renault, P., Souty, M. and Chambroy, Y. (1994) Gas exchange in modified atmosphere packaging. 1: A new theoretical approach for micro-perforated packs. *International Journal of Food Science and Technology* 29: 365–378.

Robertson, G.L. (1993) *Food Packaging: Principles and Practice.* Marcel Dekker, New York.

Rocculi, P., Romani, S. and Rosa, M.D. (2005) Effect of MAP with argon and nitrous oxide on quality maintenance of minimally processed kiwifruit. *Postharvest Biology and Technology* 35: 319–328.

Rodríguez-Hidalgo, S., Artés-Hernández, F., Gómez, P.A., Fernández, J.A. and Artés, F. (2010) Quality of fresh-cut baby spinach grown under a floating trays system as affected by nitrogen fertilization and innovative packaging treatments. *Journal of the Science of Food and Agriculture* 90: 1089–1097.

Saucier, L., Gendron, C. and Gariepy, C. (2000) Shelf life of ground poultry meat stored under modified atmosphere. *Poultry Science* 79: 1851–1856.

Sawaya, W., Elnawawy, A., Abu-Ruwaida, A., Khalafawi, S. and Dashti, B. (1995) Influence of modified atmosphere packaging on shelf-life of chicken carcasses under refrigerated storage conditions. *Journal of Food Safety* 15: 35–51.

Seyfert, M., Mancini, R.A., Hunt, M.C., Tang, J. and Faustman, C. (2007) Influence of carbon monoxide in package atmospheres containing oxygen on color, reducing activity, and oxygen consumption of bovine muscles. *Meat Science* 75: 432–442.

Simpson, R., Carevic, E. and Rojas, S. (2007) Modelling a modified atmosphere packaging system for fresh scallops (*Argopecten purpuratus*). *Packaging Technology and Science* 20: 87–97.

Sivertsvik, M., Jeksrud, W.K. and Rosnes, T. (2002) A review of modified atmosphere packaging of fish and fishery products – significance of microbial growth, activities and safety. *International Journal of Food Science and Technology* 37: 107–127

Skandamis, P.N. and Nychas, G.J.E. (2002) Preservation of fresh meat with active and modified atmosphere packaging conditions. *International Journal of Food Microbiology* 79: 35–45.

Stetzer, A.J., Wicklund, R.A., Paulson, D.D., Tucker, E.M., MacFarlane, B.J. and Brewer, M.S. (2007) Effect of carbon monoxide and high oxygen modified atmosphere packaging (MAP) on quality characteristics of beef strip steaks. *Journal of Muscle Foods* 18: 56–66.

Song, Y., Lee, D.S. and Yam, K.L. (2001) Predicting relative humidity in modified atmosphere packaging system containing blueberry and moisture absorbent. *Journal of Food Processing and Preservation* 25: 49–70.

Song, Y., Vorsa, N. and Yam, K.L. (2002) Modelling respiration-transpiration in a modified atmosphere packaging system containing blueberry. *Journal of Food Engineering* 53: 103–109.

Taniwaki, M.H., Hocking, A.D., Pitt, J.I. and Fleet, G.H. (2001) Growth of fungi and mycotoxin production on cheese under modified atmospheres. *International Journal of Food Microbiology* 68: 125–133.

Techavises, N. and Hikida, Y. (2008) Development of a mathematical model for simulating gas and water vapor exchanges in modified atmosphere packaging with macroscopic perforations. *Journal of Food Engineering* 85: 94–104.

Temiz, H., Aykut, U. and Hursit, A.K. (2009) Shelf life of Turkish whey cheese (Lor) under modified atmosphere packaging. *International Journal of Dairy Technology* 62(3): 378–386.

Thompson, A.K. (1998) *Controlled Atmosphere Storage of Fruits and Vegetables*. CAB International, Wallingford, UK.

Torrieri, E., Mahajan, P.V., Cavella, S., Gallagher, M.S., Oliveira, F.A.R. and Masi, P. (2009) Mathematical modelling of modified atmosphere package: an engineering approach to design packaging systems for fresh-cut produce. In: *Advances in Modeling Agricultural Systems* (eds P.J. Papajorgji and P.M. Pardalos). Springer Science+Business Media, New York, pp. 455–483.

Torrieri, E., Perone, N., Cavella, S. and Masi, P. (2010) Modelling the respiration rate of minimally processed broccoli (*Brassica rape* var. *sylvestris*) for modified atmosphere package design. *International Journal of Food Science and Technology* 45: 2186–2193.

Vongsawadi, P., Wongwicharn, A., Khunajakr, N. and Dejsuk, N. (2008) Shelf-life extension of precooked chicken fillets by modified atmosphere packaging. *Kasetsart Journal (Natural Science)* 42: 127–135.

Westall, S. and Filtenborg, O. (1998) Spoilage yeasts of decorated soft cheese packed in modified atmosphere. *Food Microbiology* 15: 243–249.

Yalamanchili, T. (2009) *Quality of Shell Eggs Stored under Modified Atmosphere Packaging*. Master's thesis, Texas Tech University, Lubbock, TX.

Yesudhason, P., Gopal, T.K.S., Ravishankar, C.N., Lalitha, K.V. and Kumar, K.N.A. (2009) Effect of modified atmosphere packaging on chemical, textural, microbiological and sensory quality of seer fish (*Scomberomorus commerson*) steaks packaged in thermoformed trays at 0–2C. *Journal of Food Processing and Preservation* 33: 777–797.

Yılmaz, M., Ceylan, Z.G., Kocaman, M., Kaya, M. and Yılmaz, H. (2009) The effect of vacuum and modified atmosphere packaging on growth of *Listeria* in rainbow trout (*Oncorhynchus mykiss*) fillets. *Journal of Muscle Foods* 20: 465–477.

Zhang, M., Zhan, Z.G., Wang, S.J. and Tang, J.M. 2008. Extending the shelf-life of asparagus spears with a compressed mix of argon and xenon gases. *LWT Food Science and Technology* 4: 686–691.

44

Packaging for Processed Food and the Environment

Eva Almenar, Muhammad Siddiq and Crispin Merkel

Introduction

Packaging

Packaging can be defined as "all products made of any materials of any nature to be used for the containment, protection, handling, delivery and presentation of goods, from raw materials to processed goods, from the producer to the user or the consumer" (EU, 1994). There are different types of packaging for the various stages involved in the transport of goods from the producer to the user or the consumer:

1. Primary, or sales, packaging constitutes a single sales unit to the end user or consumer at the point of purchase (e.g., jars, cans and bottles).
2. Secondary, or grouped, packaging is packaging that constitutes at the point of purchase a grouping of a certain number of sales units. It can be removed from the product without affecting its characteristics (e.g., boxes, trays and film wrap). It serves as a means to efficiently replenish the shelves at the point of sale, and the consumer can, but does not need to, purchase the goods in units of this.
3. Tertiary, or transport, packaging is packaging that facilitates the handling and transport of a number of sales units or grouped packages in order to prevent physical damage from handling and transport (e.g., large containers and pallets). Transport packaging does not include road, rail, ship and air containers.

Handbook of Food Process Design, First Edition. Edited by Jasim Ahmed and Mohammad Shafiur Rahman.

There are different categories of primary or sales packaging depending on the type of product being packaged: "food packaging" if the product being packaged is a food item, "pharmaceutical packaging" if the product being packaged is a drug, and so on.

Packaging and Processed Food

Food deteriorates owing to physicochemical and microbiological changes. These result in color, texture and flavor changes and, in some cases, possible toxicity of the product. Food is processed in order to bring quality and safety to the consumer. Packaging plays a primary role in maintaining these benefits of processed foods under specified conditions for a required period of time. In addition, it provides identification and information. This means that adequate packaging reduces food losses, improves food safety, allows larger markets and increases the consumer's choices.

In order to design an adequate packaging system for a processed food, the following particulars should be considered: (i) the characteristics and requirements of the food product: the acidity of the food, its sensitivity to light and oxygen, the amount of physical protection needed, the amount of product visibility desired, and the desired shelf-life; (ii) the food processing technology; (iii) package characteristics: size, weight, cost and availability of the packaging material, and opacity, mechanical and barrier properties, durability, chemical and corrosion resistance, and ergonomics; (iv) environmental impact, i.e., life cycle assessment; and (v) consumer attitudes about packaging.

Packaging for Processed Food and the Environment

Packaging is a crucial part of the food supply chain. It delivers the food from the point of production to the point of consumption. In order to be sustainable, packaging must balance food protection with environmental impact. The Industry Council for Packaging and the Environment (INCPEN) defines a sustainable packaging and product supply chain as "a system that enables goods to be produced, distributed, used and recovered with minimum environmental impact at the lowest social and economic cost" (INCPEN, 2010). This means: (i) optimization of the use of material, water and energy; (ii) minimizing waste both from the product and from used packaging; and (iii) maximizing the recovery of value from waste in the form of material, energy or compost. To achieve these goals, packaging has to be designed taking into account its entire life cycle and the characteristics of the product, the supply chain and the needs of consumers. In this regard, the Australian food packaging supply chain has been examined (James, 2003). The responses obtained from 27 companies demonstrated that the packaging-related environmental issues in the Australian food packaging supply chain are the management and minimization of solid waste and emissions to air, land and water. Figure 44.1 illustrates the possible different life cycles for a package in the food supply chain. The figure includes current initiatives to reduce packaging waste

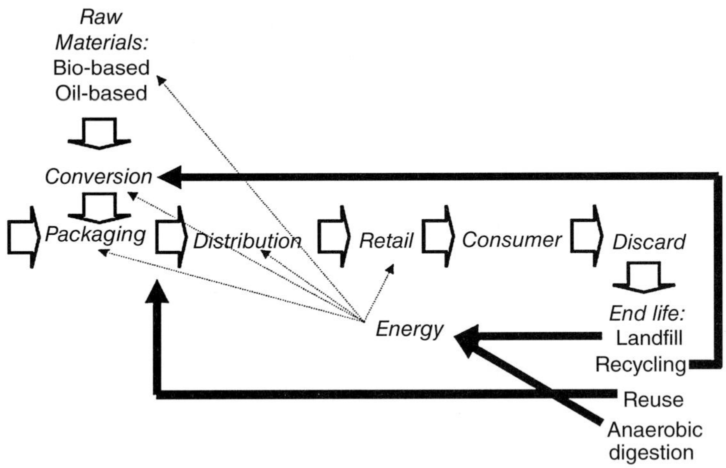

Figure 44.1 Possible different life cycles for a package in the food supply chain.

such as recycling and reuse, and emerging initiatives such as the use of renewable resources. As observed, material recovered from recycling can be used for conversion. Some packages, such as bottles and jars, can return back to the original supply chain and be reused. Energy recovered from incineration can be used for various purposes in one or several of the steps of the supply chain.

Packaging waste accounted for about 31% of the total municipal solid waste (MSW) generated in the United States in 2008 (EPA, 2009b), less than 5% of total waste by weight and volume in British landfills in that same year (INCPEN, 2010), 10% of the MSW generated in Australia in 2004 (BAO, 2004), and 25% of the MSW generated in Europe in 2005 (PIRA, 2005). Although statistics like these are difficult to compare because of the differences in definitions of MSW between countries, all amounts of packaging have remained fairly constant over recent years. For example, packaging and containers accounted for 33.1% of the US MSW in 1999 and for 30.8% in 2008. The possible methods of disposal of packaging waste (Figure 44.1) include landfill and recycling. Landfill is the dominant method. The current initiatives to reduce packaging waste focus primarily on source reduction and recovery. There are also emerging and next-generation initiatives to reduce packaging waste that focus upon compostability, anaerobic digestion and waste-to-energy incineration. Thus, there is less packaging waste than commonly expected.

Landfilling

Landfills are an important outlet for disposal of packaging waste. They can receive the packaging waste directly or receive the waste from other methods of dealing with packaging waste such as incineration. Currently, old facilities are being retired and replaced with modern ones that are designed to control the two principal environmental concerns associated with landfills:

1. *Leachate:* the mixture of water and dissolved solids produced as water passes through waste and collects at the bottom of the landfill, which can contaminate ground and surface water. Some of the dissolved solids can have toxic, polluting components. The composition of the leachate depends on the type of waste and its stage of decomposition.
2. *Gas emissions (landfill gas):* the mixture of carbon dioxide and methane, small amounts of nitrogen and oxygen, and trace amounts of a wide range of other gases produced during the biodegrading of waste resulting from bacterial action. Some components of this mixture may be toxic or explosive. The makeup of the mixture depends on the composition, temperature, moisture content and age of the waste. Landfill gas is a greenhouse gas.

Such new facilities are sited taking into consideration soil conditions, hydrology and topography, climate, and other factors (ABS, 2006). The principal environmental concerns associated with modern landfills are methane (landfill gas) production, possible long-term leachate, the energy used to transport waste, and air emissions emitted during waste transportation. Most modern landfills use a gas capture technology where the landfill gas is either flared to convert methane into carbon dioxide or collected and used as a substitute fuel or to generate energy (energy recovery). However, the methane capture rate is only about 55%. As for leachate, modern landfills are lined with impervious membrane layers to avoid leachate, but a small percentage of leachate may escape and pose an environmental risk. Despite these drawbacks, modern landfills have grown, since they generate energy from renewable resources and reduce greenhouse gas emissions. In Australia, up to 75% of landfills servicing major urban areas and capital cities use gas capture technologies (AEEMA, 2006). The quantity of MSW disposed of in landfills has declined as recycling and combustion have increased. For example, of the 88 million tons (MT) of MSW generated in the United States in 1960, 6% was recovered through recycling and 94% was landfilled, whereas of the 251 MT generated in 2006, 32.5% was recycled (including composting), 13% combusted with energy recovery, and 55% landfilled. From 1990 to 2007, the percentage of MSW generated that was sent to landfills dropped from 69 to 55%, the percentage recycled rose from 14 to 24%, the percentage composted rose from 2 to 8%, and the percentage combusted with energy recovery ranged from 13 to 15% (EPA, 2008).

Source Reduction

Source reduction includes the design, manufacture, purchase or use of products and packaging to reduce their amount before they enter the MSW management system (EPA, 2001). Source reduction includes:

1. Designing packaging to reduce the quantity of materials used. There is a constant effort in the packaging industry to reduce the amount of material used for packaging: glass containers now are, on average, 30% lighter than in 1980, the weight of

cans is now approximately 40% less than in 1970, and 2-liter polyethylene terephthalate (PET) soft-drink bottles are 25% lighter than in 1977 (INCPEN, 2010; EPA, 2009b). This means that millions of pounds of packaging materials per year are kept out of the waste stream, as well as an important reduction in the amount of raw materials used.

2. Changing the packaging format, which can also lead to overall reduction in the weight of packaging used. For example, there was a 20% reduction in the quantity of packaging (sales and transit packaging) used to pack 1-liter of soft drinks in the UK from 1997 to 2002 (PIRA, 2004).
3. Reusing existing packaging. One of the current initiatives to reduce packaging waste is reuse. Reuse can be defined as the recovery or reapplication of a package in a manner that retains its original form or identity. Reuse delays or avoids the entry of packages into the disposal system, and therefore reduces the costs of recycling, landfilling, combustion or other disposal methods and reduces the amounts of raw materials used. It has been reported that between 2 and 5% of the waste stream is potentially reusable (EPA, 2009b). Reusable and refillable packaging needs to be designed carefully and applied to appropriate products in order to be beneficial and safe. This is partially because collecting, transporting and cleaning such packages can pose logistical difficulties. Reusable and refillable packaging is currently in use, and the reused packages are typically used for the same food products as packaged originally or for similar products. Refillable glass bottles, reusable plastic food storage containers and refurbished wood pallets are common examples of reusable and refillable packaging. A considerable number of beer bottles were collected by restaurants/taverns for refilling in 2007. The Glass Packaging Institute estimates that refillable glass bottles achieve a rate of eight refillings per bottle. Over 13 million tons of wood pallets were refurbished and returned to service in 2007 (EPA, 2008).
4. Using packaging that reduces the amount of damage to or spoilage of the food product. Adequate packaging reduces the physical and chemical changes expected to happen in food during storage, reduces or inhibits the effect of environmental factors, reduces mechanical damage, retards microbial spoilage, and provides other benefits. According to the Food and Agriculture Organization of the United Nations (FAO), improved packaging and handling could immediately reduce crop loss by 5% in many countries, which would increase the world food supply by 39 MT or more each year (Burden and Wills, 1989).

One of the benefits of source reduction is the dramatic reduction in packaging waste, as well as in the amount of material recycled or sent to landfill or to combustion facilities. Another benefit includes cost reduction, and this provides economic incentives for both consumers and industry to practice source reduction. Some communities have instituted "pay-as-you-throw" programs, where citizens pay for each can or bag of trash they set out for disposal rather than through the tax base or a flat fee. Industry also reduces costs when it manufactures its products with less packaging,

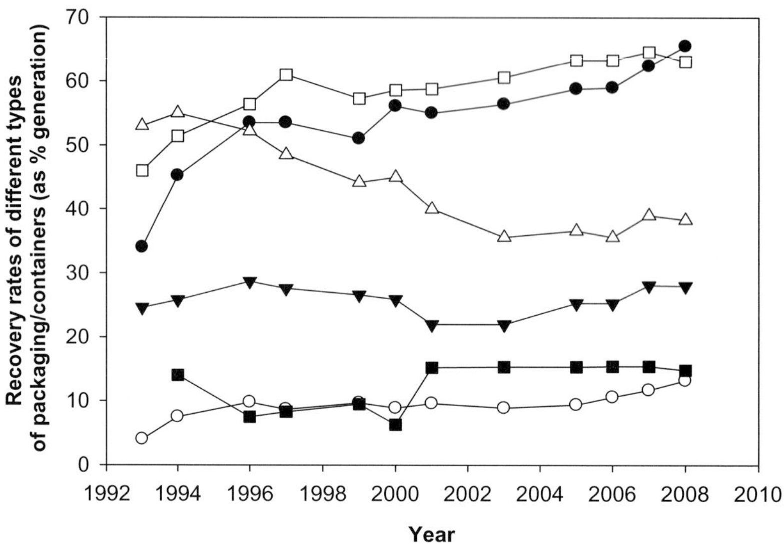

Figure 44.2 Evolution of the US recovery rates of different types of packaging/containers as percentage of generation for selected years from 1993 to 2008. ((-○-) Plastic packaging, (-●-) paper and paperboard packaging, (-△-) aluminum packaging, (-□-) steel packaging, (-▼-) glass containers, (-■-) wood.) (Source: adapted from EPA, 1994; EPA, 1995; EPA, 1997; EPA, 1998; EPA, 2001; EPA, 2002; EPA, 2003; EPA, 2005; EPA, 2006; EPA, 2007; EPA, 2008.)

since less raw material is bought. Consumers also can share the economic benefits of source reduction. Buying products in bulk, with less packaging, or with reusable (not single-use) packaging frequently means a cost saving (EPA, 2009b).

Recovery

Another current alternative method to reduce packaging waste is recovery. Materials and/or energy can be recovered from packaging waste at the same time as packaging waste is reduced. Figure 44.2 shows the evolution of the US recovery rates of different types of packaging/containers as a percentage of generation over the last 15 years. As observed, the recovery of paper and paperboard, steel packaging, plastic packaging, and wood has increased by approximately 30, 20, 10 and 7%, respectively, in the last 15 years. The recovery of glass containers has remained fairly constant (an increase of less than 5%) and that of aluminum packaging has declined significantly (by approximately 15%) during this same period of time.

Recycling

Recovery for recycling continues to be one of the most effective methods for the management of packaging waste. Recycling can be defined as the recovery of used packages or unwanted packaging materials (scrap) that would otherwise become waste, to be turned into raw materials for the manufacture of new packages or other products. Thus, recycling plays an important role at every stage in the life cycle of a package, from the raw-material state to its final disposal (EPA, 2008). Recycling includes: (i) the collection of materials that would otherwise be considered waste, (ii) the separation and processing of these materials in material recovery facilities, (iii) the sale of the processed materials, (iv) the manufacture of the processed materials into new products, and (v) the purchasing of the recycled products. This creates a loop that ensures the value of recycling. Recyclable materials can be collected from households via curbside collection or via public recycling bins, or can be delivered directly by the household to recycling depots. They will be used as feedstocks, and therefore neither the use of virgin materials nor all the upstream energy and associated environmental impacts from the extraction, transport and processing of those virgin materials are required. Some benefits associated with recycling are energy savings; reduction of greenhouse gas emissions, which contribute to global warming; reduction of air and water pollution associated with the making of new products from raw materials; and conservation of natural resources such as water (CRI, 2009). Moreover, recycling reduces the need for landfills and incineration (EPA, 2008).

Almost all packaging materials (glass, metals, paper, paperboard and plastics) are recyclable. In the United States, steel, paper products and aluminum were the most recycled materials by percentage in 2008 (EPA, 2009a). In Australia, recycling rates for beverage packaging and food packaging were about 50% and between 20% and 50%, respectively, in 2002. Metals had the highest recycling rate (82% of total metal waste generated), followed by paper (55% of total paper waste generated) and glass (38% of total glass waste generated) (Kaspura, 2006). More and more of today's products are being manufactured completely or in part from recycled content. Common household items that contain recycled materials (aluminum, plastic and glass) include steel cans, soft-drink containers and some other plastic bottles. Recycled materials made from metals and glass are considered safe for food contact because the heat used to melt and form these materials is sufficient to kill microorganisms and pyrolyze organic contaminants. This is not the case for recycled plastic materials, since their processing temperatures kill microorganisms but do not pyrolyze all organic contaminants (Marsh and Bugusu, 2007).

Waste-to-Energy Incineration

Recovery of energy from packaging waste is currently carried out, too. It is based on the collection and utilization of heat generated through controlled combustion of packaging waste. Packaging waste can also be combusted without energy recovery. Combustion with or without energy recovery is a common waste management practice in some countries where space for landfill is scarce, such as in some European and Asian countries and in countries where the annual MSW is growing rapidly, as in the case of China (ABS, 2006; Nie, 2008). Combustion with energy recovery is called waste-to-energy (WTE) incineration. WTE incinerators are installed with boilers to recover the combustion heat for energy. These WTE facilities can produce steam that can either provide heat or generate electricity, or both (Marsh and Bugusu, 2007). In the United States, this process reduces the volume of MSW by as much as 90% (when combined with recycling, otherwise the reduction is lower). Eighty-seven WTE facilities with a processing capacity of 31.4 million tons (about 87 000 tons per day) were reported in 2006, representing 12.5% of all MSW disposal (EPA, 2007; IWSA, 2007). In China, 67 WTE facilities with a total daily capacity of approximately 33 010 tons per day were operational in 2005, and accounted for about 12.9% of all the MSW being landfill-treated in China at the time (Nie, 2008). WTE facilities are designed to achieve high combustion temperatures, which helps the waste to burn cleaner and to create less ash for disposal. The major concerns about the environmental risks of MSW incinerators are the potential emission of contaminants into the air through exhaust stacks and into the water through ash leachate. These result from compounds present in the waste stream or are formed as a result of incomplete combustion. Pollution concerns include the emission of acid gases (sulfur dioxide and hydrogen chloride), heavy metals (mercury, lead and cadmium), halogens, organics (dioxins and furans) and other substances (UNEP, 2010).

In order to meet current environmental standards, modern air pollution control devices are used to remove potential harmful particulates and gases from incinerator emissions (SPI, 2010). When properly operated, the best air pollution control equipment can remove up to 99% of dioxins and furans, more than 99% of heavy metals, more than 99% of particulate matter, more than 99% of hydrogen chloride, more than 90% of sulfur dioxide, and up to 65% of nitrogen oxides (UNEP, 2010). The currently available devices for incinerators are: (i) fabric filters or baghouses, which consist of several cylindrical bags that filter emissions of metals and organic compounds that attach to fine particulates; (ii) electrostatic precipitators, which electrically charge particulate emissions and then draw the particles to oppositely charged collection plates, from where they are removed by shaking, forming fly ash; and (iii) scrubbers, which primarily control acid gases but also remove some heavy metals. There are two types of scrubbers. Wet scrubbers use an alkaline liquid solution to neutralize acids, whereas dry scrubbers use a fine alkaline spray to neutralize acids. The generally accepted air pollution control system uses dry scrubbing followed by a fabric filter. Proper control of air emissions, however, requires more than the presence of these control technologies. MSW incineration facilities must be well operated and well maintained to ensure that emissions are as low as possible. Good combustion practices can control emissions by ensuring that the temperature in the combustion chamber and the time the MSW remains in the combustion chamber are kept at optimal levels. Newer incinerators are equipped with computer control systems to help maintain a high degree of consistency in plant operations. Air pollution control equipment must also be carefully maintained to prevent the release of contaminants. These emission controls are not perfect, however. They do not contribute to removing nitrogen oxides, and actually are not very effective at removing mercury, either. Additional steps are required to remove these.

MSW incineration also generates ash, representing about 10% by volume and 25–35% by weight of the waste incinerated. Ash can be divided into bottom ash and fly ash. Bottom ash is completely or partially combusted material that passes through or is discharged from the combustion grate. Fly ash is the term for particulate matter captured from the flue gas by the air pollution control system. The total ash generated by WTE incinerators is 75–85% bottom ash and 15–25% fly ash, by weight. The main environmental concern regarding incinerator ash is that when ash is disposed of in a landfill, metals and organic compounds can leach, migrate into and contaminate water supplies. One solution to this is to stabilize and solidify the incinerator ash in order to reduce the ability of heavy metals to migrate from the ash into the environment. This can be achieved by encasing the ash in concrete, prior to disposal in an ash-only landfill known as an ash monofill. Ash monofills are often co-located with MSW incinerators or existing landfills to reduce transportation distances and siting difficulties (UNEP, 2010).

Plastics possess the highest heat content of all packaging materials, and therefore they are the most advantageous for WTE incineration. However, lead- and cadmium-based additives for plastics and colorants contribute to the heavy metal content of the ashes, and therefore the ashes are managed as potentially hazardous material (Subtitle C of the Resource Conservation and Recovery Act) (Marsh and Bugusu, 2007).

Degradation

Some emerging and next-generation initiatives to reduce packaging waste are degradation, composting and anaerobic digestion. Degradation is defined as "an irreversible process leading to a significant change of the structure of a material, typically characterized by a loss of properties (e.g., integrity, molecular weight, structure or mechanical strength) and/or fragmentation. Degradation is affected by environmental conditions and proceeds over a period of time comprising one or more steps" (ASTM, 1991). There are several different types of degradation such as thermal degradation, photodegradation, oxidative degradation, hydrolytic degradation, mechanical degradation and biodegradation, caused by heat, sunlight, oxygen, water, stress and microorganisms, respectively. Among them, biodegradation is one of the most environmentally friendly and can be defined as "the process of converting organic materials back into CO_2 and H_2O though microbial action." This process exhibits two phases: (i) disintegration (fragmentation of the material by the action of extracellular microbial enzymes) and (ii) and mineralization (microorganisms digest the water-soluble fragments of plastics and convert them to CO_2, water and cell biomass under aerobic conditions, or to CH_4, CO_2 and cell biomass under anaerobic conditions) (Krzan *et al.*, 2006). There are many environments where plastics can be biodegraded (soil, marine environments, etc.). If the environment is a compost pile, the biodegradation is known as composting.

Composting

Composting is defined as "a process where biodegradable materials, such as manure and leaves, are decomposed and transformed into a humus-like substance called compost, CO_2, water, and minerals by microorganisms through a controlled biological process" (ASTM, 2004). A material is compostable when its biodegradation is compatible with the conditions (temperature, humidity level and time) found in composting facilities. Therefore, not all biodegradable materials meet composting criteria. There are several different composting methods. The most common are backyard or onsite composting, vermicomposting, aerated windrow composting, aerated static pile composting and in-vessel composting. Composting can result in several environmental benefits, and these include the following (EPA, 2010):

1. It helps to regenerate poor soils.
2. It absorbs odors and treats semivolatile and volatile organic compounds (VOCs).
3. It avoids the production of methane and leachate formation in landfills.
4. It reduces the need for water, fertilizers and pesticides.
5. It reduces the high cost of transporting recycled materials.
6. It reduces pressure on landfill space.

Composting must be managed properly so as not to cause excessive odors, anaerobic digestion, methane, leachate production and contaminated groundwater or surface water.

Anaerobic Digestion

Anaerobic digestion is a way to recover energy from organic wastes such as biodegradable packaging waste. In this process, anaerobic conditions are created in order to transform organic waste into "bio-gas" methane. The use of anaerobic digestion for disposal of organic waste is expected to grow rapidly (Piccione, 2010). An initiative to create energy and cut landfill volumes using anaerobic digestion of food and bioplastic packaging waste was supported by the UK government in 2009. The UK government produced a report entitled *Developing an Implementation Plan for Anaerobic Digestion* that details ways in which anaerobic digestion could be boosted. This report also comments on how anaerobic digestion of renewable plastics can help to deal with food packaging waste (Harrington, 2009; DEFRA, 2009). Some of the issues related to the running of energy recovery plants using anaerobic digestion are the following: (i) there are still relatively few collection schemes, (ii) cost-effective systems are needed for the separate collection of the material, (iii) a risk-based regulatory framework is needed, and (iv) there are issues with feedstocks containing a high proportion of nitrogen (DEFRA, 2009).

Life Cycle Assessment

It is necessary to evaluate the complete life cycle of a package (production, use and disposal) to obtain a clear idea of its true environmental implications. Life cycle assessment (LCA) is a tool currently used for evaluating the environmental impact of existing and new packaging materials or packages. LCA can be defined as "a systematic process for identifying, quantifying, and assessing environmental impacts throughout the life cycle of a product, process, or activity." It is done by (i) formulating and specifying the goal and the scope of study in relation to the intended application, (ii) compiling an inventory of relevant inputs and outputs, (iii) evaluating the potential environmental impacts associated with those inputs and outputs, and (iv) interpreting the results of the inventory and impact phases in relation to the objectives of the study (ISO, 2006a). The ultimate goal of an LCA is to improve the environmental performance of the product, process or activity. An LCA considers energy and material use, and releases to the environment, from raw-material extraction through manufacturing, transportation, use and disposal. There are many different organizations and companies performing LCAs for new and existing packaging materials/packages, and there are database resources being used to generate comparative reports. The International Organization for Standardization (ISO) has been active in generating new standards on environmental management, the ISO 14000 family of standards (ISO, 2006b).

Traditional Packaging Materials and the Environment

Materials that have traditionally been used to package processed food include glass, metals (tin-plated steel, tin-free steel and aluminum), plastics, paper and paperboard, and wood. Besides their technical properties (how well the package protects a food item for the required shelf-life), their use is mostly dependent on cost and availability. Furthermore, there may be particular marketing reasons for choosing a certain type of package. Figure 44.3 shows the percentages of plastics, paper and paperboard, glass, metals, and wood in MSW in the USA from 1990 to 2008. Paper and paperboard, which have been the dominant contributors to the US MSW, have declined over the years, from approximately 38% in 1993 to 31% in 2008. This may be due to the increase in use of products made from plastics. Plastics have been a rapidly growing component of MSW. Metals have accounted for a little bit less than plastics but more than glass and wood, and have remained fairly constant as a component of MSW since 1990. Wood (around 60% of it in the form of pallets) decreased until 2006, increasing somewhat in recent years. Glass has been displaced by plastics, too. Its share in MSW has declined in 18 years from approximately 7% in 1990 to 4.9% in 2008.

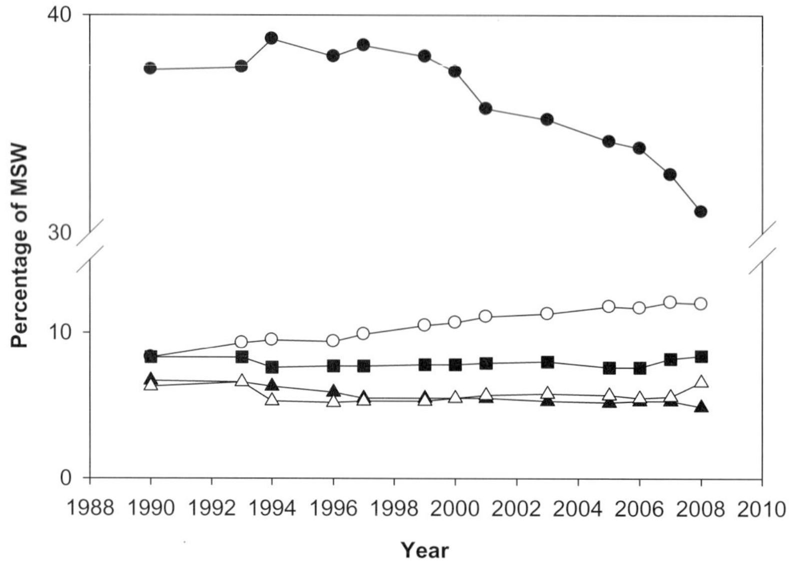

Figure 44.3 Percentages of plastics, paper and paperboard, metals, glass, and wood in the total US municipal solid waste (MSW) from 1990 to 2008. ((-○-) Plastics, (-●-) paper and paperboard, (-■-) metals, (-▲-) glass, (-△-) wood.) (Source: adapted from EPA, 1992; EPA, 1994; EPA, 1995; EPA, 1997; EPA, 1998; EPA, 2001; EPA, 2002; EPA, 2003; EPA, 2005; EPA, 2006; EPA, 2007; EPA, 2008.)

Glass

The mechanization of glass container manufacturing was introduced on a large scale in 1892 (Hanlon, 1984). Further developments have occurred, resulting in the production of a wide range of glass containers that play a significant role in the packaging of food products. Glass containers are 100% recyclable. They can be recycled indefinitely (closed-loop recycling) without any loss in purity or quality (GPI, 2010). In 2008, around 28% of the glass containers in United States were recycled (EPA, 2009a). Glass containers collected for recycling include beer and soft-drink bottles, wine and liquor bottles, and bottles and jars for foods (EPA, 2007). Recycling glass bottles back into bottles over and over again is by far the best use of secondary glass. Some benefits of recycling glass containers are as follows:

1. Recycling reduces emissions and the consumption of raw materials. One ton of carbon dioxide is saved for every six tons of cullet (furnace-ready recycled glass) used in the manufacturing process. Over a ton of natural resources are saved for every ton of glass recycled (GPI, 2010). Cullet currently accounts for up to 70% of raw materials for glassmaking (GPI, 2010). In 2010, there were 75 cullet processors operating in the USA (Cattaneo, 2010).
2. Recycling extends the life of plant equipment, such as furnaces (GPI, 2010).

3. Recycling saves energy. The use of cullet is economically desirable because somewhat less energy is required to melt cullet than raw materials (Robertson, 1993). The energy cost is reduced by about 2–3% for every 10% of cullet used in the manufacturing process (GPI, 2010). Using recycled glass to make containers saves much more energy than using recycled glass for other purposes (CRI, 2009).

In spite of all these benefits, recycled postconsumer glass from the MSW stream for use as a raw material in new glass containers is limited. Some detriments associated with the recycling of glass containers are:

1. Container glass recycling is still costly, owing to the high cost of collecting and processing (hand sorting).
2. The marketability of the recycled material depends primarily on proximity to a glass beneficiator (a place where glass cullet is prepared as a feedstock for high-end recycling markets, such as bottle manufacturing), and the beneficiator's particular quality specifications.
3. Glass breakage previous to sorting makes recycling extremely challenging. The breakage fraction is high, and is basically impossible to prevent, since it occurs during compaction in the collection truck and in the separation process at the material recovery facility. About one-third of nonrecyclable glass is broken glass that is too small to be separated for recycling, and therefore it is used for low-end uses, such as landfill cover. This is far less desirable in terms of energy conservation, emissions reductions and other high-end benefits (CRI, 2009).
4. Container glass recycling has been reduced owing to increased contamination of the glass material. Over the past few years, many recycling programs have shifted from source-separated programs to single-stream collection (all recyclable materials are placed in the same holder), and this has resulted in increased contaminated glass content and a decrease in the collection of uncontaminated glass. This means a reduction in the glass collected in single-stream systems to be used for glass bottles, one of the highest closed-loop applications. In addition, contaminated cullet reduces the life expectancy of production equipment, increases operating costs and limits how much cullet can be used in glass production (CIWMB, 2010). Since clean cullet is required to control the costs of maintenance and of production equipment, and to reduce downtime and increase usage, the growth of single-stream recycling programs and the consequent downgrading of cullet quality mean that a beneficiation facility will incur significantly higher costs for energy and cleaning, which adversely impacts the economics of building additional beneficiation capacity (CRI, 2009).
5. There is a need for improvement of the automated color sorting process (into white, green and amber) for postconsumer glass containers. The requirements for marketing recycled glass are well established and are highly dependent upon color sorting of the glass. Color-sorted glass material results in 98% being recycled and only 2% marketed as glass fines. Automated color sorting is more efficient and also more

reliable than the old manual hand sorting, and is thereby improving the recyclability of glass containers and hence the amount of recycled glass going into closed-loop applications such as new glass bottles. This would result in a reduction in the emissions and in the consumption of raw materials, as well as in energy savings.

In 2009, only 40% of glass from single-stream collection was recycled. The rest ended up in landfills (40%) and low-end uses (20%). In contrast, mixed glass from dual-stream systems yielded an average of 90% being recycled into containers and fiberglass, with 10% glass fines used for low-end uses and nearly nothing sent to landfill (CRI, 2009). Glass container manufacturers have set a goal of achieving 50% recycled material content in the manufacture of new glass bottles by 2013. Using 50% recycled glass to make new glass containers will save energy, since recycled glass can substitute for up to 70% of raw materials and save 181 550 tons of waste from landfills every month (GPI, 2010). Some improvements in recycling, while not universal, have been due to (i) container deposit legislation (which achieved a glass container recycling rate over 63%) (CRI, 2009), which requires deposits at the point of purchase for containers, and bottle deposit returns at retail outlets, and (ii) a law requiring all alcoholic-beverage permit holders to recycle their beverage containers (57 000 tons of glass bottles recovered for recycling). Currently, an estimated 80% of recovered glass containers are made into new glass bottles (GPI, 2010).

Glass containers can also be sterilized or washed with powerful detergents and reused. Refilling could make glass bottles one of the least costly packages. Factors that affect the costs and efficiency of reusing glass bottles include consumer behavior, government policy and manufacturers' attitudes. Some of these factors are (i) deposits on refillable bottles (or taxes on nonrefillable bottles), (ii) a successful program for collecting refillable bottles, (iii) adequate geographic locations for collecting and bottling operations to minimize transport distances, (iv) the willingness of retailers to collect and store empty bottles and to take on the related tasks, (v) the willingness of the industry to use standardized glass bottles (most bottles are currently manufacturer-specific) in order to minimize transport distances, and (vi) the willingness of consumers to return empty glass bottles to collection points.

The glass industry has also driven design innovations to make glass containers more environmentally friendly in recent years. The average amount of material per glass bottle has been reduced by more than 50% between 1970 and 2000 (GPI, 2010). In agreement, Girling (2003) reported that the average weight of glass containers decreased by nearly 50% from 1992 to 2002. External coatings are being used as surface treatments to increase the strength of thinner containers. This reduction in material results in less weight and is better for disposal and transportation (McKnown, 2000).

Metals

Metals have been used in food packaging for more than 200 years. New breakthroughs in metal packaging design (easy-open cans with pull-tab lids, easy-to-grasp cans, and self-heating and self-cooling cans) have guaranteed its place in the packaging market (CMI, 2010). The two metals predominantly used to package processed food are steel and aluminum. Food cans are predominantly made from steel in two- or three-piece constructions and are offered in a variety of shapes and sizes. The key market segments for metal containers are fruits, vegetables, soups, sauces, ready meals, pâté and shelf-stable canned milk products. Beverage cans are mostly made from aluminum. The latter is widely used for the packaging of soft drinks and beer. Aluminum is also used to make foil, and to laminate paper or plastic packaging materials and closures.

Steel cans (mostly cans for food products) were 1.1% of the total MSW generated in the United States in 2007 (SRI, 2010), and the recovery, including residential sources, combustion and recycling, was around 65% of the steel packaging generated (SRI, 2010). In 2008, similar amounts of metal packaging were generated and recovered, with more than 63% of steel packaging (mostly cans) recycled (EPA, 2009a). That same year, the recycling rate for aluminum packaging was 38%, including just over 48% of aluminum beverage cans (EPA, 2009a). Metal food cans are 100% recyclable and have a recycling rate 2.5 times higher than that of most other packaging options (CMI, 2010). Steel cans can be recycled again and again without losing quality. Every ton of recycled steel saves 2500 lb of iron ore, 1000 lb of coal and 40 lb of limestone (CMI, 2010). In addition, metal recycling reduces greenhouse gas emissions significantly (EPA, 2009a). The food can industry has also driven several innovations to make cans more environmentally friendly in recent years:

1. Nondetachable ring-pulls (INCPEN, 2010) make the recycling of the entire can easier.
2. A design modification to reduce the amount of material used in can manufacture has been implemented: a gradual thinning of the can walls, or "light-weighting", a reduction in the amount of tin in the tinplated steel, and a "necked-in" design for the ends (May, 2004). Aluminum cans were 26% lighter in 2005 than in 1975 (TAA, 2006). For example, the weight of aluminum cans was reduced from 19 g per can in 1983 to 13 g per can in the UK in 2003 (PIRA, 2005). Similarly, steel cans have been light-weighted, with cans now at least 40% lighter than those of 1970. For example, the weight of steel cans was reduced from 36 g per can in 1983 to 23 g per can in the UK in 2003 (PIRA, 2005). The amount of tin has been drastically reduced from pre-World War II levels of 50 lb of tin per ton of tinplate steel to a current average of 6 lb per ton (Miller, 1993). A reduction in seam dimensions has been achieved through the development of new designs for can seams, such as the Euroseam and the Kramer seam (Holdsworth and Simpson, 2007).
3. There has been a shift toward "easy-open" ends equivalent to a reduction in the thickness of the original lid, and the use of new materials such as foil instead of tinplate.
4. More water-based and high-solids coating formulations are being used, owing to environmental legislation based on new air pollution regulations (Good, 1988).

Plastics

Plastics are becoming the most important packaging material for food products because of their unique characteristics, mechanical strength, light weight, relatively low cost, and ease of processing and manufacturing. The most widely used are high-density polyethylene (HDPE), low-density polyethylene (LDPE), polypropylene (PP), polyethylene terephthalate (PET), polystyrene (PS) and polyvinyl chloride (PVC) (Komolprasert, 2000). Plastic containers and packaging have exhibited rapid growth in MSW, with generation increasing in the United States from 120 000 tons in 1960 (0.1% of MSW generation) to 13.6 million tons in 2007 (5.4% of generation) (EPA, 2008). The recovery of plastics has also increased over recent years (Figure 44.2). In 2008, more than 13% of plastic containers and packaging was recycled in the United States, mostly from soft-drink, milk and water bottles (EPA, 2009a). These higher recovery rates for plastic bottles could be associated with the use of "bottle bills," which put a deposit on beverage containers that the consumer can get back when the bottle is returned to a store or recycling facility, since there is very little collection for recycling of nonbottle plastics. For example, the state of Michigan, which has the highest deposit on beverage bottles, also has the highest recovery rate, about 97% (BBO, 2010). Every state that has implemented a bottle bill has seen an increase in bottle recovery (NAPCOR, 2010). PET has had one of the highest recycling rates of all polymers over recent years. The national recovery rate for PET bottles and jars in the United States has been 23%, 25% and 27% in 2005, 2007 and 2008, respectively (EPA, 2006; EPA, 2008; EPA, 2009a). Some American states have had much higher recovery rates because of "bottle bills," as mentioned above. A report by the National Association for PET Container Resources (NAPCOR) showed that for every pound of recycled PET used, energy use is decreased by 84% and greenhouse gas emissions are cut by 71% (NAPCOR, 2010). Once recycled, PET can be used for many things (Figure 44.4). The third highest use of postconsumer PET is for food and beverage containers (15.1%). In the United States,

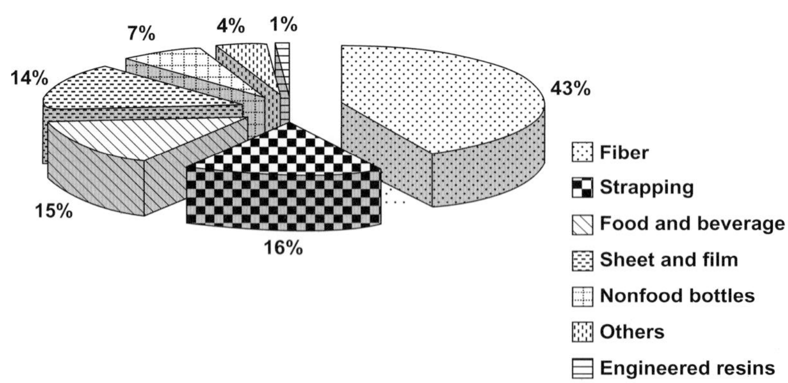

Figure 44.4 Uses of recycled PET in 2007. (Source: National Association for PET Container Resources (NAPCOR, 2010).)

1 PETE	Polyethylene terephthalate (PET): bottles (mineral water, carbonated soft drinks, salad oils and dressings), "boil-in-the-bag" products, mouth jars containing fat-rich products, and oven-ready meal trays.
2 HDPE	High-density polyethylene (HDPE): milk jugs and liquid detergent bottles.
3 V	Polyvinyl chloride (PVC): food trays, shrink wrap and bottles (mineral water and shampoo).
4 LDPE	Low-density polyethylene (LDPE): carrier bags, produce bags, shrink wrap and bin liners.
5 PP	Polypropylene (PP): lids, yoghurt caps, margarine tubs, retortable packaging and microwavable meal trays.
6 PS	Polystyrene (PS): foam produce, meat or fish trays, fast-food boxes, egg cartons and clamshell containers for produce
7 Other	"Other" plastics: any other plastics that do not fall into any of the above categories, or a mixture of the above categories.

Figure 44.5 Resin identification coding developed by the American Society of the Plastics Industry. (Source: SPI, 2010.)

the other most highly recycled polymer has been HDPE (mainly plastic milk and water bottles), with a national recovery rate of 31 and 29% in 2006 and 2007, respectively (EPA, 2007; EPA, 2008).

To facilitate the sorting of different plastics, and thus their recycling, in 1988 the American Society of the Plastics Industry (SPI) (SPI, 2010) developed a resin identification code to help consumers identify and sort the main types of plastic (Figure 44.5). However, there are some practical limitations on the separation by resin type for certain plastics and certain types of packages. While the individual compounds in laminates and metallized films are technically recyclable, the difficulty of sorting and separating the material precludes economically feasible recycling. Plastic packaging materials are often contaminated by foodstuffs and biological substances, which makes recycling of these materials impractical and, most of the time, unattractive from an economic standpoint.

Recovered and recycled plastics used for food packaging have to comply with all the rules and regulations that apply to virgin plastics. The health and safety of the consumer is of prime importance when considering the reuse and recycling of plastics in food-contact uses (ILSI, 2000). In the United States, the Food and Drug Administration (FDA), an agency within the US Department of Health and Human Services, controls the safety of recycled plastic materials used to produce plastics suitable for food-contact applications. The FDA regulates these materials as indirect food additives, and its main safety-related concerns are: (i) contamination from the postconsumer material in the final food-contact product, (ii) incorporation of nonfood-contact-approved materials from recycled postconsumer material into food-contact packaging, and (iii) incompatibility of adjuvants in the recycled plastic with the regulations for food-contact use. Currently, to address these concerns, the FDA considers each proposed use of recycled plastics on a case-by-case basis. There is no legal requirement that companies seek FDA approval, but most do so. The manufacturer is asked to provide a complete description of the recycling process, the result of migration tests to show that the recycling process removes possible contaminants, and a description of the proposed conditions of use of the plastic. Migration tests are no longer considered necessary for postconsumer recycled PET or polyethylene naphthalate (PEN) produced by a tertiary recycling process, since they are considered suitable for food contact (FDA, 2009). In addition, the FDA issues informal advice as to whether the recycling process is expected to produce plastic suitable for food-contact applications, and has prepared a document entitled *Guidance for Industry – Use of Recycled Plastics in Food Packaging: Chemistry Considerations* (FDA, 2010) that assists manufacturers of food packaging in evaluating processes for recycling plastic into food packaging.

Another alternative way to reduce the volume of plastic waste is incineration. The energy stored in plastic waste can be recovered through waste-to-energy incineration. Plastics are derived from petroleum feedstocks, giving them a stored energy value higher than that of other materials commonly found in the waste stream, and this high heat content is advantageous for WTE incineration. When plastics are processed in modern WTE facilities, they can help other waste to combust more completely, leaving less ash for disposal in landfills (SPI, 2010). In 1989, the health and safety impacts of WTE incineration were discussed and it was found that, contrary to popular misconception, there was no evidence to link the incineration of PVC with increased dioxin emissions. Similar conclusions have been presented by a number of sources, including a 1987 study for New York State Energy Research and Development (SPI, 2010). In the past, some other environmental and health concerns have been associated with specific plastics with major markets in packaging for food:

1. The production of PVC is the largest use of chlorine gas in the world, and the production of chlorine is an energy-intensive process (Thornton, 2005). The production of PVC creates many by-products, and a lot of these by-products are considered to be highly toxic, bioaccumulative and persistent. PVC by-products include dioxins, polychlorinated biphenyls and hexachlorobenzene. Dioxins are created whenever chlorine gas is burned or used for processing. Dioxins are an extremely potent

carcinogen (Selke *et al.*, 2004). Being a global pollutant, dioxins have been found even in remote ecosystems. PVC is one of the least recycled plastics. When PVC is recycled, the postconsumer resin is used in downcycled products, which means that it reduces the amount of virgin PVC produced. When it is recycled, the high temperature needed can release the toxic additives used in PVC into the air and ground. Similar to recycling, incineration can also release toxins into the air through fumes and into the ground through the ash produced. Since the additives used in PVC are not part of the chemical structure, they can leach out throughout the lifetime of the PVC. The only way to landfill PVC is to have a membrane sufficient to prevent the leakage of the additives into the soil and groundwater. In this respect, PVC is not different from other materials.

As a result of some of these environmental and health concerns, the use of PVC in food packaging is declining. Some major European supermarket chains located in Austria, Denmark and Germany have completely eliminated PVC food packaging, and others located in Switzerland and the UK are in the process of phasing it out. PVC wrapping has been eliminated by several Japanese supermarket and convenience stores, and PVC bottles have been eliminated by many big water bottling companies (Greenpeace, 2003). Plastics such as ethylene vinyl alcohol (EVOH), which also is a good barrier to oxygen, might be a viable option to reduce PVC use in some food packaging applications.

2. There are many environmental and health concerns attributed to the production of PS. In order to produce the monomer, namely styrene, benzene and ethylene are required and their production processes are wasteful processes (EPA, 1982; Neelis *et al.*, 2008). PS is found in the market in two forms, sheet or molded PS and foamed PS. The more common type is foamed PS, which can be expanded or extruded. The hydrocarbons butane and pentane and also carbon dioxide are used as foaming agents. Pentane is known to cause ground-level smog pollution, and therefore carbon dioxide is preferred, since this is neutral with respect to volatile organic compounds (VOC) emissions (ILSI, 2002). A 1986 EPA report named the production of polystyrene as the fifth largest creator of hazardous waste (ERF, 2010). However, the elimination of hydrochlorofluorocarbons (HCFCs) and chlorofluorocarbons (CFCs) (Tsai, 2002) as blowing agents in expanded PS has significantly lessened the impact of PS on the environment. Also, the late-1980s "McToxics Campaign," where consumer groups successfully stopped the fast food chain McDonalds from using PS in the packaging of its hamburgers, has also reduced the negative impact on the environment (EJnet, 2010). PS accounts for about 1% of the solid waste stream by weight (EPIC, 2008). The recycling rate of PS is less than 20%, and this is because the recycling of PS is not sustainable economically. Expanded PS consists of about 95% air, and hence large volumes of expanded PS need to be collected in order to recover a small amount of expanded PS by weight (EPS, 2010). The cost of fuel and other resources required to transport and handle the expanded PS is much larger than the value of the recovered material. Also, there is a very limited amount of products that can be made from

recycled PS. Reusing expanded PS foam may not an attractive option for open-cell foams, since food particles could remain in the holes.

Plastic waste can also reduced by using thinner gauges or by refilling plastic bottles. Light weight has been achieved in the plastics industry by the use of different materials and designs. For example, the weight of 2L PET soft-drink bottles has decreased by 25% since 1977, resulting in a saving of more than 206 million pounds of plastic material each year (APC, 2006). Similarly, the 1 gallon plastic milk jug has undergone a weight reduction of 30% in the last 20 years. Refillable plastic containers are commonly made from PET, HDPE and polycarbonate (PC), although the use of the latter is on the decline.

Paper and Paperboard

Paper and paperboard are commonly used in corrugated boxes (fruits and vegetables), cartons (milk and fresh juices), wrapping paper (meats), paper bags (sugar and flour), fiber drums (potato chips and pastes), and molded paper containers (fruits).

In the United States, corrugated boxes were the largest single product category of MSW in 2007. Other paper and paperboard packaging in the US MSW included milk cartons, and folding boxes such as cereal boxes and frozen-food boxes. Recovery of corrugated boxes is by far the largest component of paper packaging recovery (EPA, 2009b). There is increasing use of recycled material in "new" packaging. In 2008, 66% of paper and paperboard containers and packaging were recycled, including nearly 77% of all corrugated boxes (EPA, 2009b). Furthermore, some grades of paperboard now contain 100% recycled fibers, and 80% of US paper mills use recovered fiber in the production of new paper and paperboard (PP, 2010). Aggressive recycling programs are being carried out by the retail industry, municipalities and consumers. A recent report indicated that paper and paperboard topped the list of commonly recycled materials, with a share significantly higher than that of glass, metal, plastic and all other materials combined (PP, 2010). Recent data show that more than 53% of American consumers have access to paperboard recycling programs, up from 46% in 2000 (PP, 2010). Figure 44.6 shows the recycling symbol used by the Corrugated Packaging Council (CPA). It should be noted that not all corrugated packaging material can be recycled; for example, Lallanilla (2010) reported that if corrugated packaging is coated with wax or another compound, the likelihood of its being accepted is very low and, typically, it will not carry the CPA-approved recycling symbol.

The paper industry, as reported by the Paperboard Packaging Alliance (PPA), has set an aggressive goal to recover 55% of the paper and paperboard consumed in the US by 2012 (PPA, 2006). To meet this goal, the American Forest and Paper Association and its partners are working toward increasing not only the quantity but also the quality of paper recovered for recycling operations (PPA, 2006). The PPA report (PPA, 2006) also showed that besides recycling and recovery, reforestation has been the cornerstone of the forest industry's efforts to help the environment; the forest

Figure 44.6 Symbol used by the Corrugated Packaging Alliance (CPA) to show and encourage recycling efforts. (Courtesy of the Corrugated Packaging Alliance.)

products industry plants 1.7 million trees a day, which exceeds the number of trees harvested.

Paper or paperboard, when used for primary packaging, i.e., direct contact with the food, is almost always treated, coated, laminated, or impregnated with waxes, resins or lacquers to improve its functionality (Marsh and Bugusu, 2007); however, such treatments can present some challenges, such as somewhat poor biodegradability compared with plain paper. Efforts to make paper/paperboard packaging more environment-friendly have focused on using thinner-gauge paper to make it lightweight, combined with using less material. For example, Anheuser-Busch saved 7.5 million pounds of paperboard by decreasing the thickness of the packaging of its 12-bottle packs (EPA, 2004). Furthermore, the use of thinner-gauge paper and paperboard results in reduced transportation and shipping weight, thereby helping the environment indirectly.

Wood

The packaging industry is an important market for wood materials in the US; about one-third of US hardwood production is utilized in the production of pallets and containers (Bush *et al.*, 2008). The industry recovers significant volumes of pallets and containers from the waste stream for reuse, repair and recycling (used as mulch). Industry by-products (both wood and nonwood) are recycled for a variety of uses (Bush *et al.*, 2008). About 15% of wood packaging, mostly in the form of wood pallets, is recovered (EPA, 2009a).

Wood packaging constitutes only 5.5% of municipal solid waste (EPA, 2007), but it is recycled at a lower rate than paper and paperboard packaging, metal containers, and plastic packaging (EPA, 2007). According to the Waste Online Organization (WOO, 2006), wood waste has some features that warrant attention: for example, (i) it is biodegradable, and thus it can contribute to greenhouse gases if discarded in a landfill and allowed to rot; this very nature makes it subject to the European Union (EU) "EU Landfill Directive," which specifies that an increasing percentage of biological municipal waste (BMW) must be diverted from landfill (WOO, 2006); (ii) typically, wood packaging is in excellent condition, which makes it an ideal material for reuse, rather than simply trashing it (WOO, 2006); (iii) the growing, harvesting and processing of virgin timber uses energy and water, which are natural resources that are not as renewable as timber (WOO, 2006), and thus wood should be reused or recycled as much as possible; and (iv) wood waste has a disposal cost, but as with any other waste material, reusing or recycling will eventually save a company and consumers money in the long haul (WOO, 2006).

Novel Packaging Materials and the Environment

In recent years, novel packaging materials such as bio-based polymers or biopolymers, biodegradable plastics, compostable plastics, oxo-degradable plastics, and UV-degradable plastics have gained attention from industry, consumers and government. This is based on the facts that biodegradable and compostable polymers can make a significant contribution to reducing the abundance of plastic packages in landfills, since composting is an end-of-life option for single-use plastics, and that bio-based polymers can be a possible alternative to reducing dependence on declining petroleum-based resources. However, to assume that these materials always have a lower impact on the environment than conventional plastics can be an error. Life cycle assessment is an appropriate tool to evaluate the environmental impact of novel packaging plastics and to compare their environmental impact with that of conventional packaging plastics. Bohlmann (2004) used an LCA to compare the environmental impact of poly(lactic acid), or polylactide (PLA), with that of PP, and found that PLA is more energy-efficient than PP because PLA consumes no feedstock energy. However, when the uncertainty of the estimation is taken into consideration, the difference between the two polymers becomes marginal. He also found that the greenhouse gas emissions of PLA and PP are equivalent. Kijchavengkul and Auras (2008) reported that the production of the biodegradable polymers PLA and polyhydroxyalkanoate (PHA) uses less energy than the production of many nonbiodegradable polymers such as polyethylene (PE), PP, PS and PET, but that the production of the biodegradable polymer poly(ε-caprolactone) (PCL) uses more energy than does that of the aforementioned nonbiodegradable polymers. The emissions of greenhouse gases from the production of PLA and PHA are similar to those for PS, PP and PET but higher than for PCL. According to Marsh and Bugusu (2007), a switch from synthetic polymers to biopolymers would have little impact on source reduction and incineration, but recycling could be complicated by the existence of blended or modified polymers unless they are separated from the recycling stream.

Some of these bio-based and compostable polymers are starting to be used to package foods, since they provide mechanical properties similar to those of the conventional polymers and have good processability and food-contact acceptability, while leaving a smaller environmental footprint. However, they have disadvantages, such as poor thermal stability, brittleness, and poor gas and water vapor barrier properties.

Biodegradable and Bio-Based Plastics and Polymers

Biodegradable polymers are polymers that break down under the action of biological enzymes from bacteria, yeasts or fungi into CO_2, H_2O and biomass under aerobic conditions, and to hydrocarbons, methane and biomass under anaerobic conditions (Doi and Fukuda, 1994). A simplified definition has been given by ASTM International, which defines a biodegradable plastic as "a plastic in which the degradation results from the action of natural organisms such as bacteria, fungi and algae" (ASTM, 2004). Biodegradable polymers are divided into two groups based on the source of the plastic. Renewable biodegradable polymers are made from renewable resources, whereas non-renewable biodegradable polymers are not. PLA, PHAs and cellulose (Natureflex®) are examples of renewable biodegradable polymers, whereas poly(butylene adipate-*co*-terephthalate) (PBAT) (Ecoflex®) and poly(vinyl alcohol) (PVOH) are examples of biodegradable polymers produced from petroleum-based resources (Kijchavengkul and Auras, 2008). Some environmental advantages of renewable biodegradable polymers are as follows:

1. They are produced from renewable resources that are unlimited and therefore can be replaced.
2. They have a reduced environmental footprint if the end-of-life scenario is recycling or composting.
3. They help to replenish the carbon cycle (Narayan, 2010).
4. They reduce our dependence on declining petroleum-based resources.
5. They contribute to reducing the abundance of plastic packages in landfills.
6. They are important with respect to the marine environment, in which litter poses hazards to marine life.
7. They provide an alternative option to recycling.

Claims of biodegradability must be supported by verifiable scientific data from independent, third-party testing laboratories using established standard test methods and specifications developed by major standardization organizations (e.g., the International Organization for Standardization (ISO), ASTM International (ASTM) and the European Committee for Standardization (CEN) (CEN, 2010)). This certification ensures accurate communication of the nature of a material's biodegradability between governments, producers and consumers. The test methods can be divided into three categories: (i) field tests (the degradation takes place when samples are buried in soil, placed in a river or lake, or used in a full-scale composting process), (ii) simulation tests (tests performed in the laboratory, where the degradation takes place in compost, soil or seawater placed in a controlled reactor), and (iii) laboratory tests (the degradation takes place in a synthetic environment where a microbial population, screened for a particular polymer or enzymes known to depolymerize a particular polymer, is used) (Miller, 2003). These test methods establish the ambient conditions and the rate of biodegradation. The rate and extent of the biodegradability of a plastic will depend on environmental conditions (temperature, pH, moisture, UV radiation, enzymes and other factors) and the characteristics of the polymer (size, shape, weak links in the chain, conformational flexibility, crystalline content and other factors).

Biopolymers or bio-based polymers can be defined as polymers produced from renewable resources; these are mainly carbohydrates and protein substances, such as corn starch, wood pulp and soy protein (Greer, 2006). The bio-based content of a polymer can be measured according to the standard test method ASTM 6866 (ASTM, 2005) by using ^{14}C radiation, which allows one to distinguish modern carbon from fossil carbon, since the age of the latter is much greater than the half-life of ^{14}C (Kijchavengkul and Auras 2008). Not all polymers based on renewable resources are necessarily biodegradable or compostable, and not all biodegradable materials are necessarily based on renewable materials, because biodegradability is directly correlated to the chemical structure of the material rather than to its origin.

Compostable Plastics

The American standard specification for compostable plastics ASTM D 6400 (ASTM, 2003a) requires for a plastic to be compostable that it meets the following three criteria:

1. It should demonstrate inherent biodegradability at a rate and degree similar to natural biodegradable polymers, over a time period of 180 days at $58 \pm 2\,°C$. For slower-biodegrading materials, the test period can be extended to 365 days; however, appropriately ^{14}C-labeled test substrates should be used.
2. To pass disintegration requirements, the material should disintegrate during active composting, so that there are no visible, distinguishable pieces found on screens (no more than 10% of its original dry weight remains after sieving with a 2.0 mm sieve). To pass biodegradation requirements, the percentage of mineralization should be greater than 60% for a homopolymer or random copolymer, and greater than 90% for a block/graft copolymer or blend.
3. The material should have no ecotoxicity (it must not impact the ability of the resultant compost to support microbial and plant growth), one of the concerns being that the concentrations of heavy metals in plastic materials have to be less than 50% of the limits defined in 40 CFR §503.13 (CFR, 2010).

Following these criteria, many different materials have been qualified as compostable. Biodegradation of cellulose and starch during controlled composting reached 96.8% after 47 days for cellulose and 97.5% after 44 days for starch (Degli-Innocenti, 1998). PCL, poly-β-(hydroxybutyrate) (PHB) and poly-β-(hydroxybutyrate-co-β-valerate) (PHB-V) have been qualified as biodegradable in soil compost at 46 and 24 °C. PHB and PHBV were totally degraded after 104 days in soil compost at 46 °C. PHB and

Figure 44.7 BPI logo (used by permission of Biodegradable Products Institute).

PHBV were also similar to each other in terms of biodegradation at 24 °C (Lotto *et al.*, 2004). PLA and PCL showed rapid inherent degradation, since they achieved more than 60% degradation before 100 days (Pradhan *et al.*, 2010). Following the same criteria as for the above materials, various packages have been qualified as biodegradable too. PLA bottles have been qualified as biodegradable, since the mineralization was greater than 60%. This value was achieved in simulated conditions using an automatic laboratory-scale respirometric system after 63 days of exposure at 58 °C and 55% RH (Kijchavengkul *et al.*, 2006). PLA is fully degradable when composted in a large-scale operation with a temperature of 60 °C or above. Commercial PLA bottles and containers composted under real conditions (composting pile at 65 °C, 63% RH and a pH of 8.5) disintegrated in less than 30 days (Kale *et al.*, 2007).

In the United States and Canada, there are no widespread requirements to undergo certification. However, if it is preferred, all testing data required to certify a material as biodegradable can be independently reviewed by the Biodegradable Products Institute (BPI) for certification according to ASTM D 6400 and/or D 6868 (ASTM, 2003a; BPI, 2010). Plastic or paper products that will biodegrade and compost satisfactorily in actively managed compost facilities are certified, and identified with the BPI logo (Figure 44.7). These facilities are large-scale facilities and do not exist in all communities, so people need to check with municipal officials. The BPI symbol enables consumers, composters, municipal officials and waste haulers to easily differentiate compostable plastic products from those that do not have the same environmental benefits.

In Europe, polymers are defined as compostable when they have passed standardized compostability tests, which are described in the harmonized European standard EN 13432 (EN, 2000). This also includes the three criteria (biodegradation, disintegration and ecotoxicity) mentioned above. According to this standard, the degradation process takes approximately 6–12 weeks, meeting the current requirements of industrial composting facilities. This standard applies to packaging and is nearly identical to the former DIN V 54900 standard (from the German Institute of Standardization). DIN Certco is the certification organization of DIN, and it certifies and identifies compostable products made of biodegradable materials with a unique label (Figure 44.8). This compostability label is accepted in Germany, Switzerland, the Netherlands, the United Kingdom and Poland.

While specifications, criteria and requirements for compostable plastic materials and packages are described in the aforementioned standards ASTM D 6400 and EN 13432, respectively, the test procedures and methods are described in the standards

Figure 44.8 DIN Certco logo (European Bioplastics).

Figure 44.9 OK Compost HOME mark (used by permission of Vinçotte).

ASTM D 6002 (ASTM, 2002), ASTM D 5338 (ASTM, 2003b), ISO 14855-1 (ISO, 2005), ISO 14855-2 (ISO, 2006c) and ISO 16929 (ISO, 2002).

Examples of other certifying organization in the world are AIB Vinçotte International s.a./n.v. (OKCOMPOST, 2010) and GreenPla® (Japan BioPlastic Association) (JBPA, 2010). An AVI Certest certification certifies that a material or product is biodegradable in a specific natural environment (soil or fresh water) and identifies it with the OK biodegradable mark, or certifies that a material or product can be composted in an industrial composting facility or in a backyard compost heap and identifies it with the OK compost mark or the OK compost HOME mark, respectively (Figure 44.9). These marks can be assigned to any biomaterials that have passed tests and examinations carried out in accordance with the aforementioned European standard (EN 13432). GreenPla® based its certification and labeling system on the standards ISO 16929, ASTM D 5338, JIS K 6950 (ISO 14851), JIS K 6951 (ISO 14852), JIS K 6953 (ISO 14855), JIS K 6955 (ISO 17556) and ISO 14855-2 (JBPA, 2010). Products or materials that can be identified as GreenPla® are those that include in their components biodegradable organic materials that are broken down by the action of microorganisms in the natural environment and ultimately become carbon dioxide or water (JBPA, 2010).

To ensure that most biodegradable/compostable materials degrade in the appropriate environment, it is important that all of the relevant factors are understood and are applied by raw-material suppliers, product manufacturers and brand owners within the industry, and understood by governments and communities.

Oxo-Degradable and UV-Degradable Plastics and the Environment

Besides biodegradation and composting, there are other degradation mechanisms such as oxo-degradation and UV degradation, which are only effective on specially modified plastic materials. Oxo-degradable and UV-degradable plastics are basically traditional plastics (PE, PP, PVC, PET and PS) loaded with specific additives (chemical catalysts containing transition metals such as cobalt, manganese and iron) (EB, 2009) or copolymerized with active groups (as in the case of most common UV-degradable packaging materials) that allow them to degrade. Shopping bags, agricultural mulch films and plastic bottles made from oxo-degradable and UV-degradable plastics are available on the market. Plastic packaging materials with such degradation mechanisms do not meet the standards of biodegradability and compostability described above. At present there are no standards or certifications for oxo- or UV-degradable packaging materials, and therefore there are no mechanisms to ensure that such claims are reproducible and verifiable (EPR, 2009). Their degradation can be initiated by heat, physical stress or exposure to UV light. Oxo-degradable plastics seem first to be broken down into smaller molecules owing to the action of O_2 on the additive, and then converted into H_2O, CO_2 and biomass by microorganisms. For UV-degradable plastics, the exposure of the material to UV light starts a free-radical process. This process can be accelerated by heat or extended exposure to UV light, and it continues to break down the material into smaller and smaller pieces (PLASTEMART, 2010). Oxo-degradable plastics need to be stored in dry conditions and protected from UV light and from heat over 37 °C. Some of these materials claim to be biodegradable but do not meet the standards of biodegradability and compostability described above. One of the issues is that there are no overall "standards of biodegradability." This can lead to a general mistrust on the part of consumers and managers of composting plants toward the whole sector of biodegradables. The recovery of plastic materials by recycling is also affected. For example, oxo-biodegradable plastics bring their additives to the recyclates, which affect the chemical stability and cost of the recyclate (EPR, 2009).

The Future: the Role of Consumers and the Food Industry in the Impact of Packaging on the Environment

The impact of packaging on the environment has gained significant attention within the food industry, in part owing to heightened consumer concerns. These concerns focus not only on the increased total amount of waste generated, which recycling has kept from growing since the 1990s, but also on how big a carbon footprint a particular packaging material has. Hence a need for more biodegradable materials has developed. At the same time, the industry has an interest in using only the required amount of

Table 44.1 Global incidence of new food and drink products by food category with claims of environmentally friendly packaging (2006–2009)[a].

Food category	2006	2007	2008	2009	% change (2008–2009)
Cold cereal	3	22	251	487	94.0
Tea	2	44	198	351	77.3
Prepared meals	14	12	152	387	154.6
Sweet biscuits/cookies	1	33	137	338	146.7
Cakes, pastries, sweet goods	10	9	114	293	157.0
Carbonated soft drinks	1	8	74	328	343.2
Vegetables	1	13	116	261	125.0
Snack/cereal/energy bars	3	20	113	234	107.1
Juice	0	19	74	261	252.7
Wet soup	0	15	81	234	188.9

[a] *Source*: data from Mintel (2009).

packaging because this reduces costs, complies with packaging laws if required and protects the environment (INCPEN, 2010). A combination of efforts to address these concerns, on the part of consumers as well as the food processing industry, has led to more sustainable food packaging in the past few years.

Manufacturers' packaging-related efforts toward sustainability, such as reducing their carbon footprint, light-weighting, improving the raw materials used in packaging (including biodegradable ones), and recycling programs, are cited in a recent article (Anon., 2010) based on the Mintel report *Packaging Trends in Food and Drink – U.S.* (Mintel, 2009). The results of these efforts are shown in Table 44.1, as evidenced by a monumental increase in claims of environmentally friendly packaging for the top 10 food categories. Cold cereals, followed by prepared meals, led all food categories in new claims of being environmentally friendly. From 2008 to 2009, the number of claims of eco-friendly packaging for the top 10 food and beverage categories increased by 77.3–343.2%. The Mintel Global New Products Database (GNPD) reports a claim by manufacturers of increasingly launching products packaged in an environmentally friendly manner. In 2006, products in the top 10 category of the 80 global products listed in Mintel's new-products database made a claim of environmentally friendly packaging, which increased in 2009, comprising 6% of all global food and drink products (Mintel, 2009). Concerned consumers have played their due role in these changes by becoming more aware of the environmental issues and demanding environmentally friendly products; for example, in 2009, 58% of respondents to Mintel's customer survey indicated that they believed that plastic bottles are "really bad for the environment" and two-thirds reported they "usually" recycle plastic bottles and containers, which is a significant increase from the 46% who reported recycling in 2008.

Consumers are also driving some of the changes with respect to the increased use of recycled material in paper and paperboard packaging. According to a survey by the Recycled Paperboard Alliance, 61% of consumers indicated that they were more inclined to purchase products from a company that uses recycled paperboard packag-

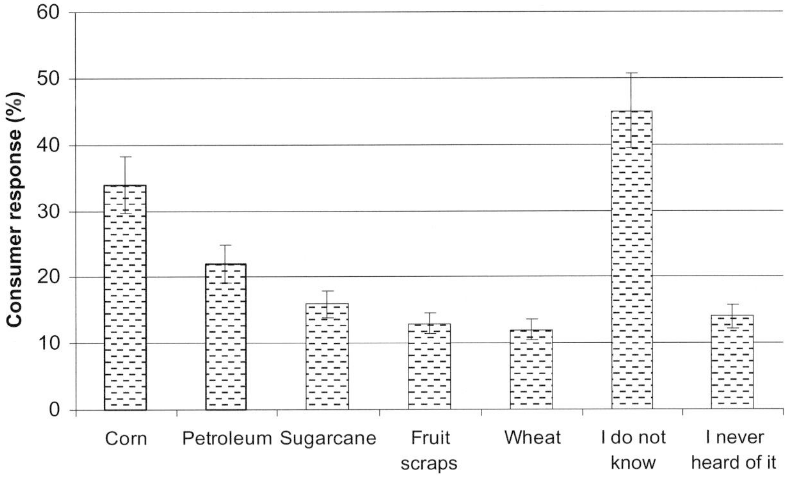

Figure 44.10 Consumers' responses to a question concerning the sources used to create bio-based plastic containers. (Based on Koutsimanis *et al.*, 2010.)

ing. This study found further that 77% of respondents felt better about a company that uses recycled paperboard and 80% felt they were "doing something good for the environment" if they purchased products having recycled paperboard packaging (Murray, 2010).

Despite the environmental concerns of consumers and their willingness to change their behavior, there are a lot of misunderstandings. When consumers see the word "biodegradable" on a package, they believe that the package will completely disappear in 12 to 18 months, no matter where it is thrown away (either in landfills, in streams or as litter) (BPI, 2010). In addition, consumers do not know the sources used to create bio-based plastic containers, although they prefer to compost food packages. In 2010, Koutsimanis *et al.* (2010) examined the perceptions, knowledge and preferences of consumers (approximately 300 people across the USA) regarding, among other topics, bio-plastics and container disposal. The findings showed that consumers do not know the sources used to create bio-based plastic containers, since the most frequent reply to the corresponding question was "I do not know." In addition, a large fraction of consumers mistakenly indicated petroleum (Figure 44.10). When asked about the future of the container after use, consumers selected the options of recycling (3.66 on a 5-point scale) and composting (3.28 on a 5-point scale) as more favorable compared with disposal in the trash bin (2.99 on a 5-point scale).

Acknowledgment

The authors thank Dr. Susan Selke for critically reviewing this chapter.

References

ABS (2006) *Solid Waste in Australia. Australia's Environment: Issues and Trends, 2006*, Cat. No. 4613.4. Australian Bureau of Statistics, Canberra. Available at http://www.abs.gov.au/ausstats/abs@.nsf/0/3B0DD93AB123A68BCA257234007B6A2F?OpenDocument (accessed October 8, 2010).

AEEMA (2006) *Waste Management*, draft report. Productivity Commission, May 2006. Available at http://www.pc.gov.au/__data/assets/pdf_file/0008/23768/subdr182.pdf (accessed October 8, 2010).

Anon. (2010) Eco-friendly food and beverage packaging. *Prepared Foods* (June), p. 10.

APC (2006) *The Many Uses of Plastics*. American Plastics Council, Arlington, VA. Available at http://www.americanplasticscouncil.org/s_apc/sec.asp?TRACKID=&SID=6&VID=86&CID=312&DID=931 (accessed September 22, 2010).

ASTM (1991) *Standard Terminology Relating to Plastics*, ASTM D 883-91. ASTM International, West Conshohocken, PA.

ASTM (2002) *Standard Guide for Assessing the Compostability of Environmentally Degradable Plastics*, ASTM D 6002-96 (reapproved 2002). ASTM International, West Conshohocken, PA.

ASTM (2003a) *Standard Specification for Biodegradable Plastics used as Coatings on Paper and Other Compostable Substrates*, ASTM D 6868-03. ASTM International, West Conshohocken, PA.

ASTM (2003b) *Standard Test Method for Determining Aerobic Biodegradation of Plastic Materials under Controlled, Composting Conditions*, ASTM D 5338-98. ASTM International, West Conshohocken, PA.

ASTM (2004) *Standard Specification for Compostable Plastics*, ASTM 6400-04. ASTM International, West Conshohocken, PA.

ASTM (2005) *Standard Test Method for Determining the Bio-based Content of Natural Range Materials using Radio Carbon and Isotope Ratio Mass Spectrometry Analysis*, ASTM 6866-05. ASTM International, West Conshohocken, PA.

BAO (2004) *Packaging Waste in Australia: An Overview, 2004*. Boomerang Alliance Organization. Available at http://www.boomerangalliance.org/000_files/28_container_deposits.pdf (accessed October 8, 2010).

BBO (2010) *Bottle Bills Promote Recycling and Reduce Waste*. Bottle Bill Organization. Available at http://www.bottlebill.org/about/benefits/waste.htm (accessed July 8, 2010).

Bohlmann, G.M. (2004) Biodegradable packaging life-cycle assessment. *Environmental Progress* 23: 342–346.

BPI (2010) *Biodegradable Additives*. Biodegradable Product Institute. Available at http://www.bpiworld.org/resources/Documents/Biodegradable%20Additives%20Fact%20Sheet%20v8%20July%2009.pdf (accessed October 15, 2010).

Burden, J. and Wills, R.B.H. (1989) *Deciding on Packaging for Produce. Prevention of Post-Harvest Food Losses: Fruits, Vegetables and Root Crops. A Training Manual*. Food and Agriculture Organization of the United Nations, Rome.

Bush, R., Araman, P. and Hager, E.B. (2008) Recovery, reuse and recycling by the United States wood packaging industry: 1993 to 2006. In: *Dean's Forum on Environment*, Virginia Tech, Blacksburg, VA.

Cattaneo, J. (2010) Glass bottles: reaching 50% recycled content. In: *Virginia Recycling Association Annual Conference*, Virginia Beach, VA, May 18, 2010.

CEN (2010) European Committee for Standardization. Available at http://www.cenorm.be (accessed October 15, 2010).

CFR (2010) *Title 40: Protection of Environment. Electronic Code of Federal Regulations Chapter I – Food and Drug Administration, Department of Health and Human Services, Subchapter B – Food For Human Consumption. Part 503: Standards for the Use or Disposal of Sewage Sludge*. 40 CFR §503.13.

CIWMB (2010) *Market Status Report: Container and Plate Glass*. California Integrated Waste Management Board. Available at http://www.ciwmb.ca.gov/markets/StatusRpts/glass.htm (accessed May 20, 2010).

CMI (2010) Can Manufacturers Institute. Available at www.cancentral.com (accessed September 7, 2010).

CRI (2009) *Understanding Economic and Environmental Impacts of Single-Stream Collection Systems*. Container Recycling Institute. Available at http://www.container-recycling.org/assets/pdfs/reports/2009-SingleStream.pdf (accessed July 19, 2010).

DEFRA (2009) Department for Environment, Food and Rural Affairs. Available at http://www.defra.gov.uk/environment/waste/ad/documents/implementation-plan.pdf (accessed August 19, 2010).

Degli-Innocenti, F., Tosin, M. and Bastioli, C. (1998) Evaluation of the biodegradation of starch and cellulose under controlled composting conditions. *Journal of Polymers and the Environment* 6: 197–202.

Doi, Y. and Fukuda, K. (1994) *Biodegradable Plastics and Polymers*, Studies in Polymer Science, No. 12. Elsevier, Amsterdam, pp. 479–497.

EB (2009) *Position Paper: "Oxo-Biodegradable" Plastics*. European Bioplastics. Available at http://www.mirelplastics.com/imagesupl/European_Bioplastics_OxoPositionPaper.pdf (accessed October 15, 2010).

EJnet (2010) *McLibel Trial Testimony on Polystyrene. 1994. Web Resources for Environmental Justice Activists*. EJnet.org. Available at http://www.ejnet.org/plastics/polystyrene/mclibel_p6.html (accessed August 10, 2010).

EN (2000) *Packaging. Requirements for Packaging Recoverable Through Composting and Biodegradation – Test Scheme and Evaluation Criteria for the Final Acceptance of Packaging*, EN 13432. European Committee for Standardization, Brussels.

EPA (1982) *Fossil Fuel Fired Industrial Boilers – Background Information Document*, Vol. 1, Chapters 1–9. Draft EIS, Report No. EPA450/3-82-006a. U.S. Environmental Protection Agency, Research Triangle Park, NC.

EPA (1992) *Solid Waste Division and Emergency Response. Characterization of Municipal Solid Waste in the United States: 1992 Update*, Report No. EPA/530-S-92-019. U.S. Environmental Protection Agency, Washington, DC.

EPA (1994) *Solid Waste Division and Emergency Response. Characterization of Municipal Solid Waste in the United States: 1994 Update*, Report No. EPA/530-S-94-042. U.S. Environmental Protection Agency, Washington, DC.

EPA (1995) *Solid Waste Division and Emergency Response. Characterization of Municipal Solid Waste in the United States: 1995 Update*, Report No. EPA/530-S-95-0. U.S. Environmental Protection Agency, Washington, DC.

EPA (1997) *Solid Municipal and Industrial Solid Waste Division. Office of Solid Waste. Characterization of Municipal Solid Waste in the United States: 1996 Update*, Report No. EPA/530-S-97-015. U.S. Environmental Protection Agency, Washington, DC.

EPA (1998) *Solid Municipal and Industrial Solid Waste Division. Office of Solid Waste. Characterization of Municipal Solid Waste in the United States: 1997 Update*, Report No. EPA530-S-98-007. U.S. Environmental Protection Agency, Washington, DC.

EPA (2001) *Solid Waste Division and Emergency Response. Municipal Solid Waste in the United States: 1999 Facts and Figures*, Report No. EPA530-R-01-014. U.S. Environmental Protection Agency, Washington, DC.

EPA (2002) *Solid Waste Division and Emergency Response. Municipal Solid Waste in the United States: 2000 Facts and Figures*, Report No. EPA530-R-02-001. U.S. Environmental Protection Agency, Washington, DC.

EPA (2003) *Solid Waste Division and Emergency Response. Municipal Solid Waste in the United States: 2001 Facts and Figures*, Report No. EPA530-R-03-011. U.S. Environmental Protection Agency, Washington, DC.

EPA (2004) *WasteWise 2004 Annual Report: Sustaining Excellence*, Report No. EPA530-R-04-035. U.S. Environmental Protection Agency, Washington, DC.

EPA (2005) *Solid Waste Division and Emergency Response. Municipal Solid Waste Generation, Recycling, and Disposal in the United States: Facts and Figures for 2003*, Report No. EPA530-F-05-003. U.S. Environmental Protection Agency, Washington, DC.

EPA (2006) *Solid Waste Division and Emergency Response. Municipal Solid Waste Generation, Recycling, and Disposal in the United States: Facts and Figures for 2005*, Report No. EPA530-F-06-039. U.S. Environmental Protection Agency, Washington, DC.

EPA (2007) *Solid Waste Division and Emergency Response. Municipal Solid Waste Generation, Recycling, and Disposal in the United States: Facts and Figures for 2006*, Report No. EPA530-F-07-030. U.S. Environmental Protection Agency, Washington, DC.

EPA (2008) *Solid Waste and Emergency Response. Municipal Solid Waste in the United States: 2007 Facts and Figures*, Report No. EPA530-R-08-010. U.S. Environmental Protection Agency, Washington, DC.

EPA (2009a) *Solid Waste and Emergency Response. Municipal Solid Waste in the United States: 2008 Facts and Figures*, Report No. EPA530-F-09-021. U.S. Environmental Protection Agency, Washington, DC.

EPA (2009b) Reuse and reduce. In: *Wastes*. U.S. Environmental Protection Agency. Available at http://www.epa.gov/epawaste/conserve/rrr/reduce.htm (accessed October 11, 2010).

EPA (2010) *Environmental Benefits*. Available at http://www.epa.gov/osw/conserve/rrr/composting/benefits.htm (accessed October 11, 2010).

EPIC (2008) *Sustainability*. Environment and Plastics Industry Council. Available at http://www.plastics.ca/EnvironmentalSustainability/index.php/ (accessed October 3, 2011).

EPR (2009) *OXO Degradables Incompatibility with Plastics Recycling*. European Plastics Recyclers. Available at www.plasticsrecyclers.eu/press (accessed June 9, 2009).

EPS (2010) *Expanded PS and the Environment*. Expanded Polystyrenes. Available at http://www.eps.co.uk/pdfs/eps_and_the_environment.pdf (accessed July 12, 2010).

ERF (2010) Earth Resource Foundation. Available at http://www.earthresource.org/campaigns/capp/capp-styrofoam.html (accessed November 5, 2010).

EU (1994) *European Parliament and Council Directive 94/62/EC of 20 December 1994 on Packaging and Packaging Waste*, OJ L 365, 31.12.1994. European Parliament, p. 10.

FDA (2009) *Recycled Plastics in Food Packaging*. United States Food and Drug Administration. Available at http://www.fda.gov/food/foodingredientspackaging/foodcontactsubstancesfcs/ucm093435.htm (accessed May 20, 2010).

FDA (2010) *Guidance for Industry – Use of Recycled Plastics in Food Packaging: Chemistry Considerations*. United States Food and Drug Administration. Available at http://www.fda.gov/food/guidancecomplianceregulatoryinformation/GuidanceDocuments/FoodIngredientsandPackaging/ucm120762.htm (accessed May 20, 2010).

Girling, P.J. (2003) Packaging of food in glass containers. In: *Food Packaging Technology* (eds R. Coles, D. McDowell and M.J. Kirwan). Blackwell Publishing Ltd., Oxford, pp. 152–173.

Good, R.H. (1988) Food and packaging interactions. In: *Recent Advances in Metal Can Interior Coatings*, ACS Symposium Series, Vol. 365, pp. 203–216.

GPI (2010) *Recycle Glass*. Glass Packaging Institute. Available at http://www.gpi.org/recycleglass/ (accessed July 19, 2010).

Greenpeace (2003) *PVC Free Solutions*. Available at http://www.greenpeace.org/international/campaigns/toxics/polyvinyl-chloride/pvc-free-solutions/ (accessed June 21, 2010).

Greer, D. (2006) A review. Aspects of compostable plastics, such as polylactic acid, starch-derived plastics, bioplastic synthetic blends, etc. are reviewed and discussed. *Biocycle* 47: 43–45.

Hanlon, J.F. (1984) *Handbook of Packaging Engineering*. McGraw-Hill, New York.

Harrington, R. (2009) *Anaerobic Digestion Scheme Key to Cutting Food and Packing in Landfill*. FOODproductionDaily.com. Available at http://www.foodproduction

daily.com/Supply-Chain/Anaerobic-digestion-scheme-key-to-cutting-food-and-packing-in-landfill (accessed October 14, 2010).

Holdsworth, S.D. and Simpson, R. (2007). Introduction. In: *Thermal Processing of Packaged Foods*, 2nd edn. Springer Science+Business Media, New York, pp. 1–13.

ILSI (2000) *Packaging Materials: 1. Polyethylene Terephthalate (PET) for Food Packaging Applications*. International Life Sciences Institute, Brussels.

ILSI (2002) *Packaging Materials: 2. Polystyrene for Food Packaging Applications*. International Life Sciences Institute, Brussels.

INCPEN (2010) The Industry Council for Packaging and the Environment. Available at www.incpen.org (accessed September 10, 2010).

ISO (2002) *Plastics – Determination of the Degree of Disintegration of Plastic Materials under Defined Composting Conditions in a Pilot-Scale Test*, ISO 16929. International Organization for Standardization, Geneva.

ISO (2005) *Determination of the Ultimate Aerobic Biodegradability and Disintegration of Plastic Materials under Controlled Composting Conditions – Method by Analysis of Evolved Carbon Dioxide*, ISO 14855-1. International Organization for Standardization, Geneva.

ISO (2006a) *Environmental Management – Life Cycle Assessment – Principles and Framework*, ISO 14040. International Organization for Standardization, Geneva.

ISO (2006b) *Environmental Management – Life Cycle Assessment – Requirements and Guidelines*, ISO 14044. International Organization for Standardization, Geneva.

ISO (2006c) *Determination of the Ultimate Aerobic Biodegradability of Plastic Materials under Controlled Composting Conditions – Method by Analysis of Evolved Carbon Dioxide. Part 2: Gravimetric Measurement of Carbon Dioxide Evolved in a Laboratory Scale Test* (under development), ISO/TC61/SC5, ISO 14855-2. International Organization for Standardization, Geneva.

IWSA (2007) *The 2007 IWSA Directory of Waste-to-Energy Plants*. Integrated Waste Services Association. Available at http://www.wte.org/docs/IWSA_2007_Directory.pdf (accessed March 4, 2008).

James, K.L. (2003) *Environmental Life Cycle Costs in the Australian Food Packaging Supply Chain*. PhD thesis, Victoria University, Melbourne, Australia.

JBPA (2010) JBPA. Available at http://www.jbpaweb.net/english/e-gp2.htm (accessed September 29, 2010).

Kale, G., Auras, R. and Singh, S.P. (2007) Comparison of the degradability of poly(lactide) packages in composting and ambient exposure conditions. *Packaging Technology and Science* 20(1): 49–70.

Kaspura, A. (2006) *Waste Generation and Resource Efficiency*. Submission to the Productivity Commission Inquiry into Waste Generation and Resource Efficiency, Department of the Environment and Heritage, February 2006. Engineers Australia.

Kijchavengkul, T, and Auras, R. (2008) Perspective compostability of polymers. *Polymer International* 57: 793–804.

Kijchavengkul, T., Auras, R., Rubin, M., Ngouajio, M. and Fernandez, T. (2006) Development of an automatic laboratory-scale respirometric system. *Polymer Testing* 25: 1006–1016.

Komolprasert, V. (2000) Food packaging: new technologies. In: *Handbook of Food Science, Technology, and Engineering*, Vol. 3 (ed. Y.H. Hui). Marcel Dekker, New York, pp. 130-1–10.

Koutsimanis, G., Harte, J., Behe, B., Bix, L., Harte, B., Whiting, M., McFerson, J. and Almenar, E. (2010) Measuring supply chain trends in the packaging of stem-free fresh sweet cherries: a sustainable approach in package design, product characteristics and expectations. *HortScience* 45(8): S71.

Krzan, A., Hemjinda, S., Miertus, S., Corti, A. and Chiellini, E. (2006) Standardization and certification in the area of environmentally degradable polymers. *Polymer Degradation and Stability* 91: 2819–2833.

Lallanilla, M. (2010) Marking symbols made easy. Available at http://greenliving.about.com/od/recyclingwaste/tp/recycling_symbols.htm (accessed October 2, 2010).

Lotto, N.T., Calil, M.R., Guedes, C.G.F. and Rosa, D.S. (2004) The effect of temperature on the biodegradation test. *Materials Science and Engineering* 245: 659–662.

May, N. (2004) Developments in packaging formats for retort processing. In: *Improving the Thermal Processing of Foods* (ed. P.S. Richardson). Woodhead/CRC, Boca Raton, FL, pp. 138–151.

Marsh, K. and Bugusu, B. (2007) Food packaging: roles, materials, and environmental issues. *Journal of Food Science* 72: R39–R55.

McKnown, C. (2000) Containers. In: *Coatings on Glass: Technology Roadmap Workshop*, report. Sandia National Laboratory, Livermore, CA, pp. 8–10.

Miller, C. (1993) Waste product profile: steel cans. *Waste Age*, January 1993, p. 69.

Miller, R.J. (2003) Biodegradability of polymers: regulations and methods for testing. In: *Biopolymers*, Vol. 10, *General Aspects and Special Applications* (ed. A. Steinbüchel). John Wiley & Sons, Berlin, pp. 365–374.

Mintel (2009) *Packaging Trends in Food and Drink – U.S. – March 2009*. Global New Products Database (GNPD). Mintel, Chicago, IL.

Murray, M. (2010) Using recycled packaging in your supply chain. Available at http://logistics.about.com/od/greensupplychain/a/recycled_pack.htm (accessed October 2, 2010).

NAPCOR (2010) *PET Recycling Links*. National Association for PET Container Resources. Available at http://www.napcor.com/ (accessed July 21, 2010).

Narayan, R. (2010) Drivers for biodegradable/compostable plastics & role of composting waste management & sustainable agriculture. Available at http://www.msu.edu/user/narayan/germanycompostingpaper.htm (accessed October 17, 2010).

Neelis, M., Worrell, E. and Masanet, E. (2008) *Energy Efficiency Improvement and Cost Saving. Opportunities for the Petrochemical Industry. An ENERGY STAR® Guide for Energy and Plant Managers*. Available at http://www.energystar.gov/ia/business/industry/Petrochemical_Industry.pdf (accessed November 5, 2010).

Nie, Y. (2008) Development and prospects of municipal solid waste (MSW) incineration in China. *Frontiers of Environmental Engineering in China* 2(1): 1–7.

OKCOMPOST (2010) Available at http://www.okcompost.be/data/pdf-document/Program-OK-10e-b-OK-Bio-SOIL.pdf (accessed September 29, 2010).

Piccione, J. (2010) *Sustainability and Bio-Based Flexible Packaging*, MSU School of Packaging Fall Seminar Series, October 8, 2010.

PIRA (2004) *Packaging's Place in Society: Resource Efficiency of Packaging in the Supply Chain for Fast Moving Consumer Goods*. Pira International and University of Brighton. Available at http://www.packagingfedn.co.uk/images/reports/Summary_26-07-04.pdf (accessed October 15, 2010).

PIRA (2005) *Study on the Implementation of Directive 94/62/EC on Packaging and Packaging Waste Options to Strengthen Prevention and Re-Use of Packaging*. Available at http://europa.eu.int/comm/environment/waste/studies/packaging/050224_final_report.pdf (accessed October 08, 2010).

PP (2010) *Sustainability*. Paperboard Packaging. Available at http://www.paperboardpackaging.org/sustainability/index.html (accessed on November 6, 2010).

PPA (2006) *A Renewable and Recyclable Resource*. Paperboard Packaging Alliance. Available at http://www.paperboardpackaging.org/ (accessed October 2, 2010).

PLASTEMART (2010) Available at http://www.plastemart.com/Upload/Ecamp/pbi/ecamp13/insidepage.htm (accessed September 29, 2010).

Pradhan, R., Reddy, M., Diebel, W., Erickson, L., Misra, M. and Mohanty, A. (2010) Comparative compostability and biodegradation studies of various components of green composites and their blends in simulated aerobic composting bioreactor. *International Journal of Plastic Technology* (online), DOI 10.1007/s12588-010-0009-z.

Robertson, G.L. (1993) Glass packaging materials. In: *Food Packaging: Principle and Practice* (ed. G.L. Robertson). Marcel Dekker, New York, pp. 232–252.

Selke, S.E.M., Cutler, J.D. and Hernandez, R.J. (2004) *Plastics Packaging*, 2nd edn. Hanser, Munich.

SPI (2010) *SPI Resin Identification Code – Guide to Correct Use*. SPI, The Plastics Industry Trade Association. Available at http://www.plasticsindustry.org/AboutPlastics/content.cfm?ItemNumber=823 (accessed September 29, 2010.

SRI (2010) Steel Recycling Institute. Available at http://www.recycle-steel.org/ (accessed October 14, 2010).

TAA (2006) *North America Aluminum Industry: A Quick Review*. The Aluminum Association. Available at http://www.aluminum.org/Content/NavigationMenu/ (accessed January 10, 2007).

Thornton, J. (2005) *PVC Production is the Largest Use of Chlorine Gas in the World: A Briefing Paper for the Healthy Building Network*. Available at http://www.healthybuilding.net/pvc/ThorntonPVCSummary.html (accessed July 01, 2010).

Tsai, W. (2002) A review of environmental hazards and adsorption recovery of cleaning solvents hydrochlorofluorocarbons (HCFCs). *Journal of Loss Prevention in the Process Industries* 15: 147–157.

UNEP (2010) *Solid Waste Management Sourcebook, Section 1.5.4, Managing Environmental Impacts of Air Emissions and Reshuffle Ash.* United Nations Environment Programme. Available at http://www.unep.or.jp/ietc/ESTdir/Pub/MSW/sp/SP5/SP5_4.asp (accessed October 10, 2010).

WOO (2006) *Wood Recycling Information Sheet.* Waste Online Organization. Available at http://www.wasteonline.org.uk/resources/InformationSheets/Wood.htm (accessed October 3, 2010).

45

Food Quality and Safety Assurance by Hazard Analysis and Critical Control Point

Tomás Norton and Brijesh Tiwari

Introduction

'Food safety' is the term used to provide the assurance that food will not cause any harm to consumers when it is prepared and/or consumed according to its intended use (FAO/WHO, 1997). Food safety requires a systematic preventive approach that addresses physical, chemical and biological hazards during food manufacturing, distribution and storage. Hazards may occur at any stage during food production, so control and remediation procedures for food safety are necessary throughout the supply chain, as well as close cooperation between countries when setting global standards. Owing to the probability of contamination at any stage in the food chain, measures to ensure food safety have to be implemented on a global scale, necessitating a global approach (Havelaar *et al.*, 2010). Therefore, ensuring food safety is a joint responsibility for all participating members, meaning that people must be aware of the potential locations where hazards may occur, and not restrict their focus to a single company or process. For example, companies that purchase raw materials, ingredients and food packaging materials for the manufacture of consumer goods must ensure that these materials are safe and fit for use. On the other hand, consumers must be vigilant of food safety and quality when making purchases, paying greatest attention to the handling, storage, preparation and final use of foods (Alli, 2004).

Handbook of Food Process Design, First Edition. Edited by Jasim Ahmed and Mohammad Shafiur Rahman.

In order to ensure that adequate food safety can be achieved, it is necessary to first conceptualise the possible effects that various hazards may have on food safety during handling, preparation, processing or packaging, and prepare responses and then devise techniques to deal with them (Doeg, 1995). In response to this, a food safety system known as Hazard Analysis and Critical Control Points (HACCP) was scientifically designed to identify specific hazards, as well as the most appropriate controlling actions, in order to ensure food safety and quality. The HACCP system is an internationally recognised method for ensuring food safety, and its principles have to be applied according to European regulations, for example EC No. 852/2004 (OJEU, 2004). Today, HACCP is used as a tool for the development, implementation and management of effective safety assurance procedures. HACCP is intended for use by individual food companies (i.e., food producers, manufacturers, distributors and retailers) as a protocol for the development of unique safety assurance procedures to meet their individual needs, and for both the food industry and health authorities in preventing food-borne diseases (Vela and Fernandez, 2003). In fact, HACCP is widely acknowledged as the best method of assuring product safety and is becoming internationally recognised as a tool for controlling food-borne safety hazards (Tsola *et al.*, 2008). In the HACCP protocol, hazards can be biological, chemical or physical agents in the food with the potential to cause an adverse health effect (FAO/WHO, 1997). While HACCP systems have become mandatory for the food industry in the European Union (ECC, 1993), the HACCP system should be optimised for every food production line by adapting its procedures to individual products and processes (da Cruz *et al.*, 2006)

In contrast to food safety, food quality is more difficult to define, understand and measure. According to Olsen *et al.* (2008), food quality transcends all steps and all actors within the food chain, but it is of an intangible nature because it is perceived individually. In fact, food quality can have several different meanings, many of which encompass aspects of the food's organoleptic characteristics, physical and functional properties, and nutrient content (Mamalis *et al.*, 2009). Food quality criteria are essentially set based on the need to achieve desirable levels of these characteristics in the final food product. 'Quality assurance' is the name given to the implementation of various programmes used to enable a company to meet these quality criteria. While it is possible to use some food quality characteristics as indicators of food safety, they are not generally considered to be food safety characteristics. For this reason, food safety principles and practices have always been integrated into activities within quality assurance or quality control programmes, such that these programmes will contemporaneously ensure that both food quality and food safety standards are met within the company (Alli, 2004).

In the past, quality assurance systems have been necessary to achieve high standards in food production; these have involved inspection, testing and monitoring of food quality, alongside additional activities dedicated to food safety hazards. Such systems have been put into place in response to the fact that high-quality products or services are the essence of a company's survival in the highly competitive global market, and customers can be confident that the food company can meet international food safety and quality regulations (Misterek *et al.*, 1990). Many of the available quality assurance systems are based on the ISO standards. In fact, one of the most widely recognised standards used for food quality was the ISO 9000 family. The earlier members of the ISO 9000 family (the 1987 and 1994 versions) had the working objectives of ensuring that products met the desired specifications, whereas the more recent version (2000) focused on the achievement of customer satisfaction by meeting customer requirements, ensuring continuous improvement of the system and preventing non-conformity (Luning *et al.*, 2006). As a result, ISO 9001:2000 is the best-known quality management system throughout many industries, and has been employed widely by the food industry. However, the fundamental difficulty associated with the application of the ISO 9001:2000 set of standards in the food industry is that the technological aspects are very global and give no concrete handles on food safety (Luning *et al.*, 2006). In 2005, ISO 22000:2005 was introduced, with flexibility to enable a tailor-made approach to food safety for all segments of the food chain, since the standards and procedures required for high-risk areas in one food sector may not be appropriate in another (Arvanitoyannis *et al.*, 2009). ISO 22000:2005 gives advice for all sorts of organisations within the food supply chain, ranging from feed providers, primary producers, food manufacturers, transport and storage operators, food and retail service outlets, and suppliers of packaging materials and cleansing agents, to suppliers of additives and ingredients. Moreover, the ISO 22000 family of standards was engineered to work in tranquillity with ISO 9001:2000 for quality management systems including continual improvement. This allows a company to integrate its own food safety management system with related management systems and establish a food safety management system that complies with ISO 22000.

From the above, it is evident that the best way to improve the performance of food businesses is by ensuring that food quality and safety are regulated in an auditable fashion, with the most recent ISO standard placing greater emphasis on this fact. Therefore, the current chapter will discuss the issues involved in the design and implementation of a HACCP system with respect to its integration within the ISO 2000:2005 quality standards. The chapter will cover the reasons of how individual aspects of ISO 22000:2005 enhance the effectiveness of a food safety and quality management system.

Introduction to Hazard Analysis Critical Control Points (HACCP)

HACCP is term that is synonymous with 'food safety' in today's vocabulary, and has grown out of consumers' desire for food which is free from microbiological, chemical and physical hazards. The need for a strict HACCP approach to food safety within all food production systems is becoming more evident as we consider the numerous reports of outbreaks of food-borne illnesses across the world. For example, in 2005 an outbreak of the *Escherichia coli* O157 bacterium occurred in Wales in the UK, resulting in 157 cases, where 31 people were hospitalised and one five-year-old child died (Pennington, 2009). More recently, in 2008, a listeriosis outbreak in Canada resulted in the death of 23 people, with 57 confirmed cases in total (CFIA, 2009). In the same year, China reported a food safety incident involving milk and infant formula, which was adulterated with melamine, resulting in an estimated 300 000 victims, of whom six infants died (Branigan, 2008). These cases bring to the forefront the need for effective HACCP procedures in the food industry with adequate integration of other practices, such as an adherence to good manufacturing practices, use of standard sanitation operating procedures and personal hygiene programmes.

HACCP was first proposed by the Pillsbury Company, in cooperation with NASA and the US Army laboratories, in a quest to make sure that food which astronauts consumed would not result in the development of illnesses. This programme came out of a NASA 'zero defects' assurance programme, which aimed to prevent food contamination through bacterial or viral pathogens and chemical or physical hazards that could cause illness or injury (Escriche *et al.*, 2006). The original form of HACCP used elements of failure, mode and effect analysis (FMEA) and had three components (Ropkins and Beck, 2000):

1. The identification and assessment of all hazards associated with the final foodstuff.
2. The identification of the steps or stages within food production at which these hazards may be controlled, reduced or eliminated: the critical control points (CCPs).
3. The implementation of monitoring procedures at these CCPs.

Since then, HACCP has evolved to meet the specific needs of food producers, manufacturers and distributors, regulators, and consumers in various different sectors of the food industry (Escriche *et al.*, 2006). Despite this development, the three components listed above remain inherent in all contemporary HACCP procedures, as the combination of practical (prevention-orientated) primary components and a flexible approach to their implementation has been identified as HACCP's greatest attribute because it allows HACCP to remain relevant, efficient and effective, despite the introduction of new food technologies (Ropkins and Beck, 2000). The essential components that make up a present day HACCP plan are shown in Figure 45.1.

Once HACCP had been recognised and developed to a point at which it could be safely used throughout the food industry, the United States National Academy of Science recommended that HACCP be employed as a food safety system in all food processing factories. In 1993, the EU implemented a directive on the hygiene of food (93/43/EEC), which established HACCP as a tool with which food manufacturers could implement and maintain control over their processes. Following this, HACCP principles were adopted worldwide, as set out by the Codex Alimentarius Commission and the National Advisory Committee on Microbiological Criteria for Foods (FAO/WHO, 1997).

Figure 45.1 All of the essential aspects that form the HACCP system.

The Advantages of Using the HACCP Approach

The HACCP system is recognised internationally as an essential tool for ensuring the safety of food. During the implementation of the HACCP system, continuous collection of essential processing variables is required, which allows auditors to gauge how well a firm is complying with food safety regulations. Besides enhancing food safety, other benefits of applying HACCP include more effective use of resources and more timely responses to food safety problems. The successful application of HACCP requires the full commitment and involvement of management and the workforce. It also requires a team approach and is compatible with the implementation of quality management systems, such as the recently introduced International Organization for Standardization's ISO 22000 series. Food safety management systems built in accordance with the principles of HACCP have a clearly defined structure:

- *They are based directly on food processes.* Each HACCP plan is specific to the food process in question. Great benefit can be drawn from each process and procedure being clearly defined.
- *Risk management.* HACCP uses a systematic and timely approach to uncovering the potential safety hazards during all aspects of food production, from raw materials, processing and distribution to consumption.

- *Consumer confidence.* A controlled operating environment and an effectively implemented and applied food safety system will improve consumer confidence in food safety.
- *Reliable record-keeping.* Record-keeping enables more efficient and effective government and customer oversight, and facilitates the inspection activities of food auditors.
- *Legal issues.* The use of the HACCP approach can help with legal cases in the event of an outbreak of food-borne disease.
- *Trading.* HACCP-based approaches are a benefit to companies seeking to meet customer and legal requirements, whether for the domestic or the export market.

Prerequisite Programmes

Hygiene standards and procedures are usually described as 'good hygienic practices' (GHPs) or 'good manufacturing practices' (GMPs), and have been in place for many years as an essential food safety tool. The following of these practices forms the foundation of HACCP, as they provide the basic environmental and operating conditions for the production of safe food. An effective HACCP system is built on a solid foundation of prerequisite programmes. Therefore, prerequisite programmes cover all GMPs and GHPs and should be documented and regularly audited; they are established and maintained separately from the HACCP plan.

Good Manufacturing Practices

The general requirements for food companies are that the premises must be kept clean and maintained in good condition at all times. The GMP rules serve as a prerequisite programme, defined as the universal procedures used to control the conditions in the plant environment, which contributes to the overall safety of the product. GMPs include aspects of plant layout, cleaning and disinfection, zone classification, waste and waste disposal, maintenance, ventilation, measures to prevent contamination by metal and pests, personal hygiene, storage, training, and record-keeping.

Good Hygienic Practices

Good hygienic practices include all practices regarding the conditions and measures necessary to ensure the safety and suitability of food at all stages of the food chain. All food handlers should have a working knowledge of personal hygiene and understand the role of food in the transmission of food-borne illnesses. GHPs include personal hygiene, operation control and hygiene, changing rooms and facilities, protective clothing, maintenance and sanitation, training, and record-keeping. Because the programmes carried out during GMPs and GHPs are interlinked, both sets of practices come under the term 'prerequisite programmes' and form the actual control measures that are needed before a HACCP plan can be put into action. Prerequisite programmes (PRPs) can be divided into 14 groups (Escriche *et al.*, 2006):

PRP 1: cleaning and disinfection;
PRP 2: pest control;
PRP 3: water and air quality;
PRP 4: temperature control and registration;
PRP 5: personnel hygiene, facilities and education;
PRP 6: structure and infrastructure;
PRP 7: technical maintenance and calibration;
PRP 8: waste management;
PRP 9: control of raw material;
PRP 10: traceability;
PRP 11: allergens;
PRP 12: contamination;
PRP 13: management of product information;
PRP 14: work methodology.

Many of these prerequisite requirements are general and ensure the minimum conditions necessary for the production of safe, wholesome foods. On the other hand, some programmes can be very specific, including aspects such as thermometer use and the type of valves used for hand washing. The planning of prerequisite programmes should be carried out as follows (Escriche *et al.*, 2006):

- Set out a programme of activities: each prerequisite should contain a description of the work that needs to be carried out. During this phase, it is important to gain an understanding of what need to be done and how and when the tasks need to be done.
- Verify the activities and control their efficiency: this is necessary to ensure that the all the work that was set out in the programme of activities is being fully carried out.
- Corrective actions: once deviations or non-compliance are detected, the most appropriate corrective actions must be planned for and the efficiency of their application should be documented.

Developing a HACCP Plan

HACCP plans will undoubtedly vary between businesses, and often the plans will vary more specifically according to the product or process under investigation. In fact, while it is common to use generic HACCP plans as guides during HACCP development, it is essential that the unique conditions within each facility be considered during the development of all components of the HACCP plan. Essential features of

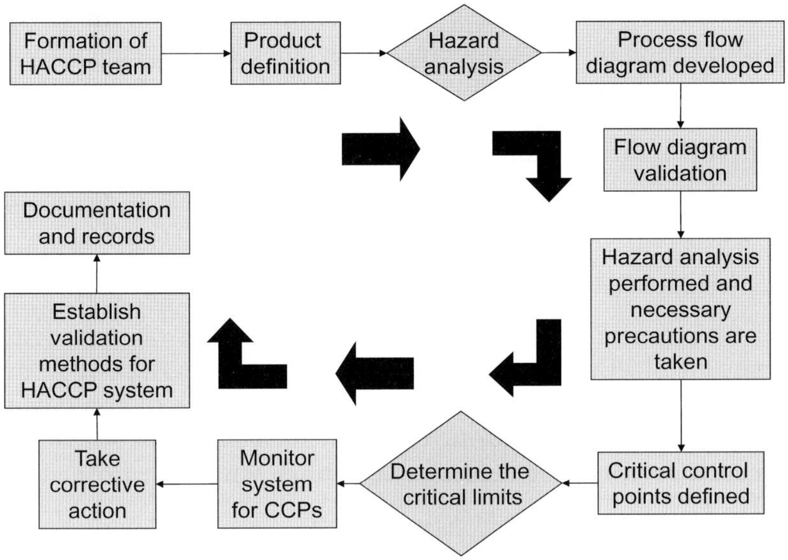

Figure 45.2 Generic process flow utilising the seven principles of HACCP.

a HACCP plan include (i) the assembly of the HACCP team, (ii) a description of the food ingredients and their distribution, (iii) a description of the intended use by the targeted consumer group, and (iv) the development and verification of a flow diagram that describes the process. Once the preliminary tasks have been done, the seven basic principles of HACCP can then be applied, in which the main goal is to establish a plan that describes each of the individuals responsible for developing, implementing and maintaining the HACCP system. Implementation of a HACCP system involves continuous monitoring and record-keeping, as well as the implementation of corrective actions, etc. Also, because maintaining an effective HACCP system depends largely on regular verification procedures, the HACCP plan should be updated and revised on a continual basis. In the following section, the seven principles of HACCP are described along with the specific stages involved in developing a HACCP plan (NACMCF, 1997). A generic process flow utilising the seven principles of HACCP is shown in Figure 45.2.

The Seven Principles of HACCP

Principle 1: Conduct a Hazard Analysis

Hazards are defined as biological, chemical or physical agents that may cause illness or injury when not effectively controlled. The purpose of a hazard analysis is to develop a list of hazards which are reasonably likely to cause injury or illness if not effectively controlled, and to develop 'process flow diagrams', which represent the various stages or steps of the food manufacturing process. The team must first identify all possible hazards and understand the measures that will control them. Through the hazard analysis, it may be possible to identify the necessary modifications to a process or product that will improve product safety, while also providing the basis for determining CCPs.

Principle 2: Identify the Critical Control Points

A critical control point is defined as a step at which control can be applied and is essential to prevent or eliminate a food safety hazard or reduce it to an acceptable level. CCPs must also address the potential hazards that are likely to cause illness or injury in the absence of control. During this identification phase, the food safety team must evaluate each potential area of risk to establish where the critical control points will exist.

Principle 3: Establish Corrective Actions To Be Taken

A critical limit is a threshold value to which a biological, chemical or physical parameter must be controlled at a CCP to prevent, eliminate or reduce the occurrence of a hazard. The HACCP team must determine measurements or values that indicate the operational limits for each critical limit, which represent the threshold at which the process falls out of control.

Principle 4: Establish CCP Monitoring Requirements

Monitoring is a planned sequence of observations and/or measurements that assess whether a CCP is under control, and it allows the production of accurate records for future use during the verification process. Establishing the monitoring requirements requires a description of the techniques and methods of measurement used to evaluate critical limits.

Principle 5: Establish Corrective Actions To Be Taken

It is inevitable that deviations from control limits may occur during food manufacturing. When such a deviation occurs, corrective actions are necessary and should include the following elements: (i) determine and correct the cause of non-compliance, (ii) determine the disposition of non-compliant product, and (c) record the corrective actions that have been taken. During the corrective-action development phase, the organisation must determine how errors can be corrected and the procedures that must be used to correct process deviations.

Principle 6: Establish Procedures for Verification

Verification procedures are put in place to check whether the HACCP plan is functioning at its best. Another important aspect of this principle is that the HACCP plan is initially validated to determine if the plan is scientifically and technically sound. This makes sure that all hazards have been identified and that these hazards will be effectively controlled.

Principle 7: Establish Documentation

This principle requires documented instructions (standard operating procedures) and accurate record-keeping for all critical control points. The records maintained for the HACCP system should include everything from a summary of the hazard analysis, through the HACCP plan documentation, to verification procedures and schedules.

Implementing the HACCP Plan

The successful implementation of a HACCP plan is facilitated by commitment from top management. Initially, the HACCP coordinator and team are selected and trained as necessary. The team is then responsible for developing the initial plan and coordinating its implementation. Product teams can be appointed to develop HACCP plans for specific products. An important aspect of developing these teams is to ensure that they have appropriate training. The workers who will be responsible for monitoring need to be adequately trained. Upon completion of the HACCP plan, operator procedures, forms, and procedures for monitoring and corrective action are developed. Often it is a good idea to develop a timeline for the activities involved in the initial implementation of the HACCP plan. An important aspect of maintaining the HACCP system is to ensure that all individuals involved are properly trained so that they understand their role and can effectively fulfil their responsibilities. The following stages of HACCP plan development are described in accordance with the advice given by Escriche *et al.* (2006) and FAO/WHO (1997).

Stage 1: Assembling the HACCP Team

Developing and implementing an effective HACCP plan requires a team with many different skills so that the most appropriate product-specific expertise and knowledge are directly at hand. The team should comprise individuals familiar with all aspects of the production process, along with engineers and microbiologists.

Stage 2: Product Description

For an effective HACCP system, it is essential that the product must be understood and described in good enough detail by the HACCP team. The product description must include all safety information, for example physical and chemical structure, processing conditions (sterilisation, freezing, chilling, blanching, etc.), shelf-life, storage conditions, and details of packaging and distribution. If a range of similar products are being developed, these can be grouped together only when their characteristics and processing steps are similar.

Stage 3: Identification of Intended Use

In this stage, the expected uses of the product by the end-user or target consumer should be considered in detail (e.g. cooking or reheating). Special care must be given to the target end-user so that different groups of users can be considered, for example children or elderly people.

Stage 4: Develop the Flow Diagram

The HACCP team should construct an accurate and detailed flow diagram of the manufacturing process being considered. The process flow diagram is a representation of how a food product or raw material is produced, prepared or manufactured. Therefore, this plan must cover every individual step of the process, while providing sufficient technical data for the HACCP plan. To ensure adequate food safety, it is necessary that the process flow diagram be constructed carefully. It should start as soon as raw materials, ingredients or packaging materials are received by the company and deal with all steps in the process through to the passing of the final product to the customer or consumer. The following points should be considered when developing a process flow diagram:

- The team responsible for this should have an intimate knowledge of the production or preparation process.
- Each processing step, including flows of ingredients, raw materials, packaging and waste materials, must be included in the drawing in a logical sequence from start to finish.
- Each step should be defined by a unit operation, in which the product undergoes a change in state, for example a change in temperature or addition or removal of ingredients.
- Each step must be named correctly so that it is easy to determine whether a critical control point needs to be put into place. To avoid confusion, the process step should be named using as few words as possible and the name should functionally describe what happens to the food, raw material, etc. at the process step.

Stage 5: Confirmation of the Flow Diagram

The HACCP team should confirm that the flow diagram matches the corresponding process in the factory. This should be done by having someone else walk through the process to check what actually happens. The process should be observed at each step or unit operation, and any discrepancies between the diagram and normal practice should be recorded. The production process should also be observed outside normal working hours, such as during night shifts, as practice may vary between shifts. It is essential that the flow diagram is accurate, since the hazard analysis and the identification of CCPs rely on the data that the diagram contains. Once the verification has been agreed, the HACCP team should then sign off the process flow diagram.

Stage 6: List All Potential Hazards Associated with Each Step, Conduct a Hazard Analysis and Identify Control Measures for Each Hazard

The HACCP team should identify all the real or potential hazards that may occur at each step in the production process. In accordance with principle 1, the HACCP team should conduct a hazard analysis to identify those hazards that can be eliminated or reduced to acceptable levels. The hazard analysis should consider the likelihood of hazards occurring, as well as their severity for public health; it should include a qualitative and/or quantitative evaluation of the presence of the hazards; and it should consider the survival or multiplication of pathogenic microorganisms, the production or persistence in foods of toxins, chemicals or physical agents, and the conditions leading to the above (FAO/WHO, 1997). Once the hazard analysis has been completed, the HACCP team should consider the control measures that can be applied to each of the identified hazards. Some hazards may require more than one control measure for adequate control, and a single control measure may act to control more than one hazard. Escriche *et al.* (2006) advises that once all potential hazards and control measures have been identified, an investigation should be made of the probability of the hazard occurring, including the possible implications for public health. Ultimately, thc hazard analysis should be itemised in a table to clearly record the analysis.

Stage 7: Determine the Critical Control Points

Determining the location of the CCPs is essential during the development of a HACCP plan. However, the identification of the CCPs can be one of the most problematic steps when developing a HACCP plan, as the previous stage will yield a great deal of control measures that may or may not be suitable as CCPs. It is vital that the HACCP team has sufficient technical data to determine the CCPs effectively, and it is also important to be aware that more than one CCP may exist for a single hazard. To assist in finding where the correct CCPs should be, a CCP decision tree can be used. A decision tree is a logical series of questions that are asked for each hazard. In the case of the CCP decision tree, this is for each hazard at each process step. The answer to each question leads the HACCP team through a particular path in the tree and to a decision as to whether or not a CCP is required at any particular step. Using a CCP decision tree promotes structured thinking and ensures a consistent approach at every process step and for each hazard identified. The Codex CCP decision tree shown in Figure 45.3 is a traditional starting point for developing a decision tree. This tool may be used as it is or may be modified to meet the requirements of the specific process in question. Using the decision tree addresses the following questions:

- Is there a control measure in place to control physical, chemical or biological hazards at this stage? This information may be easily found from the hazard analysis table developed in the previous stage.
- Is this step specifically targeted at hazards? Some examples of hazard-targeted operations are sterilisation, cleaning, distillation and polishing. If this step is in place to target hazards, then the step is most definitely a CCP.
- Is this control measure necessary? For example, if contamination with an identified hazard in excess of unacceptable levels is impossible, then there is no need for a control measure, and this step should not be a CCP.
- Are there any controls in place to render the controls at this step superfluous? This question can yield powerful insight into the necessity and rationale for control positioning.

Stage 8: Establish Critical Limits for Each CCP

Critical limits must be specified and validated for each CCP. Control criteria are used to set critical limits for measurable variables such as temperature, time, moisture level, pH and measurable organoleptic parameters. When the value of a measured variable falls outside the critical limits, the product is not acceptable.

Stage 9: Establish a Monitoring System for Each CCP

Monitoring is basically the measurement or observation of a CCP and should either be continuous or be carried out with enough frequency to control the measurable variables at the CCP. Monitoring procedures for CCPs must be rapid, so that results are available quickly enough to maintain control at the CCP. Physical and chemical on-line measurements are usually preferred, so that process deviations can be detected in enough time to make appropriate adjustments in order to maintain tight control of the process. It is essential that the data is properly documented and recorded by the organisation in case of a product recall.

Stage 10: Establish Corrective Actions

Specific corrective actions must be developed for each CCP in the HACCP plan. When developing the HACCP plan, the team should be in a position to specify the corrective measures that are required when the measurements indicate a failure or, indeed, a trend towards violation of control limits. Therefore, if deviations from the critical limits occur at a CCP, the most appropriate action can be taken to bring the process back under control.

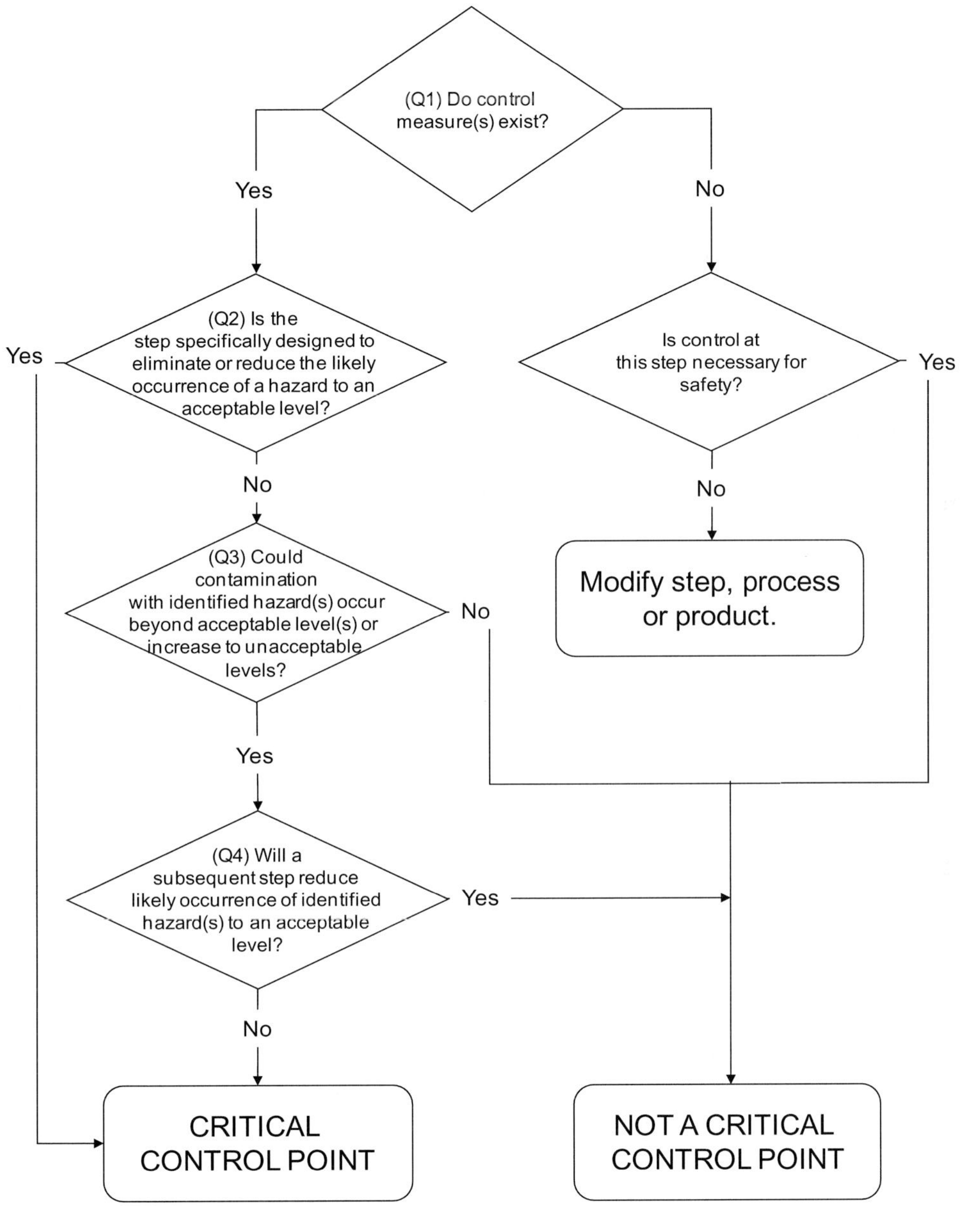

Figure 45.3 The CCP decision tree.

Stage 11: Verification and Validation

To ensure that all the above stages have been carried out effectively enough to assure food safety, the HACCP plan must be validated before being implemented. The frequency of verification depends on the type of food business, as well as the hazards involved and the number of deviations detected over time.

Stage 12: Establish Documentation and Record-Keeping

Efficient and accurate record-keeping is an essential element of the application of a HACCP system. All HACCP procedures should be documented and stored on the company's computer system. The documentation should include records arising from all the work done in the previous stages, for example hazard analysis, CCP determination and critical-limit determination. The appropriate documentation includes CCP monitoring activities and results, which in turn should include deviations from critical limits and the corrective measures used.

Reviewing the HACCP Plan

HACCP plans should change in accordance with any changes experienced by the food manufacturing process. Even small modifications to the product or process can invalidate the HACCP plan and introduce potential hazards. The implications of any such changes to the overall HACCP system must be fully considered and documented and adjustments made to procedures as necessary.

Using HACCP during Food Manufacturing

Effectiveness of HACCP

Training is essential for employees in the food industry to actively, effectively and professionally partake in the development and implementation of a HACCP system. HACCP has been characterised as the last resort of the managements of food companies, who justifiably seek some other management formula first before turning to training to solve their problems (Wallace and Williams, 2001). However, training is virtually unavoidable, particularly when staff need to know how to use machinery, understand and adopt procedures or acquire additional skills and knowledge that cannot be found in the marketplace. This is particularly true for small companies with limited access to information and often without the time or skills to interpret textbook scenarios. The fundamentals of HACCP methodology are as necessary for them as for any food company, and the typical short course must be used as an effective introduction to the concept and the jargon (Taylor, 2001).

As seen from the discussion earlier in this chapter, some of the main benefits to be drawn from HACCP are due to the preventive measures that are integrated into the system, i.e. adequate control accompanied by periodic verification and corrective actions, which supplant the need to perform simple inspections of random batches of the final product (Cormier *et al.*, 2007). However, the fact that HACCP is being employed in a manufacturing plant does not remove the need for final product monitoring. In fact, if final product monitoring is not carried out, many difficulties may be encountered during the lifetime of a food business, particularly given the fact that HACCP is simply a system for minimising the occurrence of hazards in the final end product and not removing them altogether. Take, for example, a case where a hazard has been found in a particular food product. The general reaction to this scenario is to put a lot of time, energy and of course money into recalling the product and reconstructing the HACCP system so that adequate control over food safety can be restored (Struijk, 1996). However, the actual effectiveness and performance of HACCP can be greatly improved by putting effort into the continuous monitoring of variation in the occurrence of food hazards over time. Real-time monitoring of the characteristics of the final product is not a new paradigm; it has long been known as statistical process control, which can be defined as the use of statistical methods to view and reduce process variation so that a particular process can be monitored, controlled and improved, thereby reducing defects and waste from the process, and ultimately allowing the business to continuously improve.

HACCP in Small Businesses

According to Taylor (2001), developing, installing, monitoring and verifying a successful HACCP system is dependent on a complex mix of managerial, organisational and technical hurdles such that even the largest food companies, who are equipped with significant resources of money, technical expertise and management skills, may face a difficult challenge. Therefore, it is not a surprise that small companies may feel that the difficulties of HACCP are potentially insurmountable. The factors for successful development of HACCP systems for small companies include the following (Taylor, 2001; Taylor and Kane 2005):

- Any pilot projects must be set in the context of thorough evaluation, with clearly identified indicators of success. Practitioners should only be encouraged to change when there is evidence that such change is both practical and achieving its aims.
- This commitment to high quality by the management of a small company must be channelled into the application of HACCP principles in order to secure safety across the entire food chain by making sure that the food industry, researchers, educators, enforcement agencies and governments pool resources and work towards a common goal.
- HACCP resource centres should be developed where small and medium-sized enterprises (SMEs) can access data, software, research information and advice on HACCP.
- Technology transfer could be effectively promoted by the identification of exemplary SMEs who could act as 'beacons' of good practice.

- Operation-specific guides are required which address the diversity of operations within each sector of the food industry.
- Step-by-step instructions on project-managing the HACCP process, using teams successfully, solving problems and, most importantly, managing change would be of considerable benefit to SMEs.
- A certification system is needed which will require all those offering HACCP services to have appropriate levels of knowledge of, training in and experience of HACCP.

Taylor and Kane (2005) investigated the burden of HACCP on SMEs and found that initial training had an enormous impact on the quality of the HACCP system developed, as when this training was conducted, SMEs had the confidence to (i) make decisions about the level and type of documents required for the safety of their product and (ii) negotiate with external parties in order to maintain the integrity of their system. Taylor and Kane (2005) concluded that HACCP implementation problems can be partially helped by providing simplified documents or streamlined verification methods, and by improving SMEs' basic lack of understanding of the HACCP approach.

Other consequences arising from budget limitations are operations situated in facilities that are not suitable or intended for food processing, and outdated, poorly maintained equipment not conducive to process controls or to cleaning and sanitising. Moreover, the lack of financial resources and the smaller size of the business often bring with them reduced negotiation power and an associated lack of response from suppliers and customers. Under such conditions, raw material quality, consistency, safety and availability are frequently compromised, with little room for corrective action by the manufacturer.

HACCP in a Meat Plant

HACCP was first used in processes such as canning and other cooked-foods applications where failures in food safety had the greatest consequences. It was only in the 1980s that many fresh-meat processors began to look at incorporating HACCP into their plants. Today, EU Regulation 852/2004 requires food business operators, including meat plant operators, to implement and maintain hygiene procedures based on HACCP principles. In this section, HACCP will be considered as part of meat-grinding and blending processes.

The chemical hazards that need to be considered during meat processing include antibiotics, cleaning agents and sanitising agents, and also oils, greases and wastewater. Other specific hazards arise from failure to separate meats from different species of animal, as some consumers may be allergic to certain types of meat. Also, if beef is contaminated with pork, it may not be heated enough to destroy *Trichinella spiralis* and *Toxoplasma gondii*, which can have serious health implications (Dubey *et al.*, 2006; van der Giessen *et al.*, 2007). In order to ensure that meats from specific species stay free from contamination, it is necessary to sanitise all equipment (e.g. blenders, grinders and stuffers). The scheduling of meat processing should also be done with a view to minimising cross-contamination.

A non-exhaustive list of some of the most common physical hazards includes bone, injection needles, plastic and many other foreign bodies. Bone hazards represent a serious concern when meat is ground, and this needs continuous monitoring and control to minimise the number of defects. Biological hazards are particularly threatening to the meat-processing sector, as pathogens can easily be carried in organic matter found on animal carcasses (Arvanitoyannis, 2006).

Processing Including Critical Control Points

The areas where special attention should be given to the development of critical control points will be highlighted in the following paragraphs. The plant in question deals with the processing of meat into beefburgers. More information on the process followed can be obtained from Anon. (2010a).

Meat Received From Suppliers (CCP1)

- Meat must be from an audited and approved supplier.
- All vehicles must be sealed on delivery.
- The temperature of fresh meat must be on target.
- Packaging must fully protect meat against possible contamination risks.
- Meat quality must meet the criteria set out in the plan.

If the meat passes all these checks, the container is accepted and put into chilled storage, where the preset temperature is checked regularly.

Cold Storage Monitoring (CCP2)

The storage room must operate within the desired temperature range.

Product Formulation

The meats are ground to pre-specified levels. The batch is blended and checked to make sure the fat content meets the specification. During this stage of processing, the major hazards can be either physical or chemical in nature. Physical hazards include bone fragments and other foreign material, and these hazards can be controlled by buying products from suppliers who employ good HACCP practices and by monitoring the products as they are received or used (Tompkin, 1990). Therefore, an elimination point should be included in the process, where any meat that will not go through a small bore and any pieces of bone are rejected.

Monitoring of Elimination Points (CCP3)

Defect elimination points must be working correctly in order to remove any evidence of bone or gristle.

Freezing

It is vital that correct temperatures are maintained at all times to inhibit bacterial growth and prevent dehydration of the final burgers. Therefore, sensors that read and record the temperature every hour monitor the temperature of the freezer areas in the factory. If the temperature rises, alarms are activated. The product then passes through metal detectors at the end of the freezer tunnel before packaging and then again once the boxes have been filled and sealed.

Metal Detectors (CCP4)

- In-line metal detectors fitted with automatic rejection systems are used at the freezing-tunnel exits.
- The detector sensitivity must be set appropriately.

Packaging

The beef patties are manually packed into boxes. Electronic scales are used to indicate when a box is complete. The box is sealed and coded with the date, time and production line used. This data will be stored on the computer and provides the final stage in the line of traceability.

Product Storage (CCP6)

The product is kept at a predefined storage temperature (usually −18 °C).

Inspection

Products are checked regularly by the quality assurance team and judged against the specifications. All data is recorded on hand-held data loggers before being transferred to a mainframe computer.

Cleaning

Each night, all machinery in the plant is thoroughly cleaned.

Factory Hygiene (CCP7)

- There is a daily strip-down and clean of all manufacturing equipment.
- Foam detergent is used, followed by a sanitiser.

- A daily visual inspection is performed.
- Microbiological swabbing is performed.

HACCP in a Cheese Plant

When a HACCP plan is to be developed for a cheese plant, the composition and structure of the product, and the processing, packaging systems, storage, and shelf-life should be considered, with a clear distinction being made between safety hazards and quality (Anon., 2010b).

Critical Control Points

In a cheese plant, CCPs originate from hazard areas such as the temperature of the incoming milk, and pasteurisation times and temperatures. Many hazard areas can be controlled by prerequisite programmes. Those hazards that cannot be controlled by these programmes must be identified as CCPs. Table 45.1 lists various operations that occur in a cheese plant, together with the typical locations of CCPs.

HACCP in a Fish-Smoking Plant

Fish is often smoked in order to enhance its appearance and flavour. The smoking process provides some anti-microbial properties to the treated product, but the subsequent processing must take account of the fact that any remaining microflora will multiply during subsequent refrigeration processes. The following HACCP analysis refers to a small Greek factory producing smoked trout, and has been taken from the case study conducted by Arvanitoyannis *et al.* (2009). The process is illustrated by the process flow diagram in Figure 45.4, and the issues that must be considered by the HACCP team are expanded under the following headings.

Receiving (CCP1, M, C, P)

- Fish must be from a current audited and approved supplier who ensures that the catch was taken under suitable environmental conditions and from water of suitable microbiological quality.
- Vehicles must be clean, without off-odours, fit for purpose and free from other materials. They must be sealed on delivery and temperature-monitored to within specified thresholds.
- Sensory analysis is used first to check fish quality and, if unsatisfactory, analytical testing is performed.

Table 45.1 Various procedures in the cheesemaking process, with CCPs highlighted.

Milk receiving		
Microbiological/pathogens	Temperature is monitored to prevent bacterial growth/staphylococcus toxins Each vat of milk is screened for plate loop count and soluble solid content	CCP
Chemical – animal drug residue	Milk is screened for presence of drug residues	
Physical – any physical hazards	Prerequisite programme in place to prevent contamination	
Milk storage		
Microbiological/pathogens	Temperature control is necessary to prevent bacterial growth/staphylococcus toxins in fluid milk	CCP
Chemical, physical	Prerequisite programme in place to prevent contamination	
Dry ingredient receiving		
Microbiological/pathogens, chemical, physical	Prerequisite programme in place for ingredient receiving and storage	
Dry ingredient storage		
Microbiological/pathogens	Prerequisite programme in place for ingredient receiving and storage	
Chemical, physical	Prerequisite programme in place to prevent contamination	
Cheese vat		
Microbiological/pathogens	Milk is heated to ripening temperature quickly to prevent pathogen growth. Starter is added as soon as milk is at ripening temperature	
Chemical, physical	Prerequisite programme in place to prevent contamination	
Starter		
Microbiological/pathogens, chemical, physical	Prerequisite programme in place for ingredient receiving and storage	
Rennet		
Microbiological/pathogens	Prerequisite programme in place for receiving and storage	
Chemical, physical	Prerequisite programme in place for receiving and storage	
Water		
Microbiological/pathogens	Prerequisite programme in place to prevent unwanted microbial/pathogen growth	
Forms		
Microbiological/pathogens, chemical, physical	Prerequisite programme in place to prevent unwanted microbial/pathogen growth	
Drain table		
Microbiological/pathogens	Prerequisite programme in place to prevent contamination pH monitoring to check for proper acid development before salting	CCP
Chemical, physical	Prerequisite programme in place to prevent contamination	
Sodium chloride		
Microbiological/pathogens, chemical, physical	Prerequisite programme in place for ingredient receiving and storage	
Ageing		
Microbiological/pathogens	Prerequisite programme in place for proper cheese storage	
Chemical, cedar boughs, physical, insects and rodents		
Packaging		
Microbiological/pathogens	Prerequisite programme in place for purchasing and tracking product	
Chemical, physical	Prerequisite programme in place for receiving and storage of packaging supplies	
Storage	Properly packaged product contains no hazards	
Distribution	Properly packaged product contains no hazards	

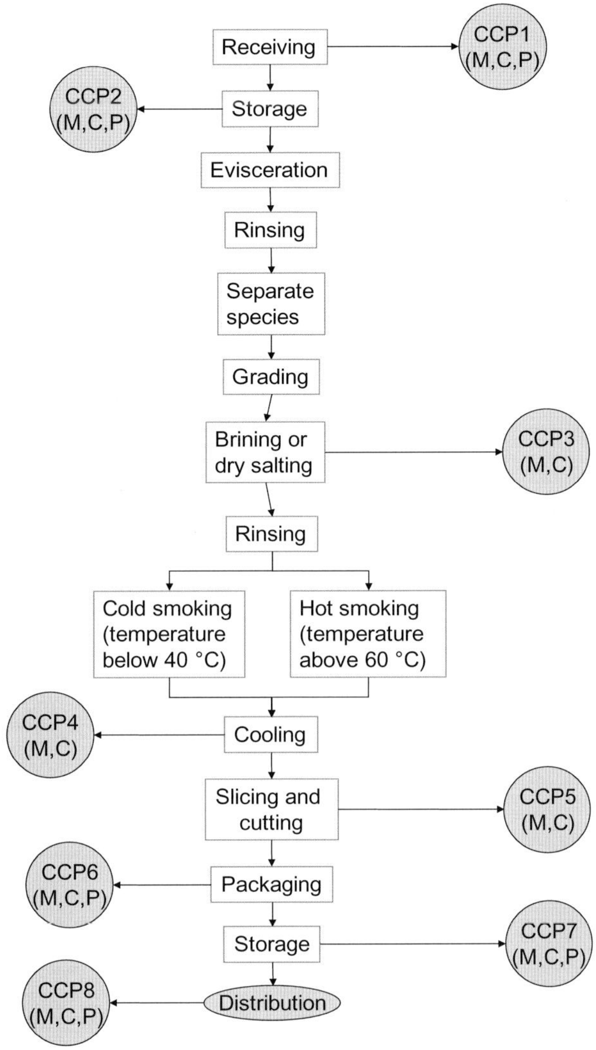

Figure 45.4 A HACCP process flow diagram of the operations during a fish-smoking process (Arvanitoyannis *et al.*, 2009). (M, microbiological contaminants; C, chemical contaminants; P, physical contaminants.) (Courtesy of Taylor and Francis.)

Storage (CCP2, M, C, P)

- Unfrozen fish must be held in a clean, sanitary environment to maintain an internal temperature of 4.4 °C or below.

Evisceration

- Evisceration should be conducted in a special area, separated from the rest of the production process.

- The process should be conducted in a manner that minimises contamination of the flesh.

Rinsing

- It must be ensured that the fish are washed and rinsed thoroughly with potable water (or chloride water) to remove debris.

Separate Species/Grading

- To achieve uniform salting, fish must be sorted by size and thickness; different species of fish should not be mixed in the same brining tank, and the weight–volume ratio, time and temperature of the process should be controlled.

Brining/Dry Salting (CCP3 M, C)

- The salt content of the brine should be monitored. The brine should be cool, and the temperature should be monitored. If the brine is recycled, a decontamination step should be performed.
- Brining tanks must be cleaned before and after each use.
- The salt concentration of the final product must be above 3% NaCl.

Rinsing

- Rinsing with potable water is necessary to remove surface brine.

Hot Smoking (CCP4 M, C)

- *Step 1: surface drying.* The drying phase should be kept short, as prolonged exposure to ambient temperature may lead to unwanted microbiological growth and to formation of histamine in susceptible species.
- *Step 2: smoking.* In the hot-smoking process, the temperature in the centre of the product should normally reach 63–72 °C for about half an hour. Time and temperature have to be monitored to ensure that meat will not coagulate on the backbone.
- *Step 3: product drying.* This step must ensure even drying of the fish to reduce moisture, i.e. raise the WPS (water per phase salt) content (and lower the water activity), thereby leading to the formation of the final texture.

Cold Smoking

- *Step 1: drying.* Smoke is introduced to dry the fish. The HACCP team must ensure that the parameters of this initial drying component are conditioned according to the type or species of fish, its fat content and the humidity

- *Step 2: smoking.* During the cold-smoking step, the fish must allow uniform smoke absorption, temperature exposure and drying.

Cooling (CCP5 M, C)

- Hot-smoked products should be cooled to below 10 °C within 2 h and to below 3 °C within 6 h. Cooling should take place as fast as possible, since pathogens can grow if the cooling period is prolonged.

Packaging (CCP6 M, C)

- Hot-smoked fish are presented to the market mostly in boxes or pre-packaged in plastic bags, possibly vacuumised or in modified-atmosphere packaging.
- The risk of recontamination is high, even if packing takes place in a high-care area, and the prevention of growth becomes imperative for safety: for example, if the products are packed in a room at ambient temperature after cooling, condensation might occur on the surface of the smoked products, thereby leading to dilution of the preservatives deposited by the smoking process. Therefore, it must be ensured that condensation of water on the surface of the smoked product is avoided and that the packaging material should be clean, sound, durable, sufficient for its intended use and of food grade material (Arvanitoyannis *et al.*, 2009).

Storage (CCP7 M, C, P)

- The products should be kept frozen or refrigerated continuously below 4 °C.
- The definition of storage temperature and shelf-life for both cold- and hot-smoked products should take into account the risk of microbiological growth during chilled storage, in particular the growth of *Listeria monocytogenes* in cold-smoked products but also in skinned hot-smoked fillets in evacuated plastic bags.

The Influence of HACCP on Hygienic Design

Figure 45.5 shows that the production of safe foods, the major concern of the food industry, stems from a thorough risk assessment in the form of HACCP. In tandem with the development and control of processes that ensure the safe production of food, it is also necessary to ensure that equipment and processes are designed to be hygienic. The hygienic design of processes and equipment deals with issues such as factory siting and construction, the design of the building, the selection of surface finishes, the flow of raw ingredients and products, the movement and control of people, the design and installation of process equipment, and the design and installation of services.

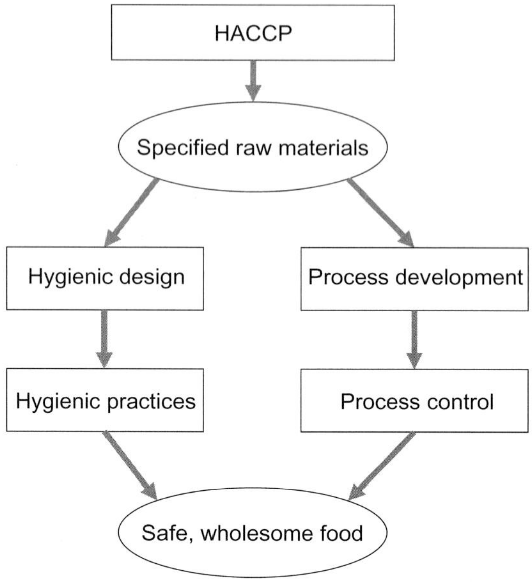

Figure 45.5 The building blocks associated with the control of a food factory to ensure that safe, wholesome food is achieved.

Advantages of Hygienic Design of Food Processing Equipment

Food Quality

Good hygienically designed equipment ensures that no product is held up within the equipment, as this could deteriorate and affect product quality on rejoining the main product flow, or cause cross-contamination.

Food Safety

Hygienic design prevents contamination of the product with substances that could adversely affect the health of the consumer.

Cost Reduction

Hygienic design reduces the time required for an item of equipment to be cleaned, which is cost-effective in the long term. There have been many examples of product recalls, lost production and, indeed, site closures due to contamination arising from poorly designed equipment.

Combining HACCP and ISO 22000:2005

ISO 22000:2005 defines the set of safety procedures needed to produce safe food, and requires the establishment and implementation of a food safety management system (FSMS) for all companies that are part of the food chain, such as feed producers, primary producers, food manufacturers, transport and storage operators, subcontractors, and retail and food service outlets. The ISO 22000 standard is fully auditable; therefore all companies using this standard must provide evidence of their ability to provide safe food. The standard promotes conformity of products and services to international standards by providing assurance about quality, safety and reliability (Tajkarimi, 2007). It has become necessary because of the significant increase in illnesses caused by infected food in both developed and developing countries. In addition to health hazards, food-borne illnesses can give rise to considerable economic costs from medical treatment, absence from work and legal compensation. A greatly beneficial aspect of ISO 22000:2005 is that is can be implemented on its own, or can be coupled with ISO 9001:2000 so that companies will find it easy to implement into current practices. In the following paragraphs, the way in which HACCP is incorporated into ISO 22000:2005 is described in more detail.

Initial Steps (ISO 22000, Sections 7.1–7.4)

According to ISO 22000:2005, the hazard analysis must be carried out in two major components, consisting of (i) the initial steps prior to developing the hazard analysis and (ii) the hazard analysis itself. The initial steps for hazard analysis (ISO 22000, Sections 7.1–7.4) involve the characterisation of the food product and the food process and ensure that the product incorporates elements of GMP as well as HACCP. Firstly, prerequisite programmes should put into place the essential practices that will enhance the role which the production environment plays in producing safe food products. Most importantly, the effectiveness of the prerequisite programmes should be audited regularly during the implementation of the HACCP plan. In order to characterise the food products and process, it is first necessary to understand the influence of the raw material on the occurrence and evaluation of hazards during food processing. Then a full description of the final food product needs to be developed, including information on the chemical properties, ingredients, packaging and delivery methods, storage conditions, and shelf-life, so that the traceability of the product can be effectively maintained. Process characterisation is carried out via flow diagrams, which details process steps, taking into account the interaction of these steps. Once all the preparation for the HACCP procedure is finished, the hazard analysis is conducted according to Section 7.4 of ISO 22000:2005, which comprises three successive steps: hazard identification, hazard assessment, and the selection and assessment of control measures.

Hazard Identification and Determination of Acceptable Levels (ISO 22000, Section 7.4.2)

The identification of food safety hazards should be based on the preliminary information about the raw product as well as on the bacterial hazards that may be relevant to the process. This step is used to define the most relevant hazards that could be expected in the food product according to (i) epidemiological data on food-borne diseases, (ii) their prevalence in the ingredients, (iii) their potential occurrence in the process during operations such as handling, and (iv) their potential elimination during the process.

Hazard Identification and Determination of Acceptable Levels (ISO 22000, Section 7.4.3)

All hazards, including potential hazards, introduced through either ingredients or handling, need to be systematically assessed for each process, in accordance with the corresponding flow diagram. During the hazard assessment, it is important to consider (i) the sources of the hazard, (ii) the probability of occurrence of the hazard, (iii) the nature of the hazard and (iv) the severity of the adverse health effects that could be caused by the hazard. To simplify things, hazards can be studied in groups defined according to similarities in (i) their growth and inactivation parameters, (ii) their spore-forming capability, (iii) their pathogenicity at a low infectious dose, and (iv) their contamination source (e.g. rhinopharyngitis or faeces). In order to make the assessment of bacterial hazards more accurate, it is necessary to quantify the hazard assessment by predicting the growth and/or inactivation potential of the microorganisms using some of the various pieces of predictive software on the market. The hazard assessment process can be time-consuming, requiring a significant amount of work.

Selection and Assessment of Control Measures (ISO 22000, Section 7.4.4.)

Control measures are selected in order to prevent or eliminate significant hazards or reduce them to an acceptable level. These control measures can be classified as either (i) hazards controlled by prerequisite programmes, i.e. measures included in good manufacturing practices and/or good hygienic practices, or (ii) critical control points and operational prerequisite programmes (OPRPs). These last measures are essential and are implemented specifically to control a defined hazard. OPRPs are essential to control the likelihood of introducing food safety hazards into a product or a processing environment, and/or contamination or proliferation that could cause a food safety hazard, i.e. any activities that can be affected by human behaviour during the production of safe food. On the other hand, a CCP is a point, step or procedure at which controls can be applied and a food safety hazard can be prevented, eliminated or reduced to an acceptable (critical) level. In order to categorise a control measure as a CCP or an OPRP, it is necessary to understand (i) the impact of that control measure on the hazard level, and (ii) the severity for consumer health of the hazard that the measure is intended to control.

Identification and Establishment of OPRPs (ISO 22000, Section 7.5)

The identification of OPRPs is based on the results of the selection of control measures and is an essential output of the hazard analysis. The determination of OPRPs sets up prevention and control measures which deal with food safety risks at a level below those which need to be included in the HACCP plan. OPRPs should include information on the hazards that need to be controlled, the control measures and monitoring methods to be applied, the monitoring records to be kept, etc.

Implementing HACCP (ISO 22000, Section 7.6)

This section of ISO 22000 is focused on putting all the actions within the HACCP plan into action. Similarly to the OPRPs, for each CCP the HACCP plan requires information on the hazards that need to be controlled, the control measures and monitoring methods to be applied, the monitoring records to be kept, and the corrections and corrective actions to be taken if critical limits are exceeded. CCPs may be located at any point in a production and manufacturing system for a food product where hazards need to be prevented, eliminated or reduced to acceptable levels, and must be chosen using the correct decision processes. They should be used only for purposes of product safety, and the CCPs and the critical limits must be established using scientific evidence based on risks to consumer health.

References

Alli, I. (2004) *Food Quality Assurance: Principles and Practices*. CRC, Boca Raton, FL.

Anon. (2010a) *HACCP During Burger Production*. Available at http://www.foodforum.org.uk/ffiles/Focus_on_HACCP-Mat+Inn+Saf+Sys+Man-Post.shtml (accessed June 10, 2010).

Anon. (2010b) *HACCP During Cheese Production*. Available at http://www.milkproduction.com/NR/rdonlyres/50B2BAA4-94F6-4FDA-857008BF39B31419/0/DevelopingaHACCPprogram.pdf (accessed June 10, 2010).

Arvanitoyannis, I.S. (2006) *HACCP and ISO 22000: Application to Foods of Animal Origin*. John Wiley & Sons, Chichester, UK.

Arvanitoyannis, I.S., Palaiokostas, C. and Panagiotaki, P. (2009) A comparative presentation of implementation of ISO 22000 versus HACCP and FMEA in a small size Greek factory producing smoked trout: a case study. *Critical Reviews in Food Science and Nutrition* 49(2): 176–201.

Branigan, T. (2008) *Chinese Figures Show Fivefold Rise in Babies Sick from Contaminated Milk*. Available at http://www.guardian.co.uk/world/2008/dec/02/china.

Cormier, R.J., Mallet, M., Chiasson, S., Magnússon, H. and Valdimarsson, G. (2007) Effectiveness and performance of HACCP-based programs. *Food Control* 18(6): 665–671.

CFIA (2009) *Listeria monocytogenes Outbreak*. Available at http://www.phac-aspc.gc.ca/alert-alerte/listeria/listeria_20100413-eng.php. (accessed March 20, 2009).

da Cruz, A.G., Cenci S.A. and Antun Maia, M.C. (2006) Good agricultural practices in a Brazilian produce plant. *Food Control* 17(10): 781–788.

Doeg, C. (1995) *Crisis Management in the Food and Drinks Industry*. Chapman and Hall, London.

Dubey, J.P., Hill, D.E., Jones, J.L., Hightower, A.W., Kirkland, E., Roberts, J.M., Marcet, P.L., Lehman, T., Vianna, M.C., Miska, K., Sreekumar C., Kwok, O.C., Shen S.K. and Gamble, H.R. (2005) Prevalence of viable *Toxoplasma gondii* in beef, chicken, and pork from retail meat stores in the United States: risk assessment to consumers. *Journal of Parasitology* 91(5): 1082–1093.

ECC (1993) European Community Council Directive 93/43/EEC of the 14th June 1993 on the Hygiene of Foods. *Official Journal of the European Communities*, No. L175/1-11.

Escriche, I., Domenech, E. and Baert, K. (2006) Design and implementation of an HACCP system. In: *Safety in the Agri-Food Chain* (eds P.A. Luning, F. Devlieghere and R. Verhe). Wageningen Academic Publishers, Wageningen, The Netherlands, pp. 303–354.

FAO/WHO (1997) Hazard Analysis and Critical Control Point (HACCP) system and guidelines for its application. Annex to CAC/RCP 1-1969, Rev. 3 (1997). In: *Food Hygiene Basic Texts*. FAO/WHO Codex Alimentarius Commission, Food and Agriculture Organization of the United Nations/World Health Organization, Rome.

Havelaar, A.H., Brul, S., de Jong, A., de Jonge, R., Zwietering, M.H. and ter Kuile, B.H. (2010) Future challenges to microbial food safety. *International Journal of Food Microbiology* 139: S79–S94.

Luning, P.A., Devlieghere, F. and Verhe, R. (eds) (2006) *Safety in the Agri-Food Chain*. Wageningen Academic Publishers, Wageningen, The Netherlands.

Mamalis, S., Kafetzopoulos, D.P. and Aggelopoulos, S. (2009) The new food safety standard ISO 22000. Assessment, comparison and correlation with HACCP and ISO 9000:2000. The practical implementation in victual business. Paper prepared for presentation at the 113th EAAE Seminar 'A resilient European food industry and food chain in a challenging world', Chania, Crete, Greece, September 3–6, 2009.

Misterek, S.A., Anderson, J.C. and Dooley, K.J. (1990) The strategic nature of process quality. Paper presented at the Annual Meeting of the Decision Sciences Institute, New Orleans, LA.

NACMCF (1997) *Hazard Analysis and Critical Control Point Principles and Application Guidelines*. National Advisory Committee on Microbiological Criteria for Foods. Available at http://www.fda.gov/food/foodsafety/HazardAnalysis CriticalControlPointsHACCP/ucm114868.htm (accessed June 8, 2010).

OJEU (2004). Regulation (EC) No. 852/2004 of the European Parliament and of the Council of 29 April 2004 on the Hygiene of Foodstuffs. *Official Journal of the European Union*, OJEU L139/1, 30.4.2004.

Olsen, J.R., Harmsen, H. and Friis, A. (2008) Linking quality goals and product development competences. *Food Quality and Preference* 19(1): 33–42.

Pennington, H. (2009) *Public Inquiry into the September 2005 Outbreak of E. coli O157 in South Wales, March 2009*. Available at http://wales.gov.uk/ecoliinquiry/?lang=en (accessed June 8, 2010).

Ropkins, K. and Beck, A.J. (2000) HACCP in the home: a framework for improving awareness of hygiene and safe food handling with respect to chemical risk. *Trends in Food Science and Technology* 11: 105–114.

Struijk, C.B. (1996) The Hamlet option in food microbiology: to analyze or not to analyze food specimens as marketed once HACCP implemented. *Acta Alimentaria* 25(1): 57–72.

Tajkarimi, M. (2007) *New Food Safety Management Systems; ISO 22000*, January 22, 2007. Available at http://www.vetmed.ucdavis.edu/PHR/PHR450/2007/45007C8T.pdf.

Taylor, E. (2001) HACCP in small companies: benefit or burden. *Food Control* 12: 217–222.

Taylor, E. (2008) A new method of HACCP for the catering and food service industry. *Food Control* 19(2): 126–134.

Taylor, E. and Kane, K. (2005) Reducing the burden of HACCP on SMEs. *Food Control* 16(10): 833–839.

Tompkin, R.B. (1990) The use of HACCP in the production of meat and poultry products. *Journal of Food Protection* 53(9): 795–803.

Tsola, E., Drosinos, E.H. and Zoiopoulos, P. (2008) Impact of poultry slaughter house modernisation and updating of food safety management systems on the microbiological quality and safety of products. *Food Control* 19(4): 423–431.

van der Giessen, J., Fonville, M., Bouwknegt, M., Langelaar, M. and Vollema, A. (2007) Seroprevalence of *Trichinella spiralis* and *Toxoplasma gondii* in pigs from different housing systems in The Netherlands. *Veterinary Parasitology* 148: 371–374.

Vela, A.R. and Fernández, J.M. (2003). Barriers for the developing and implementation of HACCP plans: results from a Spanish regional survey. *Food Control* 14(5): 333–337.

Wallace, C. and Williams, T. (2001). Pre-requisites: a help or a hindrance to HACCP. *Food Control* 12: 235–240.

46

Commercial Imperatives

Gerard La Rooy

Introduction

Dr M.S. Rahman's *Handbook of Food Preservation* (Rahman, 1999; Rahman, 2007) contains a commercially orientated chapter titled 'Managing profit and quality' (La Rooy and Perera, 1999). The chapter was included because the Editor felt it was important to cover some fundamental commercial aspects in what was otherwise a technical publication. Since that time, the world of business has moved on significantly, affecting operational staff as well as the functional specialists in all manner of ways. In particular, it is the unrelenting drive for organisations to become 'leaner' and 'flatter' which has impacted on the way the technical specialist has to function in today's business environment. Gone are the large functional departments, generally known as 'functional silos', where the managers and staff could operate largely in splendid isolation without having to worry a great deal about the commercial effects of their activities. For the technical specialist, the changed environment means that he or she can often feel as if they are being pulled in different directions. On the one hand, there are ever-increasing technical complexities to deal with, while on the other, there is the requirement to be more aware of and become more involved in commercial matters. It is because of this requirement for increased involvement that the Editor of *Handbook of Food Process Design* felt that this time too there should be a chapter

Handbook of Food Process Design, First Edition. Edited by Jasim Ahmed and Mohammad Shafiur Rahman.

dealing with relevant commercial aspects. Since I am a firm believer in the greater integration of management structures and the resulting lessening of 'functional silos', I was only too happy to contribute the chapter.

Most readers will be aware that in the majority of food companies it is the technical team that is often responsible for initiating large capital expenditures, which can have major, and hopefully beneficial, effects on business performance. It is a pity then to see, every so often, novel and promising projects not performing to expectations. If we were to look for the reasons for such failures, we would find that in quite a few cases the project's disappointing performance was not the result of any technical shortcomings but was instead caused by two deficiencies:

1. insufficient attention to commercial matters;
2. the absence of a systematic and sufficiently broad approach.

Consequently, these two deficiencies are the main focus of this chapter. It should be noted that for the sake of completeness, it was necessary to incorporate some material contained in the 1999 Handbook. In making the case for a greater understanding of commercial matters, it is worth noting that it is not only a question of improving the technical team's project development competence. There are other benefits, as an improved appreciation of the organisation's financial levers is also likely to lead to an increased identification of innovation and business opportunities.

Fundamental Financial Matters

Technical specialists should not, of course, be expected to act as financial experts. To ensure, however, that their proposals are commercially sound and therefore more likely to succeed, they should have a reasonable understanding of the financial fundamentals. Since this handbook is essentially about technical design and development, we will consider the relevant financial matters predominantly in terms of 'projects' rather than 'routine operations'.

Financial Performance

The overall financial performance of commercial organisations is generally measured by a critical performance indicator called 'return on investment' (ROI). Companies may use different indicators such as 'return on funds employed' (ROFE), which are similar in meaning and serve the same purpose. A financially attractive and sustainable ROI is essential when companies are looking for new or additional investment funds.

A company's ROI is the final outcome of a great many decisions and actions by practically all of its personnel. It is determined using three key business drivers: revenue (sales), costs and the funds invested in the business. It is calculated as follows:

$$\text{ROI\%} = \frac{\text{Revenue} - \text{Costs}}{\text{Funds invested}} \times 100\%$$

It is important to appreciate the effect of the three drivers on an organisation's ROI. Table 46.1 shows performance comparisons between two companies A and B, listing different levels of performance for the three drivers. Looking for example first at sales revenue, Company A clearly outperforms Company B. Next, because of lower costs, Company B takes the prize in terms of total profit and profit percentage. Taking into account the funds invested, however, reverses the position once more, with Company A showing the higher ROI.

Revenue

Commercial enterprises generally derive their revenue from providing products or services at an acceptable price. There are three different ways to arrive at prices in a business, depending very much on how costs are viewed – as a starting point or, more correctly, as the end point:

- *Traditional cost plus.* The traditional method of determining a selling price was to add up all direct and indirect costs (e.g. overheads) and add a suitable margin. This approach works when there is little or no competition (e.g. owing to import controls). Unfortunately, some companies will persist with this approach even when, because of changed conditions, it is no longer appropriate.
- *Market price.* In a free market, the price is set by that market and has nothing to do with costs or margins. The change from the 'traditional cost plus' to the 'market price' method took place in earnest during the latter part of the last century. The principal effect of the change to a market price (and, as a rule, a lower price) was reduced margins, since costs were generally considered as 'given'.
- *Required margins.* The squeeze on margins meant that many organisations experienced unsatisfactory returns on their invested capital. However, to attract new capital, returns and hence margins must be adequate. Since prices are largely beyond the control of the organisation and margins must be achieved, the real situation is that

$$\text{PRICE} - \text{MARGIN} \rightarrow \text{COST}$$

Table 46.1 Comparison of the performance of Companies A and B.

	Company A	Company B
Sales revenue ($M)	100	80
All costs ($M)	80	40
Profit ($M)	20	40
Profit margin (%)	20	50
Investment ($M)	40	100
ROI%	50	40

Costs

Costs can be classified in a number of ways. In the context of this handbook, the following cost classifications are the most relevant: fixed costs, variable costs, discretionary costs and intangible costs.

Fixed Costs

In brief, fixed costs are those costs not affected by changes in output. An example of a fixed cost would be the cost of monthly machine rental, which would be incurred irrespective of the level of output. Note, however, that any production beyond the machine's capacity would mean a step increase in the fixed costs, i.e. the cost of renting another machine. Also, the passage of time can lead to increases or decreases in fixed costs. Consequently, the earlier statement about fixed costs needs some qualification: 'Fixed costs are not affected by changes in output within a specified output range, and fixed costs remain constant within a specified time period'.

Management has to be constantly on guard to prevent ad hoc increases in fixed costs because:

- It is easy to increase fixed costs but usually very difficult to reduce them quickly.
- High fixed costs during a period of falling revenue are a common cause of business failure.

Variable Costs

Variable costs vary with changes in output, and the unit cost is generally deemed to be constant, although that is by no means always the case in reality. For example, bulk purchasing may bring the cost per unit down, or in other cases the cost of additional quantities may be higher than normal. When comparatively small quantity ranges are considered, however, variable costs can generally be taken as truly variable, i.e. 'The cost per unit stays constant and the total variable cost can be found by multiplying the cost per unit by the quantity used'.

Discretionary Costs

In many organisations, all 'non-variable costs' are called 'overheads' and treated as 'fixed', which often is not really correct. While for instance salaries, insurance and depreciation can be taken as truly 'fixed', some other costs should be classed as 'discretionary'. The costs of hiring, for example, technical contractors and some of the costs associated with marketing such as advertising are generally 'discretionary'.

Intangibles

Intangible costs, and benefits for that matter, can have a significant bearing on whether or not a proposal should go ahead. The point to remember is that intangibles should not be dismissed as of little importance because of our inability to put precise dollar figures on them. On the contrary, with many projects the intangibles (the 'soft' issues) are often more important than the 'hard' (dollar) issues. Examples of intangible costs include:

- adverse impact on the quality of life of people in the neighbourhood;
- waste generated by a proposed process not suitable for recycling;
- reduction in process flexibility.

Examples of intangible benefits are:

- enhanced brand image;
- improved competitive position;
- increased customer satisfaction;
- improved staff morale.

Investment in the Business

The main funds invested in a business are 'fixed assets' in the form of plant and buildings and 'current assets' in the form of cash, inventories and debtors. Technical developments often lead to companies investing in new processes and equipment. It is important that any new investments do not dilute the company's financial performance. This means that new returns should not be less than the norm for that particular industry, and preferably should be better than the company's current average return. Exceptions may occur when companies invest in new processes and facilities for strategic reasons, provided of course that this fact is understood and accepted by the decision makers. Because of the effect on the ROI, companies invariably have strict procedures in respect of proposals requiring additional capital. Most organisations will have an established 'hurdle rate' which projects need to satisfy if they are to be considered.

Working Capital

The funds tied up in inventories and debtors are known as 'working capital', which needs careful watching as it can grow rapidly. Inventories, for example, can increase in an unplanned fashion if the company does not reduce production fast enough when experiencing falling sales. Debtors, too, can rise quickly because of poorly managed credit control or because of increasing sales. What should be remembered is that working capital forms part of the funds invested in the business and any increase without an appropriate rise in earnings will adversely affect the ROI.

Cash

It is not always appreciated that 'cash' is not the same thing as 'profit'. A business may, for example, be highly profitable but still run out of cash. This may be caused by having to purchase additional equipment or increase inventories, as well as by a growing list of debtors. Conversely, a business which is losing money may be 'cash rich' owing to selling off some property, the lowering of inventory levels and a shrinking debtors list.

Cash should be seen as a company resource which must be managed carefully to avoid a cash shortage and possible insolvency. (While a proper discussion on cash management is beyond the scope of this chapter, it is important that technical professionals appreciate how technical decisions can have considerable cash impact. If you wish to pursue this topic, I suggest you talk to the financial staff in your organisation about how organisations absorb and release cash.)

Main Financial Statements

Companies tend to produce a great number of financial reports to monitor and control business performance. While much of the financial reporting is bound to be of little interest to the technical staff, it is important that they understand the main functions of the following four statements:

- *The profit and loss statement* ('P&L'). This lists revenue and costs under several headings and the total profit (or loss) for the period and year to date. Often, too, there are intermediate totals such as EBIT (earnings before interest and tax). This statement, which provides management with regular operational information against budget, is usually produced monthly or possibly weekly. It forms an essential part of the annual accounts.
- *The balance sheet.* This is a 'snapshot' of the financial position at a specified date. (Note that people sometimes refer to the 'balance sheet' when they mean the P&L.) The balance sheet lists the all the assets held by the business under the headings of fixed assets (e.g. buildings and plant) and current assets (e.g. inventories and debtors). It also shows all liabilities, such as creditors and bank loans. It is produced at least annually, although many organisations will produce a monthly balance sheet. The governing principle of the balance sheet is Assets = Liabilities + Equity. The importance of the principle becomes clear if the equation is rearranged as Assets – Liabilities = Equity. Any write-down of assets or increase in liabilities reduces shareholder equity.
- *The funds statement.* This is sometimes called the 'sources and disposition of funds' statement. It is another important statement, as it explains the changes in

assets and liabilities between two successive balance sheets. It will show where funds have been applied, for example to purchase a new building, and also the source of the new funds, say from operations or new debt.
- *The cash-flow statement.* This statement reports on the cash position usually twelve months or more into the future. As mentioned before, the idea is to treat cash as a resource. It is normally produced monthly.

It is a good idea for technical personnel to have the important financial statements explained to them by the accounting department. Better still would be to arrange for a brief course in basic financial principles.

Financial Impacts of Technical Projects

Technical projects are often the result of some initiative from marketing or operations. In addition, technical personnel also frequently initiate ideas for new processes or process enhancements to improve business performance in many areas. Turning those initiatives into successful projects is a prime responsibility of the technical specialist. Proponents of technical projects should assume responsibility for ensuring that investments deliver not only technically but also in financial terms. It is imperative then that technical staff are fully aware of how their projects impact on the three drivers and hence the ROI, as well as on cash.

Impact on Revenue

It is fair to say that it is the technical specialist who provides the lifeblood for the organisation's future, as all products and processes require constant renewal and replacement if there is to be security of future revenue. It is unfortunate that many managers do not fully appreciate the impact that technical resources have (or could have) on the business in general and on revenue generation in particular. What is more, it is quite likely that the majority of the technical team are also not completely aware of their contribution to the success of the enterprise.

When considering the impact that technical personnel have on an organisation's revenue generation, we should appreciate that there are shorter-term and longer-term impacts:

- *Shorter-term impacts (tactical or current revenue).* Projects coming under this heading are generally concerned with existing products and processes or new products within existing competencies (e.g. product line extensions, product improvements and ingredient substitutions).
- *Longer-term impacts (strategic or future revenue).* These tend to be concerned with new technology and products outside current competencies.

Impact on Costs

Technical activities and decisions have a very significant impact on costs. Frequently, the technical professionals are the ones who make the decisions that not only affect the total costs of a development but also, more particularly, determine the balance between fixed and variable operational costs. With most proposals, there are possible trade-offs between 'higher fixed costs with lower variable costs' and 'lower fixed costs with higher variable costs'. A highly automated solution is likely to incur greater fixed costs and lower variable costs. On the other hand, a less sophisticated solution is likely to mean lower fixed costs and higher variable costs. When one is designing a new food process, it is essential to be aware at all times of the cost implications of what is being requested or proposed. For example:

- *Unduly high specification levels.* Insistence on a higher specification than may be necessary is likely to push up the costs of production (higher variable costs). Such a decision can also result in higher fixed costs if, for example, the unduly close tolerances specified require a more expensive piece of equipment than would be the case with a less demanding specification.
- *Product line extensions.* Every time a new product is added to a company's product portfolio, it is likely that current production run sizes and frequencies will be affected (usually adversely).

Impact on Funds Invested

Earlier in the chapter, we looked at the importance of the assets employed when assessing a company's performance. Whenever an organisation undertakes a capital project, it increases the funds invested in the form of new fixed assets. Also, the launch of a new product invariably requires additional working capital for finished goods, new ingredients and equipment spares. In addition, further funds are required to finance the increased debtors resulting from the sales of the new product. Without any doubt, the actions and decisions of technical professionals can have a major impact on the funds invested, which should be kept in mind throughout the design process. A further point to note is that the choice of one solution over another may be decided by the size of the investment required rather than by technical considerations.

Impact on Cash

In addition to the effect on the three key drivers, the work of technical personnel also affects an organisation's cash position. Take, for example, a new item of plant costing say $1 000 000, which is to be depreciated over ten years at $100 000 per year. It is the $100 000 which would be the figure shown in the P&L as a yearly cost rather than the full million. However, all of the cash would need to be found in year one to pay for the plant. Note that besides having to apply the cash for new plant and equipment, cash is generally also needed to fund any increase in working capital.

Analytical Concepts and Techniques

Justification of food-processing projects is often the responsibility of the technical specialist. It requires a sound approach and attention to detail, since a project can be analytically correct but conceptually wrong at the same time. In other words, the arithmetic may be fine but the underlying premises are flawed. This means avoiding some common pitfalls, as well as the use of a number of practical techniques.

Inappropriate Use of Accounting Information

It is not always appreciated that traditional accounting information, if used without questioning its validity, can result in unsound economic decisions. A fairly common instance of this is the use of cost standards when preparing project cost estimates. Consider, for example, a case where a company is planning to automate a largely manual process and to justify the expenditure through expected labour savings. If the designer of the proposed process were to use the standard hourly labour rates as used by the accounting staff, the resulting labour savings would most likely be significantly overstated. This is because the standard hourly labour rates usually have built into them a considerable number of other cost elements as overhead recoveries. Now, while some of those costs will reduce in line with the lower labour usage, others are bound to be unaffected. To enable these rates to be employed in the justification of a project, it is necessary to analyse them and arrive at effective rates as shown in Table 46.2.

This example shows that without making the appropriate adjustment, the labour cost saving would be overstated by some 32%, which could have a significant impact

Table 46.2 Make-up of hourly labour cost standard.

Cost element	Standard cost ($)	Actual	Effective rate
Basic wage[a]	20.00	90%	18.00
Allowances[a]	2.50	60%	1.50
Holiday pay[a]	1.80	90%	1.62
Sick pay[a]	0.40	90%	0.36
Cafeteria use[b]	0.10	0%	0.00
Contribution to administration[b]	3.50	0%	0.00
Total	28.30		21.48

[a] The standard cost of wages, allowances and sick pay is a departmental average. The actual cost of labour units saved by the proposal is less than the average.

[b] No savings in these overheads will result from fewer labour units.

on the project's viability. It pays always to be very careful when using any regular accounting information, especially when the organisation is in the habit of allocating overheads by standard formulae. Also, when examining such information, the focus should be on the actual changes in costs resulting from the proposal: What new costs will be incurred? Which costs will no longer be incurred?

Breakeven Analysis

Although breakeven analysis is a rather unsophisticated technique, it can still be used to good effect, especially as an early 'screening tool'. The underlying idea is simple enough: the ongoing net income of a proposal is used to 'pay back' the original capital outlay. The point at which the income equals the capital outlay is known as the 'breakeven point', after which the proposal generates net profit. Consider a company planning to install some new equipment to produce a new product. The details are as follows:

Equipment cost	\$1 000 000
Selling price per unit	\$20
Variable cost per unit	\$15
Sales per annum	300 000

First, the variable margin is found by subtracting the variable unit cost from the selling price:

$$\text{Variable margin} = \$20 - \$15 = \$5$$

Now the breakeven point is found by dividing the equipment cost by the variable margin:

$$\$1\,000\,000 \div \$5 \text{ per unit} = 200\,000 \text{ units}$$

Assuming that the sales are spread evenly over the year, the time taken to reach the breakeven point is found by dividing the breakeven quantity by the annual sales volume:

$$200\,000 \text{ units} \div 300\,000 \text{ units per year} \times 12 \text{ months} = 8 \text{ months}$$

The net profit for the first year equals four months' sales at the variable margin:

$$100\,000 \text{ units} \times \$5 \text{ per unit} = \$500\,000$$

A point to remember when using breakeven analysis, or any other evaluation technique for that matter, is that we should be very clear as to which costs and benefits

Table 46.3 One-off and ongoing costs and savings in $000's[a].

Cost item	One-off			Annual ongoing		
	Current	Proposed	Net difference	Current	Proposed	Net difference
Plant	–	(50)	(50)	–	–	–
Parts	–	–	–	(4)	(2)	2
Installation	–	(10)	(10)	–	–	–
Overhaul	(15)	–	15	–	–	–
Sell existing plant	–	10	10	–	–	–
Operate	–	–	–	(20)	(12)	8
Maintenance	–	–	–	(10)	(5)	5
Totals	(15)	(50)	(35)	(34)	(19)	15

[a]Negative figures are shown in brackets.

are 'one-off' and which are 'ongoing'. On no account should they be mixed up, as that will produce an erroneous result. To ensure clarity, it is a good idea to construct a table showing 'one-off' and 'ongoing' costs and benefits for both the 'current' and the 'proposed' situation (see Table 46.3). Take the following situation. A business is planning to replace some existing equipment.

The current cost details are:

Required parts p.a.	$4 000
Maintenance p.a.	$10 000
Required overhaul	$15 000
Operating cost p.a.	$20 000

For the new plant, the details are:

Plant purchase	$50 000
Required parts p.a.	$2 000
Installation	$10 000
Maintenance p.a.	$5 000
Operating cost p.a.	$12 000
Sell existing plant	$10 000

From Table 46.3, the following can be observed:

The net difference in one-off costs is $35 000.
The business is $15 000 better off on an ongoing basis.
Payback = $35 000 : $15 000 = 2.33 years, or 2 years and 4 months.

Note that besides using breakeven analysis for evaluating proposals, it can also be used for assessing factory production shift performance. In such a situation, the cost of set-up and any start-up losses are treated as a one-off cost which is recovered during the shift (La Rooy, 1997a).

Discounted Cash-Flow (DCF) Analysis

The DCF method is a more sophisticated approach than breakeven analysis in that it takes into account the time value of money. A dollar received today is worth more than a dollar received in a year's time. A dollar invested at, say, 10% per annum would become $1.10 after one year. It can thus be said that the 'present value' of a dollar today is the same as $1.10 in year's time. Likewise, the 'present value' of a dollar in a year's time would be 90 cents and in six years 56 cents.

The concept of 'present value' can be applied to several time periods. The idea is to take into account all the cash outflows and inflows over the life of the project and express them in terms of 'present value'. By accumulating the 'discounted cash flows', the 'net present value' of a project can be calculated.

Take the following example. A business wishes to spend $100000 on new equipment to realise $30000 p.a. over the next five years. At year six, the plant is expected to be replaced and sold for $10000. Assume a discount rate of 10%. Table 46.4 shows the discounted cash outflows and inflows for the life of the project.

Looking at the accumulated DCF figure in the last column, we can say that at a 10% discount rate, the 'net present value' of the proposal is $19300. Another application of the method is to let the 'net present value' equal zero and calculate the discount factor. This figure is known as the 'internal rate of return' (IRR). Most organisations have an IRR 'hurdle rate', which proposals must equal or surpass.

In conclusion, the following should be noted. The concept of the 'time value' of money should not be confused with inflation, which is about the reduction in purchasing power of the dollar. Inflation can affect both costs and revenues, which may or may not have a bearing on our analysis. It is, however, a separate issue. Also, the analysis is concerned with cash items only. If, for example, a proposal is prepared showing the cost of new plant being depreciated over the life of the plant, then the depreciation cost needs to be added back to show the total cash outflow in the year of purchase.

Interdependent and Independent Parts of Proposals

Staff responsible for capital projects should appreciate the need for all proposals to be evaluated incrementally. Projects are often made up of a number of discrete parts,

Table 46.4 Discounted cash flow amounts in $000's.

Year	0	1	2	3	4	5	6
Cash flows	(100)	30	30	30	30	30	10
Discount factor	1.00	0.91	0.83	0.75	0.68	0.62	0.56
DCF	(100)	27.3	24.9	22.5	20.4	18.6	5.6
Accumulated DCF	(100)	(72.7)	(47.8)	(25.3)	(4.9)	13.7	19.3

Table 46.5 Payback for dependent and independent parts of a proposal.

Proposal parts	Expenditure	Savings	Payback (months)
Total dependent parts	700 000	650 000	12.9
Independent part A	200 000	80 000	30.0
Independent part B	100 000	20 000	60.0
Total	1 000 000	750 000	16.0

some of which are 'interdependent' and others of which are quite independent. For example, the installation of a new machine depends on the foundations being in place, and hence the construction of the foundations and the installation of the machine are interdependent. On the other hand, upgrading the supervisor's office as part of the same proposal would most likely be independent.

Personnel responsible for assessing capital proposals need to be aware that it is fairly common for designers not to differentiate between the interdependent and dependent parts of projects. Note also that it is not unusual for some independent 'nice to have' facility to be included because of the significant total savings of the project. Failing to treat independent parts of projects in the proper manner is likely to result in an unsound justification even though, overall, the project appears financially viable.

Take the case of a company considering a proposal to spend $1 000 000 to realise annual savings of $750 000. A quick 'payback' check shows that it will take only 16 months to recover the original outlay. The proposal looks like a winner, as the company has a two-year payback guideline. Before approving the proposal, however, it is decided to identify the dependent and independent parts and calculate the individual paybacks as shown in Table 46.5.

As can be seen, unless the independent parts A and B can be justified in some other way, it would be inappropriate for the company to approve those parts, as they both fall outside the two-year guideline. In addition to the need for taking an incremental view of the parts of a project, it is imperative that the details of all costs and benefits are examined incrementally as well. Often the information available to the designer is in the form of standard costs, which are average figures based on historical performance. It is, however, quite incorrect to assume that those figures apply also to future activities. Always be aware that the main purpose of standard costs is to provide 'after the event' control of performance. They should not, as a rule, be used for decision-making.

Assessing New Business Opportunities

A special case which requires 'incremental thinking' is when there is a request for increased production capacity due to a new sales opportunity. Consider, for example, the case where a plant manager is proposing to spend $1 000 000 on new plant to cater for new sales of 500 tonnes worth $2 000 000 p.a. The existing plant is working close to its capacity of 1500 tonnes and the new business has a very good variable

Table 46.6 Ranked annual quantities/contributions.

Product	Contribution margin	Quantity	Contribution ($000's)	Σ contributions (000's)	Σ quantities	Unused capacity
E (new)	1000	500	500	500	500	1000
D	900	200	180	680	700	800
B	850	400	340	1020	1100	400
C	700	300	210	1230	1400	100
A	600	500	300	1530	1900	(400)

contribution margin (sales price minus variable cost), meaning that the new plant would pay for itself in about two years. While on the face of it the proposal appears to be justified, more information is required before approval can be given. What is needed is an assessment of how well the existing business compares with the new. The best way to do this is by constructing a ranked 'quantity/contribution' table (starting with the highest) as shown in Table 46.6.

Note that the original justification of the $1 000 000 expenditure was based on an annual contribution of $500 000 from the new product E. Ranking the contributions as shown, however, means that the capacity is expanded by 500 tonnes to enable 400 tonnes of Product A, which has the lowest contribution, to be produced. This would provide an actual economic benefit of 400 × $600 = $ 240 000 (payback $1 000 000 : $240 000 = 4 years and 2 months).

Treatment of Intangibles

As pointed out earlier on in the chapter, intangible costs and benefits often form a significant part of process design proposals. However, while intangibles warrant due consideration, in most situations it is difficult to arrive at proper dollar estimates for them. A point to note is that although in some cases it is possible to quantify an intangible cost or benefit, it may still not be practical to assign a realistic dollar figure to it. Consider, for example, a proposed process improvement which is designed to result in a reduction of 15% in customer complaints. Assuming that the 15% figure is based on proper investigative work, the improvement can be considered 'quantified'. The next step, turning the reduction in complaints into dollars, is fraught with difficulty, as it is almost entirely subjective. It may in fact be better not to try to show dollar figures, as it often leads to much argument and debate. As an alternative approach, it is suggested that the principle of 'minimum implied value' or 'maximum implied cost' is employed (La Rooy, 1997b).

The method comprises the following steps:

- Determine the net project total by subtracting the 'hard' savings from the 'hard' costs to give the net project total.
- List all intangible ('soft') costs and benefits.

- If the net project total shows a cost balance, then the 'minimum implied value' of the combined intangibles must be considered to equal or be greater than that cost balance for the proposal to proceed.
- If the net project total shows a savings balance, then the 'maximum implied cost' of the combined intangibles must be considered to equal or be less than that savings balance.

An example may illustrate the idea more clearly. A business is considering a proposal for some new equipment which will result in some operational savings and some improvement in quality as well. The details are:

Purchase cost of new equipment	\$200 000
Expected life	4 years
Annual savings	\$40 000
Expected reduction in consumer complaints	200 p.a.
Ignore the time value of money.	

Analysis:

Cost = \$200 000
Savings = 4 × \$40 000 = \$160 000
Shortfall = \$200 000 – \$160 000 = \$40 000
Total fewer complaints = 4 × 200 = 800
Implied minimum value per complaint = \$40 000 : 800 = \$50

In this example, the implied minimum value of each complaint is \$50. If it is considered that actual value to the business is less than that figure, the proposal should probably not go ahead. If, on the other hand, it is felt that the actual value is greater than \$50, the case for the proposal is strengthened.

Some readers may feel that this method is only assigning dollar figures to intangibles by some other means. However, while to some extent they would be correct, there are some important differences. Firstly, no attempt is made to estimate the cost or benefit figures. Secondly, it places the onus for assessing the merits of the proposal squarely on the person(s) responsible for approving the proposal.

Sunk Costs

It is important not to let costs that have been incurred in the past influence our economic reasoning. Past costs are considered 'sunk', and do not affect future decisions. While in most situations sunk costs should be easy to identify, book values of plant or machine spares can cause confusion. Although any write-down of the book values of plant to be replaced does have a negative impact on the P&L, it does not affect the economic assessment. What should be remembered is that the original purchase was in the past and the amount of depreciation to date is irrelevant. Another way of looking at sunk costs is that since they are common to all the possible courses of action, they have no bearing on the evaluation of proposals and subsequent decisions.

Set-up and Changeover Considerations

Proper consideration of set-up and changeover requirements is a vital part of sound food process design, since failure to do so is likely to result in two important shortcomings. Not only will the opportunity to simplify and minimise the set-up and changeover requirements be missed, but the estimated operational costs will be understated also.

As a first step, the process designer needs to identify all the activities and requirements associated with the setting up of a typical production run, such as engineering work, sanitation and pre-production batch preparation. Next, every effort should be made to simplify and minimise the requirements, as they become a 'fixed cost' component of each future production run, largely irrespective of run size. Finally, realistic cost estimates should be made for all requirements, which should be incorporated into the output unit cost. In some cases it would be a good idea to develop a 'ready reckoner' for different run sizes, as shown in Table 46.7.

It should be appreciated that in some situations, set-up costs can be highly significant. Take, for instance, the case of a process requiring steam to be available at 6 a.m. to preheat the cooking equipment so that it is ready for production at 8 a.m. To have the right amount of steam available at the correct pressure at 6 a.m. means that the boiler needs to be fired up at 4 a.m. Ignoring any other start-up costs, there is the cost of the boiler labour together with energy costs, all of which are incurred before a single unit is produced. In cases where set-up costs are high, there is likely to be a temptation to go for larger production runs to enable the fixed set-up cost to be recovered over a greater number of units, which goes against JIT (just-in-time) thinking.

Table 46.7 Set-up costs in dollars per unit for increasing run size.

Set-up cost element	Qty 1000	Qty 2000	Qty 3000	Qty 4000
Engineering work	200	200	250[a]	250
Start-up energy	450	450	450	450
Cleaning, sanitation	150	150	300[a]	300
Wasted material	200	200	400[a]	400
Total	1000	1000	1400	1400
Set-up cost per unit	1.00	0.50	0.46	0.35

[a] After producing 2000 units, the plant needs some engineering adjustment as well as another clean.

Table 46.8 Determining development policy.

	Strategy	Implementation (or execution)
Cautious	More of the same	A step at a time
Bold	Biting the bullet	All or nothing

'Thinking Bold and Acting Cautiously'

It is important to keep in mind that in any process development proposal, we are dealing with both 'strategy' and 'implementation'. In addition, it is necessary to appreciate that there are also two different approaches to embarking on a food process design project, i.e. 'bold' and 'cautious' (La Rooy, 2010). As shown in Table 46.8, selecting a 'bold' or a 'cautious' approach provides very different 'directions' for both strategy and implementation (La Rooy, 1993a).

By combining the 'directions', four quite different policies can be developed:

- *Policy A: cautious–cautious.* Generally low in risk but unlikely to provide a long-term competitive advantage. It is likely that future expenditures will be required to update the process.
- *Policy B: cautious–bold.* As for policy A but higher risk because of an 'all or nothing' implementation.
- *Policy C: bold–bold.* The policy can deliver a good solution but can be very risky. Large up-front costs are generally unavoidable.
- *Policy D: bold–cautious.* In most situations, this is the appropriate policy as it provides for future focused gains while minimising risks. While each situation is of course unique, as an overall guide designers should, for example, go initially for lower fixed and higher variable costs to lessen the risk to the business. As throughput increases and confidence in the new process grows, more capital could be invested to lower the variable costs.

Process Variability

Process variability is a fact of life, which means that all manufacturing processes are subject to variability to a greater or lesser degree. Take, for example, a highly accurate filling process which is set to fill 500 g jars. If precise weight measurements of a number of jars were taken, differences in weights would be found, although they would be small. Since process variability can have a major impact on operational costs and consistency of output, it is important for designers of food processes to appreciate the nature and extent of the different variabilities that will affect the new process. Consideration of variabilities should form an integral part of the design stage and not be left until after the new process has been installed. This means ensuring that all sources of potential process variability are identified, properly understood and minimised where appropriate. In addition, monitoring methods and effective control of the variabilities need to be specified as part of the design.

Causes of Process Variability

It is important for designers of food processes to realise that there are two different types of causes of variability, i.e. common causes and special causes. Common-cause, or statistical, variability is the random variation in results or performance which is due to the nature of the process rather than any specific action. It is the variation which is considered 'normal' for the process. As common-cause variability is an inherent aspect of a process, reductions in variability can only come about by changing (improving) the process. Special-cause variability is due to those extraordinary and often 'one-off' events which cause a temporary increase (or decrease) in variability. A process should not be changed as the result of encountering special-cause variability.

The distinction between common and special causes is particularly relevant when improvements to processes are considered. It is, for instance, not uncommon for companies to make major and costly changes to processes in response to problems which have their origin in 'special causes'. Likewise, the reverse can also be the case: problems due to common-cause variability may be explained away as 'one-off' and unlikely to recur, and thus requiring no changes to the process.

Negative Effects of Process Variability

A high level of variability usually results in reduced customer satisfaction as well as unnecessary costs. Without any doubt, the most important of all effects is the detrimental impact that variability has on one's reputation in the marketplace. If there is one thing that annoys customers more than anything else, it is a lack of consistency. Nothing will damage a reputation more quickly and more permanently than a high level of variability in the quality of products and services. In addition to the damaging effect on the reputation of the business, there are several other negative effects of process variability, such as:

- *Increased material costs.* Quite often, high process variability means that more material may have to be used to ensure that the minimum standard in the specification is adhered to. Take, for example, a weighing machine which is expected to produce 1 kg quantities without any underweights. Assume a weighing variability of, say, ±20 g; then the machine would need to be set at 1020 g. If the weighing variability could be reduced to, say, ±10 g, some 10 g could be saved per weighing.
- *'Out-of-spec' production* and subsequent wastage or reworking, as well as the need to reschedule.

- *Reduced capacity.* Processes that are under proper control tend to operate better, with resulting increases in output.
- *Higher inventory levels.* High product variability also means that higher levels of inventory are required to service the market. Product consistency is an essential prerequisite for JIT management. Also, 'out-of-spec' products become a liability to the company as a rule, requiring 'write-downs' or additional expense, or both.
- *Staff-related costs.* These costs are associated with pride of workmanship and staff morale. High variability in processes is likely to cause many internal problems such as scheduling difficulties and rework, all of which are bound to lead to reduced job satisfaction and, as a consequence, to poorer performance.

Accumulation of Effects

Because customers experience the final result of the various processes, one needs to be aware of the potential accumulation of the effects of variability in all processes. Take, for instance, a situation where a process consists of the following sub-processes: batching, blending and cooking. Assume process performance statistics as follows: batching accuracy 90%, blending effectiveness 90% and cooking-temperature accuracy 90%. Table 46.9 shows the average effect on production batches (assuming that the performances of the three sub-processes are independent of each other). We can observe that we have, on average, perfect batches only 72.9% of the time. Also, there is one 'unlucky' batch in one thousand batches which features 'the worst with the worst with the worst' combination.

While this example shows only three sub-processes, in most practical situations there would be many more processes involved in producing a final output.

Minimising and Managing Process Variability

It is by seeking to minimise variability in the first place that the technical professional can make a significant contribution to the ultimate performance of the process. During the development phase, there are many opportunities for this but, unfortunately, the opportunities are seldom fully appreciated and acted upon. For example:

- Requiring potential equipment suppliers to furnish variability data, preferably in the form of statistical control charts.

Table 46.9 Average results of three sub-processes.

	% of batches
Perfect batches	72.9
One shortcoming	24.3
Two shortcomings	2.7
All three shortcomings	0.1

- Looking for ways to standardise materials and processes wherever appropriate. Undue and unnecessary diversity invariably leads to higher variability of output.
- Making the definition of the process control methods and the information required part of the process design and specifications.

Process Capability

For a process to be considered 'capable', the variability of the process must be no greater than a specified tolerance. There is, for instance, little point in specifying an output tolerance of, say, ±1 °C when the actual variability of the process is ±2 °C. While process capability is thus a straightforward concept, it is unfortunately not always appreciated in practice. It is, for example, not uncommon for marketing staff to require unrealistic tolerances because of a lack of appreciation of the likely variability of the process(es) being proposed.

A special area of difficulty arises when dealing with processes which involve the mixing of component materials. While it may be possible to weigh-batch the materials with reasonable accuracy, achieving a uniform distribution of components throughout the mix is much more difficult. This can be a real problem when one or more of the component materials is particularly expensive or 'quantity-critical'. In such cases, the process designer may have to opt for a 'particulate feed' solution. In conclusion, designers of food processes should keep in mind the following guiding principles:

- The specification is the voice of the customer.
- The statistical control chart is the voice of the process.

Adopting a 'Process-Based Approach'

To increase the likelihood of food process design projects meeting expectations, it is strongly recommended that a systematic, 'process-based' approach be adopted. For a start, this means looking at the overall operation under consideration 'horizontally' instead of in the traditional 'vertical' way. The usual vertical view represents the management structure (mainly who reports to whom), whereas the horizontal view represents how the operation actually works in terms of the key processes.

Charting the Operational Flow

Figure 46.1 shows an example of a horizontal view, illustrating the operational flow from incoming raw materials through to delivery of the finished goods to the customers. Note that both suppliers and customers may interface with the operation in more than one process and also that some of the production can follow alternative process routes (La Rooy, 1993b).

It needs to be appreciated that the example provided is a fairly simple one and that practical situations are likely to be considerably more complex, requiring quite a bit

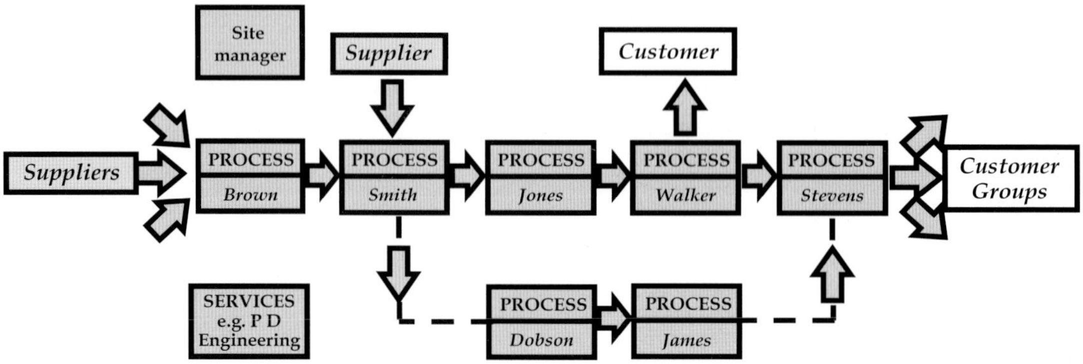

Figure 46.1 Example of a horizontal view.

of detailed investigative work. Readers should not, however, be put off by this as not only is charting a necessary first step, but drawing up the process flow chart will also lead to a greater general understanding of the overall operational process.

Developing the 'Process-Based Approach'

A system can be described as 'consisting of a number of related parts forming a connected whole'. The operational representation shown in Figure 46.1 can be considered as a 'system', with the 'parts' being the individual processes. If we were to 'drill down' into any process, however, we would find that it was also a 'system' in its own right, complete with 'parts'. This point is important when, further on, we come to the actual process of 'food process design'. Once the operational flow has been charted, a 'process-based approach' can be developed, which requires the acceptance of three fundamental principles:

1. Each individual process within the operational system is viewed as a 'mini-business', with other processes being 'internal suppliers' as well as 'internal Customers'. (As shown in Figure 46.1, some processes will also deal with external suppliers and customers.)
2. Outputs from one process become the inputs into the next process at agreed standards of acceptance.
3. There are agreed measurements to monitor and evaluate the performance of all processes in the system.

Principle 1: Processes as 'Mini-Businesses'

Figure 46.2 shows an individual process, with the main process flow running horizontally. Note that besides the main flow, there may be a significant number of other

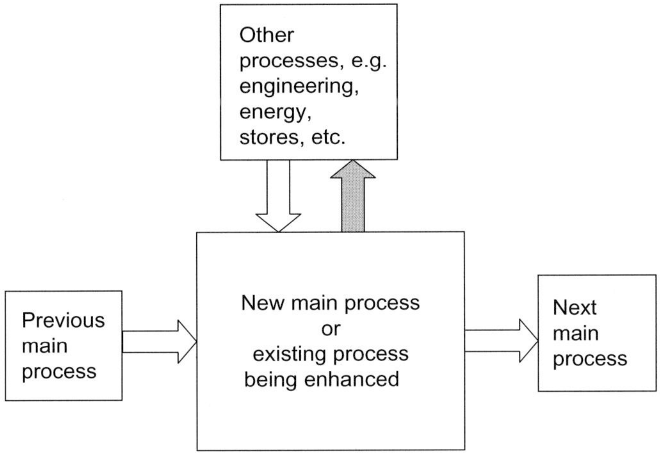

Figure 46.2 Processes as 'mini-businesses'.

processes providing input to (and possibly receiving output from) the process under consideration.

Operating a process as a 'mini-business' requires an unambiguous understanding of where the agreed boundaries are between it and other relevant processes, together with appropriate definitions of responsibilities.

Principle 2: Agreed Standards of Acceptance

Fundamental to the development of a successful 'process approach' is the need to appreciate that 'in order to meet the expectations of the ultimate external customer, we have to meet the agreed requirements of all "internal customers" in the system'. It is matter of 'a chain only being as strong as its weakest link' or, in other words, 'a sound process is unlikely to produce the required output quality if it receives sub-standard input(s)'. This is an important observation, as companies may spend large amounts of money on a new process or on enhancing existing ones, only to find that the results fall short owing to deficiencies in other processes. Improvements confined to just the 'weakest link' of a system can often result in exposing a new 'weakest link', preventing full realisation of the expected benefits. We need to consider the necessary quality standards for all inputs and outputs in order to ensure that our new or enhanced process can deliver what is expected of it. To avoid unpleasant surprises later on, this should be done not only for the process under consideration but probably for all other processes in the system as well. The required quality standards should be explored under three universal headings:

1. *Timeliness*, e.g. time of first input and time of last output.
2. *Quantity*, e.g. flow rate, quantity per batch and weight per unit.
3. *Physical attributes*, e.g. viscosity, temperature, weight and colour.

Note that the notion of 'quality', as used throughout this chapter, is about meeting the expectations of users and consumers in a consistent manner, rather than implying anything about intrinsic value. A product made from low-cost materials can still be a 'quality product' provided it meets those expectations. On the other hand, a 'classy', expensively made product can be of poor quality if it does not meet expectations such as reliability and ease of use.

In addition to the three input/output quality standards, there is also the question of costs to consider. With respect to costs, there should preferably be no tolerance, as it is managerially more sound to charge each process with the standard all-inclusive cost of inputs at that point rather than with the actual costs, the underlying idea being that each process is responsible for any cost variances, allowing each successive process to start with a 'clean sheet' cost-wise. Progressive control of costs can be provided by monitoring any cost variances incurred by each process. The total actual cost of the product can be found by adding algebraically the variances of all processes and applying the result to the standard costs (La Rooy, 1993b).

The point needs to be made that the way described here of applying standard costs and the subsequent assessment of actual final costs is by no means in general use. There can be no doubt, however, that it is the more correct method. Where designers of food processes find themselves up against more traditional methods of accounting such as allocation of overheads by some formula, they should estimate the cost of process input(s) and output(s) as realistically as possible.

Principle 3: Agreed Measurements of Performance

For the process approach to work properly, there has to be appropriate monitoring both of the output(s) and of what is happening within the process. It is essential that there be early agreement on what is to be monitored and reported. Failure to obtain prior agreement will often lead to considerable strife and difficulties later on. Instead of the team discussing the performance of the process and ways of how to improve it, the debate is about such issues as units of measure, sources of information and reporting frequency.

Most readers will be familiar with the use of KPIs (key performance indicators), which are employed by many organisations. While KPIs can be helpful to provide a measure of control, they are often used inappropriately. The trouble is that KPIs are frequently employed to report on all manner of results, irrespective of the type of result. We can in fact recognise three quite distinct classes of results:

1. *overall results*, e.g. sales volume, expenditure and profit;
2. *specific-responsibility results*, e.g. sales per representative, unit cost of production, profit per product and performance improvement;
3. *process performance*, e.g. machine uptime, weight control and waste.

Some organisations employ a large number of KPIs, perhaps at times as many as 100 or more, usually without any appreciation as to the 'result class'. Often, too, because of this lack of appreciation, the KPIs are all reported to the CEO of the organisation, who finds it difficult to absorb all this information, let alone take any required remedial action. When one thinks about it, we cannot really have a large number of *key* performance indicators. To ensure that our indicators provide some real value, we need to recognise the three different classes and adopt a suitable naming convention. For this chapter, the following has been adopted:

- *Critical overall results (CORs).* Those ten or so agreed top results which monitor the ongoing 'health' of the enterprise, for example total revenue and expenditure, market share, profit and ROI, new business, and innovation. In some organisations, these indicators are referred to as the 'cockpit indicators'. Generally speaking, CORs are the responsibility of the organisation as a whole rather than of specific individuals, as they show the amalgamated result of many responsibilities. The information is mainly for senior management and there is generally a fairly long time interval between the close-off of the reporting period and the availability of the report. (Note that CORs should, as a rule, be established as the result of a business planning exercise. Their development and use is beyond the scope of this chapter, however).
- *Key responsibility measures (KRMs).* The seven or so key results which are the specific responsibility of a manager or team, for example the process output quality standards described earlier (timeliness, quantity and physical attributes), and actual cost against standard. The information is mainly for departmental and middle management, and there is generally only a matter of days between the close-off of the reporting period and the reports being available. (Appropriate KRMs are also suitable for monitoring the progress of company-wide performance improvement programmes).
- *Process performance monitors (PPMs).* Measures which are specifically designed to monitor the performance of a number of aspects of a process, for example weighing or batching, and retort operating parameters. For complex processes, there can be many PPMs. The information is mainly for the actual operators and their supervisors and should ideally be available in 'real time' and in graphical form. Where PPMs are subject to statistical process control, it is suggested that they are referred to as process control monitors (PCMs).

While properly developed and managed KRMs and PPMs will provide focused and timely control information to operating staff and their immediate managers, there often is a requirement for more senior management to be kept informed on a regular basis as well. What this usually means is that those managers are provided with copies of the regular reports, which are likely to contain much additional information which is of little interest to them. A better idea is for the more senior managers to be able to nominate a particular KRM, PPM or PCM which is of special interest to them. For example, the Chief Technical Manager could ask for a regular PCM report on the performance of some innovative batching equipment. Likewise, the Chief Operating Officer might require, say, weekly reports on the PPM monitoring plant uptime. Once

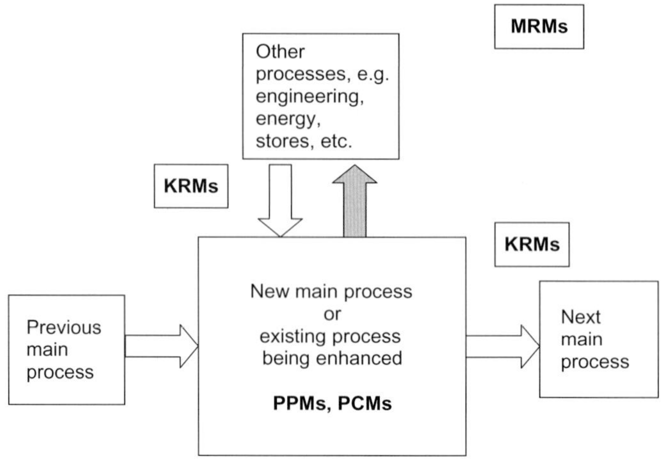

Figure 46.3 Agreed measures of performance.

selected, they should be given the supplementary label of management reporting measures (MRMs). In suggesting the use of MRMs, it should be appreciated that these are not additional measures requiring special report formats. See Figure 46.3 for the relationship between the various measures.

Some 'standard KRMs' which can be used with considerable effect in most situations are presented later in the chapter.

Separating 'Results' from 'Processes'

As stated earlier, if we were to 'drill down' into any process, we would find that it was also a 'system' in its own right complete with sub-processes, the 'parts'. Figure 46.4, which is an adaptation of the 'Ishikawa diagram' (also known as the 'cause and effect' or 'fishbone' diagram), shows how a number of sub-processes contribute to the output of a process, that is, 'the results'. While the diagram appears deceptively simple, it can be very useful in identifying the impact of the key sub-processes on the process output. In addition, the diagram illustrates the importance of separating the 'results' from the 'process'. The latter point will be particularly relevant when we come to the actual process of 'food process design' and performance improvement. At this stage it should be noted that the 'results' can be improved only by monitoring and improving the 'process' or 'processes'.

A Sound Design Process

The technical aspects of 'food process design' are undoubtedly covered most expertly by the other contributors, so we can confidently leave those matters to them. It

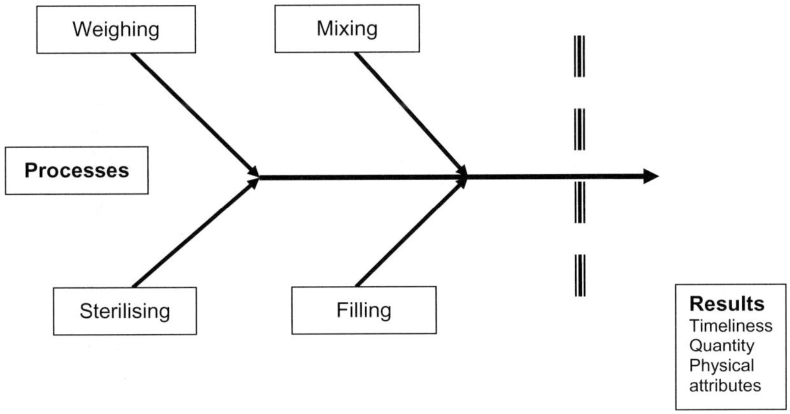

Figure 46.4 Separating 'results' from 'process'.

should be appreciated, however, that the design activity is itself also a 'process', which should be based on a sound methodology and a disciplined approach. Successful process design requires the development of a first-rate process with the following attributes:

- appropriate standardisation and predictability of outcomes;
- scope for exploration and innovation;
- a requirement for thorough analysis;
- a staged approach with clearly defined Go/No Go decision points;
- efficient use of resources;
- transparency and ease of communication;
- proper control of decision-making and progress;
- a comprehensive record of the process, e.g. options considered and decisions;
- professional preparation and presentation of the proposal;
- construction/implementation planning and progress management;
- design and approval of required changes and 'as built' recording;
- performance testing, sign-off and handover to production.

Table 46.10 outlines a standard five-stage design process template, which can be modified to suit specific organisational needs. Note that for the process to work properly, the inputs and outputs need to be defined carefully for each stage and associated decision. Once the process has been agreed to by all concerned, it would be a good idea to implement a PC-based system to manage the process. By having the system available on the organisation's network, the system could be largely paperless while at the same time allowing all relevant personnel to be kept up to date regarding progress and the decisions made.

Table 46.10 Five-stage design process template.

Key input	Stage	Key output
Ideas and suggestions captured in 'opportunity brief'	**I – Initiation** Process administration	Completed 'opportunity brief'
Completed 'opportunity brief'	First decision, by innovation team	Exploration brief
Exploration brief	**II – Exploration by technical staff**	Proposed design brief with options
Proposed design brief with options	Second decision, by innovation team	Endorsed design brief with selected/preferred option(s)
Endorsed design brief with selected/preferred option(s)	**III – Process design by technical staff**	Fully specified and costed project proposal
Presentation to Company Executive based on fully specified and costed project proposal and tentative project plan	Third decision, by Company Executive	Approved project brief Additional conditions and requirements
Approved project brief	**IV – Implementation** Management by champion	Progress reports Trial results Completion notification
Completion notification	Fourth decision, by process 'owner'	Acceptance sign-off Operational schedules
Process specifications Performance data	**V – Performance review by 'owner' and technical staff**	Project review report Required operational changes Opportunities for possible process improvements
Opportunities for possible improvements	Fifth decision, by 'owner'	Ideas and suggestions back to Stage I

Stage I: Initiation

The initiation stage is the formal starting point of the design process. It is important that all design projects have such a formal beginning, as an ill-defined and incomplete initiation is likely lead to problems later on. Stage I should be designed to handle all requests for new processes and improvements to existing processes from anywhere in the organisation.

A standard form called the 'opportunity brief' should be employed, which should allow all relevant information to be captured. For example it should include:

- what is being proposed and why (purpose and scope);
- interfaces with other processes (preferably, an 'operational flow chart' as outlined earlier should be developed at this stage);
- the 'required by date' and any other critical dates;
- the names of the proposer, the 'champion' to drive the project, and the eventual 'owner' of the new process.

The information is checked for completeness by a 'design process administrator' and the 'completed opportunity brief' is passed to the innovation team for consideration and decision. If approved, the form is suitably endorsed and now becomes the 'exploration brief', which is the key input for the next stage of the process.

Stage II: Exploration

This also needs to be a definite 'stage' to encourage the technical staff to adopt a properly broad approach from the start. All too often, a specific design is decided on early in the process without a range of possible solutions having been explored. On the other hand, the exploration stage, which can be likened to an architect producing a number of concept sketches, should not require too much effort. The idea is to have a number of potential schemes identified in sufficient detail for the assessment team to select one or more options for further development.

Besides the exploration brief, the technical staff will employ several other important inputs, such as:

- strategies and priorities;
- the business plan;
- company policies and ethics (e.g. sustainability and social responsibility);
- market information;
- financial expectations and standards;
- relevant technical reports and publications;
- equipment catalogues;
- resource information.

On completion of the exploratory work, the technical team produces a 'proposed design brief with options', together with a cost estimate for the design work, for consideration and decision by the innovation team. Following approval, and preferably with agreement on the preferred option, the form changes to the 'endorsed design brief with preferred/selected option(s), which becomes the trigger for the process design stage.

Stage III: Process Design

The endorsed design brief is the authority for the technical staff to commence work on the detailed design. It is of particular importance that all members of the process design team have a clear understanding of the 'purpose and scope' of the proposal. If not already done as part of Stage I, an 'operational flow chart' should be drawn up, together with a detailed diagram of the new process showing all inputs and outputs, as outlined earlier in the chapter. At some point during the design stage, it is also a good idea to 'drill down' into the new process to separate the 'results' from the 'processes', as shown earlier in Figure 46.4.

The inputs into this stage are much the same as those listed for Stage II, although considerably more detail will be required. Possible sources of finance may need to be identified. In addition to dealing with all the technical design matters, the team should also:

- keep a design log to record any changes requested or new requirements, changes in assumptions, etc;
- where possible, specify ingredients already in use by the organisation to avoid unnecessary complexity;
- define the three 'quality standards' (timeliness, quantity and physical attributes) for all inputs and outputs of the new process;
- define the performance measurements (KRMs, PPMs and PCMs);
- minimise set-up and changeover requirements where possible;
- work through any process capability issues;
- carry out the necessary financial analyses (breakeven and discounted cash flow);
- identify all significant intangible costs and benefits.

Proposal Presentation

The first output of Stage III is the fully specified and costed 'project proposal', which forms the basis for the presentation to the Company Executive. It is important to stress that not only should proposals be conceptually and financially sound, but they also need to be 'sold' with skill and confidence. It is unfortunately true that many a good proposal has failed to make the grade because of a poorly prepared and 'woolly' presentation. A sound proposal presentation means assessing the economies of the proposal thoroughly and presenting the justification in a logical manner.

When preparing the presentation, technical staff should be aware that presentations for commercial organisations are very different from what is normally required in an academic environment. In the main, there needs to be considerably greater emphasis on the proposal's 'purpose and scope', coupled with what is required to make things happen. On the other hand, there is usually a lesser need for advanced theory and meticulous technical explanations, the underlying thought being that the technical specialist is competent in his or her field and thus the managerial focus should be on the commercial aspects of the proposal. If it is felt necessary, more detailed technical information related to, for example, relevant research or methods of investigation could be included as appendices to the proposal. To enhance the quality of presentations, technical staff should not hesitate to seek advice, and may find it helpful to use the template shown in Table 46.11 as a guide.

The template is largely self-explanatory, but there are three points worth emphasising. Firstly, the introductory statement is most important in that it 'sets the scene' by defining the purpose and the boundaries (scope) of the proposal. To assist in getting it right, it is often useful to employ the words 'by means of' or just 'by'. Take, for example, a proposal for a new process, for which the statement may be something like this:

Table 46.11 Presentation guide template.

Presentation section	Content outline	Suggested number of slides
1 Defining the purpose and scope of the proposal	*What* is being proposed and *why* The *why* should be the single prime 'make or break' reason for the proposal, e.g. stating an important 'opportunity' or 'threat' to the business. The *what* should clearly define the boundaries of the proposal.	1
	Additional supporting reasons or benefits such as improved quality, enhanced customer satisfaction or increased production flexibility. Care should be taken not to 'scrape the barrel' by including a large number of marginal benefits.	1–3
2 Identifying what is required and who is involved	New plant and processes, additional premises and new personnel resources Impact on other departments, processes or systems	3–6
3 Justifying the proposal	New revenue, efficiency gains, production capacity increases New fixed and variable costs Breakeven analysis, discounted cash flow	4–8
	All intangible benefits and costs (drawbacks) and trade-offs, clearly identified and listed separately	2–3
4 Underlying assumptions	Such as increased sales, nature of competitor activity, improved throughput, cost of finance. If possible, assumptions should be shown in terms of both magnitude and probability.	2–4
5 Tentative project plan	Critical dates; milestones and dependencies	2–4
6 What is needed to advance the proposal	Company sign-off, regulatory approvals Securing finance, negotiations with suppliers	1– 2
7 Questions and answers	Record any new information and additional features or facilities required	

To take advantage of an opportunity for 80% additional sales, it is proposed to enhance and double the current processing capacity by:

- installing new batching and extruding equipment;
- making improvements to the existing intake facility;
- increasing the capacity of the number 2 boiler by 40%;
- implementing a new production scheduling system.

The second point that should be made is the need for conciseness. Company executives are, as a rule, very busy people and they tend to become impatient and possibly annoyed when having to sit through long-winded presentations. The template shows the suggested number of slides for each section of the presentation. Generally speaking, each slide should not have more than say six or so 'bullet points', and take 30 to 60 seconds to cover. The complete presentation should therefore take on the order of 15 to 30 minutes.

Lastly, there is the need to distinguish clearly between the 'features' and the 'benefits' of the proposal. It is not unusual for technical specialists to speak with great enthusiasm on a particular feature without identifying any associated benefits. As a general principle, features for which no benefits can be identified are better left out of the presentation. A proposal could, for example, include some new equipment employing 'the latest digital weighing technology'. However, this is likely to be of little interest to senior management unless the benefits, such as for example greater accuracy, reduced waste and improved reliability, are specifically listed and preferably quantified.

Following consideration and approval by the Company Executive, the proposal, together with any additional conditions and requirements, is the second output of Stage III and is now termed the 'approved project brief'.

Stage IV: Implementation

While in many situations the technical specialist will not be responsible for the actual construction of the new process facilities, he or she should remain actively involved to ensure that expectations will be met. As a first step, the various responsibilities need to be defined. To assist in this, a simple but effective 'brainstorm table' can be employed (see Table 46.12). The idea is to bring the relevant parties together and list

Table 46.12 Project responsibilities.

Project sub-objectives	Project champion	Technical specialist	Principal contractor	Production manager
Effective project management	Prepare detailed plan	Provide testing details	Critical construction dates	Key dates and inventory volumes
	Design and implement progress reporting	Provide input	Provide input	Provide input
Site available and prepared	Liaise with other parties		Provide temporary services	Produce buffer stock Tidy up site
Equipment purchased and on site		Inspect and certify	Take delivery and secure	
Trained staff	Assist with course development	Training and operational manuals	Assist with specialised plant	Run training courses
Etc., etc.				

their functions or names along the table's horizontal axis. The next task is to brainstorm and list all of the 'project sub-objectives' using the approved project brief together with the tentative project plan. Details of what is involved for each member of the group should be specified (light grey in the table). At the same time, prime responsibilities (dark grey in the table) are assigned also.

During the actual construction, the project champion is responsible for monitoring the project and producing regular progress reports. The matters to be covered include, for example, performance against plan, expenditure versus budget, results of tests and trials, and authorised project variations. If not already done, operational manuals should be developed and staff training material prepared. Further work may also be required to finalise the KRMs and PPMs.

Useful Standard KRMs

The kind and number of measures that should be implemented depend of course on the nature of the process and on the organisation's managerial philosophy. However, in most situations, the following 'standard' KRMs can be used to monitor and control several important aspects of the process on a daily basis (La Rooy, 1997c):

- availability (plant uptime);
- productivity;
- quality;
- overall effectiveness.

To enable the indicators to be calculated requires five kinds of data, most of which ought to be easy and quick to obtain, i.e.

- C = total crewed time (or intended operating time);
- N = total non-operating time (e.g. breakdowns and set-up time);
- O = actual output (e.g. numbers of units, tonnes or litres produced);
- T = theoretical output rate (e.g. units per hour);
- R = quantity of output rejected or outside the specification.

The formulae to calculate the indicators are straightforward enough. The availability indicator I_a is defined by

$$I_a = \frac{(C - N)}{C}$$

Assume that A = 8 hours for the day and N = 2 hours. Then

$$I_a = \frac{(8\text{ h} - 2\text{ h})}{8\text{ h}}$$
$$= 0.75 \text{ or } 75\%$$

The productivity indicator is found as follows:

$$I_p = \frac{O}{(C - N) \times T}$$

Assume that O = 10 tonnes for the day, T = 2 tonnes per hour, and C = 8 hours and N = 2 hours as before. Then

$$\begin{aligned} I_p &= \frac{10\ \text{t}}{(8\ \text{h} - 2\ \text{h}) \times 2\ \text{t/h}} \\ &= \frac{10\ \text{t}}{12\ \text{t}} \\ &= 0.83 \text{ or } 83\% \end{aligned}$$

Note that this indicator shows the productivity while the plant is actually working, as the idle time has already been allowed for in the availability indicator.

The quality indicator is found thus:

$$I_q = \frac{(O - R)}{O}$$

Assume that R = 1 tonne and O = 10 tonnes (as before). Then

$$\begin{aligned} I_q &= \frac{(10\ \text{t} - 1\ \text{t})}{10\ \text{t}} \\ &= \frac{9\ \text{t}}{10\ \text{t}} \\ &= 0.9 \text{ or } 90\% \end{aligned}$$

While the individual indicators can be very telling, it is also worthwhile to determine an overall effectiveness indicator by multiplying the three indicators. So, using the figures calculated earlier,

$$\begin{aligned} I_o &= 0.75 \times 0.83 \times 0.9 \\ &= 0.56 \text{ or } 56\% \end{aligned}$$

On completion of the project, a 'completion notification' is produced by the project champion for consideration by the project owner. Once the owner is satisfied with project, he or she will sign off the project acceptance and arrange for the production of operational schedules.

Stage V: Performance Review

Performance review is an important part of the process, although in practice it is not always carried out with sufficient vigour, if at all. What is required is comparing the original process specifications, including any approved changes, with operational performance data. In addition, the actual capital and operational costs need to be compared and justified in terms of the project proposal. A 'project review report' is the key output from this stage, together with details of any required operational changes and possible improvements to the process.

With respect to any suggested improvements to the process as the result of, say, not meeting quality standards, it is necessary to sound a word of caution. It is not uncommon for changes to processes to be requested without sufficient understanding of what is the underlying cause of the problem. When considering process performance, it is important to keep in mind the two causes of variability outlined earlier, i.e. 'common' and 'special'. How we should react to requests for changes to a process depends whether we are dealing with common-cause or special-cause variability and also whether we can or cannot control or influence the process. The importance of the ability to control or influence is well demonstrated by the lasting qualities of paint on car bodies in comparison with paint on houses. Over the years, car makers have created highly controlled environments in which the paint is applied, with the result that a very long paint life is now the norm. As shown in Table 46.13, there are four possible scenarios (La Rooy, 1995)

A: aim to avoid occurrence. An example may be where the accuracy of some weighing equipment is affected whenever a heavy forklift is travelling nearby. A possible solution is to divert the forklift traffic.

B: aim to minimise the effect. If, in the previous example, it was not possible to divert the forklift traffic, reinstalling the weighing equipment on vibration-absorbing mountings could be considered.

C: improve the process. Common-cause variability can only be reduced by improving or changing the process. In the case of the weighing equipment above, the solution could be to overhaul the machine.

D: replace the process. If the common-cause variability cannot be reduced, the process needs to be replaced.

Table 46.13 Dealing with special- and common-cause variability.

	Can control or influence	Cannot control or influence
Special-cause variability	A Aim to avoid occurrence	B Aim to minimise effect
Common-cause variability	C Change/improve the process	D Replace the process

Applying the Concepts and Techniques

In concluding the chapter, it is important to emphasise three points. Firstly, a thoroughly systematic approach should be adopted in food process design as a matter of course, as this will not only result in a sounder proposal but will also mean a more efficient use of resources. Secondly, all relevant commercial considerations and practices should form an integral part of food process design. It is definitely not sufficient for this work to be done belatedly as some 'add-on' to the technical work. Thirdly, technical specialists should not leave the commercial considerations and practices to just the accounting or other administrative staff. By all means, they should seek any necessary advice; they should not, however, abdicate their responsibilities.

Readers will appreciate that since a fairly wide range of topics had to be covered in a single chapter, there had to be some trade-offs between breadth and depth. I hope that in spite of this constraint, this chapter will nonetheless prove to be of assistance to technical specialists in their quest to contribute more effectively to the ongoing success of the enterprise in which they are employed.

References

La Rooy, G.M. (1993a) Laying the foundation for successful business systems. *NZ Business*, November: 26–31. (Republished in book form in G.M. La Rooy, *Systematic Business Success*. Profile Books, Auckland, New Zealand (1998).)

La Rooy, G.M. (1993b) Managing by work centres – a dynamic ABC application. In: *Costing Solutions & ABC*. CIMA, New Zealand.

La Rooy, G.M. (1995) Opportunities in waiting. *NZ Business*, March: 36–41. (Republished in book form in G.M. La Rooy, *Systematic Business Success*. Profile Books, Auckland, New Zealand (1998).)

La Rooy, G.M. (1997a) Breakeven revisited. *NZ Business*, August: 54–57. (Republished in book form in G.M. La Rooy, *Systematic Business Success*. Profile Books, Auckland, New Zealand (1998).)

La Rooy, G.M. (1997b) What about the intangibles. *Chartered Accountants Journal*, September: 21–24.

La Rooy, G.M. (1997c) Effective efficiency. *NZ Business*, February: 26–29. (Republished in book form in G.M. La Rooy, *Systematic Business Success*. Profile Books, Auckland, New Zealand (1998).)

La Rooy, G.M. (2010) Plan boldly act cautiously. *N Z Management*, February: 45–48.

La Rooy, G.M. and Perera, A. (1999) Managing profit and quality. In: *Handbook of Food Preservation* (ed. M.S. Rahman). Marcel Dekker, New York, pp. 1215–1242.

Rahman, M.S. (ed.) (1999) *Handbook of Food Preservation*. Marcel Dekker, New York.

Rahman, M.S. (ed.) (2007) *Handbook of Food Preservation*, 2nd edn. Marcel Dekker, New York.

Index

Note: page numbers in *italics* refer to figures and boxes, those in **bold** refer to tables

Handbook of Food Process Design, First Edition. Edited by Jasim Ahmed and Mohammad Shafiur Rahman.